SCHAUM'S SOLVED PROBLEMS SERIES

2000 SOLVED PROBLEMS IN

ELECTRONICS

by

Jimmie J. Cathey, Ph.D.
University of Kentucky

McGRAW-HILL, INC.
New York St. Louis San Francisco Auckland
Bogotá Caracas Hamburg Lisbon London Madrid
Mexico Milan Montreal New Delhi Paris San Juan
São Paulo Singapore Sydney Tokyo Toronto

Jimmie J. Cathey, Ph.D, *Professor of Electrical Engineering, Director of Graduate Studies, University of Kentucky*

Jimmie J. Cathey earned the Ph.D. from Texas A&M University and has 13 years of industrial experience in the design and development of electric drive systems. Since 1980, he has taught at the University of Kentucky, and his research and teaching interests are power electronics, electric machines, and robotics. He is a Registered Professional Engineer.

2000 SOLVED PROBLEMS IN ELECTRONICS

Library of Congress Cataloging-in-Publication Data

Cathey, Jimmie J.
  2000 solved problems in electronics / by Jimmie J. Cathey.
    p.   cm.—(Schaum's solved problems series)
  ISBN 0-07-010284-8
  1. Electronics—Problems, exercises, etc. I. Title. II. Series.
TK7862.C37 1990
621.381'076—dc20   89-29583
                                                                     CIP

2 3 4 5 6 7 8 9 0 SHP/SHP 9 4 3 2 1 0

ISBN 0-07-010284-8

Sponsoring Editor, David Beckwith
Production Supervisor, Leroy Young
Editing Supervisors, Meg Tobin, Maureen Walker

Copyright © 1990 by McGraw-Hill, Inc. All rights reserved. Printed in the United States of America. Except as permitted under the United States Copyright Act of 1976, no part of this publication may be reproduced or distributed in any form or by any means, or stored in a data base or retrieval system, without the prior written permission of the publisher.

CONTENTS

Chapter 1   PORT-BASED CIRCUIT ANALYSIS   1
One-Port Networks / Network Theorems / Two-Port Networks / Average and RMS Values / Review Problems

Chapter 2   DIODES: SIGNAL LEVEL APPLICATION   40
Semiconductor Material Properties / Terminal Characteristics / Rectifier Diode Circuit Analysis / Special Purpose Diodes / Review Problems

Chapter 3   DIODES: POWER LEVEL APPLICATIONS   75
General Circuits / Power Supply Circuits / Review Problems

Chapter 4   BIPOLAR JUNCTION TRANSISTOR FUNDAMENTALS   97
Operating Principles / Bias and Load Lines / Operating Modes / Review Problems

Chapter 5   FIELD-EFFECT TRANSISTOR FUNDAMENTALS   132
Operating Principles / Bias and Load Lines / Review Problems

Chapter 6   TRANSISTOR BIAS CONSIDERATIONS   161
β-Uncertainty and Temperature Effects / Nonlinear-Element Stabilization for BJTs / Stability-Factor Analysis / Q-Point-Bounded Bias for the FET / Review Problems

Chapter 7   SMALL-SIGNAL FREQUENCY-INDEPENDENT BJT AMPLIFIERS   192
BJT Equivalent Circuit Models / Common-Emitter Amplifiers / Common-Base Amplifiers / Common-Collector (Emitter-Follower) Amplifiers / Review Problems

Chapter 8   SMALL-SIGNAL FREQUENCY-INDEPENDENT FET AMPLIFIERS   233
FET Equivalent Circuit Models / Common-Source Amplifiers / Common-Drain Amplifiers / Common-Gate Amplifiers / Review Problems

Chapter 9   MULTIPLE-TRANSISTOR AMPLIFIERS   257
Multiple-BJT Amplifiers / Multiple-FET Amplifiers / Review Problems

Chapter 10   FREQUENCY EFFECTS IN AMPLIFIERS   281
Bode Plots and Frequency Response / Low-Frequency Effect of Bypass and Coupling Capacitors / High-Frequency BJT and FET Models / Review Problems

Chapter 11   AUDIO-FREQUENCY POWER AMPLIFIERS   313
Direct-Coupled Amplifiers / Inductor-Coupled Amplifiers / Transformer-Coupled Amplifiers / Push-Pull Amplifiers / Complementary-Symmetry Amplifiers / Distortion Considerations / Thermal Considerations / Review Problems

Chapter 12   FEEDBACK AMPLIFIERS   339
Feedback Concepts / Voltage-Series Feedback / Current-Series Feedback / Voltage-Shunt Feedback / Current-Shunt Feedback / Review Problems

Chapter 13   OPERATIONAL AMPLIFIERS   379
Op Amp Fundamentals / Linear Function Operations / Nonlinear Function Operations / Filters / Review Problems

Chapter 14   SWITCHING CIRCUITS FOR DIGITAL LOGIC   422
Binary Functions / Transistor Switching Concepts / Logic Families / Multivibrators / Review Problems

Chapter 15 BOOLEAN ALGEBRA AND LOGIC GATES 454
Numbering Systems / Theorems and Properties of Boolean Algebra / Gate Implementation of Boolean Functions / Simplification of Boolean Functions / Review Problems

Chapter 16 COMBINATIONAL AND SEQUENTIAL LOGIC 490
Combinational Logic Circuits / Sequential Logic Circuits / Review Problems

Chapter 17 VACUUM TUBES 508
Vacuum Diodes / Vacuum Triodes / Vacuum Pentodes / Review Problems

INDEX 509

To the Student

This full spectrum of potent problems ought to ready you for anything you'll be asked (plus a little bit more) in a first course in electronic circuits. In fact, the two thousand detailed solutions themselves amount to a course in electronics *practice*—in the successful construction and use of circuits, based solely on the *terminal characteristics* of the functioning control devices. So, if you want practice in practice, here is your bible.

CHAPTER 1
Port-Based Circuit Analysis

ONE-PORT NETWORKS

1.1 A *port* is a terminal pair of an electrical *network* (interconnection of two or more simple circuit elements—resistors, capacitors, inductors, voltage sources, or current sources) across which a voltage can be identified and such that the current into one terminal is the same as the current out of the other terminal. A *one-port network,* as illustrated by Fig. 1-1, has only one identified port. If the one-port network of Fig. 1-1 contains only a resistor R, determine the port voltage v_1 in terms of the port current i_1.

▌ By Ohm's law,
$$v_1 = i_1 R$$

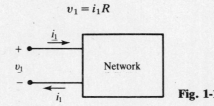

Fig. 1-1

1.2 Find the port current i_1 in terms of the port voltage v_1 for Problem 1.1.

▌ Based on Ohm's law,
$$i_1 = v_1/R$$

1.3 If the one-port network of Fig. 1-1 contains only an inductor L for which $v_1 = L\dfrac{di_1}{dt}$, determine the port current $i_1(t)$ if $i_1(-\infty) = 0$.

▌ By direct integration of the known $v - i$ relationship,
$$\int_{i_1(-\infty)}^{i_1(t)} di_1 = i_1(t) = \frac{1}{L}\int_{-\infty}^{t} v_1\, d\tau$$

1.4 If the one-port network of Fig. 1-1 contains only a capacitor C for which $i_1 = C\dfrac{dv_1}{dt}$, determine the port voltage $v_1(t)$ if the capacitor was uncharged at $t = -\infty$, that is, $v_1(-\infty) = 0$.

▌ By direct integration of the known v-i relationship,
$$\int_{v_1(-\infty)}^{v_1(t)} dv_i = v_1(t) = \frac{1}{C}\int_{-\infty}^{t} i_1\, d\tau$$

1.5 The instantaneous power supplied to the one-port network of Fig. 1-1 is given by $p(t) = v_1(t)i_1(t)$. Find the instantaneous power supplied to the one-port network of Problem 1.1.

▌
$$p(t) = v_1 i_1 = (i_1 R)(i_1) = i_1^2 R$$

1.6 Determine the instantaneous power supplied to the one-port network of Problem 1.4.

▌
$$p(t) = v_1 i_1 = (v_1)\left(C\frac{dv_1}{dt}\right) = Cv_1\frac{dv_1}{dt}$$

1.7 For the one-port network of Fig. 1-2, find the port current i_1 in terms of the port voltage v_1.

▌ The parallel-connected 2-Ω resistors can be reduced to an equivalent
$$R_{xy} = 2 \parallel 2 = \frac{(2)(2)}{2+2} = 1\ \Omega$$

2 □ CHAPTER 1

Hence, the equivalent resistance of the network is

$$R_{eq} = 1 + R_{xy} = 1 + 1 = 2\,\Omega$$

And, by Ohm's law,

$$i = v_1/R_{eq} = v_1/2$$

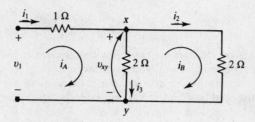

Fig. 1-2

1.8 Solve Problem 1.7 by the method of mesh currents.

▌ From Fig. 1-2

$$(1+2)i_A - 2i_B = v_1$$
$$-2i_A + (2+2)i_B = 0$$

Using Cramer's rule,

$$i_1 = \frac{\Delta_1}{\Delta} = \frac{\begin{vmatrix} v_1 & -2 \\ 0 & 4 \end{vmatrix}}{\begin{vmatrix} 3 & -2 \\ -2 & 4 \end{vmatrix}} = \frac{4v_1}{8} = \frac{v_1}{2}$$

1.9 Solve Problem 1.7 using the method of node voltages.

▌ From Fig. 1-2,

$$-i_1 + i_2 + i_3 = \frac{v_{xy} - v_1}{1} + \frac{v_{xy}}{2} + \frac{v_{xy}}{2} = 0$$

or

$$v_{xy} = v_1/2$$

Whence,

$$i_1 = \frac{v_1 - v_{xy}}{1} = \frac{v_1 - v_1/2}{1} = v_1/2$$

1.10 The one-port network of Fig. 1-3 is modeled by interconnected impedances (ratio of voltage to current) Z_1, Z_2, and Z_3. Find the Laplace domain port current I_1 in terms of the Laplace domain port voltage V_1 by use of the method of mesh currents.

▌

$$(Z_1 + Z_2)I_1 - Z_2 I_2 = V_1$$
$$-Z_2 I_1 + (Z_2 + Z_3)I_2 = 0$$

Using Cramer's rule,

$$I_1 = \frac{\Delta_1}{\Delta} = \frac{\begin{vmatrix} V_1 & -Z_2 \\ 0 & Z_2 + Z_3 \end{vmatrix}}{\begin{vmatrix} Z_1 + Z_2 & -Z_2 \\ -Z_2 & Z_2 + Z_3 \end{vmatrix}} = \frac{V_1(Z_2 + Z_3)}{Z_1 Z_2 + Z_1 Z_3 + Z_2 Z_3}$$

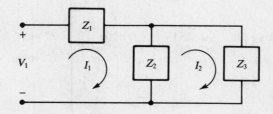

Fig. 1-3

1.11 Determine the equivalent impedance looking into the one-port network of Fig. 1-3.

▌ Based on Problem 1.10,

$$Z_{eq} = Z_{in} = \frac{V_1}{I_1} = \frac{Z_2 + Z_3}{Z_1 Z_2 + Z_1 Z_3 + Z_2 Z_3}$$

1.12 For the one-port network of Fig. 1-3 let

$Z_{11} = Z_1 + Z_2 = $ sum of impedances around mesh 1
$Z_{22} = Z_2 + Z_3 = $ sum of impedances around mesh 2
$Z_{12} (= Z_{21}) = Z_2 = $ sum of impedances in branch shared by meshes 1 and 2

Solve for port current I_1 in terms of V_1 and the newly defined impedances above.

▌ From Fig. 1-3,

$$Z_{11} I_1 - Z_{12} I_2 = V_1$$
$$-Z_{21} I_2 + Z_{22} I_2 = 0$$

By Cramer's rule,

$$I_1 = \frac{\Delta_1}{\Delta} = \frac{\begin{vmatrix} V_1 & -Z_{12} \\ 0 & Z_{22} \end{vmatrix}}{\begin{vmatrix} Z_{11} & -Z_{12} \\ -Z_{21} & Z_{22} \end{vmatrix}} = \frac{V_1 Z_{22}}{Z_{11} Z_{22} - Z_{12}^2}$$

1.13 Find current I_2 for the one-port network of Fig. 1-3 using the impedances defined in Problem 1.12.

▌ Based on Problem 1.12,

$$I_2 = \frac{\Delta_2}{\Delta} = \frac{\begin{vmatrix} Z_{11} & V_1 \\ -Z_{21} & 0 \end{vmatrix}}{Z_{11} Z_{22} - Z_{12}^2} = \frac{V_1 Z_{12}}{Z_{11} Z_{22} - Z_{12}^2}$$

1.14 Determine the port current I_1 in terms of the port voltage V_1 for the one-port network of Fig. 1-4.

▌ Applying the method of mesh currents,

$$(1 + 2 + 2)I_1 - 2I_2 - 2I_3 = V_1$$
$$-2I_1 + (2 + 2 + 3)I_2 - 2I_3 = 0$$
$$-2I_1 - 2I_2 + (1 + 2 + 2)I_3 = 0$$

By Cramer's rule,

$$I_1 = \frac{\Delta_1}{\Delta} = \frac{\begin{vmatrix} V_1 & -2 & -2 \\ 0 & 7 & -2 \\ 0 & -2 & 5 \end{vmatrix}}{\begin{vmatrix} 5 & -2 & -2 \\ -2 & 7 & -2 \\ -2 & -2 & 5 \end{vmatrix}} = \frac{31 V_1}{91} = 0.3407 V_1$$

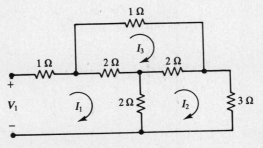

Fig. 1-4

4 □ CHAPTER 1

1.15 Find the equivalent resistance looking into the one-port network of Fig. 1-4.

▮ Based on Problem 1.14,

$$R_{eq} = \frac{V_1}{I_1} = \frac{V_1}{0.3407 V_1} = 2.935 \, \Omega$$

1.16 Define and evaluate the *generalized network impedances* for the one-port network of Fig. 1-4.

▮
$Z_{11} = 1 + 2 + 2 = 5 \, \Omega =$ sum of impedances around mesh 1
$Z_{22} = 2 + 2 + 3 = 7 \, \Omega =$ sum of impedances around mesh 2
$Z_{33} = 2 + 2 + 1 = 5 \, \Omega =$ sum of impedances around mesh 3
$Z_{12} = Z_{21} = 2 \, \Omega =$ sum of impedances in branch shared by meshes 1 and 2
$Z_{13} = Z_{31} = 2 \, \Omega =$ sum of impedances in branch shared by meshes 1 and 3
$Z_{23} = Z_{32} = 2 \, \Omega =$ sum of impedances in branch shared by meshes 2 and 3

1.17 Find the port current I_1 of the network of Fig. 1-4 in terms of the impedances defined in Problem 1.16.

▮
$$Z_{11}I_1 - Z_{12}I_2 - Z_{13}I_3 = V_1$$
$$-Z_{21}I_1 + Z_{22}I_2 - Z_{23}I_3 = 0$$
$$-Z_{31}I_1 - Z_{32}I_2 + Z_{33}I_3 = 0$$

Applying Cramer's rule,

$$I_1 = \frac{\Delta_1}{\Delta} = \frac{\begin{vmatrix} V_1 & -Z_{12} & -Z_{13} \\ 0 & Z_{22} & -Z_{23} \\ 0 & -Z_{32} & Z_{33} \end{vmatrix}}{\begin{vmatrix} Z_{11} & -Z_{12} & -Z_{13} \\ -Z_{21} & Z_{22} & -Z_{23} \\ -Z_{31} & -Z_{32} & Z_{33} \end{vmatrix}}$$

$$= \frac{V_1(Z_{22}Z_{33} - Z_{23}^2)}{Z_{11}(Z_{22}Z_{33} - Z_{23}^2) - Z_{22}Z_{13}^2 - Z_{33}Z_{12}^2 - 2Z_{12}Z_{13}Z_{23}}$$

1.18 Find a concise expression for the port current of a source-free, one-port network if the simultaneous equations are written using mesh currents and generalized network impedances.

▮ Inspection of Problems 1.12 and 1.17 leads, by induction, to

$$I_1 = \frac{V_1 \Delta_{11}}{\Delta}$$

where $\Delta =$ determinant of impedance coefficients

$\Delta_{11} =$ minor determinant of Δ formed by removal of row 1 and column 1

1.19 Determine the port current I_1 of the one-port network of Fig. 1-5.

▮ For the controlled source, $2V_c = 2(2I_2) = 4I_2$. Writing the mesh equations,

$$V_1 = (1+1)I_1 + 2V_c - (1)I_2 = 2I_1 + 3I_2$$
$$0 = -(1)I_1 + (1+2+2)I_2 - 2V_c = -I_1 + I_2$$

By Cramer's rule,

$$I_1 = \frac{\Delta_1}{\Delta} = \frac{\begin{vmatrix} V_1 & 3 \\ 0 & 1 \end{vmatrix}}{\begin{vmatrix} 2 & 3 \\ -1 & 1 \end{vmatrix}} = \frac{V_1}{5}$$

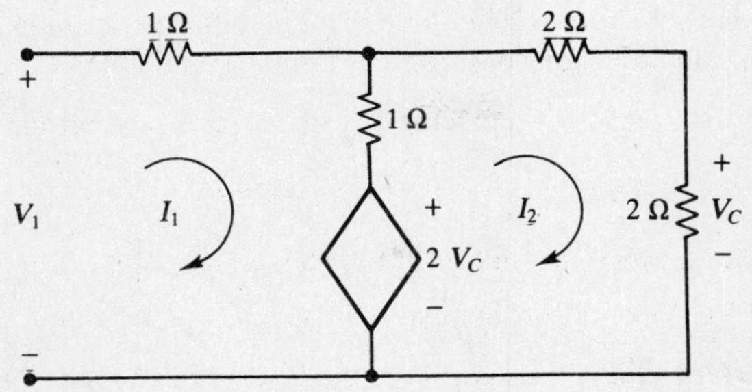

PORT-BASED CIRCUIT ANALYSIS ◻ 5

Fig. 1-5

1.20 Find the equivalent resistance looking into the one-port network of Fig. 1-5.

▮ Based on Problem 1.19,

$$R_{eq} = R_{in} = \frac{V_1}{I_1} = \frac{V_1}{V_1/5} = 5\,\Omega$$

1.21 If the one-port network of Fig. 1-1 is made up of a series connected resistor R and capacitor C, determine the Laplace domain port current $I_1(s)$ in terms of the Laplace domain port voltage $V_1(s)$.

▮ The equivalent impedance of the series-connected R and C is

$$Z_{in} = R + 1/sC = \frac{sRC + 1}{sC}$$

Hence,
$$I_1(s) = \frac{V_1(s)}{Z_{in}} = \frac{sCV_1(s)}{sRC + 1}$$

1.22 The one-port network of Fig. 1-1 is made up of a nonlinear, current dependent resistor $R = 2i_1$. Determine the port current i_1 in terms of the port voltage v_1.

▮ By Ohm's law,

$$i_1 = v_1/R = v_1/2i_1 \quad \text{or} \quad i_1 = \sqrt{v_1/2}$$

1.23 The one-port network of Fig. 1-1 is made up of a nonlinear resistor such that the current is described by $i_1 = k_1 e^{v_1/k_2}$ where k_1 and k_2 are constants. Determine the port voltage v_1.

▮ Take the logarithm of the characterizing equation and solve for v_1 to find

$$v_1 = k_2 \left[\ln(i_1) - \ln(k_1)\right]$$

1.24 Find the equivalent resistance looking into the one-port network of Problem 1.23.

▮ By Ohm's law,

$$R_{eq} = \frac{v_1}{i_1} = \frac{v_1}{k_1 e^{v_1/k_2}}$$

1.25 Determine the equivalent resistance looking into the one-port network of Problem 1.23 for large values of v_1.

▮ Applying l'Hospital rule to the result of Problem 1.24 gives

$$R_{eq} = \lim_{v_1 \to \infty} \frac{\frac{\partial}{\partial v_1}(v_1)}{\frac{\partial}{\partial v_1}(k_1 e^{v_1/k_2})} = \lim_{v_2 \to \infty} \frac{1}{\frac{k_1}{k_2} e^{v_1/k_2}} = 0$$

NETWORK THEOREMS

1.26 The *superposition theorem* states that *in a linear network containing multiple sources, the voltage across or current through any passive element may be found as the algebraic sum of the individual voltages or currents due* to each of the independent sources acting alone, with all other independent sources deactivated. Determine i_3 for the circuit of Fig. 1-6 using the superposition theorem.

▌ Deactivate current source i_s by replacing it with an open circuit and use a single prime to denote response due to v_s alone.

$$i_3' = \frac{v_s}{12 + 18} = \frac{6}{30} = 0.2 \text{ A}$$

Deactivate voltage source v_s by replacing it with a short circuit. Letting a double prime denote response due to i_s alone, current division yields

$$i_3'' = \frac{12}{12 + 18} i_s = \frac{12}{12 + 18}(2) = 0.8 \text{ A}$$

By the superposition theorem,

$$i_3 = i_3' + i_3'' = 0.2 + 0.8 = 1 \text{ A}$$

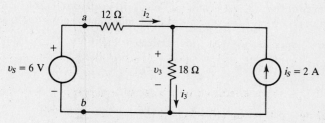

Fig. 1-6

1.27 Calculate the current i_2 for the circuit of Fig. 1-6 by use of the superposition theorem.

▌ The response due to v_s alone is

$$i_2' = \frac{v_s}{12 + 18} = \frac{6}{30} = 0.2 \text{ A}$$

The response due to i_s alone is

$$i_2'' = \frac{18}{18 + 12} i_s = \frac{18}{18 + 12}(2) = 1.2 \text{ A}$$

By superposition,

$$i_2 = i_2' + i_2'' = 0.2 + 1.2 = 1.4 \text{ A}$$

1.28 Determine voltage v_3 for the circuit of Fig. 1-6 using the superposition theorem.

▌ The response due to v_s alone is found by voltage division. Thus,

$$v_3' = \frac{18}{18 + 12} v_s = \frac{18}{18 + 12}(6) = 3.6 \text{ V}$$

The response due to i_s alone is

$$V_3'' = (12 \| 18)i_s = \frac{(12)(18)}{12 + 18}(2) = 14.4 \text{ V}$$

By the superposition theorem,

$$v_3 = v_3' + v_3'' = 3.6 + 14.4 = 18 \text{ V}$$

1.29 Verify the result of Problem 1.28 using the result of Problem 1.26.

▌ By Ohm's law,

$$v_3 = 18 i_3 = (18)(1) = 18 \text{ V}$$

1.30 For the circuit of Fig. 1-7, find current i_1 using the superposition theorem if $R_3 = 0$.

▌ Deactivate V_2 and V_3 by shorting to find the response due to V_1 alone.

$$i_1' = \frac{V_1}{2} = \frac{3}{2} = 1.5 \text{ A}$$

Short V_1 and V_3 to find the response due to V_2 alone as

$$i_1'' = -\frac{V_2}{2} = -\frac{2}{2} = -1 \text{ A}$$

Replace V_1 and V_2 by short circuits to find the response due to V_3 alone. Since shorting V_2 shunts current due to V_3 from the 2 Ω resistor, $i_1''' = 0$. By superposition,

$$i_1 = i_1' + i_1'' + i_1''' = 1.5 + (-1) + 0 = 0.5 \text{ A}$$

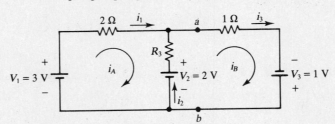

Fig. 1-7

1.31 For the circuit of Fig. 1-7, determine current i_2 using the superposition theorem if $R_3 = 0$.

▌ The response due to V_1 alone ($V_2 = V_3 = 0$) is

$$i_2' = -\frac{V_1}{2} = -\frac{3}{2} = -1.5 \text{ A}$$

The response due to V_2 alone is

$$i_2'' = \frac{V_2}{2} + \frac{V_2}{1} = \frac{2}{2} + \frac{2}{1} = 3 \text{ A}$$

The response due to V_3 alone is

$$I_2''' = \frac{V_3}{1} = \frac{1}{1} = 1 \text{ A}$$

By the superposition theorem,

$$i_2 = i_2' + i_2'' + i_2''' = -1.5 + 3 + 1 = 2.5 \text{ A}$$

1.32 For the circuit of Fig. 1-7, find current i_1 using the superposition theorem if $R_3 = 1\,\Omega$.

▌ The response due to V_1 alone is

$$i_1' = \frac{V_1}{2 + R_3 \| 1} = \frac{3}{2 + (1)(1)/(1+1)} = 1.2 \text{ A}$$

With only V_2 active,

$$i_2'' = \frac{V_2}{R_3 + 1 \| 2} = \frac{2}{1 + (1)(1)/(1+2)} = 1.2 \text{ A}$$

By current division,

$$i_1'' = -\frac{1}{1+2} i_2'' = -\frac{1}{1+2}(1.2) = -0.4 \text{ A}$$

With only V_3 active,

$$i_3''' = \frac{V_3}{1 + R_3 \| 2} = \frac{1}{1 + (1)(2)/(1+2)} = 0.6 \text{ A}$$

By current division,

$$i_1''' = \frac{R_3}{R_3 + 2} i_3''' = \frac{1}{1+2}(0.6) = 0.2 \text{ A}$$

The superposition theorem gives

$$i_1 + i_1' + i_1'' + I_1''' = 1.2 + (-0.4) + 0.2 = 1 \text{ A}$$

1.33 Verify the result of Problem 1.32 using the method of mesh currents.

$$(2 + R_3)i_A - R_3 i_B = 3i_A - i_B = V_1 - V_2 = 3 - 2 = 1$$
$$-R_3 i_A + (R_3 + 1)i_B = -i_A + 2i_B = V_2 + V_3 = 2 + 1 = 3$$

By Cramer's rule,

$$i_1 = i_A = \frac{\Delta_1}{\Delta} = \frac{\begin{vmatrix} 1 & -1 \\ 3 & 2 \end{vmatrix}}{\begin{vmatrix} 3 & -1 \\ -1 & 2 \end{vmatrix}} = \frac{5}{5} = 1 \text{ A}$$

1.34 Find current i_1 for the circuit of Fig. 1-8 by application of the superposition theorem.

Controlled sources must remain active during the superposition solution. With only V_1 active ($I = 0$), KVL yields

$$V_1 = (1 + 2)i_1' + 2i_1' \quad \text{or} \quad i_1' = \frac{V_1}{5} = \frac{10}{5} = 2 \text{ A}$$

With only I active ($V_1 = 0$), KCL requires that

$$I = 2 = -i_1'' + i_2'' = \frac{v''}{1} + \frac{v'' - 2i_1''}{2} = \frac{v''}{1} + \frac{v'' - 2(-v''/1)}{2} \quad \text{or} \quad v'' = 0.8 \text{ V}$$

Hence,

$$i_1'' = -v''/1 = -(0.8/1) = -0.8 \text{ A}$$

By the superposition theorem,

$$i_1 = i_1' + i_1'' = 2 + (-0.8) = 1.2 \text{ A}$$

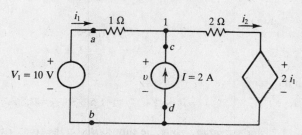

Fig. 1-8

1.35 Calculate the current i_2 for the circuit of Fig. 1-8 using the superposition theorem.

With only V_1 active, $i_2' = i_1' = 2$ A as determined in Problem 1.34. With only I active, $v'' = 0.8$ V and $i_1'' = -0.8$ A as found in Problem 1.34. By KVL,

$$i_2'' = \frac{v'' - 2i_1''}{2} = \frac{0.8 - 2(-0.8)}{2} = 1.2 \text{ A}$$

Based on the superposition theorem,

$$i_2 = i_2' + i_2'' = 2 + 1.2 = 3.2 \text{ A}$$

1.36 For the circuit of Fig. 1-9, $v_s = 10 \sin \omega t$ V, $V_b = 10$ V, $R_1 = R_2 = R_3 = 1 \, \Omega$, and $\alpha = 0$. Find current i_2 by use of the superposition theorem.

We first deactivate V_b by shorting, and use a single prime to denote a response due to v_s alone. Using the method of node voltage with unknown v_2' and summing currents at the upper node, we have

$$\frac{v_s - v_2'}{R_1} = \frac{v_2'}{R_2} + \frac{v_2'}{R_3}$$

Substituting given values and solving for v_2', we obtain
$$v_2' = \tfrac{1}{3}v_s = \tfrac{10}{3}\sin \omega t$$

Then, by Ohm's law,
$$i_2' = \frac{v_2'}{R_2} = \frac{10}{3}\sin \omega t \text{ A}$$

Now, deactivating v_s and using a double prime to denote a response due to V_b alone, we have
$$i_3'' = \frac{V_b}{R_3 + R_1 \| R_2} = \frac{V_b}{R_3 + (R_1)(R_2)/(R_1 + R_2)}$$

so that
$$i_3'' = \frac{10}{1 + 1/2} = \frac{20}{3} \text{ A}$$

Then, by current division,
$$i_2'' = \frac{R_1}{R_1 + R_2} i_3'' = \frac{1}{2} i_3'' = \frac{1}{2}\frac{20}{3} = \frac{10}{3} \text{ A}$$

Finally, by the superposition theorem,
$$i_2 = i_2' + i_2'' = \tfrac{10}{3}(1 + \sin \omega t) \text{ A}$$

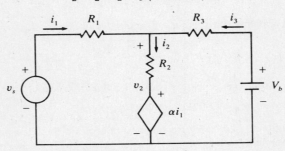

Fig. 1-9

1.37 Prove that the inductor element of Problem 1.3 is a linear element by showing that its v-i characteristic satisfies the converse of the superposition theorem.

▌ Let i_1 and i_2 be two currents that flow through the inductor. Then the voltages across the inductor for these currents are, respectively,
$$v_1 = L\frac{di_1}{dt} \quad \text{and} \quad v_2 = L\frac{di_2}{dt}$$

Now suppose $i = k_1 i_1 + k_2 i_2$, where k_1 and k_2 are distinct arbitrary constants. Then,
$$v = L\frac{d}{dt}(k_1 i_1 + k_2 i_2) = k_1 L\frac{di_1}{dt} + k_2 L\frac{di_2}{dt} = k_1 v_1 + k_2 v_2 \tag{1}$$

Since (1) holds for any pair of constants (k_1, k_2), superposition is satisfied and the element is linear.

1.38 If $R_1 = 5\,\Omega$, $R_2 = 10\,\Omega$, $V_s = 10$ V, and $I_s = 3$ A in the circuit of Fig. 1-10, find the current i by using the superposition theorem.

▌ With I_s deactivated (open-circuited), KVL and Ohm's law give the component of i due to V_s as
$$i' = \frac{V_s}{R_1 + R_2} = \frac{10}{5 + 10} = 0.667 \text{ A}$$

With V_s deactivated (short-circuited), current division determines the component of i due to I_s:
$$i'' = \frac{R_1}{R_1 + R_2} I_s = \frac{5}{5 + 10}(3) = 1 \text{ A}$$

By superposition, the total current is
$$i = i' + i'' = 0.667 + 1 = 1.667 \text{ A}$$

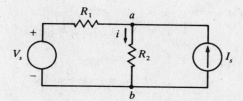

Fig. 1-10

1.39 In Fig. 1-10, assume all circuit values as in Problem 1.38 except that $R_2 = 0.25i$ Ω. Determine the current i using the method of node voltages.

▮ The voltage-current relationship for R_2 is
$$v_{ab} = R_2 i = (0.25i)(i) = 0.25i^2$$
so that
$$i = 2\sqrt{v_{ab}} \qquad (1)$$

Applying the method of node voltages at a and using (1), we get
$$\frac{v_{ab} - V_s}{R_1} + 2\sqrt{v_{ab}} - I_s = 0$$

Rearrangement and substitution of given values lead to
$$v_{ab} + 10\sqrt{v_{ab}} - 25 = 0$$

Letting $x^2 = v_{ab}$ and applying the quadratic formula, we obtain
$$x = \frac{-10 \pm \sqrt{(10)^2 - 4(-25)}}{2} = 2.071 \text{ or } -12.071$$

The negative root is extraneous, since the resulting value of v_{ab} would not satisfy KVL; thus,
$$v_{ab} = (2.071)^2 = 4.289 \text{ V} \qquad \text{and} \qquad i = 2 \times 2.071 = 4.142 \text{ A}$$

Notice that, because the resistance R_2 is a function of current, the circuit is not linear and the superposition theorem cannot be applied.

1.40 For the circuit of Fig. 1-11, find v_{ab} if $k = 0$. Do not use network theorems to simplify the circuit prior to solution.

▮ For $k = 0$, current i can be determined immediately with Ohm's law:
$$i = \frac{10}{500} = 0.02 \text{ A}$$

Since the output of the controlled current source flows through the parallel combination of two 100-Ω resistors, we have
$$v_{ab} = -(100i)(100 \parallel 100) = -100 \times 0.02 \frac{(100)(100)}{100 + 100} = -100 \text{ V} \qquad (1)$$

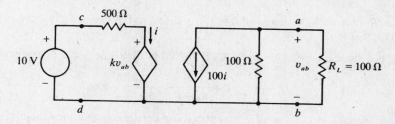

Fig. 1-11

1.41 Solve Problem 1.40 if $k = 0.01$ and all else is unchanged.

▮ With $k \neq 0$, it is necessary to solve two simultaneous equations with unknowns i and v_{ab}. Around the left loop, KVL yields
$$0.01 v_{ab} + 500i = 10 \qquad (1)$$

With i unknown, (1) of Problem 1.40 becomes

$$v_{ab} + 5000i = 0 \tag{2}$$

Solving (1) and (2) simultaneously by Cramer's rule leads to

$$v_{ab} = \frac{\begin{vmatrix} 10 & 500 \\ 0 & 5000 \end{vmatrix}}{\begin{vmatrix} 0.01 & 500 \\ 1 & 5000 \end{vmatrix}} = \frac{50{,}000}{-450} = -111.1 \text{ V}$$

1.42 For the circuit of Fig. 1-12, find i_L by the method of node voltages if $\alpha = 0.9$.

▌ With v_2 and v_{ab} as unknowns and summing currents at node c, we obtain

$$\frac{v_2 - v_s}{R_1} + \frac{v_2}{R_2} + \frac{v_2 - v_{ab}}{R_3} + \alpha i = 0 \tag{1}$$

But

$$i = \frac{v_s - v_2}{R_1} \tag{2}$$

Substituting (2) into (1) and rearranging give

$$\left(\frac{1-\alpha}{R_1} + \frac{1}{R_2} + \frac{1}{R_3}\right)v_2 - \frac{1}{R_3}v_{ab} = \frac{1-\alpha}{R_1}v_s \tag{3}$$

Now, summation of currents at node a gives

$$\frac{v_{ab} - v_2}{R_3} - \alpha i + \frac{v_{ab}}{R_L} = 0 \tag{4}$$

Substituting (2) into (4) and rearranging yield

$$-\left(\frac{1}{R_3} - \frac{\alpha}{R_1}\right)v_2 + \left(\frac{1}{R_3} + \frac{1}{R_L}\right)v_{ab} = \frac{\alpha}{R_1}v_s \tag{5}$$

Substitution of given values into (3) and (5) and application of Cramer's rule finally yield

$$v_{ab} = \frac{\begin{vmatrix} 2.1 & 0.1v_s \\ -0.1 & 0.9v_s \end{vmatrix}}{\begin{vmatrix} 2.1 & -1 \\ -0.1 & 1.1 \end{vmatrix}} = \frac{1.9v_s}{2.21} = 0.8597 v_s$$

and, by Ohm's law,

$$i_L = \frac{v_{ab}}{R_L} = \frac{0.8597 v_s}{10} = 0.08597 v_s \text{ A}$$

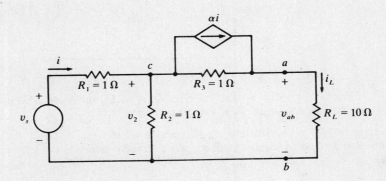

Fig. 1-12

1.43 Rework Problem 1.42 with $\alpha = 0$.

▌ With the given values (including $\alpha = 0$) substituted into (3) and (5) of Problem 1.42, Cramer's rule is used to find

$$v_{ab} = \frac{\begin{vmatrix} 3 & v_s \\ -1 & 0 \end{vmatrix}}{\begin{vmatrix} 3 & -1 \\ -1 & 1.1 \end{vmatrix}} = \frac{v_s}{2.3} = 0.4348 v_s$$

Then i_L is again found with Ohm's law:

$$i_L = \frac{v_{ab}}{R_L} = \frac{0.4348 v_s}{10} = 0.04348 v_s \text{ A}$$

1.44 *Thévenin's theorem* states that *an arbitrary linear, one-port network such as network A in Fig. 1-13(a) can be replaced at terminals 1, 2 with an equivalent series-connected voltage source V_{Th} and impedance Z_{Th} ($= R_{Th} + jX_{Th}$) as shown in Fig. 1-13(b). V_{Th} is the open-circuit voltage of network A at terminals 1, 2 and Z_{Th} is the ratio of open-circuit voltage to short-circuit current of network A determined at terminals 1, 2 with network B disconnected. If network A or B contains a controlled source, its controlling variable must be in that same network.* Alternatively, Z_{Th} is the equivalent impedance looking into network A through terminals 1, 2 with all independent sources deactivated. If network A contains a controlled source, Z_{Th} is found as the *driving-point impedance*. (See Problem 1.45).

In the circuit of Fig. 1-14, $V_A = 4$ V, $I_A = 2$ A, $R_1 = 2\,\Omega$, and $R_2 = 3\,\Omega$. Find the Thévenin equivalent voltage V_{Th} and impedance Z_{Th} for the network to the left of terminals 1, 2.

▌ With terminals 1, 2 open-circuited, no current flows through R_2; thus, by KVL,

$$V_{Th} = V_{12} = V_A + I_A R_1 = 4 + (2)(2) = 8 \text{ V}$$

The Thévenin impedance Z_{Th} is found as the equivalent impedance for the circuit to the left of terminals 1, 2 with the independent sources deactivated (that is, with V_A replaced by a short circuit and I_A replaced by an open circuit):

$$Z_{Th} = R_{Th} = R_1 + R_2 = 2 + 3 = 5\,\Omega$$

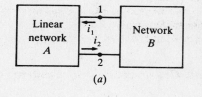 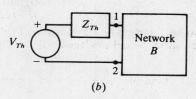 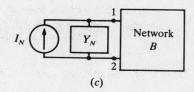

(a) (b) (c)

Fig. 1-13

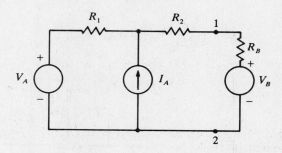

Fig. 1-14

1.45 In the circuit of Fig. 1-15(a), $V_A = 4$ V, $\alpha = 0.25$ A/V, $R_1 = 2\,\Omega$, and $R_2 = 3\,\Omega$. Find the Thévenin equivalent voltage and impedance for the network to the left of terminals 1, 2.

▌ With terminals 1, 2 open-circuited, no current flows through R_2. But the control variable V_L for the voltage-controlled dependent source is still contained in the network to the left of terminals 1, 2. Application of KVL yields

$$V_{Th} = V_L = V_A + \alpha V_{Th} R_1$$

so that

$$V_{Th} = \frac{V_A}{1 - \alpha R_1} = \frac{4}{1 - (0.25)(2)} = 8 \text{ V}$$

Since the network to the left of terminals 1, 2 contains a controlled source, Z_{Th} is found as the driving-point impedance V_{dp}/I_{dp}, with the network to the right of terminals 1, 2 in Fig. 1-15(a) replaced by the driving-point source of Fig. 1-15(b) and V_A deactivated (short-circuited). After these changes, KCL applied at node a gives

$$I_1 = \alpha V_{dp} + I_{dp} \tag{1}$$

Application of KVL around the outer loop of this circuit (with V_A still deactivated) yields

$$V_{dp} = I_{dp} R_2 + I_1 R_1 \tag{2}$$

Substitution of (1) into (2) allows solution for Z_{Th} as

$$Z_{Th} = \frac{V_{dp}}{I_{dp}} = \frac{R_1 + R_2}{1 - \alpha R_1} = \frac{2+3}{1-(0.25)(2)} = 10\,\Omega$$

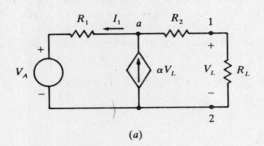 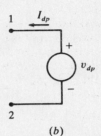

(a)　　　　　　　　　　　　　(b)　　　　**Fig. 1-15**

1.46 *Norton's theorem* states that *an arbitrary linear, one-port network such as network A in Fig. 1-13(a) can be replaced at terminals 1, 2 by an equivalent parallel-connected current source I_N and admittance Y_N as shown in Fig. 1-13(c). I_N is the short-circuit current that flows from terminal 1 to terminal 2 due to network A, and Y_N is the ratio of short-circuit current to open-circuit voltage at terminals 1, 2 with network B disconnected. If network A or B contains a controlled source, its controlling variable must be in that same network.* It is apparent that $Y_N = 1/Z_{Th}$; thus, any method for determining Z_{Th} is equally valid for finding Y_N.

Find the Norton equivalent current I_N and admittance Y_N for the circuit of Fig. 1-14 with values as given in Problem 1.44.

▌ The Norton current is found as the short-circuit current from terminal 1 to terminal 2 by superposition; it is

$$I_N = I_{12} = \text{current due to } V_A + \text{current due to } I_A = \frac{V_A}{R_1 + R_2} + \frac{R_1 I_A}{R_1 + R_2}$$

$$= \frac{4}{2+3} + \frac{(2)(2)}{2+3} = 1.6\,\text{A}$$

The Norton admittance is found from the result of Problem 1.44 as

$$Y_N = \frac{1}{Z_{Th}} = \frac{1}{5} = 0.2\,\text{S}$$

1.47 If $V_1 = 10$ V, $V_2 = 15$ V, $R_1 = 4\,\Omega$, and $R_2 = 6\,\Omega$ in the circuit of Fig. 1-16, find the Thévenin equivalent for the network to the left of terminals a, b.

▌ With terminals a, b open-circuited, only loop current I flows. Then, by KVL,

$$V_1 - IR_1 = V_2 + IR_2$$

so that

$$I = \frac{V_1 - V_2}{R_1 + R_2} = \frac{10 - 15}{4 + 6} = -0.5\,\text{A}$$

The Thévenin equivalent voltage is then

$$V_{Th} = V_{ab} = V_1 - IR_1 = 10 - (-0.5)(4) = 12\,\text{V}$$

Deactivating (shorting) the independent voltage sources V_1 and V_2 gives the Thévenin impedance to the left of terminals a, b as

$$Z_{Th} = R_{Th} = R_1 \| R_2 = \frac{R_1 R_2}{R_1 + R_2} = \frac{(4)(6)}{4+6} = 2.4 \, \Omega$$

V_{Th} and Z_{Th} are connected as in Fig. 1-13(b) to produce the Thévenin equivalent circuit.

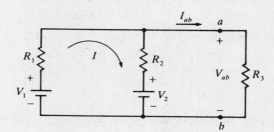

Fig. 1-16

1.48 For the circuit and values of Problem 1.47, find the Norton equivalent for the network to the left of terminals a, b.

▮ With terminals a, b shorted, the component of current I_{ab} due to V_1 alone is

$$I'_{ab} = \frac{V_1}{R_1} = \frac{10}{4} = 2.5 \text{ A}$$

Similarly, the component due to V_2 alone is

$$I''_{ab} = \frac{V_2}{R_2} = \frac{15}{6} = 2.5 \text{ A}$$

Then, by superposition,

$$I_N = I_{ab} = I'_{ab} + I''_{ab} = 2.5 + 2.5 = 5 \text{ A}$$

Now, with R_{Th} as found in Problem 1.47,

$$Y_N = \frac{1}{R_{Th}} = \frac{1}{2.4} = 0.4167 \text{ S}$$

I_N and Y_N are connected as in Fig. 1-13(c) to produce the Norton equivalent circuit.

1.49 For the circuit and values of Problems 1.47 and 1.48, find the Thévenin impedance as the ratio of open-circuit voltage to short-circuit current to illustrate the equivalence of the results.

▮ The open-circuit voltage is V_{Th} as found in Problem 1.47, and the short-circuit current is I_N from Problem 1.48. Thus,

$$Z_{Th} = \frac{V_{Th}}{I_N} = \frac{12}{5} = 2.4 \, \Omega$$

which checks with the result of Problem 1.47.

1.50 Thévenin's and Norton's theorems are applicable to other than dc steady-state circuits. For the "frequency-domain" circuit of Fig. 1-17 (where s is frequency), find the Thévenin equivalent of the circuit to the right of terminals a, b.

▮ With terminals a, b open-circuited, only loop current $I(s)$ flows; by KVL and Ohm's law, with all currents and voltages understood to be functions of s, we have

$$I = \frac{V_2 - V_1}{sL + 1/sC}$$

Now KVL gives

$$V_{Th} = V_{ab} = V_1 + sLI = V_1 + \frac{sL(V_2 - V_1)}{sL + 1/sC} = \frac{V_1 + s^2 LC V_2}{s^2 LC + 1}$$

With the independent sources deactivated, the Thévenin impedance can be determined as

$$Z_{Th} = sL \parallel \frac{1}{sC} = \frac{sL(1/sC)}{sL + 1/sC} = \frac{sL}{s^2LC + 1}$$

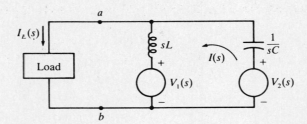

Fig. 1-17

1.51 For the circuit of Fig. 1-17, find the Norton equivalent of the circuit to the right of terminals a, b.

▌ Based on the results of Problem 1.50, the Norton current can be found as

$$I_N = \frac{V_{Th}}{Z_{Th}} = \frac{\dfrac{V_1 + s^2LCV_2}{s^2LC + 1}}{\dfrac{sL}{s^2LC + 1}} = \frac{V_1 + s^2LCV_2}{sL}$$

and the Norton admittance as

$$Y_N = \frac{1}{Z_{Th}} = \frac{s^2LC + 1}{sL}$$

1.52 Find the elements of the Thévenin's equivalent circuit looking into the one-port network of Fig. 1-5.

▌ With the port open-circuited, $I_1 = I_2 = 0$, whence $V_1 = V_{Th} = 0$. Connect a driving point voltage source so that $V_{dp} = V_1$. Then the results of Problems 1.19 and 1.20 are applicable with V_1 replaced by V_{dp} and

$$Z_{Th} = \frac{V_1}{I_1} = \frac{V_{dp}}{V_{dp}/5} = 5 \, \Omega$$

1.53 Determine the elements of the Thévenin's equivalent circuit looking to the right through terminal pair a, b of Fig. 1-6.

▌ With terminals a, b open-circuited, $i_2 = 0$ and $i_3 = i_s = 2$ A; thus,

$$V_{Th} = V_{ab} = 18i_3 = 18(2) = 36 \text{ V}$$

Deactivating i_s (open circuit),

$$Z_{Th} = 12 + 18 = 30 \, \Omega$$

1.54 Calculate the elements of a Norton's equivalent circuit looking to the right through terminals a, b of Fig. 1-6.

▌ Based on the results of Problem 1.53,

$$I_N = \frac{V_{Th}}{Z_{Th}} = \frac{36}{30} = 1.2 \text{ A} \qquad Y_N = \frac{1}{Z_{Th}} = \frac{1}{30} = 33.33 \text{ mS}$$

1.55 For the circuit of Fig. 1-7, find the elements of a Norton's equivalent circuit looking to the left through terminals a, b.

▌ With the terminals a, b shorted, the Norton current follows from use of superposition argument.

$$I_N = I_{ab} = \frac{V_1}{2} + \frac{V_2}{R_3} = \frac{3}{2} + \frac{2}{R_3} = \frac{3R_3 + 4}{2R_3}$$

Deactivating (shorting) V_1 and V_2,

$$Y_N = \frac{1}{R_3 \parallel 2} = \frac{R_3 + 2}{2R_3}$$

16 ☐ CHAPTER 1

1.56 Specify the elements of a Thévenin equivalent circuit looking to the left through terminals a, b of Fig. 1-7.

▮ Based on the results of Problem 1.55,

$$Z_{Th} = \frac{1}{Y_N} = \frac{2R_3}{R_3 + 2} \qquad V_{Th} = I_N Z_{Th} = \left(\frac{3R_3 + 4}{2R_3}\right)\left(\frac{2R_3}{R_3 + 2}\right) = \frac{3R_3 + 4}{R_3 + 2}$$

1.57 Find the elements of a Norton's equivalent circuit looking to the right through terminals a, b of Fig. 1-7.

▮ With terminals a, b shorted,

$$I_N = I_{ab} = -V_3/1 = -1/1 = -1 \text{ A}$$

Deactivating V_3, $\qquad Y_N = 1/1 = 1 \text{ S}$

The negative sign on I_N indicates that the arrow of the Norton current source must be oriented (downward) so that its current would flow from b to a if the terminal a, b were shorted.

1.58 Find the Thévenin's equivalent circuit looking to the right through terminals a, b of Fig. 1-8.

▮ With terminal a, b open-circuited, $i_1 = 0$ and the controlled source has a zero value of voltage across its terminals; thus,

$$V_{ab} = V_{Th} = I(2) = (2)(2) = 4 \text{ V}$$

Placing a driving point source such that $V_{dp} = V_{ab}$ and deactivating $I\,(=0)$, KVL yields

$$V_{dp} = (1)i_1 + (2)i_2 + 2i_1 = (1)I_{dp} + (2)I_{dp} + (2)I_{dp}$$

Whence, $\qquad Z_{Th} = \dfrac{V_{dp}}{I_{dp}} = 5 \, \Omega$

1.59 Determine the value of I_N as the short circuit current I_{ab} for a Norton's equivalent looking to the right through terminals a, b of Fig. 1-8.

▮ With terminals a, b shorted, KVL at node 1 requires that

$$I_N = I_{ab} = I - i_2 = I - \frac{v - 2i_1}{2} = I - \frac{(1)(I_{ab}) - 2(-I_{ab})}{2} = I - \frac{3}{2} I_{ab}$$

or $\qquad I_N = I_{ab} = \tfrac{2}{5} I = \tfrac{2}{5}(2) = 0.8 \text{ A}$

1.60 Use the Thévenin's equivalent circuit determined in Problem 1.58 to verify the results of Problem 1.34.

▮ With the Thévenin's circuit in place, KVL yields

$$V_1 = i_1 Z_{Th} + V_{Th} \qquad \text{or} \qquad i_1 = \frac{V_1 - V_{Th}}{Z_{Th}} = \frac{10 - 4}{5} = 1.2 \text{ A}$$

which checks with the result of Problem 1.34.

1.61 Find a Thévenin's equivalent into terminals c, d (I removed) for the circuit of Fig. 1-8.

▮ With terminals c, d open-circuited, KVL yields

$$V_1 = i_1(1) + i_2(2) + 2i_1 = i_1 + 2i_1 + 2i_1 \qquad \text{or} \qquad i_1 = \frac{V_1}{5} = \frac{10}{5} = 2 \text{ A}$$

and $\qquad V_{Th} = V_{cd} = V_1 - i_1(1) = 10 - (2)(1) = 8 \text{ V}$

With V_1 deactivated (shorted), connect a driving point source such that $V_{dp} = V_{cd}$. Then, by the method of node voltages,

$$I_{dp} = \frac{V_{dp}}{1} + \frac{V_{dp} - 2i_1}{2} = V_{dp} + \frac{V_{dp} - 2(-V_{dp}/1)}{2}$$

Whence, $\qquad Z_{Th} = \dfrac{V_{dp}}{I_{dp}} = \dfrac{2}{5} = 0.4 \, \Omega$

1.62 Since power is a nonlinear function of current, superposition of power calculations cannot be used. Illustrate this fact by first incorrectly calculating the power dissipated by the 18-Ω resistor of Fig. 1-6 as each component of i_3 due to v_s and i_s. Then correctly calculate the power dissipated by the total i_3.

▌ The results of Problem 1.26 are pertinent.

Incorrect: $\qquad P \neq (i_3')^2 18 + (i_3'')^2 18 = (0.2)^2(18) + (0.8)^2(18) = 12.24 \text{ W}$

Correct: $\qquad P = i_3^2(18) = (1)^2(18) = 18 \text{ W}$

1.63 Find the elements of the Thévenin's equivalent circuit looking to the right through terminals a, b of Fig. 1-18.

▌ With a, b open-circuited,

$$V_{Th} = V_{ab} = 6 + (3)(3) + 4V_x = 6 + 9 + 4(3)(3) = 51 \text{ V}$$

Deactivate the independent sources and connect a driving point voltage source such that $V_{dp} = V_{ab}$. Then, by KVL,

$$V_{dp} = 3I_{dp} + 4V_x = 3I_{dp} + 4(3I_{dp})$$

Whence, $\qquad Z_{Th} = \dfrac{V_{dp}}{I_{dp}} = 15 \ \Omega$

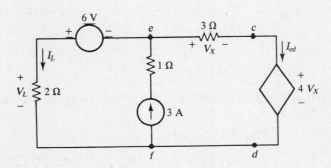

Fig. 1-18

1.64 Calculate current I_L for the circuit of Fig. 1-18.

▌ Using the elements of the Thévenin's equivalent circuit found in Problem 1.63,

$$V_{Th} = I_L(2 + Z_{Th}) \quad \text{or} \quad I_L = \frac{V_{Th}}{2 + Z_{Th}} = \frac{51}{2 + 15} = 3 \text{ A}$$

1.65 Determine the elements of the Thévenin's equivalent circuit looking to the left through terminals c, d for the circuit of Fig. 1-18.

▌ With c, d open-circuited, KVL yields

$$V_{Th} = V_{cd} = -6 + (2)(3) = 0$$

With the independent sources deactivated,

$$Z_{Th} = 3 + 2 = 5 \ \Omega$$

1.66 Find the value of current I_{cd} for the circuit of Fig. 1-18.

▌ The Thévenin's equivalent of Problem 1.65 cannot be used since the identity of V_x has been lost. Apply the method of node voltages by summing currents at node e.

$$3 = \frac{V_{ef} + 6}{2} + \frac{V_{ef} - 4V_x}{3} \qquad (1)$$

But, $\qquad V_{ef} = V_x + 4V_x \quad \text{or} \quad V_x = V_{ef}/5 \qquad (2)$

Substitute (2) into (1) to find $V_{ef} = 0$. Hence, by (2), $V_x = 0$. Thus, $I_{cd} = 0$.

18 □ CHAPTER 1

TWO-PORT NETWORKS

1.67 The network of Fig. 1-19 is a *two-port* network if $I_1 = I_1'$ and $I_2 = I_2'$. It can be characterized by the four variables V_1, V_2, I_1 and I_2, only two of which can be independent. If V_1 and V_2 are taken as independent variables and the linear network contains no independent sources, the independent and dependent variables are related by the *open-circuit impedance parameters* (or, simply, the *z parameters*) z_{11}, z_{12}, z_{21}, and z_{22} through the equation set

$$V_1 = z_{11}I_1 + z_{12}I_2 \qquad (1)$$
$$V_2 = z_{21}I_1 + z_{22}I_2 \qquad (2)$$

Establish experimental procedures to determine the z parameters of a two-port network.

▌ Each of the z parameters can be evaluated by setting the proper current to zero (or equivalently, by open-circuiting an appropriate port of the network). They are:

$$z_{11} = \frac{V_1}{I_1}\bigg|_{I_2=0} \quad (3) \qquad z_{12} = \frac{V_1}{I_2}\bigg|_{I_1=0} \quad (4)$$

$$z_{21} = \frac{V_2}{I_1}\bigg|_{I_2=0} \quad (5) \qquad z_{22} = \frac{V_2}{I_2}\bigg|_{I_1=0} \quad (6)$$

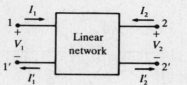

Fig. 1-19

1.68 Using the definition of the z parameters as given by (1) and (2) of Problem 1.67, find the z parameters of the network of Fig. 1-20.

▌ The loop equations for Fig. 1-20 are

$$V_1 = (Z_A + Z_C)I_1 + Z_CI_2 \qquad V_2 = Z_CI_1 + (Z_B + Z_C)I_2$$

Defining the coefficients of the currents in (1) and (2) of Problem 1.67, the z parameters are

$$z_{11} = Z_A + Z_C \qquad z_{12} = Z_C \qquad z_{21} = Z_C \qquad z_{22} = Z_B + Z_C$$

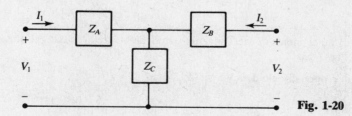

Fig. 1-20

1.69 Draw the z-parameter equivalent circuit that follows from (1) and (2) of Problem 1.67.

▌ See Fig. 1-21. Note that the z parameters are expressed in ohms and the *z-parameter model includes two controlled voltage sources*. The line joining the ports is included when a common connection is used.

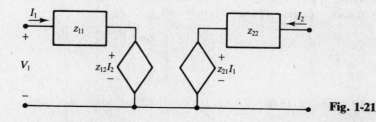

Fig. 1-21

1.70 Find the z parameters for the two-port network of Fig. 1-22.

▮ With port 2 (on the right) open-circuited, $I_2 = 0$ and the use of Problem 1.67 gives

$$z_{11} = \frac{V_1}{I_1}\bigg|_{I_2=0} = R_1 \parallel (R_2 + R_3) = \frac{R_1(R_2 + R_3)}{R_1 + R_2 + R_3}$$

Also, the current I_{R2} flowing downward through R_2 is, by current division,

$$I_{R2} = \frac{R_1}{R_1 + R_2 + R_3} I_1$$

But, by Ohm's law,

$$V_2 = I_{R2} R_2 = \frac{R_1 R_2}{R_1 + R_2 + R_3} I_1$$

Hence, by (5) of Problem 1.67,

$$z_{21} = \frac{V_2}{I_1}\bigg|_{I_2=0} = \frac{R_1 R_2}{R_1 + R_2 + R_3}$$

Similarly, with port 1 open-circuited, $I_1 = 0$ and (6) of Problem 1.67 leads to

$$z_{22} = \frac{V_2}{I_2}\bigg|_{I_1=0} = R_2 \parallel (R_1 + R_3) = \frac{R_2(R_1 + R_3)}{R_1 + R_2 + R_3}$$

The use of current division to find the current downward through R_1 yields

$$I_{R1} = \frac{R_2}{R_1 + R_2 + R_3} I_2$$

and Ohm's law gives

$$V_1 = R_1 I_{R1} = \frac{R_1 R_2}{R_1 + R_2 + R_3} I_2$$

Thus, by (4) of Problem 1.67,

$$z_{21} = \frac{V_2}{I_1}\bigg|_{I_1=0} = \frac{R_1 R_2}{R_1 + R_2 + R_3}$$

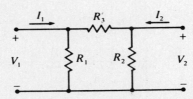

Fig. 1-22

1.71 Determine the z parameters for the two-port network of Fig. 1-23.

▮ For $I_2 = 0$, by Ohm's law,

$$I_a = \frac{V_1}{10 + 6} = \frac{V_1}{16}$$

Also, at node b, KCL gives

$$I_1 = 0.3 I_a + I_a = 1.3 I_a = 1.3 \frac{V_1}{16} \tag{1}$$

Thus, by (3) of Problem 1.67,

$$z_{11} = \frac{V_1}{I_1}\bigg|_{I_2=0} = \frac{16}{1.3} = 12.308 \; \Omega$$

Further, again by Ohm's law,

$$I_a = \frac{V_2}{6} \qquad (2)$$

Substitution of (2) into (1) yields

$$I_1 = 1.3\frac{V_2}{6}$$

so that, by (5) of Problem 1.67,

$$z_{21} = \frac{V_2}{I_1}\bigg|_{I_2=0} = \frac{6}{1.3} = 4.615\ \Omega$$

Now with $I_1 = 0$, applying KCL at node a gives us

$$I_2 = I_a = 0.3I_a = 1.3I_a \qquad (3)$$

The application of KVL then leads to

$$V_1 = V_2 - (10)(0.3I_a) = 6I_a - 3I_a = 3I_a = \frac{3I_2}{1.3}$$

so that, by (4) of Problem 1.67,

$$z_{12} = \frac{V_1}{I_2}\bigg|_{I_1=0} = \frac{3}{1.3} = 2.308\ \Omega$$

Now, substitution of (2) into (3) gives

$$I_2 = 1.3I_a = 1.3\frac{V_2}{6}$$

Hence, from (6) of Problem 1.67,

$$z_{22} = \frac{V_2}{I_2}\bigg|_{I_1=0} = \frac{6}{1.3} = 4.615\ \Omega$$

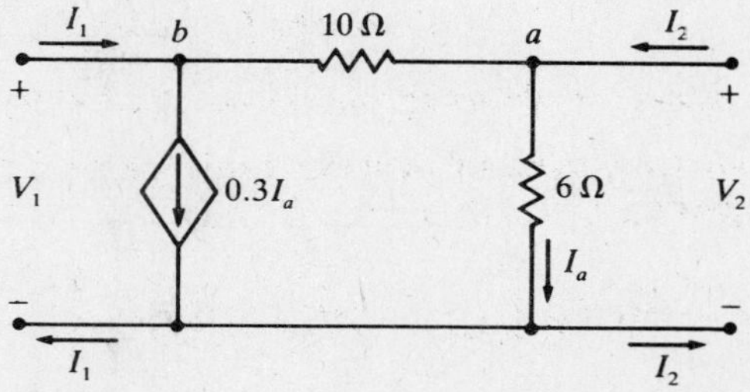

Fig. 1-23

1.72 The following open-circuit currents and voltages were determined experimentally for an unknown two-port network:

$$\left.\begin{aligned} V_1 &= 100\ \text{V} \\ V_2 &= 75\ \text{V} \\ I_1 &= 12.5\ \text{A} \end{aligned}\right|_{I_2=0} \qquad \left.\begin{aligned} V_1 &= 30\ \text{V} \\ V_2 &= 50\ \text{V} \\ I_2 &= 5\ \text{A} \end{aligned}\right|_{I_1=0}$$

Determine the z parameters.

▌

$$z_{11} = \frac{V_1}{I_1}\bigg|_{I_2=0} = \frac{100}{12.5} = 8\ \Omega \qquad z_{12} = \frac{V_1}{I_2}\bigg|_{I_1=0} = \frac{30}{5} = 6\ \Omega$$

$$z_{21} = \frac{V_2}{I_1}\bigg|_{I_2=0} = \frac{75}{12.5} = 6\ \Omega \qquad z_{22} = \frac{V_2}{I_2}\bigg|_{I_1=0} = \frac{50}{5} = 10\ \Omega$$

1.73 Draw a z-parameter model for the circuit of Problem 1.72.

▌ See Fig. 1-24.

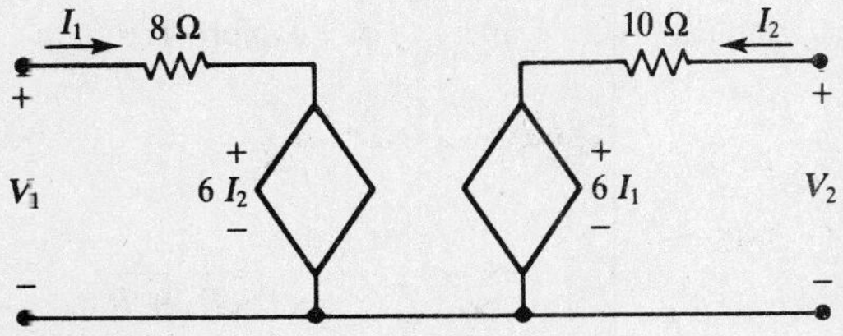

PORT-BASED CIRCUIT ANALYSIS □ 21

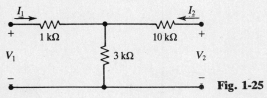

Fig. 1-24

1.74 Determine the z parameters for the network of Fig. 1-25.

▌ The loop equations become

$$V_1 = 4000I_1 + 3000I_2 \qquad V_2 = 3000I_1 + 13{,}000I_2$$

Thus, $\qquad z_{11} = 4\text{ k}\Omega \qquad z_{12} = z_{21} = 3\text{ k}\Omega \qquad z_{22} = 13\text{ k}\Omega$

Fig. 1-25

1.75 In the T network of Fig. 1-20, $Z_A = 3\,\underline{/0°}\,\Omega$, $Z_B = 4\,\underline{/90°}\,\Omega$, $Z_C = 3\,\underline{/-90°}\,\Omega$. Find the z parameters.

▌ From the results of Problem 1.68:

$$z_{11} = 3\,\underline{/0°} + 3\,\underline{/-90°} = 4.242\,\underline{/-45°}\,\Omega$$
$$z_{12} = z_{21} = Z_3 = 3\,\underline{/-90°}\,\Omega$$
$$z_{22} = 4\,\underline{/90°} + 3\,\underline{/-90°} = 1\,\underline{/90°}\,\Omega$$

1.76 In a manner similar to that of Problem 1.67, if V_1 and I_2 are taken as the independent variables, a characterization of the two-port network via the *hybrid parameters* (or, simply, the *h parameters*) results:

$$V_1 = h_{11}I_1 + h_{12}V_2 \qquad (1)$$
$$I_2 = h_{21}I_1 + h_{22}V_2 \qquad (2)$$

Establish experimental procedures to determine the h parameters of a two-port network.

▌ Two of the h parameters are determined by short-circuiting port 2, while the remaining two parameters are found by open-circuiting port 1:

$$h_{11} = \left.\frac{V_1}{I_1}\right|_{V_2=0} \quad (3) \qquad h_{12} = \left.\frac{V_1}{V_2}\right|_{I_1=0} \quad (4)$$

$$h_{21} = \left.\frac{I_2}{I_1}\right|_{V_2=0} \quad (5) \qquad h_{22} = \left.\frac{I_2}{V_2}\right|_{I_1=0} \quad (6)$$

1.77 Find the h parameters for the two-port network of Fig. 1-22.

▌ With port 2 short-circuited, $V_2 = 0$ and, by (3) of Problem 1.76,

$$h_{11} = \left.\frac{V_1}{I_1}\right|_{V_2=0} = R_1 \parallel R_3 = \frac{R_1 R_3}{R_1 + R_3}$$

By current division,

$$I_2 = -\frac{R_1}{R_1 + R_3}I_1$$

so that, by (5) of Problem 1.76,

$$h_{21} = \left.\frac{I_2}{I_1}\right|_{V_2=0} = -\frac{R_1}{R_1 + R_3}$$

If port 1 is open-circuited, voltage division and (4) of Problem 1.76 lead to

$$V_1 = \frac{R_1}{R_1 + R_3} V_2$$

and

$$h_{12} = \frac{V_1}{V_2}\bigg|_{I_1=0} = \frac{R_1}{R_1 + R_3}$$

Finally, h_{22} is the admittance looking into port 2, as given by (6) of Problem 1.76:

$$h_{22} = \frac{I_2}{V_2}\bigg|_{I_1=0} = \frac{1}{R_2 \| (R_1 + R_3)} = \frac{R_1 + R_2 + R_3}{R_2(R_1 + R_3)}$$

1.78 Draw the general h-parameter circuit model.

■ See Fig. 1-26.

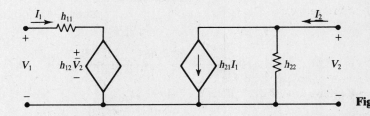

Fig. 1-26

1.79 Determine the h parameters for the two-port network of Fig. 1-23.

■ For $V_2 = 0$, $I_a \equiv 0$; thus, $I_1 = V_1/10$ and, by (3) of Problem 1.76,

$$h_{11} = \frac{V_1}{I_1}\bigg|_{V_2=0} = 10\,\Omega$$

Further, $I_2 = -I_1$ and, by (5) of Problem 1.76,

$$h_{21} = \frac{I_2}{I_1}\bigg|_{V_2=0} = -1$$

Now, $I_a = V_2/6$. With $I_1 = 0$, KVL yields

$$V_1 = V_2 - 10(0.3I_a) = V_2 - 10(0.3)\frac{V_2}{6} = \frac{1}{2}V_2$$

and, from (4) of Problem 1.76,

$$h_{12} = \frac{V_1}{V_2}\bigg|_{I_1=0} = 0.5$$

Finally, applying KCL at node a gives

$$I_2 = I_a + 0.3I_a = 1.3\frac{V_2}{6}$$

so that, by (6) of Problem 1.76,

$$h_{22} = \frac{I_2}{V_2}\bigg|_{I_1=0} = \frac{1.3}{6} = 0.2167\,\text{S}$$

1.80 Use (1) and (2) of Problem 1.67 and (3)–(6) of Problem 1.76 to find the h parameters in terms of the z parameters.

■ Setting $V_2 = 0$ in (2) of Problem 1.67 gives

$$0 = z_{21}I_1 + z_{22}I_2 \quad \text{or} \quad I_2 = -\frac{z_{21}}{z_{22}}I_1 \qquad (1)$$

from which we get

$$h_{21} = \frac{I_2}{I_1}\bigg|_{V_2=0} = -\frac{z_{21}}{z_{22}}$$

Back substitution of (1) into (1) of Problem 1.67 and use of (3) of Problem 1.76 give

$$h_{11} = \frac{V_1}{I_1}\bigg|_{V_2=0} = z_{11} - \frac{z_{12}z_{21}}{z_{22}}$$

Now, with $I_1 = 0$, (1) and (2) of Problem 1.67 become

$$V_1 = z_{12}I_2 \quad \text{and} \quad V_2 = z_{22}I_2$$

so that, from (4) of Problem 1.76,

$$h_{12} = \frac{V_1}{V_2}\bigg|_{I_1=0} = \frac{z_{12}}{z_{22}}$$

and, from (6) of Problem 1.76,

$$h_{22} = \frac{I_2}{V_2}\bigg|_{I_1=0} = \frac{I_2}{z_{22}I_2} = \frac{1}{z_{22}}$$

1.81 The h parameters of the two-port network of Fig. 1-27 are $h_{11} = 100\,\Omega$, $h_{12} = 0.0025$, $h_{21} = 20$, and $h_{22} = 1$ mS. Find the voltage-gain ratio V_2/V_1.

▎ By Ohm's law, $I_2 = -V_2/R_L$, so that (2) of Problem 1.76 may be written

$$-\frac{V_2}{R_L} = I_2 = h_{21}I_1 + h_{22}V_2$$

Solving for I_1 and substitution into (1) of Problem 1.76 give

$$V_1 = h_{11}I_1 + h_{12}V_2 = \frac{-(1/R_L + h_{22})}{h_{21}}V_2 h_{22} + h_{12}V_2$$

which can be solved for the voltage gain ratio:

$$\frac{V_2}{V_1} = \frac{1}{h_{12} - (h_{11}/h_{21})(1/R_L + h_{22})} = \frac{1}{0.0025 - (100/20)(1/2000 + 0.001)} = -200$$

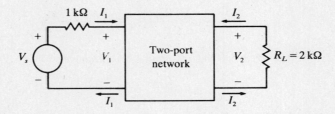

Fig. 1-27

1.82 Determine the Thévenin's equivalent voltage and impedance looking right into port 1 of the circuit of Fig. 1-27.

▎ The Thévenin voltage is V_1 of (1) of Problem 1.67 with port 1 open-circuited:

$$V_{Th} = V_1|_{I_1=0} = z_{12}I_2 \tag{1}$$

Now, by Ohm's law,

$$V_2 = -R_L I_2 \tag{2}$$

But, with $I_1 = 0$, (2) of Problem 1.67 reduces to

$$V_2 = z_{22}I_2 \tag{3}$$

Subtracting (2) from (3) leads to

$$(z_{22} + R_L)I_2 = 0 \tag{4}$$

Since, in general, $z_{22} + R_L \neq 0$, we conclude from (4) that $I_2 = 0$ and, from (1), $V_{Th} = 0$.

24 ☐ CHAPTER 1

Substituting (2) into (1) and (2) of Problem 1.67 gives

$$V_1 = z_{11}I_1 + z_{12}I_2 = z_{11}I_1 - \frac{z_{12}}{R_L}V_2 \tag{5}$$

and

$$V_2 = z_{21}I_1 + z_{22}I_2 = z_{21}I_1 - \frac{z_{22}}{R_L}V_2 \tag{6}$$

V_1 is found by solving for V_2 and substituting the result into (5):

$$V_1 = z_{11}I_1 - \frac{z_{12}z_{21}}{z_{22} + R_L}I_1$$

Then Z_{Th} is calculated as the driving-point impedance V_1/I_1:

$$Z_{Th} = \frac{V_{dp}}{I_{dp}} = \frac{V_1}{I_1} = z_{11} - \frac{z_{12}z_{21}}{z_{22} + R_L}$$

1.83 Find the Thévenin equivalent voltage and impedance looking into port 1 of the circuit of Fig. 1-27 if R_L is replaced with a current-controlled voltage source such that $V_2 = \beta I_1$, where β is a constant.

▌ As in Problem 1.82,

$$V_{Th} = V_1 \big|_{I_1 = 0} = z_{22}I_2$$

But if $I_1 = 0$, (2) of Problem 1.67 and the defining relationship for the controlled source lead to

$$V_2 = \beta I_1 = 0 = z_{22}I_2$$

from which $I_2 = 0$ and, hence, $V_{Th} = 0$.

Now we let $V_1 = V_{dp}$, so that $I_1 = I_{dp}$, and we determine Z_{Th} as the driving-point impedance. From (1) and (2) of Problem 1.67, and the defining relationship for the controlled source, we have

$$V_1 = V_{dp} = z_{11}I_{dp} + z_{12}I_2 \tag{1}$$
$$V_2 = \beta I_{dp} = z_{21}I_{dp} + z_{22}I_2 \tag{2}$$

Solving (2) for I_2 and substituting the result into (1) yields

$$V_{dp} = z_{11}I_{dp} + z_{12}\frac{\beta - z_{21}}{z_{22}}I_{dp}$$

from which Thévenin impedance is found to be

$$Z_{Th} = \frac{V_{dp}}{I_{dp}} = \frac{z_{11}z_{22} + z_{12}(\beta - z_{21})}{z_{22}}$$

1.84 Similar to the z parameters, we define the *admittance* or *y parameters* by

$$I_1 = y_{11}V_1 + y_{12}V_2 \tag{1}$$
$$I_2 = y_{21}V_1 + y_{22}V_2 \tag{2}$$

Using this definition, find the y parameters of the circuit of Fig. 1-28.

▌ From Fig. 1-28 we obtain

$$I_1 = (Y_A + Y_B)V_1 - Y_BV_2 \qquad I_2 = -Y_BV_1 + (Y_C + Y_B)V_2 \tag{3}$$

Comparing (1) and (2) with (3) yields

$$y_{11} = Y_A + Y_B \qquad y_{12} = -Y_B \qquad y_{21} = -Y_B \qquad y_{22} = Y_C + Y_B$$

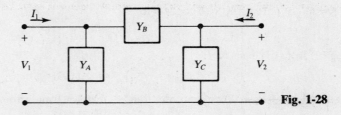

Fig. 1-28

1.85 Draw a y-parameter model for the circuit of Fig. 1-28.

⧫ See Fig. 1-29.

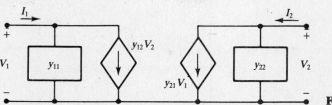

Fig. 1-29

1.86 Establish experimental procedures to determine the y parameters of a two-port network.

⧫ Inspection of (1) and (2) of Problem 1.84 shows that two of the y parameters can be determined by shorting port 1 while the other two parameters follow from shorting port 2:

$$y_{11} = \frac{I_1}{V_1}\bigg|_{V_2=0} \quad (1) \qquad y_{12} = \frac{I_1}{V_2}\bigg|_{V_1=0} \quad (2)$$

$$y_{21} = \frac{I_2}{V_1}\bigg|_{V_2=0} \quad (3) \qquad y_{22} = \frac{I_2}{V_2}\bigg|_{V_1=0} \quad (4)$$

1.87 Determine the y parameters for the circuit shown in Fig. 1-30.

⧫ The y parameters may be determined by writing the two node equations. Converting the impedances to admittances,

$$100\,\text{k}\Omega \Rightarrow 10\,\mu\text{S} \qquad 75\,\text{k}\Omega \Rightarrow 13.33\,\mu\text{S} \qquad 50\,\text{k}\Omega \Rightarrow 20\,\mu\text{S}$$

$$I_1 = (13.33 + 10)10^{-6}V_1 - (10)10^{-6}V_2 \qquad I_2 = -(10)10^{-6}V_1 + (10 + 20)10^{-6}V_2$$

Thus, the y parameters are

$$y_{11} = 23.33\,\mu\text{S} \qquad y_{12} = -10\,\mu\text{S} \qquad y_{21} = -10\,\mu\text{S} \qquad y_{22} = 30\,\mu\text{S}$$

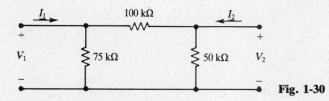

Fig. 1-30

1.88 Draw a y-parameter model of the circuit of Problem 1.87.

⧫ See Fig. 1-31.

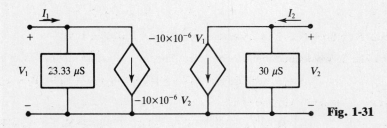

Fig. 1-31

1.89 The following short-circuit currents and voltages were determined experimentally for an unknown two-port network:

$$\begin{array}{c|c}
I_1 = 3\,\text{mA} & I_1 = -1\,\text{mA} \\
I_2 = -0.6\,\text{mA} & I_2 = 12\,\text{mA} \\
V_1 = 24\,\text{V} \quad \bigg|_{V_2=0} & V_2 = 40\,\text{V} \quad \bigg|_{V_1=0}
\end{array}$$

Determine the y parameters.

$$y_{11} = \frac{I_1}{V_1}\bigg|_{V_2=0} = \frac{0.003}{24} = 125\ \mu S \qquad y_{12} = \frac{I_1}{V_2}\bigg|_{V_1=0} = \frac{-0.001}{40} = -25\ \mu S$$

$$y_{21} = \frac{I_2}{V_1}\bigg|_{V_2=0} = \frac{-0.0006}{24} = -25\ \mu S \qquad y_{22} = \frac{I_2}{V_2}\bigg|_{V_1=0} = \frac{0.012}{40} = 300\ \mu S$$

1.90 Express the z parameters in terms of the y parameters of a two-port network.

> Solving for V_1 and V_2 in (1) and (2) of Problem 1.84 we obtain

$$V_1 = \frac{y_{22}}{\Delta_y} I_1 - \frac{y_{12}}{\Delta_y} I_2 = z_{11} I_1 + z_{12} I_2 \qquad V_2 = -\frac{y_{21}}{\Delta_y} I_1 + \frac{y_{11}}{\Delta_y} I_2 = z_{21} I_1 + z_{22} I_2$$

where
$$\Delta_y = y_{11} y_{22} - y_{12} y_{21}$$

Hence,
$$z_{11} = \frac{y_{11}}{\Delta_y} \qquad z_{12} = -\frac{y_{12}}{\Delta_y} \qquad z_{21} = -\frac{y_{21}}{\Delta_y} \qquad z_{22} = \frac{y_{22}}{\Delta_y}$$

1.91 Express the y parameters in terms of the z parameters of a two-port network.

> Solving for I_1 and I_2 in (1) and (2) of Problem 1.67, we obtain

$$I_1 = \frac{z_{22}}{\Delta_z} V_1 - \frac{z_{12}}{\Delta_z} V_2 = y_{11} V_1 + y_{12} V_2$$

$$I_2 = -\frac{z_{21}}{\Delta_z} V_1 + \frac{z_{11}}{\Delta_z} V_2 = y_{21} V_1 + y_{22} V_2$$

Hence,
$$y_{11} = \frac{z_{22}}{\Delta_z} \qquad y_{12} = -\frac{z_{12}}{\Delta_z} \qquad y_{21} = -\frac{z_{21}}{\Delta_z} \qquad y_{22} = \frac{z_{11}}{\Delta_z}$$

where
$$\Delta_z = z_{11} z_{22} - z_{12} z_{21}$$

1.92 For a two-port network show that $z_{11} y_{11} = z_{22} y_{22}$.

> From Problems 1.90 and 1.91 we obtain

$$z_{11} y_{11} = \left(\frac{y_{11}}{\Delta_y}\right)\left(\frac{z_{22}}{\Delta_z}\right) = \left(\frac{y_{11}}{\Delta_y}\right) y_{11} \tag{1}$$

$$z_{22} y_{22} = \left(\frac{y_{22}}{\Delta_y}\right)\left(\frac{z_{11}}{\Delta_z}\right) = \left(\frac{y_{22}}{\Delta_y}\right) y_{22} \tag{2}$$

Dividing (1) by (2) yields

$$\frac{z_{22}/\Delta_z}{y_{11}} = \frac{z_{11}/\Delta_z}{y_{22}}$$

Hence,
$$z_{11} y_{11} = z_{22} y_{22}$$

1.93 In Problem 1.76 we defined the h parameters. The inverse of the h parameters is the g parameters defined by

$$I_1 = g_{11} V_1 + g_{12} I_2 \tag{1}$$
$$V_2 = g_{21} V_1 + g_{22} I_2 \tag{2}$$

Establish experimental procedures to determine the g parameters.

> The g parameters are obtained from open-circuit and short-circuit tests as given by

$$g_{11} = \frac{I_1}{V_1}\bigg|_{I_2=0} \qquad g_{21} = \frac{V_2}{V_1}\bigg|_{I_2=0} \qquad g_{12} = \frac{I_1}{I_2}\bigg|_{V_1=0} \qquad g_{22} = \frac{V_2}{I_2}\bigg|_{V_1=0}$$

1.94 Draw a g-parameter model to represent (1) and (2) of Problem 1.93.

> See Fig. 1-32.

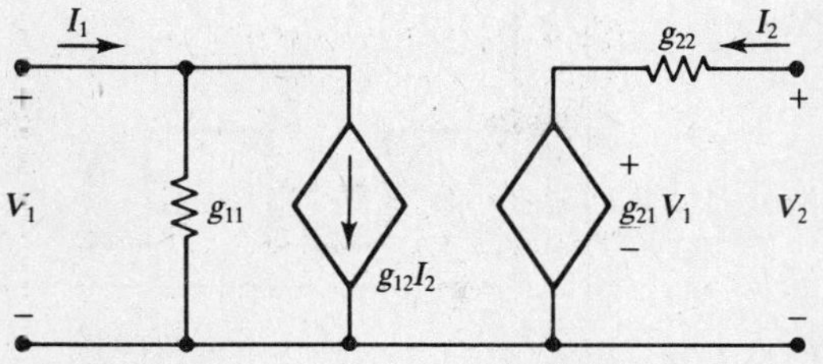

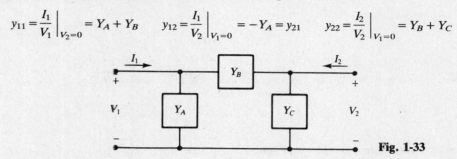

Fig. 1-32

1.95 Determine the y parameters for the general Π network shown in Fig. 1-33.

▌ Using the experimental procedure of Problem 1.86, from Fig. 1-33 we have

$$y_{11} = \frac{I_1}{V_1}\bigg|_{V_2=0} = Y_A + Y_B \qquad y_{12} = \frac{I_1}{V_2}\bigg|_{V_1=0} = -Y_A = y_{21} \qquad y_{22} = \frac{I_2}{V_2}\bigg|_{V_1=0} = Y_B + Y_C$$

Fig. 1-33

1.96 In a Π network (see Fig. 1-33), we have $Y_A = 0.2 \times 10^{-3}\,\underline{/0°}$ S, $Y_B = 0.02 \times 10^{-3}\,\underline{/-90°}$ S, and $Y_C = 0.25 \times 10^{-3}\,\underline{/90°}$ S. Find the y parameters.

▌
$$y_{11} = Y_A + Y_B = (0.2 - j0.02)10^{-3}$$
$$y_{12} = y_{21} = -Y_B = -(-j0.02)10^{-3} = (j0.02)10^{-3}$$
$$y_{22} = Y_B + Y_C = (-j0.02 + j0.25)10^{-3} = (j0.23)10^{-3}$$

1.97 Find the g-parameter values for the circuit of Fig. 1-25.

▌ Based on Problem 1.93,

$$g_{11} = \frac{I_1}{V_1}\bigg|_{I_2=0} = \frac{1}{1 \times 10^3 + 3 \times 10^3} = 2.5 \times 10^{-4}\ \text{S}$$

$$g_{12} = \frac{I_1}{I_2}\bigg|_{V_1=0} = \frac{-\dfrac{3 \times 10^3}{3 \times 10^3 + 1 \times 10^3} I_2}{I_2} = -0.75 \qquad \text{by current division}$$

$$g_{21} = \frac{V_2}{V_1}\bigg|_{I_2=0} = \frac{\dfrac{3 \times 10^3}{3 \times 10^3 + 1 \times 10^3} V_1}{V_1} = 0.75 \qquad \text{by voltage division}$$

$$g_{22} = \frac{V_2}{I_2}\bigg|_{V_1=0} = 10 \times 10^3 + \frac{(1 \times 10^3)(3 \times 10^3)}{1 \times 10^3 + 3 \times 10^3} = 1.75\ \text{k}\Omega$$

AVERAGE AND RMS VALUES

1.98 For any time-varying function $f(t)$ with period T, the *average* value over one period is given by

$$F_0 = \frac{1}{T}\int_{t_0}^{t_0+T} f(t)\, dt \tag{1}$$

where F_0 is independent of t_0. Use (1) to find the average value of the periodic waveform of Fig. 1-34.

▌ The integral of (1) is simply the area under the $v(t)$ curve for one period. We can, then, find the average current as

$$V_0 = \frac{1}{T}\left(4 \times \frac{T}{2} + 1 \times \frac{T}{2}\right) = 2.5\ \text{V}$$

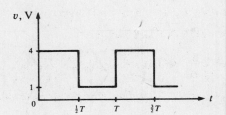

Fig. 1-34

1.99 Since the average value of a sinusoidal function of time is zero, the *half-cycle* average value, which is nonzero, is often useful. Find the half-cycle average value of the current through a resistance R connected directly across a periodic (*ac*) voltage source $v(t) = V_m \sin \omega t$.

▌ By Ohm's law,

$$i(t) = \frac{v(t)}{R} = \frac{V_m}{R} \sin \omega t$$

and from (1) of Problem 1.98, applied over the half cycle from $t_0 = 0$ to $T/2 = \pi$,

$$I_0 = \frac{1}{\pi} \int_0^\pi \frac{V_m}{R} \sin \omega t \, d(\omega t) = \frac{1}{\pi} \frac{V_m}{R} [-\cos \omega t]_{\omega t=0}^\pi = \frac{2}{\pi} \frac{V_m}{R} \qquad (1)$$

1.100 For any time-varying function $f(t)$ with period T, the *root-mean-square* (or *rms*) value is defined as

$$F = \left(\frac{1}{T} \int_{t_0}^{t_0+T} f^2(t) \, dt \right)^{1/2} \qquad (1)$$

where F is independent of t_0. Apply (1) to find the rms value of the periodic waveform of Fig. 1-34.

▌ The integral in (1) is no more than the area under the $v^2(t)$ curve. Hence,

$$V = \left[\frac{1}{T} \left(4^2 \frac{T}{2} + 1^2 \frac{T}{2} \right) \right]^{1/2} = 4.25 \text{ V}$$

1.101 The periodic waveform of Fig. 1-35 is composed of segments of a sinusoid. Find the average value of the current.

▌ Because $i(t) = 0$ for $0 \le \omega t < \alpha$, the average value of the current is, according to (1) of Problem 1.98,

$$I_0 = \frac{1}{\pi} \int_\alpha^\pi I_m \sin \omega t \, d(\omega t) = \frac{I_m}{\pi} [-\cos \omega t]_{\omega t=\alpha}^\pi = \frac{I_m}{\pi} (1 + \cos \alpha)$$

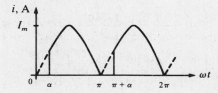

Fig. 1-35

1.102 Determine the rms value of the current depicted by Fig. 1-35.

▌ By (1) of Problem 1.100 and the identity $\sin^2 x = \frac{1}{2}(1 - \cos 2x)$,

$$I^2 = \frac{1}{\pi} \int_\alpha^\pi I_m^2 \sin^2(\omega t) \, d(\omega t) = \frac{I_m^2}{2\pi} \int_\alpha^\pi (1 - \cos 2\omega t) \, d(\omega t)$$

$$= \frac{I_m^2}{2\pi} \left[\omega t - \frac{1}{2} \sin 2\omega t \right]_{\omega t=\alpha}^\pi = \frac{I_m^2}{2\pi} \left(\pi - \alpha + \frac{1}{2} \sin 2\alpha \right)$$

so that

$$I = I_m \left(\frac{\pi - \alpha + \frac{1}{2} \sin 2\alpha}{2\pi} \right)^{1/2}$$

1.103 Calculate the average value of the current $i(t) = 4 + 10 \sin \omega t$ A.

▌ Since $i(t)$ has period 2π, (1) of Problem 1.98 gives

$$I_0 = \frac{1}{2\pi} \int_0^{2\pi} (4 + 10 \sin \omega t) \, d(\omega t) = \frac{1}{2\pi} [4\omega t - 10 \cos \omega t]_{\omega t = 0}^{2\pi} = 4 \text{ A}$$

This result was to be expected, since the average value of a sinusoidal waveform over one cycle is zero.

1.104 Find the rms value of the current of Problem 1.103.

▌ Equation (1) of Problem 1.100 and the identity $\sin^2 x = \frac{1}{2}(1 - \cos 2x)$ provide the rms value of $i(t)$:

$$I^2 = \frac{1}{2\pi} \int_0^{2\pi} (4 + 10 \sin \omega t)^2 \, d(\omega t) = \frac{1}{2\pi} \int_0^{2\pi} (16 + 80 \sin \omega t + 50 - 50 \cos 2\omega t) \, d(\omega t)$$

$$= \frac{1}{2\pi} \left[66 \omega t - 80 \cos \omega t - \frac{50}{2} \sin 2\omega t \right]_{\omega t = 0}^{2\pi} = 66$$

so that $\qquad I = \sqrt{66} = 8.125$ A

1.105 Consider a resistance R connected directly across a dc-voltage source V_{dc}. The power absorbed by R is

$$P_{dc} = \frac{V_{dc}^2}{R}$$

Now replace V_{dc} with an ac-voltage source, $v(t) = V_m \sin \omega t$, and determine the average value of power dissipated by R. Draw an association between V_{dc} and the rms value of $v(t)$.

▌ The instantaneous power is now given by

$$p(t) = \frac{v^2(t)}{R} = \frac{V_m^2}{R} \sin^2 \omega t \qquad (1)$$

Hence, the *average power* over one period is, by (1) of Problem 1.98,

$$P_0 = \frac{1}{2\pi} \int_0^{2\pi} \frac{V_m^2}{R} \sin^2 \omega t \, d(\omega t) = \frac{V_m^2}{2R} \qquad (2)$$

Comparing (1) and (2), we see that, insofar as power dissipation is concerned, an ac source of amplitude V_m is equivalent to a dc source of magnitude

$$\frac{V_m}{\sqrt{2}} = \left(\frac{1}{T} \int_0^T v^2(t) \, dt \right)^{1/2} \equiv V \qquad (3)$$

For this reason, the rms value of a sinusoid, $V = V_m/\sqrt{2}$, is also called its *effective* value.

1.106 Find the average value of the power delivered to a one-port network with *passive sign convention* (that is, the current is directed from the positive to the negative terminal) if $v(t) = V_m \cos \omega t$ and $i(t) = I_m \cos(\omega t + \theta)$.

▌ The instantaneous power flow into the port is given by

$$p(t) = v(t)i(t) = V_m I_m \cos \omega t \cos(\omega t + \theta)$$
$$= \tfrac{1}{2} V_m I_m [\cos(2\omega t + \theta) + \cos \theta]$$

By (1) of Problem 1.98,

$$P_0 = \frac{1}{2\pi} \int_0^{2\pi} p(t) \, dt = \frac{V_m}{4\pi} I_m \int_0^{2\pi} [\cos(2\omega t + \theta) + \cos \theta] \, d(\omega t)$$

After the integration is performed and its limits evaluated, the result is

$$P_0 = \frac{V_m I_m}{2} \cos \theta = \frac{V_m}{\sqrt{2}} \frac{I_m}{\sqrt{2}} \cos \theta = VI \cos \theta$$

1.107 Find the rms (or effective) value of a current consisting of the sum of two sinusoidally varying functions with frequencies whose ratio is an integer.

30 □ CHAPTER 1

▮ Without loss of generality, we may write

$$i(t) = I_1 \cos \omega t + I_2 \cos k\omega t$$

where k is an integer. Applying (1) of Problem 1.100 and recalling that $\cos^2 x = \frac{1}{2}(1 + \cos 2x)$ and $\cos x \cos y = \frac{1}{2}[\cos(x+y) + \cos(x-y)]$, we obtain

$$I^2 = \frac{1}{2\pi} \int_0^{2\pi} (I_1 \cos \omega t + I_2 \cos k\omega t)^2 \, d(\omega t)$$

$$= \frac{1}{2\pi} \int_0^{2\pi} \left\{ \frac{I_1^2}{2}(1 + \cos 2\omega t) + \frac{I_2^2}{2}(1 + \cos 2k\omega t) + I_1 I_2[\cos(k+1)\omega t + \cos(k-1)\omega t] \right\} d(\omega t)$$

Performing the indicated integration and evaluating at the limits result in

$$I = \left(\frac{I_1^2}{2} + \frac{I_2^2}{2} \right)^{1/2}$$

Review Problems

1.108 Find an expression for the instantaneous power supplied to the one-port network of Problem 1.3.

▮ $$p(t) = v_1 i_1 = \left(L \frac{di_1}{dt} \right) i_1 = L i_1 \frac{di_1}{dt}$$

1.109 Find an alternate solution to Problem 1.5 in terms of v_1 rather than i_1.

▮ $$p(t) = v_1 i_1 = (v_1)(v_1/R) = v_1^2/R$$

1.110 Determine the instantaneous power dissipated by each of the 2-Ω resistors of Problem 1.7.

▮ By KVL,

$$v_{xy} = v_1 - i_1(1) = v_1 - (v_1/2)(1) = v_1/2$$

Based on Problem 1.109,

$$p_{2\Omega}(t) = \frac{v_{xy}^2}{2} = \frac{(v_1/2)^2}{2} = v_1^2/8$$

1.111 Calculate the equivalent resistance of the one-port network of Problem 1.7 using the result of Problem 1.8.

▮ By Ohm's law,

$$R_{eq} = \frac{v_1}{i_1} = \frac{v_1}{(v_1/2)} = 2 \, \Omega$$

1.112 Find the current I_2 of Fig. 1-3 using the method of current division.

▮ Based on the result of Problem 1.10,

$$I_2 = \frac{Z_2}{Z_2 + Z_3} I_1 = \frac{Z_2}{Z_2 + Z_3} \left[\frac{V_1(Z_2 + Z_3)}{Z_1 Z_2 + Z_1 Z_3 + Z_2 Z_3} \right] = \frac{V_1}{Z_1 Z_2 + Z_1 Z_3 + Z_2 Z_3}$$

1.113 Find the input impedance of the one-port network of Fig. 1-4 in terms of the generalized network impedances defined in Problem 1.16.

▮ Based on Problem 1.17,

$$Z_{eq} = Z_{in} = \frac{V_1}{I_1} = \frac{Z_{22} Z_{33} - Z_{23}^2}{Z_{11}(Z_{22} Z_{33} - Z_{23}^2) - Z_{22} Z_{13}^2 - Z_{33} Z_{12}^2 - 2 Z_{12} Z_{13} Z_{23}}$$

1.114 Determine the current I_2 for the one-port network of Fig. 1-5.

▌ Based on Problem 1.19,

$$I_2 = \frac{\Delta_2}{\Delta} = \frac{\begin{vmatrix} 2 & V_1 \\ -1 & 0 \end{vmatrix}}{\begin{vmatrix} 2 & 3 \\ -1 & 1 \end{vmatrix}} = \frac{V_1}{5}$$

1.115 Calculate the power dissipated by the 1-Ω resistor in series with the controlled source of Fig. 1-5.

▌ Based on results of Problems 1.19 and 1.114,

$$P = (I_1 - I_2)^2(1) = (V_1/5 - V_1/5)^2(1) = 0$$

1.116 The one-port network of Fig. 1-1 is made up of a series-connected resistor R and inductor L. Determine the Laplace domain port current $I_1(s)$ in terms of the Laplace domain port voltage $V_1(s)$.

▌ The equivalent impedance of the series-connected R and L is $Z_{in} = Z_{eq} = R + sL$. Hence,

$$I_1(s) = \frac{V_1(s)}{Z_{in}} = \frac{V_1(s)}{R + sL}$$

1.117 Find the instantaneous power supplied to the nonlinear, one-port network of Problem 1.22.

▌ $$p(t) = v_1 i_1 = [(i_1)(2i_1)]i_1 = 2i_1^3$$

1.118 Find the instantaneous power supplied to the one-port network of Problem 1.23.

▌ Based on Problem 1.23 results,

$$p(t) = v_1 i_1 = [k_2 \ln(i/k_1)][k_1 e^{v_1/k_2}] = k_1 k_2 e^{v_1/k_2} \ln(1/k_1)$$

1.119 Calculate the current i_2 for the circuit of Fig. 1-7 if $R_3 = 1\,\Omega$.

▌ With only V_1 active and using i_1' from Problem 1.32, current division yields

$$i_2' = -\frac{1}{R_3 + 1} i_1' = -\frac{1}{1+1}(1.2) = -0.6\,\text{A}$$

With only V_2 active, $i_2'' = 1.2$ A, and with only V_3 active, $i_3''' = 0.6$ A as determined in Problem 1.32. By current division,

$$i_2''' = \frac{2}{2 + R_3} i_3''' = \frac{2}{2+1}(0.6) = 0.4\,\text{A}$$

Applying the superposition argument,

$$i_2 = i_2' + i_2'' + i_2''' = -0.6 + 1.2 + 0.4 = 1\,\text{A}$$

1.120 Verify the result of Problem 1.119 using the method of mesh currents.

▌ Based on the equations of Problem 1.33,

$$i_B = \frac{\Delta_2}{\Delta} = \frac{\begin{vmatrix} 3 & 1 \\ -1 & 3 \end{vmatrix}}{5} = \frac{10}{5} = 2\,\text{A}$$

and

$$i_2 = i_B - i_A = 2 - 1 = 1\,\text{A}$$

1.121 Find the voltage v across the current source I of Fig. 1-8 by use of the superposition theorem.

▌ Based on Problem 1.34 with only V_1 active, KVL requires that

$$v' = V_1 - (1)i_1' = 10 - (1)(2) = 8\,\text{V}$$

▌ Hence, by superposition,

$$v = v' + v'' = 8 + 0.8 = 8.8\,\text{V}$$

1.122 Does the controlled source of Fig. 1-8 absorb or supply power to the circuit?

▎ Based on Problems 1.34 and 1.35, the power delivered to the controlled source is determined using the passive sign convention (see Problem 1.106).

$$P = (2i_1)(i_2) = [(2)(1.2)](3.2) = 7.68 \text{ W}$$

The positive sign on P indicates power flow to the controlled source, or the controlled source absorbs power from the circuit.

1.123 Use the results of Problems 1.58 to verify the answer to Problem 1.59.

▎
$$I_N = \frac{V_{Th}}{Z_{Th}} = \frac{4}{5} = 0.8 \text{ A}$$

1.124 Find a Norton's equivalent looking right through terminals a, b for the circuit of Fig. 1-18.

▎ Based on the results of Problem 1.63,

$$I_N = \frac{V_{Th}}{Z_{Th}} = \frac{51}{15} = 3.4 \text{ A} \qquad Y_N = \frac{1}{Z_{Th}} = \frac{1}{15} = 66.67 \text{ mS}$$

1.125 Use the superposition theorem to find the current i in Fig. 1-10 if $R_1 = 5\,\Omega$, $R_2 = 10\,\Omega$, $V_s = 10\cos 2t$ V, and $I_s = 3\cos(3t + \pi/4)$ A.

▎ Although the sources are of different frequencies, the circuit is linear and the superposition theorem holds. Deactivating I_s and using phasor analysis,

$$\bar{I}' = \frac{\bar{V}_s}{R_1 + R_2} = \frac{10\,/0°}{10 + 5} = 0.667\,/0°\text{ A} \qquad \text{or} \qquad i' = 0.667\cos 2t \text{ A}$$

Deactivating V_s and applying current division,

$$\bar{I}'' = \frac{R_1}{R_1 + R_2}\bar{I}_s = \frac{5}{5 + 10}\,3\,/45° = 1\,/45° \qquad \text{or} \qquad i'' = \cos(3t + \pi/4) \text{ A}$$

By superposition,

$$i = i' + i'' = 0.667\cos 2t + \cos(3t + \pi/4) \text{ A}$$

1.126 In Fig. 1-36, find the Thévenin equivalent voltage and impedance for the network to the left of terminals a, b.

▎ With terminals a, b open-circuited, $I_L = 0$ and $V_{Th} = V_{ab} = V_1 - I_2 R_2$. Deactivating the independent sources ($V_1 = 0$, $I_2 = 0$), $R_{Th} = R_1 + R_2$.

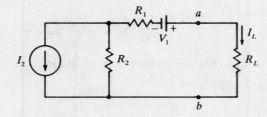

Fig. 1-36

1.127 Apply the Thévenin equivalent of Problem 1.126 to find current I_L of Fig. 1-36.

▎ With the Thévenin equivalent in place, $V_{Th} = I_L(R_{Th} + R_L)$ or

$$I_L = \frac{V_{Th}}{R_{Th} + R_L} = \frac{V_1 - I_2 R_2}{R_1 + R_2 + R_L}$$

1.128 Determine current I_L in Fig. 1-36 by use of the superposition theorem.

▎ Deactivate V_1 and apply current division.

$$I'_L = \frac{R_2}{R_1 + R_2 + R_L}I_2$$

Deactivate I_2 and use Ohm's law.

$$I''_L = \frac{V_1}{R_1 + R_2 + R_L}$$

By the superposition theorem,

$$I_L = I''_L + I'_L = \frac{V_1 - R_2 I_2}{R_1 + R_2 + R_L}$$

1.129 In the circuit of Fig. 1-17, $V_1 = 10 \cos 2t$ V, $V_2 = 20 \cos 2t$ V, $L = 1$ H, $C = 1$ F, and the load is a 1-Ω resistor. Find phasor current $\bar{I}_L$ by the method of node voltages.

▮ Since $\omega = 2$ rad/s, the impedances associated with L and C are, respectively, $Z_L = j\omega L = 2\,\underline{/90°}$ and $Z_C = -j/\omega C = 0.5\,\underline{/-90°}$. Using $\bar{V}_{ab}$ as a node voltage yields

$$\frac{\bar{V}_{ab}}{1} + \frac{\bar{V}_{ab} - 10\,\underline{/0°}}{2\,\underline{/90°}} + \frac{\bar{V}_{ab} - 20\,\underline{/0°}}{0.5\,\underline{/-90°}} = 0$$

from which $\bar{V}_{ab} = 19.4\,\underline{/33.69°}$ V; thus,

$$\bar{I}_L = \frac{\bar{V}_{ab}}{1} = 19.4\,\underline{/33.69°}\ \text{A}$$

1.130 Determine the Thévenin equivalent for the network to the right of terminals a, b of Problem 1.129.

▮ Based on the results of Problem 1.50 with $s = j\omega = j2$ rad/s,

$$\bar{V}_{Th} = \frac{\bar{V}_1 + (j\omega)^2 LC \bar{V}_2}{(j\omega)^2 LC + 1} = \frac{10\,\underline{/0°} + (j2)^2(1)(1)(20\,\underline{/0°})}{(j2)^2(1)(1) + 1} = 23.333\,\underline{/0°}$$

$$Z_{Th} = \frac{j\omega L}{(j\omega)^2 LC + 1} = \frac{(j2)(1)}{(j2)^2(1)(1) + 1} = -j0.667\ \Omega$$

1.131 Use the Thévenin equivalent of Problem 1.130 to determine $\bar{I}_L$ of Problem 1.129.

▮ By KVL,

$$\bar{V}_{Th} = \bar{I}_L(Z_{Th} + 1) \quad \text{or} \quad \bar{I}_L = \frac{\bar{V}_{Th}}{Z_{Th} + 1} = \frac{23.333\,\underline{/0°}}{-j0.667 + 1} = 19.4\,\underline{/33.69°}\ \text{A}$$

1.132 In Fig. 1-37, find the Thévenin equivalent for the bridge circuit as seen through the load resistor R_L.

▮ With terminals a, b open-circuited, $I_1 = I_2$ and $I_3 = I_4$. By KVL,

$$I_1 = \frac{V_b}{R_1 + R_2} \quad (1) \qquad I_3 = \frac{V_b}{R_3 + R_4} \quad (2)$$

Let
$$V_{ab} = V_{Th} = I_3 R_3 - I_1 R_1 \qquad (3)$$

Substitute (1) and (2) into (3) to yield

$$V_{Th} = \frac{V_b}{R_3 + R_4}(R_3) - \frac{V_b}{R_1 + R_2}(R_1) = \frac{V_b(R_2 R_3 - R_1 R_4)}{(R_1 + R_2)(R_3 + R_4)}$$

With V_b deactivated,

$$Z_{Th} = R_1 \parallel R_3 + R_2 \parallel R_4 = \frac{R_1 R_3}{R_1 + R_3} + \frac{R_2 R_4}{R_2 + R_4} = \frac{R_1 R_2 R_3 + R_1 R_3 R_4 + R_1 R_2 R_4 + R_2 R_3 R_4}{(R_1 + R_3)(R_2 + R_4)}$$

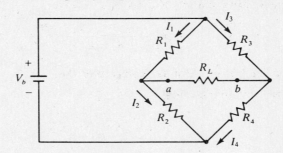

Fig. 1-37

1.133 Suppose the bridge circuit in Fig. 1-37 is balanced by letting $R_1 = R_2 = R_3 = R_4 = R$. Find the elements of the Norton equivalent circuit.

▮ Evaluating the Thévenin elements of Problem 1.132 with $R_1 = R_2 = R_3 = R_4 = R$,

$$V_{Th} = 0, \qquad Z_{Th} = \frac{R}{2} + \frac{R}{2} = R$$

Whence
$$I_N = \frac{V_{Th}}{Z_{Th}} = \frac{0}{R} = 0 \quad \text{and} \quad Y_N = \frac{1}{Z_{Th}} = \frac{1}{R}$$

1.134 For the circuit of Fig. 1-38, determine the Thévenin equivalent of the circuit to the left of terminals a, b.

▮ With terminals a, b open-circuited, $v_{ab} = V_{Th}$. By the method of node voltages,

$$\frac{30 - V_{Th}}{5} + 0.25 V_{Th} - \frac{V_{Th}}{10} = 0 \quad \text{or} \quad V_{Th} = 120 \text{ V}$$

Deactivate the 30-V source and connect a driving point source $v_{dp} = v_{ab}$ that supplies a current i_{dp}; then KCL yields

$$i_{dp} = \frac{v_{dp}}{10} - 0.25 v_{dp} + \frac{v_{dp}}{5} = 0.05 v_{dp} \quad \text{or} \quad Z_{Th} = \frac{v_{dp}}{i_{dp}} = \frac{1}{0.05} = 20 \ \Omega$$

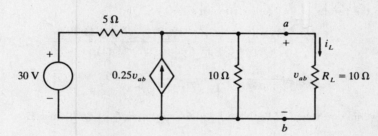

Fig. 1-38

1.135 Find current i_L of Fig. 1-38 using the Thévenin equivalent of Problem 1.135.

▮ With the Thévenin equivalent in place, KVL requires

$$i_L = \frac{V_{Th}}{Z_{Th} + R_L} = \frac{120}{20 + 10} = 4 \text{ A}$$

1.136 Verify the solution of Problem 1.135 by solving for i_L by the method of node voltages.

$$\frac{v_{ab} - 30}{5} - 0.25 v_{ab} + \frac{v_{ab}}{10} + \frac{v_{ab}}{10} = 0 = 0.15 v_{ab} - 6$$

or
$$v_{ab} = \frac{6}{0.15} = 40 \text{ V} \quad \text{and} \quad i_L = \frac{v_{ab}}{R_L} = \frac{40}{10} = 4 \text{ A}$$

1.137 In the circuit of Fig. 1-39, let $R_1 = R_2 = R_C = 1 \ \Omega$. Find the Thévenin equivalent for the circuit to the right of terminals a, b if $v_C = 0.5 i_1$.

▮ With terminals a, b open-circuited $i_1 = 0$; thus, $v_C = 0$ and $i_2 = 0$. Whence, $V_{Th} = v_C - i_2 R_2 = 0$.
Connect a driving point source such that $v_{dp} = v_{ab}$ and $i_{dp} = i_1$. Apply KVL first around the left loop and then around the outer loop to give

$$v_{dp} = R_1 i_{dp} + R_C(i_{dp} - i_2) + v_C \qquad (1)$$
$$v_{dp} = R_1 i_{dp} + R_2 i_2 \qquad (2)$$

Substitute $v_C = 0.5 i_1 = 0.5 i_{dp}$ into (1) and set $R_1 = R_2 = R_C = 1 \ \Omega$ to obtain

$$v_{dp} = 2.5 i_{dp} - i_2 \qquad (3)$$
$$v_{dp} = i_{dp} + i_2 \qquad (4)$$

Add (3) and (4) to find

$$Z_{Th} = \frac{v_{dp}}{i_{dp}} = \frac{3.5}{2} = 1.75 \, \Omega$$

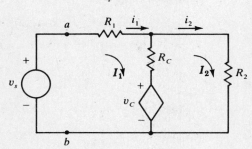

Fig. 1-39

1.138 Find the Thévenin equivalent to the right of terminals a, b for the circuit of Fig. 1-39 if $R_1 = R_2 = R_C = 1 \, \Omega$ and $v_C = 0.5i_2$.

❙ For the simultaneous equations (1) and (2) of Problem 1.137, substitute $v_C = 0.5i_2$ and set $R_1 = R_2 = R_C = 1 \, \Omega$ to get

$$v_{dp} = 2i_{dp} - 0.5i_2 \qquad (1)$$
$$v_{dp} = i_{dp} + i_2 \qquad (2)$$

Simultaneous solution of (1) and (2) gives

$$i_{dp} = \frac{\begin{vmatrix} v_{dp} & -0.5 \\ v_{dp} & 1 \end{vmatrix}}{\begin{vmatrix} 2 & -0.5 \\ 1 & 1 \end{vmatrix}} = \frac{1.5 v_{dp}}{2.5} = 0.6 v_{dp}$$

Hence,
$$Z_{Th} = \frac{v_{dp}}{i_{dp}} = \frac{1}{0.6} = 1.667 \, \Omega$$

With terminals a, b open-circuited, $i_1 = 0$. Equations (1) and (2) of Problem 1.137 become

$$V_{Th} = -i_2 R_2 + v_C = -i_2 R_2 + 0.5 i_2 = -0.5 i_2 \qquad (3)$$
$$V_{Th} = R_2 i_2 = (1) i_2 = i_2 \qquad (4)$$

Equations (1) and (2) are consistent only if $i_2 = 0$; whence, $V_{Th} = 0$.

1.139 Calculate i_1 for the circuit of Fig. 1-39 using the Thévenin equivalent of Problems 1.137 where $v_s = 7$ V, $v_C = 0.5 i_1$, and $R_1 = R_2 = R_C = 1 \, \Omega$.

❙ With the Thévenin equivalent in place, KVL yields

$$i_1 = \frac{v_s - V_{Th}}{Z_{Th}} = \frac{7 - 0}{1.75} = 4 \, \text{A}$$

1.140 Verify the result of Problem 1.139 by solving for i_1 by the method of mesh currents.

❙
$$v_s - v_C = (R_1 + R_C)I_1 - R_C I_2 \qquad (1)$$
$$v_C = -R_C I_1 + (R_C + R_2) I_2 \qquad (2)$$

Substitute $v_C = 0.5 I_1$, $v_s = 7$, and $R_1 = R_2 = R_C = 1 \, \Omega$ to obtain

$$7 = 2.5 I_1 - I_2 \qquad (3)$$
$$0 = -1.5 I_1 + 2 I_2 \qquad (4)$$

Simultaneous solution of (3) and (4) leads to

$$i_1 = I_1 = \frac{\begin{vmatrix} 7 & -1 \\ 0 & 2 \end{vmatrix}}{\begin{vmatrix} 2.5 & -1 \\ -1.5 & 2 \end{vmatrix}} = \frac{14}{3.5} = 4 \, \text{A}$$

1.141 Find the Thévenin equivalent for the network to the left of terminals a, b in Fig. 1-11 if $k = 0$.

▌ If $k = 0$, $kv_{ab} = 0$ and $i = 10/500 = 0.02$ A. Hence, the Thévenin voltage appearing at open-circuit terminals a, b is

$$V_{Th} = v_{ab} = -(100i)100 = -[(100)(0.02)]100 = -200 \text{ V}$$

Deactivate the 10-V source; thus, $i = 0$ and the controlled source has a zero value. Consequently, a driving point source connected so that $v_{dp} = v_{ab}$ sees only $Z_{Th} = 100 \ \Omega$.

1.142 Rework Problem 1.141 if $k = 0.1$.

▌ Apply KVL around the left loop and KCL at node b to find the following set of simultaneous equations if terminals a, b are open-circuited:

$$10 = 500i + 0.01v_{ab} \tag{1}$$
$$0 = 100i + v_{ab}/100 = 1 \times 10^4 i + v_{ab} \tag{2}$$

Solution yields

$$V_{Th} = v_{ab} = \frac{\begin{vmatrix} 500 & 10 \\ 1 \times 10^4 & 0 \end{vmatrix}}{\begin{vmatrix} 500 & 0.01 \\ 1 \times 10^4 & 1 \end{vmatrix}} = \frac{-1 \times 10^5}{400} = -250 \text{ V}$$

Deactivate the 10-V source and connect a driving point source such that $v_{dp} = v_{ab}$; then, by Ohm's law, the left loop yields

$$i = -\frac{0.1 \, v_{dp}}{500} = -2 \times 10^{-5} v_{dp} \text{ A} \tag{3}$$

Applying KCL at node a and using (3),

$$i_{dp} = 100i + v_{dp}/100 = 100(-2 \times 10^{-5} v_{dp}) + v_{dp}/100 \tag{4}$$

From (4),

$$Z_{Th} = \frac{v_{dp}}{i_{dp}} = \frac{1}{0.008} = 125 \ \Omega$$

1.143 An alternative solution to Problem 1.39 involves finding a Thévenin equivalent circuit which, when connected across the nonlinear $R_2 = 0.25i$, allows a quadratic equation in current i to be written using KVL. Find the elements of the Thévenin circuit and the resulting current.

▌ For terminals a, b open-circuited,

$$V_{Th} = v_{ab} = V_s + R_1 I_s = 10 + (5)(3) = 25 \text{ V}$$

With independent sources deactivated,

$$Z_{Th} = 5 \ \Omega$$

Placing the Thévenin circuit in position,

$$V_{Th} = Z_{Th}i + (0.25i)i \quad \text{or} \quad 25 = 5i + 0.25i^2 \tag{1}$$

Solve (1) for i by use of the quadratic formula and discard the extraneous negative root to find $i = 4.142$ A.

1.144 Use (1) and (2) of Problem 1.76 to find expressions for the z parameters in terms of the h parameters.

▌ Set $I_2 = 0$ in (1) and (2) of Problem 1.76 to get

$$V_1 = h_{11}I_1 + h_{12}V_2 \tag{1}$$
$$0 = h_{21}I_1 + h_{22}V_2 \tag{2}$$

From (2), $V_2 = -h_{21}I_1/h_{22}$, and substituting it into (1), find

$$V_1 = \left[h_{11} - \frac{h_{12}h_{21}}{h_{22}}\right]I_1 \quad \text{or} \quad \left.\frac{V_1}{I_1}\right|_{I_2=0} = z_{11} = h_{11} - \frac{h_{12}h_{21}}{h_{22}}$$

Directly from $V_2 = h_{21}I_1/h_{22}$,

$$z_{12} = \left.\frac{V_2}{I_1}\right|_{I_2=0} = -\frac{h_{21}}{h_{22}}$$

Set $I_1 = 0$ in (1) and (2) of Problem 1.76 to find

$$V_1 = h_{12}V_2 \tag{3}$$

$$I_2 = h_{22}V_2 \tag{4}$$

Directly from (4),

$$z_{22} = \left.\frac{V_2}{I_2}\right|_{I_1=0} = \frac{1}{h_{22}}$$

Solve (4) for $V_2 = I_2/h_{22}$ and substitute into (3) to obtain

$$V_1 = h_{12}(I_2/h_{22}) \quad \text{or} \quad z_{12} = \left.\frac{V_1}{I_2}\right|_{I_1=0} = \frac{h_{12}}{h_{22}}$$

1.145 For the two-port network of Fig. 1-27, find the voltage-gain ratio V_2/V_1 in terms of the z parameters.

▌ By Ohm's law,

$$I_2 = -V_2/R_L \tag{1}$$

Substituting (1) into (1) of Problem 1.67,

$$V_1 = z_{11}I_1 + z_{12}I_2 = z_{11}I_1 + z_{12}(-V_2/R_L) \quad \text{or} \quad I_1 = \frac{V_1 + (z_{12}/R_L)V_2}{z_{11}} \tag{2}$$

Use (1) and (2) in (2) of Problem 1.67. Thus,

$$V_2 = z_{21}I_1 + z_{22}I_2 = z_{21}\left[\frac{V_1 + (z_{12}/R_L)V_2}{z_{11}}\right] + z_{22}(-V_2/R_L) \tag{3}$$

From (3),

$$\frac{V_2}{V_1} = \frac{z_{21}R_L}{z_{11}R_L + z_{11}z_{22} - z_{12}z_{21}} \tag{4}$$

1.146 Evaluate V_2/V_1 of Problem 1.145 using the h parameters of Problem 1.81.

▌ Based on Problem 1.144,

$$z_{11} = h_{11} - \frac{h_{12}h_{21}}{h_{22}} = 100 - \frac{(0.0025)(20)}{0.001} = 50\,\Omega$$

$$z_{12} = \frac{h_{12}}{h_{22}} = \frac{0.0025}{0.001} = 2.5\,\Omega$$

$$z_{21} = -\frac{h_{21}}{h_{22}} = -\frac{20}{0.001} = -2 \times 10^4$$

$$z_{22} = \frac{1}{h_{22}} = \frac{1}{0.001} = 1 \times 10^3$$

Substitution into (4) of Problem 1.145 gives

$$\frac{V_2}{V_1} = \frac{z_{21}R_L}{z_{11}R_L + z_{11}z_{22} - z_{12}z_{21}} = \frac{(-2 \times 10^4)(2 \times 10^3)}{(50)(2 \times 10^3) + (50)(1 \times 10^3) - (2.5)(-2 \times 10^4)} = -200$$

1.147 Find the current-gain ratio I_2/I_1 for the two-port network of Fig. 1-27 in terms of the h parameters.

▌ By Ohm's law,

$$V_2 = -R_L I_2 \tag{1}$$

Use (1) in (2) of Problem 1.76. Thus,

$$I_2 = h_{21}I_1 + h_{22}V_2 = h_{21}I_1 + h_{22}(-R_L I_2)$$

Whence,

$$\frac{I_2}{I_1} = \frac{h_{21}}{1 + h_{22}R_L}$$

38 ☐ CHAPTER 1

1.148 Find the current-gain ratio I_2/I_1 for the two-port network of Fig. 1-19 in terms of the z parameters.

▌ By Ohm's law,

$$V_2 = -I_2 R_L \qquad (1)$$

Substitute (1) into (2) of Problem 1.67. So

$$V_2 = -I_2 R_L = z_{21} I_1 + z_{22} I_2$$

Whence,

$$\frac{I_2}{I_1} = -\frac{z_{21}}{z_{22} + R_L}$$

1.149 Determine the driving-point impedance (the input impedance with all independent sources deactivated) of the two-port network of Fig. 1-27.

▌ By Ohm's law,

$$V_2 = -I_2 R_L \qquad \text{or} \qquad -R_L = V_2/I_2$$

$$Z_{in} = \frac{V_1}{I_1} = \frac{T_1 V_2}{T_2 I_2} = \frac{T_1}{T_2}(-R_L)$$

Transfer functions T_1 and T_2 can be taken directly from the work of Problems 1.145 and 1.148.

$$T_1 = \frac{z_{11} R_L + z_{11} z_{22} - z_{12} z_{21}}{z_{21} R_L} \qquad T_2 = -\frac{z_{22} + R_L}{z_{21}}$$

Thus,

$$Z_{in} = \frac{z_{11} R_L + z_{11} z_{22} - z_{12} z_{21}}{z_{22} + R_L}$$

1.150 For the two-port network of Fig. 1-12, treat v_s and i as input port variables and v_{ab} and $-i_L$ as output port variables. Evaluate the z parameters.

▌ For $-i_L = 0$:

$$v_s = (R_1 + R_2)i = 2i \qquad \text{or} \qquad z_{11} = \left.\frac{v_s}{i}\right|_{-i_L = 0} = 2\,\Omega$$

$$v_{ab} = R_2 i + R_3(\alpha i) = (1 + \alpha)i \qquad \text{or} \qquad z_{21} = \left.\frac{v_{ab}}{i}\right|_{-i_L = 0} = 1 + \alpha\,\Omega$$

For $i = 0$:

$$v_s = -i_L R_3 = -i_L \qquad \text{or} \qquad z_{12} = \left.\frac{v_s}{-i_L}\right|_{i=0} = 1\,\Omega$$

$$v_{ab} = -i_L(R_2 + R_3) = -2i_L \qquad \text{or} \qquad z_{22} = \left.\frac{v_s}{-i_L}\right|_{i=0} = 2\,\Omega$$

1.151 Determine the g parameters of the network of Fig. 1-9 if $\alpha = 2$ and $R_1 = R_2 = R_3 = 1\,\Omega$. Assume v_s is a dc source.

▌ For $i_3 = 0$:

$$v_s = R_1 i_1 + R_3 i_1 + \alpha i_1 = 4i_1 \qquad \text{or} \qquad g_{11} = \left.\frac{i_1}{v_s}\right|_{i_3 = 0} = 0.25\,\text{S}$$

$$V_b = R_2 i_1 + \alpha i_1 = 3(v_s/4) \qquad \text{or} \qquad g_{21} = \left.\frac{V_b}{v_s}\right|_{i_3 = 0} = 0.75$$

For $v_s = 0$: The method of node voltages leads to

$$\frac{v_s}{R_1} + \frac{v_2 - \alpha(-v_2/R_1)}{R_2} + \frac{v_2 - V_b}{R_3} = 0$$

Use known values and solve for $v_2 = V_b/5$. Hence,

$$g_{12} = \frac{i_1}{i_3}\bigg|_{v_s=0} = \frac{-v_2/R_1}{(V_b - v_2)/R_3} = \frac{-(V_b/5)/1}{(V_b - V_b/5)/1} = -0.25$$

and

$$g_{22} = \frac{V_b}{i_3}\bigg|_{v_s=0} = \frac{V_b}{(V_b - v_2)/R_3} = \frac{V_b}{(V_b - V_b/5)/1} = 1.25\ \Omega$$

1.152 Determine the voltage gain ratio V_b/v_s for the circuit of Fig. 1-9 if $\alpha = 2$ and $R_1 = R_2 = R_3 = 1\ \Omega$. Assume v_s is a dc source.

I Based on (2) of Problem 1.93 and results of Problem 1.151,

$$\frac{V_b}{v_s} = \frac{g_{21}}{1 - g_{22}} = \frac{0.75}{1 - 1.25} = -3$$

1.153 For a one-port network with passive sign convention, $v = V_m \cos \omega t$ V and $i = I_1 + I_2 \cos(\omega t + \theta)$ A. Find the instantaneous power flowing to the network.

I $p(t) = vi = V_m \cos \omega t [I_1 + I_2 \cos(\omega t + \theta)] = V_m I_1 \cos \omega t + V_m I_2 \cos \omega t \cos(\omega t + \theta)$

Using the identity $\cos x \cos y = \tfrac{1}{2}\cos(x+y) + \tfrac{1}{2}\cos(x-y)$,

$$p(t) = V_m I_1 \cos \omega t + \tfrac{1}{2} V_m I_2 [\cos(2\omega t + \theta) + \cos \theta]$$

1.154 Find the average power delivered to the network of Problem 1.153.

I
$$P = \frac{1}{T}\int_0^T p(t)\,dt = \frac{1}{2\pi}\int_0^{2\pi} [V_m I_1 \cos \omega t + \tfrac{1}{2} V_m I_2 [\cos(2\omega t + \theta) + \cos \theta]]\,d(\omega t)$$

$$P = \tfrac{1}{2} V_m I_2 \cos \theta$$

CHAPTER 2
Diodes: Signal Level Application

SEMICONDUCTOR MATERIAL PROPERTIES

2.1 A rectangular copper conductor of length $l = 4$ mm and cross-sectional area $A = 1 \times 10^{-4}$ mm^2 conducts a current of 100 mA. Determine the voltage drop over the length of the conductor of resistivity $\rho = 1.724 \times 10^{-8}\ \Omega \cdot$ m.

$$R = \rho \frac{l}{A} = \frac{(1.724 \times 10^{-8})(4 \times 10^{-3})}{(1 \times 10^{-4})(10^{-6})} = 0.6896\ \Omega$$

By Ohm's law,

$$V = IR = (100 \times 10^{-3})(0.6896) = 0.06896\ \text{V} = 68.96\ \text{mV}$$

2.2 If the voltage measured across the rectangular copper conductor of Problem 2.1 is 100 mV, determine the current flowing through the conductor.

By Ohm's law,

$$I = \frac{V}{R} = \frac{100 \times 10^{-3}}{0.6896} = 145.01\ \text{mA}$$

2.3 Avogadro's law states that moles (atomic weight expressed in grams) of different elements contain identical number of atoms (N_a) related by Avogadro's number ($A_a = 6.023 \times 10^{23}$ cm^{-3}) and the atomic weight (W_a) as $N_a = A_a/W_a$. Find the density (n_e) of valence band electrons available to form conduction current for a copper specimen of mass density (γ) = 8.92 g/cm^3.

From a periodic table of elements, the atomic weight of copper is $W_a = 63.54$ g/mole. Since copper has one valence band electron, the electrons available for conducting equal the number of atoms.

$$n_e = N_a \gamma = \frac{A_a}{W_a} \gamma = \frac{(6.023 \times 10^{23})(8.92)}{63.54} = 8.45 \times 10^{22}\ \text{cm}^{-3}$$

2.4 The conductivity ($\Omega \cdot$ cm) of a metal conductor is given by $\sigma = n_e q \mu$ where n_e = valence band electron density (cm^{-3}), q = charge of an electron (C), and μ = mobility of electrons (cm^2/V $\cdot$ s). Find the mobility of electrons for the copper conductor of Problem 2.1.

Based on Problem 2.3,

$$\mu = \frac{\sigma}{n_e q} = \frac{1}{\rho n_e q} = \frac{1}{(1.724 \times 10^{-6})(8.45 \times 10^{22})(1.6 \times 10^{-19})} = 42.90\ \text{cm}^2/\text{V} \cdot \text{s}$$

2.5 The drift velocity (v_d), or average speed of conducting electrons past a reference point normal to the plane of current flow within a conductor, is related to the mobility (μ) and electric field intensity (E) by $v_d = \mu E$. Determine the drift velocity of the copper conductor of Problem 2.1.

Based on Problem 2.1,

$$E = \frac{V}{l} = \frac{0.06896\ \text{V}}{0.4\ \text{cm}} = 0.1724\ \text{V/cm}$$

Using the result of Problem 2.4,

$$v_d = \mu E = (4.290\ \text{cm}^2/\text{V} \cdot \text{s})(0.1724\ \text{V/cm}) = 7.396\ \text{cm/s}$$

2.6 Covalent bonded semiconductor materials (Si or Ge) contribute significantly less than one electron per atom to the current conduction process. Intrinsic (pure) Si is experimentally determined to exhibit a carrier (electron or hole) density (n_i) of 1.48×10^{10} electrons (holes)/cm^3 at 300°K. Determine the percentage of Si atoms that contribute a conduction electron (n)-hole (p) pair at 300°K if the mass density of Si is $\gamma = 2.33$ g/cm^3.

▌ The atomic density of Si is

$$N_a\gamma = \frac{A_a\gamma}{W_a} = \frac{(6.023 \times 10^{23})(2.33)}{28.09} = 4.99 \times 10^{22} \text{ atoms/cm}^3$$

$$\frac{n}{\text{atom}} = \frac{p}{\text{atom}} = \frac{n_i}{N_a\gamma}(100\%) = \frac{1.48 \times 10^{10}}{4.99 \times 10^{22}}(100\%) = 0.297 \times 10^{-10}\%$$

or approximately 1 atom in 3 trillion contributes a free electron-hole pair.

2.7 In a semiconductor specimen, the conductivity must consider the conduction process due to both electrons (n) and holes (p) and is expressed by $\gamma = q(n\mu_n + p\mu_p)$. Determine the resistivity of an intrinsic Si specimen at 300°K if the electron mobility (μ_n) = 1300 cm²/V · s and the hole mobility (μ_p) = 500 cm²/V · s.

▌ For an intrinsic specimen, $n = p$. Based on Problem 2.6,

$$\rho_i = \frac{1}{\sigma_i} = \frac{1}{q(n\mu_n + p\mu_p)} = \frac{1}{qn_i(\mu_n + \mu_p)} = \frac{1}{(1.6 \times 10^{-19})(1.48 \times 10^{10})(1300 + 500)} = 2.35 \times 10^5 \ \Omega \cdot \text{cm}$$

2.8 If a Si specimen is doped with an appropriate n-type (donor) material (pentavalent element such as Sb, As or P) to the small concentration of one part per million atoms, determine the resulting conductivity for the specimen of Problem 2.7.

▌ Each atom of the doping or impurity material will yield a free electron that has no place in the covalent bonded crystal lattice, and thus, it is available for conduction. Further, since the number of carriers contributed by the Si is small compared to the electrons contributed by the donor impurity, only the latter need be considered. Based on Problem 2.6, the donor carrier density is

$$n = \frac{N_a\gamma}{10^6} = \frac{4.99 \times 10^{22}}{10^6} = 4.99 \times 10^{16} \text{ cm}^{-3}$$

$$\sigma_d = nq\mu_n = (4.99 \times 10^{16})(1.6 \times 10^{-19})(1300) = 10.37 \ (\Omega \cdot \text{cm})^{-1}$$

2.9 What is the change in resistivity of a Si specimen due to the doping level of Problem 2.8?

▌ Based on Problems 2.7 and 2.8,

$$\frac{\rho_i}{\rho_d} = \frac{2.39 \times 10^5}{10.37} = 23{,}047$$

The resistivity has been reduced by a factor of more than 23,000 due to the small concentration of impurity material.

2.10 Repeat Problem 2.1 if the rectangular conductor is intrinsic Si material and the current is reduced to 50 μA.

▌ Based on Problem 2.7,

$$R = \rho_i \frac{l}{A} = \frac{(2.35 \times 10^3)(4 \times 10^{-3})}{(1 \times 10^{-4})(10^{-6})} = 9.40 \times 10^{10} \ \Omega$$

$$V = IR = (50 \times 10^{-6})(9.40 \times 10^{10}) = 4.7 \times 10^6 \text{ V}$$

2.11 Repeat Problem 2.1 if the rectangular conductor is the doped semiconductor material of Problem 2.8 and the current is reduced to 50 μA.

▌ Based on Problem 2.8,

$$\rho_d = \frac{1}{\sigma_d} = \frac{1}{(10.37)(10^2)} = 9.64 \times 10^{-4} \ \Omega \cdot \text{cm}$$

$$R = \frac{\rho_d l}{A} = \frac{(9.64 \times 10^{-4})(4 \times 10^{-3})}{(1 \times 10^{-4})(10^{-6})} = 3.856 = 10^4 \ \Omega$$

$$V = IR = (50 \times 10^{-6})(3.856 \times 10^4) = 1.928 \text{ V}$$

42 □ CHAPTER 2

2.12 The mobility values for Si of Problem 2.7 are valid for $T = 300°K$. The mobility varies over a temperature range from 200°K to 400°K approximately as $T^{2.6}$. Find the approximate mobility of Si carriers for $T = 400°K$.

$$\mu_n \approx 1300(400/300)^{2.6} = 2746 \text{ cm}^2/\text{V} \cdot \text{s}$$

$$\mu_p \approx 500(400/300)^{2.6} = 1056 \text{ cm}^2/\text{V} \cdot \text{s}$$

2.13 An approximate formula for intrinsic carrier concentration (cm^{-3}) of Si as a function of temperature (°K) is $n_i = 3.87 \times 10^{16} T^{3/2} e^{-7000/T}$. Determine the intrinsic carrier concentration for a Si specimen at 270°K.

▮ $$n_i = 3.87 \times 10^{16}(270)^{3/2} e^{-7000/270} = 9.446 \times 10^8 \text{ cm}^{-3}$$

2.14 The mass-action law states that for thermal equilibrium the product (np) of mobile carriers within a doped semiconductor specimen maintains the constant value of n_i^2 regardless of doping level. Further, charge neutrality requires that the positive charge density (mobile holes, p, and immobile positive donor ions, N_D) be equal to the negative charge density (mobile electrons, n, and immobile acceptor ions, N_A), i.e., $p + N_D = n + N_A$. Based on the mass-action law and charge neutrality, determine expressions for the number of mobile holes and electrons for typical levels of doping in (a) p-type semiconductor material and (b) n-type semiconductor material.

▮ (a) For a p-type semiconductor, $N_D = 0$ and at typical levels of doping, $p \gg n$. Thus, by charge neutrality, $p = N_A + n \approx N_A$. Minority carrier concentration follows from mass-action law as $n = n_i^2/p \approx n_i^2/N_A$.
  (b) For an n-type semiconductor, $N_A = 0$ and for typical doping levels, $n \gg p$. Whence, $n \approx N_D$ and $p = n_i^2/n \approx n_i^2/N_D$.

2.15 Light doping of p-type semiconductor material is defined as the case for which $p \gg n$ is not valid ($n \gg p$ is not valid for an n-type semiconductor). Derive a procedure to determine the number of mobile carriers for the case of light doping.

▮ For p-type doping, $N_D = 0$. Whence, by charge neutrality and mass-action law,

$$p + N_D = p = n + N_A = n_i^2/n \quad \text{or} \quad n^2 + N_A n - n_i^2 = 0$$

Solve the quadratic for n and discard the extraneous negative root to find

$$n = \tfrac{1}{2}(-N_A + \sqrt{N_A^2 + 4n_i^2})$$

With n known, p follows from the mass-action law as $p = n_i^2/n$.
For n-type doping, $N_A = 0$ and, by analogous procedure,

$$p = \tfrac{1}{2}(-N_D + \sqrt{N_D^2 + 4n_i^2}) \qquad n = n_i^2/p$$

2.16 Determine the electron (minority carrier) and hole (majority carrier) concentration in p-type Si with an acceptor ion concentration of 1×10^{16} cm^{-3} when operating at 270°K.

▮ From the result of Problem 2.13, it is apparent that $N_A \gg n_i$; thus, $p \approx N_A$ and the expressions of Problem 2.14 are applicable. Using n_i as calculated in Problem 2.13,

$$p \approx N_A = 1 \times 10^6 \text{ cm}^{-3} \qquad n \approx \frac{n_i^2}{N_A} = \frac{(9.446 \times 10^8)^2}{1 \times 10^{16}} = 89.2 \text{ cm}^{-3}$$

2.17 Find the electron (majority carrier) and hole (minority carrier) concentration in n-type Si with a donor ion concentration of 5×10^8 cm^{-3} when operating at 270°K.

▮ From Problem 2.13, it is seen that N_D has the same order of magnitude as n_i; hence, the expressions of Problem 2.15 are pertinent.

$$p = \tfrac{1}{2}(-N_D + \sqrt{N_D^2 + 4n_i^2}) = \tfrac{1}{2}[-5 \times 10^8 + \sqrt{(5 \times 10^8)^2 + 4(9.446 \times 10^8)^2}] = 7.27 \times 10^8 \text{ cm}^{-3}$$

$$n = \frac{n_i^2}{p} = \frac{(9.446 \times 10^8)^2}{7.27 \times 10^8} = 12.27 \times 10^8 \text{ cm}^{-3}$$

Observe that the approximation $n \approx N_D$ would introduce an appreciable error for this case of light doping.

TERMINAL CHARACTERISTICS

2.18 Metallurgical union of p and n semiconductor material as illustrated by Fig. 2-1(a) forms the diode shown schematically by Fig. 2.1(b). The static equation for junction current of the diode is given by

$$i_D = I_0(e^{v_D/\eta V_T} - 1) \text{ A} \tag{1}$$

where $V_T \equiv kT/q$, V
 $v_D \equiv$ diode terminal voltage, V
 $I_0 \equiv$ temperature-dependent saturation current, A
 $T \equiv p\text{-}n$ junction temperature, K
 $k \equiv$ Boltzmann's constant (1.38×10^{-23} J/K)
 $q \equiv$ electron charge (1.6×10^{-19} C)
 $\eta \equiv$ empirical constant, 1 for Ge and 2 for Si

Find the value of V_T in (1) at 20°C.

$$V_T = \frac{kT}{q} = \frac{(1.38 \times 10^{-23})(273 + 20)}{1.6 \times 10^{-19}} = 25.27 \text{ mV}$$

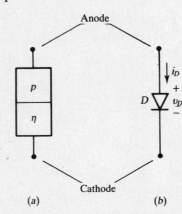

Fig. 2-1

2.19 A Si diode operating at 20°C has a saturation current of 1 μA. Sketch the static terminal characteristic (i_d vs. v_D).

Based on Problem 2.17,

$$i_D = (1 \times 10^{-6})(e^{v_D/(2)(0.02527)} - 1) = (e^{19.786 v_D} - 1)(10^{-6})$$

See Fig. 2-2 for the required sketch.

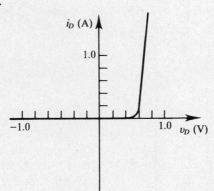

Fig. 2-2

2.20 Derive an expression for the ratio of diode currents resulting from two different forward voltage drops if $v_D \gg \eta V_T$.

If $v_D \gg \eta V_T$, (1) of Problem 2.18 yields $i_D \approx I_0 e^{v_D/V_T}$; thus,

$$\frac{i_{D2}}{i_{D1}} = \frac{I_0 e^{v_{D2}/\eta V_T}}{I_0 e^{v_{D1}/\eta V_T}} = e^{(v_{D2} - v_{D1})/\eta V_T} \tag{1}$$

44 □ CHAPTER 2

2.21 Find the per unit change in current of a Ge diode if $V_T = 26$ mV and the forward voltage drop increases by 50 mV.

▮ Based on (1) of Problem 2.20,

$$\frac{i_{D2}}{i_{D1}} = e^{(50 \times 10^{-3})/(26 \times 10^{-3})} = 6.842$$

Thus,

$$\frac{i_{D2} - i_{D1}}{i_{D1}} = \frac{6.842 i_{D1} - i_{D1}}{i_{D1}} = 5.842$$

2.22 Repeat Problem 2.21 for a Si diode.

▮

$$\frac{i_{D2}}{i_{D1}} = e^{(50 \times 10^{-3})/(2)(26 \times 10^{-3})} = 2.616$$

and

$$\frac{i_{D2} - i_{D1}}{i_{D1}} = \frac{2.616 i_{D1} - i_{D1}}{i_{D1}} = 1.616$$

2.23 Derive an expression for the change in diode forward voltage drop for a given current ratio i_{D2}/i_{D1} if $v_D \gg \eta V_T$.

▮ Take the natural logarithm of (1) from Problem 2.20 to find

$$v_{D2} - v_{D1} = \Delta v_D = \eta V_T \ln (i_{D2}/i_{D1}) \tag{1}$$

2.24 If the current of a Si diode with $V_T = 26$ mV doubles, find the increase in forward voltage drop.

▮ Based on (1) of Problem 2.23,

$$\Delta v_D = (2)(26 \times 10^{-3}) \ln (2) = 36.04 \text{ mV}$$

2.25 Find the decrease in forward voltage drop of a Ge diode for which $V_T = 26$ mV if the forward current is reduced to one-fourth of its original value.

▮ Based on (1) of Problem 2.23,

$$\Delta v_D (26 \times 10^{-3}) \ln (1/4) = -36.04 \text{ mV}$$

2.26 Repeat Problem 2.24 if the operating temperature of the diode is increased by 30 percent.

▮ Based on (1) of Problem 2.18, $V_T = 1.3(26) = 33.8$ mV. Hence,

$$\Delta v_D = (2)(33.8 \times 10^{-3}) \ln (2) = 46.86 \text{ mV}$$

2.27 For what voltage v_D will the reverse current of a Ge diode reach 99 percent of its saturation value at a temperature of 300°K?

▮ Based on (1) of Problem 2.18,

$$V_T = \frac{kT}{q} = \frac{(1.38 \times 10^{-23})(300)}{1.6 \times 10^{-19}} = 0.02587 \text{ V}$$

and

$$-0.99 I_0 = I_0 (e^{-v_D/\eta V_T} - 1)$$

Whence,

$$v_D = \eta V_T \ln (0.01) = -0.1191 \text{ V}$$

2.28 At a junction temperature of 25°C, over what range of forward voltage drop v_D can (1) of Problem 2.18 be approximated as $i_D \approx I_0 e^{v_D/V_T}$ with less than 1 percent error for a Ge diode?

▮ From (1) of Problem 2.18, with $\eta = 1$, the error will be less than 1 percent if $e^{v_D/V_T} > 101$. In that range,

$$v_D > V_T \ln (101) = \frac{kT}{q} \ln (101) = \frac{(1.38 \times 10^{-23})(25 + 273)}{1.6 \times 10^{-19}} (4.6151) = 0.1186 \text{ V}$$

2.29 A Ge diode is operated at a junction temperature of 27°C. For a forward current of 10 mA, v_D is found to be 0.3 V. If $v_D = 0.4$ V, find the forward current.

▌ We form the ratio

$$\frac{i_{D2}}{i_{D1}} = \frac{I_0(e^{v_{D2}+V_T}-1)}{I_0(e^{V_{D1}/V_T}-1)} = \frac{e^{0.4/0.02587}-1}{e^{0.3/0.02587}-1} = 47.73$$

Then $i_{D2} = (47.73)(10 \text{ mA}) = 477.3 \text{ mA}$

2.30 Find the reverse saturation current for the diode of Problem 2.29.

▌ By (1) of Problem 2.18,

$$I_0 = \frac{i_{D1}}{e^{v_{D1}/V_T}-1} = \frac{10 \times 10^{-3}}{e^{0.3/0.02587}-1} = 91 \text{ nA}$$

2.31 A Ge diode has a saturation current $I_0 = 10$ nA at $T = 300°K$. Find the forward current i_D if the forward voltage drop v_D is 0.5 V.

▌ By (1) of Problem 2.18,

$$i_D \approx I_0 e^{v_D/\eta V_T} = (10 \times 10^{-9})e^{0.5/0.02587} = 2.47 \text{ A}$$

2.32 The diode of Problem 2.31 is rated for a maximum current of 5 A. Determine the junction temperature at rated current if the forward voltage drop is 0.7 V.

▌ Assume $v_D \gg \eta V_T = \eta kT/q$ and solve (1) of Problem 2.18 for T.

$$T = \frac{qv_D}{\eta k \ln(i_D/I_0)} = \frac{(1.6 \times 10^{-19})(0.7)}{(1)(1.38 \times 10^{-23})\ln(5/10 \times 10^{-9})} = 405.2°K$$

2.33 To a close approximation, the reverse saturation current of a Si diode doubles for each increase of 10°C in temperature. Find the increase in temperature ΔT necessary to increase I_0 by a factor of 100.

▌

$$\frac{I_{02}}{I_{01}} = 100 = 2^{\Delta T/10} \quad \text{or} \quad \Delta T = \frac{10 \ln(100)}{\ln(2)} = 66.4°C$$

2.34 Laboratory data for a Si diode show that $i_D = 2$ mA when $v_D = 0.6$ V, and $i_D = 10$ mA for $v_D = 0.7$ V. Find the temperature for which the data were taken.

▌ Take the natural logarithm of (1) of Problem 2.20, let $V_T = kT/q$, and solve for T to find

$$T = \frac{q(v_{D2}-v_{D1})}{\eta k \ln(i_{D2}/i_{D1})} = \frac{(1.6 \times 10^{-19})(0.7-0.6)}{2(1.38 \times 10^{-23})\ln(10/2)} = 360.19°K = 87.19°C$$

2.35 Determine the reverse saturation current for the diode of Problem 2.34.

▌

$$I_0 = \frac{i_{D1}}{e^{qv_{D1}/\eta kT}} = \frac{2 \times 10^{-3}}{e^{(1.6 \times 10^{-19})(0.6)/(2)(1.38 \times 10^{-23})(360.19)}} = 2.397 \text{ μA}$$

2.36 A Ge diode at $T = 300°K$ conducts 5 mA at $v_D = 0.35$ V. Predict the diode current if $v_D = 0.4$ V.

▌ Based on (1) of Problem 2.20,

$$i_{D2} = i_{D1}e^{(v_{D2}-v_{D1})/V_T} = (5)e^{(0.4-0.35)/0.02587} = 34.54 \text{ mA}$$

2.37 Calculate the reverse saturation current for the diode of Problem 2.36.

▌

$$I_0 = \frac{i_{D1}}{e^{v_{D1}/\eta V_T}-1} = \frac{5 \times 10^{-3}}{e^{(0.35)/0.02587}-1} = 6.66 \text{ nA}$$

RECTIFIER DIODE CIRCUIT ANALYSIS

2.38 Find the voltage v_L in the circuit of Fig. 2-3(a), where D is an ideal diode (zero impedance to forward current and infinite impedance to reverse current).

■ The analysis is simplified if a Thévenin equivalent is found for the circuit to the left of terminals a, b; the result is

$$v_{Th} = \frac{R_1}{R_1 + R_S} v_S \quad \text{and} \quad Z_{Th} = R_{Th} = R_1 \parallel R_S = \frac{R_1 R_S}{R_1 + R_S}$$

Step 1: After replacing the network to the left of terminals a, b with the Thévenin equivalent, assume forward bias and replace diode D with a short circuit, as in Fig. 2-3(b).

Step 2: By Ohm's law,

$$i_D = \frac{v_{Th}}{R_{Th} + R_L}$$

Step 3: If $v_s \geq 0$, then $i_D \geq 0$ and

$$v_L = i_D R_L = \frac{R_L}{R_L + R_{Th}} v_{Th}$$

Step 4: If $v_S < 0$, then $i_D < 0$ and the result of step 3 is invalid. Diode D must be replaced by an open circuit as illustrated in Fig. 2-3(c), and the analysis performed again. Since now $i_D = 0$, $v_L = i_D R_L = 0$. Since $v_D = v_S < 0$, the reverse bias of the diode is verified.

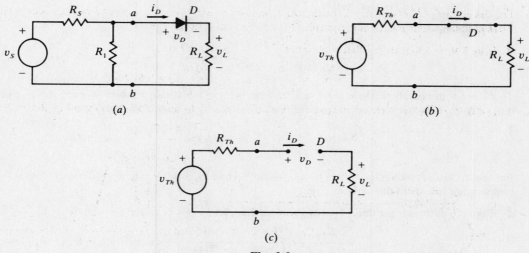

Fig. 2-3

2.39 Extend the ideal diode analysis procedure of Problem 2.38 to the case of multiple diodes by solving for the current i_L in the circuit of Fig. 2-4(a). Assume D_1 and D_2 are ideal. $R_2 = R_L = 100\ \Omega$, and v_S is a 10-V square wave of period 1 ms.

■ Step 1: Assume both diodes are forward-biased, and replace each with a short circuit as shown in Fig. 2-4(b).

Step 2: Since D_1 is "on," or in the zero-impedance state, current division requires that

$$i_{D2} = -\frac{0}{R_2 + 0} i_L = 0 \qquad (1)$$

Hence, by Ohm's law,

$$i_L = i_{D1} = \frac{v_S}{R_L} \qquad (2)$$

Step 3: Observe that when $v_S = 10 > 0$, we have, by (2), $I_{D1} = 10/100 = 0.1\ \text{A} > 0$. Also, by (1), $i_{D2} = 0$. Thus, all diode currents are greater than or equal to zero, and the analysis is valid. However, when $v_S = -10 < 0$, we have, by (2), $i_{D1} = -10/100 = -0.1\ \text{A} < 0$, and the analysis is no longer valid.

Step 4: Replace D_1 with an open circuit as illustrated in Fig. 2-4(c). Now obviously $i_{D1} = 0$ and, by Ohm's law,

$$i_L = -i_{D2} = \frac{v_S}{R_2 + R_L} = \frac{-10}{100 + 100} = -0.05\ \text{A}$$

Further, voltage division requires that

$$v_{D1} = \frac{R_2}{R_2 + R_L} v_S$$

so that $v_{D1} < 0$ if $v_S < 0$, verifying that D_1 is actually reverse-biased. Note that if D_2 had been replaced with an open circuit, we would have found that $v_{D2} = -v_S = 10 \text{ V} > 0$, so D_2 would not actually have been reverse-biased.

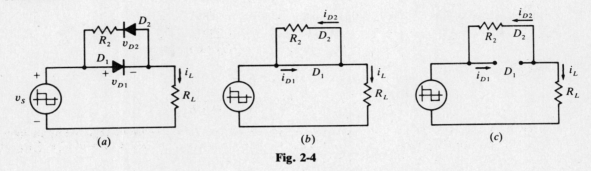

Fig. 2-4

2.40 For the circuit of Fig. 2-5(a), sketch the waveforms of v_L and v_D if the source voltage v_S is as given in Fig. 2-5(b). The diode is ideal, and $R_L = 100 \; \Omega$.

▎ If $v_S \geq 0$, D conducts, so that $v_D = 0$ and

$$v_L = \frac{R_L}{R_L + R_S} v_S = \frac{100}{100 + 10} v_S = 0.909 v_S$$

If $v_S < 0$, D blocks, so that $v_D = v_S$ and $v_L = 0$. Sketches of v_D and v_L are shown in Fig. 2-5(c).

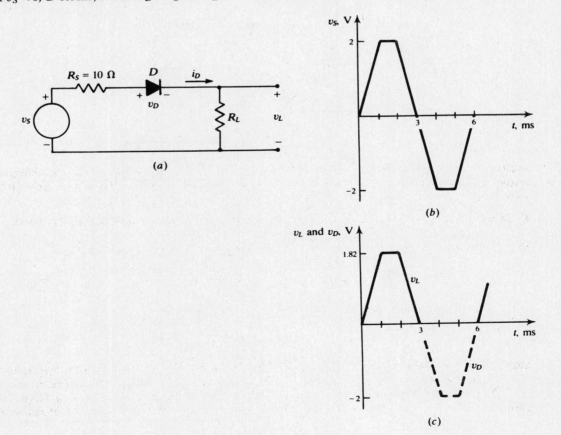

Fig. 2-5

48 □ CHAPTER 2

2.41 In the circuit of Fig. 2-6, D_1 and D_2 are ideal diodes. Find i_{D1} and i_{D2}.

▌ Because of the polarities of D_1 and D_2, it is necessary that $i_S \geq 0$. Thus, $v_{ab} \leq V_S = V_1$. But $v_{d1} = v_{ab} - V_1$; therefore, $v_D \leq 0$ and so $i_{D1} \equiv 0$, regardless of conditions in the right-hand loop. It follows that $i_{D2} = i_S$. Now using the analysis procedure of Problem 2.38, we assume D_2 is forward-biased and replace it with a short circuit. By KVL,

$$i_{D2} = \frac{V_S - V_2}{500} = \frac{5-3}{500} = 4 \text{ mA}$$

Since $i_{D2} \geq 0$, D_2 is in fact forward-biased and the analysis is valid.

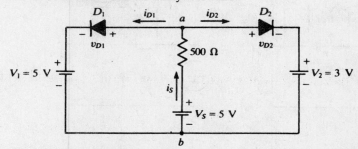

Fig. 2-6

2.42 The logic OR gate can be utilized to fabricate composite waveforms. Sketch the output v_o of the gate of Fig. 2-7(a) if the three signals of Fig. 2-7(b) are impressed on the input terminals. Assume that diodes are ideal.

▌ For this circuit, KVL gives

$$v_1 - v_2 = v_{D1} - v_{D2} \qquad v_1 - v_3 = v_{D1} - v_{D3}$$

i.e., the diode voltages have the same ordering as the input voltages. Suppose that v_1 is positive and exceeds v_2 and v_3. Then D_1 must be forward-biased, with $v_{D1} = 0$ and, consequently, $v_{D2} < 0$ and $v_{D3} < 0$. Hence, D_2 and D_3 block, while v_1 is passed as v_o. This is so in general: The logic of the OR gate is that the largest positive input signal is passed as v_o, while the remainder of the input signals are blocked. If all input signals are negative, $v_o = 0$. Application of this logic gives the sketch of v_o in Fig. 2-7(c).

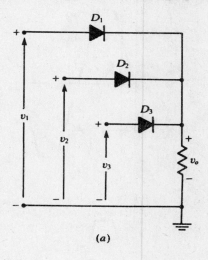
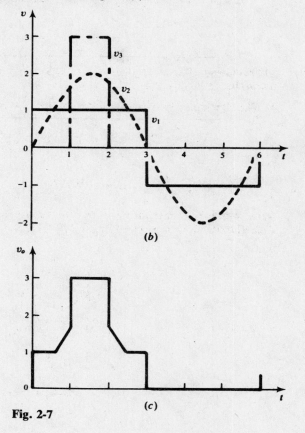

Fig. 2-7

2.43 The diode in the circuit of Fig. 2-8(a) has the nonlinear terminal characteristic of Fig. 2-8(b). Find i_D and v_D analytically, given $v_S = 0.1 \cos \omega t$ V and $V_b = 2$ V.

▌ The Thévenin equivalent circuit for the network to the left of terminals a, b in Fig. 2-8(a) has

$$V_{Th} = \frac{100}{200}(2 + 0.1 \cos \omega t) = 1 + 0.05 \cos \omega t \text{ V}$$

$$R_{Th} = \frac{(100)^2}{200} = 50 \text{ }\Omega$$

The diode can be modeled as in Fig. 2-8(b), with $V_F = 0.5$ V and

$$R_F = \frac{0.7 - 0.5}{0.004} = 50 \text{ }\Omega$$

Together, the Thévenin equivalent circuit and the diode model form the circuit in Fig. 2-8(c). Now by Ohm's law,

$$i_D = \frac{V_{Th} - V_F}{R_{Th} + R_F} = \frac{(1 + 0.05 \cos \omega t) - 0.5}{50 + 50} = 5 + 0.5 \cos \omega t \text{ mA}$$

$$v_D = V_F + R_F i_D = 0.5 + 50(0.005 + 0.0005 \cos \omega t) = 0.75 + 0.025 \cos \omega t \text{ V}$$

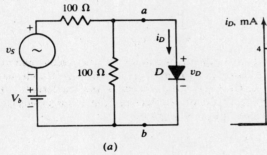
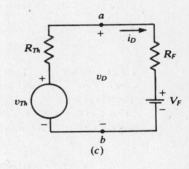

Fig. 2-8

2.44 In the circuit of Fig. 2-3(a), $v_S = 6$ V and $R_1 = R_S = R_L = 500$ Ω. Determine i_D and v_D graphically, using the diode characteristic in Fig. 2-9.

▌ The circuit may be reduced to that of Fig. 2-10, with

$$v_{Th} = \frac{R_1}{R_1 + R_S} v_S = \frac{500}{500 + 500} 6 = 3 \text{ V}$$

and

$$R_{Th} = R_1 \| R_S + R_L = \frac{(500)(500)}{500 + 500} + 500 = 750 \text{ }\Omega$$

Then, with these values the load line must be superimposed on the diode characteristic, as in Fig. 2-9. The desired solution, $i_D = 3$ mA and $v_D = 0.75$ V, is given by the point of intersection of the two plots.

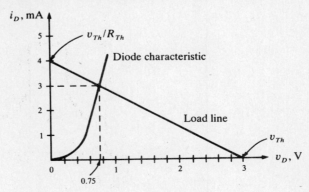

Fig. 2-9

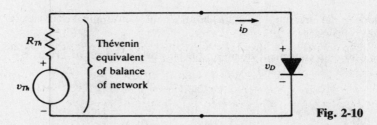

Fig. 2-10

2.45 If all sources in the original linear portion of a network vary with time, then v_{Th} is also a time-varying source. In reduced form [Fig. 2-11(a)], one such network has a Thévenin voltage that is a triangular wave with a 2-V peak. Find i_D and v_D for this network.

∎ In this case, there exists a value of i_D corresponding to each value that v_{Th} takes on. An acceptable solution for i_D may be found by considering a finite number of values of v_{Th}. Since v_{Th} is repetitive, i_D will be repetitive (with the same period), so only one cycle need be considered.

As in Fig. 2-11(b), we begin by laying out a scaled plot of v_{Th} versus time, with the v_{Th} axis parallel to the v_D axis of the diode characteristic. We then select a point on the v_{Th} plot, such as $v_{Th} = 0.5\ V$ at $t = t_1$. Considering

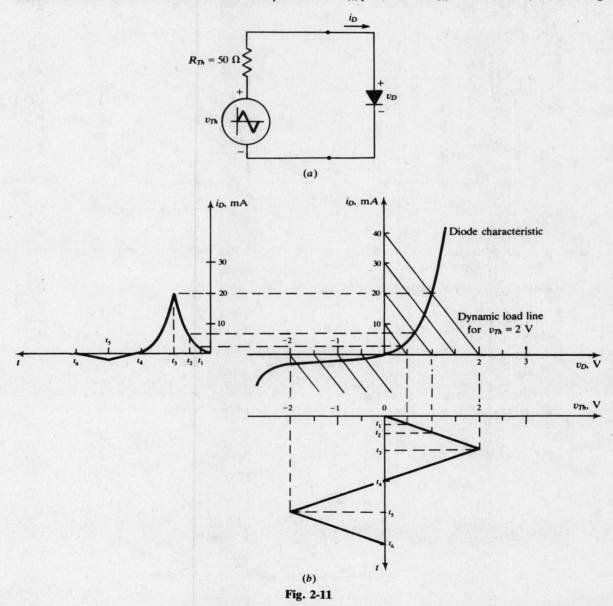

Fig. 2-11

time to be stopped at $t = t_1$, we construct a load line for this value on the diode characteristic plot; it intersects the v_D axis at $v_{Th} = 0.5$ V, and the i_D axis at $v_{Th}/R_{Th} = 0.5/50 = 10$ mA. We determine the value of i_D at which this load line intersects the characteristic, and plot the point (t_1, i_D) on a time-versus-i_D coordinate system constructed to the left of the diode characteristic curve. We then let time progress to some new value, $t = t_2$, and repeat the entire process. And we continue until one cycle of v_{Th} is completed. Since the load line is continually changing, it is referred to as a *dynamic load line*. The solution, a plot of i_D, differs drastically in form from the plot of v_{Th} because of the nonlinearity of the diode.

2.46 If both dc and time-varying sources are present in the original linear portion of a network, then v_{Th} is a series combination of a dc and a time-varying source. Suppose that the Thévenin source for a particular network combines a 0.7-V battery and a 0.1-V peak sinusoidal source, as in Fig. 2-12(a). Find i_D and v_D for the network.

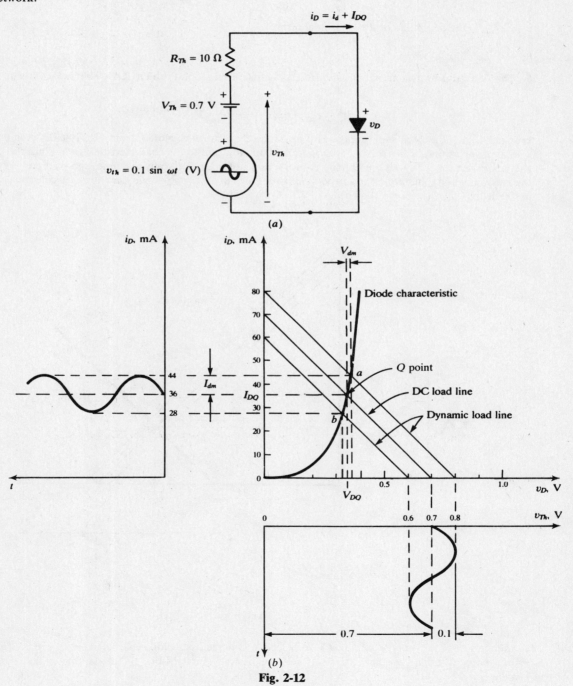

Fig. 2-12

■ We lay out a scaled plot of v_{Th}, with the v_{Th} axis parallel to the v_D axis of the diode characteristic curve. We then consider v_{th}, the ac component of v_{Th}, to be momentarily at zero ($t = 0$), and we plot a load line for this instant on the diode characteristic. This particular load line is called the *dc load line*, and its intersection with the diode characteristic curve is called the *quiescent point* or *Q point*. The values of i_D and v_D at the Q point are labeled I_{DQ} and V_{DQ}, respectively, in Fig. 2–12(b).

In general, a number of dynamic load lines are needed to complete the analysis of i_D over a cycle of v_{th}. However, for the network under study, only dynamic load lines for the maximum and minimum values of v_{th} are required. The reason is that the diode characteristic is almost a straight line near the Q point [from a to b in Fig. 2-12(b)], so that the negligible distortion of i_d, the ac component of i_D, will occur. Thus, i_d will be of the same form as v_{th} (i.e., sinusoidal), and it can easily be sketched once the extremes of variation have been determined. The solution for i_D is thus

$$i_D = I_{DQ} + i_d = I_{DQ} + I_{dm} \sin \omega t = 36 + 8 \sin \omega t \text{ mA}$$

where I_{dm} is the amplitude of the sinusoidal term.

2.47 Solve Problem 2.43 graphically for i_D.

■ The Thévenin equivalent circuit has already been determined in Problem 2.43. The dc load line is given by

$$i_D = \frac{V_{Th}}{R_{Th}} - \frac{v_D}{R_{Th}} = \frac{1}{50} - \frac{v_D}{50} = 20 - 20 v_D \text{ mA} \qquad (1)$$

In Fig. 2-13, (*1*) has been superimposed on the diode characteristic, plotted from Fig. 2-8(b). As in Problem 2.46, equivalent time scales for v_{Th} and i_D are laid out adjacent to the characteristic curve. Since the diode characteristic is linear about the Q point over the range of operation, only dynamic load lines corresponding to the maximum and minimum of v_{Th} need be drawn. Once these two dynamic load lines are constructed parallel to the dc load line, i_D can be sketched.

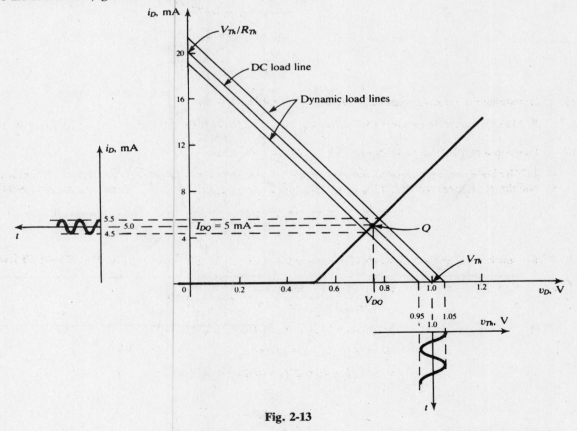

Fig. 2-13

2.48 A voltage source, $v_S = 0.4 + 0.2 \sin \omega t$ V, is placed directly across a diode characterized by Fig. 2-8(b). The source has no internal impedance and is of proper polarity to forward-bias the diode. Sketch the resulting diode current i_D.

A scaled plot of v_S has been laid adjacent to the v_D axis of the diode characteristic in Fig. 2-14. With zero resistance between the ideal voltage source and the diode, the dc load line has infinite slope and $v_D = v_S$. Thus, i_D is found by a point-by-point projection of v_S onto the diode characteristic, followed by reflection through the i_D axis. Notice that i_D is extremely distorted, bearing little resemblance to v_S.

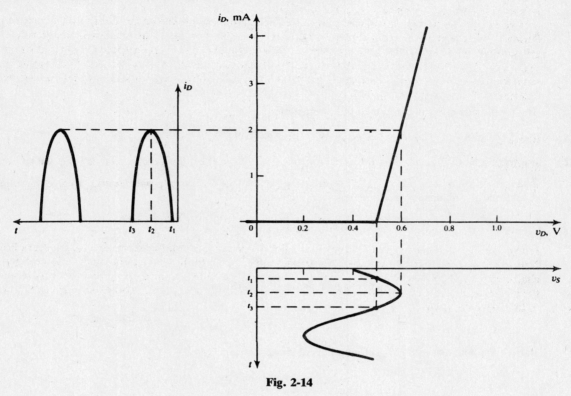

Fig. 2-14

2.49 Determine the value of quiescent current I_{DQ} in Problem 2.48.

▌ Quiescent conditions exist when the ac signal is zero. In this case, when $v_S = 0.4$ V, $i_D = I_{DQ} = 0$.

2.50 Develop a small-signal technique to find i_D and v_D in Problem 2.43.

▌ The Thévenin equivalent circuit of Problem 2.43 is valid here. Moreover, the intersection of the dc load line and the diode characteristic in Fig. 2-13 gives $i_{DQ} = 5$ mA and $V_{DQ} = 0.75$ V. Then, the *dynamic resistance* is

$$r_d = \frac{\Delta v_D}{\Delta i_D} = \frac{0.7 - 0.5}{0.004} = 50 \, \Omega$$

We now have all the values needed for analysis using the small-signal circuit of Fig. 2-15. By Ohm's law,

$$i_d = \frac{v_{th}}{R_{Th} + r_d} = \frac{0.05 \cos \omega t}{50 + 50} = 0.5 \cos \omega t \text{ mA}$$

$$v_d = r_d i_d = 50(0.0005 \cos \omega t) = 0.025 \cos \omega t \text{ V}$$

$$i_D = I_{DQ} + i_d = 5 + 0.5 \cos \omega t \text{ mA}$$

$$v_D = V_{DQ} + v_d = 0.75 + 0.025 \cos \omega t \text{ V}$$

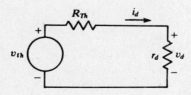

Fig. 2-15

2.51 For the circuit of Fig. 2-12, determine i_D.

▌ The Q-point current I_{DQ} has been determined as 36 mA (see Problem 2.46). The dynamic resistance of the diode at the Q point can be evaluated graphically.

$$r_d = \frac{\Delta v_D}{\Delta i_D} = \frac{0.37 - 0.33}{0.044 - 0.028} = 2.5 \, \Omega$$

Now the small-signal circuit of Fig. 2-15 can be analyzed to find i_d:

$$i_d = \frac{v_{th}}{R_{Th} + r_d} = \frac{0.1 \sin \omega t}{10 + 2.5} = 0.008 \sin \omega t \, \text{A}$$

The total diode current is obtained by superposition:

$$i_D = I_{DQ} + i_d = 36 + 8 \sin \omega t \, \text{mA}$$

2.52 For the circuit of Fig. 2-12, determine i_D if $\omega = 10^8$ rad/s and the diffusion capacitance is known to be 5000 pF.

▌ From Problem 2.51, $r_d = 2.5 \, \Omega$. The diffusion capacitance C_d acts in parallel with r_d to give the following equivalent impedance for the diode, as seen by the ac signal:

$$Z_d = r_d \parallel jx_d = r_d \parallel \left(-j\frac{1}{\omega C_d}\right) = \frac{r_d}{1 + j\omega C_d r_d} = \frac{2.5}{1 + j(10^8)(5000 \times 10^{-12})(2.5)}$$

$$= 1.56 \underline{/-51.34°} = 0.974 - j1.218$$

In the frequency domain, the small-signal circuit (Fig. 2-15) yields

$$\bar{I}_d = \frac{\bar{V}_{th}}{R_{th} + Z_d} = \frac{0.1 \underline{/-90°}}{10 + 0.974 - j1.218} = \frac{0.1 \underline{/-90°}}{11.041 \underline{/-6.33°}} = 0.0091 \underline{/-83.67°} \, \text{A}$$

In the time domain, with I_{DQ} as found in Problem 2.46, we have

$$i_D = I_{DQ} + i_d = 36 + 9.1 \cos (10^8 t - 83.67°) \, \text{mA}$$

2.53 In the circuit of Fig. 2-3(a), assume $R_S = R_1 = 200 \, \Omega$, $R_L = 50 \, \text{k}\Omega$, and $v_S = 400 \sin \omega t$ V. The diode is ideal, with reverse saturation current $I_0 = 2 \, \mu A$ and a peak inverse voltage (PIV) rating of $V_R = 100$ V. Will the diode fail in avalanche breakdown (region of negative v_D where i_D increases without bound independent of v_D)?

▌ From Problem 2.38,

$$v_{Th} = \frac{R_1}{R_1 + R_s} v_S = \frac{200}{200 + 200} (400 \sin \omega t) = 200 \sin \omega t \, \text{V}$$

$$R_{Th} = \frac{R_1 R_S}{R_1 + R_S} = \frac{(200)(200)}{200 + 200} = 100 \, \Omega$$

The circuit to be analyzed is that of Fig. 2-3(c); the instants of concern are when $\omega t = (2n + 1)\pi/2$ for $n = 1, 2, 3, \ldots$, at which times $v_{Th} = -200$ V and thus v_D is at its most negative value. An application of KVL yields

$$v_D = v_{Th} - i_D(R_{Th} + R_L) = -200 - (-2 \times 10^{-6})(100 + 50 \times 10^3) = -199.9 \, \text{V} \qquad (1)$$

Since $v_D < V_R = -100$ V, avalanche failure occurs.

2.54 Determine a value of R_L that will prevent the avalanche breakdown failure for Problem 2.53.

▌ From (1) of Problem 2.53, it is apparent that $v_D \geq -100$ V if

$$R_L \geq \frac{v_{Th} - v_D}{i_D} - R_{Th} = \frac{-200 - (-100)}{-2 \times 10^{-6}} - 100 = 50 \, \text{M}\Omega$$

2.55 Describe the performance of the *positive clipping circuit* of Fig. 2-16(a).

▌ Diode D conducts for $v_i > V_b$, and hence, removes any portion of the input signal v_i that is greater than V_b. The circuit passes as the output signal v_o any portion of v_i that is less than V_b.

DIODES: SIGNAL LEVEL APPLICATION ☐ 55

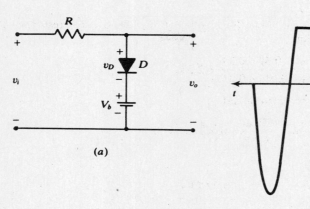
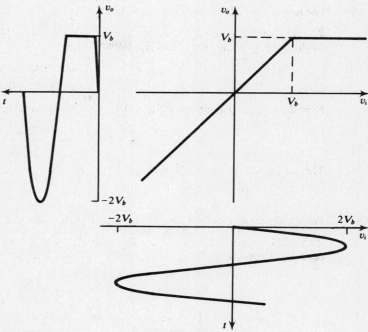

Fig. 2-16

2.56 Draw a *transfer characteristic* (a plot of v_o vs. v_i) for the clipping circuit of Fig. 2-16(a).

▌ Based on analysis of Problem 2.55, the transfer characteristic would show a one-to-one correspondence between v_o and v_i for $v_i \le V_b$. For $v_i > V_b$, v_o maintains the value of V_b. The $v_o - v_i$ plot of Fig. 2-16(b) shows the resulting transfer characteristic.

2.57 If $v_i = 2V_b \sin \omega t$ for the positive clipping circuit of Fig. 2-16(a), sketch the resulting waveform of v_o.

▌ Projection of instantaneous values of v_i from the lower sketch of Fig. 2-16(b) onto the transfer characteristic allows construction of the shown sketch of v_o.

2.58 Analyze the ideal *clamping circuit* of Fig. 2-17(b) if the triangular wave form of Fig. 2-17(a) is impressed as input.

▌ If the capacitor C is initially uncharged, the ideal diode D is forward-biased for $0 < t \le T/4$, and it acts as a short circuit while the capacitor charges to $v_C = V_p$. At $t = T/4$, D open-circuits, breaking the only possible discharge path for the capacitor. Thus, the value of $v_C = V_p$ is preserved; since v_i can never exceed V_p, D remains reverse-biased for all $t > T/4$, giving $v_o = v_D = v_i - V_p$. The function v_o is sketched in Fig. 2-17(c); all positive peaks are clamped at zero, and the average value is shifted from 0 to $-V_p$.

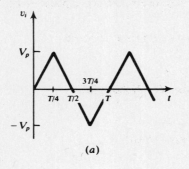
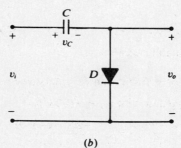
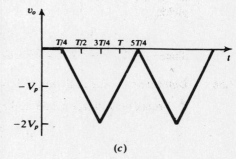

Fig. 2-17

2.59 In the circuit of Fig. 2-18, v_S is a 10-V square wave of period 4 ms, $R = 100\ \Omega$, and $C = 20\ \mu F$. Sketch v_C for the first two cycles of v_S if the capacitor is initially uncharged and the diode is ideal.

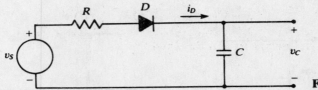
Fig. 2-18

■ In the interval $0 \leq t < 2$ ms,

$$v_C(t) = v_S(1 - e^{t/RC}) = 10(1 - e^{-500t}) \text{ V}$$

For $2 \leq t < 4$ ms, D blocks and the capacitor voltage remains at

$$v_C(2 \text{ ms}) = 10(1 - e^{-500(0.002)}) = 6.32 \text{ V}$$

For $4 \leq t < 6$ ms,

$$v_C(t) = v_S - (v_S - 6.32)e^{-(t-0.004)/RC} = 10 - (10 - 6.32)e^{-500(t-0.004)} \text{ V}$$

And for $6 \leq t < 8$ ms, D again blocks and the capacitor voltage remains at

$$v_C(6 \text{ ms}) = 10 - (10 - 6.32)e^{-500(0.002)} = 8.654 \text{ V}$$

The waveforms of v_S and v_C are sketched in Fig. 2-19.

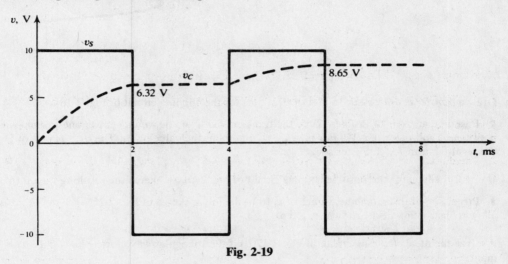
Fig. 2-19

2.60   If v_i is a 10-V triangular wave of period T and $V_b = 6$ V for the positive clipping circuit of Fig. 2-16(a), sketch one cycle of the output voltage v_o.

■ The diode blocks for $v_i < 6$ V, giving $v_o = v_i$. For $v_i \geq 6$ V, the diode is in forward conduction to yield $v_o = 6$ V. See Fig. 2-20 for resulting sketch.

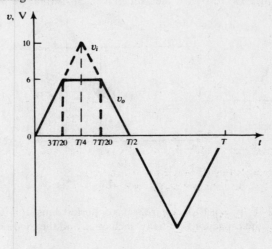
Fig. 2-20

2.61 Draw a transfer characteristic relating v_o to v_i for the positive clipping circuit of Problem 2.60.

▌ The diode blocks for $v_i < 6$ V and conducts for $v_i \geq 6$ V. Thus, $v_o = v_i$ for $v_i < 6$ V, and $v_o = 6$ V for $v_i \geq 6$ V. See Fig. 2-21(a) for resulting transfer characteristic.

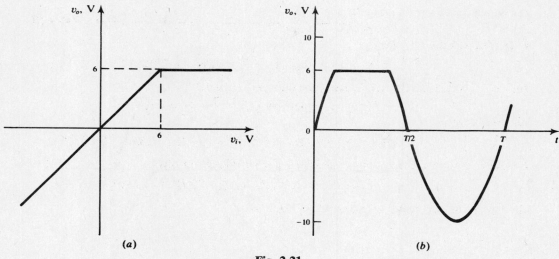

Fig. 2-21

2.62 Sketch one cycle of the output waveform for the circuit of Problem 2.60 if $v_i = 10 \sin \omega t$ V.

▌ The output v_o is a sine wave with the positive peak clipped at 6 V as shown by Fig. 2-21(b).

2.63 Reverse the diode in Fig. 2-16(a) to create a *negative clipping circuit*. Let $V_b = 6$ V, and draw the network transfer characteristic.

▌ The diode conducts for $v_i \leq 6$ V and blocks for $v_i > 6$ V. Consequently, $v_o = v_i$ for $v_i > 6$ V, and $v_o = 6$ V for $v_i \leq 6$ V. The transfer characteristic is drawn in Fig. 2-22(a).

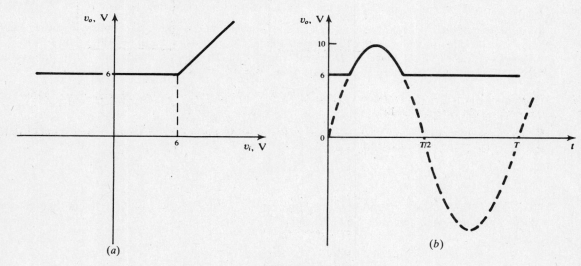

Fig. 2-22

2.64 For the negative clipping circuit of Problem 2.63, sketch one cycle of the output v_o if $v_i = 10 \sin \omega t$ V.

▌ With negative clipping, the output is made up of the positive peaks of $10 \sin \omega t$ above 6 V and is 6 V otherwise. Figure 2-22(b) displays the output waveform.

2.65 A signal, $v_i = 10 \sin \omega t$ V, is applied to the negative clamping circuit of Fig. 2-17(b). Treating the diode as ideal, sketch the output waveform for $1\frac{1}{2}$ cycles of v_i. The capacitor is initially uncharged.

■ For $0 \leq t \leq T/4$, the diode is forward-biased, giving $v_o = 0$ as the capacitor charges to $v_C = +10$ V. For $t > T/4$, $v_o \leq 0$, and thus the diode remains in the blocking mode, resulting in

$$v_o = -v_C + v_i = -10 + v_i = -10(1 - \sin \omega t) \text{ V}$$

The output waveform is sketched in Fig. 2-23.

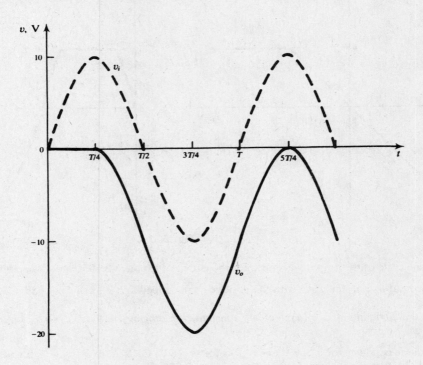

Fig. 2-23

2.66 The diodes in the circuit of Fig. 2-24 are ideal. Sketch the transfer characteristic for $-20 \text{ V} \leq V_1 \leq 20 \text{ V}$.

■ Inspection of the circuit shows that I_2 can have no component due to the 10-V battery because of the one-way conduction property of D_2. Therefore, D_1 is "off" for $V_1 < 0$; then $v_{D2} = -10$ V and $V_2 = 0$.

Now D_1 is "on" if $V_1 \geq 0$; however, D_2 is "off" for $V_2 < 10$. The onset of conduction for D_2 occurs when $V_{ab} = 10$ V with $I_2 = 0$, or when, by voltage division,

$$V_{ab} = V_2 = 10 = \frac{R_2}{R_1 + R_2} V_1$$

Hence,

$$V_1 = \frac{R_1 + R_2}{R_2}(10) = \frac{5 + 10}{10}(10) = 15 \text{ V} \tag{1}$$

Thus, if $V_1 \geq 15$ V, D_2 is "on" and $V_2 = 10$ V. But, for $0 \leq V_1 < 15$ V, D_2 is "off," $I_2 = 0$, and V_2 is given as a function of V_1 by (1). Figure 2-25 shows the composite result.

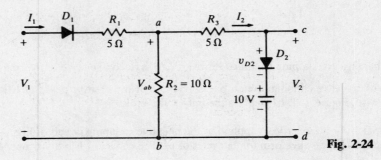

Fig. 2-24

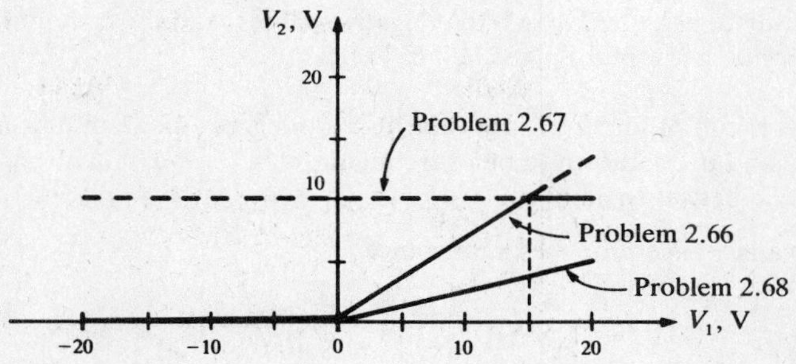

Fig. 2-25

2.67 Suppose diode D_2 is reversed in the circuit of Fig. 2-24. Sketch the resulting transfer characteristic for $-20\text{ V} \leq V_1 \leq 20\text{ V}$.

▮ Diode D_2 is now "on" and $V_2 = 10\text{ V}$ until V_1 increases enough so that $V_{ab} = 10\text{ V}$, at which point $I_2 = 0$. That is, $V_2 = 10\text{ V}$ until

$$V_2 = V_{ab} = 10 = \frac{R_2}{R_1 + R_2} V_1 = \frac{10}{5 + 10} V_1 = \frac{2}{3} V_1 \tag{1}$$

or until

$$V_1 = \tfrac{3}{2} V_2 = 15\text{ V}$$

For $V_1 > 15\text{ V}$, $I_2 = 0$ and (1) remains valid. The resulting transfer characteristic is shown dashed in Fig. 2-25.

2.68 Suppose a resistor $R_4 = 5\ \Omega$ is added across terminals c, d of the circuit of Fig. 2-24. Describe the changes that result in the transfer characteristic of Problem 2.66.

▮ There is no change in the transfer characteristic for $V_1 \leq 0$. However, D_2 remains "off" until $V_1 > 0$ increases to where $V_2 = 10\text{ V}$. At the onset of conduction for D_2, the current through D_2 is zero; thus,

$$I_1 = \frac{V_1}{R_1 + R_2 \| (R_3 + R_4)} = \frac{V_1}{10} \quad \text{and} \quad I_2 = \frac{R_2}{R_2 + R_3 + R_4} I_1 = \frac{I_1}{2}$$

Hence, by Ohm's law,

$$V_2 = I_2 R_4 = \frac{I_1 R_4}{2} = \frac{V_1 R_4}{20} = \frac{V_1}{4}$$

Thus, $V_1 = 40\text{ V}$ when $V_2 = 10\text{ V}$, and it is apparent that the breakpoint of Problem 2.66 at $V_1 = 15\text{ V}$ has moved to $V_1 = 40\text{ V}$. The transfer characteristic for $-20\text{ V} \leq V_1 \leq 20\text{ V}$ is sketched in Fig. 2-25.

2.69 Sketch the i–v input characteristic of the network of Fig. 2-26(a) when the switch is open.

▮ The solution is more easily found if the current source and resistor are replaced with the Thévenin equivalents $V_{Th} = IR$ and $R_{Th} = R$. KVL gives $v = iR_{Th} + IR$, which is the equation of a straight line intersecting the i axis at $-I$ and the v axis at IR. The slope of the line is $1/R$. The characteristic is sketched in Fig. 2-26(b).

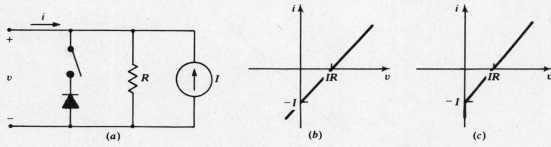

Fig. 2-26

2.70 Sketch the i–v input characteristic of the network of Fig. 2-26(a) when the switch is closed.

▮ The diode is reverse-biased and acts as an open circuit when $v > 0$. It follows that the i–v characteristic here is identical to that with the switch open if $v > 0$. But if $v \leq 0$, the diode is forward-biased, acting as a short

60 ☐ CHAPTER 2

circuit. Consequently, v can never reach the negative values, and the current i can increase negatively without limit. The corresponding i–v plot is sketched in Fig. 2-26(c).

2.71 In the small-signal circuit of Fig. 2-27, the capacitor models the diode diffusion capacitance, so that $C = C_d = 0.02\ \mu\text{F}$, and v_{th} is known to be of frequency $\omega = 10^7$ rad/s. Also, $r_d = 2.5\ \Omega$ and $Z_{Th} = R_{Th} = 10\ \Omega$. Find the phase angle between i_d and v_d.

▌ The diffusion capacitance produces a reactance

$$x_d = \frac{1}{\omega C_d} = \frac{1}{(10^7)(0.02 \times 10^{-6})} = 5\ \Omega$$

so that

$$Z_d = r_d \parallel (-jx_d) = \frac{(2.5)(5\ \underline{/-90°})}{2.5 - j5} = 2.236\ \underline{/-26.57°} = 2 - j1\ \Omega$$

Thus, i_d leads v_d by a phase angle of 26.57°.

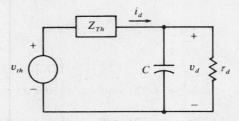

Fig. 2-27

2.72 Determine the phase angle between v_d and v_{th} in Prob. 2.71.

▌ Let Z_{eq} be the impedance looking to the right from v_{th}; then

$$Z_{eq} = Z_{Th} + Z_d = 10 + (2 - j1) = 12 - j1 = 12.04\ \underline{/-4.76°}\ \Omega$$

Hence, v_{th} leads v_d by an angle of $26.57° - 4.76° = 21.81°$.

2.73 Using ideal diodes, resistors, and batteries, synthesize a function-generator circuit that will yield the i–v characteristic of Fig. 2-28(a).

▌ Since the i–v characteristic has two breakpoints, two diodes are required. Both diodes must be oriented so that no current flows for $v < -5$ V. Further, one diode must move into forward bias at the first breakpoint, $v = -5$ V, and the second diode must begin conduction at $v = +10$ V. Note also that the slope of the i–v plot is the reciprocal of the Thévenin equivalent resistance of the active portion of the network.

The circuit of Fig. 2-28(b) will produce the given i–v plot if $R_1 = 6\ \text{k}\Omega$, $R_2 = 3\ \text{k}\Omega$, $V_1 = 5$ V, and $V_2 = 10$ V. These values are arrived at as follows:
1. If $v < -5$ V, both v_{D1} and v_{D2} are negative, both diodes block, and no current flows.
2. If $-5\ \text{V} \le v < 10\ \text{V}$, D_1 is forward-biased and acts as a short circuit, whereas v_{D2} is negative causing D_2 to

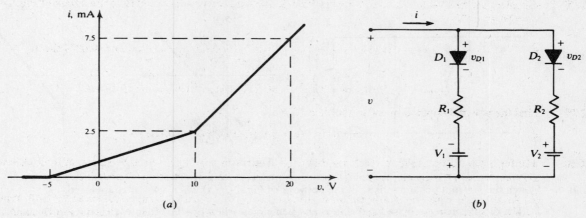

Fig. 2-28

act as an open circuit. R_1 is found as the reciprocal of the slope in that range:

$$R_1 = \frac{10-(-5)}{0.0025} = 6 \text{ k}\Omega$$

3. If $v \geq 10$ V, both diodes are forward-biased,

$$R_{Th} = \frac{R_1 R_2}{R_1 + R_2} = \frac{\Delta v}{\Delta i} = \frac{20-10}{(7.5-2.5) \times 10^{-3}} = 2 \text{ k}\Omega$$

and

$$R_2 = \frac{R_1 R_{Th}}{R_1 - R_{Th}} = \frac{(6 \times 10^3)(2 \times 10^3)}{4 \times 10^3} = 3 \text{ k}\Omega$$

2.74 The diode of Problem 2.31 is operating in a circuit where it has dynamic resistance $r_d = 100\,\Omega$. What must be the quiescent voltage V_{DQ}?

❙ Based on (1) of Problem 2.18,

$$\frac{1}{r_d} = \frac{\partial i_D}{\partial v_D}\bigg|_Q = \frac{\partial}{\partial v_D}[I_0(e^{v_D/\eta V_T}-1)]\bigg|_Q = \frac{I_0}{\eta V_T}e^{V_{DQ}/\eta V_T}$$

or

$$V_{DQ} = \eta V_T \ln\left(\frac{\eta V_T}{r_d I_0}\right) = (1)(0.02587)\ln\left[\frac{(1)(0.02587)}{(100)(10 \times 10^{-9})}\right] = 0.263 \text{ V}$$

2.75 Calculate the quiescent current I_{DQ} for Problem 2.74.

❙ $$I_{DQ} = i_D|_Q = I_0(e^{v_D/\eta V_T}-1)|_Q = (10 \times 10^{-9})(e^{0.263/(1)(0.02587)}-1) = 0.260 \text{ mA}$$

2.76 The diode of Problem 2.31 has a forward current $i_D = 2 + 0.004 \sin \omega t$ mA. Find the quiescent voltage V_{DQ}.

❙ $$I_{DQ} = 0.002 = i_D|_Q \approx I_o e^{V_{DQ}/\eta V_T} = (10 \times 10^{-9})e^{V_{DQ}/(1)(0.02587)}$$

Solution gives $V_{DQ} = 0.3395$ V.

2.77 Find the total voltage $v_D = V_{DQ} + v_d$ across the diode of Problem 2.76.

❙ Based on Problem 2.74,

$$r_d = \frac{\eta V_T}{I_o}e^{-V_{DQ}/\eta V_T} = \frac{(1)(0.02587)}{10 \times 10^{-9}}e^{-0.3395/(1)(0.02587)} = 5.17\,\Omega$$

and
$$v_d = r_d i_d = (5.17)(0.004 \times 10^{-3} \sin \omega t) = 0.0207 \sin \omega t \text{ mV}$$

Using the result of Problem 2.76,

$$v_D = 339.5 + 0.0207 \sin \omega t \text{ mV}$$

2.78 Calculate the rms value of current flowing through the load resistor $R_L = 100\,\Omega$ of the circuit of Fig. 2-5(a) if $v_S = 10 \sin \omega t$ V.

❙ $$i_D = 0 \text{ for } v_S < 0.\text{ For } v_S \geq 0,\ i_D = \frac{v_S}{R_L + R_S} = \frac{10}{100+10}\sin \omega t = 0.0909 \sin \omega t.$$

$$I_{Drms} = \left(\frac{1}{T}\int_0^T i_D^2\,dt\right)^{1/2} = \left[\frac{1}{2\pi}\int_0^\pi (0.0909 \sin \omega t)\,d(\omega t)\right]^{1/2} = 45.45 \text{ mA}$$

2.79 Find the power dissipated by R_L in Problem 2.78.

❙ $$P_L = I_{Drms}^2 R_L = (0.04545)^2(100) = 206.6 \text{ mW}$$

2.80 The logic AND gate of Fig. 2-29(a) has trains of input pulses arriving at the gate inputs, as indicated by Fig. 2-29(b). Signal v_2 is erratic, dropping below nominal logic level on occasion. Determine v_o.

❙ At all points where v_1 or v_2 equal zero ($0 < t < 1$ ms, $2 < t < 3$ ms) a diode conducts and $v_o = 0$. When $v_1 = v_2 = 10$ V ($1 \leq t \leq 2$ ms), both diodes block and $v_o = 10$. For $4 \leq t \leq 5$ ms, D_1 blocks and D_2 conducts, requiring $v_o = v_2 = 5$ V.

62 ☐ **CHAPTER 2**

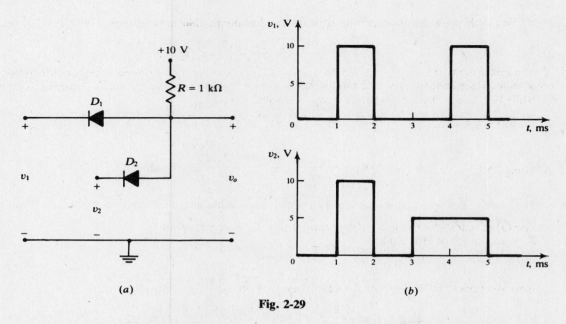

(a) (b)

Fig. 2-29

2.81 The logic AND gate of Fig. 2-29(a) is to be used to generate a crude pulse train by letting $v_1 = 5 + 5 \sin \omega t$ V and $v_2 = 5$ V. Determine (a) the amplitude and (b) the period of the pulse train appearing as v_o.

▌ (a) The circuit simply passes as the output v_o the lesser of v_1 and v_2. Thus, for $0 \leq \omega t < \pi$, $v_o = v_2 = 5$ V. Over the interval $\pi \leq \omega t < 2\pi$, $v_o = v_1 = 5 + 5 \sin \omega t$.
  (b) The pulse train is of period $T = 2\pi/\omega$.

2.82 The forward voltage across a Ge diode is $v_D = 0.3 + 0.060 \cos \omega t$ V. Find the ac component of the diode current i_d if $I_0 = 10$ nA.

▌ Based on Prob. 2.77,

$$r_d = \frac{\eta V_T}{I_0} e^{-V_{DQ}/\eta V_T} = \frac{(1)(0.02587)}{(10 \times 10^{-9})} e^{-0.3/(1)(0.02587)} = 23.8 \ \Omega$$

$$i_d = \frac{v_d}{r_d} = \frac{0.06}{23.8} \cos \omega t = 2.52 \cos \omega t \ \text{mA}$$

SPECIAL PURPOSE DIODES

2.83 Find the voltage v_Z across the Zener diode of Fig. 2-30(a) if $i_Z = 10$ mA and it is known that $V_Z = 5.6$ V, $I_Z = 25$ mA, and $R_Z = 10 \ \Omega$.

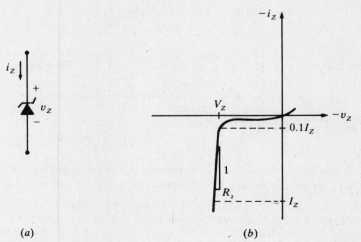

(a) (b) **Fig. 2-30**

▌ Since $0.1I_z \le i_Z \le I_Z$, operation is along the safe and predictable region of Zener operation. Consequently,
$$v_Z \approx V_Z + i_Z R_Z = 5.6 + (10 \times 10^{-3})(10) = 5.7 \text{ V}$$

2.84 The Zener diode in the voltage regulator circuit of Fig. 2-31 has a constant reverse breakdown voltage $V_Z = 8.2$ V for 75 mA $\le i_Z \le 1$ A. If $R_L = 9\,\Omega$, size R_S so that $v_L = V_Z$ is regulated to (maintained at) 8.2 V while V_b varies by ± 10 percent from its nominal value of 12 V.

▌ By Ohm's law,
$$i_L = \frac{v_L}{R_L} = \frac{V_Z}{R_L} = \frac{8.2}{9} = 0.911 \text{ A}$$

Using KVL,
$$R_S = \frac{V_b - V_Z}{i_Z + i_L} \tag{1}$$

Use (1) to size R_S for maximum Zener current I_Z at the largest value of V_b:
$$R_S = \frac{(1.1)(12) - 8.2}{1 + 0.911} = 2.62\,\Omega$$

Now we check to see if $i_Z \ge 75$ mA at the lowest value of V_b:
$$i_Z = \frac{V_b - v_Z}{R_S} - i_L = \frac{(0.9)(12) - 8.2}{2.62} - 0.911 = 81.3 \text{ mA}$$

Since $i_Z > 75$ mA, $v_Z = V_Z = 8.2$ V and regulation is preserved.

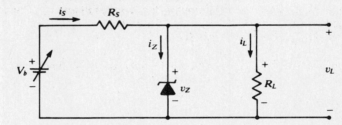

Fig. 2-31

2.85 A Zener diode has the specifications $V_Z = 5.2$ V and $P_{D\,\text{max}} = 260$ mW. Assume $R_Z = 0$. Find the maximum allowable current i_Z when the Zener diode is acting as a regulator.

▌
$$i_{z\,\text{max}} = I_Z = \frac{P_{D\,\text{max}}}{V_Z} + \frac{260 \times 10^{-3}}{5.2} = 50 \text{ mA}$$

2.86 If a single-loop circuit consists of an ideal 15-V dc source V_S, a variable resistor R, and the Zener diode of Problem 2.84, find the range of values of R for which the Zener diode remains in constant reverse voltage breakdown with no danger of failure.

▌ By KVL,
$$V_S = Ri_Z + V_Z \quad \text{so that} \quad R = \frac{V_S - V_Z}{i_Z}$$

From Problem 2.83 we know that regulation is preserved if
$$R \le \frac{V_S - V_Z}{0.1 I_{Z\,\text{max}}} = \frac{15 - 5.2}{(0.1)(50 \times 10^{-3})} = 1.96 \text{ k}\Omega$$

Overcurrent failure is avoided if
$$R \ge \frac{V_S - V_Z}{I_{Z\,\text{max}}} = \frac{15 - 5.2}{50 \times 10^{-3}} = 196\,\Omega$$

Thus, we need $196\,\Omega \le R \le 196$ kΩ.

2.87 The Zener diode in the voltage-regulator circuit of Fig. 2-31 has $v_Z = V_Z = 18.6$ V at a minimum i_Z of 15 mA. If $V_b = 24 \pm 3$ V and R_L varies from 250 Ω to 2 kΩ, find the value of R_S to maintain regulation.

64 ☐ CHAPTER 2

▮ With $v_L = v_Z = V_Z$, the range on i_L as R_L varies is

$$\frac{V_Z}{R_{L\,max}} \leq i_L \leq \frac{V_Z}{R_{L\,min}} \quad \text{or} \quad 9.3\,\text{mA} \leq i_L \leq 74.4\,\text{mA}$$

Minimum Zener current exists when i_L is maximum and V_b is minimum. Hence, application of KVL leads to

$$\frac{V_{b\,min} - V_Z}{R_S} - i_{L\,max} \geq i_{Z\,min} \quad \text{or} \quad R_S < \frac{V_{b\,min} - V_Z}{i_{Z\,min} + i_{L\,max}} = \frac{21 - 18.6}{0.015 + 0.0744} = 26.8\,\Omega$$

2.88 Specify the minimum power rating of the Zener diode of Problem 2.87 if $R_S = 20\,\Omega$.

▮ Maximum Zener current occurs when V_b is maximum and i_L is minimum. Based on Problem 2.87,

$$i_{Z\,max} = \frac{V_{b\,max} - V_Z}{R_S} - i_{L\,min} = \frac{27 - 18.6}{20} - 0.0093 = 410.7\,\text{mA}$$

$$P_Z \geq V_Z i_{Z\,max} = (18.6)(0.4107) = 7.64\,\text{W}$$

2.89 The regulator circuit of Fig. 2-31 is modified by replacing the Zener diode with two Zener diodes in series to obtain a regulated voltage of 20 V. The characteristics of the two Zeners are:

Zener 1: $V_Z = 9.2\,\text{V}$ for $15 \leq i_Z \leq 300\,\text{mA}$
Zener 2: $V_Z = 10.8\,\text{V}$ for $12 \leq i_Z \leq 240\,\text{mA}$

If i_L varies from 10 mA to 90 mA and V_b varies from 22 V to 26 V, size R_S so that regulation is preserved.

▮ Since the two Zener diodes carry the same current, design must assure that $i_Z = i_{Z1} = i_{Z2} \geq 12\,\text{mA}$ to maintain regulation of Zener diode 2. Based on Problem 2.87, with V_Z replaced by $V_{Z1} + V_{Z2}$,

$$R_S < \frac{V_{b\,min} - V_{Z1} - V_{Z2}}{i_{Z\,min} + i_{L\,max}} = \frac{26 - 9.2 - 10.8}{0.012 + 0.090} = 19.6\,\Omega$$

2.90 Will either Zener diode of Problem 2.89 exceed its rated value at light load conditions if $R_S = 19.6\,\Omega$?

▮ Based on Problem 2.88,

$$i_{Z1\,max} = i_{Z2\,max} = \frac{V_{b\,max} - V_{Z1} - V_{Z2}}{R_S} - i_{L\,min} = \frac{26 - 9.2 - 10.8}{19.6} - 0.010 = 296.1\,\text{mA}$$

Zener 1 is properly sized, but Zener 2 will exceed its rated current.

2.91 Determine the maximum value of i_L for the circuit of Problem 2.89 to preserve regulation if $R_S = 25\,\Omega$.

▮ Based on Problem 2.87,

$$i_{L\,max} = \frac{V_{b\,min} - V_{Z1} - V_{Z2}}{R_S} - i_{z\,min} = \frac{22 - 9.2 - 10.8}{25} - 0.012 = 92\,\text{mA}$$

2.92 Calculate the minumum value of i_L for the circuit of Problem 2.89 to preserve regulation if $R_S = 25\,\Omega$.

▮ Based on Problem 2.87,

$$i_{L\,min} = \frac{V_{b\,max} - V_{Z1} - V_{Z2}}{R_S} - i_{z\,max} = \frac{26 - 9.2 - 10.8}{25} - 0.240 = 0$$

2.93 Find the range of R_L for which the circuit of Problem 2.89 will preserve regulation for $R_S = 25\,\Omega$.

▮ Based on Problems 2.91 and 2.92, application of Ohm's law requires that

$$R_{L\,max} \leq \frac{V_{Z1} + V_{Z2}}{i_{L\,min}} = \frac{9.2 + 10.8}{0} \to \infty \quad R_{L\,min} \geq \frac{V_{Z1} + V_{Z2}}{i_{L\,max}} = \frac{9.2 + 10.8}{0.092} = 217.4\,\Omega$$

2.94 The two Zener diodes of Fig. 2-32 have negligible forward voltage drops, and both regulate at constant V_Z for $50\,\text{mA} \leq i_Z \leq 500\,\text{mA}$. If $R_1 = R_L = 10\,\Omega$, $V_{Z1} = 8\,\text{V}$, and $V_{Z2} = 5\,\text{V}$, is regulation of the Zener diodes preserved for v_i being a 10-V square wave?

■ For $v_i = 10$ V,

$$i_{Z2} = \frac{v_i - V_{Z2}}{R_1 + R_L} = \frac{10-5}{10+10} = 250 \text{ mA}$$

For $v_i = -10$ V,

$$i_{Z1} = -\frac{v_i + V_{Z1}}{R_1 + R_L} = -\frac{-10+8}{10+10} = 100 \text{ mA}$$

Hence, regulation is maintained for both Zener diodes.

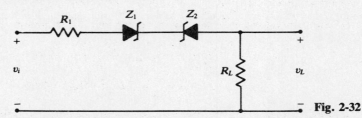

Fig. 2-32

2.95 Calculate the value of output voltage v_L for Problem 2.94.

■ By voltage division, for $v_i = 10$ V,

$$v_L = (v_i - V_{Z2})\left(\frac{R_L}{R_L + R_1}\right) = \frac{(10-5)(10)}{10+10} = 2.5 \text{ V}$$

For $v_i = -10$ V,

$$v_L = -(v_i + V_{Z1})\left(\frac{R_L}{R_L + R_1}\right) = -\frac{(-10+8)(10)}{10+10} = -1 \text{ V}$$

$$V_{L \text{ ave}} = \frac{1}{T}\int_0^T v_L \, dt = \frac{1}{T}[2.5(T/2) - 1(T/2)] = 0.75 \text{ V}$$

2.96 The Zener diode of Problem 2.85 is used in a simple series circuit consisting of a variable dc voltage source V_S, the Zener diode, and current limiting resistor $R = 1$ kΩ. Find the allowable range of V_S for which the Zener diode is safe and regulation is preserved.

■ To preserve regulation ($i_Z \geq 0.1 I_Z$), KVL requires

$$V_S \geq R(0.1 I_Z) + V_Z = (1 \times 10^3)(50 \times 10^{-3}) + 5.2 = 10.2 \text{ V}$$

To assure $P_{D \text{ max}}$ is not exceeded, KVL gives

$$V_S \leq R I_Z + V_Z = (1 \times 10^3)(50 \times 10^{-3}) + 5.2 = 55.2 \text{ V}$$

2.97 Derive an expression for the power dissipated by the Zener diode of the circuit described by Problem 2.96.

■ By KVL,
$$i_Z = \frac{V_S - V_Z}{R}$$

and
$$P_D = V_Z i_Z = V_Z(V_S - V_Z)/R$$

2.98 Calculate the range of power dissipated by the circuit of Problem 2.96 over the range of V_S for which regulation is preserved.

■ Based on Problem 2.97,

$$P_{D \text{ max}} = \frac{V_Z(V_{S \text{ max}} - V_Z)}{R} = \frac{5.2(55.2 - 5.2)}{1000} = 260 \text{ mW}$$

$$P_{D \text{ min}} = \frac{V_Z(V_{S \text{ min}} - V_Z)}{R} = \frac{5.2(10.2 - 5.2)}{1000} = 26 \text{ mW}$$

2.99 A *light-emitting diode* (LED) has a greater forward voltage drop than does a common signal diode. A typical LED can be modeled as a constant forward voltage drop $v_D = 1.6$ V. Its *luminous intensity* I_v varies directly with forward current and is described by

$$I_v = 40i_D \approx \text{millicandela(mcd)} \qquad (1)$$

A series circuit consists of such an LED, a current-limiting resistor R, and a 5-V dc source V_S. Find the value of LED current i_D such that the luminous intensity is 1 mcd.

▌ By (1), we must have

$$i_D = \frac{I_v}{40} = \frac{1}{40} = 25 \text{ mA}$$

2.100 Determine the value of R for the circuit of Problem 2.99 such that the luminous intensity is 1 mcd.

▌ From KVL, we have

$$V_S = Ri_D + 1.6$$

so that

$$R = \frac{V_S - 1.6}{i_D} = \frac{5 - 1.6}{25 \times 10^{-3}} = 136 \, \Omega$$

2.101 The reverse breakdown voltage V_R of the LED of Problem 2.99 is guaranteed by the manufacturer to be no lower than 3 V. Knowing that the 5-V dc source may be inadvertently applied so as to reverse-bias the LED, we wish to add a Zener diode to ensure that reverse breakdown of the LED can never occur. A Zener diode is available with $V_Z = 4.2$ V, $I_Z = 30$ mA, and a forward drop of 0.6 V. Describe the proper connection of the Zener in the circuit to protect the LED, and find the value of the luminous intensity that will result if R is unchanged from Problem 2.99.

▌ The Zener diode and LED should be connected in series so that the anode of one device connects to the cathode of the other. Then, even if the 5-V source is connected in reverse, the reverse voltage across the LED will be less that $5 - 4.2 = 0.8$ V < 3 V. When the dc source is connected to forward-bias the LED, we will have

$$i_D = \frac{V_S - V_{F\,LED} - V_{FZ}}{R} = \frac{5 - 1.6 - 0.6}{136} = 20.6 \text{ mA}$$

so that

$$I_v = 40i_D = (40)(20.6 \times 10^{-3}) = 0.824 \text{ mcd}$$

2.102 An LED with luminous intensity described by (1) of Problem 2.99 is modeled by the piecewise-linear function of Fig. 2-33, with $R_F = 3 \, \Omega$ and $V_F = 1.5$ V. Find the maximum and minimum diode current that result if the LED is used in a series circuit consisting of the LED, a current-limiting resistor $R = 125 \, \Omega$, and a source $v_S = 5 + 1.13 \sin 0.1t$ V.

▌ Application of KVL yields

$$i_{D\,\max} = \frac{v_{S\,\max} - V_F}{R + R_F} = \frac{(5 + 1.13) - 1.5}{100 + 3} = 44.95 \text{ mA}$$

$$i_{D\,\min} = \frac{v_{S\,\min} - V_F}{R + R_F} = \frac{(5 - 1.13) - 1.5}{100 + 3} = 23.01 \text{ mA}$$

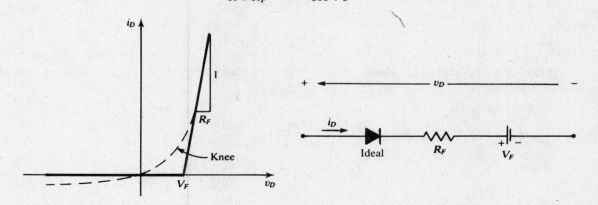

Fig. 2-33

2.103 Calculate the maximum and minimum luminous intensity of the LED of Problem 2.102.

▌ Since the period of v_S exceeds 1 minute, it is logical to assume that the luminous intensity follows i_D without the necessity to consider the physics of the light-emitting process.

$$I_{v\,max} = 40i_{D\,max} = 40(0.04495) = 1.798 \text{ mcd}$$

$$I_{v\,min} = 40i_{D\,min} = 40(0.02301) = 0.9204 \text{ mcd}$$

2.104 The *varactor diode* is designed to operate reverse-biased and is manufactured by a process that increases the voltage-dependent depletion capacitance or junction capacitance C_j. A varactor diode is frequently connected in parallel with an inductor L to form a resonant circuit for which the resonant frequency, $f_R = 1/2\pi\sqrt{LC_j}$, is voltage-dependent. Such a circuit can form the basis of a *frequency modulation* (FM) transmitter. If a varactor diode whose depletion capacitance is $C_j = 10^{-11}/(1 - 0.75v_D)^{1/2}$ F is reverse-biased with a voltage magnitude of 10 V, find C_j.

▌ $$C_j = 10^{-11}/[1 - 0.75v_D]^{1/2} = 10^{-11}/[1 - 0.75(-10)]^{1/2} = 3.244 \text{ pF}$$

2.105 If the varactor diode of Problem 2.104 is connected in parallel with a 0.8-μH inductor, find the value of v_D to establish resonance at a frequency of 100 MHz.

▌ The required depletion capacitance is

$$C_j = \frac{1}{(2\pi f_R)^2 L} = \frac{1}{[(2\pi)(100 \times 10^6)]^2(0.8 \times 10^{-6})} = 3.1663 \text{ pF}$$

Solving the depletion capacitance expression for v_D gives

$$v_D = [1 - 10^{-22}/C_j^2]/0.75 = [1 - 10^{-22}/(3.1663 \times 10^{-12})^2]/0.75 = -11.966 \text{ V}$$

Review Problems

2.106 If the resistivity of Problem 2.1 is given for a temperature of 20°C, determine the voltage drop across the rectangular copper conductor if the temperature of the bar is increased to 100°C.

▌ $$R_2 = \left[\frac{T_2 + 234.5}{T_1 + 234.5}\right]R_1 = \left[\frac{100 + 234.5}{20 + 234.5}\right](0.6896) = 0.906 \text{ }\Omega$$

$$V = IR_2 = (100 \times 10^{-3})(0.906) = 0.0906 \text{ V} = 90.6 \text{ mV}$$

2.107 If the material of the conductor of Problem 2.1 were changed to steel having a conductivity (σ) 10 per cent that of copper, find the voltage drop across the conductor length.

▌ $$\sigma = 1/\rho = 0.1\frac{1}{1.724 \times 10^{-8}} = 5.8 \times 10^6 \text{ S}$$

$$R = \frac{l}{\sigma A} = \frac{4 \times 10^{-3}}{(5.8 \times 10^6)(1 \times 10^{-4})(10^{-6})} = 6.896 \text{ }\Omega$$

$$V = IR = (100 \times 10^{-3})(6.896) = 0.6896 \text{ V} = 689.6 \text{ V}$$

2.108 Determine the drift velocity of conducting electrons for the copper conductor of Problem 2.1 if the current is reduced to 10 mA.

▌ Resistance is unchanged from value of Problem 2.1. By Ohm's law,

$$E = \frac{V}{l} = \frac{IR}{l} = \frac{(10 \times 10^{-3})(0.6896)}{0.4} = 0.01724 \text{ V/cm}$$

The mobility of Problem 2.4 is applicable; hence,

$$v_d = \mu E = (42.90)(0.01724) = 0.7396 \text{ cm/s}$$

2.109 If the carrier density (n_i) of intrinsic Ge is 2.4×10^{13} cm^{-3}, determine the percentage of Ge atoms that contribute an electron-hole pair to conduction if the mass density of Ge is 5.32 g/cm^3.

■ Referring to Problem 2.6 and finding $W_A = 72.6$ from a periodic table,

$$N_a\gamma = \frac{A_a\gamma}{W_a} = \frac{(6.023 \times 10^{23})(5.32)}{72.6} = 4.42 \times 10^{22} \text{ atoms/cm}^3$$

$$\text{Carriers per atom} = \frac{n_i}{N_a\gamma}(100\%) = \frac{2.4 \times 10^{13}}{4.42 \times 10^{22}}(100\%) = 0.543 \times 10^{-7}\%$$

or approximately 1 of every 2 billion atoms contributes a carrier.

2.110 Find the density (n_e) of valence band electrons available for conduction of current in an aluminum specimen of mass density (γ) = 2.77 g/cm^3.

■ From a periodic table of elements find the atomic weight of aluminum at 26.98 g/mole and with 3 valence band electrons available for conduction.

$$n_e = 3N_a\gamma = \frac{3A_a\gamma}{W_a} = \frac{(3)(6.023 \times 10^{23})(2.77)}{26.98} = 1.854 \times 10^{23} \text{ cm}^{-3}$$

2.111 Determine the mobility of electrons in an aluminum specimen with conductivity $\sigma = 2.91 \times 10^5$ ($\Omega \cdot$ cm)$^{-1}$.

■ Based on Problem 2.110,

$$\mu = \frac{\sigma}{n_e q} = \frac{2.91 \times 10^5}{(1.854 \times 10^{23})(1.6 \times 10^{-19})} = 9.81 \text{ cm}^2/\text{V} \cdot \text{s}$$

2.112 Calculate the conductivity of intrinsic Ge if electron mobility μ_n = 3900 cm^2/V · s, hole mobility μ_p = 1900 cm^2/V · s, and the intrinsic carrier density $n_i = 2.4 \times 10^{13}$ cm^{-3}.

■ $\sigma_i = q(n\mu_p + p\mu_p) = qn_i(\mu_n + \mu_p) = (1.6 \times 10^{-19})(2.4 \times 10^{13})(3900 + 1900) = 0.0223$ ($\Omega \cdot$ cm)$^{-1}$

2.113 If a Ge specimen is doped with an appropriate p-type (acceptor) material (trivalent element such as Al, B or Ga) to the small concentration of one part per half-million atoms, determine the resulting conductivity for the specimen of Problem 2.109.

■ Each atom of the impurity will yield an excess hole. Based on Problem 2.109,

$$p = \frac{N_a\gamma}{0.5 \times 10^6} = \frac{4.42 \times 10^{22}}{0.5 \times 10^6} = 8.84 \times 10^{16} \text{ cm}^{-3}$$

Neglecting conduction due to intrinsic carriers (see Problem 2.109) and based on data of Problem 2.112,

$$\sigma_d = pq\mu_p = (8.84 \times 10^{16})(1.6 \times 10^{-19})(1900) = 26.87 \quad (\Omega \cdot \text{cm})^{-1}$$

2.114 If an electric field with intensity of 5 V/m is applied across an aluminum conductor, find the drift velocity of free electrons.

■ Based on Problem 2.111,

$$v_d = \mu E = (9.81)(5 \times 10^{-2}) = 0.49 \text{ cm/s}$$

2.115 For the lightly doped semiconductor material of Problem 2.17, find the ratio of donor atoms to Si atoms.

■ Based on Problems 2.6 and 2.7,

$$\frac{N_D}{N_a\gamma} = \frac{5 \times 10^8}{4.99 \times 10^{22}} \approx \frac{1}{10^{14}}$$

or approximately 1 carrier per 100 trillion atoms.

2.116 Find the intrinsic carrier density for a Si specimen at 340°K.

■ Based on Problem 2.13,

$$n_i = (3.87 \times 10^{16})(340)^{3/2} e^{-7000/340} = 2.78 \times 10^{11} \text{ cm}^{-3}$$

DIODES: SIGNAL LEVEL APPLICATION ☐ 69

2.117 Repeat Problem 2.17 if the semiconductor is operating at 340°K.

▌ n_i of Problem 2.116 is pertinent.

$$p = \tfrac{1}{2}[-5 \times 10^8 + \sqrt{(5 \times 10^8)^2 + 4(2.78 \times 10^{11})^2}] = 2.78 \times 10^{11} \text{ cm}^{-3}$$

$$n = \frac{n_i^2}{p} = \frac{(2.78 \times 10^{11})^2}{2.78 \times 10^{11}} = 2.78 \times 10^{11} \text{ cm}^{-3}$$

Since $n \approx p \approx n_i$, this lightly doped semiconductor is acting as intrinsic Si.

2.118 Show that the typically doped semiconductor of Problem 2.16 does not appear intrinsic ($n \approx p \approx n_i$) at the elevated temperature of 340°K.

▌ n_i of Problem 2.116 is applicable. Since $N_D \gg n_i$,

$$n \approx N_D = 1 \times 10^{16} \text{ cm}^{-3}$$

$$p \approx \frac{n_i^2}{N_D} = \frac{(2.78 \times 10^{11})^2}{1 \times 10^{16}} = 7.73 \times 10^{22} \text{ cm}^{-3}$$

2.119 Find the resistivity of Si at 340°K and compare the effect of temperature on a semiconductor versus a metal conductor.

▌ Based on Problems 2.12, 2.116, and 2.7,

$$\mu_n = 1300(340/300)^{2.6} = 1800.0 \text{ cm}^2/\text{V} \cdot \text{s}$$

$$\mu_p = 500(340/300)^{2.6} = 692.3 \text{ cm}^2/\text{V} \cdot \text{s}$$

$$\rho = \frac{1}{qn_i(\mu_n + \mu_p)} = \frac{1}{(1.6 \times 10^{-19})(2.78 \times 10^{11})(1800.0 + 692.3)} = 9020.8 \text{ }\Omega \cdot \text{cm}$$

Compare the result with Problem 2.7 to see that the resistivity of a semiconductor decreases with increase in temperature whereas the resistivity of a metal conductor increases with increase in temperature.

2.120 To a close approximation, the reverse saturation current of a Ge diode doubles for each increase of 7°C in temperature. Find the increase in temperature ΔT necessary to increase I_o by a factor of 100.

▌ $$\frac{I_{02}}{I_{01}} = 100 = 2^{\Delta T/7} \quad \text{or} \quad \Delta T = 7 \ln(100)/\ln(2) = 46.5°\text{C}$$

2.121 Laboratory data for a Ge diode show that $i_D = 2$ mA when $v_D = 0.36$ V, and $i_D = 10$ mA for $i_D = 0.4$ V. Find the temperature for which the data were recorded.

▌ Based on Problem 2.34,

$$T = \frac{q(v_{D2} - v_{D1})}{\eta k \ln(i_{D2}/i_{D1})} = \frac{(1.6 \times 10^{-19})(0.4 - 0.36)}{(1.38 \times 10^{-23}) \ln(10/2)} = 288.2°\text{K}$$

2.122 Determine the reverse saturation current for the diode of Problem 2.121.

▌ $$I_0 = \frac{i_{D2}}{e^{qv_{D2}/\eta kT}} = \frac{10 \times 10^{-3}}{e^{(1.6 \times 10^{-19})(0.4)/(1.38 \times 10^{-23})(288.2)}} = 1.027 \text{ nA}$$

2.123 Repeat Problem 2.27 for a Si diode.

▌ Based on Problem 2.27,

$$v_D = \eta V_T \ln(0.01) = (2)(0.02587) \ln(0.01) = -0.2383 \text{ V}$$

2.124 Repeat Problem 2.27 if $T = 425°\text{K}$.

▌ By (1) of Problem 2.18,

$$V_T = \frac{kT}{q} = \frac{(1.38 \times 10^{-23})(425)}{1.6 \times 10^{-19}} = 0.0366 \text{ V}$$

Based on Problem 2.27,
$$v_D = \eta V_T \ln(0.01) = (1)(0.0366)\ln(0.01) = -0.1685 \text{ V}$$

2.125 Repeat Problem 2.29 for a Si diode.

$$\frac{i_{D2}}{i_{D1}} = \frac{e^{v_{D2}/\eta V_T} - 1}{e^{V_{D1}/\eta V_T} - 1} = \frac{e^{0.4/(2)(0.02587)} - 1}{e^{0.3/(2)(0.02587)} - 1} = 6.926$$

Hence,
$$i_{D2} = (6.926)(10 \times 10^{-3}) = 69.26 \text{ mA}$$

2.126 Find the reverse saturation current for the Si diode of Problem 2.125.

By (1) of Problem 2.18,
$$I_o = \frac{i_{D1}}{e^{v_{D1}/\eta V_T} - 1} = \frac{10 \times 10^{-3}}{e^{0.3/(2)(0.02587)} - 1} = 30.42 \text{ mA}$$

2.127 Find the per unit change in current of a Ge diode for which $V_T = 26$ mV if the forward voltage drop v_D decreases by 50 mV.

Based on (1) of Problem 2.20,
$$\frac{i_{D2}}{i_{D1}} = e^{(-50 \times 10^{-3})/(26 \times 10^{-3})} = 0.1462$$

Whence,
$$\frac{i_{D2} - i_{D1}}{i_{D1}} = \frac{0.1462 i_{D1} - i_{D1}}{i_{D1}} = -0.8538$$

2.128 Find the increase in forward voltage drop of a Ge diode for which $V_T = 26$ mV if the forward current doubles.

Based on (1) of Problem 2.21,
$$\Delta v_D = (26 \times 10^{-3}) \ln(2) = 18.02 \text{ mV}$$

2.129 Repeat Problem 2.128 if the operating temperature is 30 percent greater.

Based on (1) of Problem 2.18,
$$V_T = 1.3(26) = 33.8 \text{ mV}$$

By (1) of Problem 2.21,
$$\Delta v_d = (33.8 \times 10^{-3}) \ln(2) = 23.43 \text{ mV}$$

2.130 A series circuit consists of a 12-V battery and two identical Si diodes with PIV rating of 15 V and a reverse saturation current of 15 nA. The diodes are connected anode-to-anode or series opposing. Find the current flowing in the circuit.

One diode is forward biased and the other diode is reverse biased. Hence, current is limited by the reverse-biased diode to $I_0 = 15$ nA.

2.131 Determine the voltage across each diode of Problem 2.130 at 300°K.

Based on (1) of Problem 2.18, the current relationship for the forward-biased diode is
$$i_{DF} = I_0 = I_0(e^{v_{DF}/\eta V_T} - 1)$$
or
$$v_{DF} = \eta V_T \ln(2) = (2)(0.02587)\ln(2) = 0.0359 \text{ V}$$

By KVL, the voltage across the reverse-biased diode is
$$v_{DR} = 12 - v_{DF} = 12 - 0.0359 = 11.964 \text{ V}$$

2.132 Find the power dissipated by each diode of Problem 2.130.

Based on Problems 2.130 and 2.131,
$$P_F = v_{DF} i_{DF} = v_{DF} I_0 = (11.964)(15 \times 10^{-9}) = 0.1795 \text{ }\mu\text{W}$$
$$P_R = v_{DR} i_{DR} = v_{DR} I_0 (0.0359)(15 \times 10^{-9}) = 538.5 \text{ pW}$$

2.133 Calculate the power dissipated by the Ge diode of Problem 2.31 at $T = 300°K$.

$$P = v_D i_D = (0.5)(2.47) = 1.235 \text{ W}$$

2.134 Find the power dissipated by the Ge diode of Problem 2.31 at $i_D = 5$ A and $T = 405.2°K$.

Based on Problem 2.32,

$$P = v_D i_D = (0.7)(5) = 3.5 \text{ W}$$

2.135 Find the output voltage v_L for the circuit of Fig. 2-3(a) if $v_S = 10$ V, $R_S = 10 \, \Omega$, and $R_1 = R_L = 1 \, \text{k}\Omega$. Assume the diode is ideal.

Based on Problem 2.38,

$$v_{Th} = \frac{R_1}{R_1 + R_S} v_S = \frac{1 \times 10^3}{1 \times 10^3 + 10}(10) = 9.9 \text{ V}$$

$$R_{Th} = \frac{R_1 R_S}{R_1 + R_S} = \frac{(1 \times 10^3)(10)}{1 \times 10^3 + 10} = 9.9 \, \Omega$$

$$v_L = \frac{R_L}{R_L + R_{Th}} v_{Th} = \frac{1 \times 10^3}{1 \times 10^3 + 9.9}(9.9) = 9.803 \text{ V}$$

2.136 Solve Problem 2.38 if diode D is represented by the piecewise linear model of Fig. 2-33.

The Thévenin equivalent circuit looking to the left from R_L is found as

$$v_{Th} = \frac{R_1}{R_1 + R_S} v_S - V_F = \frac{R_1 v_S - (R_1 + R_S) V_F}{R_1 + R_S}$$

$$R_{Th} = R_1 \| R_S + R_F = \frac{R_1 R_S}{R_1 + R_S} + R_F = \frac{R_1 R_S + R_F(R_1 + R_S)}{R_1 + R_S}$$

$$v_L = i_D R_L = \frac{v_{Th}}{R_{Th} + R_L} R_L = \frac{[R_1 v_S - (R_1 + R_S) V_F] R_L}{R_1 R_S + (R_F + R_L)(R_1 + R_S)}$$

2.137 Determine the output voltage for the circuit of Fig. 2-3(a) if $v_S = 10$ V, $R_S = 10 \, \Omega$, $R_L = R_1 = 1 \, \text{k}\Omega$, and the nonideal Si diode is modeled as in Fig. 2-33 where $V_F = 0.6$ V and $R_F = 25 \, \Omega$.

Based on Problem 2.136,

$$v_L = \frac{[R_1 v_S - (R_1 + R_S) V_F] R_L}{R_1 R_S + (R_F + R_L)(R_1 + R_S)} = \frac{[(1000)(10) - (1010)(0.6)](1000)}{(1000)(10) + (1025)(1010)} = 8.987 \text{ V}$$

2.138 Rework Problem 2.137 if the nonideal diode is a Ge device with $V_F = 0.3$, $R_F = 10 \, \Omega$, and all else unchanged.

Based on Problem 2.136,

$$v_L = \frac{[R_1 v_S - (R_1 + R_S) V_F] R_L}{R_1 R_S + (R_F + R_L)(R_1 + R_S)} = \frac{[(1000)(10) - (1010)(0.3)](1000)}{(1000)(10) + (1010)(1010)} = 9.414 \text{ V}$$

2.139 Find the power supplied by the source v_S for Problem 2.135.

Based on Problems 2.38 and 2.135,

$$i_D = \frac{v_{Th}}{R_{Th} + R_L} = \frac{9.9}{9.9 + 1 \times 10^3} = 9.803 \text{ mA}$$

$$P_S = \frac{v_L^2}{R_L} + \frac{v_L^2}{R_1} + (2 i_D)^2 R_S = \frac{(9.803)^2}{1 \times 10^3} + \frac{(9.803)^2}{1 \times 10^3} + [(2)(9.803 \times 10^{-3})]^2 (10) = 195.5 \text{ mW}$$

2.140 Rework Problem 2.74 if the diode is a Si device with $I_0 = 8$ nA, $r_d = 125 \, \Omega$, and $T = 300°K$.

$$V_{DQ} = \eta V_T \ln \left[\frac{\eta V_T}{r_d I_0} \right] = (2)(0.02587) \ln \left[\frac{(2)(0.02587)}{(125)(8 \times 10^{-9})} \right] = 0.562 \text{ V}$$

2.141 Find the quiescent current I_{DQ} for Problem 2.140.

▮ $\quad I_{DQ} = i_D|_Q = I_0(e^{v_D/\eta V_T} - 1)|_Q = (8 \times 10^{-9})(e^{0.562/(2)(0.02587)} - 1) = 0.417 \text{ mA}$

2.142 If the forward current flowing through the diode of Problem 2.140 is given by $i_D = 2 + 0.004 \sin \omega t$ mA, find the quiescent voltage V_{DQ}.

▮ $\quad I_{DQ} = i_D|_Q = 0.002 \approx I_0 e^{V_{DQ}/\eta V_T} = (8 \times 10^{-9}) e^{V_{DQ}/(2)(0.02587)}$

Whence, $\quad V_{DQ} = 0.643 \text{ V}$

2.143 Determine the total voltage $v_D = V_{DQ} + v_d$ across the diode of Problem 2.142.

▮ Based on Problem 2.74,

$$r_d = \frac{\eta V_T}{I_0} e^{-V_{DQ}/\eta V_T} = \frac{(2)(0.02587)}{8 \times 10^{-9}} e^{-0.643/(2)(0.02587)} = 25.91 \ \Omega$$

Thus, $\quad v_d = r_d i_d = (25.91)(0.004 \times 10^{-3} \sin \omega t) = 0.104 \text{ mV}$

Using the result of Problem 2.142,

$$v_D = V_{DQ} + v_d = 643 + 0.104 \sin \omega t \text{ mV}$$

2.144 Calculate the rms value of the current flowing through the diode of Problem 2.142.

$$I_{D \text{ rms}} = \left[\frac{1}{T} \int_0^T i_D^2 \, dt\right]^{1/2} = \left[\frac{1}{2\pi} \int_0^{2\pi} (2 + 0.004 \sin \omega t)^2 \, d(\omega t)\right]^{1/2} = \left[(2)^2 + \left(\frac{0.004}{\sqrt{2}}\right)^2\right]^{1/2} \approx 2 \text{ mA}$$

2.145 Find the value of average power dissipated by the diode of Problem 2.142.

▮ $\quad P = \frac{1}{T} \int_0^T v_D i_D \, dt = \frac{10^{-6}}{2\pi} \int_0^{2\pi} (2 + 0.004 \sin \omega t)(643 + 0.104 \sin \omega t) \, d(\omega t) \approx 1.286 \text{ mW}$

2.146 Find v_o for the circuit of Fig. 2-34 if $v_1 = 10$ V and $v_2 = 0$ V.

▮ Assume D_1 ON and D_2 OFF; then,

$$v_o = \frac{R_L}{R_L + R_1} v_1 = \frac{9}{9+1}(10) = 9 \text{ V}$$

Check assumptions:

$$i_{D1} = \frac{v_1 - v_o}{R_1} = \frac{10 - 9}{1} = 1 \text{ A} > 0, \quad \text{therefore } D_1 \text{ ON}$$

$$v_{D2} = v_2 - v_o = 0 - 9 = -9 < 0, \quad \text{therefore } D_2 \text{ OFF}$$

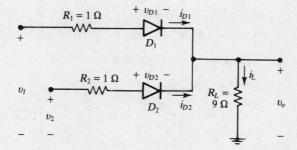

Fig. 2-34

2.147 Repeat Problem 2.146 if $v_1 = 10$ V and $v_2 = 5$ V.

▮ Assume D_1 ON and D_2 OFF; then,

$$v_o = \frac{R_L}{R_L + R_1} v_1 = \frac{9}{9+1}(10) = 9 \text{ V}$$

Check assumptions:

$$i_{D1} = \frac{v_1 - v_o}{R_1} = \frac{10-9}{1} = 1 \text{ A} > 0, \quad \text{therefore } D_1 \text{ ON}$$

$$v_{D2} = v_2 - v_0 = 5 - 10 = -5 \text{ V} < 0, \quad \text{therefore } D_2 \text{ OFF}$$

2.148 Repeat Problem 2.146 if $v_1 = v_2 = 10$ V.

▌ Assume D_1 and D_2 ON; then,

$$i_{D1} = \frac{v_1 - v_o}{R_1} \quad i_{D2} = \frac{v_2 - v_o}{R_2}$$

$$v_o = i_L R_L = (i_{D1} + i_{D2})R_L = \left[\frac{v_1 - v_o}{R_1} + \frac{v_2 - v_o}{R_2}\right]R_L = \frac{[v_1 + v_2 - 2v_o]}{1} \quad (9)$$

or

$$v_o = 9.474 \text{ V}$$

Check assumptions:

$$i_{D1} = i_{D2} = \frac{v_1 - v_o}{R_1} = \frac{10 - 9.474}{1} = 0.526 \text{ A} > 0, \quad \text{therefore } D_1 \text{ and } D_2 \text{ ON}$$

2.149 Repeat Problem 2.146 if $v_1 = -5$ V and $v_2 = 5$ V.

▌ Assume D_1 OFF and D_2 ON; then,

$$v_o = \frac{R_L}{R_L + R_2} v_2 = \frac{9}{9+1}(5) = 4.5 \text{ V}$$

Check assumptions:

$$i_{D2} = \frac{v_2 - v_o}{R_2} = \frac{5 - 4.5}{1} = 0.5 \text{ A} > 0, \quad \text{therefore } D_2 \text{ ON}$$

$$v_{D1} = v_1 - v_o = -5 - 4.5 = -9.5 < 0, \quad \text{therefore } D_1 \text{ OFF}$$

2.150 Find v_o for the circuit of Fig. 2-35 if $v_1 = 10$ V and $v_2 = 0$ V.

▌ Assume D_1 ON and D_2 OFF; then,

$$i_L = \frac{v_1 - V_B}{R_1 + R_L} = \frac{10 - 5}{1 + 9} = 0.5 \text{ A}$$

$$v_o = V_B + i_L R_L = 5 + (0.5)(9) = 9.5 \text{ V}$$

Check assumptions:

$$i_{D1} = i_L = 0.5 \text{ A} > 0, \quad \text{therefore } D_1 \text{ ON}$$

$$v_{D2} = v_2 - v_o - 9.5 = -9.5 \text{ V} < 0, \quad \text{therefore } D_2 \text{ OFF}$$

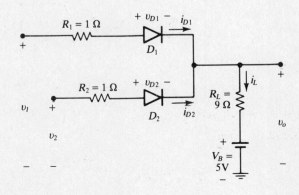

Fig. 2-35

2.151 Repeat Problem 2.150 if $v_1 = v_2 = 10$ V.

∎ Assume D_1 and D_2 ON; then,

$$i_{D1} = \frac{v_1 - v_o}{R_1} \qquad i_{D2} = \frac{v_2 - v_o}{R_2}$$

$$v_o = V_B + i_L R_L = V_B + (i_{D1} + i_{D2})R_L = V_B + \left[\frac{v_1 - v_o}{R_1} + \frac{v_2 - v_o}{R_2}\right]R_L$$

$$= 5 + \left[\frac{10 - v_o}{1} + \frac{10 - v_o}{1}\right](9)$$

or

$$v_o = 9.737 \text{ V}$$

Check assumptions:

$$i_{D1} = i_{D2} = \frac{v_1 - v_o}{R_1} = \frac{10 - 9.737}{1} = 0.263 \text{ A} > 0, \quad \text{therefore } D_1 \text{ and } D_2 \text{ ON}$$

2.152 Repeat Problem 2.150 if $v_1 = -5$ V and $v_2 = 10$ V.

∎ Assume D_1 OFF and D_2 ON; then,

$$i_L = \frac{v_2 - V_B}{R_2 + R_L} = \frac{10 - 5}{1 + 9} = 0.5 \text{ A}$$

$$v_o = V_B + i_L R_L = 5 + (0.5)(9) = 9.5 \text{ V}$$

Check assumptions:

$$i_{D1} = i_L = 0.5 \text{ A} > 0, \quad \text{therefore } D_1 \text{ ON}$$

$$v_{D2} = v_2 - v_o = -5 - 9.5 = -14.5 \text{ V} < 0, \quad \text{therefore } D_2 \text{ OFF}$$

CHAPTER 3
Diodes: Power Level Applications

GENERAL CIRCUITS

3.1 If $v_S = 115\sqrt{2} \sin t$ V, $V_B = 50$ V, and $R = 2\Omega$, describe the current i_o and voltage v_o for the half-wave rectifier circuit of Fig. 3-1(a).

▌ Diode D_{FW} remains reverse-biased, and thus, never conducts in this circuit arrangement. Diode D_R conducts only for $v_S > V_B$ giving $v_o = v_S$ over the interval $\alpha \leq \omega t \leq \beta$ where $\alpha = \sin^{-1}(150/115\sqrt{2}) = 17.91°$ and $\beta = 180° - \alpha = 162.1°$. Otherwise, D_R blocks requiring $v_o = V_B$. The current is described by

$$i_o = \frac{v_S - V_B}{R} = \frac{115\sqrt{2}}{2} \sin \omega t - \frac{50}{2} = 81.32 \sin \omega t - 25 \text{ A}, \qquad \alpha \leq \omega t \leq \beta$$

$$i_o = \frac{v_o - V_B}{R} = \frac{50 - 50}{2} = 0, \quad \text{otherwise}$$

The result is sketched by Fig. 3-1(b).

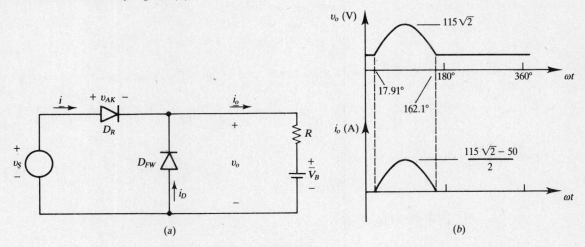

Fig. 3-1

3.2 Find the average value of current i through D_R (I_{ave}) and the average value of current i_o through R ($I_{o\text{ ave}}$) in Problem 3.1.

▌ Since i_D is everywhere zero, $i = i_o$ for all t and

$$I_{\text{ave}} = I_{o\text{ ave}} = \frac{1}{2\pi} \int_\alpha^\beta \left(\frac{115\sqrt{2}}{2} \sin \omega t - \frac{50}{2}\right) d(\omega t) = 14.62 \text{ A}$$

3.3 If $v_S = 115\sqrt{2} \sin \omega t$ V, $V_B = -50$ V, and $R = 2\ \Omega$ for the circuit of Fig. 3-1(a), describe and sketch the current i_o and voltage v_o.

▌ For $0 \leq \omega t \leq \pi$, D_R conducts and D_{FW} blocks. D_{FW} conducts and D_R blocks for $\pi < \omega t < 2\pi$; hence,

$$i_o = \frac{v_S - V_B}{R} = \frac{115\sqrt{2}}{2} \sin \omega t - \frac{(-50)}{2} = 81.32 \sin \omega t + 25 \text{ A}, \qquad 0 \leq \omega t \leq \pi$$

$$i_o = \frac{v_o - V_B}{R} = \frac{0 - (-50)}{2} = 25 \text{ A}, \qquad \pi < \omega t < 2\pi$$

See Fig. 3-2 for sketches of v_o and i_o.

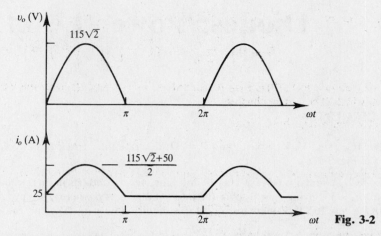

Fig. 3-2

3.4 Determine the average values of current i (I_{ave}) and current i_o ($I_{o\,ave}$) for the circuit of Problem 3.3.

▌ Referring to Fig. 3-2,

$$I_{o\,ave} = \frac{1}{2\pi}\left[\int_0^\pi \left(\frac{115\sqrt{2}}{2}\sin\omega t + \frac{50}{2}\right)d(\omega t) + \int_\pi^{2\pi}\left(\frac{50}{2}\right)d(\omega t)\right] = 50.88\text{ A}$$

Now $i = i_o$ for $0 \le \omega t \le \pi$ and $i = 0$ over $\pi < \omega < 2\pi$; hence,

$$I_{ave} = \frac{1}{2\pi}\int_0^\pi \left(\frac{115\sqrt{2}}{2}\sin\omega t + \frac{50}{2}\right)d(\omega t) = 38.38\text{ A}$$

3.5 Find expressions to describe $i(t)$ for the half-wave rectifier circuit of Fig. 3-3 if $i(0) = 0$ and v_S is a square wave of amplitude V_p and period T.

▌ Assume D conducts; then, by KVL,

$$v_S = L\frac{di_o}{dt} \qquad (1)$$

For $0 \le t \le T/2$, $v_S = V_p$ and direct integration of (1) yields

$$\int_0^{i_o(t)} di_o = \frac{V_p}{L}\int_0^t d\tau \quad\text{or}\quad i_o(t) = \frac{V_p}{L}t$$

For $T/2 < t < T$, $v_S = -V_p$; thus by (1),

$$\int_{i_o(0)}^{i_o(t')} di_o = \frac{-V_p}{L}\int_0^{t'} d\tau$$

where $i_o(t'=0) = i_o(t=T/2) = (V_p/L)(T/2)$ and $t' = t - T/2$.
Integrating and substituting for $i_o(0)$ and t' gives

$$i_o(t) = \frac{V_p}{L}\frac{T}{2} - \frac{V_p}{L}(t - T/2) = \frac{-V_p}{L}(t - T)$$

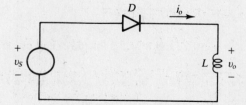

Fig. 3-3

3.6 Determine expressions for the average value ($I_{o\,ave}$) and rms value ($I_{o\,rms}$) of current i_o of Problem 3.5.

▌ Since i_o is a triangular waveform, the average value is simply $\frac{1}{2}$(base)(height) or

$$I_{o\,ave} = \frac{1}{2}(T)\left(\frac{V_p}{L}\frac{T}{2}\right) = \frac{1}{4}\frac{V_p}{L}T$$

Due to symmetry, the rms value can be found over $0 \le t \le T/2$ as

$$I_{o\,\text{rms}} = \left[\frac{1}{T/2}\int_0^{T/2} i_o^2\, dt\right]^{1/2} = \left[\frac{2}{T}\int_0^{T/2}\left(\frac{V_p}{L}t\right)^2 dt\right]^{1/2} = \frac{\sqrt{3}}{6}\frac{V_p}{L}T$$

3.7 If $v_S = V_m \sin \omega t$ and $V_C(0) = 0$ for the circuit of Fig. 3-4, find $i(t)$ over the interval $0 \le \omega t \le \beta$ where β is the point at which D ceases conduction.

▌ If D is conducting, KVL yields

$$Ri + \frac{1}{C}\int i\, dt = V_m \sin \omega t \tag{1}$$

The forced response of (1) follows from sinusoidal steady-state methods as

$$i_f = \frac{\omega C V_m}{\sqrt{1 + (\omega RC)^2}} \sin(\omega t - \phi) \qquad \text{where} \qquad \phi = \tan^{-1}\left(-\frac{1}{\omega RC}\right)$$

The form of the natural response of (1) is $i_n = Ae^{-t/RC}$. Thus,

$$i(t) = i_f + i_n = \frac{\omega C V_m}{\sqrt{1 + \omega RC)^2}} \sin(\omega t - \phi) + Ae^{-t/RC} \tag{2}$$

Since $v_C(0) = v_S(0) = 0$, (1) requires that $i(0) = 0$. Hence, (2) gives

$$i(0) = 0 = \frac{\omega C V_m}{\sqrt{1 + (\omega RC)^2}} \sin(-\phi) + A \qquad \text{or} \qquad A = \frac{\omega C V_m \sin \phi}{\sqrt{1 + (\omega RC)^2}}$$

and

$$i(t) = \frac{\omega C V_m}{\sqrt{1 + (\omega RC)^2}}[\sin(\omega t - \phi) + \sin \phi\, e^{-t/RC}], \qquad 0 \le \omega t \le \beta \tag{3}$$

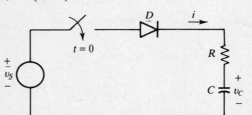

Fig. 3-4

3.8 Determine the transcendental equation that must be solved to find β, the point at which D ceases the conduction in Problem 3.7.

▌ Substitute $\omega t = \beta$ or $t = \beta/\omega$ into (3) of Problem 3.7 and set $i(\beta/\omega) = 0$ to get

$$0 = \sin(\beta - \phi) + \sin \phi\, e^{-\beta/\omega RC}$$

3.9 Let $v_S = V_m \sin \omega t$ for the circuit of Fig. 3-5 and find (a) $i_o(t)$ and (b) $v_o(t)$.

▌ (a) Assuming D forward-biased, KVL gives

$$L\frac{di_o}{dt} = v_S = V_m \sin \omega t \tag{1}$$

Direct integration of (1) with $i_o(0) = 0$ leads to

$$i_o(t) = \frac{V_m}{L}\int_0^t \sin \omega \tau\, d\tau = \frac{-V_m}{\omega L}\cos \omega \tau\Big|_{\tau=0}^t = \frac{V_m}{\omega L}(1 - \cos \omega t) \tag{2}$$

(b) Since (2) shows that $i_o \ge 0$ for all t, D remains continuously in conduction; thus,

$$v_o(t) = v_S(t) = V_m \sin \omega t$$

3.10 Find expressions to describe $i(t)$ for the circuit of Fig. 3-5 if $v_S = V_m \cos \omega t$.

▌ D is forward-biased upon closing the switch. By KVL,

$$L\frac{di_o}{dt} = V_m \cos \omega t \tag{1}$$

78 □ CHAPTER 3

Direct integration of (1) using $i_o(0) = 0$ gives

$$i_o(t) = \frac{V_m}{L} \int_0^t \cos \omega\tau \, d\tau = \frac{V_m}{\omega L} \sin \omega t \qquad (2)$$

Equation (2) is valid until $\omega t = \pi$, at which point $i_o(t)$ reaches zero. For $\pi < \omega t < 3\pi/2$, the solution is similar to that of Problem 3.9. Using a primed coordinate frame ($\omega t' = \omega t - 3\pi/2$) for convenience,

$$i_o(t') = \frac{V_m}{L} \int_0^{t'} \sin \omega\tau \, d\tau = \frac{V_m}{\omega L}(1 - \cos \omega t') \qquad (3)$$

Using the unit step function with characteristic $U(x) = 0$ for $x < 0$ and $U(x) = 1$ for $x \geq 0$, (2) can be written in terms of the unprimed coordinate frame. In summary,

$$i_o(t) = \frac{V_m}{\omega L} \sin \omega t, \qquad 0 \leq \omega t \leq \pi$$

$$i_o(t) = 0, \qquad \pi < \omega t < 3\pi/2$$

$$i_o(t) = \frac{V_m}{\omega L}[1 - \cos(\omega t - 3\pi/2)]U(\omega t - 3\pi/2), \qquad \omega t \geq 3\pi/2$$

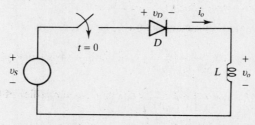

Fig. 3-5

3.11 If $v_S = V_m \sin(\omega t - \theta)$, where $0 \leq \theta \leq \pi$, find expressions to describe $i_o(t)$ for the half-wave rectifier circuit of Fig. 3-5.

▌ The diode is reverse-biased until $\omega t = \theta$, or $i_o(t) = 0$, $0 \leq \omega t \leq \theta$. For $\omega t \geq \theta$, $v_o = v_S$ and

$$i_o(t) = \frac{V_m}{L} \int_{\theta/\omega}^t \sin(\omega\tau - \theta) \, d\tau = \frac{V_m}{\omega L}[1 - \cos(\omega t - \theta)]U(\omega t - \theta)$$

3.12 If $v_S = V_m \sin \omega t$ and $i_o(0) = 0$ for the half-wave rectifier circuit of Fig. 3-6, find an expression for current $i(t)$.

▌ By KVL, if D is conducting,

$$L\frac{di}{dt} + Ri = V_m \sin \omega t \qquad (1)$$

The complete response of $i(t)$ is found as the sum of the forced response and natural response.

$$i(t) = i_f + i_n = \frac{V_m}{Z} \sin(\omega t - \theta) + Ae^{-tR/L}$$

where $Z = \sqrt{R^2 + (\omega L)^2}$ and $\phi = \tan^{-1}(\omega L/R)$.

Applying the given initial condition

$$i(0) = 0 = \frac{V_m}{Z}\sin(-\phi) + A \qquad \text{or} \qquad A = \frac{V_m}{Z}\sin\phi$$

Hence, $\qquad i(t) = \frac{V_m}{Z}[\sin(\omega t - \phi) + \sin\phi \, e^{-tR/L}], \qquad i \geq 0$

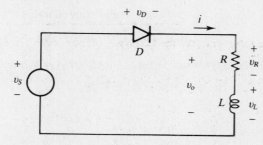

Fig. 3-6

3.13 If $v_S = 100\sqrt{2}\sin 377t$ V, $R = 10\,\Omega$, and $L = 26.53$ mH in the half-wave rectifier circuit of Fig. 3-6, find an expression for $i(t)$ valid over $0 \leq \omega t \leq \beta$ where β is the point at which D ceases conduction.

▌ The results of Problem 3.12 are applicable.

$$Z = [R^2 + (\omega L)^2]^{1/2} = [(10)^2 + ((377)(0.02653))^2]^{1/2} = 14.142\,\Omega$$

$$\phi = \tan^{-1}(\omega L/R) = \tan^{-1}\left[\frac{(377)(0.02653)}{10}\right] = 45°$$

$$i(t) = \frac{100\sqrt{2}}{14.142}[\sin(377t - 45°) + \sin(45°)e^{-10t/0.02653}] = 10[\sin(377t - 45°) + 0.707e^{-377t}] \quad (1)$$

3.14 Calculate the point $\beta = \omega t$ at which D ceases conduction in Problem 3.13.

▌ Substitute $\beta = \omega t$ in (1) of Problem 3.13 and equate the result to zero to yield the transcendental equation

$$i(\beta/\omega) = 0 = \sin(\beta - 45°) + 0.707e^{-\beta}$$

A trial-and-error solution finds $\beta \approx 225.8°$.

3.15 Find the average value of current i of Problem 3.13.

▌ $\beta \approx 225.8°$ as determined in Problem 3.14 is used.

$$I_{\text{ave}} = \frac{1}{2\pi}\int_0^{2\pi} i_o\,d(\omega t) = \frac{1}{2\pi}\int_0^{\beta} 10[\sin(\omega t - 45°) + 0.707e^{-377t}]\,d(\omega t) = 2.695\text{ A}$$

3.16 Show that the average value of voltage v_o for the half-wave rectifier of Fig. 3-6 is given by $V_{o\,\text{ave}} = RI_{\text{ave}}$.

▌
$$V_{o\,\text{ave}} = \frac{1}{2\pi}\int_0^{2\pi} v_o\,d(\omega t) = \frac{1}{2\pi}\int_0^{2\pi}(v_R + v_L)\,d(\omega t)$$

$$= R\left[\frac{1}{\pi}\int_1^{2\pi} i\,d(\omega t)\right] + \frac{\omega L}{2\pi}\int_{i(0)}^{i(2\pi)} di$$

Since $i(0) = i(2\pi)$, the right integral is zero. The expression within the brackets is recognized as I_{ave}. Thus, $V_{o\,\text{ave}} = RI_{\text{ave}}$.

3.17 Calculate the conduction angle (γ) of D in Fig. 3-7 if $v_S = 20\sin\omega t$ V and $V_B = 10$ V.

▌ Diode D begins conduction for $\omega t = \alpha$ at which point v_s becomes greater than V_B, or

$$20\sin\alpha = 10 \quad \text{or} \quad \alpha = \sin^{-1}(10/20) = 30°$$

Conduction of D ceases for $\omega t = \beta$ at which point v_S decreases to less than V_B. By symmetry,

$$\beta = 180° - \alpha = 180° - 30° = 150°$$

Whence, $$\gamma = \beta - \alpha = 150° - 30° = 120°$$

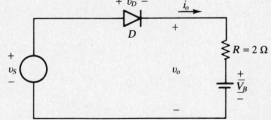

Fig. 3-7

3.18 Determine the average value of v_o for the half-wave rectifier circuit of Fig. 3-7 if $V_B = 10$ V and $v_S = 20\sin\omega t$ V.

▌ The values of α and β of Problem 3.17 are pertinent. For $\alpha \leq \omega t \leq \beta$, D conducts and $v_o = v_S$. Otherwise $v_o = V_B$; hence

$$V_{o\,\text{ave}} = \frac{1}{2\pi}\int_0^{2\pi} v_o\,d(\omega t) = \frac{1}{2\pi}\left[\int_0^\alpha 10\,d(\omega t) + \int_\alpha^\beta 20\sin\omega t\,d(\omega t) + \int_\beta^{2\pi} 10\,d(\omega t)\right] = 12.12\text{ V}$$

80 □ CHAPTER 3

3.19 Find the average value of current i_o for Problem 3.17.

▮ $i_o \neq 0$ only during conduction interval of D; thus,

$$i_o = \frac{v_S - V_B}{R} = 10 \sin \omega t - 5 \text{ A}, \qquad \alpha \leq \omega t \leq \beta$$

$$I_{o \text{ ave}} = \frac{1}{2\pi} \int_\alpha^\beta [10 \sin \omega t - 5] \, d(\omega t) = 1.09 \text{ A}$$

3.20 Calculate the average value of power supplied by source v_S in Problem 3.17.

▮ Using i_o determined in Problem 3.19,

$$P = \frac{1}{2\pi} \int_0^{2\pi} v_S i_o \, d(\omega t) = \frac{1}{2\pi} \int_\alpha^\beta (20 \sin \omega t)(10 \sin \omega t - 5) \, d(\omega t) = 26.44 \text{ W}$$

3.21 Determine $i(t)$ for the circuit of Fig. 3-8 if the capacitor is initially uncharged.

▮ Upon closing the switch, D is forward biased and, by KVL,

$$L\frac{di}{dt} + \frac{1}{C} \int i \, dt = V_B \qquad (1)$$

The forced response of $i(t)$ is zero. Set $V_B = 0$ and differentiate once to obtain the homogeneous differential equation

$$\frac{d^2 i}{dt^2} + \frac{1}{LC} i = 0 \qquad (2)$$

If $\omega = 1/\sqrt{LC}$, the solution of (2) is

$$i(t) = A_1 \cos \omega t + A_2 \sin \omega t \qquad (3)$$

Applying the initial condition $i(0) = 0$ to (3) gives $A_1 = 0$. Since $v_C(0) = 0$ and $i(0) = 0$, (1) requires that

$$L\frac{di}{dt}\bigg|_0 = V_B \qquad \text{or} \qquad \frac{di}{dt}\bigg|_0 = \frac{V_B}{L} \qquad (4)$$

By (2) and (4),

$$\frac{di}{dt}\bigg|_0 = \frac{V_B}{L} = \omega A_2 \cos \omega t|_0 \qquad \text{or} \qquad A_2 = \frac{V_B}{\omega L}$$

Thus,

$$i(t) = \frac{V_B}{\omega L} \sin \omega t, \qquad 0 \leq \omega t \leq \pi$$

$$i(t) = 0, \qquad \omega t > \pi$$

3.22 Use the method of Laplace transforms to find the current $i(t)$ over the interval $0 \leq t \leq \pi/\omega$ for the circuit of Fig. 3-8 if $v_C(0) = V_o$.

▮ By KVL,

$$L\frac{di}{dt} + \frac{1}{C} \int i \, dt = V_B$$

Applying Laplace transform,

$$LsI + \frac{1}{Cs} I + \frac{V_o}{s} = \frac{V_B}{s}$$

or

$$I = \frac{(V_B - V_o)/L}{s^2 + 1/LC} = \frac{V_B - V_o}{\omega L} \frac{\omega}{s^2 + \omega^2} \qquad \text{where} \qquad \omega = 1/\sqrt{LC}$$

Inverse transforming,

$$i(t) = \frac{V_B - V_o}{\omega L} \sin \omega t, \qquad 0 \leq t \leq \pi/\omega$$

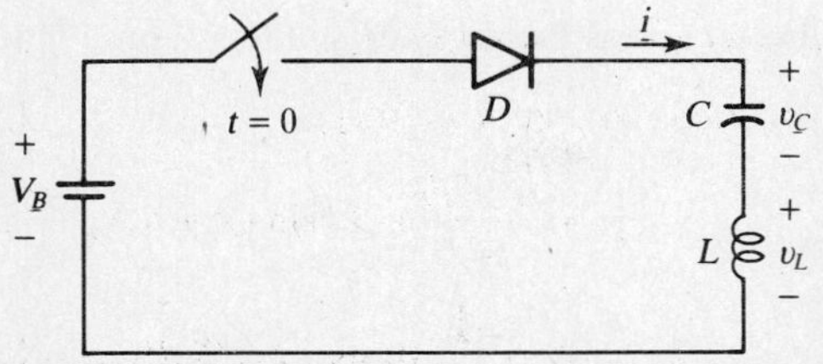

DIODES: POWER LEVEL APPLICATIONS

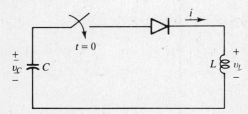

Fig. 3-8

3.23 Determine an expression for the voltage at $t = \pi/\omega$ across the capacitor of Fig. 3-8 if the capacitor has an initial voltage of V_o.

▌ The pertinent current is known from Problem 3.22. Then,

$$v_L = L\frac{di}{dt} = (V_B - V_o)\cos \omega t$$

By KVL,

$$v_C(\pi/\omega) = V_B - v_L(\pi/\omega) = V_B + (V_B - V_o) = 2V_B - V_o$$

3.24 For the source-free circuit of Fig. 3-9, find $i(t)$ if $v_C(0) = V_o$.

▌ By KVL,

$$L\frac{di}{dt} + \frac{1}{C}\int i\,dt = 0 \qquad (1)$$

By direct analogy with Problem 3.21,

$$i(t) = \frac{V_o}{\omega L}\sin \omega t, \qquad 0 \le \omega t \le \pi$$

Fig. 3-9

3.25 Find the voltages v_L and v_C for Problem 3.24.

▌ Current $i(t)$ over $0 \le \omega t \le \pi$ is known from Problem 3.24; thus,

$$v_C(t) = v_L(t) = L\frac{di}{dt} = V_o \cos \omega t, \qquad 0 \le t \le \pi/\omega$$

For $\omega t > \pi$, D blocks and $i(t)$ remains zero. Consequently,

$$v_L(t) = 0, \qquad v_C(t) = v_C(\pi/\omega) = V_0, \qquad t > \pi/\omega$$

3.26 Solve Problem 3.24 by method of Laplace transforms.

▌ By KVL,

$$L\frac{di}{dt} + \frac{1}{C}\int i\,dt = 0$$

Transforming,

$$LsI + \frac{1}{Cs}I - \frac{V_o}{s} = 0$$

or

$$I = \frac{V_o/L}{s^2 + 1/LC} = \frac{V_o}{\omega L}\frac{\omega}{s^2 + \omega^2} \qquad \text{where} \qquad \omega = 1/\sqrt{LC}$$

Inverse transforming,

$$i(t) = \frac{V_o}{\omega L}\sin \omega t, \qquad 0 \le \omega t \le \pi$$

3.27 If $v_S = V_m \sin \omega t$, determine $i(t)$ over the conduction interval of D in the circuit of Fig. 3-10. Assume $i(0) = 0$ and $V_m > V_B$.

▌ By KVL, and with D conducting,

$$L\frac{di}{dt} + Ri = V_m \sin \omega t - V_B$$

The complete response of the current follows from addition of the forced response, formed by superposition of sinusoidal steady-state and dc steady-state responses, and the natural response.

$$i(t) = i_f + i_n = \frac{V_m}{Z} \sin (\omega t - \phi) - \frac{V_B}{R} + Ae^{-tR/L} \quad (1)$$

where $\quad \phi = \tan^{-1}(\omega L/R) \quad$ and $\quad Z = \sqrt{R^2 + (\omega L)^2}$

Current initially remains zero until the point $\omega t = \alpha$, at which v_S becomes greater than V_B; thus, from (1),

$$i(\alpha/\omega) = 0 = \frac{V_m}{Z} \sin (\alpha - \phi) - \frac{V_B}{R} + Ae^{-(R/L)(\alpha/\omega)} \quad (2)$$

Solution of (2) for A and substitution of result into (1) leads to

$$i(t) = \frac{V_m}{Z} \sin (\omega t - \phi) - \frac{V_B}{R} + \left[\frac{V_B}{R} - \frac{V_m}{Z} \sin (\alpha - \phi)\right] e^{-(R/L)(t - \alpha/\omega)}, \quad i(t) \geq 0$$

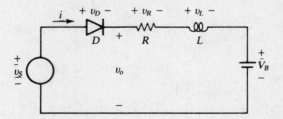

Fig. 3-10

3.28 Sketch v_o and v_{AK} for the circuit of Fig. 3-10 with $i(t)$ as found in Problem 3.27.

▌ When D conducts, $v_o = v_S = V_m \sin \omega t$ and $v_D = 0$. Otherwise, $v_o = V_B$ and $v_D = v_S - v_o = V_m \sin \omega t - V_B$. See Fig. 3-11 for sketches of v_o and v_D.

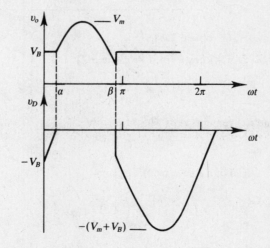

Fig. 3-11

3.29 Show that by use of impedance triangle relationships the current resulting from solution can be written as

$$i(t) = \frac{V_m}{R} \{\cos \phi \sin (\omega t - \phi) - \sin \alpha + [\sin \alpha - \sin (\alpha - \phi)]e^{-\cot \phi (\omega t - \alpha)}\}$$

▌ D becomes forward biased for $V_B = V_m \sin \alpha$. From the impedance triangle, $R = Z \sin \phi$. Thus, the ratio $V_B/R = V_m \sin \alpha / Z \cos \phi$ is formed. Further, the impedance triangle requires $R/\omega L = \cot \phi$. Direct substitution into the expression for $i(t)$ from Problem 3.27 gives

$$i(t) = \frac{V_m}{Z}\left\{\sin(\omega t - \phi) - \frac{\sin\alpha}{\cos\phi} + \left[\frac{\sin\alpha}{\cos\phi} - \sin(\alpha - \phi)\right]e^{-\cos\phi(\omega t - \alpha)}\right\}$$

Factoring out $1/\cos\phi$ and using $Z\cos\phi = R$ yield the desired expression.

3.30 Find v_R and v_L for the circuit of Fig. 3-10 if $v_S = V_m \sin\omega t$.

▎ Current $i(t)$ as found in Problem 3.27 is applicable.

$$v_R = Ri = \frac{V_m R}{Z}\sin(\omega t - \phi) - V_B + \left[V_B - \frac{V_m R}{Z}\sin(\alpha - \phi)\right]e^{-(R/L)(t - \alpha/\omega)}, \qquad i(t) > 0$$

$$v_L = L\frac{di}{dt} = \frac{\omega V_m}{Z}\sin(\omega t - \phi) - \frac{R}{L}\left[\frac{V_B}{R} - \frac{V_m}{Z}\sin(\alpha - \phi)\right]e^{-(R/L)(t - \alpha/\omega)}, \qquad i(t) > 0$$

For $i(t) = 0$, $v_R = v_L = 0$.

3.31 Form the transcendental equation that must be solved to determine the point $\omega t = \beta$ where D ceases conduction for Problem 3.27.

▎ Use the form of $i(t)$ as verified in Problem 3.29 and substitute $\beta = \omega t$ to yield the transcendental equation.

$$i(\beta/\omega) = 0 = \cos\phi \sin(\beta - \phi) - \sin\alpha + [\sin\alpha - \sin(\alpha - \phi)]e^{-\cot\phi(\beta - \alpha)}$$

3.32 Find $i(t)$ for the circuit of Fig. 3-10 if $v_S = V_{dc}$ (a dc source). Assume $V_{dc} > V_B$ and $i(0) = 0$.

▎ Since D is forward biased, KVL gives

$$L\frac{di}{dt} + Ri = V_{dc} - V_B$$

Adding the forced and natural response,

$$i(t) = \frac{V_{dc} - V_B}{R} + Ae^{-tR/L}$$

Using $i(0) = 0$ finds $A = -(V_{dc} - V_B)/R$. Hence,

$$i(t) = \frac{V_{dc} - V_B}{R}(1 - e^{-tR/L})$$

3.33 For the source-free circuit of Fig. 3-12, find $i(t)$ if $v_{C1}(0) = V_o$ and $v_{C2}(0) = 0$.

▎ By KVL,

$$L\frac{di}{dt} + \frac{1}{C_1}\int i\, dt + \frac{1}{C_2}\int i\, dt = 0 \tag{1}$$

Differentiate (1) once and define $C_{eq} = C_1 C_2/(C_1 + C_2)$ to get

$$\frac{d^2 i}{dt^2} + \frac{1}{LC_{eq}} i = 0 \tag{2}$$

If $\omega = 1/\sqrt{LC_{eq}}$, the solution of (2) for $i(t)$ gives

$$i(t) = A_1 \cos\omega t + A_2 \sin\omega t \tag{3}$$

Use of $i(0) = 0$ in (3) finds $A_1 = 0$. With $v_{C1}(0) = V_o$, $v_{C2}(0) = 0$, and $i(0) = 0$ in (1) leads to $(di/dt)|_0 = V_o/L$; thus, from (3),

$$\left.\frac{di}{dt}\right|_0 = \frac{V_o}{L} = \omega A_2 \cos\omega t|_0 \qquad \text{or} \qquad A_2 = \frac{V_o}{\omega L}$$

and

$$i(t) = \frac{V_o}{\omega L}\sin\omega t, \qquad 0 \le \omega \le \pi \tag{4}$$

84 □ CHAPTER 3

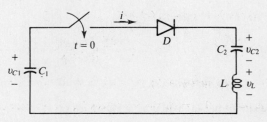

Fig. 3-12

3.34 Find v_{C1} at the point that D ceases conduction in Problem 3.33.

▌ Using $i(t)$ as given by (4) of Problem 3.33 and recalling that $\omega = 1/\sqrt{LC_{eq}}$

$$v_{C1} = -\frac{1}{C_1}\int i\, dt = V_o - \frac{1}{C_1}\int_0^{\pi/\omega} \frac{V_o}{\omega L}\sin\omega t\, dt = V_o\left(1 - \frac{2}{\omega^2 LC_1}\right) = V_o\left(1 - \frac{2C_1}{C_1 + C_2}\right)$$

3.35 Find v_{C2} at the point that D ceases conduction in Problem 3.33.

▌ From use of (4) of Problem 3.33,

$$v_{C2} = \frac{1}{C_2}\int i\, dt = \frac{1}{C_2}\int_0^{\pi/\omega} \frac{V_o}{\omega L}\sin\omega t\, dt = \frac{2V_o}{C_2\omega^2 L} = \frac{2V_o C_1}{C_1 + C_2}$$

3.36 Find v_L at $t = \pi/\omega - \epsilon$ where $\epsilon \to 0$ in Problem 3.33; that is, find v_L as D approaches the point at which conduction ceases.

▌

$$v_L(t) = L\frac{di}{dt} = V_o\cos\omega t, \qquad v_L(\pi/\omega) = -V_o$$

3.37 For the diode circuit of Fig. 3-12, find $i(t)$ if $v_{C1}(0) = V_1$ and $v_{C2}(0) = V_2$ where $V_1 > V_2$.

▌ The solution is identical to that of Problem 3.33 up to the point of equation (3), when $\omega = 1/\sqrt{LC_{eq}}$:

$$i(t) = A_1\cos\omega t + A_2\sin\omega t \qquad (1)$$

Use of $i(0) = 0$ in (1) gives $A_1 = 0$. Since $v_{C1}(0) = V_1$, $v_{C2}(0) = V_2$, and $i(0) = 0$, (1) of Problem 3.33 leads to $(di/dt)|_0 = (V_1 - V_2)/L$; hence, from (1)

$$\left.\frac{di}{dt}\right|_0 = \frac{V_1 - V_2}{L} = \omega A_2\cos\omega t \qquad \text{or} \qquad A_2 = \frac{V_1 - V_2}{\omega L}$$

Thus,
$$i(t) = \frac{V_1 - V_2}{\omega L}\sin\omega t, \qquad 0 \le \omega t \le \pi$$

3.38 Let $v_S = V_B$ (an ideal battery) for the circuit of Fig. 3-13. Find $i(0^+)$ and $(di/dt)|_{0^+}$ if (a) $v_C(0) = V_o$ and (b) $v_C(0) = 0$.

▌ (a) Since current $i(t)$ through the inductor cannot be discontinuous, KVL requires that

$$v_S(0^+) = V_B = L\left.\frac{di}{dt}\right|_{0^+} + Ri(0^+) + \frac{1}{C}\int_{-\infty}^{0^-} i\, dt + \frac{1}{C}\int_{0^-}^{0^+} i\, dt = L\left.\frac{di}{dt}\right|_{0^+} + 0 + V_o + 0$$

or
$$\left.\frac{di}{dt}\right|_{0^+} = \frac{V_B - V_o}{L}$$

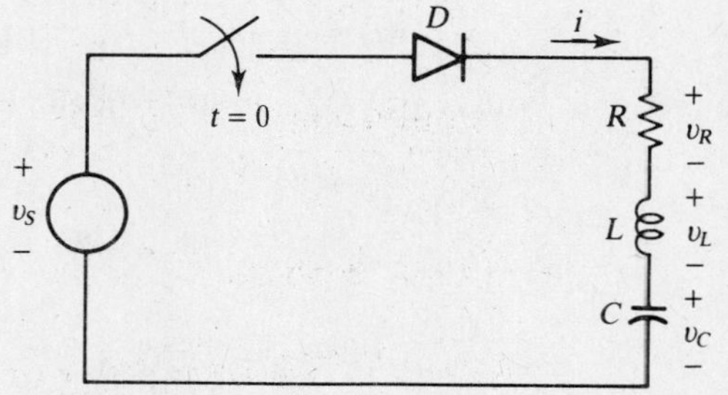

Fig. 3-13

(b) Solution for $v_C(0) = 0$ follows from (a) with $V_o = 0$.

$$\left.\frac{di}{dt}\right|_{0^+} = \frac{V_B}{L}$$

3.39 Rework Problem 3.38 if $v_S = V_m \sin(\omega t + \theta)$.

▌ The only difference is that $v_S(0^+) = V_m \sin\theta$ replaces V_B everywhere or

(a) $\left.\dfrac{di}{dt}\right|_{0^+} = \dfrac{V_m \sin\theta - V_o}{L}$ (b) $\left.\dfrac{di}{dt}\right|_{0^+} = \dfrac{V_m \sin\theta}{L}$

3.40 Let $v_S = V_B$, $v_C(0) = V_o$, and $L = 0$ for the circuit of Fig. 3-13. Find (a) $i(0^+)$ and (b) $(di/dt)|_0$.

▌ (a) By KVL applied at $t = 0^+$,

$$V_B = Ri(0^+) + V_o \quad \text{or} \quad i(0^+) = \frac{V_B - V_o}{R}$$

(b) Noting that $i(0^-) = 0$ and using the result of part (a),

$$\left.\frac{di}{dt}\right|_0 = \lim_{\Delta t \to 0}\left[\frac{i(0+\Delta t) - i(0-\Delta t)}{\Delta t}\right] = \lim_{\Delta t \to 0}\frac{i(0^+)}{\Delta t} \to \infty$$

3.41 For the circuit of Fig. 3-13, let $C = 1000\ \mu F$ and $L = 1$ mH. Find the form of the natural response of $i(t)$ over the conduction interval of D if (a) $R = 4\ \Omega$, (b) $R = 2\ \Omega$, and (c) $R = 1\ \Omega$.

▌ By KVL with $v_S = 0$,

$$L\frac{di}{dt} + Ri + \frac{1}{C}\int i\,dt = 0$$

Differentiate once to obtain

$$\frac{d^2 i}{dt^2} + \frac{R}{L}\frac{di}{dt} + \frac{1}{LC}i = 0$$

Try $i(t) = Ae^{st}$ and find the roots of the characteristic equation as

$$s_{1,2} = \zeta\omega_n \pm \omega_n\sqrt{\zeta^2 - 1} \quad \text{where} \quad \zeta = (R/2)\sqrt{C/L},\quad \omega_n = \sqrt{1/LC}$$

(a) For $R = 4\ \Omega$,

$$\zeta = \frac{4}{2}\left[\frac{1000 \times 10^{-6}}{1 \times 10^{-3}}\right]^{1/2} = 2 \qquad \omega_n = \left[\frac{1}{(1 \times 10^{-3})(1 \times 10^{-6})}\right]^{1/2} = 1000\ \text{rad/s}$$

$$s_{1,2} = -(2)(1000) \pm 1000\sqrt{2^2 - 1}, \quad s_1 = -268, \quad s_2 = -3732$$

$$i_n(t) = A_1 e^{-268t} + A_2 e^{-3732t}$$

(b) For $R = 2\ \Omega$,

$$\zeta = \frac{2}{2}\left[\frac{1000 \times 10^{-6}}{1 \times 10^{-3}}\right]^{1/2} = 1 \qquad \omega_n = 1000\ \text{rad/s}$$

$$s_{1,2} = -1000$$

$$i_n(t) = A_1 e^{-1000t} + A_2 t e^{-1000t}$$

(c) For $R = 1\ \Omega$,

$$\zeta = \frac{1}{2}\left[\frac{1000 \times 10^{-6}}{1 \times 10^{-3}}\right]^{1/2} = \frac{1}{2} \qquad \omega_n = 1000\ \text{rad/s}$$

$$s_{1,2} = -\frac{1}{2}(1000) \pm 1000\sqrt{(1/2)^2 - 1} = -500 \pm j866$$

$$i_n(t) = e^{-500t}(A_1 \cos\omega t + A_2 \sin\omega t)$$

3.42 If $v_S = V_m \sin \omega t$ for the circuit of Fig. 3-14, find expressions to describe $i(t)$ over $0 \leq \omega t \leq 4\pi$.

▌ If $v_S \geq 0$, D_R is ON, D_{FW} is OFF, and

$$v_D = v_s = V_m \sin \omega t = L \frac{di_o}{dt} \tag{1}$$

If $v_S < 0$, D_R is OFF, D_{FW} is ON, and

$$0 = L \frac{di_o}{dt} \tag{2}$$

For $0 \leq \omega t \leq \pi$, direct integration of (1) yields

$$\int_0^{i_o(t)} di_o = \frac{V_m}{L} \int_0^t \sin \omega \tau \, d\tau \quad \text{or} \quad i_o(t) = \frac{V_m}{\omega L} (1 - \cos \omega t) \tag{3}$$

For $\pi \leq \omega t \leq 2\pi$, direct integration of (2) requires that $i_o =$ constant. Since $i_o(t)$ must be continuous at $t = \pi/\omega$, the particular constant value of i_o follows from evaluation of (3) at $t = \pi/\omega$, or

$$i_o(t) = \frac{2V_m}{\omega L} \tag{4}$$

For $2\pi \leq \omega t \leq 3\pi$, direct integration of (2) gives

$$\int_{i_o(2\pi/\omega)}^{i_o(t)} di_o = \frac{V_m}{L} \int_{2\pi/\omega}^t \sin \omega \tau \, d\tau$$

or

$$i_o(t) = i_o(2\pi/\omega) + \frac{V_m}{\omega L}(1 - \cos \omega t) = \frac{V_m}{\omega L}(3 - \cos \omega t) \tag{5}$$

For $3\pi \leq \omega t \leq 4\pi$, $i_o =$ constant where the particular value is given by evaluation of (5) at $t = 3\pi/\omega$, or

$$i_o(t) = \frac{4V_m}{\omega L} \tag{6}$$

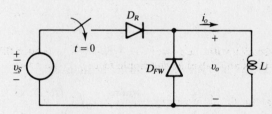

Fig. 3-14

POWER SUPPLY CIRCUITS

3.43 The source v_S with internal resistance R_S and half-wave rectifier of Fig. 3-15 act as a power supply for load R_L. If $v_S = V_m \sin \omega t$, calculate v_L and the average value of v_L.

▌ Only one cycle of v_S need be considered. For the positive half-cycle, $i_L > 0$ and, by voltage division,

$$v_L = \frac{R_L}{R_L + R_S}(V_m \sin \omega t) \equiv V_{Lm} \sin \omega t$$

For the negative half-cycle, the diode is reverse-biased, $i_L = 0$, and $v_L = 0$. Hence,

$$V_{L \text{ ave}} = \frac{1}{2\pi} \int_0^{2\pi} v_L(\omega t) \, d(\omega t) = \frac{1}{2\pi} \int_0^\pi V_{Lm} \sin \omega t \, d(\omega t) = \frac{V_{Lm}}{\pi}$$

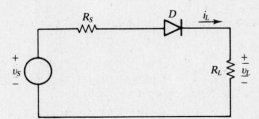

Fig. 3-15

DIODES: POWER LEVEL APPLICATIONS ◊ 87

3.44 The voltage regulation (ideal value of zero) of a power supply is defined as

$$\text{Reg} \equiv \frac{(\text{no-load } V_{L\,\text{ave}}) - (\text{full-load } V_{L\,\text{ave}})}{\text{full-load } V_{L\,\text{ave}}} \tag{1}$$

Find the percentage voltage regulation of the power supply of Fig. 3-15.

▎ From Problem 3.43, it is known that

$$\text{Full-load } V_{L\,\text{ave}} = \frac{V_{Lm}}{\pi} = \frac{R_L}{\pi(R_L + R_S)} V_m$$

Realizing that $R_L \to \infty$ for no load, then

$$\text{No-load } V_{L\,\text{ave}} = \lim_{R_L \to \infty} \left| \frac{R_L}{\pi(R_L + R_S)} V_m \right| = \frac{V_m}{\pi}$$

Thus, the voltage regulation is

$$\text{Reg} = \frac{\dfrac{V_m}{\pi} - \dfrac{R_L}{\pi(R_L + R_S)} V_m}{\dfrac{R_L}{\pi(R_L + R_S)} V_m} = \frac{R_S}{R_L} = \frac{100 R_S}{R_L} \%$$

3.45 As a measure of goodness of a power supply circuit, the ripple factor (ideal value of zero) is introduced.

$$F_r \equiv \frac{\text{maximum variation in output voltage}}{\text{average value of output voltage}} = \frac{\Delta v_L}{V_{L\,\text{ave}}} \tag{1}$$

Calculate the ripple factor for the half-wave rectifier of Problem 3.43.

▎ For the circuit of Problem 3.43,

$$F_r = \frac{\Delta v_L}{V_{L\,\text{ave}}} = \frac{V_{Lm}}{V_{Lm}/\pi} = \pi \approx 3.14$$

3.46 Commonly a filter is cascaded with the power supply rectifier to reduce the output voltage ripple. Make appropriate approximations and calculate the ripple factor for the power supply of Fig. 3-16(a) with the filter capacitor present.

▎ The capacitor in Fig. 3-16(a) stores energy while the diode allows current to flow, and delivers energy to the load when current flow is blocked. The actual load voltage v_L that results with the filter inserted is sketched in Fig. 3-16(b), for which it is assumed that $v_S = V_{Sm} \sin \omega t$ and D is an ideal diode. For $0 < t \le t_1$, D is forward-biased and capacitor C charges to the value V_{Sm}. For $t_1 < t \le t_2$, v_S is less than v_L, reverse-biasing D and causing it to act as an open circuit. During this interval the capacitor is discharging through the load R_L, giving

$$v_L = V_{Sm} e^{-(t-t_1)/R_L C} \qquad (t_1 < t \le t_2) \tag{1}$$

Over the interval $t_2 < t \le t_2 + \delta$, v_S forward-biases diode D and again charges the capacitor to V_{Sm}. Then v_S falls below the value of v_L and another discharge cycle identical to the first occurs.

Obviously, if the time constant $R_L C$ is large enough compared to T to result in a decay like that indicated in Fig. 3-16(b), a major reduction in Δv_L and a major increase in $V_{L\,\text{ave}}$ will have been achieved, relative to the unfiltered rectifier. The introduction of two quite reasonable approximations leads to simple formulas for Δv_L and $V_{L\,\text{ave}}$, and hence for F_r, that are sufficiently accurate for design and analysis work:
1. If Δv_L is to be small, then $\delta \to 0$ in Fig. 3-16(b) and $t_2 - t_1 \approx T$.
2. If Δv_L is small enough, then (1) can be represented over the interval $t_1 < t \le t_2$ by a straight line with a slope of magnitude $V_{Sm}/R_L C$.

The dashed line labeled "Approximate v_L" in Fig. 3-16(b) implements these two approximations. From right triangle abc,

$$\frac{\Delta v_L}{T} = \frac{V_{Sm}}{R_L C} \qquad \text{or} \qquad \Delta v_L = \frac{V_{Sm}}{f R_L C}$$

where f is the frequency of v_S. Since, under this approximation,

$$V_{L\,\text{ave}} = V_{Sm} - \tfrac{1}{2} \Delta v_L$$

and $R_L C/T = fR_L C$ is presumed large,

$$F_r = \frac{\Delta v_L}{V_{L\,\text{ave}}} = \frac{2}{2fR_L C - 1} \approx \frac{1}{fR_L C} \qquad (2)$$

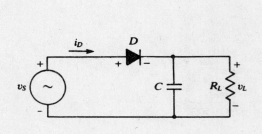
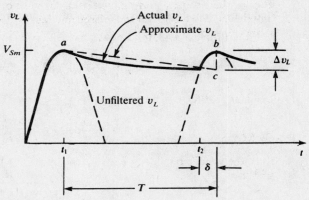

Fig. 3-16

3.47 The circuit of Fig. 3-15 is to be used as a dc power supply for a load R_L that varies from 10 Ω to 1 kΩ; v_S is a 10-V square wave. Find the percentage change in the average value of v_L over the range of load variation, and comment on the quality of regulation exhibited by this circuit.

▌ Let T denote the period of v_S. For $R_L = 10\,\Omega$,

$$v_L = \begin{cases} \dfrac{R_L}{R_L + R_S} v_S = \dfrac{10}{10+10}(10) = 5\,\text{V} & 0 \leq t < T/2 \\ 0 \text{ (diode blocks)} & T/2 \leq t < T \end{cases}$$

and so

$$V_{L\,\text{ave}} = \frac{5(T/2) + 0(T/2)}{T} = 2.5\,\text{V}$$

For $R_L = 1\,\text{k}\Omega$,

$$v_L = \begin{cases} \dfrac{R_L}{R_L + R_S} v_S = \dfrac{1000}{1010}\,10 = 9.9\,\text{V} & 0 \leq t < T/2 \\ 0 \text{ (diode blocks)} & T/2 \leq t < T \end{cases}$$

and so

$$V_{L\,\text{ave}} = \frac{9.9(T/2) + 0}{T} = 4.95\,\text{V}$$

Then, by (1) of Problem 3.44 and using $R_L = 10\,\Omega$ as full load, we have

$$\text{Reg} = \frac{4.95 - 2.5}{2.5}(100\%) = 98\%$$

This large value of regulation is prohibitive for most applications. Either another circuit or a filter network would be necessary to make this power supply useful.

3.48 Size the filter capacitor in the rectifier circuit of Fig. 3-16(a) so that the ripple voltage is approximately 5 percent of the average value of the output voltage. The diode is ideal, $R_L = 1\,\text{k}\Omega$, and $v_S = 90 \sin 2000t$ V.

▌ With $F_r = 0.05$, (2) of Problem 3.46 gives

$$C \approx \frac{1}{fR_L(0.05)} = \frac{1}{(2000/2\pi)(1 \times 10^3)(0.05)} = 62.83\,\mu\text{F}$$

3.49 Find the average value of v_L in Problem 3.48 with the filter capacitor in place.

▌ Using the approximations that led to (2) of Problem 3.46,

$$V_{L\,\text{ave}} = V_{Sm} - \frac{1}{2}\Delta v_L = V_{Sm} - \frac{V_{Sm}}{2fR_L C} \approx V_{Sm}\left(1 - \frac{0.05}{2}\right) = (90)(0.975) = 87.75\,\text{V}$$

3.50 Find v_L for the full-wave rectifier circuit of Fig. 3-17(a), treating the transformer and diodes as ideal.

▌ The two voltages labeled v_2 in Fig. 3-17(a) are identical in magnitude and phase. The ideal transformer and the voltage source v_S can therefore be replaced with two identical voltage sources, as in Fig. 3-17(b), without altering the electrical performance of the balance of the network. When v_S/n is positive, D_1 is forward-biased and conducts but D_2 is reverse-biased and blocks. Conversely, when v_S/n is negative, D_2 conducts and D_1 blocks. In short,

$$i_{D1} = \begin{cases} \dfrac{v_S/n}{R_L} & \dfrac{v_S}{n} \geq 0 \\ 0 & \dfrac{v_S}{n} < 0 \end{cases} \quad \text{and} \quad i_{D2} = \begin{cases} 0 & \dfrac{v_S}{n} > 0 \\ -\dfrac{v_S/n}{R_L} & \dfrac{v_S}{n} \leq 0 \end{cases}$$

By KCL,
$$i_L = i_{D1} + i_{D2} = \frac{|v_S/n|}{R_L}$$

and so $v_L = R_L i_L = |v_S/n|$.

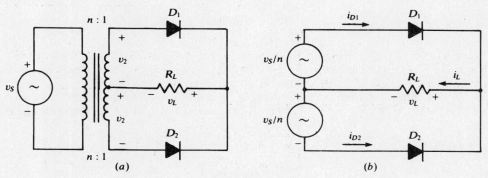

Fig. 3-17

3.51 Find the average value of output voltage v_L for the full-wave rectifier power supply of Fig. 3-17 if $v_S = V_m \sin \omega t$.

▌ Based on the results of Problem 3.50,
$$v_L = \left|\frac{V_m}{n} \sin \omega t\right|$$

Thus,
$$V_{L\,\text{ave}} = \frac{1}{2\pi} \int_0^{2\pi} \left|\frac{V_m}{n} \sin \omega t\right| d(\omega t) = \frac{V_m}{n\pi} \int_0^{\pi} \sin \omega t\, d(\omega t) = \frac{2V_m}{n\pi}$$

3.52 Determine the ripple factor for the power supply of Fig. 3-17.

▌ Using the results of Problem 3.51,
$$F_r = \frac{\Delta v_L}{V_{L\,\text{ave}}} = \frac{V_m/n}{2V_m/n\pi} = \frac{\pi}{2} \approx 1.57$$

3.53 Find the ripple factor for the full-wave rectifier power supply of Fig. 3-17 if a filter capacitor C is connected in parallel with R_L. Use analogous approximations to those introduced in Problem 3.46.

▌ The periodic voltage has period $T/2$ where T is the period of v_S. Hence, the derivation would be changed only in that T is replaced by $T/2$ and f by $2f$. Hence,
$$F_r = \frac{\Delta v_L}{V_{L\,\text{ave}}} = \frac{2}{2(2f)R_L C - 1} \approx \frac{1}{2fR_L C} \tag{1}$$

3.54 Size a filter capacitor that could be connected in parallel with R_L of Fig. 3-17 so that the ripple voltage is approximately 3 percent of the average value if $R_L = 500\,\Omega$ and the source frequency is 60 Hz.

I By (1) of Problem 3.53,

$$C \approx \frac{1}{2fR_LF_r} = \frac{1}{(2)(60)(500)(0.03)} = 555.5 \ \mu F$$

3.55 The identical diodes of Fig. 3-17 are nonideal and can be modeled by $R_F = 5 \ \Omega$ in the forward conduction state and an infinite resistance in the reverse direction. If $R_L = 500 \ \Omega$, $V_{L \ ave} = 80$ V, and $v_S = 115\sqrt{2} \sin(120\pi t)$ V, find the turns ratio of the ideal transformer.

I Because of symmetry, it is only necessary to analyze one leg of the rectifier circuit. Based on Problem 3.51,

$$v_L(t) = \frac{80}{2/\pi} \sin(120\pi t) = 125.66 \sin(120\pi t), \qquad 0 \le t \le T/2$$

$$i_L(t) = \frac{v_L(t)}{R_L} = \frac{125.66}{500} \sin(120\pi t) = 0.251 \sin(120\pi t), \qquad 0 \le t \le T/2$$

$$v_{D1}(t) = R_F i_L(t) = (5)(0.251) \sin(120\pi t) = 1.255 \sin(120\pi t), \qquad 0 \le t \le T/2$$

$$\frac{v_S}{n} = v_{D1} + v_L = 126.92 \sin(120\pi t), \qquad 0 \le t \le T/2$$

Thus, $\dfrac{v_S}{v_S/n} = \dfrac{115\sqrt{2}}{126.92} \approx 1.28$

3.56 Diodes of the full-wave bridge rectifier circuit of Fig. 3-18 conduct in pairs. Pair D_1-D_2 conduct for $0 \le \omega t \le \pi$ and pair D_3-D_4 conduct for $\pi \le \omega t \le 2\pi$. If i_o is continuous and $i_o(0) = i_o(\pi) = i_o(2\pi)$, find the average value of $i_o(t)$.

I By KVL,

$$v_o = Ri_o + L\frac{di_o}{dt} + V_B \tag{1}$$

Since i_o is periodic with period π, the average value need only consider the interval $0 \le t \le \pi/\omega$. Multiply (1) by dt, divide by π/ω, and integrate to form

$$\frac{\omega}{\pi} \int_0^{\pi/\omega} v_o \, dt = R\frac{\omega}{\pi} \int_0^{\pi/\omega} i_o \, dt + L_a \frac{\omega}{\pi} \int_{i_o(0)}^{i_o(\pi/\omega)} di_o + V_B \frac{\omega}{\pi} \int_0^{\pi/\omega} dt \tag{2}$$

Recognize the left-hand term as $V_{o \ ave}$ and the first term on the right of the equal sign as $RI_{o \ ave}$. Further, since the limits of integration of the second term on the right of the equal sign are equal, the term vanishes. Whence,

$$V_{o \ ave} = RI_{o \ ave} + 0 + V_B \qquad \text{or} \qquad I_{o \ ave} = \frac{V_{o \ ave} - V_B}{R}$$

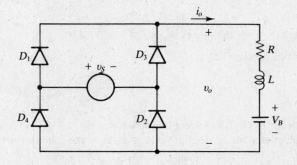

Fig. 3-18

Review Problems

3.57 Let v_S be a 10-V square wave of 500-Hz frequency and $L = 5$ mH for the circuit of Fig. 3-3. Find the average and rms values of $i_o(t)$.

▮ $T = 1/f = 1/500 = 0.002$ s. From the result of Problem 3.6,

$$I_{o\ \text{ave}} = \frac{1}{4}\frac{V_p}{L}T = \frac{1}{4}\left(\frac{10}{5 \times 10^{-3}}\right)(0.002) = 1 \text{ A}$$

$$I_{o\ \text{rms}} = \frac{\sqrt{3}}{6}\frac{V_p}{L}T = \frac{\sqrt{3}}{6}\left(\frac{10}{5 \times 10^{-3}}\right)(0.002) = 1.155 \text{ A}$$

3.58 Solve Problem 3.5 by methods of Laplace transforms.

▮ By KVL, with D conducting,

$$v_0 = v_S = L\frac{di_o}{dt}$$

Take the Laplace transform with $i(0) = 0$.

$$V_p = \frac{1 - 2e^{-Ts/2}}{s} = LsI$$

$$I = \frac{V_p}{L}\frac{1 - 2e^{-Ts/2}}{s^2} = \frac{V_p}{L}\left[\frac{1}{s^2} - \frac{2e^{-Ts/2}}{s^2}\right]$$

After inverse transformation,

$$i_o(t) = \frac{V_p}{L}[tU(t) - 2(t - T/2)U(t - T/2)]$$

3.59 Assume that the capacitor of Fig. 3-4 is initially uncharged. If $v_S = 100 \sin(120\pi t)$ V, $R = 20\ \Omega$, and $C = 500\ \mu\text{F}$, find $i(t)$ over the conduction interval of D.

▮ Based on results of Problem 3.7,

$$\phi = \tan^{-1}\left[-\frac{1}{\omega RC}\right] = \tan^{-1}\left[-\frac{1}{(120\pi)(20)(500 \times 10^{-6})}\right] = -14.85°$$

$$\frac{\omega CV_m}{[1 + (\omega RC)^2]^{1/2}} = \frac{(120\pi)(500 \times 10^{-6})(100)}{[1 + ((120\pi)(20)(500 \times 10^{-6}))^2]^{1/2}} = 4.833$$

$$1/RC = 1/(20)(500 \times 10^{-6}) = 100, \quad \sin\phi = \sin(-14.85°) = -0.256$$

$$i(t) = 4.833[\sin(120\pi t + 14.85°) - 0.256e^{-100t}], \quad 0 \le \omega t \le \beta$$

3.60 Determine $\beta = \omega t$, the point at which conduction of D ceases in Problem 3.59.

▮ The appropriate transcendental equation is given by the results of Problem 3.8. By trial-and-error solution, using $\phi = -14.85°$ gives $\beta \approx 158.1°$.

3.61 Rework Problem 3.7 if $v_S = V_m \cos \omega t$ and all else is unchanged.

▮ The complete response is given by

$$i(t) = i_f + i_n = \frac{\omega CV_m}{[1 + (\omega RC)^2]^{1/2}}\cos(\omega t - \phi) + Ae^{-t/RC}$$

where $\phi = \tan^{-1}(-1/\omega RC)$.
With $v_C(0) = 0$, $i(0)v_S(0)/R = V_m/R$; thus,

$$i(0) = \frac{V_m}{R} = \frac{\omega CV_m}{[1 + (\omega RC)^2]^{1/2}}\cos(-\phi) + A$$

And, after solving for A,

$$i(t) = \frac{\omega CV_m}{[1 + (\omega RC)^2]^{1/2}}\cos(\omega t - \phi) + \left[\frac{V_m}{R} - \frac{\omega CV_m}{[1 + (\omega RC)^2]^{1/2}}\right]e^{-t/RC}, \quad 0 \le \omega t \le \beta$$

3.62 Find the average value of i_o for Problem 3.9.

▌ Based on Problem 3.9,

$$I_{o\,\text{ave}} = \frac{1}{2\pi}\int_0^{2\pi} i_o\,d(\omega t) = \frac{1}{2\pi}\int_0^{2\pi} \frac{V_m}{\omega L}(1-\cos\omega t)\,d(\omega t) = \frac{V_m}{\omega L}$$

3.63 For the circuit of Fig. 3-5, let $v_S = 100\sin(120\pi t)$ V and $L = 10$ mH. Find (a) $i_o(t)$ and (b) $I_{o\,\text{ave}}$.

▌ (a) By the result of Problem 3.9,

$$i_o(t) = \frac{V_m}{\omega L}[1-\cos\omega t] = \frac{100}{(120\pi)(10\times 10^{-3})}[1-\cos(120\pi t)] = 26.53[1-\cos(120\pi t)]\text{ A}$$

(b) By the result of Problem 3.62,

$$I_{o\,\text{ave}} = \frac{V_m}{\omega L} = \frac{100}{(120\pi)(10\times 10^{-3})} = 26.53\text{ A}$$

3.64 Determine the voltage across R and L for the half-wave rectifier circuit of Fig. 3-6.

▌ The solution for $i(t)$ of Problem 3.12 is applicable. By Ohm's law,

$$v_R = Ri = \frac{RV_m}{Z}[\sin(\omega t - \phi) + e^{-tR/L}], \qquad i \geq 0$$

and

$$v_L = L\frac{di}{dt} = \frac{\omega L}{Z}\left[\cos(\omega t - \phi) - \frac{R}{L}e^{-tR/L}\right], \qquad i \geq 0$$

3.65 Let $V_B = 10$ V and v_S be a 20-V square wave of period T for the half-wave rectifier circuit of Fig. 3-7. Find (a) $V_{o\,\text{ave}}$ and (b) $I_{o\,\text{ave}}$.

▌ (a) For $0 \leq t \leq T/2$, $v_S = 20$ V; thus, d is forward-biased and $v_o = v_S = 20$ V. For $T/2 < t < T$, D is reverse-biased and $v_o = V_B = 10$ V. Hence,

$$V_{o\,\text{ave}} = \frac{1}{T}\left[\int_0^{T/2} 20\,dt + \int_{T/2}^T 10\,dt\right] = 15\text{ V}$$

(b) Current i_o is nonzero for $0 \leq t \leq T/2$ with value

$$i_o = \frac{v_S - V_B}{R} = \frac{20-10}{2} = 5\text{ A}$$

$$I_{o\,\text{ave}} = \frac{1}{T}\int_0^{T/2} 5\,dt = \frac{5}{2} = 2.5\text{ A}$$

3.66 Find the rms value of i_o for Problem 3.65.

▌

$$I_{o\,\text{rms}} = \left[\frac{1}{T}\int_0^T i_o\,dt\right]^{1/2} = \left[\frac{1}{T}\int_0^{T/2}(5)^2\,dt\right]^{1/2} = 3.535\text{ A}$$

3.67 Calculate the average values of power delivered to R and to V_B in Problem 3.65.

▌ Using the rms value of i_o from Problem 3.66,

$$P_R = I_{o\,\text{rms}}^2 R = (3.535)^2(2) = 25\text{ W}$$

$$P_{V_B} = \frac{1}{T}\int_0^T V_B i_o\,dt = \frac{1}{T}\int_0^{T/2}(10)(5)\,dt = 25\text{ W}$$

3.68 If C is initially uncharged, $V_B = 15$ V, $L = 10$ mH, and $C = 1000$ μF in the circuit of Fig. 3-8, find and sketch $i(t)$.

▌ By the results of Problem 3.21,

$$\omega = \sqrt{1/LC} = \left[\frac{1}{(10 \times 10^{-3})(1000 \times 10^{-6})}\right]^{1/2} = 316.23 \text{ rad/s}$$

$$i(t) = \frac{V_B}{\omega L} \sin \omega t = \frac{15}{(316.23)(1000 \times 10^{-3})} \sin(316.23t) = 4.74 \sin(316.23t), \quad 0 \leq \omega t \leq \pi$$

$$i(t) = 0, \quad \pi < \omega t$$

Figure 3-19 shows a sketch of $i(t)$.

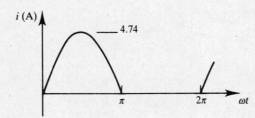

Fig. 3-19

3.69 Find the voltage across L and C of Problem 3.21.

▌ Using the results of Problem 3.21,

$$v_L = L\, di/dt = V_B \cos \omega t, \quad 0 \leq \omega t \leq \pi$$

$$v_L = 0, \quad \omega t > \pi$$

By KVL,

$$v_C = V_B - v_L = V_B(1 - \cos \omega t), \quad 0 \leq \omega t \leq \pi$$

and

$$v_C = v_C(\pi/\omega) = 2V_B, \quad \omega t > \pi$$

3.70 Assume the capacitor initially uncharged in the circuit of Fig. 3-8 and find $i(t)$ by the method of Laplace transforms.

▌ By KVL, with D conducting,

$$L\frac{di}{dt} + \frac{1}{C}\int i\, dt = V_B$$

Transforming,

$$LsI + \frac{1}{Cs}I = \frac{V_B}{s}$$

$$I = \frac{V_B/L}{s^2 + 1/LC} = \frac{V_B/L}{s^2 + \omega^2} = \frac{V_B}{\omega L}\left(\frac{\omega}{s^2 + \omega^2}\right) \quad \text{where} \quad \omega = \sqrt{1/LC}$$

Inverse transform to find

$$i(t) = \frac{V_B}{\omega L} \sin \omega t, \quad i \geq 0$$

3.71 Solve Problem 3.31 using the method of Laplace transforms.

▌ By KVL,

$$\frac{di}{dt} + \frac{R}{L}i = \frac{V_{dc} - V_B}{L}$$

Transforming with $i(0) = 0$,

$$sI + \frac{R}{L}I = \frac{V_{dc} - V_B}{sL} \quad \text{or} \quad I = \left(\frac{V_{dc} - V_B}{L}\right)\frac{1}{s(s + R/L)} = \frac{A}{s} + \frac{B}{s + R/L}$$

By residue theory,

$$A = \frac{V_{dc} - V_B}{L}\left[\frac{1}{s + R/L}\right]_{s=0} = \frac{V_{dc} - V_B}{R}, \quad B = \frac{V_{dc} - V_B}{L}\left[\frac{1}{s}\right]_{s=-R/L} = -\frac{(V_{dc} - V_B)}{R}$$

Whence, inverse transformation yields

$$i(t) = \frac{V_{dc} - V_B}{R}(1 - e^{-tR/L})$$

3.72 Find current $i(t)$ for the circuit of Fig. 3-13 if $v_S = 10$ V, $v_C(0) = 5$ V, $R = 1\,\Omega$, $L = 1$ mH, and $C = 1000\,\mu$F.

▌ In the dc steady-state, the current-forced response is zero; thus, the complete response is given by part (c) of Problem 3.41.

$$i(t) = i_f + i_n = 0 + e^{-500t}(A_1 \cos 866t + A_2 \sin 866t)$$

For $i(0) = 0$, $A_1 = 0$. Based on the results of part (a) of Problem 3.38,

$$\left.\frac{di}{dt}\right|_0 = \frac{v_S - v_C(0)}{L} = \frac{10-5}{1\times 10^{-3}} = 5000 = 866 A_2 \quad \text{or} \quad A_2 = 5.77$$

Hence, $$i(t) = 5.77 e^{-500t} \sin(866t), \quad i \geq 0$$

3.73 For the circuit of Fig. 3-13, let $v_S = 10$ V, $v_C(0) = 0$, $R = 2\,\Omega$, $L = 1$ mH, and $C = 1000\,\mu$F. Find $i(t)$.

▌ In the dc steady-state $i_f = 0$ and the complete response of $i(t)$ is given by the results of part (b) of Problem 3.41.

$$i(t) = i_f + i_n = 0 + A_1 e^{-1000t} + A_2 t e^{-1000t}$$

With $i(0) = 0$, $A_1 = 0$. Using the results of part (b) of Problem 3.38,

$$\left.\frac{di}{dt}\right|_0 = \frac{v_S}{L} = \frac{10}{1\times 10^{-3}} = 1 \times 10^4 = A_2$$

Hence, $$i(t) = 1 \times 10^4 t e^{-1000t}$$

3.74 If $v_S = 10$ V, $v_C(0) = 0$, $R = 4\,\Omega$, $L = 1$ mH, and $C = 1000\,\mu$F for the circuit of Fig. 3-13, find $i(t)$.

▌ Since $i_f = 0$ in the dc steady-state, the complete response of $i(t)$ is given by part (a) of Problem 3.41.

$$i(t) = i_f + i_n = 0 + A_1 e^{-268t} + A_2 e^{-3732t}$$

$$i(0) = 0 = A_1 + A_2 \tag{1}$$

Based on part (b) of Problem 3.38,

$$\left.\frac{di}{dt}\right|_0 = \frac{v_S}{L} = \frac{10}{1\times 10^{-3}} = -268 A_1 - 3732 A_2 \tag{2}$$

Simultaneous solution of (1) and (2) yields $A_1 = -A_2 = 2.89$, and

$$i(t) = 2.89(e^{-268t} - e^{-3732t})$$

3.75 Calculate $I_{o\,\text{ave}}$ for the circuit of Fig. 3-18 if v_S is a square wave of magnitude V_p and period T. Assume i_o is continuous and $i_o(0) = i_o(T/2) = i_o(T)$.

▌ Equation (2) of Problem 3.56 becomes

$$\frac{2}{T}\int_0^{T/2} v_o\, dt = R\frac{2}{T}\int_0^{T/2} i_0\, dt + L\frac{2}{T}\int_{i_o(0)}^{i_o(T/2)} di_o + V_B \frac{2}{T}\int_0^{T/2} dt$$

or $$V_p = R I_{o\,\text{ave}} + 0 + V_B \quad \text{and} \quad I_{o\,\text{ave}} = \frac{V_p - V_B}{R}$$

3.76 If i_o is continuous for the circuit of Fig. 3-18, $v_S = 120\sqrt{2}\sin(120\pi t)$ V, $R = 3\,\Omega$, and $V_B = 96$ V, find $I_{o\,\text{ave}}$.

▌ Based on Problem 3.56,

$$I_{o\,\text{ave}} = \frac{(2/\pi)(120\sqrt{2}) - 96}{3} = 3.98\text{ A}$$

3.77 The nonideal diodes of the full-wave rectifier of Fig. 3-17(a) can be modeled by $R_F = 4\,\Omega$ in forward conduction and an infinite resistance in blocking. If $v_S = 120\sqrt{2}\sin \omega t$ V, $R_L = 400\,\Omega$, and $n = 2$, find $I_{L\,\text{ave}}$.

$$\frac{v_S}{n} = \frac{120\sqrt{2}}{2}\sin \omega t = 60\sqrt{2}\sin \omega t \text{ V}$$

$$i_L(t) = \frac{v_S/n}{R_F + R_L} = \frac{60\sqrt{2}}{4 + 400}\sin \omega t = 0.210\sin \omega t \text{ A}$$

$$I_{L\,\text{ave}} = \frac{2}{\pi}(0.210) = 0.134 \text{ A}$$

3.78 Predict $i(t)$ over $n\pi \leq \omega t \leq (n+2)\pi$ if $v_S = V_m \sin \omega t$ for the circuit of Fig. 3-14. Assume that n is an even integer.

By induction from the results of Problem 3.42,

$$i_o(t) - i_o(n\pi/\omega) = i_o(t) - \frac{nV_m}{\omega L} = \frac{V_m}{\omega L}(1 + \cos \omega t)$$

or

$$i_o(t) = \frac{V_m}{\omega L}[(n+1) - \cos \omega t], \qquad n\pi \leq \omega t \leq (n+1)\pi$$

Since $di_o/dt = 0$, $i_o(t) =$ constant between odd and even values of π, or

$$i_o(t) = \frac{(n+2)V_m}{\omega L}, \qquad (n+1)\pi \leq \omega t \leq (n+2)\pi$$

3.79 If $v_1 = 100 \sin \omega t$ V and $v_2 = 50 \sin \omega t$ V for the rectifier circuit of Fig. 3-20, sketch v_o and i_o and label maximum and minimum values.

Diode D conducts for $v_1 > v_2$ or $0 \leq \omega t \leq \pi$. Over $\pi < \omega t < 2\pi$, $v_1 < v_2$ and D blocks. Thus,

$0 \leq \omega t \leq \pi$:
$$v_o = v_1 = 100 \sin \omega t$$
$$i_o = \frac{v_1 - v_2}{R} = 25 \sin \omega t$$

$\pi < \omega t < 2\pi$:
$$v_o = v_2 = 50 \sin \omega t$$
$$i_o = 0$$

Figure 3-21 displays the sketches of v_o and i_o.

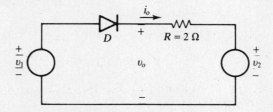

Fig. 3-20

3.80 Determine the average value of v_o for Problem 3.79.

$$V_{o\,\text{ave}} = \frac{1}{2\pi}\left[\int_0^\pi 100 \sin \omega t \, d(\omega t) + \int_\pi^{2\pi} 50 \sin \omega t \, d(\omega t)\right] = \frac{50}{\pi} \text{ V}$$

3.81 Calculate the average value of i_o for Problem 3.79.

$$I_{o\,\text{ave}} = \frac{1}{2\pi}\int_0^{2\pi} 25 \sin \omega t \, d(\omega t) = \frac{25}{\pi} \text{ A}$$

3.82 Find the average value of power delivered to v_2 for Problem 3.79.

$$P_2 = \frac{1}{2\pi}\int_0^{2\pi} v_2 i_o \, d(\omega t) = \frac{1}{2\pi}\int_0^\pi (50 \sin \omega t)(25 \sin \omega t) \, d(\omega t) = \frac{1250}{2\pi}\int_0^\pi \sin^2 \omega t \, d(\omega t) = 312.5 \text{ W}$$

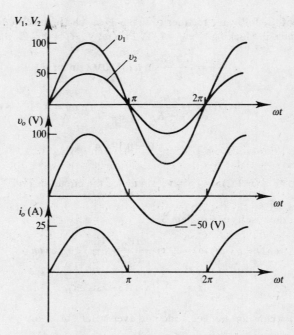

Fig. 3-21

3.83 Find the average value of power dissipated by the 2-Ω resistor for Problem 3.79.

$$P_R = \frac{1}{2\pi} \int_0^{2\pi} 2i_o^2 \, d(\omega t) = \frac{1}{\pi} \int_0^{\pi} (25 \sin \omega t)^2 \, d(\omega t) = 312.5 \text{ W}$$

3.84 Show that the average value of power delivered by v_1 to the circuit for Problem 3.79 is exactly the average power absorbed by v_2 and the 2-Ω resistor.

$$P_1 = \frac{1}{2\pi} = \int_0^{2\pi} v_1 i_o \, d(\omega t) = \frac{1}{2\pi} \int_0^{\pi} (100 \sin \omega t)(25 \sin \omega t) \, d(\omega t) = \frac{1250}{\pi} \int_0^{\pi} \sin^2 \omega t \, d(\omega t) = 625 \text{ W}$$

Based on Problems 3.82 and 3.83, $P_2 + P_R = 312.5 + 312.5 = 625 = P_1$.

CHAPTER 4
Bipolar Junction Transistor Fundamentals

OPERATING PRINCIPLES

4.1 The *bipolar junction transistor* (BJT) is a three-element [*emitter* (E), *base* (B), and *collector* (C)] device made up of alternating layers of *n*- and *p*-type semiconductor materials joined metallurgically. Figure 4-1 displays the schematic representations and defines positive current directions of the *pnp* type (principal conduction by holes) and the *npn* (principal conduction by electrons) transistors. Since electronic signals commonly consist of superimposed components, a set of notations given by Table 4-1 is adopted to clarify discussion of signal components.

If base current i_B of the *npn* transistor of Fig. 4-1(a) consists of a dc component of 10 μA and a sinusoidal component of 2 μA peak value, relate these values through use of Table 4-1.

$$I_B = 10\ \mu\text{A} \quad \text{and} \quad i_b = I_{bm} \sin \omega t = 2 \sin \omega t\ \mu\text{A}$$

Hence, the total base current is

$$i_B = I_B + i_b = 10 + 2 \sin \omega t\ \mu\text{A}$$

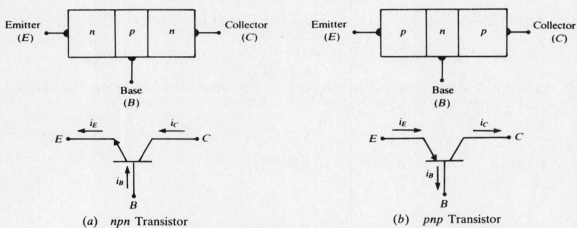

Fig. 4-1

TABLE 4-1

type of value	symbol		examples
	variable	subscript	
Total instantaneous	Lowercase	Uppercase	i_B, v_{BE}
DC	Uppercase	Uppercase	I_B, V_{BE}
Quiescent-point	Uppercase	Uppercase plus Q	I_{BQ}, V_{BEQ}
AC instantaneous	Lowercase	Lowercase	i_b, v_{be}
RMS	Uppercase	Lowercase	I_b, V_{be}
Maximum (sinusoid)	Uppercase	Lowercase plus m	I_{bm}, V_{bem}
Average	Uppercase	Uppercase plus ave	$I_{B\text{ ave}}, V_{BE\text{ ave}}$

4.2 For the *pnp* transistor of Fig. 4-1(b), v_{CE} (increase in potential from emitter to collector) is given by $10 + 2 \cos \omega t$ V. Determine the components of v_{CE} consistent with Table 4-1.

$$v_{CE} = V_{CE} + v_{ce} = V_{CE} + V_{cem} \cos \omega t$$

where $V_{CE} = 10$ V and $V_{cem} = 2$ V.

4.3 Determine the rms value for v_{CE} of Problem 4.2 using notation consistent with Table 4-1.

$$V_{ce}^2 = \frac{1}{T}\int_0^T v_{CE}^2\, dt = \frac{1}{2\pi}\int_0^{2\pi}(V_{CE} + V_{cem}\cos\omega t)^2\, d(\omega t) = V_{CE}^2 + \frac{V_{cem}^2}{2}$$

Whence,
$$V_{ce} = \left[(10)^2 + \frac{(2)^2}{2}\right]^{1/2} = 10.099\text{ V}$$

4.4 Calculate the average value for i_B of Problem 4.1.

$$I_{B\text{ ave}} = \frac{1}{T}\int_0^T i_B\, dt = \frac{1}{2\pi}\int_0^{2\pi}(I_B + I_{bm}\sin\omega t)\, d(\omega t) = I_B = 10\ \mu\text{A}$$

4.5 In the *npn* transistor of Fig. 4-1(a), 10^8 holes/μs move from the base to the emitter region while 10^{10} electrons/μs move from the emitter to the base region. An ammeter reads the base current as $i_B = 16\ \mu$A. Determine the emitter current i_E.

The emitter current is found as the net rate of flow of positive charge into the emitter region:

$$i_E = (1.602\times 10^{-19}\text{ C/hole})(10^{14}\text{ holes/s}) - (-1.602\times 10^{-19}\text{ C/electron})(10^{16}\text{ electron/s})$$
$$= 1.602\times 10^{-5} + 1.602\times 10^{-3} = 1.618\text{ mA}$$

4.6 Determine the collector current i_C for the *npn* transistor of Problem 4.5.

By KCL,
$$i_C = i_E - i_B = 1.618\times 10^{-3} - 16\times 10^{-6} = 1.602\text{ mA}$$

4.7 The collector current for each of the transistors of Fig. 4-1 is made up of two components:
1. $\alpha i_E = h_{FB}i_E$ = per unit portion of i_E that crosses the base region.
2. I_{CBO} = leakage current (similar to I_o of a diode) that flows across the reverse-biased collector-base junction with the emitter open-circuited plus surface leakage around the junction and avalanche multiplication current (negligible at low voltage) due to collisions of carriers in the collector transition region.

Hence,
$$i_C = \alpha i_E + I_{CBO} \tag{1}$$

Find an expression for the transistor base current that includes saturation current.

Using KCL and (1),
$$i_B = i_E - i_C = i_E - (\alpha i_E + I_{CBO}) = (1-\alpha)i_E - I_{CBO} \tag{2}$$

4.8 Express the base current of Problem 4.7 in terms of collector current rather than emitter current.

Solve (1) of Problem 4.7 for i_E and use KCL to give
$$i_E = i_C/\alpha - I_{CBO}/\alpha$$

But,
$$i_B = i_E - i_C = (i_C/\alpha - I_{CBO}/\alpha) - i_C = \frac{1-\alpha}{\alpha}i_C - \frac{I_{CBO}}{\alpha} = i_C/\beta - I_{CBO}/\alpha \tag{1}$$

where
$$\beta \equiv \frac{\alpha}{1-\alpha} \tag{2}$$

4.9 Collector current of the BJT can also be expressed as

$$i_C = \beta i_B + I_{CEO} = \frac{\alpha}{1-\alpha}i_B + I_{CEO} = h_{FE}i_B + I_{CEO} \tag{1}$$

where I_{CEO} is the saturation plus surface leakage current that flows from the collector to emitter with the base open-circuited. Subtract (1) of Problem 4.8 from (1) above to find a relationship between I_{CBO} and I_{CEO}.

$$\beta i_B = i_C - I_{CEO}$$
$$\beta i_B = i_C - (\beta/\alpha)I_{CBO}$$
$$\overline{\phantom{0 = -I_{CEO} + (\beta/\alpha)I_{CBO}}}$$
$$0 = -I_{CEO} + (\beta/\alpha)I_{CBO} \tag{2}$$

But, $\beta = \alpha/(1-\alpha)$; whence,
$$\alpha = \beta/(\beta+1) \tag{3}$$
Substitute (3) into (2) to find
$$I_{CEO} = (\beta+1)I_{CBO} \tag{4}$$

4.10 Determine α for the transistor of Problem 4.5 if leakage currents (flow due to holes) are negligible and the described charge flow is constant.

▍ If we assume $I_{CBO} = I_{CEO} = 0$, then
$$\alpha = \frac{i_C}{i_E} = \frac{i_E - i_B}{i_E} = \frac{1.602 - 0.016}{1.602} = 0.99$$

4.11 Neglect leakage currents and calculate β for the transistor of Problem 4.5.

▍ With $I_{CBO} = I_{CEO} = 0$,
$$\beta = \frac{i_C}{i_B} = \frac{i_E - i_B}{i_B} = \frac{1.602 - 0.016}{0.016} = 99.125$$

4.12 A BJT has $\alpha = 0.99$, $i_B = I_B = 25\ \mu A$, and $I_{CBO} = 200$ nA. Find the dc collector current.

▍ With $\alpha = 0.99$, (2) of Problem 4.8 gives
$$\beta = \frac{\alpha}{1-\alpha} = 99$$

Using (3) of Problem 4.9 in (1) of Problem 4.8 then gives
$$I_C = \beta I_B + (\beta+1)I_{CBO} = 99(25 \times 10^{-6}) + (99+1)(200 \times 10^{-9}) = 2.495\ \text{mA}$$

4.13 Determine the dc emitter current for the BJT of Problem 4.12.

▍ The dc emitter current follows from (1) of Problem 4.7:
$$I_E = \frac{I_C - I_{CBO}}{\alpha} = \frac{2.495 \times 10^{-3} - 200 \times 10^{-9}}{0.99} = 2.518\ \text{mA}$$

4.14 Calculate the percentage error in emitter current when the leakage current is neglected for the BJT of Problem 4.13.

▍ Neglecting the leakage current, we have
$$I_C = \beta I_B = 99(25 \times 10^{-6}) = 2.475\ \text{mA} \qquad \text{so} \qquad I_E = \frac{I_C}{\alpha} = \frac{2.475}{0.99} = 2.5\ \text{mA}$$

giving an emitter-current error of
$$\frac{2.518 - 2.5}{2.518}(100\%) = 0.71\%$$

4.15 For a certain BJT, $\beta = 50$, $I_{CEO} = 3\ \mu A$, and $I_C = 1.2$ mA. Find I_B.

▍ By (1) of Problem 4.9,
$$I_B = \frac{I_C - I_{CEO}}{\beta} = \frac{1.2 \times 10^{-3} - 3 \times 10^{-6}}{50} = 23.94\ \mu A$$

4.16 Determine the emitter current for the BJT of Problem 4.15.

▍ By KCL,
$$I_E = I_C + I_B = 1.2 \times 10^{-3} - 23.94 \times 10^{-6} = 1.224\ \text{mA}$$

4.17 A Ge transistor with $\beta = 100$ has a base-to-collector leakage current I_{CBO} of $5\ \mu A$. If the transistor is connected for common-emitter operation, find the collector current for $I_B = 0$.

▮ With $I_B = 0$, only emitter-to-collector leakage flows, and by (4) of Problem 4.9,
$$I_{CEO} = (\beta + 1)I_{CBO} = (100 + 1)(5 \times 10^{-6}) = 505 \, \mu A$$

4.18 If $I_B = 40 \, \mu A$ for the transistor of Problem 4.17, calculate the collector current.

▮ If we substitute (4) of Problem 4.9 into (1) of Problem 4.8 and solve for I_C, we get
$$I_C = \beta I_B + (\beta + 1)I_{CBO} = (100)(40 \times 10^{-6}) + (101)(5 \times 10^{-6}) = 4.505 \text{ mA}$$

4.19 The *Ebers–Moll transistor model* accounts for the exponential nature of the transistor junctions regardless of polarities of applied terminal voltages describing the currents of an *npn* transistor by

$$i_E = i_{EF} - i_{ER} = i_{EF} - \alpha_R i_{CR} \tag{1}$$
$$i_C = i_{CF} - i_{CR} = \alpha_F i_{EF} - i_{CR} \tag{2}$$
$$i_B = i_{BF} + i_{BR} \tag{3}$$

where
$$i_{EF} = I_{EO}(e^{v_{BE}/\eta V_T} - 1) \tag{4}$$
$$i_{CR} = I_{CO}(e^{v_{BC}/\eta V_T} - 1) \tag{5}$$
$$\alpha_R I_{CO} = \alpha_F I_{EO} \tag{6}$$

The subscripts *F* and *R* denote conduction in the forward and reverse directions, respectively. I_{EO} and I_{CO} are the saturation currents, respectively, of the emitter-base and collector-base junctions. Based on (1)–(3), draw the Ebers–Moll model equivalent circuit.

▮ See Fig. 4-2.

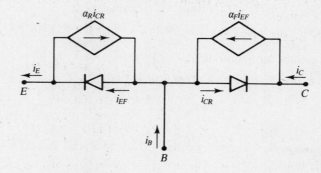

Fig. 4-2

4.20 Using back substitution, rewrite (1) and (2) of Problem 4.19 in terms of terminal currents and voltages to give a more useful form of the Ebers–Moll equations.

▮ Solve (2) for i_{CR}, substitute the result into (1), and use (4).
$$i_E = i_{EF} - \alpha_R(\alpha_F i_{EF} - i_C) = \alpha_R i_C + (1 - \alpha_R \alpha_F)i_{EF}$$
or
$$i_E = \alpha_R i_C + (1 - \alpha_F \alpha_R)I_{EO}(e^{v_{BE}/\eta V_T} - 1) \tag{1'}$$

Solve (1) for i_{EF}, substitute the result into (2), and use (5).
$$i_C = \alpha_F(i_E + \alpha_R i_{CR}) - i_{CR} = \alpha_F i_E - (1 - \alpha_F \alpha_R)i_{CR}$$
or
$$i_C = \alpha_F i_E - (1 - \alpha_F \alpha_R)I_{CO}(e^{v_{BC}/\eta V_T} - 1) \tag{2'}$$

4.21 Devise a test to determine the parameter α_R (*common base reverse short circuit current gain*) of the Ebers–Moll model.

▮ Short the base-emitter terminals so that $v_{BE} = 0$ and apply $v_{BC} > 0$; then by (1') of Problem 4.20,
$$\alpha_R = \frac{i_E}{i_C}\bigg|_{v_{BE}=0, v_{BC}>0} \tag{1}$$

4.22 Propose a test to measure the parameter α_F (*common base forward short circuit current gain*) of the Ebers–Moll model.

■ Short the base-collector terminals so that $v_{BC}=0$ and apply $v_{BE}>0$; then, by (2') of Problem 4.20,
$$\alpha_F = \left.\frac{i_C}{i_E}\right|_{v_{BC}=0,\,v_{BE}>0} \tag{1}$$

4.23 Find an expression for the base-emitter voltage of a transistor modeled by the Ebers–Moll equations.

■ Direct solution of (1') of Problem 4.20 for v_{BE} yields
$$v_{BE} = \eta V_T \ln\left[1 + \frac{i_E - \alpha_R i_C}{(1-\alpha_F\alpha_R)I_{EO}}\right] \tag{1}$$

4.24 If $i_E = 0$ (*cutoff*) exists for a transistor, find an expression for the base-emitter voltage v_{BE}.

■ With $i_E = 0$, $i_C \approx -I_{CO}$. Based on (1) of Problem 4.23 and using (6) of Problem 4.19,
$$v_{BE} \approx \eta V_T \ln\left[1 + \frac{0 - \alpha_R(-I_{CO})}{(1-\alpha_F\alpha_R)I_{EO}}\right] = \eta V_T \ln\left[1 + \frac{\alpha_F}{1-\alpha_F\alpha_R}\right] \tag{1}$$

4.25 For a certain Si *npn* transistor, $I_{CO} = 3\,\mu\text{A}$, $V_T = 26\,\text{mV}$, $\alpha_F = 0.99$, and $\alpha_R = 0.2$. Calculate the value of v_{BE} at cutoff.

■ By (1) of Problem 4.24,
$$v_{BE} = \eta V_T \ln\left[1 + \frac{\alpha_F}{1-\alpha_F\alpha_R}\right] = 2(0.026)\ln\left[1 + \frac{0.99}{(1-0.99(0.2))}\right] = 41.8\,\text{mV}$$

4.26 If the transistor of Problem 4.25 is operating at a point where $i_E = 10\,\text{mA}$ and $i_C = 9.99\,\text{mA}$, determine the base-emitter voltage v_{BE}.

■ From (6) of Problem 4.19,
$$I_{EO} = \frac{\alpha_R}{\alpha_F}I_{CO} = \frac{0.2}{0.99}(3\times 10^{-6}) = 0.606\,\mu\text{A}$$

By (1) of Problem 4.23,
$$v_{BE} = \eta V_T \ln\left[1 + \frac{i_E - \alpha_R i_C}{(1-\alpha_F\alpha_R)I_{EO}}\right] = 2(0.026)\ln\left[1 + \frac{10\times 10^{-3} - 0.2(9.99\times 10^{-3})}{(1-0.99(0.2))0.606\times 10^{-9}}\right] = 0.505\,\text{V}$$

4.27 If the transistor of Problem 4.25 were operating with $v_{BE} = 0.6\,\text{V}$ and $v_{BC} = 0.35\,\text{V}$, find the value of collector current i_C.

■ By (2), (4), and (5) of Problem 4.19 and the result of Problem 4.26,
$$i_{EF} = I_{EO}(e^{v_{BE}/\eta V_T} - 1) = 0.606\times 10^{-6}(e^{0.6/(2)(0.026)} - 1) = 62.17\,\text{mA}$$
$$i_{CR} = I_{CO}(e^{v_{BC}/\eta V_T} - 1) = 3\times 10^{-6}(e^{0.35/(2)(0.026)} - 1) = 2.51\,\text{mA}$$
$$i_C = \alpha_F i_{EF} - i_{CR} = 0.99(62.17) - 2.51 = 59.04\,\text{mA}$$

4.28 Determine the value of collector-emitter voltage v_{CE} for the transistor of Problem 4.27.

■ Applying KVL to Fig. 4-1(*a*),
$$v_{CE} = v_{BE} - v_{BC} = 0.6 - 0.35 = 0.25\,\text{V}$$

BIAS AND LOAD LINES

4.29 The *common-base* (CB) connection is a two-port transistor arrangement in which the base shares a common point with the input and output terminals. The independent input variables are emitter current i_E and base-to-emitter voltage v_{EB}. The corresponding independent output variables are collector current i_C and base-to-collector voltage v_{CB}. Practical CB transistor analysis is based on two experimentally determined sets of curves:

1. *Input* or *transfer characteristics* relate i_E and v_{EB} (port input variables), with v_{CB} (port output variable)

held constant. The method of laboratory measurement is indicated in Fig. 4-3(a), and the typical form of the resulting family of curves is depicted in Fig. 4-3(b).

2. *Output* or *collector characteristics* give i_C as a function of v_{CB} (port output variables) for constant values of i_E (port input variable), measured as in Fig. 4-3(a). Figure 4-3(c) shows the typical form of the resulting family of curves.

For the CB circuit of Fig. 4-3(a), let $v_s = V_{sm} \sin \omega t$ where $V_{sm} = 1$ V. If $V_{EE} = 2$ V, $V_{CC} = 25$ V, $R_E = 1$ kΩ, $R_C = 10$ kΩ, $I_{CBO} \approx 0$, $\alpha = 0.99$, and $v_{EB} = 0.7$ V, find i_E.

▌ Use KVL around the left-hand mesh to give

$$i_E = \frac{V_{EE} - v_{BE} + v_s}{R_E} = \frac{2 - 0.7 + \sin \omega t}{1 \times 10^3} = 1.3 + \sin \omega t \text{ mA} \qquad (1)$$

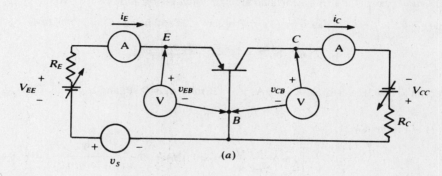

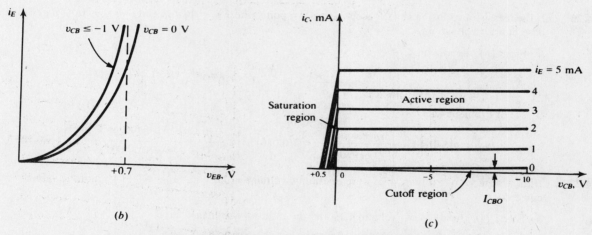

Fig. 4-3 Common-base characteristics (*pnp*, Si device).

4.30 Determine the collector-base voltage v_{CB} for the CB circuit of Problem 4.29.

▌ Apply KVL around the collector-base mesh, use $i_C = \alpha i_E$, and substitute i_E from Problem 4.29 to yield

$$v_{CB} = -V_{CC} + i_C R_C = -V_{CC} + \alpha i_E R_C = -25 + 0.99(1.3 + \sin \omega t) \times 10^{-3}(10 \times 10^3)$$
$$= -12.13 + 10 \sin \omega t \text{ V}$$

4.31 Find the ac voltage gain ratio A_v for the CB circuit of Problem 4.29.

▌ Based on the result of Problem 4.30,

$$A_v = \frac{v_{cb}}{v_s} = \frac{V_{cbm}}{V_{sm}} = \frac{10}{1} = 10$$

4.32 The *common-emitter* (CE) connection is a two-port transistor arrangement (widely used because of its high current amplification) in which the emitter shares a common point with the input and output terminals. The independent port input variables are base current i_B and emitter-to-base voltage v_{BE}, and the independent port output variables are collector current i_C and emitter-to-collector voltage v_{CE}. Like CB analysis, CE analysis is

based on:
1. *Input* or *transfer characteristics* that relate the port input variables i_B and v_{BE}, with v_{CE} held constant. Figure 4-4(a) shows the measurement setup, and Fig. 4-4(b) the resulting input characteristics.
2. *Output* or *collector characteristics* that show the functional relationship between port output variables i_C and v_{CE} for constant i_B, measured as in Fig. 4-4(a). Typical collector characteristics are displayed in Fig. 4-4(c).

For the CE circuit of Fig. 4-4(a), let $v_s = V_{sm} \cos \omega t$ where $V_{sm} = 1$ V. If $V_{BB} = 2$ V, $V_{CC} = 15$ V, $R_B = 100$ kΩ, $R_C = 10$ kΩ, $I_{CBO} \approx 0$, $v_{BE} = 0.7$ V, and $\beta = 50$, find i_B.

▌ Use KVL around the left-hand mesh to give

$$i_B = \frac{V_{BB} - v_{BE} + V_{sm} \cos \omega t}{R_B} = \frac{2 - 0.7 + \cos \omega t}{100 \times 10^3} = 13 + 10 \cos \omega t \ \mu A$$

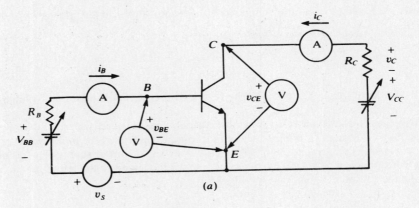

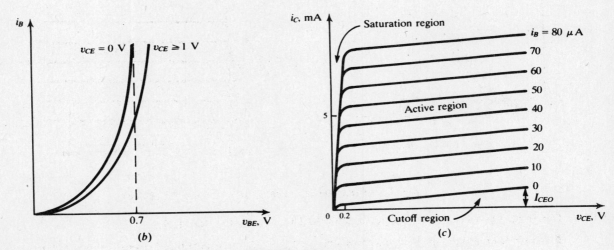

Fig. 4-4 Common-emitter characteristics (*npn*, Si device).

4.33 Determine the collector current i_C for the CE circuit of Problem 4.32.

▌ From (1) of Problem 4.8 with $I_{CBO} = 0$ and the result of Problem 4.32,

$$i_C = \beta i_B = 50(13 + 10 \cos \omega t) \times 10^{-6} = 0.65 + 0.5 \cos \omega t \ \text{mA}$$

4.34 Find the emitter current i_E for the CE circuit of Problem 4.32.

▌ $$\alpha = \beta/(\beta + 1) = 50/(50 + 1) = 0.9804$$

From (2) of Problem 4.7 with $I_{CBO} = 0$ and the result of Problem 4.32,

$$i_E = i_B/(1 - \alpha) = (13 + 10 \cos \omega t) \times 10^{-6}/(1 - 0.9804)$$

$$= 0.663 + 0.51 \cos \omega t \ \text{mA}$$

4.35 Calculate the ac current gain ratio A_i for the CB circuit of Problem 4.32.

▎ Based on the results of Problems 4.32 and 4.33,

$$A_i = \frac{i_c}{i_b} = \frac{I_{cm}}{I_{bm}} = \frac{0.5 \times 10^{-3}}{10 \times 10^{-6}} = 50 = \beta$$

4.36 Determine the collector-emitter voltage v_{CB} for the CB circuit of Problem 4.32.

▎ Apply KVL around the collector-emitter mesh and use the result of Problem 4.33 to find

$$v_{CE} = V_{CC} - i_C R_C = 15 - (0.65 + 0.5 \cos \omega t) \times 10^{-3}(10 \times 10^3)$$
$$= 8.5 + 5 \cos \omega t \text{ V}$$

4.37 Evaluate the ac voltage gain ratio A_v for the CE circuit of Problem 4.32 if v_C is considered the output voltage.

▎ Based on the result of Problem 4.33,

$$A_v = \frac{-i_c R_C}{v_s} = \frac{-I_{cm} R_C}{V_{sm}} = \frac{-(0.5 \times 10^{-3})(10 \times 10^3)}{1} = -5$$

where the negative sign indicates a polarity reversal between input and output voltages.

4.38 Supply voltages and resistors *bias* a transistor; that is, they establish a specific set of dc terminal voltages and currents, thus determining a point of active-mode operation (called the *quiescent point* or *Q point*).

With universal bias arrangement of Fig. 4-5(a), only one dc power supply (V_{CC}) is needed to establish active-mode operation. Apply Thévenin's theorem for the circuit to the left of the points *a*, *b* of Fig. 4-5(a) to determine the values of R_B and V_{BB} of Fig. 4-5(b).

$$R_{Th} = R_B = \frac{R_1 R_2}{R_1 + R_2} \qquad V_{Th} = V_{BB} = \frac{R_1}{R_1 + R_2} V_{CC} \qquad (1)$$

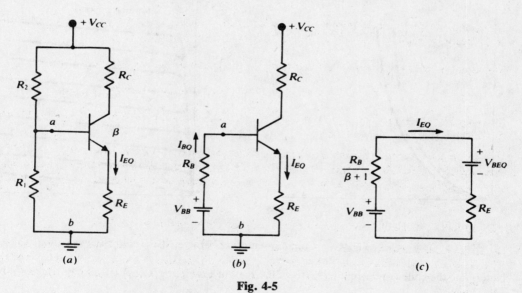

Fig. 4-5

4.39 Neglecting leakage current so that $I_{EQ} = (\beta + 1)I_{BQ}$ and assuming the quiescent base-emitter voltage V_{BEQ} is constant (0.7 V and 0.3 V for Si and Ge, respectively) for the transistor of Fig. 4-5(b), show that the bias circuit of Fig. 4-5(c) is valid.

▎ Application of KVL around the base-emitter loop yields

$$V_{BB} = \frac{I_{EQ}}{\beta + 1} R_B + V_{BEQ} + I_{EQ} R_E \qquad (1)$$

which can be represented by the emitter-loop equivalent bias circuit of Fig. 4-5(c).

4.40 From the emitter-loop equivalent bias circuit of Fig. 4-5(c), determine the conditions under which the quiescent current I_{CQ} is independent of the transistor parameter β.

▎ Solving (1) of Problem 4.39 for I_{EQ} and noting that $I_{EQ} = I_{CQ}/\alpha \approx I_{CQ}$, we obtain

$$I_{CQ} \approx I_{EQ} = \frac{V_{BB} - V_{BEQ}}{R_B/(\beta+1) + R_E} \tag{1}$$

If component values and the worst-case β value are such that

$$\frac{R_B}{\beta+1} \approx \frac{R_B}{\beta} \ll R_E \tag{2}$$

then I_{EQ} (and thus I_{CQ}) is nearly constant, regardless of changes in β; the circuit then has β-independent bias.

4.41 Define the *dc load line*.

▎ From Fig. 4-5(c) it is apparent that the family of collector characteristics is described by the mathematical relationship $i_C = f(v_{CE}, i_B)$ with independent variable v_{CE} and the parameter i_B. We assume that the collector circuit can be biased so as to place the Q point anywhere in the active region. A typical setup is shown in Fig. 4-6(a), from which

$$I_{CQ} = -\frac{V_{CEQ}}{R_{dc}} + \frac{V_{CC}}{R_{dc}}$$

Thus, if the *dc load line*

$$i_C = -\frac{v_{CE}}{R_{dc}} + \frac{V_{CC}}{R_{dc}} \tag{1}$$

and the specification

$$i_B = I_{BQ}$$

are combined with the relationship for the collector characteristics, the resulting system can be solved (analytically or graphically) for the collector quiescent quantities I_{CQ} and V_{CEQ}.

4.42 The signal source switch of Fig. 4-6(a) is closed, and the transistor base current becomes

$$i_B = I_{BQ} + i_b = 40 + 20 \sin \omega t \ \mu A$$

The collector characteristics of the transistor are those displayed in Fig. 4-6(b). If $V_{CC} = 12$ V and $R_{dc} = 1$ kΩ, graphically determine I_{CQ} and V_{CEQ}.

▎ The dc load line has ordinate intercept $V_{CC}/R_{dc} = 12$ mA and abscissa intercept $V_{CC} = 12$ V and is constructed on Fig. 4-6(b). The Q point is the intersection of the load line with the characteristic curve $i_B = i_{BQ} = 40 \ \mu A$. The collector quiescent quantities may be read from the axes as $I_{CQ} = 4.9$ mA and $V_{CEQ} = 7.2$ V.

4.43 Graphically determine i_c and v_{ce} for the transistor amplifier of Problem 4.42.

▎ A time scale is constructed perpendicular to the load line at the Q point, and a scaled sketch of $i_b = 20 \sin \omega t \ \mu A$ is drawn [see Fig. 4-6(b)] and translated through the load line to sketches of i_c and v_{ce}. As i_b swings $\pm 20 \ \mu A$ along the load line from point a to b, the ac components of collector current and voltage take on the values

$$i_c = 2.25 \sin \omega t \ \text{mA} \quad \text{and} \quad v_{ce} = -2.37 \sin \omega t \ \text{V}$$

The negative sign on v_{ce} signifies a 180° phase shift.

4.44 Calculate $h_{FE} (=\beta)$ at the Q point for the transistor amplifier of Problem 4.42.

▎ From (1) of Problem 4.8 with $I_{CEO} = 0$ [the $i_B = 0$ curve coincides with the v_{CE} axis in Fig. 4-6(b)],

$$h_{FE} = \frac{I_{CQ}}{I_{BQ}} = \frac{4.9 \times 10^{-3}}{40 \times 10^{-6}} = 122.5$$

106 □ CHAPTER 4

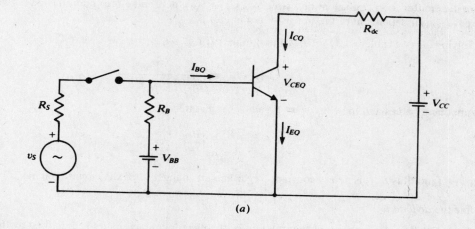

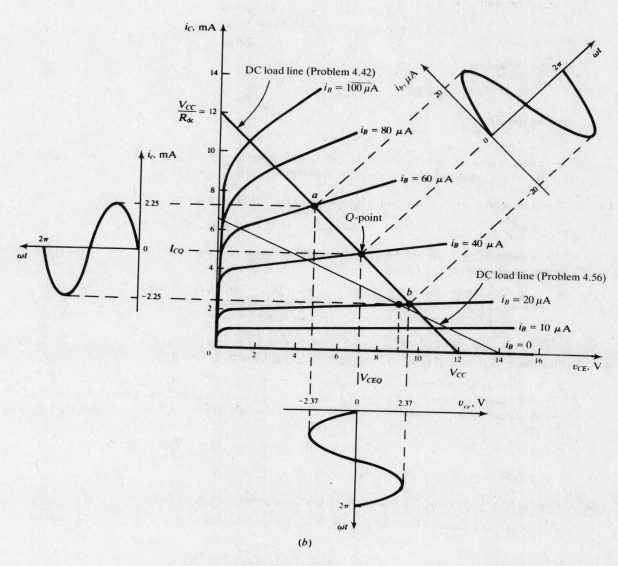

Fig. 4-6

4.45 A transistor with $\alpha = 0.98$ and $I_{CBO} = 5\ \mu A$ is biased so that $I_{BQ} = 100\ \mu A$. Find I_{CQ} and I_{EQ}.

By (2) of Problem 4.8 and (4) of Problem 4.9,

$$\beta = \frac{\alpha}{1-\alpha} = \frac{0.98}{1-0.98} = 49$$

so that
$$I_{CEO} = (\beta + 1)I_{CBO} = (49+1)(5 \times 10^{-6}) = 0.25\ \text{mA}$$

and, from (1) of Problem 4.9 and KCL,

$$I_{CQ} = \beta I_{BQ} + I_{CEO} = 49(100 \times 10^{-6}) + 0.25 \times 10^{-3} = 5.15\ \text{mA}$$

$$I_{EQ} = I_{CQ} + I_{BQ} = 5.15 \times 10^{-3} + 100 \times 10^{-6} = 5.25\ \text{mA}$$

4.46 The transistor of Fig. 4-7 has $\alpha = 0.98$ and a base current of $30\ \mu A$. Find β if leakage current is negligible.

$$\beta = \frac{\alpha}{1-\alpha} = \frac{0.98}{1-0.98} = 49$$

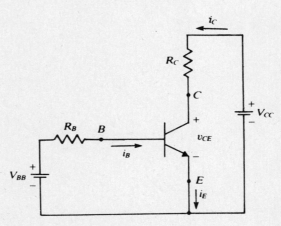

Fig. 4-7

4.47 Determine quiescent collector current I_{CQ} for the transistor of Problem 4.46.

From (1) of Problem 4.9 with $I_{CEO} = 0$, we have $I_{CQ} = \beta I_{BQ} = (49)(30 \times 10^{-6}) = 1.47\ \text{mA}$.

4.48 Calculate the quiescent emitter current I_{EQ} for the transistor of Problem 4.46.

From (1) of Problem 4.8 with $I_{CBO} = 0$,

$$I_{EQ} = \frac{I_{CQ}}{\alpha} = \frac{1.47}{0.98} = 1.50\ \text{mA}$$

4.49 The transistor circuit of Fig. 4-7 is to be operated with a base current of $40\ \mu A$ and $V_{BB} = 6\ V$. The Si transistor ($V_{BEQ} = 0.7\ V$) has negligible leakage current. Find the required value of R_B.

By KVL around the base-emitter loop,

$$V_{BB} = I_{BQ}R_B + V_{BEQ} \quad \text{so that} \quad R_B = \frac{V_{BB} - V_{BEQ}}{I_{BQ}} = \frac{6 - 0.7}{40 \times 10^{-6}} = 132.5\ \text{k}\Omega$$

4.50 In the circuit of Fig. 4-7, $\beta = 100$, $I_{BQ} = 20\ \mu A$, $V_{CC} = 15\ V$, and $R_C = 3\ \text{k}\Omega$. If $I_{CBO} = 0$, find I_{EQ}.

$$\alpha = \frac{\beta}{\beta + 1} = \frac{100}{101} = 0.9901$$

Now, using (1) of Problem 4.9 and (1) of Problem 4.8 with $I_{CBO} = I_{CEO} = 0$, we get

$$I_{CQ} = \beta I_{BQ} = 100(20 \times 10^{-6}) = 2\ \text{mA}$$

and
$$i_{EQ} = \frac{I_{CQ}}{\alpha} = \frac{2 \times 10^{-3}}{0.9901} = 2.02\ \text{mA}$$

4.51 Determine V_{CEQ} for the circuit of Problem 4.50.

▌ From an application of KVL around the collector circuit,
$$V_{CEQ} = V_{CC} - I_{CQ}R_C = 15 - 2(3) = 9 \text{ V}$$

4.52 For the circuit of Problem 4.50, change R_C to 6 kΩ and all else remains the same. Find the resulting V_{CEQ}.

▌ If I_{BQ} is unchanged, then I_{CQ} is unchanged. The solution proceeds as in Problem 4.51:
$$V_{CEQ} = V_{CC} - I_{CQ}R_C = 15 - 2(6) = 3 \text{ V}$$

4.53 The transistor of Fig. 4-8 is a Si device with a base current of 40 μA and $I_{CBO} = 0$. If $V_{BB} = 6$ V, $R_E = 1$ kΩ, and $\beta = 80$, find I_{EQ}.

▌
$$\alpha = \frac{\beta}{\beta + 1} = \frac{80}{81} = 0.9876$$

Then, combining (1) of Problem 4.8 and (1) of Problem 4.9 with $I_{CBO} = I_{CEO} = 0$ gives
$$I_{EQ} = \frac{I_{BQ}}{1 - \alpha} = \frac{40 \times 10^{-6}}{1 - 0.9876} = 3.226 \text{ mA}$$

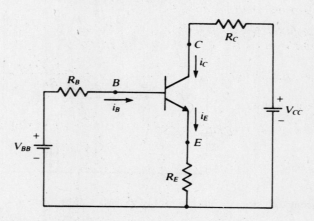

Fig. 4-8

4.54 Determine the necessary value of R_B for Problem 4.53.

▌ Applying KVL around the base-emitter loop gives
$$V_{BB} = I_{BQ}R_B + V_{BEQ} + I_{EQ}R_E$$

or (with V_{BEQ} equal to the usual 0.7 V for a Si device)
$$R_B = \frac{V_{BB} - V_{BEQ} - I_{EQ}R_E}{I_{BQ}} = \frac{6 - 0.7 - 3.226(1)}{40 \times 10^{-6}} = 51.85 \text{ k}\Omega$$

4.55 If $V_{CC} = 15$ V and $R_C = 3$ kΩ, find V_{CEQ} for the transistor circuit of Problem 4.53.

▌ From (1) of Problem 4.9 with $I_{CEO} = 0$,
$$I_{CQ} = \beta I_{BQ} = 80(40 \times 10^{-6}) = 3.2 \text{ mA}$$

Then, by KVL around the collector circuit,
$$V_{CEQ} = V_{CC} - I_{EQ}R_E - I_{CQ}R_C = 15 - 3.226(1) - 3.2(3) = 2.174 \text{ V}$$

4.56 Assume that the CE collector characteristics of Fig. 4-6(b) apply to the transistor of Fig. 4-7. If $I_{BQ} = 20$ μA, $V_{CEQ} = 9$ V, and $V_{CC} = 14$ V, find I_{CQ} graphically.

▌ The Q point is the intersection of $i_B = I_{BQ} = 20$ μA and $v_{CE} = V_{CEQ} = 9$ V. The dc load line must pass through the Q point and intersect the v_{CE} axis at $V_{CC} = 14$ V. Thus, the dc load line can be drawn on Fig. 4-6(b), and $I_{CQ} = 2.25$ mA can be read as the i_C coordinate of the Q point.

4.57 Determine the necessary value of R_C for Problem 4.56.

▎ The i_C intercept of the dc load line is $V_{CC}/R_{dc} = V_{CC}/R_C$, which, from Fig. 4-6(b), has the value 6.5 mA; thus,

$$R_C = \frac{V_{CC}}{6.5 \times 10^{-3}} = \frac{14}{6.5 \times 10^{-3}} = 2.15 \text{ k}\Omega$$

4.58 Evaluate I_{EQ} for the transistor amplifier of Problem 4.56.

▎ By KVL,

$$I_{EQ} = I_{CQ} + I_{BQ} = 2.25 \times 10^{-3} + 20 \times 10^{-6} = 2.27 \text{ mA}$$

4.59 Determine the numerical value of β for Problem 4.56 if leakage current is negligible.

▎ With $I_{CEO} = 0$, (1) of Problem 4.9 yields

$$\beta = \frac{I_{CQ}}{I_{BQ}} = \frac{2.25 \times 10^{-3}}{20 \times 10^{-6}} = 112.5$$

4.60 In the *pnp* Si transistor circuit of Fig. 4-9, $R_B = 500$ kΩ, $R_C = 2$ kΩ, $R_E = 0$, $V_{CC} = 15$ V, $I_{CBO} = 20$ μA, and $\beta = 70$. Find the Q-point collector current I_{CQ}.

▎ By (4) of Problem 4.9, $I_{CEO} = (\beta + 1)I_{CBO} = (70 + 1)(20 \times 10^{-6}) = 1.42$ mA. Now, application of the KVL around the loop that includes V_{CC}, R_B, R_E (=0), and ground

$$V_{CC} = V_{BEQ} + I_{BQ}R_B \quad \text{so that} \quad I_{BQ} = \frac{V_{CC} - V_{BEQ}}{R_B} = \frac{15 - 0.7}{500 \times 10^3} = 28.6 \text{ }\mu\text{A}$$

Thus, by (1) of Problem 4.9,

$$I_{CQ} = \beta I_{BQ} + I_{CEO} = 70(28.6 \times 10^{-6}) + 1.42 \times 10^{-3} = 3.42 \text{ mA}$$

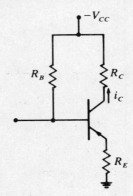

Fig. 4-9

4.61 The Si transistor of Fig. 4-10 is biased for constant base current. If $\beta = 80$, $V_{CEQ} = 8$ V, $R_C = 3$ kΩ, and $V_{CC} = 15$ V, find I_{CQ}.

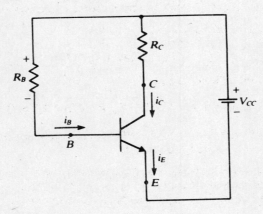

Fig. 4-10

■ By KVL around the collector-emitter circuit,

$$I_{CQ} = \frac{V_{CC} - V_{CEQ}}{R_C} = \frac{15 - 8}{3 \times 10^3} = 2.333 \text{ mA}$$

4.62 Determine the value of R_B required by the operating conditions of Problem 4.61.

■ If leakage current is neglected, (1) of Problem 4.9 gives

$$I_{BQ} = \frac{I_{CQ}}{\beta} = \frac{2.333 \times 10^{-3}}{80} = 29.16 \text{ } \mu\text{A}$$

Since the transistor is a Si device, $V_{BEQ} = 0.7$ V and, by KVL around the outer loop,

$$R_B = \frac{V_{CC} - V_{BEQ}}{I_{BQ}} = \frac{15 - 0.7}{29.16 \times 10^{-6}} = 490.4 \text{ k}\Omega$$

4.63 Repeat Problem 4.62 if the transistor were a Ge device.

■ The only difference here is that $V_{BEQ} = 0.3$ V; thus,

$$R_B = \frac{15 - 0.3}{29.16 \times 10^{-6}} = 504.1 \text{ k}\Omega$$

4.64 The Si transistor of Fig. 4-11 has $\beta = 50$ and negligible leakage current. Let $V_{CC} = 18$ V, $V_{EE} = 4$ V, $R_E = 200 \ \Omega$, and $R_C = 4 \text{ k}\Omega$. Find R_B so that $I_{CQ} = 2$ mA.

■ KVL around the base-emitter-ground loop gives

$$V_{EE} = I_{BQ}R_B + V_{BEQ} + I_{EQ}R_E \qquad (1)$$

Also, from (1) of Problem 4.7 and (3) of Problem 4.9,

$$I_{EQ} = \frac{\beta + 1}{\beta} I_{CQ} \qquad (2)$$

Now, using (1) of Problem 4.8, substituting (2) in (1), and solving for R_B yields

$$R_B = \frac{\beta(V_{EE} - V_{BEQ})}{I_{CQ}} - (\beta + 1)R_E = \frac{50(4 - 0.7)}{2 \times 10^{-3}} - (50 + 1)(200) = 72.3 \text{ k}\Omega$$

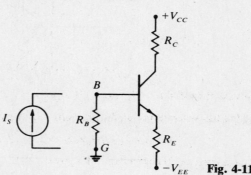

Fig. 4-11

4.65 Determine the value of V_{CEQ} for Problem 4.64.

■ KVL around the collector-emitter-ground loop gives

$$V_{CEQ} = V_{CC} + V_{EE} - \left(R_C + \frac{\beta + 1}{\beta} R_E\right) I_{CQ}$$

$$= 18 + 4 - \left(4 \times 10^3 + \frac{50 + 1}{50}(200)\right)(2 \times 10^{-3}) = 13.59 \text{ V}$$

4.66 The dc current source $I_S = 10 \ \mu\text{A}$ of Fig. 4-11 is connected from G to node B. The Si transistor has negligible leakage current and $\beta = 50$. If $R_B = 75 \text{ k}\Omega$, $R_E = 200 \ \Omega$, and $R_C = 4 \text{ k}\Omega$, find the dc current-gain ratio I_{CQ}/I_S for $V_{CC} = 18$ V and $V_{EE} = 4$ V.

▌ A Thévenin equivalent for the network to the left of terminals B, G has $V_{Th} = R_B I_S$ and $R_{Th} = R_B$. With the Thévenin equivalent circuit in place, KVL around the base-emitter loop yields

$$R_B I_S + V_{EE} = I_{BQ} R_B + V_{BEQ} + I_{EQ} R_E \qquad (1)$$

Using (1) of Problem 4.8 and (2) of Problem 4.64 in (1), solving for I_{CQ}, and then dividing by I_S result in the desired ratio:

$$\frac{I_{CQ}}{I_S} = \frac{R_B I_S + V_{EE} - V_{BEQ}}{I_S\left(\frac{R_B}{\beta} + \frac{\beta+1}{\beta} R_E\right)} = \frac{(75 \times 10^3)(10 \times 10^{-6}) + 4 - 0.7}{(10 \times 10^{-6})\left(\frac{75 \times 10^3}{50} + \frac{50+1}{50}(200)\right)} = 237.67 \qquad (2)$$

Note that the value of V_{CC} must be large enough so that cutoff does not occur, but otherwise it does not affect the value of I_{CQ}.

4.67 Repeat Problem 4.66 if $V_{CC} = 22$ V and $V_{EE} = 0$ V and all else is unchanged.

▌ $V_{EE} = 0$ in (2) of Problem 4.66 directly gives

$$\frac{I_{CQ}}{I_S} = \frac{(75 \times 10^3)(10 \times 10^{-6}) - 0.7}{(10 \times 10^{-6})\left[\frac{75 \times 10^3}{50} + \frac{50+1}{50}\right](200)} = 2.93$$

Obviously, V_{EE} strongly controls the dc current gain of this amplifier.

4.68 The Si transistor of Fig. 4-12 has $\alpha = 0.99$ and $I_{CEO} = 0$. Also, $V_{EE} = 4$ V and $V_{CC} = 12$ V. If $I_{EQ} = 1.1$ mA, find R_E.

▌ By KVL around the emitter-base loop,

$$R_E = \frac{V_{EE} + V_{BEQ}}{I_{EQ}} = \frac{4 + (-0.7)}{1.1 \times 10^{-3}} = 3\ \Omega$$

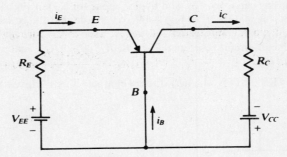

Fig. 4-12

4.69 For the transistor circuit of Problem 4.68, find R_C if $V_{CEQ} = -7$ V.

▌ By KVL around the transistor terminals (which constitute a closed path),

$$V_{CBQ} = V_{CEQ} - V_{BEQ} = -7 - (-0.7) = -6.3\ \text{V}$$

With negligible leakage current, (1) of Problem 4.7 gives

$$I_{CQ} = \alpha I_{EQ} = 0.99(1.1 \times 10^{-3}) = 1.089\ \text{mA}$$

Finally, by KVL around the base-collector loop,

$$R_C = \frac{V_{CC} + V_{CBQ}}{I_{CQ}} = \frac{12 - 6.3}{1.089 \times 10^{-3}} = 5.234\ \text{k}\Omega$$

4.70 Collector characteristics for the Ge transistor of Fig. 4-12 are given in Fig. 4-13. If $V_{EE} = 2$ V, $V_{CC} = 12$ V, and $R_C = 2$ kΩ, size R_E so that $V_{CEQ} = -6.4$ V.

▌ We construct, on Fig. 4-13, a dc load line having v_{CB} intercept $-V_{CC} = -12$ V and i_C intercept $V_{CC}/R_C = 6$ mA. The abscissa of the Q point is given by KVL around the transistor terminals:

$$V_{CBQ} = V_{CEQ} - V_{BEQ} = -6.4 - (-0.3) = -6.1\ \text{V}$$

With the Q point defined, we read $I_{EQ} = 3$ mA from the graph. Now KVL around the emitter-base loop leads to
$$R_E = \frac{V_{EE} + V_{BEQ}}{I_{EQ}} = \frac{2 + (-0.3)}{3 \times 10^{-3}} = 566.7 \ \Omega$$

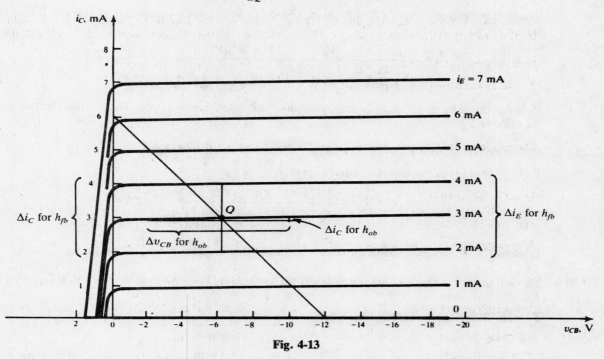

Fig. 4-13

4.71 The capacitors (coupling capacitors C_C and bypass capacitor C_E) of Fig. 4-14(a) are shorted in the circuit as it appears to ac signals [Fig. 4-14(b)]. In Fig. 4-14(a), we note that the collector-circuit resistance seen by the dc bias current I_{CQ} ($\approx I_{EQ}$) is $R_{dc} = R_C + R_E$. However, from Fig. 4-14(b) it is apparent that the collector signal current i_c sees a collector-circuit resistance $R_{ac} = R_C R_L / (R_C + R_L)$. Since $R_{ac} \neq R_{dc}$ in general, define the concept of an *ac load line*.

▌ By application of KVL to Fig. 4-14(b), the v-i characteristic of the external signal circuitry is found to be

$$v_{ce} = i_c R_{ac} \tag{1}$$

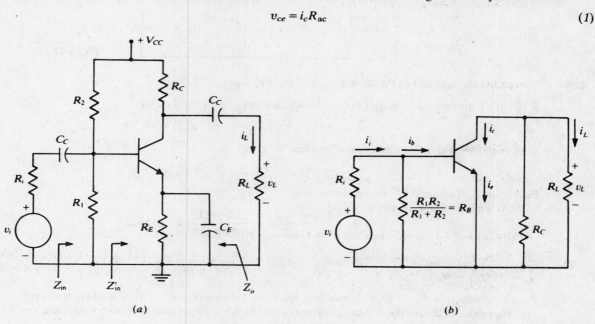

Fig. 4-14

BIPOLAR JUNCTION TRANSISTOR FUNDAMENTALS □ 113

Since $i_c = i_C - I_{CQ}$ and $v_{ce} = v_{CE} - V_{CEQ}$, (1) can be written analogously to (1) of Problem 4.41 as

$$i_C = -\frac{v_{CE}}{R_{ac}} + \frac{V_{CEQ}}{R_{ac}} + I_{CQ} \qquad (2)$$

All excursions of the ac signals i_c and v_{ce} are represented by points on the ac load line, (2). If the value $i_C = I_{CQ}$ is substituted into (2), we find that $v_{CE} = V_{CEQ}$; thus, the ac load line intersects the dc load line at the Q point.

4.72 Find the points at which the ac load line intersects the axes of the collector characteristic.

▌ The i_C intercept ($i_{C\,max}$) is found by setting $v_{CE} = 0$ in (2) of Problem 4.71:

$$i_{C\,max} = \frac{V_{CEQ}}{R_{ac}} + I_{CQ} \qquad (1)$$

The v_{CE} intercept is found by setting $i_C = 0$ in (2) of Problem 4.71:

$$v_{CE\,max} = V_{CEQ} + I_{CQ}R_{ac} \qquad (2)$$

4.73 In the *common-collector* (CC) or *emitter-follower* (EF) amplifier of Fig. 4-15(a), $V_{CC} = 12$ V, $R_E = 1$ kΩ, $R_L = 3$ kΩ, and $C_C \rightarrow \infty$. The Si transistor is biased so that $V_{CEQ} = 5.7$ V and has the collector characteristic of Fig. 4-15(b). Construct the dc load line.

▌ The dc load line must intercept the v_{CE} axis at $V_{CC} = 12$ V. It intercepts the i_C axis at

$$\frac{V_{CC}}{R_{dc}} = \frac{V_{CC}}{R_E} = \frac{12}{1 \times 10^3} = 12 \text{ mA}$$

The intercepts are connected to form the dc load line shown on Fig. 4-15(b).

4.74 Calculate the value of β for the amplifier of Problem 4.73.

▌ I_{BQ} is determined by entering Fig. 4-15(b) at $V_{CEQ} = 5.7$ V and interpolating between i_B curves to find $I_{BQ} \approx 50$ μA. I_{CQ} is then read as ≈ 6.3 mA. Thus

$$\beta = \frac{I_{CQ}}{I_{BQ}} = \frac{6.3 \times 10^{-3}}{50 \times 10^{-6}} = 126$$

4.75 Determine the value of R_B to satisfy the operating point of Problem 4.7.

▌ By KVL and using β of Problem 4.74,

$$R_B = \frac{V_{CC} - V_{BEQ} - \frac{\beta+1}{\beta}I_{CQ}R_E}{I_{BQ}} = \frac{12 - 0.7 - \frac{126+1}{126}(6 \times 10^{-3})(1 \times 10^3)}{50 \times 10^{-6}} = 105.05 \text{ kΩ}$$

4.76 Construct the ac load line for the CC amplifier of Problem 4.73.

$$R_{ac} = R_E \parallel R_L = \frac{R_E R_L}{R_E + R_L} = \frac{(1 \times 10^3)(3 \times 10^3)}{1 \times 10^3 + 3 \times 10^3} = 750 \text{ Ω}$$

v_{CE} intercept: $\qquad v_{CEQ} + I_{CQ}R_{ac} = 5.7 + (6.3 \times 10^{-3})(750) = 10.43$ V

i_C intercept: $\qquad \dfrac{V_{CEQ}}{R_{ac}} + I_{CQ} = \dfrac{5.7}{750} + (6.3 \times 10^{-3}) = 13.9$ mA

The ac load line is shown on Fig. 4-15(b).

4.77 Determine v_{CE} for the CC amplifier of Problem 4.76 if $i_S = 10 \sin \omega t$ μA.

▌ A time scale is constructed perpendicular to the ac load line at the Q point of Fig. 4-15(b). A scaled sketch of i_S is drawn and translated through the ac load line to a sketch of v_{CE}. As i_S swings ± 10 μA along the ac load line, the collector-emitter voltage takes on the value $v_{CE} = 5.7 - \sin \omega t$ V.

4.78 Find the value of v_L for the CC amplifier of Fig. 4-15(a) if v_{CE} is as determined in Problem 4.77.

▌ The dc component of v_{CE} (5.7 V) appears across C_C and $v_L = -v_{ce} = \sin \omega t$ V.

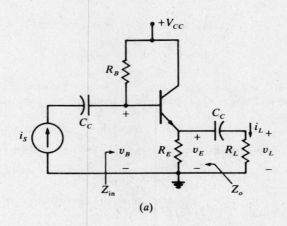

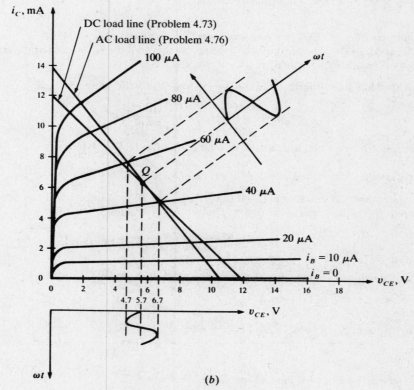

Fig. 4-15

4.79 The Si *Darlington transistor pair* of Fig. 4-16 has negligible leakage current, and $\beta_1 = \beta_2 = 50$. Let $V_{CC} = 12$ V, $R_E = 1$ kΩ, and $R_2 \to \infty$. Find the value of R_1 needed to bias the circuit so that $V_{CEQ2} = 6$ V.

∎ Since $R_2 \to \infty$, $I_{R2} = 0$ and $I_{BQ1} = I_{R1}$. By KVL,

$$I_{EQ2} = \frac{V_{CC} - V_{CEQ2}}{R_E} = \frac{12 - 6}{1 \times 10^3} = 6 \text{ mA}$$

Now
$$I_{BQ2} = \frac{I_{EQ2}}{\beta_2 + 1} = I_{EQ1}$$

and
$$I_{R1} = I_{BQ1} = \frac{I_{EQ1}}{\beta_1 + 1} = \frac{I_{EQ2}}{(\beta_1 + 1)(\beta_2 + 1)} = \frac{6 \times 10^{-3}}{(50 + 1)(50 + 1)} = 2.31 \text{ μA}$$

By KVL (around a path that includes R_1, both transistors, and R_E) and Ohm's law,

$$R_1 = \frac{V_{R1}}{I_{R1}} = \frac{V_{CC} - V_{BEQ1} - V_{BEQ2} - I_{EQ2}R_E}{I_{R1}} = \frac{12 - 0.7 - 0.7 - (6 \times 10^{-3})(1 \times 10^3)}{2.31 \times 10^{-6}} = 1.99 \text{ MΩ}$$

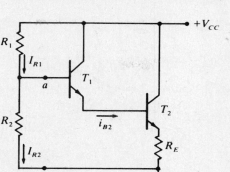

Fig. 4-16

4.80 Determine the value of V_{CEQ1} for the Darlington transistor amplifier of Problem 4.79.

▮ Applying KVL around a path including both transistors and R_E, we have

$$V_{CEQ1} = V_{CC} - V_{BEQ2} - I_{EQ2}R_E = 12 - 0.7 - (6 \times 10^{-3})(1 \times 10^3) = 5.3 \text{ V}$$

4.81 The Si Darlington transistor pair of Fig. 4-16 has negligible leakage current, and $\beta_1 = \beta_2 = 60$. Let $R_1 = R_2 = 1 \text{ M}\Omega$, $R_E = 500 \text{ }\Omega$, and $V_{CC} = 12$ V and find I_{EQ2}.

▮ A Thévenin equivalent for the circuit to the left of terminals a, b has

$$V_{Th} = \frac{R_2}{R_1 + R_2} V_{CC} = \frac{1 \times 10^6}{1 \times 10^6 + 1 \times 10^6} 12 = 6 \text{ V}$$

and

$$R_{Th} = \frac{R_1 R_2}{R_1 + R_2} = \frac{(1 \times 10^6)(1 \times 10^6)}{1 \times 10^6 + 1 \times 10^6} = 500 \text{ k}\Omega$$

With the Thévenin circuit in place, KVL gives

$$V_{Th} = I_{BQ1}R_{Th} + V_{BEQ1} + V_{BEQ2} + I_{EQ2}R_E \tag{1}$$

Realizing that

$$I_{EQ2} = (\beta_2 + 1)I_{BQ2} = (\beta_2 + 1)(\beta_1 + 1)I_{BQ1}$$

we can substitute for I_{BQ1} in (1) and solve for I_{EQ2}, obtaining

$$I_{EQ2} = \frac{(\beta_1 + 1)(\beta_2 + 1)(V_{Th} - V_{BEQ1} - V_{BEQ2})}{R_{Th} + (\beta_1 + 1)(\beta_2 + 1)R_E} = \frac{(60 + 1)(60 + 1)(6 - 0.7 - 0.7)}{500 \times 10^3 + (60 + 1)(60 + 1)(500)} = 7.25 \text{ mA}$$

4.82 Determine the value of V_{CEQ2} for the amplifier of Problem 4.81.

▮ By KVL,

$$V_{CEQ2} = V_{CC} - I_{EQ2}R_E = 12 - (7.25 \times 10^{-3})(500) = 8.375 \text{ V}$$

4.83 Calculate the value of I_{CQ1} for the amplifier of Problem 4.81.

▮ From (1) of Problem 4.7 and (1) of Problem 4.9,

$$I_{CQ1} = \frac{\beta_1}{\beta_1 + 1} I_{EQ1} = \frac{\beta_1}{\beta_1 + 1} I_{BQ2} = \frac{\beta_1}{\beta_1 + 1} \frac{I_{EQ2}}{\beta_2 + 1} = \frac{60}{60 + 1} \frac{7.25 \times 10^{-3}}{60 + 1} = 116.9 \text{ }\mu\text{A}$$

4.84 The Si transistors in the differential amplifier circuit of Fig. 4-17 have negligible leakage current, and $\beta_1 = \beta_2 = 60$. Also, $R_C = 6.8 \text{ k}\Omega$, $R_B = 10 \text{ k}\Omega$, $V_{CC} = V_{EE} = 15$ V. Find the value of I_{EQ1} if the amplifier is biased such that $V_{CEQ1} = V_{CEQ2} = 8$ V.

▮ By symmetry, $I_{EQ1} = I_{EQ2}$. Then, by KCL,

$$i_E = I_{EQ1} + I_{EQ2} = 2I_{EQ1} \tag{1}$$

Using (1) and (2) of Problem 4.64 (which apply to the T_1 circuit here), along with KVL around the left collector loop, gives

$$V_{CC} + V_{EE} = \frac{\beta_1}{\beta_1 + 1} I_{EQ1}R_C + V_{CEQ1} + 2I_{EQ1}R_E \tag{2}$$

Applying KVL around the left base loop gives

$$V_{EE} = I_{BQ1}R_B + V_{BEQ1} + i_E R_E = \frac{I_{EQ1}}{\beta_1 + 1} R_B + V_{BEQ1} + 2I_{EQ1}R_E \qquad (3)$$

Solving (3) for $2I_{EQ1}R_E$, substituting the result into (2), and solving for I_{EQ1} yield

$$I_{EQ1} = \frac{(\beta_1 + 1)(V_{CC} - V_{CEQ1} + V_{BEQ1})}{\beta_1 R_C - R_B} = \frac{(60 + 1)(15 - 8 + 0.7)}{60(6.8 \times 10^3) - 10 \times 10^3} = 1.18 \text{ mA}$$

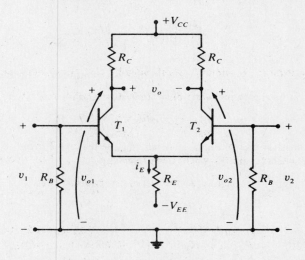

Fig. 4-17

4.85 Determine the value of R_E necessary to attain the Q point for the differential amplifier of Problem 4.84.

▌ Solving (3) of Problem 4.84 for R_E yields

$$R_E = \frac{V_{EE} - V_{BEQ1} - \dfrac{R_B}{\beta_1 + 1} I_{EQ1}}{2I_{EQ1}} = \frac{15 - 0.7 - \dfrac{10 \times 10^3}{60 + 1}(1.18 \times 10^{-3})}{2(1.18 \times 10^{-3})} = 5.97 \text{ k}\Omega$$

OPERATING MODES

4.86 The *pn* junctions of the BJT can be independently forward- or reverse-biased, resulting in four possible *operating modes*. Describe these possible operating modes and comment on their characteristics.

▌ See Table 4-2.

TABLE 4-2

emitter-base bias	collector-base bias	operating mode
Forward	Forward	Saturation
Reverse	Reverse	Cutoff
Reverse	Forward	Inverse
Forward	Reverse	Linear or active

Saturation denotes operation (with $|v_{CE}| \approx 0.2$ V and $|v_{BC}| \approx 0.5$ V for Si devices) such that maximum collector current flows and the transistor acts much like a closed switch from collector to emitter terminals. [See Figs. 4-3(c) and 4-4(c).]

Cutoff denotes operation near the voltage axis of the collector characteristics, where the transistor acts much like an open switch. Only leakage current (similar to I_o of the diode) flows in this mode of operation; thus, $i_C = I_{CBO} \approx 0$ for CB connection, and $i_C = I_{CEO} \approx 0$ for CE connection. Figures 4-3(c) and 4-4(c) indicate these leakage currents.

The *inverse* mode is a little-used, inefficient active mode with the emitter and collector interchanged.

The *active* or *linear* mode describes transistor operation in the region to the right of saturation and above cutoff in Figs. 4-3(c) and 4-4(c); here, near-linear relationships exist between terminal currents as described by (1) of Problem 4.7 and (1) of Problem 4.9.

4.87 In the circuit of Fig. 4-8, $I_{BQ} = 30\,\mu\text{A}$, $R_E = 1\,\text{k}\Omega$, $V_{CC} = 15\,\text{V}$, and $\beta = 80$. Find the minimum value of R_C that will maintain the transistor quiescent point at saturation, if $V_{CE\,\text{sat}} = 0.2\,\text{V}$, β is constant, and leakage current is negligible.

▮ We first find

$$\alpha = \frac{\beta}{\beta+1} = \frac{80}{81} = 0.9876$$

Then the use of (1) of Problem 4.7 and (1) of Problem 4.8 with negligible leakage current yields

$$I_{CQ} = \beta I_{BQ} = 80(30 \times 10^{-6}) = 2.4\,\text{mA}$$

and

$$I_{EQ} = \frac{I_{CQ}}{\alpha} = \frac{2.4 \times 10^{-3}}{0.9876} = 2.43\,\text{mA}$$

Now KVL around the collector circuit leads to the minimum value of R_C to ensure saturation:

$$R_C = \frac{V_{CC} - V_{CE\,\text{sat}} - I_{EQ}R_E}{I_{CQ}} = \frac{15 - 0.2 - 2.43(1)}{2.4 \times 10^{-3}} = 5.154\,\text{k}\Omega$$

4.88 In the circuit of Fig. 4-18, $V_{CC} = 12\,\text{V}$, $V_S = 2\,\text{V}$, $R_C = 4\,\text{k}\Omega$, and $R_S = 100\,\text{k}\Omega$. The Ge transistor is characterized by $\beta = 50$, $I_{CEO} = 0$, and $V_{CE\,\text{sat}} = 0.2\,\text{V}$. Find the value of R_B that just results in saturation.

▮ Application of KVL around the collector loop gives the collector current at the onset of saturation as

$$I_{CQ} = \frac{V_{CC} - V_{CE\,\text{sat}}}{R_C} = \frac{12 - 0.2}{4 \times 10^3} = 2.95\,\text{mA}$$

With C blocking, $I_S = 0$; hence the use of KVL leads to

$$R_B = \frac{V_{CC} - V_{BEQ}}{I_{BQ}} = \frac{V_{CC} - V_{BEQ}}{I_{CQ}/\beta} = \frac{12 - 0.3}{(2.95 \times 10^{-3})/50} = 198.3\,\text{k}\Omega$$

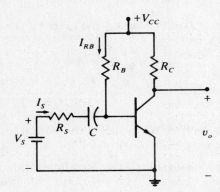

Fig. 4-18

4.89 Solve Problem 4.88 if the capacitor is replaced by a short circuit and all else is unchanged.

▮ With C shorted, the application of (1) of Problem 4.9, KCL, and KVL results in

$$I_{BQ} = \frac{I_{CQ}}{\beta} = I_S + I_{RB} = \frac{V_S - V_{BEQ}}{R_S} + \frac{V_{CC} - V_{BEQ}}{R_B}$$

so that

$$R_B = \frac{V_{CC} - V_{BEQ}}{\dfrac{I_{CQ}}{\beta} - \dfrac{V_S - V_{BEQ}}{R_S}} = \frac{12 - 0.3}{\dfrac{2.95 \times 10^{-3}}{50} - \dfrac{2 - 0.3}{100 \times 10^3}} = 278.6\,\text{k}\Omega$$

4.90 The circuit of Fig. 4-19 uses current- (or shunt-) feedback bias. The Si transistor has $I_{CEO} \approx 0$, $V_{CE\,sat} \approx 0$, and $h_{FE} = 100$. If $R_C = 2\,k\Omega$ and $V_{CC} = 12\,V$, size R_F for ideal *maximum symmetrical swing* in the active operating mode (that is, location of the quiescent point such that $V_{CEQ} = V_{CC}/2$).

▌ Application of KVL to the collector-emitter bias circuit gives

$$(I_{BQ} + I_{CQ})R_C = V_{CC} - V_{CEQ}$$

With $I_{CQ} = h_{FE}I_{BQ}$, this leads to

$$I_{BQ} = \frac{V_{CC} - V_{CEQ}}{(h_{FE} + 1)R_C} = \frac{12 - 6}{(100 + 1)(2 \times 10^3)} = 29.7\,\mu A$$

Then, by KVL around the transistor terminals,

$$R_F = \frac{V_{CEQ} - V_{BEQ}}{I_{BQ}} = \frac{6 - 0.7}{29.7 \times 10^{-6}} = 178.5\,k\Omega$$

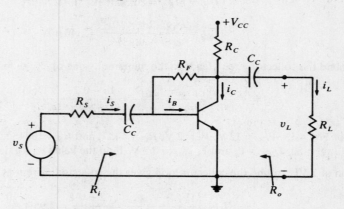

Fig. 4-19

4.91 Find the value of the emitter resistor R_E that, when added to the Si transistor circuit of Fig. 4-19, would bias for operation about $V_{CEQ} = 5\,V$. Let $I_{CEO} = 0$, $\beta = 80$, $R_F = 220\,k\Omega$, $R_C = 2\,k\Omega$, and $V_{CC} = 12\,V$.

▌ Application of KVL around the transistor terminals yields

$$I_{BQ} = \frac{V_{CEQ} - V_{BEQ}}{R_F} = \frac{5 - 0.7}{220 \times 10^3} = 19.545\,\mu A$$

Since leakage current is zero, (2) of Problem 4.7 and (3) of Problem 4.9 give $I_{EQ} = (\beta + 1)I_{BQ}$; thus KVL around the collector circuit gives

$$(I_{BQ} + \beta I_{BQ})R_C + (\beta + 1)I_{BQ}R_E = V_{CC} - V_{CEQ}$$

so

$$R_E = \frac{V_{CC} - V_{CEQ} - (\beta + 1)I_{BQ}R_C}{(\beta + 1)I_{BQ}} = \frac{12 - 5 - (80 + 1)(19.545 \times 10^{-6})(2 \times 10^3)}{(80 + 1)(19.545 \times 10^{-6})} = 2.42\,k\Omega$$

4.92 Find the proper collector current bias for maximum symmetrical (or undistorted) swing along the ac load line of a transistor amplifier for which $V_{CE\,sat} = I_{CEO} = 0$.

▌ For maximum symmetrical swing, the Q point must be set at the midpoint of the ac load line. Hence, from (1) of Problem 4.72, we want

$$I_{CQ} = \frac{1}{2}i_{C\,max} = \frac{1}{2}\left(\frac{V_{CEQ}}{R_{ac}} + I_{CQ}\right) \qquad (1)$$

But for a circuit such as that in Fig. 4-6(a), KVL gives

$$V_{CEQ} \approx V_{CC} - I_{CQ}R_{dc} \qquad (2)$$

which becomes an equality if no emitter resistor is present. Substituting (2) into (1), assuming equality, and solving for I_{CQ} yield the desired result:

$$I_{CQ} = \frac{V_{CC}}{R_{ac} + R_{dc}} \qquad (3)$$

4.93 In the circuit of Fig. 4-5(a), $R_E = 300\ \Omega$, $R_C = 500\ \Omega$, $V_{CC} = 15$ V, $\beta = 100$, and the Si transistor has β-independent bias. Size R_1 and R_2 for maximum symmetrical swing if $V_{CE\ \text{sat}} \approx 0$.

■ For maximum symmetrical swing, the quiescent collector current is

$$I_{CQ} = \frac{1}{2}\frac{V_{CC}}{R_E + R_C} = \frac{15}{2(300 + 500)} = 9.375\ \text{mA}$$

Standard practice is to use a factor of 10 as the margin of inequality for β independence in (2) of Problem 4.40. Then,

$$R_B = \frac{\beta R_E}{10} = \frac{100(300)}{10} = 3\ \text{k}\Omega$$

and, from (1) of Problem 4.40,

$$V_{BB} \approx V_{BEQ} + I_{CQ}(1.1R_E) = 0.7 + (9.375 \times 10^{-3})(330) = 3.794\ \text{V}$$

Equations (1) of Problem 4.38 may now be solved simultaneously to obtain

$$R_1 = \frac{R_B}{1 - V_{BB}/V_{CC}} = \frac{3 \times 10^3}{1 - 3.794/15} = 4.02\ \text{k}\Omega$$

and

$$R_2 = R_B \frac{V_{CC}}{V_{BB}} = 3 \times 10^3 \frac{15}{3.794} = 11.86\ \text{k}\Omega$$

4.94 In the circuit of Fig. 4-14(a), the transistor is a Si device, $R_E = 200\ \Omega$, $R_2 = 10R_1 = 10\ \text{k}\Omega$, $R_L = R_C = 2\ \text{k}\Omega$, $\beta = 100$, and $V_{CC} = 15$ V. Assume that C_C and C_E are very large, that $C_{CE\ \text{sat}} \approx 0$, and that $i_C = 0$ at cutoff. Find I_{CQ}.

■ Equations (1) of Problem 4.38 and (1) of Problem 4.40 give

$$R_B = \frac{(1 \times 10^3)(10 \times 10^3)}{11 \times 10^3} = 909\ \Omega \quad \text{and} \quad V_{BB} = \frac{1 \times 10^3}{11 \times 10^3}(15) = 1.364\ \text{V}$$

so

$$I_{CQ} \approx \frac{V_{BB} - V_{BEQ}}{R_B/(\beta + 1) + R_E} = \frac{1.364 - 0.7}{(909/101) + 200} = 3.177\ \text{mA}$$

4.95 Determine the value of V_{CEQ} for the amplifier of Problem 4.94.

■ KVL around the collector-emitter circuit, with $I_{CQ} \approx I_{EQ}$, gives

$$V_{CEQ} = V_{CC} - I_{CQ}(R_E + R_C) = 15 - (3.177 \times 10^{-3})(2.2 \times 10^3) = 8.01\ \text{V}$$

4.96 Calculate the slope of the dc load line for the amplifier of Problem 4.94.

■
$$\text{Slope} = \frac{1}{R_{\text{dc}}} = \frac{1}{R_C + R_E} = \frac{1}{2.2 \times 10^3} = 0.454\ \text{mS}$$

4.97 Find the slope of the ac load line for the amplifier of Problem 4.94.

■
$$\text{Slope} = \frac{1}{R_{\text{ac}}} = \frac{1}{R_C} + \frac{1}{R_L} = 2\frac{1}{2 \times 10^3} = 1\ \text{mS}$$

4.98 Determine the peak value of undistorted i_L for the amplifier of Problem 4.94.

■ From (2) of Problem 4.72, the ac load line intersects the v_{CE} axis at

$$v_{CE\ \text{max}} = V_{CEQ} + I_{CQ}R_{\text{ac}} = 8.01 + (3.177 \times 10^{-3})(1 \times 10^3) = 11.187\ \text{V}$$

Since $v_{CE\ \text{max}} < 2V_{CEQ}$, cutoff occurs before saturation and thus sets V_{cem}. With the large capacitor appearing as ac shorts,

$$i_L = \frac{v_L}{R_L} = \frac{v_{ce}}{R_L}$$

or, in terms of peak values,

$$I_{Lm} = \frac{V_{cem}}{R_L} = \frac{v_{CE\ \text{max}} - V_{CEQ}}{R_L} = \frac{11.187 - 8.01}{2 \times 10^3} = 1.588\ \text{mA}$$

4.99 In the circuit of Fig. 4-5(a), $R_C = 300\ \Omega$, $R_E = 200\ \Omega$, $R_1 = 2\ k\Omega$, $R_2 = 15\ k\Omega$, $V_{CC} = 15$ V, and $\beta = 110$ for the Si transistor. Assume that $I_{CQ} \approx I_{EQ}$ and $V_{CE\,sat} \approx 0$. Find the maximum symmetrical swing in collector current if an ac base current is injected.

▌ From (1) of Problem 4.38 and (1) of Problem 4.40,

$$R_B = \frac{(2 \times 10^3)(15 \times 10^3)}{17 \times 10^3} = 1.765\ k\Omega \quad \text{and} \quad V_{BB} = \frac{2 \times 10^3}{17 \times 10^3}(15) = 1.765\ V$$

so
$$I_{CQ} \approx I_{EQ} = \frac{V_{BB} - V_{BEQ}}{R_B/(\beta + 1) + R_E} = \frac{1.765 - 0.7}{1765/111 + 200} = 4.93\ \text{mA}$$

By KVL around the collector-emitter circuit with $I_{CQ} \approx I_{EQ}$,

$$V_{CEQ} = V_{CC} - I_{CQ}(R_C + R_E) = 15 - (4.93 \times 10^{-3})(200 + 300) = 12.535\ V$$

Since $V_{CEQ} > V_{CC}/2 = 7.5$ V, cutoff occurs before saturation, and i_C can swing ± 4.93 mA about I_{CQ} and remain in the active region.

4.100 Solve Problem 4.99 if $V_{CC} = 10$ V and all else is unchanged.

▌
$$V_{BB} = \frac{R_1}{R_1 + R_2} V_{CC} = \frac{2 \times 10^3}{17 \times 10^3}(10) = 1.1765\ V$$

so that
$$I_{CQ} \approx I_{EQ} = \frac{V_{BB} - V_{BEQ}}{R_B/(\beta + 1) + R_E} = \frac{1.1765 - 0.7}{1765/111 + 200} = 2.206\ \text{mA}$$

and
$$V_{CEQ} = V_{CC} - I_{CQ}(R_C + R_E) = 10 - (2.206 \times 10^{-3})(0.5) = 8.79\ V$$

Since $V_{CEQ} > V_{CC}/2 = 5$ V, cutoff again occurs before saturation, and i_C can swing ± 2.206 mA about I_{CQ} and remain in the active region of operation. Here, the 33.3 percent reduction in power supply voltage has resulted in a reduction of over 50 percent in symmetrical collector-current swing.

4.101 If a Si transistor were removed from the circuit of Fig. 4-5(a) and a Ge transistor of identical β were substituted, would the Q point move in the direction of saturation or of cutoff?

▌ Since R_1, R_2, and V_{CC} are unchanged, R_B and V_{BB} would remain unchanged. However, owing to the different emitter-to-base forward drops for Si (0.7 V) and Ge (0.3 V) transistors,

$$I_{CQ} \approx \frac{V_{BB} - V_{BEQ}}{R_B/(\beta + 1) + R_E}$$

would be higher for the Ge transistor. Thus, the Q point would move in the direction of saturation.

4.102 In the circuit of Fig. 4-14(a), $V_{CC} = 12$ V, $R_C = R_L = 1\ k\Omega$, $R_E = 100\ \Omega$, and $C_C = C_E \to \infty$. The Si transistor has negligible leakage current, and $\beta = 100$. If $V_{CE\,sat} = 0$ and the transistor is to have β-independent bias (by having $R_1 \| R_2 = \beta R_E/10$), size R_1 and R_2 for maximum symmetrical swing.

▌ Evaluating R_{ac} and R_{dc}, we find

$$R_{ac} = R_L \| R_C = \frac{(1 \times 10^3)(1 \times 10^3)}{1 \times 10^3 + 1 \times 10^3} = 500\ \Omega \qquad R_{dc} = R_C + R_E = 1 \times 10^3 + 100 = 1100\ \Omega$$

Thus, according to (3) of Problem 4.92, maximum symmetrical swing requires that

$$I_{CQ} = \frac{V_{CC}}{R_{ac} + R_{dc}} = \frac{12}{500 + 1100} = 7.5\ \text{mA}$$

Now,
$$R_B = R_1 \| R_2 = \frac{\beta R_E}{10} = \frac{100(100)}{10} = 1\ k\Omega$$

and, by (1) of Problem 4.39 and (2) of Problem 4.64,

$$V_{BB} = \left(\frac{R_B}{\beta} + \frac{\beta + 1}{\beta} R_E\right) I_{CQ} + V_{BEQ} = \left(\frac{1 \times 10^3}{100} + \frac{100 + 1}{100} 100\right) 7.5 \times 10^{-3} + 0.7 = 1.53\ V$$

Finally, from (1) of Problem 4.38,

$$R_1 = \frac{R_B}{1 - V_{BB}/V_{CC}} = \frac{1 \times 10^3}{1 - 1.53/12} = 1.34 \text{ k}\Omega \quad \text{and} \quad R_2 = \frac{R_B V_{CC}}{V_{BB}} = \frac{(1 \times 10^3)(12)}{1.53} = 10.53 \text{ k}\Omega$$

4.103 The Si transistor of Fig. 4-14(a) has $V_{CE\,\text{sat}} = I_{CBO} = 0$ and $\beta = 75$. C_E is removed from the circuit, and $C_C \to \infty$. Also, $R_1 = 1 \text{ k}\Omega$, $R_2 = 9 \text{ k}\Omega$, $R_E = R_L = R_C = 1 \text{ k}\Omega$, and $V_{CC} = 15$ V. Sketch the ac load line for this amplifier on a set of $i_C - v_{CE}$ axes.

$$R_{\text{dc}} = R_C + R_E = 1 \times 10^3 + 1 \times 10^3 \times 10^3 = 2 \text{ k}\Omega$$

and

$$R_{\text{ac}} = R_E + R_C \parallel R_L = 1 \times 10^3 + \frac{(1 \times 10^3)(1 \times 10^3)}{1 \times 10^3 + 1 \times 10^3} = 1.5 \text{ k}\Omega$$

By (1) of Problem 4.38,

$$V_{BB} = \frac{R_1}{R_2} V_{CC} = \frac{1 \times 10^3}{9 \times 10^3}(15) = 1.667 \text{ V} \quad \text{and} \quad R_B = R_1 \parallel R_2 = \frac{(1 \times 10^3)(9 \times 10^3)}{1 \times 10^3 + 9 \times 10^3} = 900 \text{ }\Omega$$

and from (1) of Problem 4.40,

$$I_{CQ} = \frac{(\beta + 1)(V_{BB} - V_{BEQ})}{R_B + (\beta + 1)R_E} = \frac{(75 + 1)(1.667 - 0.7)}{900 + (75 + 1)(1 \times 10^3)} = 0.956 \text{ mA}$$

By KVL around the collector loop and (2) of Problem 4.64,

$$V_{CEQ} = V_{CC} - \left(R_C + \frac{\beta + 1}{\beta} R_E\right) I_{CQ} = 15 - \left(1 \times 10^3 + \frac{75 + 1}{75}(1 \times 10^3)\right) 0.956 \times 10^{-3} = 13.07 \text{ V}$$

The ac load-line intercepts now follow directly from (1) and (2) of Problem 4.72:

$$i_{C\,\text{max}} = \frac{V_{CEQ}}{R_{\text{ac}}} + I_{CQ} = \frac{13.07}{1.5 \times 10^3} + 0.956 \times 10^{-3} = 9.67 \text{ mA}$$

$$v_{CE\,\text{max}} = V_{CEQ} + I_{CQ}R_{\text{ac}} = 13.07 + (0.956 \times 10^{-3})(1.5 \times 10^3) = 14.5 \text{ V}$$

The required ac load line is sketched in Fig. 4-20.

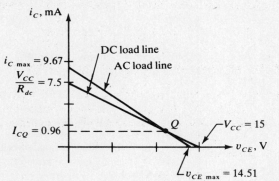

Fig. 4-20

4.104 Sketch the dc load line for the amplifier of Problem 4.103.

The dc load-line intercepts follow from (1) of Problem 4.41:

$$i_C - \text{axis intercept} = \frac{V_{CC}}{R_{\text{dc}}} = \frac{15}{2 \times 10^3} = 7.5 \text{ mA}$$

$$v_{CE} - \text{axis intercept} = V_{CC} = 15 \text{ V}$$

The required dc load line is sketched in Fig. 4-20.

4.105 For the amplifier of Problem 4.103, find the maximum undistorted value of i_L and determine whether cutoff or saturation limits i_L swing.

Since $I_{CQ} < \frac{1}{2} i_{C\,\text{max}}$, it is apparent that cutoff limits the undistorted swing of i_c to $\pm I_{CQ} = \pm 0.956$ mA. By current division,

$$i_L = \frac{R_E}{R_E + R_L} i_c = \frac{1 \times 10^3}{1 \times 10^3 + 1 \times 10^3}(\pm 0.956 \text{ mA}) = \pm 0.478 \text{ mA}$$

4.106 The Si transistor of Fig. 4-21 has negligible leakage current, and $\beta = 100$. If $V_{CC} = 15$ V, $V_{EE} = 4$ V, $R_E = 3.3$ kΩ, and $R_C = 7.1$ kΩ, find I_{BQ}.

▮ By KVL around the base-emitter loop,

$$I_{EQ} = \frac{V_{EE} - V_{BEQ}}{R_E} = \frac{4 - 0.7}{3.3 \times 10^3} = 1 \text{ mA}$$

Then, by (2) of Problem 4.7 and (2) of Problem 4.8,

$$I_{BQ} = \frac{I_{EQ}}{\beta + 1} = \frac{1 \times 10^{-3}}{100 + 1} = 9.9 \text{ } \mu\text{A}$$

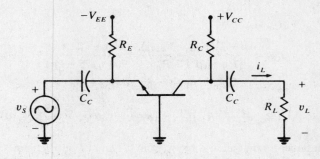

Fig. 4-21

4.107 Determine V_{CEQ} for the amplifier of Problem 4.106.

▮ KVL and (2) of Problem 4.64 yield

$$V_{CEQ} = V_{CC} + V_{EE} - I_{EQ}R_E - I_{CQ}R_C = V_{CC} + V_{EE} - \left(R_E + \frac{\beta}{\beta + 1}R_C\right)I_{EQ}$$

$$= 15 + 4 - \left(3.3 \times 10^3 + \frac{100}{100 + 1}(7.1 \times 10^3)\right)(1 \times 10^{-3}) = 8.67 \text{ V}$$

4.108 It is clear that amplifiers can be biased for operation at any point along the dc load line. Classify amplifiers based on the percentage of the signal cycle over which they operate in the linear or active region.

▮ See Table 4-3.

TABLE 4-3

class	percentage of active-region signal excursion
A	100
AB	Between 50 and 100
B	50
C	Less than 50

4.109 The amplifier of Fig. 4-22 uses a Si transistor for which $V_{BEQ} = 0.7$ V. Assuming that the collector-emitter bias does not limit voltage excursion, classify the amplifier according to Table 4-3 if $V_B = 1.0$ V and $v_S = 0.25 \cos \omega t$ V.

▮ As long as $v_S + V_B > 0.7$ V, the emitter-base junction is forward-biased; thus, classification becomes a matter of determining the portion of the period of v_S over which the above inequality holds.

$v_S + V_B \geq 0.75$ V through the complete cycle; thus, the transistor is always in the active region, and the amplifier is of class A.

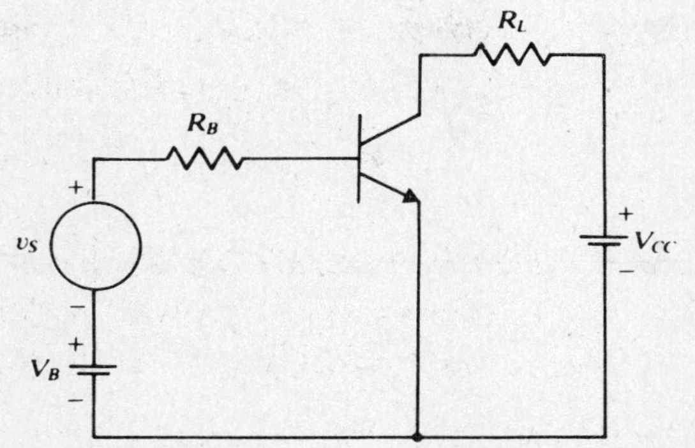

BIPOLAR JUNCTION TRANSISTOR FUNDAMENTALS □ 123

Fig. 4-22

4.110 Classify the amplifier of Problem 4.109 according to Table 4-3 if $V_B = 1.0$ V and $v_S = 0.5 \cos \omega t$ V.

▮ $0.5 \leq v_S + V_B \leq 1.5$ V; thus the transistor is cut off for a portion of the negative excursion v_S. Since cutoff occurs during less than 180°, the amplifier is of class AB.

4.111 Classify the amplifier of Problem 4.109 according to Table 4-3 if $V_B = 0.5$ V and $v_S = 0.6 \cos \omega t$ V.

▮ $-0.1 \leq v_S + V_B \leq 1.1$ V, which gives conduction for less than 180° of the period of v_S, for class C operation.

4.112 Classify the amplifier of Problem 4.109 according to Table 4-3 if $V_B = 0.7$ V and $v_S = 0.5 \cos \omega t$ V.

▮ $v_S + V_B \geq 0.7$ V over exactly 180° of the period of v_S, for class B operation.

Review Problems

4.113 Prove the leakage current relationship $I_{CEO} = I_{CBO}/(1 - \alpha)$.

▮ Proof follows directly from substitution of (2) from Problem 4.8 into (4) of Problem 4.9.

$$I_{CEO} = (\beta + 1)I_{CBO} = \left(\frac{\alpha}{1 - \alpha} + 1\right)I_{CBO} = I_{CBO}/(1 - \alpha)$$

4.114 Find an expression for the base-collector voltage v_{BC} of a transistor represented by the Ebers–Moll model.

▮ Solving (2′) of Problem 4.20 for v_{BC} yields

$$v_{BC} = \eta V_T \ln\left[1 - \frac{i_C - \alpha_F i_E}{(1 - \alpha_F \alpha_R)I_{CO}}\right] \tag{1}$$

4.115 Determine the base current i_B for the transistor of Problem 4.27.

▮ Use (1) of Problem 4.19 and the result of Problem 4.27.

$$i_E = i_{EF} - \alpha_R i_{CR} = 62.17 - 0.2(2.51) = 61.67 \text{ mA}$$

By KCL, $\qquad i_B = i_E - i_C = 61.67 - 59.04 = 2.63$ mA

4.116 For a Ge npn transistor, $I_{EO} = 10$ μA, $V_T = 26$ mV, $\alpha_F = 0.99$, $\alpha_R = 0.45$. Determine the value of v_{BE} at cutoff.

▮ By (1) of Problem 4.24,

$$v_{BE} = \eta V_T \ln\left[1 + \frac{\alpha_F}{1 - \alpha_F \alpha_R}\right] = (1)(0.026) \ln\left[1 + \frac{0.99}{1 - 0.99(0.45)}\right] = 26.63 \text{ mV}$$

4.117 If the transistor of Problem 4.116 is operating at a point where $i_E = 15$ mA and $i_B = 200$ μA, determine the base-emitter voltage v_{BE}.

▮ By KCL,

$$i_C = i_E - i_B = 15 \times 10^{-3} - 200 \times 10^{-6} = 14.8 \text{ mA}$$

Based on (1) of Problem 4.23,

$$v_{BE} = \eta V_T \ln\left[1 + \frac{i_E - \alpha_R i_C}{(1 - \alpha_F \alpha_R)I_{EO}}\right] = 1(0.026)\ln\left[1 + \frac{15 \times 10^{-3} - 0.45(14.8 \times 10^{-3})}{(1 - 0.99(0.45))(10 \times 10^{-6})}\right] = 0.1902 \text{ V}$$

4.118 Determine the base-collector voltage v_{BC} for the transistor of Problem 4.117.

▌ Based on (6) of Problem 4.19, the results of Problem 4.5, and (1) of Problem 4.114,

$$I_{CO} = \frac{\alpha_F}{\alpha_R} I_{EO} = \frac{0.99}{0.45}(10 \times 10^{-6}) = 22 \text{ } \mu A$$

$$v_{BC} = \eta V_T \ln\left[1 - \frac{i_C - \alpha_F i_E}{(1 - \alpha_F \alpha_R)I_{CO}}\right] = (1)(0.026)\ln\left[1 - \frac{14.8 \times 10^{-3} - 0.99(15 \times 10^{-3})}{(1 - (0.99)(0.45))(22 \times 10^{-6})}\right] = 42.3 \text{ mV}$$

4.119 If the transistor of Problem 4.116 were operating with $v_{BE} = 0.3$ V and $v_{BC} = 0.15$ V, find the collector current i_C.

▌ By (2), (4), and (5) of Problem 4.19 and the result of Problem 4.118,

$$i_{EF} = I_{EO}(e^{v_{BE}/\eta V_T} - 1) = (10 \times 10^{-6})(e^{0.3/(1)(0.026)} - 1) = 1.026 \text{ A}$$
$$i_{CR} = I_{CO}(e^{v_{BC}/\eta V_T} - 1) = (22 \times 10^{-6})(e^{0.15/(1)(0.026)} - 1) = 7.02 \text{ mA}$$
$$i_C = \alpha_F i_{EF} - i_{CR} = 0.99(1.026) - 7.02 \times 10^{-3} = 1.009 \text{ A}$$

4.120 Calculate the collector-emitter voltage v_{CE} for the transistor of Problem 4.117.

▌ Using KVL and the results of Problems 4.117 and 4.118,

$$v_{CE} = v_{BC} - v_{BE} = 0.1902 - 0.0423 = 0.148 \text{ V}$$

4.121 If the temperature of the transistor of Problem 4.117 were doubled from 25°C to 50°C, predict the resulting value of v_{BE}. Assume α_F and α_R unchanged.

▌ Let subscripts 1 and 2 denote old and new conditions, respectively. Saturation current (as in the case of a diode) doubles for each increase of 10°C in temperature, or by a procedure similar to that of Problem 2.33,

$$I_{EO2} = I_{EO1} 2^{\Delta T/10} = (10 \times 10^{-6})2^{25/10} = 56.57 \text{ } \mu A$$

Based on (1) of Problem 2.18,

$$V_{T2} = \frac{T_2}{T_1} V_{T1} = \frac{50}{25}(0.026) = 52 \text{ mV}$$

By (1) of Problem 4.23,

$$v_{BE2} = \eta V_T \ln\left[1 + \frac{i_E - \alpha_R i_C}{(1 - \alpha_F \alpha_R)I_{EO2}}\right] = 1(0.052)\ln\left[1 + \frac{15 \times 10^{-3} - 0.45(14.8 \times 10^{-3})}{(1 - (0.99)(0.45))(56.57 \times 10^{-6})}\right] = 0.287 \text{ V}$$

4.122 The leakage currents of a transistor are $I_{CBO} = 5$ μA and $I_{CEO} = 0.4$ mA, and $I_B = 30$ μA. Determine the value of β.

▌ From (4) of Problem 4.9,

$$\beta = \frac{I_{CEO}}{I_{CBO}} - 1 = \frac{0.4 \times 10^{-3}}{5 \times 10^{-6}} - 1 = 79$$

4.123 Determine the value of I_C for the transistor of Problem 4.122.

▌ By (1) of Problem 4.9,

$$I_C = \beta I_B + I_{CEO} = 79(30 \times 10^{-6}) + 0.4 \times 10^{-3} = 2.77 \text{ mA}$$

BIPOLAR JUNCTION TRANSISTOR FUNDAMENTALS ◻ 125

1.124 For a BJT, $I_C = 5.2$ mA, $I_B = 50\ \mu$A, and $I_{CBO} = 0.5\ \mu$A. Find β.

▌ By (1) and (4) of Problem 4.9,

$$I_C = \beta I_B + I_{CEO} = \beta I_B + (\beta + 1)I_{CBO}$$

or

$$\beta = \frac{I_C - I_{CBO}}{I_B + I_{CBO}} = \frac{5.2 \times 10^{-3} - 0.5 \times 10^{-6}}{50 \times 10^{-6} + 0.5 \times 10^{-6}} = 102.96$$

4.125 Determine I_E for the transistor of Problem 4.124.

▌ By (3) of Problem 4.9,

$$\alpha = \beta/(\beta + 1) = \frac{102.96}{102.96 + 1} = 0.9904$$

Using (1) of Problem 4.7,

$$I_E = (I_C - I_{CBO})/\alpha = \frac{5.2 \times 10^{-3} - 0.5 \times 10^{-6}}{0.9904} = 5.25\ \text{mA}$$

4.126 For the transistor of Problem 4.124, what is the percentage error in the calculation of β if leakage current is neglected?

▌ Neglecting leakage current, (1) of Problem 4.9 gives

$$\beta = \frac{I_C}{I_B} = \frac{5.2 \times 10^{-3}}{50 \times 10^{-6}} = 104$$

Thus,

$$\text{percent error} = \frac{104 - 102.96}{102.96}(100\%) = 1.01\%$$

4.127 Collector-to-base leakage current can be modeled by a current source as in Fig. 4-23, with the understanding that transistor action relates currents I'_C, I'_B, and I_E ($I'_C = \alpha I_E$, and $I'_C = \beta I'_B$). Prove that $I_C = \beta I_B + (\beta + 1)I_{CBO}$.

▌ Due to transistor action,

$$I'_C = \beta I'_B \tag{1}$$

By KCL,

$$I'_C = I_C - I_{CBO} \tag{2}$$

$$I'_B = I_B + I_{CBO} \tag{3}$$

Substitute (2) and (3) into (1) to find

$$I_C - I_{CBO} = \beta(I_B + I_{CBO}) \quad\text{or}\quad I_C = \beta I_B + (\beta + 1)I_{CBO}$$

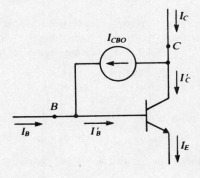

Fig. 4-23

4.128 For the transistor model of Fig. 4-23, prove that $I_B = I_E/(\beta + 1) - I_{CBO}$.

▌ Apply KCL at base node.

$$I_B = I'_B - I_{CBO} \tag{1}$$

Use $I'_C = \alpha I_E$, (1) of Problem 4.127, and (3) of Problem 4.9 in (1) to yield

$$I_B = \frac{I'_C}{\beta} - I_{CBO} = \frac{\alpha I_E}{\beta} - I_{CBO} = I_E/(\beta + 1) - I_{CBO}$$

4.129 For the transistor model of Fig. 4-23, prove that $I_E = \frac{\beta + 1}{\beta}(I_C - I_{CBO})$.

▌ From (2) of Problem 4.127 and use of $I'_C = \alpha I_E$,

$$I_C = I'_C + I_{CBO} = \alpha I_E + I_{CBO} \quad (1)$$

Rearrange (1) and use (3) of Problem 4.9 to yield

$$I_E = (I_C - I_{CBO})/\alpha = \frac{\beta + 1}{\beta}(I_C - I_{CBO})$$

4.130 If the transistor of Problem 4.46 were replaced by a new transistor with 1 percent greater α, what would be the percentage change in emitter current? Neglect leakage currents.

▌ Let subscripts 1 and 2 denote the original and new transistors, respectively. Based on results of Problem 4.47, $I_{E1} = I_{C1}/\alpha_1 = (1.47 \times 10^{-3})/0.98 = 1.50$ mA. But,

$$\alpha_2 = 1.01\alpha_1 = (1.01)(0.98) = 0.9898$$

$$\beta_2 = \frac{\alpha_2}{1 - \alpha_2} = \frac{0.9898}{1 - 0.9898} = 97.04$$

$$I_{E2} = (\beta_2 + 1)I_B = (97.04 + 1)(30 \times 10^{-6}) = 2.941 \text{ mA}$$

$$\text{percent } \Delta I_E = \frac{I_{E2} - I_{E1}}{I_{E1}}(100\%) = \frac{2.941 - 1.50}{1.50}(100\%) = 96.07\% \text{ increase}$$

4.131 In the circuit of Fig. 4-7, $V_{CE\,\text{sat}} = 0.2$ V, $\alpha = 0.99$, $I_{BQ} = 20\,\mu\text{A}$, $V_{CC} = 15$ V, and $R_C = 15$ kΩ. What is the value of V_{CEQ}?

▌ Assume linear or active mode operation, then

$$\beta = \frac{\alpha}{1 - \alpha} = \frac{0.99}{1 - 0.99} = 99$$

$$I_{CQ} = \beta I_{BQ} = 99(20 \times 10^{-6}) = 1.98 \text{ mA}$$

By KVL,

$$V_{CEQ} = V_{CC} - I_{CQ}R_C = 15 - (1.98 \times 10^{-3})(15 \times 10^3) = -14.7 \text{ V}$$

However, V_{CEQ} must be positive for an *npn* transistor, and thus, the solution is invalid. V_{CEQ} is constrained by the collector characteristic curve to $V_{CEQ} = V_{CE\,\text{sat}} = 0.2$ V.

4.132 In many switching applications, the transistor may be utilized without a heat sink, since $P_C \approx 0$ in cutoff and P_C is small in saturation. Support this statement by calculating the collector power dissipated in (*a*) Problem 4.50 (active-region bias) and (*b*) Problem 4.131 (saturation-region bias).

▌ (*a*) Based on results of Problems 4.50 and 4.51,

$$P_C = V_{CEQ}I_{CQ} = 9(2 \times 10^{-3}) = 18 \text{ mW}$$

(*b*) Based on results of Problem 4.131,

$$P_C = V_{CEQ}I_{CQ} = 0.2(1.98 \times 10^{-3}) = 0.396 \text{ mW}$$

The collector power dissipation in the saturation mode is less by a factor of 46 in this particular case.

4.133 The collector characteristics of the transistor of Fig. 4-7 are given in Fig. 4-6(*b*). If $I_{BQ} = 40\,\mu\text{A}$, $V_{CC} = 12$ V, and $R_C = 1$ kΩ, specify the minimum power rating of the transistor to ensure there is no danger of thermal damage.

▌ The dc load line with ordinate intercept $V_{CC}/R_C = 12$ mA and abscissa intercept $V_{CC} = 12$ V is identical to that of Problem 4.42. Whence, for $I_{BQ} = 40\ \mu A$, $V_{CEQ} = 7.2\ V$ and $I_{CQ} = 4.9\ mA$. Thus,

$$P_C = V_{CEQ}I_{CQ} = 7.2(4.9 \times 10^{-3}) = 35.28\ mV$$

4.134 In the circuit of Fig. 4-9, $V_{CC} = 20\ V$, $R_C = 5\ k\Omega$, $R_E = 4\ k\Omega$, and $R_B = 300\ k\Omega$. The Si transistor has $I_{CBO} = 0$ and $\beta = 50$. Find I_{CQ}.

▌ By KVL and using $I_{EQ} = (\beta + 1)I_{BQ}$,

$$V_{CC} - V_{BEQ} = I_{BQ}R_B + I_{EQ}R_E = I_{BQ}R_B + (\beta + 1)I_{BQ}R_E$$

or

$$I_{BQ} = \frac{V_{CC} - V_{BEQ}}{R_B + (\beta + 1)R_E} = \frac{20 - 0.7}{300 \times 10^3 + (50 + 1)(4 \times 10^3)} = 38.3\ \mu A$$

Thus, $\quad I_{CQ} = \beta I_{BQ} = 50(38.3 \times 10^{-6}) = 1.91\ mA$

4.135 Calculate V_{CEQ} for the transistor circuit of Problem 4.134.

▌ Use KVL, the result of Problem 4.134, and recall that $I_{EQ} = (\beta + 1)I_{BQ}$.

$$V_{CEQ} = V_{CC} - I_{EQ}R_E - I_{CQ}R_C = V_{CC} - (\beta + 1)I_{BQ}R_E - I_{CQ}R_C \tag{1}$$

$$V_{CEQ} = 20 - (50 + 1)(38.3 \times 10^{-6})(4 \times 10^3) - (1.91 \times 10^{-3})(5 \times 10^3) = 2.64\ V$$

4.136 The transistor of Problem 4.134 failed and was replaced with a new transistor with $I_{CBO} = 0$ and $\beta = 75$. Is the transistor still biased for active-region operation?

▌ Assume active region bias. As in Problems 4.134 and 4.135,

$$I_{BQ} = \frac{V_{CC} - V_{BEQ}}{R_B + (\beta + 1)R_E} = \frac{20 - 0.7}{300 \times 10^3 + (75 + 1)(4 \times 10^3)} = 31.9\ \mu A$$

$$I_{CQ} = \beta I_{BQ} = 75(31.9 \times 10^{-6}) = 2.39\ mA$$

$$V_{CEQ} = V_{CC} - (\beta + 1)I_{BQ}R_E - I_{CQ}R_C = 20 - (75 + 1)(31.9 \times 10^{-6})(4 \times 10^3) - (2.39 \times 10^{-3})(5 \times 10^3) = -1.65$$

Since the calculated $V_{CEQ} = -1.65 < 0$, the *npn* transistor is not in the active region.

4.137 What value of R_B will result in saturation of the Si transistor of Fig. 4-9 if $V_{CC} = 20\ V$, $R_C = 5\ k\Omega$, $R_E = 0$, $\beta = 50$, and $V_{CE\ sat} = 0.2\ V$?

▌ By (1) of Problem 4.135 with $I_{CQ} = \beta I_{BQ}$, $R_E = 0$, and $V_{CEQ} = V_{CE\ sat}$,

$$I_{BQ} = \frac{V_{CC} - V_{CE\ sat}}{\beta R_C} = \frac{20 - 0.2}{50(5 \times 10^3)} = 79.2\ \mu A$$

By KVL, $\quad R_B = \dfrac{V_{CC} - V_{BEQ}}{I_{BQ}} = \dfrac{20 - 0.7}{79.2 \times 10^{-6}} = 243.7\ k\Omega$

Hence, the transistor operates in saturation for $R_B \leq 243.7\ k\Omega$.

4.138 The circuit of Fig. 4-24 illustrates a method for biasing a CB transistor using a single dc source. The transistor is a Si device ($V_{BEQ} = 0.7\ V$), $\beta = 99$, and $I_{BQ} = 30\ \mu A$. Find the necessary value of R_2.

$$I_{CQ} = \beta I_{BQ} = 99(30 \times 10^{-6}) = 2.97\ mA$$

$$I_{EQ} = (\beta + 1)I_{BQ} = (99 + 1)(30 \times 10^{-6}) = 3\ mA$$

$$V_2 = I_{EQ}R_E + V_{BEQ} = (3 \times 10^{-3})(1 \times 10^3) + 0.7 = 3.7\ V$$

$$V_1 = V_{CC} - V_2 = 15 - 3.7 = 11.3\ V$$

$$I_1 = V_1/R_1 = 11.3/(10 \times 10^3) = 1.13\ mA$$

$$I_2 = I_1 - I_{BQ} = 1.13 \times 10^{-3} - 30 \times 10^{-6} = 1.1\ mA$$

$$R_2 = V_2/I_2 = 3.7/(1.1 \times 10^{-3}) = 3.36\ k\Omega$$

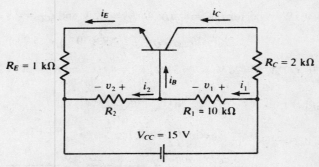

Fig. 4-24

4.139 Determine V_{CEQ} for the CB transistor amplifier of Fig. 4-24.

▮ By KVL and the results of Problem 4.138,

$$V_{CEQ} = V_{CC} - I_{CQ}R_C - I_{EQ}R_E = 15 - (2.97 \times 10^{-3})(2 \times 10^3) - (3 \times 10^{-3})(1 \times 10^{-3}) = 6.06 \text{ V}$$

4.140 Rework Problem 4.99 with $R_2 = 5$ kΩ and all else unchanged.

$$R_B = \frac{R_1 R_2}{R_1 + R_2} = \frac{(2 \times 10^3)(5 \times 10^3)}{2 \times 10^3 + 5 \times 10^3} = 1.429 \text{ k}\Omega$$

$$V_{BB} = \frac{R_1}{R_1 + R_2} V_{CC} = \frac{2 \times 10^3}{2 \times 10^3 + 5 \times 10^3}(15) = 4.286 \text{ V}$$

$$I_{CQ} \approx I_{EQ} = \frac{V_{BB} - V_{BEQ}}{R_E + R_B/(\beta + 1)} = \frac{4.286 - 0.7}{200 + (1.429 \times 10^3)/(110 + 1)} = 16.84 \text{ mA}$$

$$V_{CEQ} \approx V_{CC} - I_{CQ}(R_C + R_E) = 15 - (16.84 \times 10^{-3})(300 + 200) = 6.58 \text{ V}$$

Since $V_{CEQ} < V_{CC}/2$, saturation occurs before cutoff, and i_C at saturation must be calculated to determine maximum swing.

$$i_{C \text{ sat}} = \frac{V_{CC} - V_{CE \text{ sat}}}{R_C + R_E} = \frac{15 - 0}{300 + 200} = 30 \text{ mA}$$

Maximum swing of i_C is $\pm(i_{C \text{ sat}} - I_{CQ}) = \pm(30 - 16.84) = \pm 13.16$ mA about $I_{CQ} = 16.84$ mA.

4.141 Because of a poor solder joint, resistor R_1 of Problem 4.99 becomes open-circuited. Calculate the percentage change in I_{CQ} that will be observed.

▮ With $R_1 = 2$ kΩ, $I_{CQ} = 4.93$ mA from Problem 4.99. With $R_1 \to \infty$,

$$I_{CQ} \approx I_{EQ} = \frac{V_{CC} - V_{BEQ}}{R_2/(\beta + 1) + R_E} = \frac{15 - 0.7}{(15 \times 10^3)/(110 + 1) + 200} = 42.67 \text{ mA}$$

Check for saturation.

$$I_{C \text{ sat}} = \frac{V_{CC} - V_{CE \text{ sat}}}{R_C + R_E} = \frac{15 - 0}{300 + 200} = 30 \text{ mA}$$

Whence, I_{CQ} cannot reach 42.67 mA, but rather is limited by saturation to 30 mA. And

$$\Delta I_{CQ} = \frac{30 - 4.93}{4.93}(100\%) = 508.5\%$$

4.142 The circuit of Problem 4.99 has β-independent bias ($R_E \geq 10 R_B/\beta$). Find the allowable range of β if I_{CQ} can change at most ± 2 percent from its value for $\beta = 110$.

▮

$$R_B = \frac{R_1 R_2}{R_1 + R_2} = \frac{(2 \times 10^3)(15 \times 10^3)}{2 \times 10^3 + 15 \times 10^3} = 1.765 \text{ k}\Omega$$

$$V_{BB} = \frac{R_1}{R_1 + R_2} V_{CC} = \frac{2 \times 10^3}{2 \times 10^3 + 15 \times 10^3}(15) = 1.765 \text{ V}$$

$$I_{CQ \text{ nom}} \approx I_{EQ \text{ nom}} = \frac{V_{BB} - V_{BEQ}}{R_E + R_B/(\beta + 1)} = \frac{1.765 - 0.7}{200 + (1.765 \times 10^{-3})/(110 + 1)} = 4.93 \text{ mA}$$

By KVL,
$$I_{CQ}(R_E + R_B/\beta) \approx V_{BB} - V_{BEQ}$$

or
$$\beta \approx \frac{R_B I_{CQ}}{V_{BB} - V_{BEQ} - I_{CQ} R_E} \tag{1}$$

For $I_{CQ} = 1.02 I_{CQ \text{ nom}} = 1.02(4.93 \times 10^{-3}) = 5.0286$ mA, (1) yields
$$\beta_1 = \frac{1.765(5.0286 \times 10^{-3})}{1.765 - 0.7 - (5.0286 \times 10^{-3})(200)} = 149.7$$

For $I_{CQ} = 0.98 I_{CQ \text{ nom}} = 0.98(4.93 \times 10^{-3}) = 4.8314$ mA, (1) gives
$$\beta_2 = \frac{1.765(4.8314 \times 10^{-3})}{1.765 - 0.7 - (4.8314 \times 10^{-3})(200)} = 86.4$$

Hence, the range on β is $86.4 \leq \beta \leq 149.7$.

4.143 For the circuit of Fig. 4-22, $v_S = 0.25 \cos \omega t$ V, $R_B = 30$ kΩ, $V_B = 1$ V, and $V_{CC} = 12$ V. The transistor is a Si device with negligible base-to-emitter resistance and $\beta = 100$. Assume that $V_{CE \text{ sat}} \approx 0$ and $I_{CBO} = 0$. Find the range of R_L for class A operation.

▌ Cutoff can only be accomplished by reverse bias of the base-emitter junction and is unaffected by R_L. However, saturation can be controlled by choice of value for R_L.

$$I_{BQ} = \frac{V_B - V_{BEQ}}{R_B} = \frac{1 - 0.7}{30 \times 10^3} = 10 \ \mu A$$

$$i_b = \frac{v_S}{R_B} = \frac{0.25 \cos \omega t}{30 \times 10^3} = 8.333 \cos \omega t \ \mu A$$

$$i_C = I_{CQ} + i_c = \beta(I_{BQ} + i_b) = 100(10 \times 10^{-6} + 8.333 \times 10^{-6} \cos \omega t)$$

$$i_C = 1 + 0.8333 \cos \omega t \text{ mA}$$

To avoid saturation, and thus, maintain class A operation,
$$\frac{V_{CC}}{R_L} \geq I_{CQ} + I_{cm}$$

or
$$R_L \leq \frac{V_{CC}}{I_{CQ} + I_{cm}} = \frac{12}{(1 + 0.8333) \times 10^{-3}} = 6.545 \text{ k}\Omega$$

4.144 If an emitter resistor is added to the circuit of Fig. 4-19, find the value of R_F needed to bias for maximum symmetrical swing. Let $V_{CC} = 15$ V, $R_E = 1.5$ kΩ, and $R_C = 5$ kΩ. Assume the transistor is a Si device with $I_{CEO} = V_{CE \text{ sat}} = 0$ and $\beta = 80$.

▌ By KVL, and using $I_{CQ} = \beta I_{BQ}$ and $I_{EQ} = (\beta + 1)I_{BQ}$,
$$(I_{BQ} + I_{CQ})R_C + I_{EQ}R_E = (I_{BQ} + \beta I_{BQ})R_C + (\beta + 1)I_{BQ} = V_{CC} - V_{CEQ}$$

Whence,
$$I_{BQ} = \frac{V_{CC} - V_{CEQ}}{(\beta + 1)(R_C + R_E)} = \frac{15 - 15/2}{(80 + 1)(5 \times 10^3 + 1.5 \times 10^3)} = 14.245 \ \mu A$$

where for maximum symmetrical swing, $V_{CEQ} = V_{CC}/2$. From application of KVL over the path around transistor terminals,
$$R_F = \frac{V_{CEQ} - V_{BEQ}}{I_{BQ}} = \frac{15/2 - 0.7}{14.245 \times 10^{-6}} = 477.4 \text{ k}\Omega$$

4.145 In the circuit of Fig. 4-18, the Ge transistor has $I_{CEO} = 0$ and $\beta = 50$. Assume the capacitor is replaced with a short circuit. Let $V_S = 2$ V, $V_{CC} = 12$ V, $R_C = 4$ kΩ, $R_S = 100$ kΩ, and $R_B = 330$ kΩ. Find the ratio I_{CQ}/I_S.

▌
$$I_S = \frac{V_S - V_{BEQ}}{R_S} = \frac{2 - 0.3}{100 \times 10^3} = 17 \ \mu A$$

$$I_B = \frac{V_{CC} - V_{BEQ}}{R_B} = \frac{12 - 0.3}{330 \times 10^3} = 35.45 \ \mu A$$

By KCL, $I_{BQ} = I_B + I_S = (35.45 + 17) \times 10^{-6} = 52.45\ \mu\text{A}$

$$\frac{I_{CQ}}{I_S} = \frac{\beta I_{BQ}}{I_S} = \frac{50(52.45 \times 10^{-6})}{17 \times 10^{-6}} = 374.6$$

4.146 Determine the ratio V_{CEQ}/V_S for the transistor circuit of Problem 4.145.

▎ By KVL and the result of Problem 4.145,

$$V_{CEQ} = V_{CC} - I_{CQ}R_C = V_{CC} - \beta I_{BQ}R_C = 12 - 50(52.45 \times 10^{-6})(4 \times 10^3) = 1.51\ \text{V}$$

Hence, $\quad\dfrac{V_{CEQ}}{V_S} = \dfrac{1.51}{2} = 0.755$

4.147 In the differential amplifier of Fig. 4-17, the Si transistors have negligible leakage current, and $\beta_1 = \beta_2 = 75$. If $R_B = 10\ \text{k}\Omega$, $R_E = R_C = 6.8\ \text{k}\Omega$, and $V_{EE} = V_{CC} = 15\ \text{V}$, find V_{CEQ1}.

▎ By KVL around left base loop with identical transistor currents,

$$V_{EE} = \frac{I_{EQ1}}{\beta_1 + 1} R_B + V_{BEQ1} + 2I_{EQ1}R_E$$

or $\quad I_{EQ1} = \dfrac{V_{EE} - V_{BEQ1}}{R_B/(\beta_1 + 1) + 2R_E} = \dfrac{15 - 0.7}{(10 \times 10^3)/(75 + 1) + 2(6.8 \times 10^3)} = 1.041\ \text{mA}$

Apply KVL around collector-emitter path and use $I_{CQ1} = \dfrac{\beta_1}{\beta_1 + 1} I_{EQ1}$.

$$V_{CC} + V_{EE} = I_{CQ1}R_C + V_{CEQ1} + 2I_{EQ1}R_E$$

or $\quad V_{CEQ1} = V_{CC} + V_{EE} - \left[\dfrac{\beta_1}{\beta_1 + 1} R_C + 2R_E\right] I_{EQ1}$

$$= 15 + 15 - \left[\frac{75}{75 + 1}(6.8 \times 10^3) + 2(6.8 \times 10^3)\right](1.041 \times 10^{-3}) = 8.85\ \text{V}$$

4.148 Find the voltage $v_{o1} = v_{o2}$ for Problem 4.147.

▎ By KVL and the results of Problem 4.147,

$$v_{o1} = V_{CEQ1} + 2I_{EQ1}R_E - V_{EE} = 8.85 + 2(1.041 \times 10^{-3})(6.8 \times 10^3) - 15 = 8.01\ \text{V}$$

4.149 If in Problem 4.103, R_1 is changed to $9\ \text{k}\Omega$ and all else remains unchanged, determine the maximum undistorted swing of i_c.

▎ $R_{dc} = 2\ \text{k}\Omega$ and $R_{ac} = 1.5\ \text{k}\Omega$ of Problem 4.103 are valid.

$$V_{BB} = \frac{R_1}{R_1 + R_2} V_{CC} = \frac{9 \times 10^3}{9 \times 10^3 + 9 \times 10^3}(15) = 7.5\ \text{V}$$

$$R_B = R_1 \parallel R_2 = \frac{(9 \times 10^3)(9 \times 10^3)}{9 \times 10^3 + 9 \times 10^3} = 4.5\ \text{k}\Omega$$

$$I_{CQ} = \frac{\beta(V_{BB} - V_{BEQ})}{R_B + (\beta + 1)R_E} = \frac{75(7.5 - 0.7)}{4.5 \times 10^3 + (75 + 1)(1 \times 10^3)} = 6.33\ \text{mA}$$

$$V_{CEQ} = V_{CC} - \left[R_C\left(\frac{\beta + 1}{\beta}\right)R_E\right]I_{EQ} = 15 - \left[1 \times 10^3\left(\frac{75 + 1}{75}\right)(1 \times 10^3)\right](6.33 \times 10^{-3}) = 2.256\ \text{V}$$

The i_C intercept of the ac load line is

$$i_{C\,\text{max}} = \frac{V_{CEQ}}{R_{ac}} + I_{CQ} = \frac{2.256}{1.5 \times 10^3} + 6.33 \times 10^{-3} = 7.83\ \text{mA}$$

Since $i_{C\,\text{max}} < 2I_{CQ}$, saturation will limit the undistorted swing. Signal i_c swings symmetrically about $I_{CQ} = 6.33\ \text{mA}$ by $\pm(i_{C\,\text{max}} - I_{CQ}) = \pm(7.83 - 6.33) \times 10^{-3} = \pm 1.5\ \text{mA}$.

4.150 For the circuit of Fig. 4-5(a), $R_C = 1\,\text{k}\Omega$, $R_E = 750\,\Omega$, $V_{CC} = 15\,\text{V}$, and $\beta = 75$. The transistor is a Si device. Use β-independent bias, size R_1 and R_2 such that $V_{CEQ} = 10\,\text{V}$.

$$R_B = \frac{\beta R_E}{10} = \frac{75(750)}{10} = 5.625\,\text{k}\Omega$$

$$I_{CQ} \approx \frac{V_{CC} - V_{CEQ}}{R_E + R_C} = \frac{15 - 10}{750 + 1 \times 10^3} = 2.86\,\text{mA}$$

$$V_{BB} = I_{CQ}(R_E + R_B/\beta) + V_{BEQ} = (2.86 \times 10^{-3})(750 + 5.625 \times 10^3/75) + 0.7 = 3.0595\,\text{V}$$

$$R_1 = \frac{R_B}{1 - V_{BB}/V_{CC}} = \frac{5.625 \times 10^3}{1 - 3.0595/15} = 7.07\,\text{k}\Omega$$

$$R_2 = R_B \frac{V_{CC}}{V_{BB}} = 5.625 \times 10^3 \left(\frac{15}{3.0595}\right) = 27.58\,\text{k}\Omega$$

4.151 If the transistor circuit of Fig. 4-14(a) has $\beta = 100$, $V_{CE\,\text{sat}} = 0$, $R_L = R_C = 2\,\text{k}\Omega$, and $R_E = 1\,\text{k}\Omega$, size R_1 and R_2 for maximum symmetrical swing.

▌ For β-independent bias,

$$R_B = \frac{\beta R_E}{10} = \frac{100(1 \times 10^3)}{10} = 10\,\text{k}\Omega$$

For maximum symmetrical swing,

$$I_{CQ} = \frac{V_{CC}}{R_{ac} + R_{dc}} = \frac{15}{1 \times 10^3 + 3 \times 10^3} = 3.75\,\text{mA}$$

where $R_{ac} = R_L \parallel R_C$ and $R_{dc} = R_E + R_C$. Since

$$V_{BB} \approx V_{BEQ} + I_{CQ}(R_E + R_B/\beta) = 0.7 + (3.75 \times 10^{-3})(1 \times 10^3 + 10 \times 10^3/100) = 4.825\,\text{V}$$

we have

$$R_1 = \frac{R_B}{1 - V_{BB}/V_{CC}} = \frac{10 \times 10^3}{1 - 4.825/15} = 14.74\,\text{k}\Omega$$

$$R_2 = R_B \frac{V_{CC}}{V_{BB}} = (10 \times 10^3)\left(\frac{15}{4.825}\right) = 31.09\,\text{k}\Omega$$

CHAPTER 5
Field-Effect Transistor Fundamentals

OPERATING PRINCIPLES

5.1 The operation of the *field-effect transistor* (FET) can be explained in terms of only majority-carrier charge flow from *source* (S) to *drain* (D) through a *channel* (*p*- or *n*-type material) wherein the conductivity is regulated by the electric field resulting from application of a voltage between the source and *gate* (G) terminals. Figure 5-1 displays the physical arrangement of, and schematics for, the *junction field-effect transistor* (JFET); conduction mechanism is by electrons for the *n*-channel device, while conduction by holes characterizes the *p*-channel device. Current and voltage symbology for FETs parallels that given in Table 4-1. Evaluate the drain current i_D of a FET if it consists of a dc component of 5 mA and a sinusoidal component of 10-μA peak value.

$$I_D = 5 \text{ mA} \quad \text{and} \quad i_d = I_{dm} \sin \omega t = 0.01 \sin \omega t \text{ mA}$$

Hence, the total drain current is

$$i_D = I_D + i_d = 5 + 0.01 \sin \omega t \text{ mA}$$

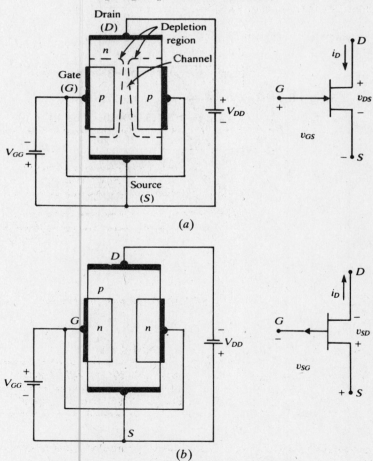

Fig. 5-1 (*a*) *n*-channel JFET, (*b*) *p*-channel JFET.

5.2 For the *n*-channel JFET of Fig. 5-1(*a*), v_{DS} (increase in potential from source to drain) is given by $12 + 2 \cos \omega t$ V. Determine the components of v_{DS} consistent with Table 4-1.

$$v_{DS} = V_{DS} + v_{ds} = V_{DS} + V_{dsm} \cos \omega t$$

where $V_{DS} = 12$ V and $V_{dsm} = 2$ V.

5.3 Determine the rms value of v_{DS} of Problem 5.2 using notation consistent with Table 4-1.

$$V_{ds}^2 = \frac{1}{T}\int_0^T v_{DS}^2\, dt = \frac{1}{2\pi}\int_0^{2\pi}(V_{DS}+V_{dsm}\cos\omega t)^2\, d(\omega t) = V_{DS}^2 + \frac{V_{dsm}^2}{2}$$

Whence,
$$V_{ds} = \left[(12)^2 + \frac{(2)^2}{2}\right]^{1/2} = 12.083\text{ V}$$

5.4 Calculate the average value of i_D for Problem 5.1.

$$I_{D\text{ ave}} = \frac{1}{T}\int_0^T i_D\, dt = \frac{1}{2\pi}\int_0^{2\pi}(I_D+I_{dm}\sin\omega t)\, d(\omega t) = I_D = 5\text{ mA}$$

5.5 Typical *output* or *drain characteristics for an* n-channel JFET in *common-source* (CS) connection with $v_{GS} \leq 0$ are given by Fig. 5-2. For a constant value of v_{GS}, the JFET acts as a linear resistive device (in the *ohmic region*) until the depletion region of the reverse-biased gate-source junction extends the width of the channel (a condition called *pinchoff*) at the value of source-to-drain voltage V_p (*pinchoff voltage*). For specification purposes, *shorted-gate parameters* ($v_{GS}=0$) are defined on Fig. 5-2. Sketch a normalized *transfer characteristic*, that is, plot i_D/I_{DSS} vs. v_{GS}/V_{p0}, from the drain characteristic of Fig. 5-2 for $v_{DS} > V_{p0}$.

▌ Establishing a vertical scale on Fig. 5-2 where $i_D = I_{DSS} = I_{p0}$, the values of Table 5-1 are found and the results are plotted by the solid line of Fig. 5-3.

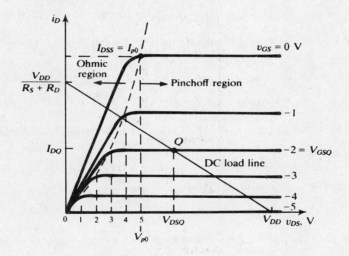

Fig. 5-2 JFET drain characteristics.

TABLE 5-1

i_D/I_{DSS}	1.0	0.633	0.390	0.207	0.083	0.0
v_{GS}/V_{p0}	0.0	−0.2	−0.4	−0.6	−0.8	−1.0

5.6 The drain current of the JFET shows an approximate square-law dependence on source-to-gate voltage for constant values of v_{DS} in the pinchoff region given by

$$\frac{i_D}{I_{DSS}} = \left(1 + \frac{v_{GS}}{V_{p0}}\right)^2 \tag{1}$$

Based on (1), plot a normalized transfer characteristic for a JFET on the same axes as the result of Problem 5.5.

▌ See dashed line plot of Fig. 5-3.

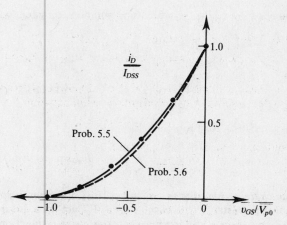

Fig. 5-3 Transfer characteristics.

5.7 For a JFET with a pinchoff voltage $V_{p0} = 4$ V and a saturation current $I_{DSS} = 10$ mA, predict the value of drain current if $v_{GS} = -2$ V. Assume $v_{DS} > V_{p0}$.

▌ By (1) of Problem 5.6,

$$i_D = I_{DSS}(1 + v_{GS}/V_{p0})^2 = 10 \times 10^{-3}[1 + (-2)/4]^2 = 5 \text{ mA}$$

5.8 Predict the value of source-to-gate voltage for the JFET of Problem 5.7 if $i_D = 7$ mA and the device is operating with $v_{DS} > V_{p0}$.

▌ Based on (1) of Problem 5.6,

$$v_{GS} = V_{p0}\left[(i_D/I_{DSS})^{1/2} - 1\right] = 4\left[(7 \times 10^{-3}/10 \times 10^{-3})^{1/2} - 1\right] = -0.837 \text{ V}$$

5.9 The shorted-gate saturation current of a JFET is known to be inversely proportional to the 3/2 power of temperature, or

$$I_{DSS} = k_I T^{-3/2} \tag{1}$$

Evaluate the constant k_I of a particular JFET for which $I_{DSS} = 10$ mA at 25°C.

▌ $$k_I = I_{DSS} T^{3/2} = (10 \times 10^{-3})(273 + 25)^{3/2} = 51.443 \text{ A} \cdot {}^\circ K^{3/2}$$

5.10 Predict the saturation current for the JFET of Problem 5.9 at an operating temperature of 100°C.

▌ Based on the results of Problem 5.9,

$$I_{DSS} = k_I T^{-3/2} = 51.443(273 + 100)^{-3/2} = 7.14 \text{ mA}$$

5.11 At what temperature will the saturation current for the JFET of Problem 5.9 reduce to 5 mA?

▌ Based on (1) of Problem 5.9,

$$T_2 = \left(\frac{I_{DSS1}}{I_{DSS2}}\right)^{2/3} T_1 = \left(\frac{10 \times 10^{-3}}{5 \times 10^{-3}}\right)^{2/3}(273 + 25) = 473^\circ K = 200^\circ C$$

5.12 Similar to the base-emitter voltage of the BJT, the temperature dependence of pinchoff voltage for the JFET is described by

$$\Delta V_{p0} = -k_V \Delta T \tag{1}$$

For a particular JFET, the manufacturer's specification sheet gives $V_{p0} = 5$ V at 25°C and $V_{p0} = 5.15$ V at −45°C. Determine the temperature coefficient k_V for the device.

▌ Based on (1),

$$k_V = -\frac{V_{p01} - V_{p02}}{T_1 - T_2} = -\frac{5 - 5.15}{(273 + 25) - (273 - 45)} = 2.14 \text{ mV/}^\circ C$$

5.13 Predict the value of pinchoff voltage for the JFET of Problem 5.12 at an operating temperature of 100°C.

▌ Using (1) of Problem 5.12 and manufacturer's data for 25°C,

$$V_{p0} = 5 + \Delta V_{p0} = 5 - k_v \, \Delta T = 5 - (2.14 \times 10^{-3})(100 - 25) = 4.84 \text{ V}$$

5.14 Typically the short-gate pinchoff voltage V_{p0} of the JFET is between 4 and 5 V. As gate potential decreases, the pinchoff voltage V_p also decreases, approximately obeying the equation

$$V_p = V_{p0} + v_{GS} \qquad (1)$$

Predict the pinchoff voltage for the JFET of Problem 5.8.

▌ $$V_p = V_{p0} + v_{GS} = 4 + (-0.837) = 3.163 \text{ V}$$

5.15 Figure 5-4 displays the physical arrangement of, and schematic for, an n-channel *metal-oxide-semiconductor field-effect transistor* (MOSFET). The gate (positive plate), metal oxide film (dielectric), and substrate (negative plate) form a capacitor, the electric field of which controls channel resistance. When the positive potential of the gate reaches a *threshold voltage* V_T (typically 2 to 4 V), sufficient free electrons are attracted to the region immediately beside the metal oxide film (this is called *enhancement-mode* operation) to induce a conducting channel of low resistivity. If the source-to-drain voltage is increased, the enhanced channel is depleted of free charge carriers in the area near the drain, and pinchoff occurs as in the JFET. Typical drain characteristics are displayed in Fig. 5-5, where $V_T = 2$ V is used for illustration. Commonly, the manufacturer specifies V_T and a value of pinchoff current $I_{D \text{ on}}$; the corresponding value of source-to-gate voltage is $V_{GS \text{ on}}$. Sketch a normalized transfer characteristic, that is, plot $i_D/I_{D \text{ on}}$ vs. $v_{GS}/V_{GS \text{ on}}$, from the drain characteristic of Fig. 5-5 for $v_{DS} > V_{GS \text{ on}} - V_T$.

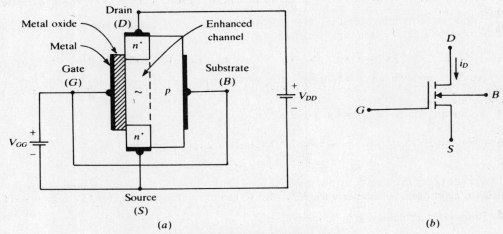

Fig. 5-4 n-channel MOSFET.

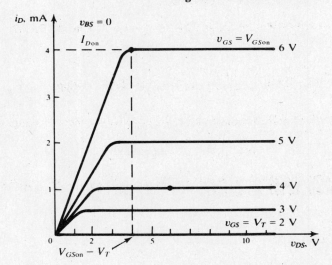

Fig. 5-5 MOSFET drain characteristics.

▎ Values of $i_D/I_{D\,on}$ and $v_{GS}/V_{GS\,on}$ determined from Fig. 5-5 are given by Table 5-2 and the results are plotted by the solid line of Fig. 5-6.

TABLE 5-2

$i_D/I_{D\,on}$	1.0	0.5	0.25	0.125	0.0
$v_{GS}/V_{GS\,on}$	1.0	0.833	0.667	0.50	0.333

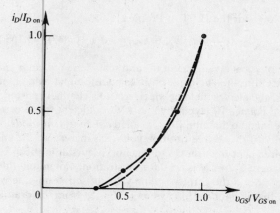

Fig. 5-6

5.16 The drain current of the MOSFET shows an approximate square-law dependence on the source-to-gate voltage for constant value of v_{DS} in the pinchoff region given by

$$i_D = k_0(v_{GS} - V_T)^2 \qquad (1)$$

where k_0 depends on the channel characteristics. Determine the value of k_0 for MOSFET characterized by Fig. 5-5.

▎ Evaluating for $i_D = I_{D\,on} = 4$ mA and $v_{GS} = V_{GS\,on} = 6$ V,

$$k_0 = \frac{i_D}{(v_{GS} - V_T)^2} = \frac{4 \times 10^{-3}}{(6-2)^2} = 0.25 \times 10^{-3}\ \text{A/V}^2$$

5.17 Rearrange (1) of Problem 5.16 into the form of (1) of Problem 5.6 to show the similarity of the JFET and MOSFET for operation above pinchoff.

▎ From (1) of Problem 5.16,

$$i_D = k_0 V_T^2\left(1 - \frac{v_{GS}}{V_T}\right)^2$$

Whence,

$$\frac{i_D}{I_{D0}} = \left(1 - \frac{v_{GS}}{V_T}\right)^2 \qquad (1)$$

where $I_{D0} \equiv k_0 V_T^2$.

5.18 Based on (1) of Problem 5.17, plot a normalized transfer characteristic for the MOSFET of Fig. 5-5 on the same axes as the result of Problem 5.15.

▎ Normalize (1) of Problem 5.17 through dividing both sides by $I_{D\,on}$.

$$\frac{i_D}{I_{D\,on}} = \frac{I_{D0}}{I_{D\,on}}\left(1 - \frac{v_{GS}}{V_T}\right)^2 = \frac{k_0 V_T^2}{I_{D\,on}}\left(1 - \frac{v_{GS}}{V_T}\right)^2$$

Using $k_0 = 0.25 \times 10^{-3}$, $I_{D\,on} = 4$ mA, and $V_T = 2$ V, the results are plotted by the dashed line of Fig. 5-6.

5.19 The constant k_0 from (1) of Problem 5.16 is given by

$$k_0 = \frac{\mu \epsilon W}{2tL} \qquad (1)$$

where μ = mobility of channel carriers
ϵ = permittivity of metal oxide
t = thickness of metal oxide
W = channel width
L = channel length

For a particular MOSFET, $\mu = 1350 \text{ cm}^2/\text{V} \cdot \text{s}$, $\epsilon = 1.4 \times 10^{-10}$ F/m, $t = 0.1$ μm, $W = 300$ μm, and $L = 25$ μm. Evaluate k_0 for this MOSFET.

$$k_0 = \frac{\mu \epsilon W}{2tL} = \frac{0.1350(1.4 \times 10^{-10})(300 \times 10^{-6})}{2(0.1 \times 10^{-6})(25 \times 10^{-6})} = 1.134 \times 10^{-3} \text{ A/V}^2$$

5.20 If the MOSFET of Problem 5.19 is characterized by $V_T = 2$ V and $V_{GS \text{ on}} = 7$ V, predict the saturation current $I_{D \text{ on}}$.

I Using the result of Problem 5.19 and (*1*) of Problem 5.16,

$$I_{D \text{ on}} = k_0(V_{GS \text{ on}} - V_T)^2 = (1.134 \times 10^{-3})(7 - 2)^2 = 28.35 \text{ mA}$$

5.21 What geometric characteristic would you expect to observe for a high current rated MOSFET?

I A large value of k_0 is necessary. Based on (*1*) of Problem 5.19, one would expect a large W/L (channel width to length ratio).

5.22 If, instead of depending on the enhanced channel for conduction, the region between the two heavily doped n^+ regions of the MOSFET is made up of lightly doped n material, a *depletion-enhancement-mode* MOSFET can be formed with drain characteristics as displayed by Fig. 5-7, where v_{GS} may be either positive or negative.

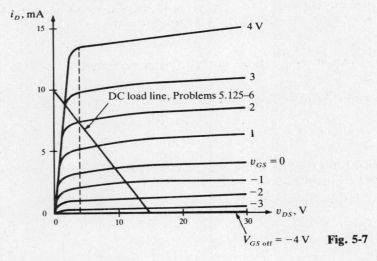

Fig. 5-7

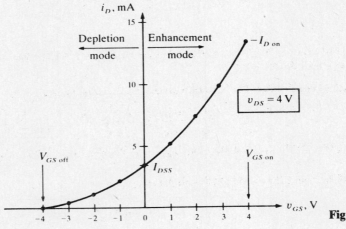

Fig. 5-8

Construct a transfer characteristic for the drain characteristics of Fig. 5-7, and clearly label the regions of depletion-mode and enhancement-mode operation.

▎ If a constant value of $v_{DS} = V_{GS\,\text{on}} = 4$ V is taken as indicated by the broken line on Fig. 5-7, the transfer characteristic of Fig. 5-8 results. $v_{GS} = 0$ is the dividing line between depletion- and enhancement-mode operation.

BIAS AND LOAD LINES

5.23 The commonly used *voltage-divider* bias arrangement of Fig. 5-9 can be reduced to its equivalent of Fig. 5-10. Justify this statement.

▎ Only dc signals are of interest in bias calculations; thus, the capacitor C_C can be treated as an open circuit. Whence, the Thévenin voltage and current looking to the right through the port formed by G and ground of Fig. 5-9 are

$$R_{Th} = R_G = \frac{R_1 R_2}{R_1 + R_2} \quad \text{and} \quad V_{Th} = V_{GG} = \frac{R_1}{R_1 + R_2} V_{DD} \qquad (1)$$

The circuit of Fig. 5-10 follows from placement of the Thévenin equivalent circuit.

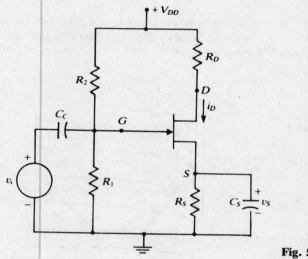

Fig. 5-9

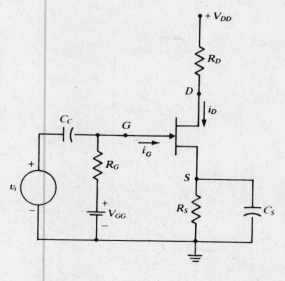

Fig. 5-10

5.24 Assume $i_G = 0$ and apply KVL around the gate-source loop of Fig. 5-10 to find the *transfer bias line*, an equation expressing i_D as a function of v_{GS}. Comment on how the transfer bias line can be used to determine the quiescent variables I_{DQ} and V_{GSQ}.

▍ By KVL,

$$i_D = \frac{V_{GG}}{R_S} - \frac{v_{GS}}{R_S} \quad (1)$$

Equation (1) can be solved simultaneously with (1) of Problem 5.6 or plotted as indicated on Fig. 5-11 to yield I_{DQ} and V_{GSQ}.

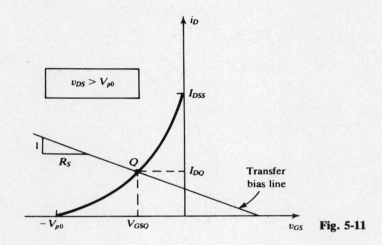

Fig. 5-11

5.25 Apply KVL around the drain-source loop of Fig. 5-10 to find the *dc load line*, an equation expressing i_D as a function of v_{DS}. Comment on how the dc load line can be used to determine the quiescent value V_{DSQ}.

▍ By KVL,

$$i_D = \frac{V_{DD}}{R_S + R_D} - \frac{v_{DS}}{R_S + R_D} \quad (1)$$

Equation (1) can be plotted on the drain characteristics (see Fig. 5-2) to yield V_{DSQ} using V_{GSQ} as determined in Problem 5.24. Alternatively, with I_{DQ} already determined in Problem 5.24,

$$V_{DSQ} = V_{DD} - (R_S + R_D)I_{DQ} \quad (2)$$

5.26 In the amplifier of Fig. 5-9, $V_{DD} = 20$ V, $R_1 = 1$ MΩ, $R_2 = 15.7$ MΩ, $R_D = 3$ kΩ, and $R_S = 2$ kΩ. If the JFET characteristics are given by Fig. 5-12, find I_{DQ}.

▍ By (1) of Problem 5.23,

$$V_{GG} = \frac{R_1}{R_1 + R_2} V_{DD} = \frac{1 \times 10^6}{16.7 \times 10^6} 20 = 1.2 \text{ V}$$

On Fig. 5-12(a), we construct the transfer bias line given by (1) of Problem 5.24; it intersects the transfer characteristic at the Q point, giving $I_{DQ} = 1.5$ mA.

5.27 Determine V_{GSQ} for the amplifier of Problem 5.26.

▍ The Q point of Fig. 5-12(a) also gives $V_{GSQ} = -2$ V.

5.28 Find the quiescent value V_{DSQ} for the amplifier of Problem 5.26.

▍ We construct the dc load line on the drain characteristics, making use of the v_{DS} intercept of $V_{DD} = 20$ V and the i_D intercept of $V_{DD}/(R_S + R_D) = 4$ mA. The Q point was established at $I_{DQ} = 1.5$ mA in Problem 5.26 (and at $V_{GSQ} = -2$ V in Problem 5.27); its abscissa is $V_{DSQ} = 12.5$ V. Analytically,

$$V_{DSQ} = V_{DD} - (R_S + R_D)I_{DQ} = 20 - (5 \times 10^3)(1.5 \times 10^{-3}) = 12.5 \text{ V}$$

140 ▢ CHAPTER 5

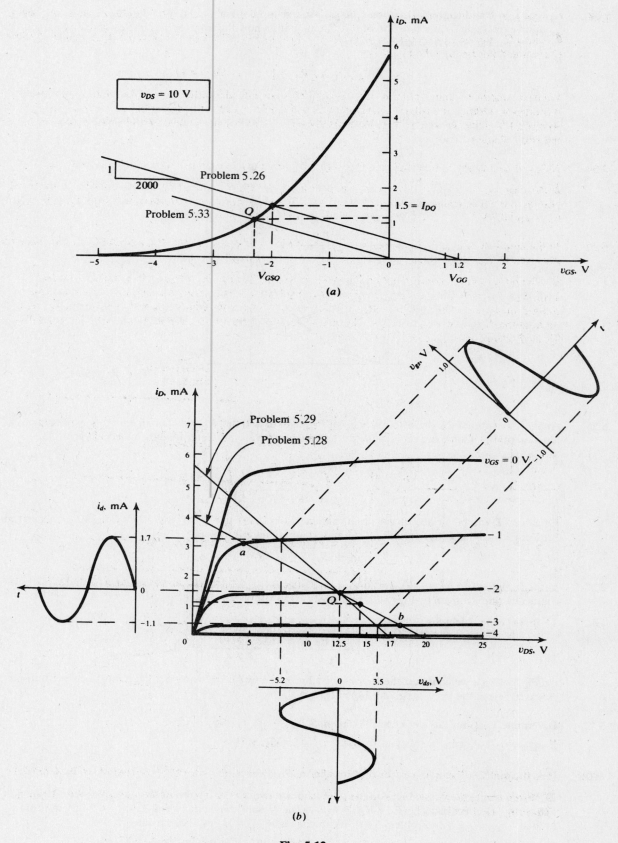

Fig. 5-12

5.29 For the amplifier of Problem 5.26, let $v_i = \sin t(\omega = 1 \text{ rad/s})$ and $C_S \to \infty$. Graphically determine v_{ds} and i_d.

▌ Since C_s appears as a short to ac signals, an ac load line must be added to Fig. 5-12(b), passing through the Q point and intersecting the v_{DS} axis at

$$V_{DSQ} + I_{DQ}R_{ac} = 12.5 + (1.5)(3) = 17 \text{ V}$$

We next construct an auxiliary time axis through Q, perpendicular to the ac load line, for the purpose of showing, on additional auxiliary axes as constructed in Fig. 5-12(b), the excursions of i_d and v_{ds} as $v_{gs} = v_i$ swings ± 1 V along the ac load line. Note the distortion in both signals, introduced by the square-law behavior of the JFET characteristics.

5.30 If $C_S = 0$ and all else is unchanged in Problem 5.26, find the extremes between which v_S swings.

▌ Voltage v_{gs} will swing along the dc load line of Fig. 5-12(b) (which is now identical to the ac load line) from point a to point b, giving, as extremes of i_D, 3.1 mA and 0.4 mA. The corresponding extremes of $v_s = i_D R_S$ are 6.2 V and 0.8 V.

5.31 In the amplifier of Fig. 5-13(a), $V_{DD} = 15$ V, $R_L = 3$ kΩ, and $R_F = 50$ MΩ. If the MOSFET drain characteristics are given by Fig. 5-13(b), determine the values of the quiescent quantities.

▌ The dc load line is constructed on Fig. 5-13(b) with v_{DS} intercept of $V_{DD} = 15$ V and i_D intercept of $V_{DD}/R_L = 5$ mA. With gate current negligible, no voltage appears across R_F, and so $V_{GS} = V_{DS}$. The *drain-feedback bias line* of Fig. 5-13(b) is the locus of all points for which $V_{GS} = V_{DS}$. Since the Q point must lie on both the dc load line and the drain-feedback bias line, their intersection is the Q point. From Fig. 5-13(b), $I_{DQ} \approx 2.65$ mA and $V_{DSQ} = V_{GSQ} \approx 6.90$ V.

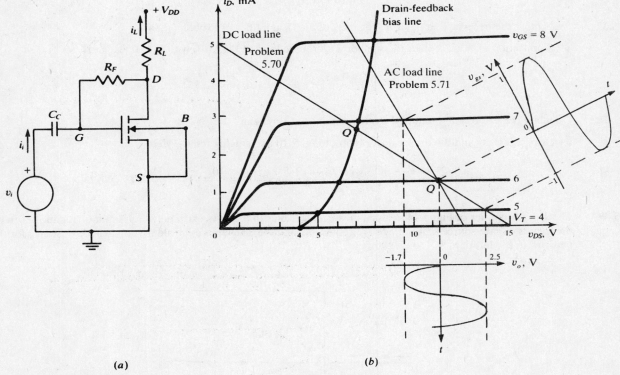

Fig. 5-13

5.32 For the MOSFET amplifier of Problem 5.31, let $V_{GSQ} = 6.90$ V. Calculate I_{DQ} from (1) of Problem 5.17.

▌ From the drain characteristics of Fig. 5-13(b), we see that $V_T = 4$ V and that $I_{D \text{ on}} = 5$ mA at $V_{GS \text{ on}} = 8$ V. Thus, by (1) of Problem 5.16,

$$k_0 = \frac{I_{D \text{ on}}}{(V_{GS \text{ on}} - V_T)^2} = \frac{5 \times 10^{-3}}{(8-4)^2} = 0.3125 \times 10^{-3} \text{ A/V}^2$$

And by (1) of Problem 5.17,

$$I_{DQ} = k_0 V_T^2 \left(1 - \frac{V_{GSQ}}{V_T}\right)^2 = (0.3125 \times 10^{-3})(4)^2 \left(1 - \frac{6.90}{4}\right)^2 = 2.63 \text{ mA}$$

5.33 By a method called *self-bias*, the Q point of a JFET amplifier may be established using only a single resistor from gate to ground (Fig. 5-10 with $V_{GG} = 0$). If $R_D = 3$ kΩ, $R_S = 2$ kΩ, $R_G = 5$ MΩ, and $V_{DD} = 20$ V in Fig. 5-10, and the JFET characteristics are given by Fig. 5-12, find I_{DQ}.

▌ On Fig. 5-12(a) we construct a transfer bias line having a v_{GS} intercept of $V_{GG} = 0$ and a slope of $-1/R_S = -0.5$ mS; the ordinate of its intersection with the transfer characteristic is $I_{DQ} = 1.15$ mA.

5.34 Determine V_{GSQ} for the JFET amplifier of Problem 5.33.

▌ The abscissa of the Q point of Fig. 5-12(a) is $V_{GSQ} = -2.3$ V.

5.35 Find the quiescent value V_{DSQ} for the JFET amplifier of Problem 5.33.

▌ The dc load line from Problem 5.26, already constructed on Fig. 5-12(b), is applicable here. The Q point was established at $I_{DQ} = 1.15$ mA in Problem 5.33; the corresponding abscissa is $V_{DSQ} \approx 14.2$ V.

5.36 Replace the JFET of Fig. 5-9 with an *n*-channel enhancement-mode MOSFET characterized by Fig. 5-5. Let $V_{DD} = 8$ V, $V_{GSQ} = 4$ V, $V_{DSQ} = 6$ V, $I_{DQ} = 1$ mA, $R_1 = 5$ MΩ, and $R_2 = 3$ MΩ. Find V_{GG}.

▌ By (1) of Problem 5.23,

$$V_{GG} = R_1 V_{DD}/(R_1 + R_2) = 5 \text{ V}$$

5.37 Determine the necessary value of R_S for the MOSFET amplifier of Problem 5.36.

▌ Application of KVL around the smaller gate-source loop of Fig. 5-10 with $i_G = 0$ leads to

$$R_S = \frac{V_{GG} - V_{GSQ}}{I_{DQ}} = \frac{5 - 4}{1 \times 10^{-3}} = 1 \text{ k}\Omega$$

5.38 Calculate the value of R_D required by the amplifier of Problem 5.36.

▌ Using KVL around the drain-source loop of Fig. 5-10 and solving for R_D yield

$$R_D = \frac{V_{DD} - V_{DSQ} - I_{DQ} R_S}{I_{DQ}} = \frac{8 - 6 - (1 \times 10^{-3})(1 \times 10^3)}{1 \times 10^{-3}} = 1 \text{ k}\Omega$$

5.39 The JFET amplifier of Fig. 5-14 shows a means of self-bias that allows extremely high input impedance even if low values of gate-source bias voltage are required. Find the Thévenin equivalent voltage and resistance for the network to the left of *a, b*.

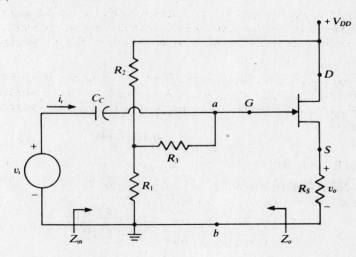

Fig. 5-14

With a, b open there is no voltage drop across R_3, and the voltage at the open-circuited terminals is determined by the R_1-R_2 voltage divider:

$$v_{Th} = V_{GG} = \frac{R_1}{R_1 + R_2} V_{DD}$$

With V_{DD} deactivated (shorted), the resistance to the left of a, b is

$$R_{Th} = R_G = R_3 + \frac{R_1 R_2}{R_1 + R_2}$$

It is apparent that if R_3 is made large, then $R_G = Z_{in}$ is large regardless of the values of R_1 and R_2.

5.40 The manufacturer's specification sheet for a certain kind of n-channel JFET has nominal and worst-case shorted-gate parameters as follows:

value	I_{DSS}, mA	V_{p0}, V
Maximum	7	4.2
Nominal	6	3.6
Minimum	5	3.0

Sketch the nominal and worst-case transfer characteristics that can be expected from a large sample of the device.

▌ Values can be calculated for the nominal, maximum, and minimum transfer characteristics using (1) of Problem 5.6 over the range $-V_{p0} \leq v_{GS} \leq 0$. The results are plotted in Fig. 5-15.

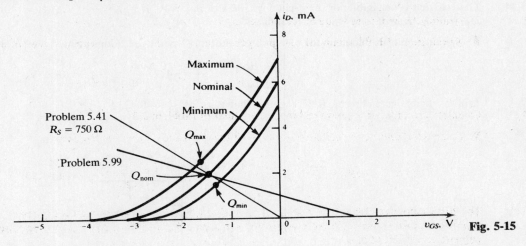

Fig. 5-15

5.41 A self-biased JFET amplifier (Fig. 5-14) is to be designed with $V_{DSQ} = 15$ V and $V_{DD} = 24$ V, using a device as described in Problem 5.40. For the control of gain variation, the quiescent drain current must satisfy $I_{DQ} = 2 \pm 0.4$ mA regardless of the particular parameters of the JFET utilized. Determine appropriate values of R_S and R_D.

▌ Quiescent points are first established on the transfer characteristics of Fig. 5-15: Q_{max} at $I_{DQ} = 2.4$ mA, Q_{nom} at $I_{DQ} = 2.0$ mA, and Q_{min} at $I_{DQ} = 1.6$ mA. A transfer bias line is then constructed to pass through the origin (i.e., we choose $V_{GG} = 0$) and Q_{nom}. Since its slope is $-1/R_S$, the source resistor value may be determined as

$$R_S = \frac{0 - (-3)}{(4 - 0) \times 10^{-3}} = 750 \ \Omega$$

The drain resistor value is found by applying KVL around the drain-source loop and solving for R_D:

$$R_D = \frac{V_{DD} - V_{DSQ} - I_{DQ} R_S}{I_{DQ}} = \frac{24 - 15 - 0.002(750)}{0.002} = 3.75 \ k\Omega$$

When R_S and R_D have these values, the condition on I_{DQ} is satisfied.

5.42 An n-channel JFET has worst-case shorted-gate parameters given by the manufacturer as follows:

value	I_{DSS}, mA	V_{p0}, V
Maximum	8	6
Minimum	4	3

If the JFET is used in the circuit of Fig. 5-10, where $R_S = 0$, $R_G = 1$ MΩ, $R_D = 2.2$ kΩ, $V_{GG} = -1$ V, and $V_{DD} = 15$ V, find the maximum and minimum values of I_{DQ}. Assume (1) of Problem 5.6 describes the JFET.

▌ By (1) of Problem 5.6,

$$I_{DQ\,max} = I_{DSS\,max}\left(1 + \frac{V_{GSQ}}{V_{p0\,max}}\right)^2 = (8 \times 10^{-3})\left(1 + \frac{-1}{6}\right)^2 = 5.55 \text{ mA}$$

$$I_{DQ\,min} = I_{DSS\,min}\left(1 + \frac{V_{GSQ}}{V_{p0\,min}}\right)^2 = (4 \times 10^{-3})\left(1 + \frac{-1}{3}\right)^2 = 1.77 \text{ mA}$$

5.43 Determine the maximum and minimum values of V_{DSQ} that could be expected for the amplifier design of Problem 5.42.

▌ Application of KVL around the drain-source loop gives $V_{DSQ} = V_{DD} - I_{DQ}R_D$. Hence,

$$V_{DSQ\,max} = 15 - (5.55 \times 10^{-3})(2.2 \times 10^3) = 2.78 \text{ V}$$

$$V_{DSQ\,min} = 15 - (1.77 \times 10^{-3})(2.2 \times 10^3) = 11.08 \text{ V}$$

5.44 Gate current is negligible for the p-channel JFET of Fig. 5-16. If $V_{DD} = -20$ V, $I_{DSS} = -10$ mA, $I_{DQ} = -8$ mA, $V_{p0} = -4$ V, $R_S = 0$, and $R_D = 1.5$ kΩ, find V_{GG}.

▌ Solving (1) of Problem 5.6 for v_{GS} and substituting Q-point conditions yield

$$V_{GSQ} = V_{p0}\left[\left(\frac{I_{DQ}}{I_{DSS}}\right)^{1/2} - 1\right] = -4\left[\left(\frac{-8}{-10}\right)^{1/2} - 1\right] = 0.422 \text{ V}$$

With negligible gate current, KVL requires that $V_{GG} = V_{GSQ} = 0.422$ V.

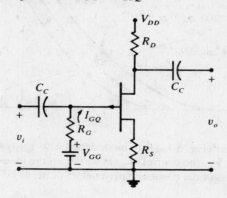

Fig. 5-16

5.45 Determine the value of V_{DSQ} for the amplifier of Problem 5.44.

▌ Applying KVL around the drain-source loop gives

$$V_{DSQ} = V_{DD} - I_{DQ}R_D = (-20) - (-8 \times 10^{-3})(1.5 \times 10^3) = -8 \text{ V}$$

5.46 The n-channel enhancement-mode MOSFET of Fig. 5-17 is characterized by $V_T = 4$ V, $V_{GS\,on} = 6$ V, and $I_{D\,on} = 10$ mA. Assume negligible gate current, $R_1 = 50$ kΩ, $R_2 = 0.4$ MΩ, $R_S = 0$, $R_D = 1$ kΩ, and $V_{DD} = 15$ V. Find V_{GSQ}.

▌ With negligible gate current, (1) of Problem 5.23 leads to

$$V_{GSQ} = V_{GG} = \frac{R_2}{R_2 + R_1}V_{DD} = \frac{50 \times 10^3}{50 \times 10^3 + 0.4 \times 10^6}(15) = 1.67 \text{ V}$$

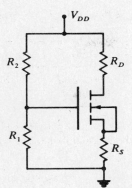

Fig. 5-17

5.47 Calculate the quiescent value of I_{DQ} for the MOSFET amplifier of Problem 5.46.

▌ By (1) of Problem 5.16 and (1) of Problem 5.17,

$$k_0 = \frac{I_{D\,on}}{(V_{GS\,on} - V_T)^2} = \frac{10 \times 10^{-3}}{(6-4)^2} = 2.5 \times 10^{-3}\,\text{A/V}^2$$

$$I_{DQ} = k_0 V_T^2 \left(1 - \frac{V_{GSQ}}{V_T}\right)^2 = (2.5 \times 10^{-3})(4)^2 \left(1 - \frac{1.67}{4}\right)^2 = 13.57\,\text{mA}$$

5.48 Determine V_{DSQ} for the amplifier of Problem 5.46.

▌ By KVL around the drain-source loop,

$$V_{DSQ} = V_{DD} - I_{DQ}R_D = 15 - (13.57 \times 10^{-3})(1 \times 10^3) = 1.43\,\text{V}$$

5.49 For the n-channel enhancement-mode MOSFET of Fig. 5-17, gate current is negligible, $V_{GS\,on} = 8\,\text{V}$, $I_{D\,on} = 10\,\text{mA}$, and $V_T = 4\,\text{V}$. If $R_S = 0$, $R_1 = 50\,\text{k}\Omega$, $V_{DD} = 15\,\text{V}$, $V_{GSQ} = 3\,\text{V}$, and $V_{DSQ} = 9\,\text{V}$, determine the value of R_1.

▌ Since $i_G = 0$, $V_{GSQ} = V_{GG}$ in (1) of Problem 5.23. Solving for R_2 gives

$$R_2 = R_1\left(\frac{V_{DD}}{V_{GSQ}} - 1\right) = 50 \times 10^3 \left(\frac{15}{3} - 1\right) = 200\,\text{k}\Omega$$

5.50 Determine the value of R_D necessary for the amplifier operating point of Problem 5.49.

▌ Using (1) of Problem 5.16 and (1) of Problem 5.17,

$$k_0 = \frac{I_{D\,on}}{(V_{GS\,on} - V_T)^2} = \frac{10 \times 10^{-3}}{(8-4)^2} = 0.625 \times 10^{-3}$$

$$I_{DQ} = k_0 V_T^2 \left(1 - \frac{V_{GSQ}}{V_T}\right)^2 = (0.625 \times 10^{-3})(4)^2 \left(1 - \frac{3}{4}\right)^2 = 0.625\,\text{mA}$$

Then KVL around the drain-source loop requires that

$$R_D = \frac{V_{DD} - V_{DSQ}}{I_{DQ}} = \frac{15 - 9}{0.625 \times 10^{-3}} = 9.6\,\text{k}\Omega$$

5.51 A p-channel MOSFET operating in the enhancement mode is characterized by $V_T = -3\,\text{V}$ and $I_{DQ} = -8\,\text{mA}$ when $V_{GSQ} = -4.5\,\text{V}$. Find V_{GSQ} if $I_{DQ} = -16\,\text{mA}$.

▌ By (1) of Problem 5.16,

$$k_0 = \frac{I_{DQ}}{(V_{GSQ} - V_T)^2} = \frac{-8 \times 10^{-3}}{[-4.5 - (-3)]^2} = -3.555 \times 10^{-3}\,\text{A/V}^2$$

Rearranging (1) of Problem 5.16,

$$V_{GSQ} = (I_{DQ}/k_0)^{1/2} + V_T = \left[\frac{-16 \times 10^{-3}}{-3.555 \times 10^{-3}}\right]^{1/2} + (-3) = -0.88\,\text{V}$$

5.52 For the MOSFET of Problem 5.51, find I_{DQ} if $V_{GSQ} = -5$ V.

▮ Using the result of Problem 5.51 and (1) of Problem 5.16,
$$I_{DQ} = k_0(V_{GSQ} - V_T)^2 = (-3.555 \times 10^{-3})[-5 - (-3)]^2 = -14.22 \text{ mA}$$

5.53 The n-channel JFET circuit of Fig. 5-18 employs one of several methods of self-bias. Assume negligible gate leakage current ($i_G \approx 0$), and show that if $V_{DD} > 0$, then $V_{GSQ} < 0$ and, hence, the device is properly biased.

▮ By KVL,
$$I_{DQ} = \frac{V_{DD} - V_{DSQ}}{R_S + R_D} \tag{1}$$

Now $V_{DSQ} < V_{DD}$, so it is apparent that $I_{DQ} > 0$. Since $i_G \approx 0$, KVL around the gate-source loop gives
$$V_{GSQ} = -I_{DQ}R_S < 0 \tag{2}$$

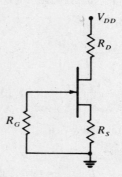

Fig. 5-18

5.54 For the circuit described by Problem 5.53, let $R_D = 3$ kΩ, $R_S = 1$ kΩ, $V_{DD} = 15$ V, and $V_{DSQ} = 7$ V. Find I_{DQ} and V_{GSQ}.

▮ By (1) of Problem 5.53,
$$I_{DQ} = \frac{15 - 7}{3 \times 10^3 + 1 \times 10^3} = 2 \text{ mA}$$

and (2) of Problem 5.53,
$$V_{GSQ} = -(2 \times 10^{-3})(1 \times 10^3) = -2 \text{ V}$$

5.55 The n-channel JFET of Fig. 5-19 is characterized by $I_{DSS} = 5$ mA and $V_{p0} = 3$ V. Let $R_D = 3$ kΩ, $R_S = 8$ kΩ, $V_{DD} = 15$ V, and $V_{SS} = -8$ V. Find V_{GSQ} and V_0 if $V_G = 0$.

▮ Applying KVL around the gate-source loop yields
$$V_G = V_{GSQ} + R_S I_{DQ} + V_{SS} \tag{1}$$

Solving (1) for I_{DQ} and equating the result to the right side of (1) from Problem 5.6 gives
$$\frac{V_G - V_{GSQ} - V_{SS}}{R_S} = I_{DSS}\left(1 + \frac{V_{GSQ}}{V_{p0}}\right)^2 \tag{2}$$

Rearranging (2) leads to the following quadratic in V_{GSQ}:
$$V_{GSQ}^2 + V_{p0}\frac{V_{p0} + 2I_{DSS}R_S}{I_{DSS}R_S}V_{GSQ} + \frac{V_{p0}^2}{I_{DSS}R_S}(I_{DSS}R_S - V_G + V_{SS}) = 0 \tag{3}$$

Substituting known values into (3) and solving for V_{GSQ} with the quadratic formula lead to
$$V_{GSQ}^2 + 3\frac{3 + (2)(5 \times 10^{-3})(8 \times 10^{-3})}{(5 \times 10^{-3})(8 \times 10^3)}V_{GSQ} + \frac{(3)^2}{(5 \times 10^{-3})(8 \times 10^3)}[(5 \times 10^{-3})(8 \times 10^3) - 0 - 8] = 0$$

so that
$$V_{GSQ}^2 + 6.225V_{GSQ} + 7.2 = 0$$

and $V_{GSQ} = -4.69$ V or -1.53 V. Since $V_{GSQ} = -4.69$ V $< -V_{p0}$, this value must be considered extraneous as it will result in $i_D = 0$. Hence, $V_{GSQ} = -1.53$ V. Now, from (1) of Problem 5.6,

$$I_{DQ} = I_{DSS}\left(1 + \frac{V_{GSQ}}{V_{p0}}\right)^2 = (5 \times 10^{-3})\left(1 + \frac{-1.53}{3}\right)^2 = 1.2 \text{ mA}$$

and, by KVL,

$$V_0 = I_{DQ}R_S + V_{SS} = (1.2 \times 10^{-3})(8 \times 10^3) + (-8) = 1.6 \text{ V}$$

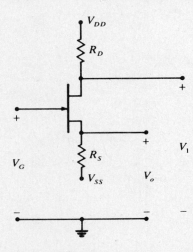

Fig. 5-19

5.56 Rework Problem 5.55 if $V_G = 10$ V and all else remains unchanged.

▌ Substitution of known values into (3) of Problem 5.55 leads to

$$V_{GSQ}^2 + 6.225V_{GSQ} + 4.95 = 0$$

which, after elimination of the extraneous root, results in $V_{GSQ} = -0.936$ V. Then, as in Problem 5.55,

$$I_{DQ} = I_{DSS}\left(1 + \frac{V_{GSQ}}{V_{p0}}\right)^2 = (5 \times 10^{-3})\left(1 + \frac{-0.936}{4}\right)^2 = 2.37 \text{ mA}$$

and

$$V_0 = I_{DQ}R_S + V_{SS} = (2.37 \times 10^{-3})(8 \times 10^3) + (-8) = 10.96 \text{ V}$$

5.57 Find the equivalent of the two identical n-channel JFETs connected in parallel in Fig. 5-20.

▌ Assume the devices are described by (1) of Problem 5.6; then

$$i_D = i_{D1} + i_{D2} = I_{DSS}\left(1 + \frac{v_{GS}}{V_{p0}}\right)^2 + I_{DSS}\left(1 + \frac{v_{GS}}{V_{p0}}\right)^2 = 2I_{DSS}\left(1 + \frac{v_{GS}}{V_{p0}}\right)^2$$

Because the two devices are identical and connected in parallel, the equivalent JFET has the same pinchoff voltage as the individual devices. However, it has a value of shorted-gate current I_{DSS} equal to twice that of the individual devices.

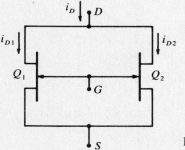

Fig. 5-20

5.58 The differential amplifier of Fig. 5-21 includes identical JFETs with $I_{DSS} = 10$ mA and $V_{p0} = 4$ V. Let $V_{DD} = 15$ V, $V_{SS} = 5$ V, and $R_S = 3\ k\Omega$. If the JFETs are described by (1) of Problem 5.6, find the value of R_D required to bias the amplifier such that $V_{DSQ1} = V_{DSQ2} = 7$ V.

▌ By symmetry, $I_{DQ1} = I_{DQ2}$. KCL at the source node requires that

$$I_{SQ} = I_{DQ1} + I_{DQ2} = 2I_{DQ1} \tag{1}$$

With $i_{G1} = 0$, KVL around the left gate-source loop gives

$$V_{GSQ1} = V_{SS} - I_{SQ}R_S = V_{SS} - 2I_{DQ1}R_S \tag{2}$$

Solving (1) of Problem 5.6 for V_{GSQ} and equating the result to the right side of (2) give

$$V_{p0}\left[\left(\frac{I_{DQ1}}{I_{DSS}}\right)^{1/2} - 1\right] = V_{SS} - 2I_{DQ1}R_S \tag{3}$$

Rearranging (3) results in a quadratic in I_{DQ}:

$$I_{DQ1}^2 - \left[\frac{V_{SS} + V_{p0}}{R_S} + \left(\frac{V_{p0}}{2R_S}\right)^2 \frac{1}{I_{DSS}}\right]I_{DQ1} + \left(\frac{V_{SS} + V_{p0}}{2R_S}\right)^2 = 0 \tag{4}$$

Substituting known values into (4) yields

$$I_{DQ1}^2 - 3.04 \times 10^{-3} I_{DQ1} + 2.25 \times 10^{-6} = 0 \tag{5}$$

Applying the quadratic formula to (5) and disregarding the extraneous root yields $I_{DQ1} = 1.27$ mA. Now the use of KVL around the left drain-source loop gives

$$V_{DD} + V_{SS} - V_{DSQ1} = I_{DQ1}R_D + I_{SQ}R_S \tag{6}$$

Substituting (1) into (6) and solving the result for R_D leads to

$$R_D = \frac{V_{DD} + V_{SS} - V_{DSQ1} - 2I_{DQ1}R_S}{I_{DQ1}} = \frac{15 + 5 - 7 - 2(1.27 \times 10^{-3})(3 \times 10^3)}{1.27 \times 10^{-3}} = 4.20 \text{ k}\Omega$$

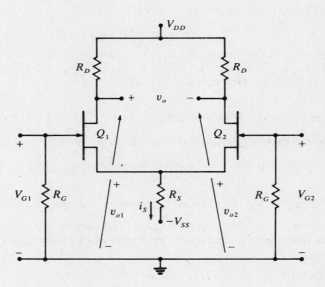

Fig. 5-21

5.59 For the series-connected identical JFETs of Fig. 5-22, $I_{DSS} = 8$ mA and $V_{p0} = 4$ V. If $V_{DD} = 15$ V, $R_D = 5$ kΩ, $R_S = 2$ kΩ, and $R_G = 1$ MΩ, find V_{DSQ1}.

▌ By KVL,

$$V_{GSQ1} = V_{GSQ2} + V_{DSQ1} \tag{1}$$

But, since $I_{DQ1} \equiv I_{DQ2}$, (1) of Problem 5.6 leads to

$$I_{DSS}\left(1 + \frac{V_{GSQ1}}{V_{p0}}\right)^2 = I_{DSS}\left(1 + \frac{V_{GSQ2}}{V_{p0}}\right)^2$$

or

$$V_{GSQ1} = V_{GSQ2} \tag{2}$$

Substitution of (2) into (1) yields $V_{DSQ1} = 0$.

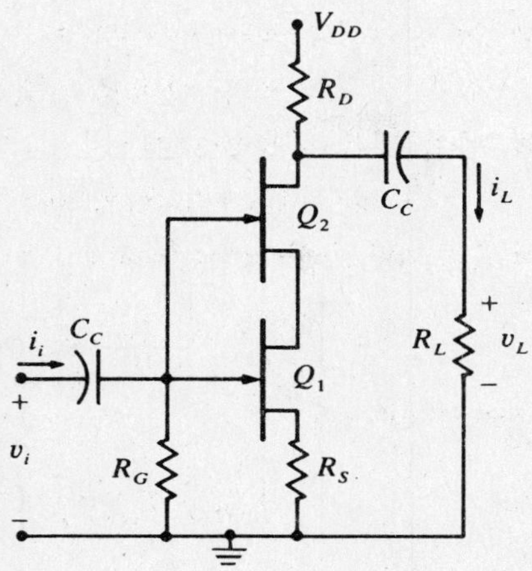

FIELD-EFFECT TRANSISTOR FUNDAMENTALS □ 149

Fig. 5-22

5.60 Determine the value of I_{DQ1} for the series-connected JFETs of Problem 5.59.

▮ With negligible gate current, KVL applied around the lower gate-source loop requires that $V_{GSQ1} = -I_{DQ1}R_S$. Substituting into (1) of Problem 5.6 and rearranging now give a quadratic in I_{DQ1}:

$$I_{DQ1}^2 - \left(\frac{V_{p0}}{R_S}\right)^2 \left(\frac{1}{I_{DSS}} + \frac{2R_S}{V_{p0}}\right) I_{DQ1} + \left(\frac{V_{p0}}{R_S}\right)^2 = 0 \qquad (1)$$

Substitution of known values gives

$$I_{DQ1}^2 - 4.5 \times 10^{-3} I_{DQ1} + 4 \times 10^{-6} = 0$$

from which we obtain $I_{DQ1} = 3.28$ mA and 1.22 mA. The value $I_{DQ1} = 3.28$ mA would result in $V_{GSQ1} < -V_{p0}$, so that the value is extraneous. Hence, $I_{DQ1} = 1.22$ mA.

5.61 Calculate the quiescent source-to-gate voltage of Q_1 for the combined JFETs of Problem 5.59.

▮ Using the result of Problem 5.60,

$$V_{GSQ1} = -I_{DQ1}R_S = -(1.22 \times 10^{-3})(2 \times 10^3) = -2.44 \text{ V}$$

5.62 Evaluate the quiescent source-to-gate voltage of Q_2 for JFET connected and operated as described by Problem 5.59.

▮ Based on (2) of Problem 5.59 and the result of Problem 5.61, we have $V_{GSQ2} = V_{GSQ1} = -2.44$ V.

5.63 Determine the value of V_{DSQ2} for the JFET circuit of Problem 5.59.

▮ By KVL, and using the results of Problems 5.59 and 5.60,

$$V_{DSQ2} = V_{DD} - V_{DSQ1} - I_{DQ1}(R_S + R_D) = 15 - 0 - (1.22 \times 10^{-3})(2 \times 10^3 + 5 \times 10^3) = 6.46 \text{ V}$$

5.64 Identical JFETs characterized by $i_G = 0$, $I_{DSS} = 10$ mA, and $V_{p0} = 4$ V are connected as shown in Fig. 5-23. Let $R_D = 1$ kΩ, $R_S = 2$ kΩ, and $V_{DD} = 15$ V, and find V_{GSQ1}.

▮ With negligible gate current, (1) of Problem 5.6 gives

$$i_{G2} = I_{DQ1} = 0 = I_{DSS}\left(1 + \frac{V_{GSQ1}}{V_{p0}}\right)^2$$

so $$V_{GSQ1} = -V_{p0} = -4 \text{ V}$$

150 □ CHAPTER 5

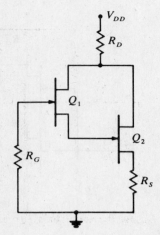

Fig. 5-23

5.65 Determine the quiescent drain current for Q_2 of Problem 5.64.

▮ With negligible gate current, KVL applied around the lower left-hand loop of Fig. 5-23 yields

$$V_{GSQ2} = -V_{GSQ1} - I_{DQ2}R_S \qquad (1)$$

Substituting (1) into (1) of Problem 5.6 and rearranging give

$$I_{DQ2}^2 - \left(\frac{V_{p0}}{R_S}\right)^2 \left[\frac{1}{I_{DSS}} + 2\left(1 - \frac{V_{GSQ1}}{V_{p0}}\right)\frac{R_S}{V_{p0}}\right]I_{DQ2} + \left(\frac{V_{p0} - V_{GSQ1}}{R_S}\right)^2 = 0$$

which becomes, with known values substituted,

$$I_{DQ2}^2 - 8.4 \times 10^{-3} I_{DQ2} + 1.6 \times 10^{-5} = 0$$

The quadratic formula may be used to find the relevant root $I_{DQ2} = 2.92$ mA.

5.66 Find the value of V_{GSQ2} for the amplifier of Problem 5.64.

▮ With negligible gate current and results of Problems 5.64 and 5.65, KVL leads to

$$V_{GSQ2} = -V_{GSQ1} - I_{DQ2}R_S = -(-4) - (2.92 \times 10^{-3})(2 \times 10^3) = -1.84 \text{ V}$$

5.67 Calculate the value of quiescent source-to-drain voltage for the cascaded JFET arrangement of Problem 5.64.

▮ By KVL and results of Problems 5.64 and 5.66,

$$V_{DSQ1} = V_{DD} - (I_{DQ1} + I_{DQ2})R_D - I_{DQ2}R_S - V_{GSQ2}$$
$$= 15 - (0 + 2.92 \times 10^{-3})(1 \times 10^3) - (2.92 \times 10^{-3})(2 \times 10^3) - (-1.84) = 8.08 \text{ V}$$

5.68 What is the value of V_{DSQ2} for the amplifier of Problem 5.64?

▮ By KVL and results of Problems 5.64 and 5.65,

$$V_{DSQ2} = V_{DD} - (I_{DQ1} + I_{DQ2})R_D - I_{DQ2}R_S$$
$$= 15 - (0 + 2.92 \times 10^{-3})(1 \times 10^3) - (2.92 \times 10^{-3})(2 \times 10^3) = 6.24 \text{ V}$$

5.69 Fixed bias can also be utilized for the enhancement-mode MOSFET, as is illustrated by the circuit of Fig. 5-24. The MOSFET is described by the drain characteristic of Fig. 5-13. Let $R_1 = 60$ kΩ, $R_2 = 40$ kΩ, $R_D = 3$ kΩ, $R_L = 1$ kΩ, $V_{DD} = 15$ V, and $C_C \to \infty$. Find V_{GSQ}.

▮ Assume $i_G = 0$. Then, by (1) of Problem 5.23,

$$V_{GSQ} = V_{GG} = \frac{R_2}{R_2 + R_1}V_{DD} = \frac{40 \times 10^3}{40 \times 10^3 + 60 \times 10^3}(15) = 6 \text{ V}$$

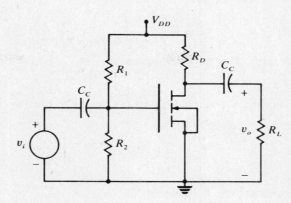

Fig. 5-24

5.70 Graphically determine V_{DSQ} and I_{DQ} for the amplifier of Problem 5.69.

▮ The dc load line is constructed on Fig. 5-13 with v_{DS} intercept $V_{DD} = 15$ V and i_D intercept $V_{DD}/R_L = 5$ mA. The Q-point quantities can be read directly from projections back to the i_D and v_{DS} axes; they are $V_{DSQ} \approx 11.3$ V and $I_{DQ} \approx 1.4$ mA.

5.71 For the enhancement-mode MOSFET amplifier of Problems 5.69 and 5.70, let $v_i = \sin \omega t$ and graphically determine v_o.

▮ We have, first,

$$R_{ac} = R_D \parallel R_L = \frac{(3 \times 10^3)(1 \times 10^3)}{3 \times 10^3 + 1 \times 10^3} = 0.75 \text{ k}\Omega$$

An ac load line must be added to Fig. 5-13; it passes through the Q point and intersects the v_{DS} axis at

$$V_{DSQ} + I_{DQ}R_{ac} = 11.3 + (1.4 \times 10^{-3})(0.75 \times 10^3) = 12.35 \text{ V}$$

Now we construct an auxiliary time axis through the Q point and perpendicular to the ac load line; on it, we construct the waveform $v_{gs} = v_i$ as it swings ± 1 V along the ac load line about the Q point. An additional auxiliary time axis is constructed perpendicular to the v_{DS} axis, to display the output voltage $v_o = v_{ds}$ as v_{gs} swings along the ac load line.

5.72 A common-gate JFET amplifier is shown in Fig. 5-25. The JFET obeys (1) of Problem 5.6. If $I_{DSS} = 10$ mA, $V_{p0} = 4$ V, $V_{DD} = 15$ V, $R_1 = R_2 = 10$ kΩ, $R_D = 500$ Ω, and $R_S = 2$ kΩ, determine V_{GSQ}. Assume $i_G = 0$.

▮ By KVL,

$$V_{GSQ} = \frac{R_2}{R_1 + R_2} V_{DD} - I_{DQ}R_S \tag{1}$$

Solving (1) for I_{DQ} and equating the result to the right side of (1) from Problem 5.6 yield

$$\frac{\frac{R_2}{R_1 + R_2} V_{DD} - V_{GSQ}}{R_S} = I_{DSS}\left(1 + \frac{V_{GSQ}}{V_{p0}}\right)^2 \tag{2}$$

Rearranging leads to a quadratic in V_{GSQ},

$$V_{GSQ}^2 + \left(2V_{p0} + \frac{V_{p0}^2}{I_{DSS}R_S}\right)V_{GSQ} + V_{p0}^2\left[1 - \frac{R_2 V_{DD}}{(R_1 + R_2)I_{DSS}R_S}\right] = 0 \tag{3}$$

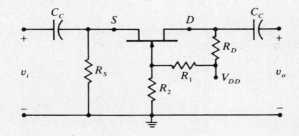

Fig. 5-25

or, with known values substituted,

$$V_{GSQ}^2 + 8.8V_{GSQ} + 10 = 0 \qquad (4)$$

Solving for V_{GSQ} and disregarding the extraneous root $V_{GSQ} = -7.46 < -V_{p0}$, we determine that $V_{GSQ} = -1.34$ V.

5.73 Find the value of quiescent drain current for the CG amplifier of Problem 5.72.

▌ By (1) of Problem 5.6 and result of Problem 5.72,

$$I_{DQ} = I_{DSS}\left(1 + \frac{V_{GSQ}}{V_{p0}}\right)^2 = (10 \times 10^{-3})\left(1 + \frac{-1.34}{4}\right)^2 = 4.42 \text{ mA}$$

5.74 Determine V_{DSQ} for the amplifier of Problem 5.72.

▌ By KVL and result of Problem 5.73,

$$V_{DSQ} = V_{DD} - I_{DQ}(R_S + R_D) = 15 - (4.42 \times 10^{-3})(2 \times 10^3 + 500) = 3.95 \text{ V}$$

5.75 The method of *source bias*, illustrated in Fig. 5-26, can be employed for both JFETs and MOSFETs. For a JFET with characteristics given by Fig. 5-12 and with $R_D = 1$ kΩ, $R_S = 4$ kΩ, and $R_G = 10$ MΩ, determine V_{DD} and V_{SS} so that the amplifier has the same quiescent conditions as the amplifier of Problem 5.26.

▌ Using results of Problem 5.26, assuming $i_G = 0$, and applying KVL around the gate-source loop yield

$$V_{SS} = V_{GSQ} + I_{DQ}R_S = -2 + (1.5 \times 10^{-3})(4 \times 10^3) = 4 \text{ V}$$

Applying KVL around the drain-source loop and using result of Problem 5-28 give

$$V_{DD} = I_{DQ}(R_S + R_D) + V_{DSQ} - V_{SS} = (1.5 \times 10^{-3})(4 \times 10^3 + 1 \times 10^3) + 12.5 - 4 = 16 \text{ V}$$

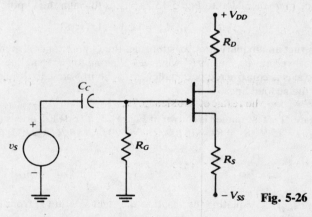

Fig. 5-26

5.76 In the circuit of Fig. 5-27, $R_G \gg R_{S1}, R_{S2}$. The JFET is described by (1) of Problem 5.6, $I_{DSS} = 10$ mA, $V_{p0} = 4$ V, $V_{DD} = 15$ V, $V_{DSQ} = 10$ V, and $V_{GSQ} = -2$ a V. Find R_{S1}.

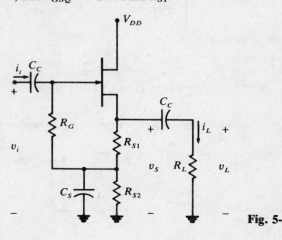

Fig. 5-27

▌ By (1) of Problem 5.26,

$$I_{DQ} = I_{DSS}\left(1 + \frac{V_{GSQ}}{V_{p0}}\right)^2 = (10 \times 10^{-3})\left(1 + \frac{-2}{4}\right)^2 = 2.5 \text{ mA}$$

Apply KVL around the gate-source loop, assuming $i_G = 0$ leads to

$$V_{GSQ} = -I_{DQ}R_{S1} \quad \text{or} \quad R_{S1} = -\frac{V_{GSQ}}{I_{DQ}} = -\frac{-2}{2.5 \times 10^{-3}} = 800 \ \Omega$$

5.77 Determine R_{S2} for the JFET amplifier of Problem 5.76.

▌ Applying KVL around the drain-source loop and using the results of Problem 5.76 yield

$$V_{DSQ} = V_{DD} - I_{DQ}(R_{S1} + R_{S2})$$

Whence,
$$R_{S2} = \frac{V_{DD} - V_{DSQ}}{I_{DQ}} - R_{S1} = \frac{15 - 10}{2.5 \times 10^{-3}} - 800 = 1.2 \text{ k}\Omega$$

5.78 Calculate the value of v_S at quiescent conditions for the amplifier of Problem 5.76.

▌ By KVL around the drain-source loop,

$$v_S = V_{DD} - V_{DSQ} = 15 - 10 = 5 \text{ V}$$

Alternatively, Ohm's law and the results of Problems 5.76 and 5.77 give

$$v_S = I_{DQ}(R_{S1} + R_{S2}) = (2.5 \times 10^{-3})(800 + 1.2 \times 10^3) = 5 \text{ V}$$

5.79 Replace the JFET of Fig. 5-18 with a p-channel device characterized by $V_{p0} = -5$ V and $I_{DSS} = -12$ mA. If $I_{DQ} = -4$ mA and $V_{DSQ} = -6$ V, determine the value of V_{GSQ}.

▌ By (1) of Problem 5.6,

$$V_{GSQ} = V_{p0}[(I_{DQ}/I_{DSS})^{1/2} - 1] = (-5)[(-4 \times 10^{-3}/-12 \times 10^{-3})^{1/2} - 1] = 2.11 \text{ V}$$

5.80 If $V_{DD} = -15$ V, determine the necessary values of R_D and R_S for the circuit of Problem 5.79.

▌ By Ohm's law and the result of Problem 5.79,

$$R_S = -\frac{V_{GSQ}}{I_{DQ}} = -\frac{2.11}{-4 \times 10^{-3}} = 527.5 \ \Omega$$

Applying KVL around the drain-source loop leads to

$$R_D = \frac{V_{DD} - V_{DSQ} - I_{DQ}R_S}{I_{DQ}} = \frac{-15 - (-6) - (-4 \times 10^{-3})(527.5)}{-4 \times 10^{-3}} = 1.723 \text{ k}\Omega$$

5.81 The n-channel JFET of Fig. 5-9 has $V_{p0} = 5$ V and $I_{DSS} = 12$ mA. If $V_{DD} = 15$ V, $R_S = R_D = 2$ kΩ, $R_1 = 90$ kΩ, and $R_2 = 400$ kΩ, determine V_{GSQ}.

▌ By (1) of Problem 5.23,

$$V_{GG} = \frac{R_1}{R_1 + R_2}V_{DD} = \frac{90 \times 10^3}{90 \times 10^3 + 400 \times 10^3}(15) = 2.755 \text{ V}$$

Apply KVL around the gate-source loop and use (1) of Problem 5.6.

$$V_{GG} = V_{GSQ} + I_{DQ}R_S = V_{GSQ} + I_{DSS}\left(1 + \frac{V_{GSQ}}{V_{p0}}\right)^2 R_S$$

Substitute known values and rearrange to find

$$V_{GSQ}^2 + 10.04V_{GSQ} + 24.88 = 0$$

Solving for V_{GSQ} and disregarding the extraneous root $V_{GSQ} = -5.59 < -V_{p0}$, we determine that $V_{GSQ} = -4.45$ V.

5.82 Find the value of I_{DQ} for the JFET circuit of Problem 5.81.

▌ Based on the result of Problem 5.81 and (1) of Problem 5.6,

$$I_{DQ} = I_{DSS}\left(1 + \frac{V_{GSQ}}{V_{p0}}\right)^2 = (12 \times 10^{-3})\left(1 + \frac{-4.45}{5}\right)^2 = 0.145 \text{ mA}$$

5.83 Calculate the value of quiescent source-to-drain voltage for the JFET amplifier of Problem 5.81.

▌ Apply KVL around the drain-source loop and use the result of Problem 5.82.

$$V_{DSQ} = V_{DD} - I_{DQ}(R_D + R_S) = 15 - (0.145 \times 10^{-3})(2 \times 10^3 + 2 \times 10^3) = 14.42 \text{ V}$$

5.84 Would you expect saturation or cutoff to limit the symmetrical signal swing for the amplifier of Problem 5.81?

▌ Based on the result of Problem 5.83, cutoff would limit the symmetrical signal swing to a value of $v_{ds} = \pm(V_{DD} - V_{DSQ}) = \pm(15 - 14.42) = \pm 0.58$ V.

5.85 Since the swing of signal v_{DS} is limited by cutoff in Problem 5.81, it is desired to change R_1 so that $I_{DQ} = 2$ mA. Find the necessary value of R_1.

▌ From (1) of Problem 5.6,

$$V_{GSQ} = V_{p0}[(I_{DQ}/I_{DSS})^{1/2} - 1] = 5[(2 \times 10^{-3}/12 \times 10^{-3})^{1/2} - 1] = -2.96 \text{ V}$$

Apply KVL around the gate-source loop and use (1) of Problem 5.23 to find

$$V_{GSQ} = V_{GG} - I_{DQ}R_S = \frac{R_1}{R_1 + R_2}V_{DD} - I_{DQ}R_S$$

Substitute known numerical values and solve for R_1.

$$-2.96 = \frac{R_1}{R_1 + 400 \times 10^3}(15) - (2 \times 10^{-3})(2 \times 10^3)$$

or

$$R_1 = 27.8 \text{ k}\Omega$$

5.86 For the n-channel MOSFET (NMOS) amplifier of Fig. 5-17, $k_0 = 2 \times 10^{-3}$ A/V^2, $V_T = 4$ V, $V_{DD} = 15$ V, $R_D = R_S = 2$ kΩ, $R_1 = 300$ kΩ, and $R_2 = 200$ kΩ. Find V_{GSQ}.

▌ By (1) of Problem 5.23,

$$V_{GG} = \frac{R_1}{R_1 + R_2}V_{DD} = \frac{300 \times 10^3}{300 \times 10^3 + 200 \times 10^3}(15) = 9 \text{ V}$$

Apply KVL around the gate-source loop and use (1) of Problem 5.16.

$$V_{GSQ} = V_{GG} - I_{DQ}R_S = V_{GG} - k_0(V_{GSQ} - V_T)^2 R_S$$

Substitute known numerical values and rearrange to find

$$V_{GSQ}^2 - 7.75V_{GSQ} + 13.75 = 0$$

Solving for V_{GSQ} and disregarding the extraneous root $V_{GSQ} = 2.75 < V_T$, we determine $V_{GSQ} = 5$ V.

5.87 Determine I_{DQ} and V_{DSQ} for the NMOS amplifier of Problem 5.86.

▌ Based on the result of Problem 5.86 and (1) of Problem 5.16,

$$I_{DQ} = k_0(V_{GSQ} - V_T)^2 = (2 \times 10^{-3})(5 - 4)^2 = 2 \text{ mA}$$

Application of KVL around the drain-source loop yields

$$V_{DSQ} = V_{DD} - I_{DQ}(R_S + R_D) = 15 - (2 \times 10^{-3})(2 \times 10^3 + 2 \times 10^3) = 7 \text{ V}$$

5.88 If the NMOS transistor of Problem 5.86 is replaced with a device differing only in having twice the channel length (L), determine the new value of R_S to maintain $I_{DQ} = 2$ mA.

▌ Based on (1) of Problem 5.19,

$$k_0 = 2 \times 10^{-3}/2 = 1 \times 10^{-3} \text{ A/V}^2$$

FIELD-EFFECT TRANSISTOR FUNDAMENTALS □ 155

By (*1*) of Problem 5.16,

$$V_{GSQ} = (I_{DQ}/k_0)^{1/2} + V_T = (2 \times 10^{-3}/1 \times 10^{-3})^{1/2} + 4 = 5.41 \text{ V}$$

From application of KVL around the gate-source loop,

$$R_S = \frac{V_{GG} - V_{GSQ}}{I_{DQ}} = \frac{9 - 5.41}{2 \times 10^{-3}} = 1.795 \text{ k}\Omega$$

5.89 The NMOS device of Fig. 5-17 is replaced by a PMOS transistor (*p*-channel MOSFET) with $V_T = -3$ V and $k_0 = -0.4 \times 10^{-3}$ A/V^2. If $V_{DD} = -15$ V, $R_1 = 120$ kΩ, and $R_2 = 240$ kΩ, determine R_S so that $V_{GSQ} = -2$ V.

▌ By (*1*) of Problem 5.23,

$$V_{GG} = \frac{R_1}{R_1 + R_2} V_{DD} = \frac{120 \times 10^3}{120 \times 10^3 + 240 \times 10^3}(-15) = -5 \text{ V}$$

From (*1*) of Problem 5.16,

$$I_{DQ} = k_0(V_{GSQ} - V_T)^2 = (-0.4 \times 10^{-3})[-2 - (-3)]^2 = -0.4 \text{ mA}$$

Application of KVL around the gate-source loop gives

$$R_S = \frac{V_{GG} - V_{GSQ}}{I_{DQ}} = \frac{-5 - (-2)}{-0.4 \times 10^{-3}} = 7.5 \text{ k}\Omega$$

5.90 With R_S as determined in Problem 5.89, find the value of R_D necessary to maintain $V_{DSQ} = -7$ V for the circuit.

▌ By application of KVL around the drain-source loop,

$$R_D = \frac{V_{DD} - V_{DSQ} - I_{DQ}R_S}{I_{DQ}} = \frac{-15 - (-5) - (-0.4 \times 10^{-3})(7.5 \times 10^3)}{-0.4 \times 10^{-3}} = 12.5 \text{ k}\Omega$$

Review Problems

5.91 Predict the shorted-gate saturation current for the JFET of Problem 5.9 at an operating temperature of $-50°C$.

▌ Based on the results of Problem 5.9,

$$I_{DSS} = k_I T^{-3/2} = 51.443(273 - 50)^{-3/2} = 15.45 \text{ mA}$$

5.92 If the pinchoff voltage is measured as 4.65 V for the FET of Problem 5.12, determine the temperature of operation.

▌ Based on (*1*) and the manufacturer's specification of Problem 5.12,

$$T_2 = T_1 + \frac{V_{p01} - V_{p02}}{k_V} = 25 + \frac{5 - 4.65}{2.14 \times 10^{-3}} = 188.6°C$$

5.93 Replace Q_1 and Q_2 of Fig. 5-20 with identical MOSFETs and determine an equivalent of the two parallel connected devices.

▌ Assume the MOSFETs are described by (*1*) of Problem 5.17; then,

$$i_D = i_{D1} + i_{D2} = k_0 V_T^2 \left(1 - \frac{v_{GS}}{V_T}\right)^2 + k_0 V_T^2 \left(1 - \frac{v_{GS}}{V_T}\right)^2 = 2k_0 V_T^2 \left(1 - \frac{v_{GS}}{V_T}\right)^2$$

where, from (*1*) of Problem 5.16, $k_0 = I_{D \text{ on}}/(V_{GSQ} - V_T)^2$.

5.94 In the JFET amplifier of Problem 5.26, R_1 is changed to 2 MΩ to increase the input impedance. If R_D, R_S, and V_{DD} are unchanged, what value of R_2 is needed to maintain the original Q point?

▌ The transfer bias line of Fig. 5-12(*a*) must remain unchanged; hence, $V_{GG} = 1.2$ V. Solving (*1*) of Problem 5.23 for R_2 gives

$$R_2 = R_1(V_{DD}/V_{GG} - 1) = (1 \times 10^6)(20/1.2 - 1) = 15.67 \text{ M}\Omega$$

5.95 Find the voltage across R_S in Problem 5.26.

▌ Based on result of Problem 5.26 and Ohm's law with $i_G = 0$,
$$V_{R_s} = I_{DQ}R_S = (1.5 \times 10^{-3})(2 \times 10^3) = 3 \text{ V}$$

5.96 Find the input impedance as seen by source v_i of Problem 5.26 if C_C is large.

▌ Since $i_G = 0$,
$$Z_{\text{in}} = R_G = R_1 \parallel R_2 = \frac{(1 \times 10^6)(15.7 \times 10^6)}{1 \times 10^6 + 15.7 \times 10^6} = 940 \text{ k}\Omega$$

5.97 In the drain-feedback-biased amplifier of Fig. 5-13(a), $V_{DD} = 15$ V, $R_F = 5$ MΩ, $I_{DQ} = 0.75$ mA, and $V_{GSQ} = 4.5$ V. Find V_{DSQ}.

▌ Since $i_G = 0$, the voltage drop across R_F is zero. Hence, $V_{DSQ} = V_{GSQ} = 4.5$ V.

5.98 Determine the necessary value of R_L for the NMOS amplifier of Problem 5.97.

▌ Application of KVL around the drain-source loop yields
$$R_L = \frac{V_{DD} - V_{DSQ}}{I_{DQ}} = \frac{15 - 4.5}{0.75 \times 10^{-3}} = 14 \text{ k}\Omega$$

5.99 A JFET amplifier with the circuit arrangement of Fig. 5-9 is to be manufactured using devices as described in Problem 5.40. For the design, assume a nominal device and use $V_{DD} = 24$ V, $V_{DSQ} = 15$ V, $I_{DQ} = 2$ mA, $R_1 = 2$ MΩ, and $R_2 = 30$ MΩ. Determine the values of R_S and R_D for the amplifier.

▌ From (1) of Problem 5.23,
$$V_{GG} = \frac{R_1}{R_1 + R_2} V_{DD} = \frac{2 \times 10^6}{2 \times 10^6 + 30 \times 10^6}(24) = 1.5 \text{ V}$$

Construct a transfer bias line on Fig. 5-15 through $v_{GS} = V_{GG} = 1.5$ V and Q_{nom} at $I_{DQ} = 2$ mA. Since the transfer bias line has a slope of $-1/R_S$, selection of convenient values gives
$$R_S = \frac{1.5 - (-1.45)}{2 \times 10^{-3} - 0} = 1.475 \text{ k}\Omega$$

Application of KVL around the drain-source loop yields
$$R_D = \frac{V_{DD} - V_{DSQ} - I_{DQ}R_S}{I_{DQ}} = \frac{24 - 15 - (2 \times 10^{-3})(1.475 \times 10^3)}{2 \times 10^{-3}} = 3.03 \text{ k}\Omega$$

5.100 Predict the range of I_{DQ} that can be expected for the JFET amplifier of Problem 5.99.

▌ The range of I_{DQ} can be determined by intersections of maximum and minimum transfer characteristics with the transfer bias line on Fig. 5-15: $I_{DQ \text{ max}} \approx 2.2$ mA; $I_{DQ \text{ min}} \approx 1.8$ mA.

5.101 To see the effect of a source resistor on Q-point conditions, solve Problem 5.44 with $R_S = 500$ Ω and all else unchanged.

▌ V_{GSQ} is unchanged from the value of Problem 5.44. Applying KVL around the gate-source loop,
$$V_{GG} = V_{GSQ} + I_{DQ}R_S = 0.422 + (-8 \times 10^{-3})(500) = -3.58 \text{ V}$$

5.102 Also, rework Problem 5.45 with $R_S = 500$ Ω and all else unchanged.

▌ Applying KVL around the drain-source loop,
$$V_{DSQ} = V_{DD} - I_{DQ}(R_S + R_D) = -20 - (-8 \times 10^{-3})(500 + 1.5 \times 10^3) = -4 \text{ V}$$

5.103 Solve Problem 5.49 with a 200-Ω resistor R_S added to the circuit and all else unchanged.

▌ With $i_G = 0$, $V_{GG} = V_{GSQ} + I_{DQ}R_S$ where $I_{DQ} = 0.625$ mA from Problem 5.50. Solving (1) of Problem 5.23 for R_2,
$$R_2 = R_1\left[\frac{V_{DD}}{V_{GSQ} + I_{DQ}R_S} - 1\right] = (50 \times 10^3)\left[\frac{15}{3 + (0.625 \times 10^{-3})(200)} - 1\right] = 190 \text{ k}\Omega$$

FIELD-EFFECT TRANSISTOR FUNDAMENTALS   157

5.104 Solve Problem 5.50 with a 200-Ω source resistor R_S added to the circuit and all else unchanged.

▌ By application of KVL to the drain-source loop,

$$R_D = \frac{V_{DD} - V_{DSQ}}{I_{DQ}} - R_S = \frac{15 - 9}{0.625 \times 10^{-3}} - 200 = 9.4 \text{ k}\Omega$$

5.105 For the n-channel JFET circuit of Fig. 5-19, $I_{DSS} = 6$ mA, $V_{p0} = 4$ V, $R_D = 5$ kΩ, $R_S = 10$ kΩ, $V_{DD} = 15$ V, and $V_{SS} = 10$ V. The JFET is described by (1) of Problem 5.6. Find the value of V_G that renders $V_o = 0$.

▌ By KVL with $V_o = 0$,

$$I_{DQ} = \frac{V_{SS}}{R_S} = \frac{10}{10 \times 10^3} = 1 \text{ mA}$$

From (1) of Problem 5.6,

$$V_{GSQ} = V_{p0}[(I_{DQ}/I_{DSS})^{1/2} - 1] = 4[(1 \times 10^{-3}/6 \times 10^{-3})^{1/2} - 1] = -2.37 \text{ V}$$

Apply KVL around the gate-source loop.

$$V_G = V_{GSQ} + I_{DQ}R_S + V_{SS} = -2.37 + (1 \times 10^{-3})(10 \times 10^3) + 10 = 17.63 \text{ V}$$

5.106 Determine V_{DSQ} for the JFET circuit of Problem 5.105 with $V_o = 0$.

▌ Applying KVL around the drain-source loop gives

$$V_{DSQ} = V_{DD} - V_o - I_{DQ}R_D = 15 - 0 - (1 \times 10^{-3})(5 \times 10^3) = 10 \text{ V}$$

5.107 In the differential amplifier of Fig. 5-21, the identical JFETs are characterized by $I_{DSS} = 10$ mA, $V_{p0} = 4$ V, and $i_G = 0$. If $V_{DD} = 15$ V, $V_{SS} = 5$ V, $R_S = 3$ kΩ, and $R_D = 5$ kΩ, find I_{DQ1} and V_{DSQ1}.

▌ The value of $I_{DQ1} = 1.27$ mA is independent of R_D and identical to that found in Problem 5.58. Apply KVL around the left drain-source loop and use $I_{SQ} = 2I_{DQ1}$ to find

$$V_{DSQ1} = V_{DD} + V_{SS} - I_{DQ1}(R_D + 2R_S) = 15 + 5 - (1.27 \times 10^{-3})[(5 \times 10^3 + 2(5 \times 10^3)] = 6.03 \text{ V}$$

5.108 Find the voltage $v_{o1} = v_{o2}$ for the differential amplifier of Problem 5.107.

▌ Apply KVL around a left drain-source loop, use the results of Problem 5.107, and realize that $I_{SQ} = 2I_{DQ1}$ to obtain

$$v_{o1} = V_{DSQ1} + I_{SQ}R_S - V_{SS} = 6.03 + 2(1.27 \times 10^{-3})(3 \times 10^3) - 5 = 8.65 \text{ V}$$

5.109 A voltage source is connected to the differential amplifier of Fig. 5-21 such that $V_{G1} = 0.5$ V. Let $V_{DD} = 15$ V, $V_{SS} = 2$ V, $I_{DSS} = 10$ mA, $V_{p0} = 4$ V for the identical JFETs, $R_D = 6$ kΩ, and $R_S = 1$ kΩ. Find v_{o1}.

▌ Apply KVL around the left gate-source loop.

$$V_{G1} = V_{GSQ1} + I_{SQ}R_S - V_{SS} \tag{1}$$

Since $i_{G2} = 0$, $v_{G2} = i_{G2}R_G = 0$. Apply KVL around a loop that includes both gate-source voltages to obtain

$$V_{GSQ2} = V_{GSQ1} - V_{G1} \tag{2}$$

Apply KVL at the node above R_S, use (1) of Problem 5.6, and substitute (2) to find

$$I_{SQ} = I_{DQ1} + I_{DQ2} = I_{DSS}\left(1 + \frac{V_{GSQ1}}{V_{p0}}\right)^2 + I_{DSS}\left(1 + \frac{V_{GSQ1} - V_{G1}}{V_{p0}}\right)^2 \tag{3}$$

Substitute (3) into (1) to obtain the following quadratic in V_{GSQ1}:

$$V_{GSQ1}^2 + \left(2V_{p0} - V_{G1} + \frac{V_{p0}^2}{2I_{DSS}R_S}\right)V_{GSQ1} + \frac{1}{2}\left[\left(1 - \frac{V_{G1} + V_{SS}}{I_{DSS}R_S}\right)V_{p0}^2 + (V_{p0} - V_{G1})^2\right] = 0 \tag{4}$$

Place the known values into (4) to find

$$V_{GSQ1}^2 + 8.3V_{GSQ1} + 13.325 = 0$$

Solving for V_{GSQ1} and disregarding the extraneous root $V_{GSQ1} = -6.124 < -V_{p0}$, we find $V_{GSQ1} = -2.176$ V.

By (3),

$$I_{DQ1} = I_{DSS}\left(1 + \frac{V_{GSQ1}}{V_{p0}}\right)^2 = (10 \times 10^{-3})\left(1 + \frac{-2.176}{4}\right)^2 = 2.079 \text{ mA}$$

$$I_{DQ2} = I_{DSS}\left(1 + \frac{V_{GSQ1} - V_{G1}}{V_{p0}}\right)^2 = (10 \times 10^{-3})\left(1 + \frac{-2.176 - 0.5}{4}\right)^2 = 1.096 \text{ mA}$$

$$I_{SQ} = I_{DQ1} + I_{DQ2} = 2.079 \times 10^{-3} + 1.096 \times 10^{-3} = 3.175 \text{ mA}$$

Apply KVL around the left drain-source loop and use known values.

$$V_{DSQ1} = V_{DD} + V_{SS} - I_{DQ1}R_D - I_{SQ}R_S$$

$$V_{DSQ1} = 15 + 2 - (2.079 \times 10^{-3})(6 \times 10^3) - (3.175 \times 10^{-3})(1 \times 10^3) = 1.351 \text{ V}$$

and

$$v_{o1} = V_{DSQ1} + I_{SQ}R_S - V_{SS} = 1.351 + (3.175 \times 10^{-3})(1 \times 10^3) - 2 = 2.53 \text{ V}$$

5.110 Find the value of v_{o2} for the differential amplifier of Problem 5.109.

▎ Applying KVL around the right drain-source loop and using results of Problem 5.109,

$$V_{DSQ2} = V_{DD} + V_{SS} - I_{DQ2}R_D - I_{SQ}R_S$$

$$V_{DSQ2} = 15 + 2 - (1.096 \times 10^{-3})(6 \times 10^3) - (3.175 \times 10^{-3})(1 \times 10^3) = 7.249 \text{ V}$$

and

$$v_{o2} = V_{DSQ2} + I_{SQ}R_S - V_{SS} = 7.249 + (3.175 \times 10^{-3})(1 \times 10^3) - 2 = 8.42 \text{ V}$$

5.111 For the series-connected, nonidentical JFETs of Fig. 5-22, $i_{G1} = i_{G2} = 0$, $I_{DSS1} = 8$ mA, $I_{DSS2} = 10$ mA, and $V_{p01} = V_{p02} = 4$ V. Let $V_{DD} = 15$ V, $R_G = 1$ MΩ, $R_D = 5$ kΩ, and $R_S = 2$ kΩ. Find I_{DQ1} and V_{GSQ1}.

▎ Since parameters of JFET Q_1 are unchanged from those of Problems 5.60 and 5.61, the quiescent quantities remain $I_{DQ1} = 1.22$ mA and $V_{GSQ1} = -2.44$ V.

5.112 Determine the value of V_{GSQ2} for the JFET amplifier of Problem 5.111.

▎ Due to series connection of JFETs, $I_{DQ1} = I_{DQ2}$. Hence, by (1) of Problem 5.6,

$$I_{DSS1}\left(1 + \frac{V_{GSQ1}}{V_{p01}}\right)^2 = I_{DSS2}\left(1 + \frac{V_{GSQ2}}{V_{p02}}\right)^2$$

Solving for V_{GSQ2} and using known values give

$$V_{GSQ2} = V_{p02}\left[\left(\frac{I_{DSS1}}{I_{DSS2}}\right)^{1/2}\left(1 + \frac{V_{GSQ1}}{V_{p01}}\right) - 1\right] = 4\left[\left(\frac{8 \times 10^{-3}}{10 \times 10^{-3}}\right)^{1/2}\left(1 + \frac{-2.44}{4}\right) - 1\right] = -2.605 \text{ V}$$

5.113 Find V_{DSQ1} for the JFET circuit of Problem 5.111.

▎ Applying KVL around a path that includes the drain-source of Q_1 and both gate-source voltages, and using the results of Problems 5.111 and 5.112 give

$$V_{DSQ1} = V_{GSQ1} - V_{GSQ2}$$
$$= -2.44 - (-2.605) = 0.165 \text{ V} \tag{1}$$

5.114 Calculate the value of V_{DSQ2} for the circuit of Problem 5.111.

▎ By KVL applied around a path that includes both drain-source voltages, and using the results of Problems 5.111 and 5.113 yield

$$V_{DSQ2} = V_{DD} - V_{DSQ1} - I_{DQ}(R_D + R_S)$$
$$= 15 - 0.165 - (1.22 \times 10^{-3})(5 \times 10^3 + 2 \times 10^3) = 6.295 \text{ V} \tag{1}$$

5.115 The series-connected, identical JFETs of Fig. 5-22 are characterized by $I_{DSS} = 8$ mA, $V_{p0} = 4$ V, and $i_G = 0.5$ μA. If $V_{DD} = 15$ V, $R_D = 5$ kΩ, $R_S = 2$ kΩ, and $R_G = 1$ MΩ, find v_{GSQ_1}.

▎ $I_{DQ1} = I_{DQ2} = 1.22$ mA as determined in Problem 5.60. Since both gate currents flow through R_G, KVL around lower gate-source loop gives

$$V_{GSQ1} = -I_{DQ1}R_S - (I_{GQ1} + I_{GQ2})R_G = -(1.22 \times 10^{-3})(2 \times 10^3) - 2(1 \times 10^{-6})(1 \times 10^6) = -3.44 \text{ V}$$

5.116 Determine V_{GSQ2} for amplifier of Problem 5.115.

▌ By (2) of Problem 5.59 and the result of Problem 5.115,
$$V_{GSQ2} = V_{GSQ1} = -3.44 \text{ V}$$

5.117 Find the value of V_{DSQ1} for the JFET circuit of Problem 5.115.

▌ Based on (1) of Problem 5.113 and the results of Problem 5.116,
$$V_{DSQ1} = V_{GSQ1} - V_{DSQ2} = -3.44 - (-3.44) = 0$$

5.118 Calculate the value of V_{DSQ2} for the series-connected JFETs of Problem 5.115.

▌ Based on (1) of Problem 5.114 and the results of Problems 5.115 and 5.117,
$$V_{DSQ2} = V_{DD} - V_{DSQ1} - I_{DQ}(R_D + R_S) = 15 - 0 - (1.22 \times 10^{-3})(5 \times 10^{-3} + 2 \times 10^{-3}) = 6.46 \text{ V}$$

5.119 In the circuit of Fig. 5-23, the identical JFETs are described by $I_{DSS} = 8$ mA, $V_{p0} = 4$ V, and $i_{G1} = i_{G2} = 0.1$ μA. If $R_D = 1$ kΩ, $R_S = 2$ kΩ, $R_G = 1$ MΩ, and $V_{DD} = 15$ V, find V_{GSQ1}.

▌ Since $I_{DQ1} = i_{G2} = 0.1$ μA, (1) of Problem 5.6 requires that
$$V_{GSQ1} = V_{p0}[(I_{GQ2}/I_{DSS})^{1/2} - 1] = 4[(0.1 \times 10^{-6}/8 \times 10^{-3})^{1/2} - 1] = -3.986 \text{ V}$$

5.120 Determine V_{GSQ2} for the cascaded JFET amplifier of Problem 5.119.

▌ Apply KVL around a loop that includes both gate-source voltages to find
$$V_{GSQ1} + V_{GSQ2} + I_{GQ1}R_G + I_{DQ2}R_S = 0 \tag{1}$$

Solving (1) for I_{DQ2} and using (1) of Problem 5.6 leads to
$$\frac{-V_{GSQ1} - V_{GSQ2} - I_{GQ1}R_G}{R_S} = I_{DSS}\left(1 + \frac{V_{GSQ2}}{V_{p0}}\right)^2 \tag{2}$$

Rearranging (2) and substituting known values give a quadratic in V_{GSQ2}.
$$V_{GSQ2}^2 + 9V_{GSQ2} + 12.114 = 0 \tag{3}$$

Solving (3) for V_{GSQ2} and discarding extraneous root $V_{GSQ2} = -7.35 < -V_{p0}$, we determine $V_{GSQ2} = -1.65$ V.

5.121 Find the value of I_{DQ2} for the circuit of Problem 5.119.

▌ From the result of Problem 5.120 and (1) of Problem 5.6,
$$I_{DQ2} = I_{DSS}\left(1 + \frac{V_{GSQ2}}{V_{p0}}\right)^2 = (8 \times 10^{-3})\left(1 + \frac{-1.65}{4}\right)^2 = 2.76 \text{ mA}$$

5.122 Evaluate the quiescent source-to-drain voltage of Q_2 for the cascaded JFETs of Problem 5.119.

▌ Applying KVL around the drain-source loop of Q_2, realizing that $I_{SQ2} = I_{GQ2} + I_{DQ2} = I_{DQ1} + I_{DQ2}$, and using the result of Problem 5.121 yield
$$V_{DSQ2} = V_{DD} - (I_{DQ1} + I_{DQ2})(R_D + R_S) = 15 - (0.1 \times 10^{-6} + 2.76 \times 10^{-3})(1 \times 10^3 + 2 \times 10^3) = 6.72 \text{ V}$$

5.123 Calculate the value of V_{DSQ1} that exists for the JFET amplifier of Problem 5.119.

▌ Apply KVL over a path around the terminals of the two JFETs and use the result of Problems 5.120 and 5.122 to produce
$$V_{DSQ1} = V_{DSQ2} - V_{GSQ2} = 6.72 - (-1.65) = 8.37 \text{ V}$$

5.124 For the enhancement-mode MOSFET of Problem 5.69, determine the value of k_0.

▌ From Fig. 5-13, $I_{D\,on} = 5$ mA, $V_T = 4$ V, and $V_{GS\,on} = 8$ V. By (1) of Problem 5.16,
$$k_0 = \frac{I_{D\,on}}{(V_{G\,on} - V_T)^2} = \frac{5 \times 10^{-3}}{(8-4)^2} = 0.3125 \text{ A/V}^2$$

5.125 The drain characteristic of Fig. 5-7 describes the MOSFET of Fig. 5-17. Let $V_{DD} = 15$ V, $R_D = 1$ kΩ, $R_S = 500$ Ω, and $R_1 = 10$ kΩ. Find R_2 such that the MOSFET is biased for depletion-mode operation.

▌ The dc load line is constructed on Fig. 5-8 with v_{DS}-axis intercept at $v_{DD} = 15$ V and i_D-axis intercept at $V_{DD}/(R_D + R_S) = 15/1500 = 10$ mA. Depletion-mode operation occurs for $v_{GS} < 0$. From the intersection of the dc load line with the $v_{GS} = 0$ curve, it is seen that any value of $V_{DSQ} > 9.2$ V and $I_{DQ} < 3.7$ mA leads to depletion-mode operation. Finding a Thévenin equivalent by (1) of Problem 5.23 and applying KVL around the gate-source loop give

$$V_{GSQ} = \frac{R_1}{R_1 + R_2} V_{DD} - I_{DQ} R_S \tag{1}$$

Setting $V_{GSQ} = 0$ and solving for R_2 of (1) leads to

$$R_2 = \left[\frac{V_{DD}}{I_{DQ} R_S} - 1 \right] R_1 = \left[\frac{15}{(3.7 \times 10^{-3})(500)} - 1 \right] (10 \times 10^3) = 71.08 \text{ k}\Omega$$

Hence, selection of $R_2 > 71.08$ kΩ results in $V_{GSQ} < 0$ and depletion-mode operation occurs.

5.126 Select R_2 for the amplifier of Problem 5.125 such that the MOSFET is biased for enhancement-mode operation.

▌ For enhancement-mode operation, $v_G > 0$. Based on work of Problem 5.125, it is seen that if $R_2 < 71.08$ kΩ, $V_{GSQ} < 0$, and enhancement-mode operation results.

5.127 The common-gate JFET amplifier of Problem 5.72 is not biased for maximum symmetrical swing. Shift the bias point by letting $R_1 = 10$ kΩ and $R_2 = 5$ kΩ while all else is unchanged. Does the amplifier bias point move closer to the condition of maximum symmetrical swing?

▌ Substituting known values into (3) of Problem 5.72, the resulting quadratic in V_{GSQ} is

$$V_{GSQ}^2 + 8.8 V_{GSQ} + 12 = 0$$

Whence, disregarding the extraneous root $V_{GSQ} = -7.11 < -V_{p0}$, we determine $V_{GSQ} = -1.68$ V.
From (1) of Problem 5.6,

$$I_{DQ} = I_{DSS} \left(1 + \frac{V_{GSQ}}{V_{p0}} \right)^2 = (10 \times 10^{-3}) \left(1 + \frac{-1.68}{4} \right)^2 = 3.364 \text{ mA}$$

Apply KVL around the drain-source loop to find

$$V_{DSQ} = V_{DD} - I_{DQ}(R_S + R_D) = 15 - (3.364 \times 10^{-3})(2 \times 10^3 + 500) = 6.59 \text{ V}$$

Since $V_{DSQ} = 6.59$ is nearer to $V_{DD}/2 = 15/2 = 7.5$ than is the value of $V_{DSQ} = 3.95$ V as determined in Problem 5.74, the amplifier now operates closer to the point of maximum symmetrical swing.

CHAPTER 6
Transistor Bias Considerations

β-UNCERTAINTY AND TEMPERATURE EFFECTS

6.1 The *constant-base-current bias* arrangement of Fig. 4-10 has the advantage of high current gain; however, the sensitivity of its Q point to changes in β limits its usage. If the Si transistor is biased for constant base current, neglect leakage current I_{CBO}, and let $V_{CC} = 15$ V, $R_B = 500$ kΩ, and $R_C = 5$ kΩ. Find I_{CQ} and V_{CEQ} for $\beta = 50$.

By KVL,
$$V_{CC} = V_{BEQ} + I_{BQ}R_B \tag{1}$$

Since $I_{BQ} = I_{CQ}/\beta$, we may write, using (1),
$$I_{CQ} = \beta I_{BQ} = \frac{\beta(V_{CC} - V_{BEQ})}{R_B} = \frac{50(15 - 0.7)}{500 \times 10^3} = 1.43 \text{ mA} \tag{2}$$

so that, by KVL,
$$V_{CEQ} = V_{CC} - I_{CQ}R_C = 15 - 1.43(5) = 7.85 \text{ V} \tag{3}$$

6.2 Repeat Problem 6.1 if $\beta = 100$ and all else is unchanged. Compare the results with those of Problem 6.1.

With β changed to 100, (2) of Problem 6.1 gives
$$I_{CQ} = \frac{100(15 - 0.7)}{500 \times 10^3} = 2.86 \text{ mA}$$

and from (3) of Problem 6.1,
$$V_{CEQ} = 15 - 2.86(5) = 0.7 \text{ V}$$

Comparing Problems 6.1 and 6.2, the collector current I_{CQ} doubled with the doubling of β, and the Q point moved from near the middle of the dc load line to near the saturation region.

6.3 Show that, in the circuit of Fig. 4-10, I_{CQ} varies linearly with β even if leakage current is not neglected, provided $\beta \gg 1$.

Using the result of Problem 4.127 and KVL, we have
$$I_{BQ}R_B = \frac{I_{CQ} - (\beta + 1)I_{CBO}}{\beta} R_B = V_{CC} - V_{BEQ}$$

Rearranging and assuming $\beta \gg 1$ lead to the desired result:
$$I_{CQ} = \frac{\beta(V_{CC} - V_{BEQ})}{R_B} + \frac{\beta + 1}{\beta} I_{CBO} \approx \frac{\beta(V_{CC} - V_{BEQ})}{R_B} + I_{CBO}$$

6.4 Use the *constant-emitter-current bias* circuit of Fig. 6-1 to show that (2) of Problem 4.40 is the condition for β-independent bias even when leakage current is not neglected.

By KVL,
$$V_{BB} = I_{BQ}R_B + V_{BEQ} + I_{EQ}R_E \tag{1}$$

Using the results of Problems 4.129 and 4.127, and assuming that $\beta \gg 1$, we may write
$$I_{EQ} = \frac{\beta + 1}{\beta}(I_{CQ} - I_{CBO}) \approx I_{CQ} - I_{CBO} \tag{2}$$

and
$$I_{BQ} = \frac{I_{CQ}}{\beta} - \frac{\beta + 1}{\beta} I_{CBO} \approx \frac{I_{CQ}}{\beta} - I_{CBO} \tag{3}$$

Substituting (2) and (3) into (1) and rearranging then give

$$I_{CQ} = \frac{V_{BB} - V_{BEQ} + I_{CBO}(R_B + R_E)}{R_B/\beta + R_E} \quad (4)$$

From (4) it is apparent that leakage current I_{CBO} increases I_{CQ}. However, I_{CQ} is relatively independent of β only when $R_B/\beta \ll R_E$.

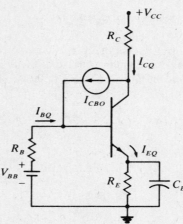

Fig. 6-1

6.5 In the *shunt-feedback bias* circuit of Fig. 4-19, $V_{CC} = 15$ V, $R_C = 2$ kΩ, $R_F = 150$ kΩ, and $I_{CBO} \approx 0$. The transistor is a Si device. Find I_{CQ} and V_{CEQ} if $\beta = 50$.

▌ By KVL,

$$V_{CC} = (I_{CQ} + I_{BQ})R_C + I_{BQ}R_F + V_{BEQ} = \left(I_{CQ} + \frac{I_{CQ}}{\beta}\right)R_C + \frac{I_{CQ}}{\beta}R_F + V_{BEQ}$$

so that

$$I_{CQ} = \frac{\beta(V_{CC} - V_{BEQ})}{R_F + (\beta+1)R_C} = \frac{50(15 - 0.7)}{150 \times 10^3 + 51(2 \times 10^3)} = 2.84 \text{ mA}$$

Now KVL gives

$$V_{CEQ} = V_{CC} - (I_{BQ} + I_{CQ})R_C = V_{CC} - [(1/\beta) + 1]I_{CQ}R_C = 15 - [(1/50) + 1](2.84 \times 10^{-3})(2 \times 10^3) = 9.21 \text{ V}$$

6.6 Rework Problem 6.5 with $\beta = 100$ and all else unchanged. Compare the shunt-feedback bias with the constant-base-current bias and the constant-emitter-current bias.

▌ For $\beta = 100$,

$$I_{CQ} = \frac{100(15 - 0.7)}{150 \times 10^3 + 101(2 \times 10^3)} = 4.06 \text{ mA}$$

and

$$V_{CEQ} = 15 - [(1/100) + 1](4.06 \times 10^{-3})(2 \times 10^3) = 6.80 \text{ V}$$

With shunt-feedback bias the increase in I_{CQ} is appreciable (here, 43 percent); this case lies between the β-insensitive case of constant-emitter-current bias and the directly sensitive case of constant-base-current bias.

6.7 Neglecting leakage current in the shunt-feedback-bias amplifier of Fig. 4-19, find a set of conditions that will render the collector current I_{CQ} insensitive to small variations in β. Is the condition practical?

▌ From Problem 6.5, if $\beta \gg 1$,

$$I_{CQ} = \frac{V_{CC} - V_{BEQ}}{\frac{R_F}{\beta} + \frac{\beta+1}{\beta}R_c} \approx \frac{V_{CC} - V_{BEQ}}{\frac{R_F}{\beta} + R_C}$$

The circuit would be insensitive to β variations if $R_F/\beta \ll R_C$. However, since $0.3 \leq V_{BEQ} \leq 0.7$, that would lead to $I_{CQ}R_C \to V_{CC}$; hence, V_{CEQ} would come close to 0 and the transistor would operate near the saturation region.

6.8 In the circuit of Fig. 4-9, a transistor that has $\beta = \beta_1$ is replaced with a transistor that has $\beta = \beta_2$. Find an expression for the percentage change in collector current. Neglect leakage current.

■ By KVL,
$$V_{CC} = I_{BQ}R_B + V_{BEQ} + I_{EQ}R_E \tag{1}$$

Using (1) of Problem 4.9 and (2) of Problem 4.7 in (1) and rearranging lead to
$$V_{CC} - V_{BEQ} = (R_B + R_E)\frac{I_{CQ}}{\beta} + R_E I_{CQ} \tag{2}$$

This equation may be written for the original transistor (with $\beta = \beta_1$ and $I_{CQ} = I_{CQ1}$) and for the replacement transistor (with β_2 and I_{CQ2}). Subtracting the former from the latter then gives
$$0 = (R_B + R_E)\left(\frac{I_{CQ2}}{\beta_2} - \frac{I_{CQ1}}{\beta_1}\right) + R_E(I_{CQ2} - I_{CQ1}) \tag{3}$$

If we define $I_{CQ2} = I_{CQ1} + \Delta I_{CQ}$, then (3) can be rewritten as
$$0 = (R_B + R_E)\frac{\beta_1(I_{CQ1} + \Delta I_{CQ}) - \beta_2 I_{CQ1}}{\beta_1\beta_2} + R_E \Delta I_{CQ}$$

which, when rearranged, gives the desired ratio:
$$\frac{\Delta I_{CQ}}{I_{CQ1}} = \frac{(\beta_2 - \beta_1)(R_B + R_E)}{\beta_1[R_B + (\beta_2 + 1)R_E]}(100\%) \tag{4}$$

6.9 Will the collector current increase or decrease in magnitude if $\beta_2 > \beta_1$ for the amplifier of Problem 6.8?

■ By inspection of (4) of Problem 6.8, it is apparent that ΔI_{CQ} is positive for an increase in β ($\beta_2 > \beta_1$).

6.10 The transistor in the circuit of Fig. 4-11 is a Si device with $I_{CEO} \approx 0$. Let $V_{CC} = 18$ V, $V_{EE} = 4$ V, $R_E = 2$ kΩ, $R_C = 6$ kΩ, and $R_B = 25$ kΩ. Find I_{CQ} and V_{CEQ} for $\beta = 50$.

■ By KVL around base-emitter loop,
$$V_{EE} - V_{BEQ} = I_{BQ}R_B + I_{EQ}R_E \tag{1}$$

We let $I_{BQ} = I_{CQ}/\beta$ and $I_{EQ} = I_{CQ}(\beta + 1)/\beta$ in (1) and rearrange to obtain
$$I_{CQ} = \frac{V_{EE} - V_{BEQ}}{\dfrac{R_B}{\beta} + \dfrac{\beta+1}{\beta}R_E} = \frac{4 - 0.7}{\dfrac{25 \times 10^3}{50} + \dfrac{51}{50}(2 \times 10^3)} = 1.3 \text{ mA} \tag{2}$$

Then KVL around the collector loop with $I_{EQ} = I_{CQ}(\beta + 1)/\beta$ yields
$$V_{CEQ} = V_{CC} + V_{EE} - \left(R_C + \frac{\beta+1}{\beta}R_E\right)I_{CQ} = 18 + 4 - \left(6 + \frac{51}{50}(2)\right)(1.3) = 11.55 \text{ V} \tag{3}$$

6.11 Rework Problem 6.10 if $\beta = 100$ and all else is unchanged. Comment on the β-sensitivity of this circuit.

■ By (2) and (3) of Problem 6.10,
$$I_{CQ} = \frac{4 - 0.7}{(25 \times 10^3)/100 + (101/100)(2 \times 10^3)} = 1.45 \text{ mA}$$
$$V_{CEQ} = 18 + 4 - [6 + (101/100)(2)](1.45) = 10.37 \text{ V}$$

When β is doubled, I_{CQ} only increased by 11.5%. Thus, the circuit is reasonably insensitive to changes in β.

6.12 In the circuit of Fig. 4-11, under what condition will the bias current I_{CQ} be practically independent of β if $I_{CEO} \approx 0$?

■ With $\beta \gg 1$, the expression for I_{CQ} from Problem 6.10 gives
$$I_{CQ} = \frac{V_{EE} - V_{BEQ}}{\dfrac{R_B}{\beta} + \dfrac{\beta+1}{\beta}R_E} \approx \frac{V_{EE} - V_{BEQ}}{\dfrac{R_B}{\beta} + R_E}$$

It is apparent that I_{CQ} is practically independent of β if $R_B/\beta \ll R_E$. The inequality is generally considered to be satisfied if $R_B \leq \beta R_E/10$.

6.13 In the circuit of Fig. 4-21, the Si transistor has negligible leakage current, $V_{CC} = 15$ V, $V_{EE} = 5$ V, $R_E = 3$ kΩ, and $R_C = 7$ kΩ. Find I_{CQ}, I_{BQ}, and V_{CEQ} if $\beta = 50$.

▌ KVL around the base loop yields

$$I_{EQ} = \frac{V_{EE} - V_{BEQ}}{R_E} = \frac{4 - 0.7}{3 \times 10^3} = 1.1 \text{ mA}$$

Now,
$$I_{CQ} = \frac{\beta}{\beta + 1} I_{EQ} = \frac{50}{51}(1.1) = 1.078 \text{ mA}$$

and
$$I_{BQ} = \frac{I_{CQ}}{\beta} = \frac{1.078 \times 10^{-3}}{50} = 21.56 \text{ μA}$$

and KVL around the collector loop gives

$$V_{CEQ} = V_{CC} + V_{EE} - I_{EQ}R_E - I_{CQ}R_C = 15 + 5 - 1.1(3) - 1.078(7) = 9.154 \text{ V}$$

6.14 Rework Problem 6.13 if $\beta = 100$ and all else is unchanged.

▌ For $\beta = 100$, I_{EQ} is unchanged. However,

$$I_{CQ} = \frac{100}{101}1.1 = 1.089 \text{ mA} \qquad I_{BQ} = \frac{1.089 \times 10^{-3}}{100} = 10.89 \text{ μA}$$

and
$$V_{CEQ} = 15 + 5 - 1.1(3) - 1.089(7) = 9.077 \text{ V}$$

It is seen that the circuit is nearly independent of β.

6.15 Leakage current approximately doubles for every 10°C increase in the temperature of a Si transistor. If a Si transistor has $I_{CBO} = 500$ nA at 25°C, find its leakage current at 90°C.

▌ $I_{CBO} = (500 \times 10^{-9})2^{(90-25)/10} = (500 \times 10^{-9})(90.51) = 45.25 \text{ μA}$

6.16 Sketch a set of common-emitter output characteristics for each of two different temperatures, indicating which set is for the higher temperature.

▌ The CE collector characteristics of Fig. 4-4(c) are obtained as sets of points (i_C, v_{CE}) from the ammeter and voltmeter readings of Fig. 4-4(a). For each fixed value of i_B, $i_C = \beta i_B + (\beta + 1)I_{CBO}$ must increase with temperature, since I_{CBO} increases with temperature (Problem 6.15) and β is much less temperature sensitive than I_{CBO}. The resultant shift in the collector characteristics is shown in Fig. 6-2.

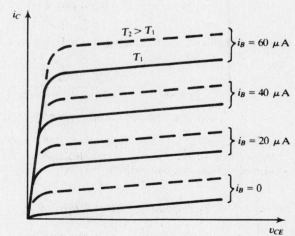

Fig. 6-2

6.17 Temperature variations can shift the quiescent point by affecting leakage current and base-to-emitter voltage. In the circuit of Fig. 6-1, $V_{BB} = 6$ V, $R_B = 50$ kΩ, $R_E = 1$ kΩ, $R_C = 3$ kΩ, $\beta = 75$, $V_{CC} = 15$ V, and the transistor is a Si device. Initially, $I_{CBO} = 0.5$ μA and $V_{BEQ} = 0.7$ V, but the temperature of the device increases by 20°C. Find the exact change in I_{CQ}.

▌ Let the subscript 1 denote quantities at the original temperature T_1, and 2 denote quantities at $T_1 + 20°C = T_2$. By (4) of Problem 6.4,

$$I_{CQ1} = \frac{V_{BB} - V_{BEQ1} + I_{CBO1}(R_B + R_E)}{R_B/\beta + R_E} = \frac{6 - 0.7 + (0.5 \times 10^{-6})(51 \times 10^3)}{50 \times 10^3/75 + 1 \times 10^3} = 3.1953 \text{ mA}$$

Now, according to Problem 6.15 and realizing that base-emitter voltage decreases approximately 2 mV for each 1°C temperature increase in a Si device,

$$I_{CBO2} = I_{CBO1} 2^{\Delta T/10} = 0.5 \times 10^{-6} 2^{20/10} = 2 \text{ μA}$$

$$\Delta V_{BEQ} = -2 \times 10^{-3} \Delta T = (-2 \times 10^{-3})(20) = -0.04 \text{ V}$$

so
$$V_{BEQ2} = V_{BEQ1} + \Delta V_{BEQ} = 0.7 - 0.04 = 0.66 \text{ V}$$

Again, by (4) of Problem 6.4,

$$I_{CQ2} = \frac{V_{BB} - V_{BEQ2} + I_{CBO2}(R_B + R_E)}{R_B/\beta + R_E} = \frac{6 - 0.66 + (2 \times 10^{-6})(51 \times 10^3)}{50 \times 10^3/75 + 1 \times 10^3} = 3.2652 \text{ mA}$$

Thus,
$$\Delta I_{CQ} = I_{CQ2} - I_{CQ1} = 3.2652 - 3.1953 = 0.0699 \text{ mA}$$

6.18 In the constant-base-current-bias-circuit arrangement of Fig. 6-3, the leakage current is explicitly modeled as a current source I_{CBO}. Find I_{CQ} as a function of I_{CBO}, V_{BEQ}, and β.

▌ By KVL,
$$V_{CC} = I_{BQ}R_b + I_{EQ}R_E \tag{1}$$

Substitution of (2) and (3) of Problem 6.4 into (1) and rearrangement give

$$I_{CQ} \approx \frac{V_{CC} - V_{BEQ} + I_{CBO}(R_b + R_E)}{R_b/\beta + R_E} \tag{2}$$

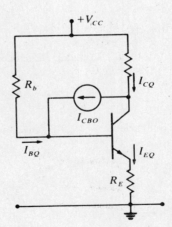

Fig. 6-3

6.19 In the shunt-feedback-biased arrangement of Fig. 6-4, the leakage current is explicitly shown as a current source I_{CBO}. Find I_{CQ} as a function of I_{CBO}, V_{BEQ}, and β.

▌ By KVL,
$$V_{CC} = I_{CQ}R_C + I_{BQ}(R_C + R_F) + V_{BEQ} + I_{EQ}R_E \tag{1}$$

Substituting (2) and (3) of Problem 6.4 into (1), rearranging, and then assuming $\beta \gg 1$, we obtain

$$I_{CQ} \approx \frac{V_{CC} - V_{BEQ} + I_{CBO}(R_C + R_F + R_E)}{\dfrac{\beta+1}{\beta}R_C + \dfrac{R_F}{\beta} + R_E} \approx \frac{V_{CC} - V_{BEQ} + I_{CBO}(R_C + R_F + R_E)}{R_F/\beta + R_C + R_E} \tag{2}$$

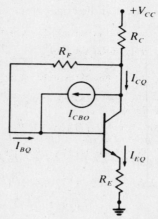

Fig. 6-4

6.20 In the CB amplifier of Fig. 6-5, the transistor leakage current is shown explicitly as a current source I_{CBO}. Find I_{CQ} as a function of I_{CBO}, V_{BEQ}, and β.

▌ By KVL,
$$V_{EE} = V_{BEQ} + I_{EQ}R_E \tag{1}$$

Substituting (2) of Problem 6.4 into (1) and rearranging yield

$$I_{CQ} = \frac{\beta+1}{\beta}\frac{V_{EE}-V_{BEQ}}{R_E} + I_{CBO} \tag{2}$$

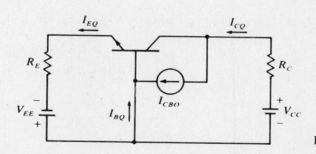

Fig. 6-5

6.21 The CB amplifier of Fig. 6-5 has $V_{CC} = 15$ V, $V_{EE} = 5$ V, $R_E = 3$ kΩ, $R_C = 7$ kΩ, and $\beta = 50$. At a temperature of 25°C, the Si transistor has $V_{BEQ} = 0.7$ V and $I_{CBO} = 0.5$ μA. Find an expression for I_{CQ} at any temperature.

▌ Let the subscript 1 denote quantities at $T_1 = 25$°C, and 2 denote them at any other temperature T_2. Then, similar to Problem 6.17,

$$I_{CBO2} = 2^{(T_2-25)/10}I_{CBO1}$$

$$V_{BEQ2} = V_{BEQ1} + \Delta V_{BEQ} = V_{BEQ1} - 0.002(T_2 - 25)$$

Hence, by (2) of Problem 6.20.

$$I_{CQ2} = \frac{\beta+1}{\beta}\frac{V_{EE}-V_{BEQ1}+0.002(T_2-25)}{R_E} + 2^{(T_2-25)/10}I_{CB01} \tag{1}$$

6.22 Evaluate the expression for I_{CQ2} of Problem 6.21 at $T_2 = 125$°C.

▌ At $T_2 = 125$°C, (1) of Problem 6.21 gives

$$I_{CQ2} = \frac{51}{50}\frac{5 - 0.7 + 0.002(125 - 25)}{3 \times 10^3} + (2^{(125-25)/10})(0.5 \times 10^{-6}) = 1.53 + 0.512 = 2.042 \text{ mA}$$

6.23 For the Darlington-pair emitter-follower of Fig. 6-6, find I_{CQ1} as a function of the six temperature-sensitive variables I_{CBO1}, I_{CBO2}, V_{BEQ1}, V_{BEQ2}, β_1, and β_2.

■ By KVL,
$$V_{CC} = I_{BQ1}R_F + V_{BEQ1} + V_{BEQ2} + I_{EQ2}R_E \quad (1)$$
By KCL,
$$I_{EQ2} = I_{EQ1} + I_{CQ2} \quad (2)$$

Using the result of Problem 4.127 in (2) and then substituting $I_{BQ2} = I_{EQ1}$, we obtain
$$I_{EQ2} = I_{EQ1} + \beta_2 I_{BQ2} + (\beta_2 + 1)I_{CBO2} = (\beta_2 + 1)I_{EQ1} + (\beta_2 + 2)I_{CBO2}$$

Assuming $\beta_1, \beta_2 \gg 1$ and substituting for I_{EQ1} according to (2) of Problem 6.4, we obtain
$$I_{EQ2} \approx (\beta_2 + 1)I_{CQ1} + (\beta_2 + 1)(I_{CBO2} - I_{CBO1}) \quad (3)$$

Also, from (3) of Problem 6.4,
$$I_{BQ1} \approx \frac{I_{CB1}}{\beta_1} - I_{CBO1} \quad (4)$$

Now we substitute (3) and (4) into (1) and rearrange to get
$$I_{CQ1} = \frac{V_{CC} - V_{BEQ1} - V_{BEQ2} + I_{CBO1}(R_F + \beta_2 R_E) - I_{CBO2}\beta_2 R_E}{R_F/\beta_1 + \beta_2 R_E} \quad (5)$$

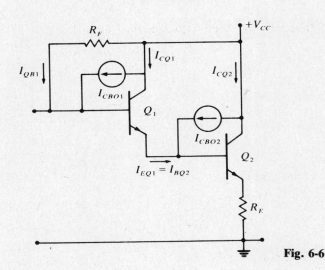

Fig. 6-6

6.24 In the constant-base-current-biased amplifier of Fig. 4-9, $V_{CC} = 15$ V, $R_C = 2.5$ kΩ, $R_E = 500$ Ω, and $R_B = 500$ kΩ. $I_{CBO} \approx 0$ for the Si device. Find I_{CQ} and V_{CEQ} if $\beta = 100$.

■ By KVL,
$$V_{CC} = V_{BEQ} + I_{BQ}R_B + I_{EQ}R_E \quad (1)$$

Substituting $I_{EQ} = (\beta + 1)I_{BQ}$ into (1) and rearranging give
$$I_{BQ} = \frac{V_{CC} - V_{BEQ}}{R_B + (\beta + 1)R_E} = \frac{15 - 0.7}{500 \times 10^3 + (100 + 1)(500)} = 25.98 \ \mu A \quad (2)$$

But
$$I_{CQ} = \beta I_{BQ} = 100(25.98 \times 10^{-6}) = 2.6 \text{ mA} \quad (3)$$

By KVL,
$$V_{CEQ} = V_{CC} - I_{EQ}R_E - I_{CQ}R_C \quad (4)$$

Using $I_{EQ} = (\beta + 1)I_{BQ}$ and $I_{CQ} = \beta I_{BQ}$ in (4) leads to
$$V_{CEQ} = V_{CC} - [(\beta + 1)R_E + \beta R_C]I_{BQ}$$
$$= 15 - [(100 + 1)(500) + (100)(2.5 \times 10^3)](25.98 \times 10^{-6}) = 7.19 \text{ V} \quad (5)$$

6.25 Rework Problem 6.24 if $\beta = 50$ and all else is unchanged.

▌ Based on (2), (3) and (5) of Problem 6.24,

$$I_{BQ} = \frac{V_{CC} - V_{BEQ}}{R_B + (\beta + 1)R_E} = \frac{15 - 0.7}{500 \times 10^3 + (50 + 1)(500)} = 27.21 \ \mu A$$

$$I_{CQ} = \beta I_{BQ} = 50(27.21 \times 10^{-6}) = 1.36 \ \text{mA}$$

$$V_{CEQ} = V_{CC} - [(\beta + 1)R_E + \beta R_C]I_{BQ} = 15 - [(50 + 1)(500) + 50(2.5 \times 10^3)](27.21 \times 10^{-6}) = 10.9 \ \text{V}$$

6.26 Under what condition will the bias current I_{CQ} of the amplifier in Fig. 4-10 be practically independent of β? Is this condition practical?

▌ By KVL,

$$V_{CC} = V_{CEQ} + I_{BQ}[R_B + (\beta + 1)R_E] \tag{1}$$

Solving (1) for I_{BQ}, using $I_{CQ} = \beta I_{BQ}$, and realizing that $\beta \gg 1$ give

$$I_{CQ} = \beta I_{BQ} = \frac{V_{CC} - V_{CEQ}}{\dfrac{R_B}{\beta} + \dfrac{\beta + 1}{\beta}R_E} \approx \frac{V_{CC} - V_{CEQ}}{R_B/\beta + R_E}$$

Now, I_{CQ} would be independent of β if $R_B/\beta \ll R_E$. The condition is not practical as a value of R_B large enough to properly limit I_{BQ} leads to a value of R_E so large that it forces cutoff.

NONLINEAR-ELEMENT STABILIZATION FOR BJTs

6.27 In the CE amplifier circuit of Fig. 6-1, assume that the Si device has negligible leakage current and (2) of Problem 4.40 holds to the point that R_B/β can be neglected. Also, V_{BEQ} decreases by 2 mV/°C from its value of 0.7 V at 25°C. Find the change in I_{CQ} as the temperature increases from 25°C to 125°C.

▌ Let the subscript 1 denote "at $T = 25$°C," and 2 denote "at $T = 125$°C." Under the given assumptions, (4) of Problem 6.4 reduces to

$$I_{CQ} = \frac{V_{BB} - V_{BEQ}}{R_E}$$

The change in I_{CQ} is then

$$\Delta I_{CQ} = I_{CQ2} - I_{CQ1} = \frac{0.002(T_2 - T_1)}{R_E} = \frac{0.2}{R_E}$$

6.28 Assume that the amplifier circuit of Fig. 6-7 has been designed so it is totally insensitive to variations of β. Further, $R_B \gg R_D$. As in Problem 6.27, V_{BEQ} is equal to 0.7 V at 25°C and decreases by 2 mV/°C. Also assume that V_D varies with temperature exactly as V_{BEQ} does. Find an expression for I_{CQ}.

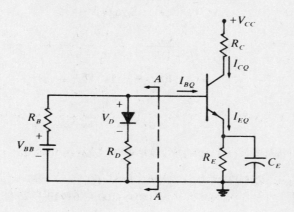

Fig. 6-7

▌ A Thévenin equivalent circuit can be found for the network to the left of terminals A, A, under the assumption that the diode can be modeled by a voltage source V_D. The result is

$$R_{Th} = R_D \| R_B \approx R_D$$

$$V_{Th} = V_D + \frac{V_{BB} - V_D}{R_B + R_D} R_D = \frac{V_{BB} R_D + V_D R_B}{R_D + R_B}$$

With the Thévenin equivalent in place, KVL and the assumption $I_{BQ} = I_{CQ}/\beta \approx I_{EQ}/\beta$ give

$$I_{CQ} \approx I_{EQ} = \frac{(V_{BB} R_D + V_D R_B)/(R_D + R_B) - V_{BEQ}}{R_D/\beta + R_E} \tag{1}$$

6.29 Beginning with (1) of Problem 6.28, find an expression for the change in I_{CQ} of the amplifier of Fig. 6-7 as the temperature increases from 25°C to 125°C.

▌ Now if there is total independence of β, then R_D/β must be negligible compared with R_E. Further, since only V_D and V_{BEQ} are dependent on temperature,

$$\frac{\partial I_{CQ}}{\partial T} \approx \frac{\dfrac{R_B}{R_B + R_D} \dfrac{\partial V_D}{\partial T} - \dfrac{\partial V_{BEQ}}{\partial T}}{R_E} = \frac{0.002 R_D}{R_E(R_B + R_D)} \approx \frac{0.002}{R_E} \frac{R_D}{R_B}$$

Hence,

$$\Delta I_{CQ} \approx \frac{\partial I_{CQ}}{\partial T} \Delta T = \frac{0.002}{R_E} \frac{R_D}{R_B}(100) = \frac{0.2}{R_E} \frac{R_D}{R_B}$$

Because $R_D \ll R_B$ here, the change in I_{CQ} has been reduced appreciably from what it was in the circuit of Problem 6.27.

6.30 The circuit of Fig. 6-8 includes nonlinear diode compensation for variations in V_{BEQ}. Neglecting I_{CBO}, find an expression for I_{CQ} that is a function of the temperature sensitive variables β, V_{BEQ}, and V_D.

▌ The usual Thévenin equivalent can be used to replace the R_1-R_2 voltage divider. Then, by KVL,

$$V_{BB} = R_B I_{BQ} + V_{BEQ} + I_{EQ} R_E - V_D \tag{1}$$

Substitution of $I_{BQ} = I_{CQ}/\beta$ and $I_{EQ} = I_{CQ}(\beta + 1)/\beta$ into (1) and rearranging yield

$$I_{CQ} = \frac{\beta[V_{BB} - (V_{BEQ} - V_D)]}{R_B + (\beta + 1)R_E} \tag{2}$$

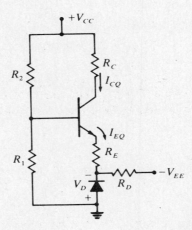

Fig. 6-8

6.31 Show that if V_{BEQ} and V_D are equal for the amplifier of Fig. 6-8, then the sensitivity of I_{CQ} to changes in V_{BEQ} is zero.

▌ From (2) of Problem 6.30, it is apparent that if $V_D = V_{BEQ}$, then I_{DQ} is independent of variations in V_{BEQ}.

6.32 Show for the amplifier of Fig. 6-8 it is not necessary that $V_{BEQ} = V_D$, but only (and less restrictively) that $dV_{EQ}/dT = dV_D/dT$, to ensure the insensitivity of I_{CQ} to temperature T.

■ If β is independent of temperature, differentiation of (2) from Problem 6.30 with respect to T results in

$$\frac{dI_{CQ}}{dT} = \frac{\beta}{R_B + (\beta + 1)R_E}\left(\frac{dV_D}{dT} - \frac{dV_{BEQ}}{dT}\right)$$

Hence, if $dV_D/dT = dV_{BEQ}/dT$, I_{CQ} is insensitive to temperature.

6.33 The circuit of Fig. 6-9 includes nonlinear diode compensation for variations in I_{CBO}. Find an expression for I_{CQ} as a function of the temperature-sensitive variables V_{BEQ}, β, I_{CBO}, and V_D.

■ By KVL,

$$V_{BB} = (I_{BQ} + I_D)R_B + V_{BEQ} + I_{EQ}R_E$$

Substitution for I_{EQ} and I_{BQ} via (2) and (3) of Problem 6.4 and rearranging give

$$I_{CQ} = \frac{V_{BB} - V_{BEQ} + I_{CBO}(R_B + R_E) - I_D R_B}{R_B/\beta + R_E} \tag{1}$$

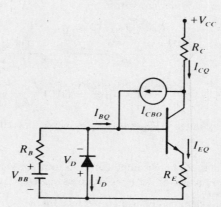

Fig. 6-9

6.34 What conditions will render I_{CQ} for the amplifier of Fig. 6-9 insensitive to changes in I_{CBO}?

■ According to (1) of Problem 6.33, if $R_B \gg R_E$ and $I_D = I_{CBO}$, then I_{CQ} is, in essence, independent of I_{CBO}.

6.35 Show that if a second identical diode is placed in series with the diode of Problem 6.28 (see Fig. 6-7), and if R_D is made equal in value to R_B, then the collector current ($I_{CQ} \approx I_{EQ}$) displays zero sensitivity to temperature changes that affect V_{BEQ}. Make the reasonable assumption that $\partial V_D/\partial T = \partial V_{BEQ}/\partial T$.

■ Equation (1) of Problem 6.28 describes I_{CQ} in this problem if V_D is replaced by $2V_D$; that gives

$$I_{CQ} \approx \frac{(V_{BB}R_D + 2V_D R_B)/(R_D + R_B) - V_{BEQ}}{(2R_D \parallel R_B)/\beta + R_E} \tag{1}$$

Assuming that only V_{BEQ} and V_D are temperature dependent, we have

$$\frac{\partial I_{CQ}}{\partial T} = \frac{\dfrac{2R_B}{R_D + R_B}\dfrac{\partial V_D}{\partial T} - \dfrac{\partial V_{BEQ}}{\partial T}}{(2R_D \parallel R_B)/\beta + R_E} \tag{2}$$

With $\partial V_D/\partial T = \partial V_{BEQ}/\partial T$ and $R_B = R_D$, (2) reduces to zero, indicating that I_{CQ} is not a function of temperature.

STABILITY-FACTOR ANALYSIS

6.36 *Stability-factor* or *sensitivity* analysis is based on the assumption that, for small changes, the variable of interest is a linear function of the other variables, and thus its differential can be replaced by its increment. In a study of BJT Q-point stability, we examine changes in quiescent collector current I_{CQ} due to variations in transistor quantities and/or elements of the surrounding circuit. Specifically, if

$$I_{CQ} = f(\beta, I_{CBO}, V_{BEQ}, \ldots) \tag{1}$$

take the total differential of I_{CQ}, define suitable sensitivity factors, replace the differentials with increments, and find a first-order approximation to the total change in I_{CQ}.

▌ By application of the chain rule, the total differential is

$$dI_{CQ} = \frac{\partial I_{CQ}}{\partial \beta} d\beta + \frac{\partial I_{CQ}}{\partial I_{CBO}} dI_{CBO} + \frac{\partial I_{CQ}}{\partial V_{BEQ}} dV_{BEQ} + \cdots \tag{2}$$

We may define a set of *stability factors* or *sensitivity factors* as follows:

$$S_\beta = \frac{\Delta I_{CQ}}{\Delta \beta}\bigg|_Q \approx \frac{\partial I_{CQ}}{\partial \beta}\bigg|_Q \tag{3}$$

$$S_I = \frac{\Delta I_{CQ}}{\Delta I_{CBO}}\bigg|_Q \approx \frac{\partial I_{CQ}}{\partial I_{CBO}}\bigg|_Q \tag{4}$$

$$S_V = \frac{\Delta I_{CQ}}{\Delta V_{BEQ}}\bigg|_Q \approx \frac{\partial I_{CQ}}{\partial V_{BEQ}}\bigg|_Q \tag{5}$$

and so on. Then replacing differentials with increments in (2) yields a first-order approximation to the total change in I_{CQ}:

$$\Delta I_{CQ} \approx S_\beta \Delta\beta + S_I \Delta I_{CBO} + S_V \Delta V_{BEQ} + \cdots \tag{6}$$

6.37 For the CE amplifier of Fig. 6-1, use stability-factor analysis to find an expression for the change in I_{CQ} due to variations in β, I_{CBO}, and V_{BEQ}.

▌ The quiescent collector current I_{CQ} is expressed as a function of β, I_{CBO}, and V_{BEQ} in (4) of Problem 6.4. Thus, by (6) of Problem 6.36,

$$\Delta I_{CQ} \approx S_\beta \Delta\beta + S_I \Delta I_{CBO} + S_V \Delta V_{BEQ} \tag{1}$$

where the stability factors, according to (3)–(5) of Problem 6.36, are

$$S_\beta = \frac{\partial I_{CQ}}{\partial \beta} = \frac{\partial}{\partial \beta}\left\{\frac{\beta[V_{BB} - V_{BEQ} + I_{CBO}(R_B + R_E)]}{R_B + \beta R_E}\right\} = \frac{R_B[V_{BB} - V_{BEQ} + I_{CBO}(R_B + R_E)]}{(R_B + \beta R_E)^2} \tag{2}$$

$$S_I = \frac{\partial I_{CQ}}{\partial I_{CBO}} = \frac{R_B + R_E}{R_B/\beta + R_E} \tag{3}$$

$$S_V = \frac{\partial I_{CQ}}{\partial V_{BEQ}} = -\frac{\beta}{R_B + \beta R_E} \tag{4}$$

6.38 In the circuit of Fig. 4-10, let $V_{CC} = 15$ V, $R_B = 500$ kΩ, and $R_C = 5$ kΩ. Assume a Si transistor with $I_{CBO} \approx 0$. Find the β sensitivity factor S_β and use it to calculate the change in I_{CQ} when β changes from 50 to 100.

▌ By KVL,

$$V_{CC} = V_{BEQ} + I_{BQ}R_B = V_{BEQ} + \frac{I_{CQ}}{\beta}R_B$$

so that

$$I_{CQ} = \frac{\beta(V_{CC} - V_{BEQ})}{R_B}$$

and by (3) of Problem 6.36,

$$S_\beta = \frac{\partial I_{CQ}}{\partial \beta} = \frac{V_{CC} - V_{BEQ}}{R_B} = \frac{15 - 0.7}{500 \times 10^3} = 28.6 \times 10^{-6}$$

According to (6) of Problem 6.36, the change in I_{CQ} due to β alone is

$$\Delta I_{CQ} \approx S_\beta \Delta\beta = (28.6 \times 10^{-6})(100 - 50) = 1.43 \text{ mA}$$

6.39 Compare the result of Problem 6.38 with those of Problems 6.1 and 6.2.

▌ From Problems 6.1 and 6.2, we have

$$\Delta I_{CQ} = I_{CQ}|_{\beta=100} - I_{CQ}|_{\beta=50} = 2.86 - 1.43 = 1.43 \text{ mA}$$

Because I_{CQ} is of the first degree in β, (6) of Problem 6.36 produces the exact change.

6.40 Find the β sensitivity factor for the amplifier of Fig. 4-5.

▌ Since
$$I_{EQ} = \frac{I_{CQ}}{\alpha} = \frac{\beta+1}{\beta} I_{CQ}$$

we have, from (1) of Problem 4.39,
$$V_{BB} = I_{CQ}\frac{R_B}{\beta} + V_{BEQ} + \frac{\beta+1}{\beta} I_{CQ} R_E$$

Rearranging gives
$$I_{CQ} = \frac{V_{BB} - V_{BEQ}}{\frac{R_B}{\beta} + \frac{\beta+1}{\beta}R_E} = \frac{\beta(V_{BB} - V_{BEQ})}{R_B + (\beta+1)R_E} \qquad (1)$$

and, from (3) of Problem 6.36,
$$S_\beta = \frac{\partial I_{CQ}}{\partial \beta} = \frac{(R_B + R_E)(V_{BB} - V_{BEQ})}{[R_B + (\beta+1)R_E]^2} \qquad (2)$$

6.41 Show that the condition under which the β sensitivity factor of Problem 6.40 is reduced to zero is identical to the condition under which the emitter current bias is constant.

▌ Note in (2) of Problem 6.40 that $\lim_{\beta \to \infty} S_\beta = 0$. Now if $\beta \to \infty$ in (1) of Problem 6.40, then
$$I_{CQ} \approx (V_{BB} - V_{BEQ})/R_E = \text{constant}$$

6.42 Predict the new value of I_{CQ} for Problem 6.17 using stability-factor analysis.

▌ By (3) and (4) of Problem 6.37,
$$S_I = \frac{R_B + R_E}{R_B/\beta + R_E} = \frac{50+1}{50/75+1} = 30.6$$

$$S_V = \frac{-\beta}{R_B + \beta R_E} = \frac{-75}{50 \times 10^3 + 75(1 \times 10^3)} = -0.6 \times 10^{-3}$$

▌ Then, according to (6) of Problem 6.36,
$$\Delta I_{CQ} = S_I \Delta I_{CBO} + S_V \Delta V_{BEQ} = 30.6(1.5 \times 10^{-6}) + (-0.6 \times 10^{-3})(-0.04) = 0.0699 \text{ mA}$$
and
$$I_{CQ2} = I_{CQ1} + \Delta I_{CQ} = 3.1953 + 0.0699 = 3.2652 \text{ mA}$$

6.43 In Problem 6.17, assume that the given values of I_{CBO} and V_{BEQ} are valid at 25°C (that is, that $T_1 = 25$°C). Use stability-factor analysis to find an expression for the change in collector current resulting from a change to any temperature T_2.

▌ Recalling that leakage current I_{CBO} doubles for each 10°C rise in temperature, we have
$$\Delta I_{CBO} = I_{CBO}|_{T_2} - I_{CBO}|_{T_1} = I_{CBO}|_{25°C}(2^{(T_2-25)/10} - 1)$$

Since V_{BEQ} for a Si device decreases by 2 mV/°C, we have
$$\Delta V_{BEQ} = -0.002(T_2 - 25)$$

Now, substituting S_I and S_V as determined in Problem 6.17 into (6) of Problem 6.36, we obtain
$$\Delta I_{CQ} = S_I \Delta I_{CBO} + S_V \Delta V_{BEQ} = \frac{\beta(R_B + R_E)}{R_B + \beta R_E} I_{CBO}|_{25°C}(2^{(T_2-25)/10} - 1) + \frac{\beta}{R_B + \beta R_E}(0.002)(T_2 - 25) \qquad (1)$$

6.44 Use (1) of Problem 6.43 to find ΔI_{CQ} for the amplifier of Problem 6.17 when $T_2 = 125$°C.

▌ At $T_2 = 125$°C, with the values of Problem 6.17, this expression for ΔI_{CQ} gives
$$\Delta I_{CQ} = 30.6(0.5 \times 10^{-6})(2^{(125-25)/10} - 1) + 0.0006(0.002)(125 - 25)$$
$$= 15.65 \text{ mA} + 0.12 \text{ mA} = 15.77 \text{ mA}$$

6.45 What percentage of the change in I_{CQ} of Problem 6.44 is attributable to a change in leakage current?

▌ From Problem 6.44, the percentage of ΔI_{CQ} due to I_{CBO} is $(15.65/15.77)(100) = 99.24$ percent.

6.46 Determine the stability factors that should be used in (6) of Problem 6.36 to express the influence of I_{CBO}, V_{BEQ}, and β on I_{CQ} for the amplifier of Fig. 6-3 as given by (2) of Problem 6.18.

▌ Based on the symmetry between (2) of Problem 6.18 and (4) of Problem 6.4 we have, from Problem 6.37,

$$S_I = \frac{R_b + R_E}{R_b/\beta + R_E} \qquad S_V = \frac{-\beta}{R_b + \beta R_E} \qquad S_\beta = \frac{R_b[V_{CC} - V_{BEQ} + I_{CBO}(R_b + R_E)]}{(R_b + \beta R_E)^2}$$

6.47 Find the stability factors that should be used in (6) of Problem 6.36 to express the influence of I_{CBO}, V_{BEQ}, and β on I_{CQ} for the amplifier of Fig. 6-4 as given by (2) of Problem 6.19.

▌ Based on the symmetry between (2) of Problem 6.19 and (4) of Problem 6.4 we have, from Problem 6.37,

$$S_I = \frac{R_C + R_F + R_E}{R_F/\beta + R_C + R_E} \qquad S_V = \frac{-\beta}{R_F + \beta(R_C + R_E)}$$

$$S_\beta = \frac{R_F[V_{CC} - V_{BEQ} + I_{CBO}(R_C + R_F + R_E)]}{[R_F + \beta(R_C + R_E)]^2}$$

6.48 Determine the stability factors that should be used in (6) of Problem 6.36 to express the influence of I_{CBO}, V_{BEQ}, and β on I_{CQ} for the amplifier of Fig. 6-5 as given by (2) of Problem 6.20.

▌ Direct application of (3)–(5) of Problem 6.36 to (2) of Problem 6.20 gives the desired stability factors as

$$S_\beta = \frac{\partial I_{CQ}}{\partial \beta} = -\frac{1}{\beta^2} \frac{V_{EE} - V_{BEQ}}{R_E} \qquad S_I = \frac{\partial I_{CQ}}{\partial I_{CBO}} = 1 \qquad S_V = \frac{\partial I_{CQ}}{\partial V_{BEQ}} = -\frac{\beta}{(\beta + 1)R_E}$$

6.49 Determine a first-order approximation for the change in I_{CQ1} in the circuit of Fig. 6-6, in terms of the six variables I_{CBO1}, I_{CBO2}, V_{BEQ1}, V_{BEQ2}, β_1, and β_2.

▌ Since $I_{CQ1} = f(I_{CBO1}, I_{CBO2}, V_{BEQ1}, V_{BEQ2}, \beta_1, \beta_2)$, its total differential is given by

$$dI_{CQ1} = \frac{\partial I_{CQ1}}{\partial I_{CBO1}} dI_{CBO1} + \frac{\partial I_{CQ1}}{\partial I_{CBO2}} dI_{CBO2} + \frac{\partial I_{CQ1}}{\partial V_{BEQ1}} dV_{BEQ1}$$
$$+ \frac{\partial I_{CQ1}}{\partial V_{BEQ2}} dV_{BEQ2} + \frac{\partial I_{CQ1}}{\partial \beta_1} d\beta_1 + \frac{\partial I_{CQ1}}{\partial \beta_2} d\beta_2 \tag{1}$$

Using the method of Problem 6.36, we may write this as

$$\Delta I_{CQ1} \approx S_{I1} \Delta I_{CBO1} + S_{I2} \Delta I_{CBO2} + S_{V1} \Delta V_{BEQ1} + S_{V2} \Delta V_{BEQ2} + S_{\beta 1} \Delta \beta_1 + S_{\beta 2} \Delta \beta_2 \tag{2}$$

6.50 Use I_{CQ1} as found in Problem 6.23 to evaluate the sensitivity factors in the expression determined in Problem 6.49.

▌ The sensitivity factors in (1) of Problem 6.49 may be evaluated with the use of (5) of Problem 6.23:

$$S_{I1} = \frac{\partial I_{CQ1}}{\partial I_{CBO1}} = \frac{R_F + \beta_2 R_E}{R_F/\beta_1 + \beta_2 R_E}$$

$$S_{I2} = \frac{\partial I_{CQ1}}{\partial I_{CBO2}} = \frac{-\beta_2 R_E}{R_F/\beta_1 + \beta_2 R_E}$$

$$S_{V1} = \frac{\partial I_{CO1}}{\partial V_{BEQ1}} = \frac{-1}{R_F/\beta_1 + \beta_2 R_E} = S_{V2} = \frac{\partial I_{CQ1}}{\partial V_{BEQ2}}$$

$$S_{\beta 1} = \frac{\partial I_{CQ1}}{\partial \beta_1} = \frac{R_F[V_{CC} - V_{BEQ1} - V_{BEQ2} + I_{CBO1}(R_F + \beta_2 R_E) - I_{CBO2}\beta_2 R_E]}{(R_F + \beta_1\beta_2 R_E)^2}$$

$$S_{\beta 2} = \frac{\partial I_{CQ2}}{\partial \beta_2} = \frac{\beta_1 R_E[R_F(I_{CBO1} - I_{CBO2}) - \beta_1(V_{CC} - V_{BEQ1} - V_{BEQ2} + I_{CBO1}R_F)]}{(R_F + \beta_1\beta_2 R_E)^2}$$

174 □ CHAPTER 6

6.51 It is possible that variations in passive components will have an effect on transistor bias. In the circuit of Fig. 4-5(a), let $R_1 = R_C = 500\ \Omega$, $R_2 = 5\ \text{k}\Omega$, $R_E = 100 \pm 10\ \Omega$, $\beta = 75$, $I_{CBO} = 0.2\ \mu A$, $V_{CC} = 20\ \text{V}$. Find an expression for the change in I_{CQ} due to a change in R_E alone.

▌ We seek a stability factor

$$S_{RE} = \frac{\partial I_{CQ}}{\partial R_E} \quad \text{such that} \quad \Delta I_{CQ} \approx S_{RE}\, \Delta R_E$$

Starting with I_{CQ} as given by (4) of Problem 6.4, we find

$$S_{RE} = \frac{\partial I_{CQ}}{\partial R_E} = \frac{\beta(R_B + \beta R_E)I_{CBO} - \beta^2[V_{BB} - V_{BEQ} + I_{CBO}(R_B + R_E)]}{(R_B + \beta R_E)^2}$$

$$= \frac{\beta R_B I_{CBO} - \beta^2(V_{BB} - V_{BEQ} + I_{CBO}R_B)}{(R_B + \beta R_E)^2}$$

6.52 Use the stability factor of Problem 6.51 to predict the change that will occur in I_{CQ} as R_E changes from the minimum to the maximum allowable value for the described circuit.

▌ We first need to evaluate S_{RE} at $R_E = 100 - 10 = 90\ \Omega$:

$$R_B = R_1 \parallel R_2 = 454.5\ \Omega$$

$$V_{BB} = \frac{R_1}{R_1 + R_2} V_{CC} = \frac{500}{5500}(20) = 1.818\ \text{V}$$

and

$$S_{RE} = \frac{75(454.5)(0.2 \times 10^{-6}) - (75)^2[1.818 - 0.7 + (0.2 \times 10^{-6})(454.5)]}{(454.5 + 75 \times 90)^2} = -1.212 \times 10^{-4}\ \text{A}/\Omega$$

Then $\quad \Delta I_{CQ} = S_{RE}\, \Delta R_E = (-1.212 \times 10^{-4})(110 - 90) = -2.424\ \text{mA}$

6.53 The amplifier of Fig. 4-9 uses a Si transistor for which $I_{CBO} \approx 0$. Let $V_{CC} = 15\ \text{V}$, $R_C = 2.5\ \text{k}\Omega$, $R_E = 500\ \Omega$, and $R_B = 500\ \text{k}\Omega$. Find the value of the β sensitivity factor $S_\beta = \partial I_{CQ}/\partial \beta$ for $\beta = 50$.

▌ Based on the result of Problem 4.60, use $I_{CQ} = \beta I_{BQ}$ to find

$$I_{CQ} = \frac{\beta(V_{CC} - V_{BEQ})}{R_B + (\beta + 1)R_E} \tag{1}$$

Whence,

$$S_\beta = \frac{\partial I_{CQ}}{\partial \beta} = \frac{(R_B + R_E)(V_{CC} - V_{BEQ})}{[R_B + (\beta + 1)R_E]^2} = \frac{(500 \times 10^3 + 500)(15 - 0.07)}{[500 \times 10^3 + (50 + 1)(500)]^2} = 2.59 \times 10^{-5}$$

6.54 Use S_β of Problem 6.53 to predict I_{CQ} for the amplifier when $\beta = 100$. Compare the result with the exact answer of Problem 6.24.

▌ Using the result of Problem 6.26 for I_{CQ1},

$$I_{CQ2} = I_{CQ1} + \Delta I_{CQ} = I_{CQ1} + S_\beta \Delta_\beta = 1.36 \times 10^{-3} + (2.59 \times 10^{-5})(100 - 50) = 2.65\ \text{mA}$$

The answer is within 2% of the exact answer.

6.55 Find I_{CQ} of Problem 4.99 if $\beta = 75$ and all else is unchanged.

▌ Using the result of Problem 4.99,

$$I_{CQ} \approx I_{EQ} = \frac{V_{BB} - V_{BEQ}}{R_B/(\beta + 1) + R_E} = \frac{1.765 - 0.7}{1.765 \times 10^3/(75 + 1) + 200} = 4.77\ \text{mA}$$

6.56 Use the β sensitivity factor found in Problem 6.40 to predict the change in I_{CQ} when β changes from 110 to 75.

▌ Evaluating (2) of Problem 6.40 for $\beta = 110$ and values from Problem 4.99 yield

$$S_\beta = \frac{(R_B + R_E)(V_{BB} - V_{BEQ})}{[R_B + (\beta + 1)R_E]^2} = \frac{(1.765 \times 10^3 + 200)(1.765 - 0.7)}{[1.765 \times 10^3 + (110 + 1)(200)]^2} = 3.643 \times 10^{-6}$$

Hence, by (6) of Problem 6.36,

$$\Delta I_{CQ} \approx S_\beta \Delta_\beta = (3.643 \times 10^{-6})(75 - 110) = -0.127 \text{ mA}$$

6.57 In the shunt-feedback-biased amplifier of Fig. 4-19, $V_{CC} = 15$ V, $R_C = 2$ kΩ, $R_F = 150$ kΩ, $I_{CEO} \approx 0$, and the transistor is a Si device. Find an expression for the β sensitivity factor S_β.

▌ By KVL,
$$V_{CC} = (I_{CQ} + I_{BQ})R_C + I_{BQ}R_F + V_{BEQ}$$

Substituting $I_{BQ} = I_{CQ}/\beta$ and rearranging lead to

$$I_{CQ} = \frac{\beta(V_{CC} - V_{BEQ})}{R_F + (\beta + 1)R_C}$$

By use of (3) from Problem 6.36,

$$S_\beta = \frac{\partial I_{CQ}}{\partial \beta} = \frac{(R_F + R_C)(V_{CC} - V_{BEQ})}{[R_F + (\beta + 1)R_C]^2} \tag{1}$$

6.58 Apply S_β of Problem 6.57 to predict the change in quiescent collector current due to a change in β from 50 to 100.

▌ Evaluating (1) of Problem 6.57 for $\beta = 50$,

$$S_\beta = \frac{(150 \times 10^3 + 2 \times 10^3)(15 - 0.7)}{[150 \times 10^3 + (50 + 1)(2 \times 10^3)]} = 3.423 \times 10^{-5}$$

Hence,
$$\Delta I_{CQ} \approx S_\beta \Delta_\beta = (3.423 \times 10^{-5})(100 - 50) = 1.71 \text{ mA}$$

6.59 In the CB amplifier of Fig. 4-21, $V_{CC} = 15$ V, $V_{EE} = 5$ V, $R_E = 3$ kΩ, $R_C = 7$ kΩ, and $\beta = 50$. Find an expression for the β sensitivity factor S_β.

▌ By KVL,
$$V_{EE} = V_{BEQ} + I_{EQ}R_E$$

Solve for I_{EQ} and recall that $I_{CQ} = I_{EQ}\beta/(\beta + 1)$ to find

$$I_{CQ} = \frac{\beta}{\beta + 1} I_{EQ} = \frac{\beta(V_{EE} - V_{BEQ})}{(\beta + 1)R_E}$$

By (3) of Problem 6.36,
$$S_\beta = \frac{\partial I_{CQ}}{\partial \beta} = \frac{V_{EE} - V_{BEQ}}{(\beta + 1)^2 R_E} \tag{1}$$

6.60 Evaluate S_β for the amplifier of Problem 6.59 if the transistor is a Si device.

▌
$$S_\beta = \frac{V_{EE} - V_{BEQ}}{(\beta + 1)^2 R_E} = \frac{5 - 0.7}{(50 + 1)^2(3 \times 10^3)} = 5.51 \times 10^{-7}$$

6.61 The circuit of Fig. 6-1 has the values given in Problem 6.17 except the transistor is a Ge device with $I_{CBO} = 1$ μA and $V_{BEQ} = 0.3$ V. Assume that the initial values of I_{CBO} and V_{BEQ} are for 25°C. Find an expression for the value of I_{CQ} at any temperature $T_2 \geq 25$°C.

▌ By procedure similar to that of Problem 6.17,

$$I_{CBO2} = I_{CBO1} 2^{(T_2 - 25)/10}$$

$$V_{BEQ2} = V_{BEQ1} + \Delta V_{BEQ} = V_{BEQ1} - k(T_2 - 25)$$

$$I_{CQ2} = \frac{V_{BB} - V_{BEQ1} + k(T_2 - 25) + I_{CBO1}(R_B + R_E)2^{(T_2 - 25)/10}}{R_B/\beta + R_E}$$

6.62 Evaluate the expression for I_{CQ} determined in Problem 6.61 at a temperature $T_2 = 125$°C. For a Ge transistor, $k = 1.8$ mV/°C.

▌
$$I_{CQ2} = \frac{6 - 0.3 + (1.8 \times 10^{-3})(125 - 25) + (1 \times 10^{-6})(50 \times 10^3 + 1 \times 10^3)2^{(125-25)/10}}{(50 \times 10^3)/(75) + 1 \times 10^3} = 34.86 \text{ mA}$$

6.63 The constant-base-current-biased amplifier of Fig. 6-3 contains a Si transistor. Let $V_{CC} = 15$ V, $R_C = 2.5$ kΩ, $R_E = 500$ Ω, $R_b = 500$ kΩ, and $\beta = 100$. At 25°C, $I_{CBO} = 0.5$ μA and $V_{BEQ} = 0.7$ V. Find the exact change in I_{CQ} if the temperature changes to 100°C.

▌ The temperature dependencies of I_{CBO} and V_{BEQ} are established in Problem 6.17. Let the subscripts 1 and 2 denote $T_1 = 25$°C and $T_2 = 100$°C, respectively. Then, by (2) of Problem 6.18,

$$I_{CQ1} = \frac{V_{CC} - V_{BEQ1} + I_{CBO1}(R_b + R_E)}{R_b/\beta + R_E} = \frac{15 - 0.7 + (0.5 \times 10^{-6})(500 \times 10^3 + 500)}{(500 \times 10^3)/100 + 500} = 2.645 \text{ mA}$$

Following the procedure of Problem 6.17,

$$V_{BEQ2} = V_{BEQ1} + \Delta V_{BEQ} = 0.7 - (2 \times 10^{-3})(100 - 25) = 0.55 \text{ V}$$

$$I_{CBO2} = I_{CBO1} 2^{\Delta T/10} = (0.5 \times 10^{-6}) 2^{(100-25)/10} = 0.09051 \text{ mA}$$

$$I_{CQ2} = \frac{V_{CC} - V_{BEQ2} + I_{CBO2}(R_b + R_E)}{R_b/\beta + R_E}$$

$$= \frac{15 - 0.55 + (0.09051 \times 10^{-6})(500 \times 10^3 + 500)}{(500 \times 10^3)/100 + 500} = 10.864 \text{ mA}$$

Whence, $\Delta I_{CQ} = I_{CQ2} - I_{CQ1} = 10.864 - 2.645 = 8.219$ mA

6.64 Use the stability factors developed in Problem 6.46 to predict ΔI_{CQ} for the amplifier of Problem 6.63 with the temperature increase of 100°C.

▌ Based on the results of Problem 6.46,

$$S_I = \frac{R_b + R_E}{R_b/\beta + R_E} = \frac{500 \times 10^3 + 500}{(500 \times 10^3)/100 + 500} = 91$$

$$S_V = \frac{-\beta}{R_b + \beta R_E} = \frac{-100}{500 \times 10^3 + 100(500)} = -0.182 \times 10^{-3}$$

Using (6) of Problem 6.36,

$$\Delta I_{CQ} \approx S_I \Delta I_{CBO} + S_V \Delta V_{BEQ} \approx 91(90.51 \times 10^{-6} - 0.5 \times 10^{-6}) + (-0.182 \times 10^3)(0.55 - 0.7) = 8.218 \text{ mA}$$

6.65 In the constant-base-current-biased amplifier of Fig. 6-3, the Si transistor is characterized by $I_{CBO} = 0.5$ μA and $V_{BEQ} = 0.7$ V at 25°C. Find an expression for I_{CQ} at any temperature $T_2 \geq 25$°C.

▌ By procedure similar to that of Problem 6.17 where I_{CQ} is given by (2) of Problem 6.18, we find

$$I_{CQ2} = \frac{V_{CC} - 0.7 + (2 \times 10^{-3})(T_2 - 25) + (0.5 \times 10^{-6})(R_b + R_E) 2^{(T_2-25)/10}}{R_b/\beta + R_E}$$

6.66 For the amplifier of Fig. 6-3, the Si transistor is described at 25°C by $I_{CBO} = 0.5$ μA and $V_{BEQ} = 0.7$ V. Let $V_{CC} = 15$ V, $R_C = 2.5$ kΩ, $R_E = 500$ Ω, $R_b = 500$ kΩ, and $\beta = 100$. Find I_{CQ} at 100°C.

▌ Using the result of Problem 6.65 with $T_2 = 100$°C,

$$I_{CQ} = \frac{15 - 0.7 + (2 \times 10^{-3})(100 - 25) + (0.5 \times 10^{-6})(500 \times 10^3 + 500) 2^{(100-25)/10}}{(500 \times 10^3)/100 + 500} = 10.864 \text{ mA}$$

6.67 In the current-feedback-biased amplifier of Fig. 6-4, $V_{CC} = 15$ V, $R_C = 1.5$ kΩ, $R_F = 150$ kΩ, $R_E = 500$ Ω, and $\beta = 100$. $I_{CBO} = 0.2$ μA and $V_{BEQ} = 0.7$ V at 25°C for this Si transistor. Find the exact change in I_{CQ} when the temperature changes to 125°C.

▌ Using (2) of Problem 6.19 and a procedure similar to that of Problem 6.17, we have

$$I_{CQ1} = \frac{V_{CC} - V_{BEQ1} + I_{CBO1}(R_C + R_F + R_E)}{R_F/\beta + R_C + R_E}$$

$$= \frac{15 - 0.7 + (0.2 + 10^{-6})(1.5 \times 10^3 + 150 \times 10^3 + 500)}{(150 \times 10^3)/100 + 1.5 \times 10^3 + 500} = 4.094 \text{ mA}$$

$$V_{BEQ2} = V_{BEQ1} + \Delta V_{BEQ} = 0.7 - (2 \times 10^{-3})(125 - 25) = 0.5 \text{ V}$$

$$I_{CBO2} = I_{CBO1} 2^{\Delta T/10} = (0.2 \times 10^{-6}) 2^{(125-25)/10} = 0.2048 \text{ mA}$$

$$I_{CQ2} = \frac{V_{CC} - V_{BEQ2} + I_{CBO2}(R_C + R_F + R_E)}{R_F/\beta + R_C + R_E}$$

$$= \frac{15 - 0.5 + (0.2048 \times 10^{-3})(1.5 \times 10^3 + 150 \times 10^3 + 500)}{(150 \times 10^3)/100 + 1.5 \times 10^3 + 500} = 13.037 \text{ mA}$$

Whence, $\quad \Delta I_{CQ} = I_{CQ2} - I_{CQ1} = 13.037 - 4.094 = 8.943 \text{ mA}$

6.68 Apply the stability factors developed in Problem 6.47 to predict ΔI_{CQ} for the amplifier of Problem 6.67 when the operating temperature increases from 25°C to 125°C.

▌ Based on the results of Problem 6.47,

$$S_I = \frac{R_C + R_F + R_E}{R_F/\beta + R_C + R_E} = \frac{1.5 \times 10^3 + 150 \times 10^3 + 500}{(150 \times 10^3)/100 + 1.5 \times 10^3 + 500} = 43.43$$

$$S_V = \frac{-\beta}{R_F + \beta(R_C + R_E)} = \frac{-100}{150 \times 10^3 + (100)(1.5 \times 10^3 + 500)} = -0.2857 \times 10^{-3}$$

Using (6) of Problem 6.36 and results of Problem 6.67,

$$\Delta I_{CQ} \approx S_I \Delta I_{CBO} + S_V \Delta V_{BEQ}$$

$$\approx 43.43(0.2048 \times 10^{-3} - 0.2 \times 10^{-6}) + (-0.2857 \times 10^{-3})(0.5 - 0.7) = 8.943 \text{ mA}$$

6.69 The shunt-feedback-biased amplifier of Fig. 6-4 uses a Si transistor for which $I_{CBO} = 0.2 \ \mu\text{A}$ and $V_{BEQ} = 0.7 \text{ V}$ at 25°C. Find an expression for I_{CQ} at any temperature $T_2 \geq 25°C$.

▌ I_{CQ} is given by (2) of Problem 6.19. Based on symmetry between (2) of Problem 6.18 and (2) of Problem 6.19, the result of Problem 6.65 can be used to write

$$I_{CQ2} = \frac{V_{CC} - V_{BEQ1} + (2 \times 10^{-3})(T_2 - 25) + (0.2 \times 10^{-6})(R_C + R_F + R_E) 2^{(T_2-25)/10}}{R_F/\beta + R_C + R_E}$$

6.70 If the transistor for the amplifier of Fig. 6-4 is described in Problem 6.69, $V_{CC} = 15 \text{ V}$, $R_C = 1.5 \text{ k}\Omega$, $R_F = 150 \text{ k}\Omega$, $R_E = 500 \ \Omega$, and $\beta = 100$, find I_{CQ} at $T_2 = 125°C$.

▌ Based on the result of Problem 6.69,

$$I_{CQ2} = \frac{15 - 0.7 + (2 \times 10^{-3})(125 - 25) + (0.2 \times 10^{-6})(1.5 \times 10^3 + 150 \times 10^3 + 500) 2^{(125-25)/10}}{(150 \times 10^3)/100 + 1.5 \times 10^3 + 500}$$

$$= 13.037 \text{ mA}$$

6.71 In the CB amplifier of Fig. 6-5, $V_{CC} = 15 \text{ V}$, $V_{EE} = 5 \text{ V}$, $R_E = 3 \text{ k}\Omega$, $R_C = 7 \text{ k}\Omega$, and $\beta = 50$. For the Si transistor, $I_{CBO} = 0.5 \ \mu\text{A}$ and $V_{BEQ} = 0.7 \text{ V}$ at 25°C. Find the exact change in I_{CQ} when the temperature changes to 125°C.

▌ Based on (2) of Problem 6.20 and letting subscripts 1 and 2 denote $T_1 = 25°C$ and $T_2 = 125°C$, respectively,

$$I_{CQ1} = \frac{\beta + 1}{\beta} \frac{V_{EE} - V_{BEQ1}}{R_E} + I_{CBO1} = \frac{50 + 1}{5} \frac{5 - 0.7}{3 \times 10^3} + 0.5 \times 10^{-6} = 1.4625 \text{ mA}$$

Proceeding as in Problem 6.17,

$$V_{BEQ2} = V_{BEQ1} + \Delta V_{BEQ} = 0.7 - (2 \times 10^{-3})(125 - 25) = 0.5 \text{ V}$$

$$I_{CBO2} = I_{CBO1} 2^{\Delta T/10} = (0.5 \times 10^{-6}) 2^{(125-25)/10} = 0.512 \text{ mA}$$

And

$$I_{CQ2} = \frac{\beta + 1}{\beta} \frac{V_{EE} - V_{BEQ2}}{R_E} + I_{CBO2} = \frac{50 + 1}{50} \frac{5 - 0.5}{3 \times 10^3} + 0.512 \times 10^{-3} = 2.042 \text{ mA}$$

Whence, $\quad \Delta I_{CQ} = I_{CQ2} - I_{CQ1} = 2.042 - 1.4625 = 0.5795 \text{ mA}$

178 ◻ CHAPTER 6

6.72 Apply the stability factors developed in Problem 6.48 to predict ΔI_{CQ} for the amplifier of Problem 6.71.

▮ Based on results of Problem 6.48,

$$S_I = 1 \qquad S_V = -\frac{\beta}{(\beta+1)R_E} = -\frac{50}{(50+1)(3 \times 10^3)} = -0.327 \times 10^{-3}$$

From (6) of Problem 6.36 and results of Problem 6.71,

$$\Delta I_{CQ} \approx S_I \Delta I_{CBO} + S_V \Delta V_{BEQ} \approx 1(0.512 \times 10^{-3} - 0.5 \times 10^{-6}) + (-0.327 \times 10^{-3})(0.5 - 0.7) = 0.5769 \text{ mA}$$

6.73 Sensitivity analysis can be extended to handle uncertainties in power-supply voltage. In the circuit of Fig. 4-5(a), let $R_1 = R_C = 500 \, \Omega$, $R_2 = 5 \, \text{k}\Omega$, $R_E = 100 \, \Omega$, $\beta = 75$, $V_{BEQ} = 0.7 \, \text{V}$, $I_{CBO} = 0.2 \, \mu\text{A}$, and $V_{CC} = 20 \pm 2 \, \text{V}$. Find an expression for the change in I_{CQ} due to changes in V_{CC} alone.

▮ If I_{CBO} is not neglected and $\beta \gg 1$, the amplifier is modeled by Fig. 6-1, and I_{CQ} is given by (4) of Problem 6.4. Proceeding similar to (2)–(6) of Problem 6.36 obtains

$$\Delta I_{CQ} \approx S_{VCC} \Delta V_{CC} \qquad \text{where} \qquad S_{VCC} = \frac{\partial I_{CQ}}{\partial V_{CC}}$$

Based on (4) of Problem 6.4, where V_{BB} and R_B are given by (1) of Problem 4.38,

$$S_{VCC} = \frac{\partial I_{CQ}}{\partial V_{CC}} = \frac{\partial}{\partial V_{CC}} \left\{ \frac{\beta[V_{CC}R_1/(R_1+R_2) - V_{BEQ} + I_{CBO}(R_B+R_E)]}{R_B + (\beta+1)R_E} \right\}$$

$$= \frac{\beta[R_1/(R_1+R_2)]}{R_B + (\beta+1)R_E} \approx \frac{R_1}{(R_1+R_2)(R_B/\beta + R_E)}$$

where $R_B = R_1 R_2/(R_1 + R_2)$.

6.74 Predict the change in I_{CQ} for the amplifier of Problem 6.73 as V_{CC} changes from its minimum to its maximum value.

▮ Based on results of Problem 6.73,

$$R_B = R_1 R_2/(R_1 + R_2) = (500)(5 \times 10^3)/(500 + 5 \times 10^3) = 454.5 \, \Omega$$

$$S_{VCC} \approx \frac{R_1}{(R_1+R_2)(R_B/\beta + R_E)} = \frac{500}{(500 + 5 \times 10^3)(454.5/75 + 100)} = 8.571 \times 10^{-4}$$

$$\Delta I_{CQ} \approx S_{VCC} \Delta V_{CC} \approx (8.571 \times 10^{-4})(22 - 18) = 3.428 \text{ mA}$$

6.75 In the circuit of Fig. 6-8, $R_1 = R_C = 500 \, \Omega$, $R_2 = 5 \, \text{k}\Omega$, $R_E = 100 \, \Omega$, $\beta = 75$, and $V_{CC} = 20 \, \text{V}$. Leakage current is negligible. At 25°C, $V_{BEQ} = 0.7 \, \text{V}$ and $V_D = 0.65 \, \text{V}$; however, both change at a rate of $-2 \, \text{mV}/°C$. Find the exact change in I_{CQ} due to an increase in temperature to 125°C.

▮ By (1) of Problem 4.38,

$$R_B = R_1 R_2/(R_1 + R_2) = 500(5 \times 10^3)/(500 + 5 \times 10^3) = 454.5 \, \Omega$$

$$V_{BB} = \frac{R_1}{R_1+R_2} V_{CC} = \frac{500}{500 + 5 \times 10^3}(20) = 1.818 \text{ V}$$

As usual, let subscripts 1 and 2 denote $T_1 = 25°C$ and $T_2 = 125°C$. Then, using (2) of Problem 6.30 with a procedure similar to that of Problem 6.17 leads to

$$I_{CQ1} = \frac{\beta[V_{BB} - (V_{BEQ1} - V_{D1})]}{R_B + (\beta+1)R_E} = \frac{75[1.818 - (0.7 - 0.65)]}{454.5 + (75+1)(100)} = 9.5455 \text{ mA}$$

$$V_{BEQ2} = V_{BEQ1} + \Delta V_{BEQ} = 0.7 - (2 \times 10^{-3})(125 - 25) = 0.5 \text{ V}$$

$$V_{D2} = V_{D1} + \Delta V_D = 0.65 - (2 \times 10^{-3})(125 - 25) = 0.45 \text{ V}$$

$$I_{CQ2} = \frac{\beta[V_{BB} - (V_{BEQ2} - V_{D2})]}{R_B + (\beta+1)R_E} = \frac{75[1.818 - (0.5 - 0.45)]}{454.5 + (75+1)(100)} = 9.5455 \text{ mA}$$

Whence, $\qquad \Delta I_{CQ} = I_{CQ2} - I_{CQ1} = 9.5455 - 9.5455 = 0$

6.76 By sensitivity analysis, predict the change in I_{CQ} of Problem 6.75 when the temperature increases to 125°C.

▮ Letting $\Delta I_{CBO} = \Delta \beta = 0$, and replacing V_{BEQ} with $V_{BEQ} - V_D$, extension of the work of Problem 6.37 leads to

$$\Delta I_{CQ} \approx S_V \Delta V_{BEQ} + S_{VD} \Delta V_D$$

where
$$S_V = -S_{VD} = -\frac{\beta}{R_B + \beta R_E} = -\frac{75}{454.5 + 75(100)} = -9.43 \times 10^{-3}$$

Thus,
$$\Delta I_{CQ} \approx (-9.43 \times 10^{-3})(0.5 - 0.7) + (9.43 \times 10^{-3})(0.45 - 0.65) = 0$$

Q-POINT-BOUNDED BIAS FOR THE FET

6.77 Just as β may vary in the BJT, the FET shorted-gate parameters I_{DSS} and V_{p0} can vary widely within devices of the same classification. If the extremes of FET parameter variation are known and (1) of Problem 5.6 has been used to establish the upper and lower (worst-case) transfer characteristics of Fig. 6-10, develop a procedure to set the gate-source bias for the amplifier of Fig. 5-10 so that the Q-point (and, hence, the quiescent drain current) remains within fixed limits of $I_{DQ\,max}$ and $I_{DQ\,min}$.

▮ A dc load line is established by choosing $R_D + R_S$ in the circuit of Fig. 6-10 so that v_{DS} remains within a desired region of the nominal drain characteristics. If now a value of R_S is selected such that

$$R_S \geq \frac{|V_{GSQ\,max} - V_{GSQ\,min}|}{I_{DQ\,max} - I_{DQ\,min}} \qquad (1)$$

Then the transfer bias line with slope $-1/R_S$ and v_{GS} intercept $V_{GG} \geq 0$ is located as shown in Fig. 6-10, and the nominal Q-point is forced to lie beneath Q_{max} and above Q_{min}, so that, as desired,

$$I_{DQ\,min} \leq I_{DQ} \leq I_{DQ\,max}$$

With R_S, R_D, and V_{GG} already assigned, R_G is chosen large enough to give a satisfactory input impedance, and then R_1 and R_2 are determined from (1) of Problem 5.23.

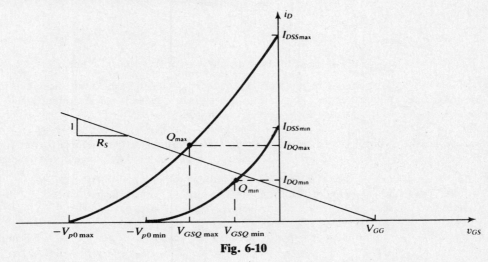

Fig. 6-10

6.78 The JFET of Fig. 5-10(b) is said to have *fixed bias* if $R_S = 0$. The worst-case shorted-gate parameters are given by the manufacturer of the device as

value	I_{DSS}, mA	V_{p0}, V
Maximum	8	6
Minimum	4	3

Let $V_{DD} = 15$ V, $V_{GG} = -1$ V, and $R_D = 2.5$ kΩ. Find the range of values of I_{DQ} that could be expected in using this FET.

▌ The maximum and minimum transfer characteristics are plotted in Fig. 6-11, based on (1) of Problem 5.6. Because $V_{GSQ} = V_{GG} = -1$ V is a fixed quantity unaffected by I_{DQ} and V_{DSQ}, the transfer bias line extends vertically at $V_{GS} = -1$, as shown. Its intersections with the two transfer characteristics give $I_{DQ\,max} \approx 5.5$ mA and $I_{DQ\,min} \approx 1.3$ mA.

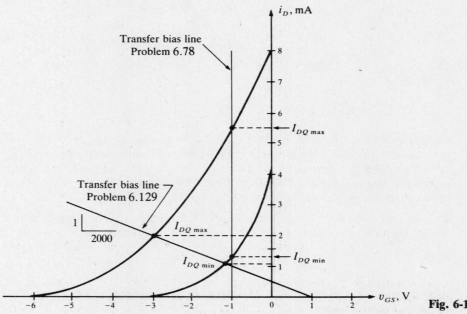

Fig. 6-11

6.79 Find the range of V_{DSQ} corresponding to the range of I_{DQ} determined for the JFET amplifier of Problem 6.78.

▌ For $I_{DQ} = I_{DQ\,max}$, KVL requires that
$$V_{DSQ\,max} = V_{DD} - I_{DQ\,max}R_D = 15 - 5.5(2.5) = 1.25 \text{ V}$$
And, for $I_{DQ\,min}$,
$$V_{DSQ\,min} = V_{DD} - I_{DQ\,min}R_D = 15 - 1.3(2.5) = 11.75 \text{ V}$$

6.80 Comment on the desirability of the JFET bias arrangement of Problems 6.78 and 6.79.

▌ The spread in FET parameters (and thus in transfer characteristics) makes the fixed-bias technique an undesirable one. The value of the Q-point drain current can vary from near the ohmic region to near the cutoff region, as is apparent from range on V_{DSQ} of Problem 6.79.

6.81 The self-biased JFET of Fig. 5-18 has a set of worst-case shorted-gate parameters that yield the plots of Fig. 6-12. Let $V_{DD} = 24$ V, $R_D = 3$ kΩ, $R_S = 1$ kΩ, and $R_G = 10$ MΩ. Find the range of I_{DQ} that can be expected.

▌ Since $V_{GG} = 0$, the transfer bias line must pass through the origin of the transfer characteristics plot, and its slope is $-1/R_S$ (solid line in Fig. 6-12). From the intersections of the transfer bias line and the transfer characteristics, we see that $I_{DQ\,max} \approx 2.5$ mA and $I_{DQ\,min} \approx 1.2$ mA.

6.82 Determine the range of drain-source quiescent voltage that can be expected for the JFET amplifier of Problem 6.81.

▌ For $I_{DQ} = I_{DQ\,max}$, KVL requires that
$$V_{DSQ\,max} = V_{DD} - I_{DQ\,max}(R_S + R_D) = 24 - 2.5(1 + 3) = 14 \text{ V}$$
And, for $I_{DQ\,min}$,
$$V_{DSQ\,min} = V_{DD} - I_{DQ\,min}(R_S + R_D) = 24 - 1.2(1 + 3) = 19.2 \text{ V}$$

6.83 Discuss the idea of reducing the I_{DQ} variation by increasing the value of R_S for the self-bias JFET amplifier of Problem 6.81.

▌ The transfer bias lines for $R_S = 2$ kΩ and 3 kΩ are also plotted on Fig. 6-12 (dashed lines). An increase in R_S obviously does decrease the difference between $I_{DQ\,max}$ and $I_{DQ\,min}$; however, in the process I_{DQ} is reduced to

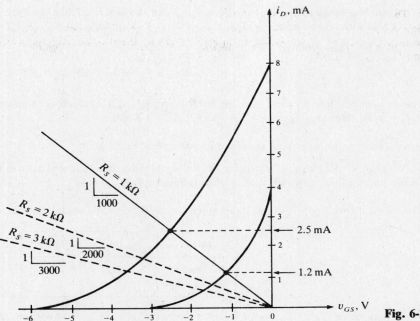

Fig. 6-12

quite low values, so that operation is on the nonlinear portion of the drain characteristics near the ohmic region where appreciable signal distortion results. But if self-bias with an external source is utilized (see Problem 6.86), the transfer bias line can be given a small negative slope without forcing I_{DQ} to approach zero.

6.84 In the JFET circuit of Fig. 5-10(a), using self-bias with an external source, $V_{DD} = 24$ V and $R_S = 3$ kΩ. The JFET is characterized by worst-case shorted-gate parameters that result in the transfer characteristics of Fig. 6-13. Find the range of I_{DQ} that can be expected if $R_1 = 1$ MΩ and $R_2 = 3$ MΩ.

▌ By (1) of Problem 5.23,

$$V_{GG} = \frac{R_1}{R_1 + R_2} V_{DD} = \frac{1}{1+3}(24) = 6 \text{ V}$$

In this case the transfer bias line, shown on Fig. 6-13, has abscissa intercept $v_{GS} = V_{GG} = 6$ V and slope $-1/R_S$. The range of I_{DQ} is determined by the intersections of the transfer bias line and the transfer characteristics: $I_{DQ\,\text{max}} \approx 2.8$ mA and $I_{DQ\,\text{min}} \approx 2.2$ mA.

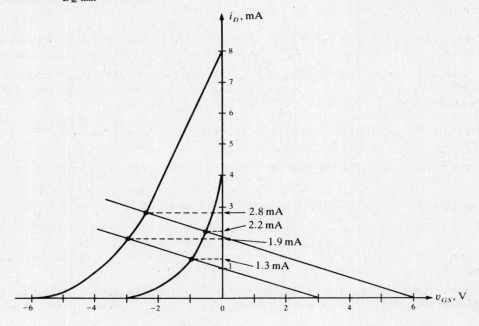

Fig. 6-13

6.85 Determine the range of I_{DQ} that can be expected for the JFET amplifier of Problem 6.84 if $R_1 = 1\ \text{M}\Omega$ and $R_2 = 7\ \text{M}\Omega$ with all else unchanged.

▌ Again by (1) of Problem 5.23,

$$V_{GG} = \frac{1}{1+7}(24) = 3\ \text{V}$$

The transfer bias line for this case is also drawn on Fig. 6-13; it has abscissa intercept $v_{GS} = V_{GG} = 3\ \text{V}$ and slope $-1/R_S$. Here $I_{DQ\ \text{max}} \approx 1.9\ \text{mA}$ and $I_{DQ\ \text{min}} \approx 1.3\ \text{mA}$.

6.86 Discuss the significance of the results of Problems 6.84 and 6.85.

▌ We changed V_{GG} by altering the R_1-R_2 voltage divider. This allowed us to maintain a small negative slope on the transfer bias line (and, thus, a small difference $I_{DQ\ \text{max}} - I_{DQ\ \text{min}}$) while shifting the range of I_{DQ}.

6.87 The MOSFET of Fig. 5-17 is an enhancement-mode device with worst-case shorted-gate parameters as follows:

value	I_{D0}, mA	V_T, V
Maximum	8	4
Minimum	4	2

These parameter values lead to the transfer characteristics of Fig. 6-14 because the device may be assumed to obey (1) of Problem 5.17. Let $V_{DD} = 24\ \text{V}$, $R_1 = 2\ \text{M}\Omega$, $R_2 = 2\ \text{M}\Omega$, $R_D = 1\ \text{k}\Omega$, and $R_S = 2\ \text{k}\Omega$. Find the range of I_{DQ} that can be expected.

▌ By (1) of Problem 5.23,

$$V_{GG} = \frac{R_1}{R_1 + R_2} V_{DD} = \frac{2}{2+2}(24) = 12\ \text{V}$$

The transfer bias line, with abscissa intercept $v_{GS} = V_{GG} = 12\ \text{V}$ and slope $-1/R_S$, is drawn on Fig. 6-14. From the intersections of the transfer bias line with the transfer characteristics, we see that $I_{DQ\ \text{max}} \approx 4\ \text{mA}$ and $I_{DQ\ \text{min}} \approx 2.8\ \text{mA}$.

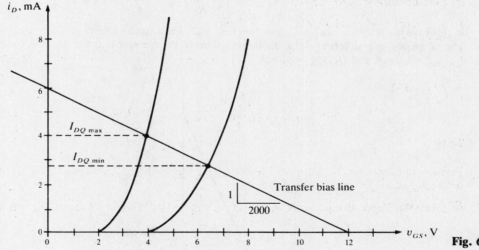

Fig. 6-14

6.88 Determine the range of V_{DSQ} that can be expected for the MOSFET amplifier of Problem 6.87.

▌ By KVL,

$$V_{DSQ\ \text{max}} = V_{DD} - I_{DQ\ \text{max}}(R_S + R_D) = 24 - 4(2+1) = 12\ \text{V}$$

$$V_{DSQ\ \text{min}} = V_{DD} - I_{DQ\ \text{min}}(R_S + R_D) = 24 - 2.8(2+1) = 15.6\ \text{V}$$

6.89 Discuss a technique, suggested by Problems 6.87 and 6.88, for minimizing the range of I_{DQ} for the MOSFET amplifier of Fig. 5-17.

TRANSISTOR BIAS CONSIDERATIONS ◻ 183

▌ As in the case of the JFET (see Problem 6.83), the range of I_{DQ} can be decreased by increasing R_S. However, to avoid undesirably small values for I_{DQ}, it is also necessary to increase V_{GG} by altering the R_1-R_2 voltage-divider ratio.

Review Problems

6.90 For the BJT amplifier of Fig. 6-15, find an expression for quiescent collector current I_{CQ} if leakage current I_{CBO} is neglected.

▌ Based on (1) of Problem 4.38, the Thévenin equivalent is found by letting $R_F = R_2$ and $R_1 \to \infty$, giving $R_B = R_F$ and $V_{BB} = V_{CC}$. Hence, I_{CQ} follows from (1) of Problem 4.40 as

$$I_{CQ} \approx \frac{V_{CC} - V_{BEQ}}{R_F/(\beta + 1) + R_E} \tag{1}$$

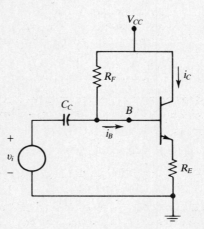

Fig. 6-15

6.91 For the amplifier of Fig. 6-15, $V_{CC} = 15$ V, $R_F = 50$ kΩ, $R_E = 1$ kΩ, and $I_{CBO} \approx 0$. Find the range of I_{CQ} if β of the Si transistor varies from 50 to 150.

▌ Applying (1) of Problem 6.90,
For $\beta = 50$:

$$I_{CQ} \approx \frac{V_{CC} - V_{BEQ}}{R_F/(\beta + 1) + R_E} = \frac{15 - 0.7}{(50 \times 10^3)/(50 + 1) + 1 \times 10^3} = 7.22 \text{ mA}$$

For $\beta = 150$:

$$I_{CQ} \approx \frac{15 - 0.7}{(50 \times 10^3)/(150 + 1) + 1 \times 10^3} = 10.74 \text{ mA}$$

Thus, $\qquad 7.22 \leq I_{CQ} \leq 10.74 \text{ mA}$

6.92 Find an expression for the quiescent collector current I_{CQ} of the amplifier in Fig. 6-15 if the leakage current is not negligible.

▌ Using the Thévenin equivalent described in Problem 6.90 and referring to Fig. 4-5(c), KVL gives

$$V_{CC} = \left(\frac{R_F}{\beta + 1} + R_E\right) I_{EQ} + V_{BEQ} \tag{1}$$

Substitution of I_{EQ} as determined by Problem 4.129 into (1) leads to

$$V_{CC} = \left(\frac{R_F}{\beta + 1} + R_E\right)\left[\frac{\beta + 1}{\beta}(I_{CQ} - I_{CBO})\right] + V_{BEQ} \tag{2}$$

Rearranging (2) and assuming $\beta \gg 1$ yield

$$I_{CQ} = \frac{V_{CC} - V_{BEQ}}{R_F/\beta + R_E(\beta + 1)/\beta} + I_{CBO} \approx \frac{V_{CC} - V_{BEQ}}{R_F/\beta + R_E} + I_{CBO} \tag{3}$$

184 CHAPTER 6

6.93 For the amplifier of Fig. 6-15, use stability-factor analysis to find an approximate expression for the total change in I_{CQ} due to variations in β, I_{CBO}, and V_{BEQ}.

The stability factors are found according to (3)–(5) of Problem 6.36 where I_{CQ} is given by (3) of Problem 6.92.

$$S_\beta = \frac{\partial I_{CQ}}{\partial \beta} = \frac{\partial}{\partial \beta}\left[\frac{V_{CC} - V_{BEQ}}{R_F/\beta + R_E} + I_{CBO}\right] = \frac{R_F(V_{CC} - V_{BEQ})}{(R_F + \beta R_E)^2} \quad (1)$$

$$S_I = \frac{\partial I_{CQ}}{\partial I_{CBO}} = 1 \quad (2)$$

$$S_V = \frac{\partial I_{CQ}}{\partial V_{BEQ}} = \frac{-\beta}{R_F + \beta R_E} \quad (3)$$

Whence, by (6) of Problem 6.36,

$$\Delta I_{CQ} \approx S_\beta \Delta_\beta + S_I \Delta I_{CBO} + S_V \Delta V_{BEQ} \quad (4)$$

6.94 It is determined that $\Delta I_{CQ} = 10.74 - 7.22 = 3.52$ mA as β changes from 50 to 150 for the amplifier of Problem 6.92. Find the approximate change in I_{CQ} predicted by sensitivity analysis.

Based on (1) and (4) of Problem 6.93,

$$S_\beta = \frac{R_F(V_{CC} - V_{BEQ})}{(R_F + \beta R_E)^2} = \frac{(50 \times 10^3)(15 - 0.7)}{[50 \times 10^3 + (50)(1 \times 10^3)]^2} = 7.15 \times 10^{-5}$$

$$\Delta I_{CQ} \approx S_\beta \Delta_\beta = (7.15 \times 10^{-5})(150 - 50) = 7.15 \text{ mA}$$

Note that the error is appreciable. Hence, in this case a first order approximation to the change in I_{CQ} is not acceptable.

6.95 Repeat Problem 6.94 if β varies from 50 to 60.

Based on (1) of Problem 6.90 with $\beta = 60$,

$$I_{CQ} = \frac{V_{CC} - V_{BEQ}}{R_F/(\beta + 1) + R_E} = \frac{15 - 0.7}{(50 \times 10^3)/(60 + 1) + 1 \times 10^3} = 8.08 \text{ mA}$$

Thus, the exact value of $\Delta I_{CQ} = 8.08 - 7.22 = 0.86$ mA. By sensitivity analysis,

$$\Delta I_{CQ} \approx S_\beta \Delta_\beta = (7.15 \times 10^{-5})(60 - 50) = 0.715 \text{ mA}$$

Note that the percentage error in ΔI_{CQ} is reduced compared with the result of Problem 6.94 for a smaller change in β.

6.96 For the amplifier of Problem 6.91, let $I_{CBO} = 0.1$ μA and $V_{BEQ} = 0.7$ V at a temperature of 20°C. If the temperature increases to 60°C, find the exact change in I_{CQ} for $\beta = 150$.

Let subscripts 1 and 2 denote temperatures of 20°C and 60°C, respectively; then, by (3) of Problem 6.92,

$$I_{CQ1} = \frac{V_{CC} - V_{BEQ1}}{R_F/\beta + R_E} + I_{CBO1} = \frac{15 - 0.7}{(50 \times 10^3)/150 + 1 \times 10^3} + 0.5 \times 10^{-6} = 10.7255 \text{ mA}$$

Now following the procedure of Problem 6.17,

$$I_{CBO2} = I_{CBO1} 2^{\Delta T/10} = (0.1 \times 10^{-6}) 2^{(60-20)/10} = 1.6 \ \mu\text{A}$$

$$\Delta V_{BEQ} = -2 \times 10^{-3} \Delta T = (-2 \times 10^{-3})(60 - 40) = -0.08 \text{ V}$$

$$V_{BEQ2} = V_{BEQ1} + \Delta T = 0.7 + (-0.08) = 0.62 \text{ V}$$

Again, by (3) of Problem 6.92,

$$I_{CQ2} = \frac{V_{CC} - V_{BEQ2}}{R_F/\beta + R_E} + I_{CBO2} = \frac{15 - 0.62}{(50 \times 10^3)/150 + 1 \times 10^3} + 1.6 \times 10^{-6} = 10.7866 \text{ mA}$$

Hence, $$\Delta I_{CQ} = I_{CQ2} - I_{CQ1} = 10.7866 - 10.7255 = 0.0611 \text{ mA}$$

6.97 Repeat Problem 6.96 except predict the change of I_{CQ} using stability-factor analysis.

▌ By (3) and (4) of Problem 6.37,

$$S_I = 1 \qquad S_V = \frac{-\beta}{R_F + \beta R_E} = \frac{-150}{50 \times 10^3 + (150)(1 \times 10^3)} = -0.75 \times 10^{-3}$$

Thus, according to (6) of Problem 6.36 and using the results of Problem 6.96,

$$\Delta I_{CQ} \approx S_I \Delta I_{CBO} + S_V \Delta V_{BEQ} \approx 1(1.6 \times 10^{-6} - 0.1 \times 10^{-6}) + (-0.75 \times 10^{-3})(-0.08) = 0.0615 \text{ mA}$$

6.98 For the amplifier of Problem 6.91 with $\beta = 100$, find the change in I_{CQ} if R_E changes from 1 kΩ to 900 Ω.

▌ Based on (1) of Problem 6.90,
For $R_E = 1$ kΩ:

$$I_{CQ} = \frac{V_{CC} - V_{BEQ}}{R_F/(\beta + 1) + R_E} = \frac{15 - 0.7}{(50 \times 10^3)/(100 + 1) + 1 \times 10^3} = 9.565 \text{ mA}$$

For $R_E = 900$ Ω:

$$I_{CQ} = \frac{15 - 0.7}{(50 \times 10^3)/(100 + 1) + 900} = 10.250 \text{ mA}$$

Whence, $\qquad \Delta I_{CQ} = 10.250 - 9.565 = 0.685$ mA

6.99 Repeat Problem 6.98 using stability-factor analysis similar to that of Problem 6.51.

▌ Using (3) of Problem 6.92,

$$S_{RE} = \frac{\partial I_{CQ}}{\partial R_E} = \frac{\partial}{\partial R_E}\left[\frac{V_{CC} - V_{BEQ}}{R_F/\beta + R_E} + I_{CBO}\right] = \frac{-\beta^2(V_{CC} - V_{BEQ})}{(R_F + \beta R_E)^2}$$

Substituting known values,

$$S_{RE} = \frac{-(100)^2(15 - 0.7)}{[150 \times 10^3 + (100)(1 \times 10^3)]^2} = -6.355 \times 10^{-6}$$

Thus, $\qquad \Delta I_{CQ} \approx S_{RE} \Delta R_E = (-6.355 \times 10^{-6})(900 - 1 \times 10^3) = 0.635$ mA

6.100 Develop a stability-factor analysis procedure to predict changes in I_{CQ} for the amplifier of Fig. 6-15 due to changes in power supply voltage V_{CC}.

▌ Similar to the procedure of Problem 6.36,

$$\Delta I_{CQ} \approx S_{VCC} \Delta V_{CC} \qquad (1)$$

where $S_{VCC} = \partial I_{CQ}/\partial V_{CC}$. Using I_{CQ} as given by (3) of Problem 6.92,

$$S_{VCC} = \frac{\partial}{\partial V_{CC}}\left[\frac{V_{CC} - V_{BEQ}}{R_F/\beta + R_E}\right] = \frac{\beta}{R_F + \beta R_E} \qquad (2)$$

6.101 Apply the sensitivity-factor analysis procedure developed in Problem 6.100 to predict the change in I_{CQ} for the amplifier of Problem 6.91 if V_{CC} were increased to 17 V and $\beta = 100$.

▌ Based on (1) and (2) of Problem 6.100,

$$S_{VCC} = \frac{\beta}{R_F + \beta R_E} = \frac{100}{50 \times 10^3 + 100(1 \times 10^3)} = 0.667 \times 10^{-3}$$

$$\Delta I_{CQ} \approx S_{VCC} \Delta V_{CC} = (0.667 \times 10^{-3})(17 - 15) = 1.333 \text{ mA}$$

6.102 The constant-base-current-biased transistor amplifier of Fig. 4-10 has values of Problem 6.1 except $\beta = 100$. Find the exact change in I_{CQ} if R_B increases by 10 percent to 550 kΩ.

▌ By (2) of Problem 6.1 for $R_B = 550$ kΩ,

$$I_{CQ} = \frac{\beta(V_{CC} - V_{BEQ})}{R_B} = \frac{100(15 - 0.7)}{550 \times 10^3} = 2.6 \text{ mA}$$

Using the result of Problem 6.2 for I_{CQ} with $R_B = 500\ \text{k}\Omega$,
$$\Delta I_{CQ} = 2.6 - 2.86 = -0.26\ \text{mA}$$

6.103 Develop a sensitivity-factor analysis procedure to predict changes in I_{CQ} for the BJT amplifier of Fig. 4-10.

I Similar to the procedure of Problem 6.36,
$$\Delta I_{CQ} \approx S_{RB}\, \Delta R_B \qquad (1)$$
where $S_{RB} = \partial I_{CQ}/\partial R_B$. Based on I_{CQ} as given by (2) of Problem 6.1,
$$S_{RB} = \frac{\partial}{\partial R_B}\left[\frac{\beta(V_{CC} - V_{BEQ})}{R_B}\right] = \frac{-\beta(V_{CC} - V_{BEQ})}{R_B^2} \qquad (2)$$

6.104 Predict the change in I_{CQ} for the amplifier of Problem 6.102 by application of the sensitivity-factor procedure developed in Problem 6.103.

I Based on (1) and (2) of Problem 6.103,
$$S_{RB} = \frac{-\beta(V_{CC} - V_{BEQ})}{R_B^2} = \frac{-100(15 - 0.7)}{(500 \times 10^3)^2} = -5.72 \times 10^{-9}$$
$$\Delta I_{CQ} \approx S_{RB}\, \Delta R_B = (-5.72 \times 10^{-9})(550 \times 10^3 - 500 \times 10^3) = -0.286\ \text{mA}$$

6.105 For the constant-emitter-current-biased circuit of Fig. 6-1, neglect I_{CBO} of the Si transistor and let $V_{BB} = 6\ \text{V}$, $V_{CC} = 15\ \text{V}$, $R_B = 50\ \text{k}\Omega$, $R_E = 1\ \text{k}\Omega$, $R_C = 3\ \text{k}\Omega$, and $\beta = 75$. Find the exact change in I_{CQ} if R_B increases by 10 percent to $55\ \text{k}\Omega$.

I By (4) of Problem 6.4 with $I_{CBO} = 0$,
For $R_B = 50\ \text{k}\Omega$:
$$I_{CQ} = \frac{V_{BB} - V_{BEQ}}{R_B/\beta + R_E} = \frac{6 - 0.7}{(50 \times 10^3)/75 + 1 \times 10^3} = 3.18\ \text{mA}$$
For $R_B = 55\ \text{k}\Omega$:
$$I_{CQ} = \frac{6 - 0.7}{(55 \times 10^3)/75 + 1 \times 10^3} = 3.058$$
Hence, $\quad \Delta I_{CQ} = 3.058 - 3.18 = -0.122\ \text{mA}$

6.106 For the amplifier described by Problem 6.105, find the exact change in I_{CQ} if R_E increases by 10 percent to $1.1\ \text{k}\Omega$.

I By (4) of Problem 6.4 with $I_{CBO} = 0$,
For $R_E = 1\ \text{k}\Omega$:
$$I_{CQ} = \frac{V_{BB} - V_{BEQ}}{R_B/\beta + R_E} = \frac{6 - 0.7}{(50 \times 10^3)/75 + 1 \times 10^3} = 3.18\ \text{mA}$$
For $R_E = 1.1\ \text{k}\Omega$:
$$I_{CQ} = \frac{6 - 0.7}{(50 \times 10^3)/75 + 1.1 \times 10^3} = 3\ \text{mA}$$
Hence, $\quad \Delta I_{CQ} = 3 - 3.18 = -0.18\ \text{mA}$

6.107 Develop a sensitivity-factor analysis procedure to predict changes in I_{CQ} for the amplifier of Fig. 6-1 due to variations in R_B.

I Similar to the procedure of Problem 6.36,
$$\Delta I_{CQ} \approx S_{RB}\, \Delta R_B \qquad (1)$$
where $S_{RB} = \partial I_{CQ}/\partial R_B$. For I_{CQ} given by (4) of Problem 6.4 with $I_{CBO} = 0$,
$$S_{RB} = \frac{\partial}{\partial R_B}\left[\frac{\beta(V_{BB} - V_{BEQ})}{R_B + \beta R_E}\right] = \frac{-\beta(V_{BB} - V_{BEQ})}{(R_B + \beta R_E)^2} \qquad (2)$$

6.108 Predict the change in I_{CQ} for the amplifier of Problem 6.105 by application of the sensitivity-factor analysis procedure of Problem 6.107.

▌ Based on (1) and (2) of Problem 6.107,
$$S_{RB} = \frac{-\beta(V_{BB} - V_{BEQ})}{(R_B + \beta R_E)^2} = \frac{-75(6 - 0.7)}{[50 \times 10^3 + 75(1 \times 10^3)]^2} = -2.544 \times 10^{-8}$$
$$\Delta I_{CQ} \approx S_{RB} \Delta R_B = (-2.544 \times 10^{-8})(55 \times 10^3 - 50 \times 10^3) = -0.1272 \text{ mA}$$

6.109 Develop a sensitivity analysis procedure to predict changes in I_{CQ} for the amplifier of Fig. 6-1 due to variations in R_E.

▌ By method of Problem 6.36,
$$\Delta I_{CQ} \approx S_{RE} \Delta R_E \qquad (1)$$
where $S_{RE} = \partial I_{CQ} / \partial R_E$. Using I_{CB} from (4) of Problem 6.4 with $I_{CBO} = 0$,
$$S_{RE} = \frac{\partial}{\partial R_E}\left[\frac{\beta(V_{BB} - V_{BEQ})}{R_B + \beta R_E}\right] = \frac{-\beta^2(V_{BB} - V_{BEQ})}{(R_B + \beta R_E)^2} \qquad (2)$$

6.110 Predict the change in I_{CQ} for the amplifier described by Problem 6.106 by use of the sensitivity-factor analysis procedure developed in Problem 6.109.

▌ Directly applying (1) and (2) of Problem 6.109,
$$S_{RE} = \frac{-\beta^2(V_{BB} - V_{BEQ})}{(R_B + \beta R_E)^2} = \frac{-(75)^2(6 - 0.7)}{[50 \times 10^3 + (75)(1 \times 10^3)]^2} = -1.908 \times 10^{-6}$$
$$\Delta I_{CQ} \approx S_{RE} \Delta R_E = (-1.908 \times 10^{-6})(1.1 \times 10^3 - 1 \times 10^3) = -0.1908 \text{ mA}$$

6.111 The shunt-feedback-bias circuit of Fig. 4-19 uses a Si transistor. If $I_{CBO} = 0$, $V_{CC} = 15$ V, $R_C = 1.5$ kΩ, $R_F = 150$ kΩ, and $\beta = 75$, find the exact change in I_{CQ} if R_C decreases to 1.35 kΩ.

▌ Using I_{CQ} as determined in Problem 6.5,
For $R_C = 1.5$ kΩ:
$$I_{CQ} = \frac{\beta(V_{CC} - V_{BEQ})}{R_F + (\beta + 1)R_C} = \frac{75(15 - 0.7)}{150 \times 10^3 + (75 + 1)(1.5 \times 10^3)} = 4.063 \text{ mA}$$

For $R_C = 1.35$ kΩ:
$$I_{CQ} = \frac{75(15 - 0.7)}{150 \times 10^3 + (75 + 1)(1.35 \times 10^3)} = 4.246 \text{ mA}$$
and $\Delta I_{CQ} = 4.246 - 4.063 = 0.183$ mA

6.112 For the amplifier described by Problem 6.111, find the exact change in I_{CQ} if R_F decreases to 135 kΩ.

▌ Based on the result of Problem 6.5,
For $R_F = 150$ kΩ:
$$I_{CQ} = \frac{\beta(V_{CC} - V_{BEQ})}{R_F + (\beta + 1)R_C} = \frac{75(15 - 0.7)}{150 \times 10^3 + (75 + 1)(1.5 \times 10^3)} = 4.063 \text{ mA}$$

For $R_F = 135$ kΩ:
$$I_{CQ} = \frac{75(15 - 0.7)}{135 \times 10^3 + (75 + 1)(1.5 \times 10^3)} = 4.333 \text{ mA}$$
Hence, $\Delta I_{CQ} = 4.333 - 4.063 = 0.27$ mA

6.113 Develop a sensitivity-factor analysis procedure to predict changes in I_{CQ} for the amplifier of Fig. 4-19 due to variations in R_C.

▌ Following the method of Problem 6.36,
$$\Delta I_{CQ} \approx S_{RC} \Delta R_C \qquad (1)$$

where $S_{RC} = \partial I_{CQ}/\partial R_C$. With I_{CQ} as given by Problem 6.5,

$$S_{RC} = \frac{\partial}{\partial R_C}\left[\frac{\beta(V_{CC} - V_{BEQ})}{R_F + (\beta + 1)R_C}\right] = \frac{-\beta(\beta + 1)(V_{CC} - V_{BEQ})}{[R_F + (\beta + 1)R_C]^2} \quad (2)$$

6.114 Predict the change in I_{CQ} for the amplifier of Problem 6.111 by use of the sensitivity-factor analysis procedure developed in Problem 6.113.

▌ Applying (1) and (2) of Problem 6.113,

$$S_{RC} = \frac{-\beta(\beta + 1)(V_{CC} - V_{BEQ})}{[R_F + (\beta + 1)R_C]^2} = \frac{-75(75 + 1)(15 - 0.7)}{[150 \times 10^3 + (75 + 1)(1.5 \times 10^3)]^2} = -1.169 \times 10^{-6}$$

$$\Delta I_{CQ} \approx S_{RC}\,\Delta R_C = (-1.169 \times 10^{-6})(1.35 \times 10^3 - 1.5 \times 10^3) = 0.175 \text{ mA}$$

6.115 Develop a sensitivity-factor analysis procedure to predict changes in I_{CQ} for the amplifier of Fig. 4-19 due to variations of R_F.

▌ Proceeding as in Problem 6.36,

$$\Delta I_{CQ} \approx S_{RF}\,\Delta R_F \quad (1)$$

where $S_{RF} = \partial I_{CQ}/\partial R_F$. Based on the expression for I_{CQ} of Problem 6.5,

$$S_{RF} = \frac{\partial}{\partial R_F}\left[\frac{\beta(V_{CC} - V_{BEQ})}{R_F + (\beta + 1)R_C}\right] = \frac{-\beta(V_{CC} - V_{BEQ})}{[R_F + (\beta + 1)R_C]^2} \quad (2)$$

6.116 Predict the change in I_{CQ} for the amplifier of Problem 6.112 by application of the sensitivity-factor analysis procedure developed in Problem 6.115.

▌ By (1) and (2) of Problem 6.115,

$$S_{RF} = \frac{-\beta(V_{CC} - V_{BEQ})}{[R_F + (\beta + 1)R_C]^2} = \frac{-75(15 - 0.7)}{[150 \times 10^3 + (75 + 1)(1.5 \times 10^3)]^2} = -1.539 \times 10^{-8}$$

$$\Delta I_{CQ} \approx S_{RF}\,\Delta R_F = (-1.539 \times 10^{-8})(135 \times 10^3 - 150 \times 10^3) = 0.231 \text{ mA}$$

6.117 For the shunt-feedback-bias amplifier of Fig. 4-19 with parameters as described in Problem 6.111, use sensitivity analysis to predict the value of I_{CQ} if both R_C and R_F decrease by 10 percent.

▌ Proceeding as in Problem 6.36, and using the results of Problems 6.114 and 6.116,

$$\Delta I_{CQ} \approx S_{RC}\,\Delta R_C + S_{RF}\,\Delta R_F = 0.175 + 0.231 = 0.406 \text{ mA}$$

and

$$I_{CQ} \approx 4.063 + \Delta I_{CQ} = 4.063 + 0.406 = 4.469 \text{ mA}$$

6.118 Find an exact solution to Problem 6.117.

▌ Based on the result of Problem 6.4,

$$I_{CQ} = \frac{\beta(V_{CC} - V_{BEQ})}{R_F + (\beta + 1)R_C} = \frac{75(15 - 0.7)}{135 \times 10^3 + (75 + 1)(1.35 \times 10^3)} = 4.514 \text{ mA}$$

6.119 A JFET for which (1) of Problem 5.6 holds is biased by the voltage-divider arrangement of Fig. 5-10. Find I_{DQ} as a function of I_{DSS}, V_{p0}, and V_{GG}.

▌ We use (1) of Problem 5.24 to find an expression for V_{GSQ} and then use (1) of Problem 5.6 to obtain

$$I_{DQ} = I_{DSS}\left(1 + \frac{V_{GG} - I_{DQ}R_S}{V_{p0}}\right)^2 \quad (1)$$

which we can solve for I_{DQ}:

$$I_{DQ} = \frac{V_{GG} + V_{p0}}{R_S} + \frac{V_{p0}^2}{2R_S^2 I_{DSS}} \pm \frac{V_{p0}}{2R_S}\left[\left(\frac{V_{p0}}{I_{DSS}}\right)^2 + \frac{4(V_{GG} + V_{p0})R_S}{I_{DSS}}\right]^{1/2} \quad (2)$$

6.120 For a JFET described by (1) of Problem 5.6 and biased by the voltage-divider arrangement of Fig. 5-10, find the total differential of I_{DQ}. Make reasonable linearity assumptions that allow us to replace differentials with increments so as to find an expression for the JFET analogous to (6) of Problem 6.36.

▌ Since V_{SQ} depends upon the bias network chosen, our result will have more general application if we take the differential of (1) of Problem 5.6 and then specialize it to the case at hand, instead of taking the differential of (2) of Problem 6.119. Assuming that I_{DSS}, V_{p0}, and V_{GSQ} are the independent variables, we have, for the total differential of (1) of Problem 5.6,

$$dI_{DQ} = \frac{\partial I_{DQ}}{\partial I_{DSS}} dI_{DSS} + \frac{\partial I_{DQ}}{\partial V_{p0}} dV_{p0} + \frac{\partial I_{DQ}}{\partial V_{GSQ}} dV_{GSQ} \tag{1}$$

For the case at hand, V_{GSQ} is given by (1) of Problem 5.24, from which

$$dV_{GSQ} = -R_S \, dI_{DQ} \tag{2}$$

Substituting (2) into (1) and rearranging, we find

$$dI_{DQ} = \frac{\partial I_{DQ}/\partial I_{DSS}}{1 + R_S \, \partial I_{DQ}/\partial V_{GSQ}} dI_{DSS} + \frac{\partial I_{DQ}/\partial V_{p0}}{1 + R_S \, \partial I_{DQ}/\partial V_{GSQ}} dV_{p0} \tag{3}$$

The assumption of linearity allows us to replace the differentials in (3) with increments and define appropriate stability factors:

$$\Delta I_{DQ} \approx S_I \, \Delta I_{DSS} + S_V \, \Delta V_{p0} \tag{4}$$

where

$$S_I = \frac{\partial I_{DQ}/\partial I_{DSS}}{1 + R_S \, \partial I_{DQ}/\partial V_{GSQ}} = \frac{(1 + V_{GSQ}/V_{p0})^2}{1 + (2R_S I_{DSS}/V_{p0})(1 + V_{GSQ}/V_{p0})} \tag{5}$$

$$S_V = \frac{\partial I_{DQ}/\partial I_{DSS}}{1 + R_S \, \partial I_{DQ}/\partial V_{GSQ}} = \frac{-2I_{DSS}(1 + V_{GSQ}/V_{p0})(V_{GSQ}/V_{p0}^2)}{1 + (2R_S I_{DSS}/V_{p0})(1 + V_{GSQ}/V_{p0})} \tag{6}$$

6.121 In Problem 6.120, it was assumed that V_{GG}, and hence V_{DD}, was constant. Supposing now that the power-supply voltage does vary, find an expression for ΔI_{DQ} using stability factors.

▌ Taking the total differential of (1) from Problem 5.24, we have

$$dV_{GSQ} = dV_{GG} - R_S \, dI_{DQ} \tag{1}$$

Substitute (1) into (1) of Problem 6.120 to give

$$dI_{DQ} = \frac{\partial I_{DQ}}{\partial I_{DSS}} dI_{DSS} + \frac{\partial I_{DQ}}{\partial V_{p0}} dV_{p0} + \frac{\partial I_{DQ}}{\partial V_{GSQ}} (dV_{GG} - R_S \, dI_{DQ}) \tag{2}$$

After rearranging,

$$dI_{DQ} = \frac{\partial I_{DQ}/\partial I_{DSS}}{1 + R_S \, \partial I_{DQ}/\partial V_{GSQ}} dI_{DSS} + \frac{\partial I_{DQ}/\partial V_{p0}}{1 + R_S \, \partial I_{DQ}/\partial V_{GSQ}} dV_{p0} + \frac{\partial I_{DQ}/\partial V_{GSQ}}{1 + R_S \, \partial I_{DQ}/\partial V_{GSQ}} dV_{GG} \tag{3}$$

Using (5) and (6) of Problem 6.120, defining

$$S_{VGG} = \frac{\partial I_{DQ}/\partial V_{GSQ}}{1 + R_S \, \partial I_{DQ}/\partial V_{GSQ}} = \frac{(2I_{DSS}/V_{p0})(1 + V_{GSQ}/V_{p0})}{1 + (2R_S I_{DSS}/V_{p0})(1 + V_{GSQ}/V_{p0})} \tag{4}$$

and replacing differentials in (3) with increments lead to

$$\Delta I_{DQ} \approx S_I \, \Delta I_{DSS} + S_V \, \Delta V_{p0} + S_{VGG} \, \Delta V_{GG} \tag{5}$$

6.122 The MOSFET of Fig. 5-17 is characterized by $V_T = 4$ V and $I_{D0} = 10$ mA. The device obeys (1) of Problem 5.17. Let $i_G \approx 0$, $R_2 = 0.4$ MΩ, $R_1 = 5$ kΩ, $R_S = 0$, $R_D = 2$ kΩ, and $V_{DD} = 20$ V. Find the exact change in I_{DQ} when the MOSFET is replaced with a new device characterized by $V_T = 3.8$ V and $I_{D0} = 9$ mA.

▌ Let subscripts 1 and 2 denote the original and the new device, respectively. Since $R_S = 0$, KVL requires that

$$V_{GSQ} = V_{GG} = \frac{R_1}{R_1 + R_2} V_{DD} = \frac{50 \times 10^3}{50 \times 10^3 + 0.4 \times 10^6} (15) = 1.667 \text{ V}$$

By (1) of Problem 5.17,

$$I_{DQ1} = I_{D01}\left(1 - \frac{V_{GSQ1}}{V_{T1}}\right)^2 = (10 \times 10^{-3})\left(1 - \frac{1.667}{4}\right)^2 = 3.402 \text{ mA}$$

$$I_{DQ2} = I_{D02}\left(1 - \frac{V_{GSQ2}}{V_{T2}}\right)^2 = (9 \times 10^{-3})\left(1 - \frac{1.667}{3.8}\right)^2 = 2.836 \text{ mA}$$

and

$$\Delta I_{DQ} = I_{DQ2} - I_{DQ1} = 2.836 - 3.402 = -0.566 \text{ mA}$$

6.123 Solve Problem 6.122 by sensitivity-analysis methods.

▌ Equations (5) and (6) of Problem 6.120 are applicable if V_{p0} and I_{DSS} were replaced by $-V_T$ and I_{D0}, respectively. Recalling that $R_S = 0$, we find

$$S_I = \frac{(1 - V_{GSQ}/V_T)^2}{1 + (2R_S I_{D0}/V_{p0})(1 - V_{GSQ}/V_{p0})} = \frac{(1 - 1.667/4)^2}{1 + 0} = 0.3402$$

$$S_V = \frac{-2I_{DSS}(1 - V_{GSQ}/V_T)(V_{GSQ}/V_{p0}^2)}{1 + (2R_S I_{D0}/V_{p0})(1 - V_{GSQ}/V_{p0})} = \frac{-2(10 \times 10^{-3})(1 - 1.667/4)(1.667/4^2)}{1 + 0} = 0.00122$$

Analogous to (4) of Problem 6.120,

$$\Delta I_{DQ} \approx S_I \Delta I_{D0} + S_V \Delta V_T \approx 0.3402(9 \times 10^{-3} - 10 \times 10^{-3}) + 0.00122(3.8 - 4) = -0.548 \text{ mA}$$

6.124 In the JFET amplifier of Fig. 5-10, $V_{DD} = 20$ V, $R_1 = 1$ MΩ, $R_2 = 15.7$ MΩ, $R_D = 3$ kΩ, $R_S = 2$ kΩ, and $i_G \approx 0$. The JFET obeys (1) of Problem 5.6 and is characterized by $I_{DSS} = 5$ mA and $V_{p0} = 5$ V. Due to aging, the resistance of R_1 increases by 20 percent. Find the value of quiescent source-to-gate voltage for this amplifier.

▌ Applying KVL around the gate-source loop with $i_G = 0$ yields

$$V_{GSQ} = \frac{R_1}{R_1 + R_2} V_{DD} - I_{DQ} R_S \tag{1}$$

Solving (1) for I_{DQ}, equating the result to (1) of Problem 5.6, and rearranging lead to

$$V_{GSQ}^2 + \left(2V_{p0} + \frac{V_{p0}^2}{I_{DSS} R_S}\right) V_{GSQ} + V_{p0}^2 \left[1 - \frac{R_1 V_{DD}}{(R_1 + R_2) I_{DSS} R_S}\right] = 0 \tag{2}$$

Using known values in (2) gives

$$V_{GSQ}^2 + 12.5 V_{GSQ} + 22.006 = 0 \tag{3}$$

Solving (3) for V_{GSQ} and disregarding the extraneous root $V_{GSQ} = -10.38 < -V_{p0}$, it is determined that $V_{GSQ} = -2.12$ V.

6.125 For the amplifier of Problem 6.124, find the exact change in I_{DQ} due to the increase in R_1.

▌ Let subscripts 1 and 2 denote the old and new cases, respectively. Then, by (1) of Problem 5.6,

$$I_{DQ1} = I_{DSS}(1 + V_{GSQ1}/V_{p0})^2 = (5 \times 10^{-3})(1 + -2.12/5)^2 = 1.658 \text{ mA}$$

Substituting known values, including $R_1 = 1.2$ MΩ, into (2) of Problem 6.124 gives the quadratic in V_{GSQ}.

$$V_{GSQ}^2 + 12.5 V_{GSQ} + 21.45 = 0 \tag{1}$$

Solving (1) for V_{GSQ} and disregarding the extraneous root $V_{GSQ} = -10.455 < -V_{p0}$, it is determined that $V_{GSQ} = -2.055$ V. Hence,

$$I_{DQ2} = I_{DSS}(1 + V_{GSQ2}/V_{p0})^2 = (5 \times 10^{-3})(1 + -2.055/5)^2 = 1.735 \text{ mA}$$

and

$$\Delta I_{DQ} = I_{DQ2} - I_{DQ1} = 1.735 - 1.658 = 0.077 \text{ mA}$$

6.126 Solve Problem 6.125 using sensitivity analysis.

▌ Equation (5) of Problem 6.121 is applicable where $\Delta I_{DSS} = \Delta V_{p0} = 0$. Since $V_{GG} = [R_1/(R_1 + R_2)]V_{DD}$,

$$\Delta V_{GG} = V_{GG2} - V_{GG1} = \left[\frac{1.2 \times 10^6}{1.2 \times 10^6 + 15.7 \times 10^6} - \frac{1 \times 10^6}{1 \times 10^6 + 15.7 \times 10^6}\right](20) = 0.2225 \text{ V}$$

By (4) of Problem 6.121,

$$S_{VGG} = \frac{(2I_{DSS}/V_{p0})(1 + V_{GSQ}/V_{p0})}{1 + (2R_S I_{DSS}/V_{p0})(1 + V_{GSQ}/V_{p0})} = \frac{[2(5 \times 10^{-3})/5](1 - 2.12/5)}{1 + [2(2 \times 10^3)(5 \times 10^{-3})/5](1 - 2.12/5)} = 3.48 \times 10^{-4}$$

and

$$\Delta I_{DQ} \approx S_{VGG} \Delta V_{GG} = (3.48 \times 10^{-4})(0.2225) = 0.0776 \text{ mA}$$

6.127 For a FET, the temperature dependence of V_{GSQ} is very small when I_{DQ} is held constant. Moreover, for constant V_{DSQ}, the temperature dependency of V_{GSQ} is primarily due to changes in the shorted-gate current;

those changes are given by

$$I_{DSS} = I_{DSSO}(k\,\Delta T + 1.1) \tag{1}$$

where I_{DSSO} = value of I_{DSS} at 0°C
ΔT = change in temperature from 0°C
k = constant (typically $0.003°\text{C}^{-1}$)

For the JFET of Fig. 5-10, $V_{DD} = 20\,\text{V}$, $R_1 = 1\,\text{M}\Omega$, $R_2 = 15.7\,\text{M}\Omega$, $R_D = 3\,\text{k}\Omega$, $R_S = 2\,\text{k}\Omega$, $i_G \approx 0$, $I_{DSSO} = 5\,\text{mA}$, and $V_{p0} = 5\,\text{V}$ (and is temperature-independent). Find the exact value of I_{DQ} at 100°C.

I Using the typical value of k in (1) gives

$$I_{DSS} = (5 \times 10^{-3})[0.003(100 - 0) + 1.1] = 7\,\text{mA}$$

Substitute known values into (2) of Problem 6.124 to find

$$V_{GSQ}^2 + 11.786 V_{GSQ} + 22.86 = 0 \tag{2}$$

Solve (2) for V_{GSQ}, discard the extraneous root $V_{GSQ} = -9.38 < -V_{p0}$, and determine $V_{GSQ} = -2.448\,\text{V}$. Then, by (1) of Problem 5.6,

$$I_{DQ} = I_{DSS}(1 + V_{GSQ}/V_{p0})^2 = (7 \times 10^{-3})(1 + -2.448/5)^2 = 1.82\,\text{mA}$$

6.128 Use sensitivity analysis to predict I_{DQ} at 100°C for the JFET of Problem 6.127.

I For $T = 0°\text{C}$, (1) of Problem 6.127 where $\Delta T = 0$ gives

$$I_{DSS} = I_{DSSO}(1.1) = 5 \times 10^{-3}(1.1) = 5.5\,\text{mA}$$

Substitute known values, including $I_{DSS} = 5.5\,\text{mA}$, into (2) of Problem 6.124 to find

$$V_{GSQ}^2 + 12.273 V_{GSQ} + 22.278 = 0 \tag{1}$$

Solving (1) for V_{GSQ} and discarding the extraneous root $V_{GSQ} = -10.06 < -V_{p0}$ find $V_{GSQ} = -2.215\,\text{V}$. By (5) of Problem 6.120,

$$S_I = \frac{\left(1 + \dfrac{V_{GSQ}}{V_{p0}}\right)^2}{1 + \dfrac{2R_S I_{DSS}}{V_{p0}}\left(1 + \dfrac{V_{GSQ}}{V_{p0}}\right)} = \frac{\left(1 + \dfrac{-2.215}{5}\right)^2}{1 + \dfrac{2(2\times 10^3)(5.5\times 10^{-3})}{5}\left(1 + \dfrac{-2.215}{5}\right)} = 0.0899$$

By (4) of Problem 6.120,

$$\Delta I_{DQ} \approx S_I\,\Delta I_{DSS} = 0.0899(7 \times 10^{-3} - 5.5 \times 10^{-3}) = 0.135\,\text{mA}$$

From (1) of Problem 5.6, I_{DQ} at $T = 0°$ is given by

$$I_{DQ}|_{0°\text{C}} = I_{DSS}(1 + V_{GSQ}/V_{p0})^2 = (5.5 \times 10^{-3})(1 + -2.215/5)^2 = 1.706\,\text{mA}$$

Hence, $I_{DQ}|_{100°\text{C}} = I_{DQ}|_{0°\text{C}} + \Delta I_{DQ} = 1.706 + 0.135 = 1.841\,\text{mA}$

6.129 Solve Problem 6.78 if $R_S = 2\,\text{k}\Omega$, $V_{GG} = 1\,\text{V}$, and all else remains unchanged.

I The transfer bias line is drawn on Fig. 6-11 with slope of $-1/R_S = -1/2000$ and abscissa intercept at $v_{GS} = V_{GG} = 1\,\text{V}$. The two intersections of the transfer bias line with the maximum and minimum transfer characteristics give $I_{DQ\,\text{max}} \approx 2\,\text{mA}$ and $I_{DQ\,\text{min}} \approx 1.1\,\text{mA}$.

6.130 Find the range of V_{DSQ} that could be expected for the amplifier of Problem 6.129.

I For $I_{DQ} = I_{DQ\,\text{max}}$, KVL requires that

$$V_{DSQ\,\text{max}} = V_{DD} - I_{DQ\,\text{max}}(R_D + R_S) = 15 - (2 \times 10^{-3})(2.5 \times 10^3 + 2 \times 10^3) = 6\,\text{V}$$

and, for $I_{DQ\,\text{min}}$,

$$V_{DSQ\,\text{min}} = V_{DD} - I_{DQ\,\text{min}}(R_D + R_S) = 15 - (1.1 \times 10^{-3})(2.5 \times 10^3 + 2 \times 10^3) = 10.05\,\text{V}$$

CHAPTER 7
Small-Signal Frequency-Independent BJT Amplifiers

BJT EQUIVALENT CIRCUIT MODELS

7.1 For sufficiently small emitter-collector voltage and current excursions about the quiescent point (*small signals*), the BJT may be replaced with any of several two-port *small-signal equivalent-circuit models*. Of particular application interest is a range of frequencies (*midfrequency range*) over which the amplifier can be considered independent of capacitive reactances. For a common-emitter (CE) transistor connection, assume that the total emitter-to-base voltage v_{BE} goes through only small excursions (ac signals) about the Q point so that $\Delta v_{BE} = v_{be}$ and $\Delta i_C = i_c$ and determine a linear system of equations to relate the small-signal quantities v_{be}, v_{ce}, i_b, and i_c.

▌ From Fig. 4-4(*b*) and (*c*), we see that if i_C and v_{BE} are taken as dependent variables in the CE transistor configuration, then

$$v_{BE} = f_1(i_B, v_{CE}) \qquad (1)$$

$$i_C = f_2(i_B, v_{CE}) \qquad (2)$$

Since $\Delta v_{BE} = v_{be}$ and $\Delta i_C = i_c$, application of the chain rule to (*1*) and (*2*), respectively, leads to the set of linear equations

$$v_{be} = \Delta v_{BE} \approx dv_{BE} = \left.\frac{\partial v_{BE}}{\partial i_B}\right|_Q i_b + \left.\frac{\partial v_{BE}}{\partial v_{CE}}\right|_Q v_{ce} \qquad (3)$$

$$i_c = \Delta i_C \approx di_C = \left.\frac{\partial i_C}{\partial i_B}\right|_Q i_b + \left.\frac{\partial i_C}{\partial v_{CE}}\right|_Q v_{ce} \qquad (4)$$

The four partial derivatives, evaluated at the Q point, that occur in (*3*) and (*4*) are called *CE hybrid parameters* and are denoted as follows:

$$\text{Input resistance} \qquad h_{ie} \equiv \left.\frac{\partial v_{BE}}{\partial i_B}\right|_Q \approx \left.\frac{\Delta v_{BE}}{\Delta i_B}\right|_Q \qquad (5)$$

$$\text{Reverse voltage ratio} \qquad h_{re} \equiv \left.\frac{\partial v_{BE}}{\partial v_{CE}}\right|_Q \approx \left.\frac{\Delta v_{BE}}{\Delta v_{CE}}\right|_Q \qquad (6)$$

$$\text{Forward current gain} \qquad h_{fe} \equiv \left.\frac{\partial i_C}{\partial i_B}\right|_Q \approx \left.\frac{\Delta i_C}{\Delta i_B}\right|_Q \qquad (7)$$

$$\text{Output admittance} \qquad h_{oe} \equiv \left.\frac{\partial i_C}{\partial v_{CE}}\right|_Q \approx \left.\frac{\Delta i_C}{\Delta v_{CE}}\right|_Q \qquad (8)$$

7.2 Draw a small-signal equivalent circuit based on (*3*) and (*4*) of Problem 7.1 valid for use with a CE-connected BJT whose signal excursions about the Q point are sufficiently small for the h parameters to be treated as constants.

▌ See Fig. 7-1.

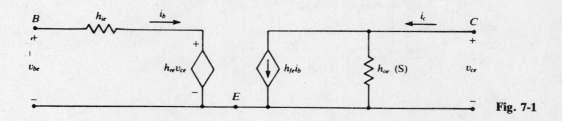

Fig. 7-1

7.3 In a manner analogous to the procedure of Problem 7.1, determine a system of linear equations to relate the small-signal quantities v_{eb}, v_{cb}, i_e, and i_c for the common-base (CB) connected BJT. Specifically, assume $\Delta v_{EB} = v_{eb}$ and $\Delta i_C = i_c$.

❙ If v_{EB} and i_C are taken as the dependent variables for the CB transistor characteristics of Fig. 4-3(b) and (c), then, as in the CE case of Problem 7.1, equations can be found specifically for small excursions about the Q point. The results are

$$v_{eb} = h_{ib} i_e + h_{rb} v_{cb} \tag{1}$$

$$i_c = h_{fb} i_e + h_{ob} v_{cb} \tag{2}$$

The partial-derivative definitions of the CB h parameters are:

$$\text{Input resistance} \qquad h_{ib} \equiv \left.\frac{\partial v_{EB}}{\partial i_E}\right|_Q \approx \left.\frac{\Delta v_{EB}}{\Delta i_E}\right|_Q \tag{3}$$

$$\text{Reverse voltage ratio} \qquad h_{rb} \equiv \left.\frac{\partial v_{EB}}{\partial v_{CB}}\right|_Q \approx \left.\frac{\Delta v_{EB}}{\Delta v_{CB}}\right|_Q \tag{4}$$

$$\text{Forward current gain} \qquad h_{fb} \equiv \left.\frac{\partial i_C}{\partial i_E}\right|_Q \approx \left.\frac{\Delta i_C}{\Delta i_E}\right|_Q \tag{5}$$

$$\text{Output admittance} \qquad h_{ob} \equiv \left.\frac{\partial i_C}{\partial v_{CB}}\right|_Q \approx \left.\frac{\Delta i_C}{\Delta v_{CB}}\right|_Q \tag{6}$$

7.4 Draw a small-signal equivalent circuit based on (*1*) and (*2*) of Problem 7.3 valid for use with a CB-connected BJT whose signal excursions about the Q point are sufficiently small for the h parameters to be treated as constants.

❙ See Fig. 7-2.

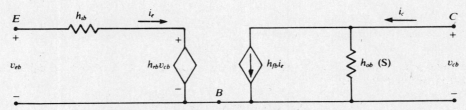

Fig. 7-2 CB small-signal equivalent circuit.

7.5 For the *common-collector* (CC) or *emitter-follower* (EF) amplifier, with the universal-bias circuitry of Fig. 7-3(a), replace the CE-connected transistor with its h-parameter model of Fig. 7-1 to establish a model valid for small-signal analysis.

❙ See Fig. 7-3(b).

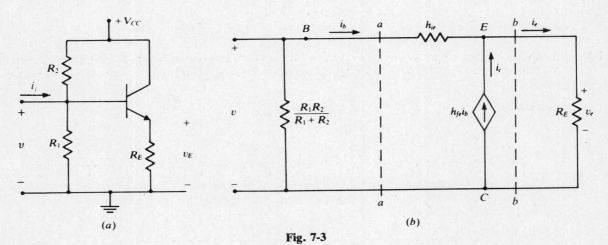

Fig. 7-3

7.6 Find a more simple small-signal model of the CC amplifier by obtaining a Thévenin equivalent for the circuit to the right of a, a in Fig. 7-3(b).

▌ Application of KVL around the outer loop of Fig. 7-3(b) gives

$$v = i_b h_{ie} + i_e R_E = i_b h_{ie} + (h_{fe} + 1) i_b R_E \qquad (1)$$

The Thévenin impedance is the driving-point impedance:

$$R_{Th} = \frac{v}{i_b} = h_{ie} + (h_{fe} + 1) R_E \qquad (2)$$

The Thévenin voltage is zero (computed with terminals a, a open); thus, the equivalent circuit consists only of R_{Th}. This is shown, in a base-current frame of reference, in Fig. 7-4.

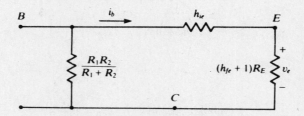

Fig. 7-4 CC amplifier.

7.7 The CE h-parameter transistor model (with $h_{re} = h_{oe} = 0$) was applied to the CC amplifier in Problem 7.6. Taking i_B and v_{EC} as independent variables, develop a linear system of equations to relate the small-signal quantities $v_{bc}, v_{ec}, i_b,$ and i_e, thus allowing for more accurate representation of the transistor than the circuit of Fig. 7-4.

▌ CC characteristics are not commonly given by transistor manufacturers, but they would be plots of i_B vs. v_{BC} with v_{EC} as parameter (input characteristics) and plots of i_E vs. v_{EC} with i_B as parameter (output or emitter characteristics). With i_B and v_{EC} as independent variables, we have

$$v_{BC} = f_1(i_B, v_{EC}) \qquad (1)$$

$$i_E = f_2(i_B, v_{EC}) \qquad (2)$$

Next we apply the chain rule to form the total differentials of (1) and (2), assuming that $v_{bc} = \Delta v_{BC} \approx dv_{BC}$, and similarly for i_e:

$$v_{bc} = \Delta v_{BC} \approx dv_{BC} = \left.\frac{\partial v_{BC}}{\partial i_B}\right|_Q i_b + \left.\frac{\partial v_{BC}}{\partial v_{EC}}\right|_Q v_{ec} \qquad (3)$$

$$i_e = \Delta i_E \approx di_E = \left.\frac{\partial i_E}{\partial i_B}\right|_Q i_b + \left.\frac{\partial i_E}{\partial v_{EC}}\right|_Q v_{ec} \qquad (4)$$

Finally, we define

$$\text{Input resistance} \quad h_{ic} \equiv \left.\frac{\partial v_{BC}}{\partial i_B}\right|_Q \approx \left.\frac{\Delta v_{BC}}{\Delta i_B}\right|_Q \qquad (5)$$

$$\text{Reverse voltage ratio} \quad h_{rc} \equiv \left.\frac{\partial v_{BC}}{\partial v_{EC}}\right|_Q \approx \left.\frac{\Delta v_{BC}}{\Delta v_{EC}}\right|_Q \qquad (6)$$

$$\text{Forward current gain} \quad h_{fc} \equiv \left.\frac{\partial i_E}{\partial i_B}\right|_Q \approx \left.\frac{\Delta i_E}{\Delta i_B}\right|_Q \qquad (7)$$

$$\text{Output admittance} \quad h_{oc} \equiv \left.\frac{\partial i_E}{\partial v_{EC}}\right|_Q \approx \left.\frac{\Delta i_E}{\Delta v_{EC}}\right|_Q \qquad (8)$$

7.8 Draw a small-signal equivalent circuit based on (3) and (4) of Problem 7.7 valid for use with a CC-connected BJT whose signal excursions about the Q point are sufficiently small for the h parameters as defined by (5)–(8) to be treated as constants.

▌ See Fig. 7-5.

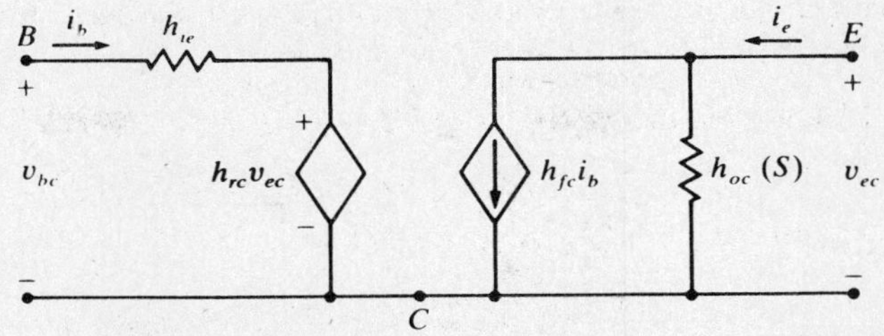

Fig. 7-5 CC small-signal equivalent circuit.

7.9 Apply the z-parameter definitions of (3)–(6) of Problem 1.67 to the CB small-signal equivalent circuit of Fig. 7-2 to write the linear system of equations given by (1) and (2) of Problem 1.67 so as to describe small-signal operation of the CB-connected BJT in terms of the CB h parameters.

▌ By (3)–(6) of Problem 1.67,

$$z_{11} = h_{ib} - \frac{h_{rb}h_{fb}}{h_{ob}} \tag{1}$$

$$z_{12} = \frac{h_{rb}}{h_{ob}} \tag{2}$$

$$z_{21} = -\frac{h_{fb}}{h_{ob}} \tag{3}$$

$$z_{22} = \frac{1}{h_{ob}} \tag{4}$$

Substitution of these z parameters into (1) and (2) of Problem 1.67 yields

$$v_{eb} = \left(h_{ib} - \frac{h_{rb}h_{fb}}{h_{ob}}\right)i_e + \frac{h_{rb}}{h_{ob}}(-i_c) \tag{5}$$

$$v_{cb} = -\frac{h_{fb}}{h_{ob}}i_e + \frac{1}{h_{ob}}(-i_c) \tag{6}$$

7.10 Define an appropriate set of parameters with units of Ohms (called *r parameters*) based on (5) and (6) of Problem 7.9 and rewrite the linear system of equations for small-signal description of the CB-connected BJT.

▌ If we now define

$$r_b = \frac{h_{rb}}{h_{ob}} \tag{1}$$

$$r_e = h_{ib} - \frac{h_{rb}}{h_{ob}}(1 + h_{fb}) \tag{2}$$

$$r_c = \frac{1 - h_{rb}}{h_{ob}} \tag{3}$$

$$\alpha' = -\frac{h_{fb} + h_{rb}}{1 - h_{rb}} \tag{4}$$

then (5) and (6) of Problem 7.9 can be written

$$v_{eb} = (r_e + r_b)i_e - r_b i_c \tag{5}$$

and

$$v_{cb} = (\alpha' r_c + r_b)i_e - (r_b + r_c)i_c \tag{6}$$

Typically, $-0.9 > h_{fb} > -1$ and $0 \le h_{rb} \ll 1$. Letting $h_{rb} \approx 0$ in (4), comparing (5) of Problem 7.3 with (1) of Problem 4.7 while neglecting thermally generated leakage currents, and assuming that $h_{FB} = h_{fb}$ (which is a valid assumption *except* near the boundary of active-region operation) result in

$$\alpha' \approx -h_{fb} = \alpha \tag{7}$$

7.11 Draw a small-signal equivalent circuit (*tee-equivalent circuit* or *r-parameter model*) based on (5) and (6) of Problem 7.10 suitable for analysis of a CB-connected BJT.

▌ See Fig. 7-6.

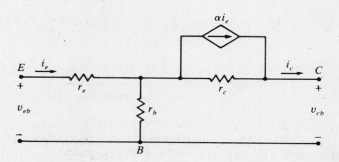

Fig. 7-6

7.12 The transistor of a CE amplifier can be modeled with the tee-equivalent circuit of Fig. 7-6 if the base and emitter terminals are interchanged, as shown by Fig. 7-7(a); however, the controlled source is no longer given in terms of a port current—an analytical disadvantage. Show that the circuit of Fig. 7-7(b), where the controlled variable of the dependent source is the input current i_b, can be obtained by application of Thévenin's theorem to the circuit of Fig. 7-7(a).

▌ The Thévenin equivalent for the circuit above terminals 1, 2 of Fig. 7-7(a) has

$$v_{th} = \alpha r_c i_e \qquad Z_{th} = r_c$$

By KCL, $i_e = i_c + i_b$, so that

$$v_{th} = \alpha r_c i_c + \alpha r_c i_b \qquad (1)$$

We recognize that if the Thévenin elements are placed in the network, the first term on the right side of (1) must be modeled by using a "negative resistance." The second term represents a controlled voltage source. Thus, a modified Thévenin equivalent can be introduced, in which the "negative resistance" is combined with Z_{th} to give

$$v'_{th} = \alpha r_c i_b = r_m i_b \qquad Z'_{th} = (1 - \alpha) r_c \qquad (2)$$

With the modified Thévenin elements of (2) in position, we obtain Fig. 7-7(b).

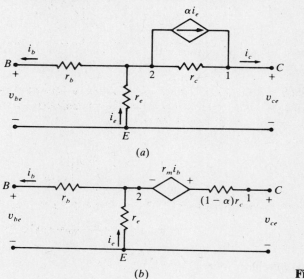

Fig. 7-7

7.13 Show that the *r*-parameter circuit for the CE-connected BJT of Fig. 7-8 can be obtained by application of Norton's theorem to the circuit of Fig. 7-7(a).

▌ The elements of the Norton equivalent circuit can be determined directly from (2) of Problem 7.12 as

$$Z_N = \frac{1}{Y_N} = Z'_{th} = (1 - \alpha) r_c \qquad I_N = \frac{v'_{th}}{Z'_{th}} = \frac{\alpha r_c i_b}{(1 - \alpha) r_c} = \beta i_b \qquad (1)$$

The elements of (1) give the circuit of Fig. 7-8.

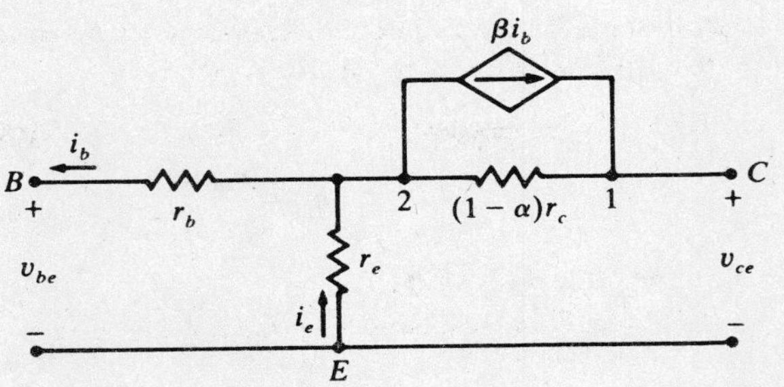

SMALL-SIGNAL FREQUENCY-INDEPENDENT BJT AMPLIFIERS ◻ 197

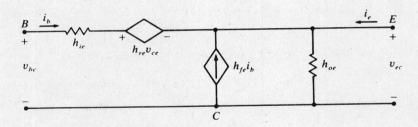

Fig. 7-8

7.14 Redraw the CE small-signal equivalent circuit of Fig. 7-1 so that the collector C is common to the input and output ports. Then apply KVL at the input port and KCL at the output port to find a set of equations that can be compared with (3) and (4) of Problem 7.7 to determine the CC h parameters in terms of the CE h parameters.

▮ Figure 7-1 is rearranged, to make the collector common, in Fig. 7-9. Applying KVL around the B, C loop, with $v_{ce} = -v_{ec}$, results in

$$v_{bc} = h_{ie}i_b + h_{re}v_{ce} + v_{ec} = h_{ie}i_b + (1 - h_{re})v_{ec} \tag{1}$$

Applying KCL at node E gives

$$i_e = -i_b - h_{fe}i_b + h_{oe}v_{ec} = -(h_{fe} + 1)i_b + h_{oe}v_{ec} \tag{2}$$

Comparison of (1) and (2) above with (3) and (4) of Problem 7.7 yields, by direct analogy,

$$h_{ic} = h_{ie} \qquad h_{rc} = 1 - h_{re} \qquad h_{fc} = -(h_{fe} + 1) \qquad h_{oc} = h_{oe} \tag{3}$$

Fig. 7-9

7.15 Apply the definitions of the general h parameters given by (3)–(6) of Problem 1.76 to the circuit of Fig. 7-2 to determine the CE parameter h_{ie} in terms of the CB h parameters. Use the typically good approximations $h_{rb} \ll 1$ and $h_{ob}h_{ib} \ll 1 + h_{fb}$ to simplify the results.

▮ By (3) of Problem 1.76,

$$h_{ie} = \left.\frac{v_{be}}{i_b}\right|_{v_{ce}=0} \tag{1}$$

If $v_{ce} = 0$ (short-circuited) in the network of Fig. 7-2, then $v_{cb} = -v_{be}$, so that, by KVL around the E, B loop,

$$v_{be} = -h_{ib}i_e - h_{rb}v_{cb} = -h_{ib}i_e + h_{rb}v_{be}$$

which gives

$$i_e = \frac{h_{rb} - 1}{h_{ib}} v_{be} \tag{2}$$

KCL at node b then gives

$$i_b = -(1 + h_{fb})i_e - h_{ob}v_{cb} = \left[\frac{(1 + h_{fb})(1 - h_{rb})}{h_{ib}} + h_{ob}\right]v_{be}$$

Now, by (1) and the given approximations,

$$h_{ie} = \frac{h_{ib}}{h_{ib}h_{ob} + (1 + h_{fb})(1 - h_{rb})} \approx \frac{h_{ib}}{1 + h_{fb}}$$

7.16 Based on (3)–(6) of Problem 1.76, determine the CE parameter h_{re} in terms of the CB parameters for the circuit of Fig. 7-2.

▌ By (4) of Problem 1.76,
$$h_{re} = \left.\frac{v_{be}}{v_{ce}}\right|_{i_b=0} \tag{1}$$

If $i_b = 0$, then $i_c = -i_e$ in Fig. 7-2. By KVL,
$$v_{ce} = v_{cb} - h_{rb}v_{cb} - h_{ib}i_e = (1 - h_{rb})v_{cb} - h_{ib}i_e \tag{2}$$

KCL at node C then gives
$$i_c = -i_e = h_{fb}i_e + h_{ob}v_{cb}$$

so that
$$i_e = -\frac{h_{ob}}{1 + h_{fb}} \tag{3}$$

Substituting (3) into (2) with $v_{cb} = v_{ce} - v_{be}$ gives
$$v_{ce} = (1 - h_{rb})(v_{ce} - v_{be}) + \frac{h_{ib}h_{ob}}{1 + h_{fb}}(v_{ce} - v_{be})$$

After rearranging, (1) and the approximations suggested by Problem 7.15 lead to
$$h_{re} = \frac{h_{rb}(1 + h_{fb}) - h_{ib}h_{ob}}{-h_{ib}h_{ob} + (h_{rb} - 1)(1 + h_{fb})} \approx \frac{h_{ib}h_{ob}}{1 + h_{fb}} - h_{rb}$$

7.17 Based on (3)–(6) of Problem 1.76, find an expression for the CE parameter h_{fe} in terms of the CB parameters for the circuit of Fig. 7-2.

▌ By (5) of Problem 1.76,
$$h_{fe} = \left.\frac{i_c}{i_b}\right|_{v_{ce}=0} \tag{1}$$

By KCL at node B of Fig. 7-2, with $v_{ce} = 0$ (and thus $v_{cb} = v_{eb} = -v_{be}$),
$$i_b = -(1 + h_{fb})i_e - h_{ob}v_{cb} = -(1 + h_{fb})i_e + h_{ob}v_{be}$$

Solving (2) of Problem 7.15 for v_{be}, with $i_e = -i_b - i_c$, and substituting now give
$$i_b = (1 + h_{fb})(i_b + i_c) + \frac{h_{ib}h_{ob}}{1 - h_{rb}}(i_b + i_c)$$

After rearranging, (1) and the approximations suggested by Problem 7.15 lead to
$$h_{fe} = \frac{-h_{fb}(1 - h_{rb}) - h_{ib}h_{ob}}{(1 + h_{fb})(1 - h_{rb}) + h_{ib}h_{ob}} \approx \frac{-h_{fb}}{1 + h_{fb}}$$

7.18 With (3)–(6) of Problem 1.76 as a basis, determine the CE parameter h_{oe} in terms of the CB parameters for the circuit of Fig. 7-2.

▌ By (6) of Problem 1.76,
$$h_{oe} = \left.\frac{i_c}{v_{ce}}\right|_{i_b=0} \tag{1}$$

If $i_b = 0$, then $-i_c = i_e$. Replacing i_e with $-i_c$ in (2) and (3) of Problem 7.16, solving (2) of Problem 7.16 for v_{cb}, and substituting into (3) of Problem 7.16 give
$$i_c = \frac{h_{ob}}{1 + h_{fb}}\left(\frac{v_{ce}}{1 - h_{rb}} - \frac{h_{ib}}{1 - h_{rb}}i_c\right)$$

After rearranging, (1) and the approximations suggested by Problem 7.15 lead to
$$h_{oe} = \frac{h_{ob}}{(1 - h_{fb})(1 + h_{rb}) + h_{ib}h_{ob}} \approx \frac{h_{ob}}{1 + h_{fb}}$$

7.19 Apply the definitions of the z parameters given by (3)–(6) of Problem 1.67 to the CB h-parameter circuit of Fig. 7-2 to find values for the z parameters in terms of the CB h parameters.

▎The circuit of Fig. 7-2 is described by the linear system of equations

$$\begin{bmatrix} h_{ib} & h_{rb} \\ h_{fb} & h_{ob} \end{bmatrix} \begin{bmatrix} i_e \\ v_{cb} \end{bmatrix} = \begin{bmatrix} v_{eb} \\ i_c \end{bmatrix} \qquad (1)$$

By (3) of Problem 1.67 and Fig. 1-19,

$$z_{11} = \left. \frac{v_{eb}}{i_e} \right|_{i_c=0} \qquad (2)$$

Setting $i_c = 0$ in (1) yields

$$v_{cb} = -\frac{h_{fb}}{h_{ob}} i_e \qquad (3)$$

Substituting (3) into the first equation of (1) and applying (2) yield

$$z_{11} = h_{ib} - \frac{h_{rb}h_{fb}}{h_{ob}}$$

By (5) of Problem 1.67 and (3),

$$z_{21} = \left. \frac{v_{cb}}{i_e} \right|_{i_c=0} = -\frac{h_{fb}}{h_{ob}}$$

By (4) of Problem 1.67,

$$z_{12} = \left. \frac{v_{eb}}{i_c} \right|_{i_e=0} \qquad (4)$$

Setting $i_e = 0$ in (1), solving the two equations for v_{cb}, and equating the results give

$$\frac{v_{eb}}{h_{rb}} = \frac{i_c}{h_{ob}} \qquad \text{from which} \qquad z_{12} = \frac{h_{rb}}{h_{ob}}$$

Finally, by (6) of Problem 1.67,

$$z_{22} = \left. \frac{v_{cb}}{i_c} \right|_{i_e=0} \qquad (5)$$

Letting $i_e = 0$ in the second equation of (1) and applying (5) yield $z_{22} = 1/h_{ob}$ directly.

7.20 Apply KVL and KCL to Fig. 7-1 to obtain $v_{eb} = g_1(i_e, v_{cb})$ and $i_c = g_2(i_e, v_{eb})$. Compare these equations with (1) and (2) of Problem 7.3 to find the CB h parameters in terms of the CE h parameters. Use the typically reasonable approximations $h_{re} \ll 1$ and $h_{fe} + 1 \gg h_{ie}h_{oe}$ to simplify the computations and results.

▎KVL around the E, B loop of Fig. 7-1 (with assumed current directions reversed) yields

$$v_{eb} = -h_{ie}i_b - h_{re}v_{ce} \qquad (1)$$

But KCL at node E requires that

$$i_b = -i_e - i_c = -i_e - h_{fe}i_b - h_{oe}v_{ce}$$

or

$$-i_b = \frac{1}{h_{fe}+1} i_e + \frac{h_{oe}}{h_{fe}+1} v_{ce} \qquad (2)$$

In addition, KVL requires that

$$v_{ce} = v_{cb} - v_{eb} \qquad (3)$$

Substituting (2) and (3) into (1) and rearranging give

$$\frac{(1-h_{re})(h_{fe}+1)+h_{ie}h_{oe}}{h_{fe}+1} v_{eb} = \frac{h_{ie}}{h_{fe}+1} i_e + \left(\frac{h_{ie}h_{oe}}{h_{fe}+1} - h_{re} \right) v_{cb} \qquad (4)$$

Use of the given approximations reduces the coefficient of v_{eb} in (4) to unity, so that

$$v_{eb} \approx \frac{h_{ie}}{h_{fe}+1} i_e + \left(\frac{h_{ie}h_{oe}}{h_{fe}+1} - h_{re} \right) v_{cb} \qquad (5)$$

Now KCL at node C of Fig. 7-1 (again with assumed current directions reversed) yields

$$i_c = h_{fe}i_b + h_{oe}v_{ce} \tag{6}$$

Substituting (2), (3), and (5) into (6) and solving for i_c give

$$i_c = -\left[\frac{h_{fe}}{h_{fe}+1} + \frac{h_{oe}h_{ie}}{(h_{fe}+1)^2}\right]i_e - h_{oe}\left[\frac{h_{ie}h_{oe}}{(h_{fe}+1)^2} - \frac{h_{re}+1}{h_{fe}+1}\right]v_{cb} \tag{7}$$

Use of the given approximations then leads to

$$i_c \approx \frac{h_{fe}}{h_{fe}+1}i_e + \frac{h_{oe}}{h_{fe}+1}v_{cb} \tag{8}$$

Comparing (5) with (1) of Problem 7.3 and (8) with (2) of Problem 7.3, we see that

$$h_{ib} = \frac{h_{ie}}{h_{fe}+1} \tag{9}$$

$$h_{rb} = \frac{h_{ie}h_{oe}}{h_{fe}+1} - h_{re} \tag{10}$$

$$h_{fb} = -\frac{h_{fe}}{h_{fe}+1} \tag{11}$$

$$h_{ob} = \frac{h_{oe}}{h_{fe}+1} \tag{12}$$

COMMON-EMITTER AMPLIFIERS

7.21 Draw a small-signal h-parameter equivalent circuit for the simplified (bias network omitted) CE amplifier of Fig. 7-10(a).

▮ See Fig. 7-10(b).

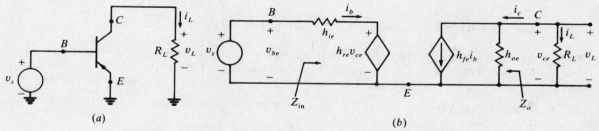

Fig. 7-10

7.22 In the CE amplifier of Fig. 7-10(b), let $h_{ie} = 1\text{ k}\Omega$, $h_{re} = 10^{-4}$, $h_{fe} = 100$, $h_{oe} = 12\ \mu\text{S}$, and $R_L = 2\text{ k}\Omega$. (These are typical CE amplifier values.) Find an expression for the current-gain ratio A_i.

▮ By current division at node C,

$$i_L = \frac{1/h_{oe}}{1/h_{oe} + R_L}(-h_{fe}i_b) \tag{1}$$

and

$$A_i = \frac{i_L}{i_b} = -\frac{h_{fe}}{1 + h_{oe}R_L} = -\frac{100}{1 + (12 \times 10^{-6})(2 \times 10^3)} = -97.7 \tag{2}$$

Note that $A_i \approx -h_{fe}$, where the minus sign indicates a 180° phase shift between input and output currents.

7.23 Derive an expression for the voltage-gain ratio A_v of the CE amplifier in Fig. 7-10(b) if described by the parameters of Problem 7.22.

▮ By KVL around B, E mesh,

$$v_s = v_{be} = h_{ie}i_b + h_{re}v_{ce} \tag{1}$$

Ohm's law applied to the output network requires that

$$v_{ce} = -h_{fe}i_b\left(\frac{1}{h_{oe}} \middle\| R_L\right) = \frac{-h_{fe}R_L i_b}{1 + h_{oe}R_L} \tag{2}$$

Solving (2) for i_b, substituting the result into (1), and rearranging yield

$$A_v = \frac{v_s}{v_{ce}} = -\frac{h_{fe}R_L}{h_{ie} + R_L(h_{ie}h_{oe} - h_{fe}h_{re})}$$

$$= -\frac{100(2 \times 10^3)}{1 \times 10^3 + (2 \times 10^3)[(1 \times 10^3)(12 \times 10^{-6}) - 100(1 \times 10^{-4})]} = -199.2 \tag{3}$$

Observe that $A_v \approx -h_{fe}R_L/h_{ie}$, where the minus sign indicates a 180° phase shift between input and output voltages.

7.24 Find an expression for the input impedance Z_{in} of the CE amplifier in Fig. 7-10(b) if described by the parameters of Problem 7.22.

▎ Substituting (2) into (1) of Problem 7.23 and rearranging yield

$$Z_{in} = \frac{v_s}{i_b} = h_{ie} - \frac{h_{re}h_{fe}R_L}{1 + h_{oe}R_L} = 1 \times 10^3 - \frac{(1 \times 10^{-4})(100)(2 \times 10^3)}{1 + (12 \times 10^{-6})(2 \times 10^3)} = 980.5\ \Omega \tag{1}$$

Note that for typical CE amplifier values, $Z_{in} \approx h_{ie}$.

7.25 Calculate the output impedance Z_o for the CE amplifier of Fig. 7-10(b) if described by the parameters of Problem 7.22.

▎ We deactivate (short) v_s and replace R_L with a driving-point source so that $v_{dp} = v_{ce}$. Then, for the input mesh, Ohm's law requires that

$$i_b = -\frac{h_{re}}{h_{ie}}v_{dp} \tag{1}$$

However, at node C (with, now, $i_c = i_{dp}$), KCL yields

$$i_c = i_{dp} = h_{fe}i_b + h_{oe}v_{dp} \tag{2}$$

Using (1) in (2) and rearranging then yield

$$Z_o = \frac{v_{dp}}{i_{dp}} = \frac{1}{h_{oe} - h_{fe}h_{re}/h_{ie}} = \frac{1}{12 \times 10^{-6} - 100(1 \times 10^{-4})/(1 \times 10^3)} = 500\ k\Omega \tag{3}$$

The output impedance is increased by feedback due to the presence of the controlled source $h_{re}v_{ce}$.

7.26 Evaluate a typical CE amplifier from the results of Problems 7.22–7.25.

▎ Based on the typical values of Problems 7.22–7.25, the characteristics of the CE amplifier can be summarized as follows:
1. Large current gain
2. Large voltage gain
3. Large power gain ($A_i A_v$)
4. Current and voltage phase shifts of 180°
5. Moderate input impedance
6. Moderate output impedance

7.27 Utilize the r-parameter model of Fig. 7-7(b) to draw a small-signal equivalent circuit for the CE amplifier of Fig. 4-14.

▎ See Fig. 7-11.

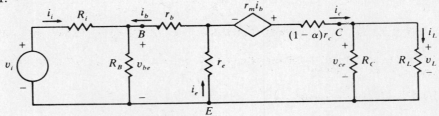

Fig. 7-11

7.28 Using the small-signal circuit of Fig. 7-11, find an expression for the voltage gain ratio $A_v = v_L/v_i$ of the amplifier of Fig. 4-14.

▎ After finding the Thévenin equivalent for the network to the left of terminals B, E in Fig. 7-11, we may write

$$v_{be} = \frac{R_B}{R_B + R_i} v_i + \frac{R_B R_i}{R_B + R_i} i_b \tag{1}$$

Ohm's law at the output requires that

$$v_{ce} = v_L = \frac{R_C R_L}{R_C + R_L} i_c \tag{2}$$

Applying KVL around the B, E mesh and around the C, E mesh while noting that $i_e = i_c + i_b$ yields, respectively,

$$v_{be} = -r_b i_b - r_e i_e = -(r_b + r_e) i_b - r_e i_c \tag{3}$$

and

$$v_{ce} = -r_e i_e + r_m i_b - (1-\alpha) r_c i_c = -(r_e - r_m) i_b - [(1-\alpha) r_c + r_e] i_c \tag{4}$$

Equating (1) to (3) and (2) to (4) allows formulation of the system of linear equations

$$\begin{bmatrix} -\left(r_b + r_e + \dfrac{R_B R_i}{R_B + R_i}\right)\dfrac{R_B + R_i}{R_B} & -\dfrac{r_e(R_B + R_i)}{R_B} \\ -(r_e - r_m) & -\left[(1-\alpha)r_c + r_e + \dfrac{R_C R_L}{R_C + R_L}\right] \end{bmatrix} \begin{bmatrix} i_b \\ i_c \end{bmatrix} = \begin{bmatrix} v_i \\ 0 \end{bmatrix} \tag{5}$$

from which, by Cramer's rule, $i_c = \Delta_2/\Delta$, where

$$\Delta = \frac{R_B + R_i}{R_B}\left\{\left(r_b + r_e + \frac{R_B R_i}{R_B + R_i}\right)\left[(1-\alpha)r_c + r_e + \frac{R_C R_L}{R_C + R_L}\right] - r_e(r_e - r_m)\right\} \tag{6}$$

$$\Delta_2 = (r_e - r_m) v_i$$

Then

$$A_v = \frac{v_L}{v_i} = \frac{(R_L \parallel R_C) i_c}{v_i} = \frac{R_L R_C}{R_L + R_C} \frac{r_e - r_m}{\Delta} \tag{7}$$

7.29 For the CE amplifier of Fig. 4-14, let $R_C = R_L = 800\ \Omega$, $R_i = 0$, $R_1 = 1.2$ kΩ, $R_2 = 2.7$ kΩ, $r_b = 99\ \Omega$, $r_e = 1\ \Omega$, $r_c = 1$ MΩ, and $\alpha = 0.99$. Find the voltage gain $A_v = v_L/v_i$.

▎ Recall from Problem 7.12 that $r_m = \alpha r_c = (0.99)(1 \times 10^6) = 0.99$ MΩ. And, by (1) of Problem 4.38,

$$R_B = \frac{R_1 R_2}{R_1 + R_2} = \frac{(1.2 \times 10^3)(2.7 \times 10^3)}{1.2 \times 10^3 + 2.7 \times 10^3} = 830\ \Omega$$

Direct application of (6) and (7) of Problem 7.28 leads to the desired result.

$$\Delta = \frac{830 + 0}{830}\left\{\left(99 + 1 + \frac{830(0)}{830 + 0}\right)\left[(1 - 0.99)(1 \times 10^6) + 1 + \frac{800(800)}{800 + 800}\right] - 1(1 - 0.99 \times 10^6)\right\}$$

$$= 2.03 \times 10^6$$

$$A_v = \frac{R_L R_C}{R_L + R_C} \frac{r_e - r_m}{\Delta} = \frac{800(800)}{800 + 800} \frac{1 - 0.99 \times 10^6}{2.03 \times 10^6} = -195.07$$

7.30 Repeat Problem 7.29 if $R_i = 50\ \Omega$ and all else is unchanged.

▎ Applying (6) and (7) of Problem 7.28,

$$\Delta = \frac{830 + 50}{830}\left\{\left(99 + 1 + \frac{830(50)}{830 + 50}\right)\left[(1 - 0.99)(1 \times 10^6) + 1 + \frac{800(800)}{800 + 800}\right] - 1(1 - 0.99 \times 10^6)\right\}$$

$$= 2.67 \times 10^6$$

$$A_v = \frac{800(800)}{800 + 800} \frac{1 - 0.99 \times 10^6}{2.67 \times 10^6} = -148.31$$

7.31 Use the *r*-parameter model of Fig. 7-7(*b*) to find the current gain ratio $A_i' = i_L/i_b$.

▎ The equivalent circuit of Fig. 7-11 is applicable. Current i_b can be found from (5) of Problem 7.28 by use of Cramer's rule as $i_b = \Delta_1/\Delta$. Δ is given by (6) of Problem 7.28, and

$$\Delta_1 = -v_i\left[(1-\alpha)r_c + r_e + \frac{R_C R_L}{R_C + R_L}\right] \quad (1)$$

Apply current division at node *C* and use i_c as determined in Problem 7.28 to find

$$i_L = \frac{R_C}{R_C + R_L}i_c = \frac{R_C}{R_C + R_L}\frac{\Delta_2}{\Delta} \quad (2)$$

Thus,

$$A_i' = \frac{i_L}{i_b} = \frac{\left(\frac{R_C}{R_C + R_L}\frac{\Delta_2}{\Delta}\right)}{\Delta_1/\Delta} = \frac{R_C}{R_C + R_L}\frac{\Delta_2}{\Delta_1} \quad (3)$$

Substitution of known expressions for Δ_1 and Δ_2 into (3) yields

$$A_i' = \frac{R_C}{R_C + R_L}\left[\frac{r_e - r_m}{(1-\alpha)r_c + r_e + R_C R_L/(R_C + R_L)}\right] \quad (4)$$

7.32 If the CE amplifier of Fig. 4-14 is described by the parameters of Problem 7.29, find the current gain $A_i' = i_L/i_b$.

▎ Based on (4) of Problem 7.31,

$$A_1' = \frac{R_C}{R_C + R_L}\left[\frac{r_e - r_m}{(1-\alpha)r_c + r_e + R_C R_L/(R_C + R_L)}\right]$$

$$= \frac{800}{800 + 800}\left[\frac{1 - 0.99 \times 10^6}{(1 - 0.99)(1 \times 10^6) + 1 + 800(800)/(800 + 800)}\right] = -38.1 \times 10^3$$

7.33 Determine the current gain ratio $A_i = i_L/i_i$ for the amplifier of Fig. 4-14 if $R_i = 0$.

▎ With $R_i = 0$, KCL applied at node *B* leads to

$$i_i = \frac{v_i}{R_B} - i_b = \frac{v_i}{R_B} - \frac{\Delta_1}{\Delta} \quad (1)$$

where Δ and Δ_1 are given by (1) of Problem 7.31 and (6) of Problem 7.28, respectively. Based on (7) of Problem 7.28, and realizing that $v_L = R_L i_L$, we find

$$v_i = \frac{v_L}{A_v} = \frac{R_L i_L}{\left[\frac{R_L R_C(r_e - r_m)}{(R_L + R_C)\Delta}\right]} = \frac{\Delta(R_L + R_C)i_L}{R_C(r_e - r_m)} \quad (2)$$

Substitute (2) into (1), use Δ_1 as given by (1) of Problem 7.31, and rearrange to yield

$$A_i = \frac{i_L}{i_i} = \frac{R_B R_C(r_e - r_m)}{(R_L + R_C)\{\Delta + R_B[(1-\alpha)r_c + r_e + R_C R_L/(R_C + R_L)]\}} \quad (3)$$

7.34 If the CE amplifier of Fig. 4-14 is described by the values of Problem 7.29, find the current gain $A_i = i_L/i_i$.

▎ By (3) of Problem 7.33,

$$A_i = \frac{830(800)(1 - 0.99 \times 10^6)}{(800 + 800)\{2.03 \times 10^6 + 830[(1 - 0.99)(1 \times 10^6) + 1 + 800(800)/(800 + 800)]\}} = -38.53$$

7.35 For a CE-connected transistor, how are the input characteristics (i_B vs. v_{BE}) affected if there is negligible feedback of v_{CE}?

▎ The family of input characteristics degenerates to a single-curve—one that is frequently used to approximate the family.

7.36 For a CE-connected transistor, what might be the effect of too-small emitter-base junction bias?

▮ If I_{BQ} were so small that operation occurred near the knee of an input characteristic curve, distortion would result.

7.37 Suppose that a CE-connected transistor has an infinite output impedance; how would that affect the output characteristic?

▮ The slope of the output characteristic curves would be zero in the active region.

7.38 With reference to Fig. 4-6(b), does the current gain of a CE-connected transistor increase or decrease as the mode of operation approaches saturation from the active region?

▮ Δi_C decreases for constant Δi_B; hence, the current gain decreases.

7.39 Draw a small-signal h-parameter equivalent circuit suitable to use for analysis of the CE amplifier of Fig. 4-14(a).

▮ Application of the h-parameter model of Fig. 7-1 leads to the circuit of Fig. 7-12.

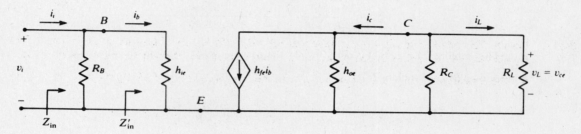

Fig. 7-12

7.40 Find an expression for the voltage gain ratio $A_v = v_L/v_i$ in terms of the CE h parameters for the amplifier of Fig. 4-14 if $R_i = 0$.

▮ The small-signal circuit is shown in Fig. 7-12, where $R_B = R_1 R_2/(R_1 + R_2)$. By current division in the collector circuit,

$$-i_L = \frac{R_C(1/h_{oe})}{R_C(1/h_{oe}) + R_L(1/h_{oe}) + R_L R_C} h_{fe} i_b$$

The voltage gain is then

$$A_v \equiv \frac{v_L}{v_i} = \frac{R_L i_L}{h_{ie} i_b} = -\frac{h_{fe} R_L R_C}{h_{ie}(R_C + R_L + h_{oe} R_L R_C)} \tag{1}$$

7.41 For the amplifier of Fig. 4-14, let $R_C = R_L = 800\,\Omega$, $R_i = 0$, $R_1 = 1.2\,k\Omega$, $R_2 = 2.7\,k\Omega$, $h_{re} \approx 0$, $h_{oe} = 100\,\mu S$, $h_{fe} = 90$, and $h_{ie} = 200\,\Omega$. Calculate the voltage gain $A_v = v_L/v_i$.

▮ $$R_B = R_1 R_2/(R_1 + R_2) = (1.2 \times 10^3)(2.7 \times 10^3)/(1.2 \times 10^3 + 2.7 \times 10^3) = 831\,\Omega$$

By (1) of Problem 7.40,

$$A_v = -\frac{90(800)(800)}{200[800 + 800 + (100 \times 10^{-6})(800)(800)]} = -173.08$$

7.42 Find an expression for the current gain ratio $A_i = i_L/i_i$ in terms of the CE h parameters for the amplifier of Fig. 4-14 if $R_i = 0$. The circuit of Fig. 7-12 is applicable.

▮ By current division,

$$i_b = \frac{R_B}{R_B + h_{ie}} i_i$$

so $$A_i \equiv \frac{i_L}{i_i} = \frac{R_B}{R_B + h_{ie}} \frac{i_L}{i_b} = \frac{R_B h_{ie}}{R_L(R_B + h_{ie})} A_v = -\frac{h_{fe} R_C R_B}{(R_B + h_{ie})(R_C + R_L + h_{oe} R_L R_C)} \tag{1}$$

7.43 If the CE amplifier of Fig. 4-14 is described by the parameters of Problem 7.41, calculate the current gain $A_i = i_L/i_i$.

By (1) of Problem 7.42,
$$A_i = -\frac{90(800)(831)}{(831+200)[800+800+(100\times 10^{-6})(800)(800)]} = -34.87$$

7.44 Draw a small-signal h-parameter equivalent circuit suitable to use for analysis of the CE amplifier of Fig. 4-19.

Application of the h-parameter model of Fig. 7-1 leads to the circuit of Fig. 7-13.

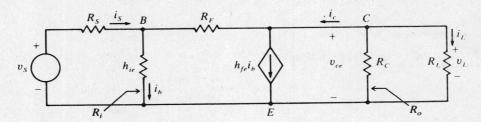

Fig. 7-13

7.45 For the CE amplifier of Fig. 4-19, assume that $h_{re} = h_{oe} \approx 0$, $h_{ie} = 1.1 \text{ k}\Omega$, $h_{fe} = 50$, $C_C \to \infty$, $R_F = 100 \text{ k}\Omega$, $R_S = 5 \text{ k}\Omega$, and $R_C = R_L + 20 \text{ k}\Omega$. Using CE h parameters, find and evaluate an expression for $A_i = i_L/i_S$.

The small-signal equivalent circuit for the amplifier is given in Fig. 7-13. By the method of node voltages,

$$\frac{v_s - v_{be}}{R_S} + \frac{v_{ce} - v_{be}}{R_F} - \frac{v_{be}}{h_{ie}} = 0 \tag{1}$$

$$\frac{v_{be} - v_{ce}}{R_F} - h_{fe} i_b - \frac{R_C + R_L}{R_C R_L} v_{ce} = 0 \tag{2}$$

Rearranging (1) and (2) and substituting $i_b = v_{be}/h_{ie}$ lead to

$$\begin{bmatrix} \dfrac{1}{R_S} + \dfrac{1}{R_F} + \dfrac{1}{h_{ie}} & -\dfrac{1}{R_F} \\ \dfrac{h_{fe}}{h_{ie}} - \dfrac{1}{R_F} & \dfrac{1}{R_F} + \dfrac{R_C + R_L}{R_C R_L} \end{bmatrix} \begin{bmatrix} v_{be} \\ v_{ce} \end{bmatrix} = \begin{bmatrix} \dfrac{v_S}{R_S} \\ 0 \end{bmatrix}$$

The determinant of coefficients is then

$$\Delta = \left(\frac{1}{R_S} + \frac{1}{R_F} + \frac{1}{h_{ie}}\right)\left(\frac{1}{R_F} + \frac{R_C + R_L}{R_C R_L}\right) + \frac{1}{R_F}\left(\frac{h_{fe}}{h_{ie}} - \frac{1}{R_F}\right)$$

$$= \left(\frac{1}{5} + \frac{1}{100} + \frac{1}{1.1}\right)\left(\frac{1}{100} + \frac{10+10}{10\times 10}\right)(10^{-6}) + \frac{1}{100}\left(\frac{50}{1.1} - \frac{1}{100}\right)(10^{-6}) = 4.557 \times 10^{-6}$$

By Cramer's rule,

$$v_{be} = \frac{\Delta_1}{\Delta} = \frac{\left(\dfrac{1}{R_F} + \dfrac{R_C+R_L}{R_C R_L}\right) v_S}{R_S \Delta} = \frac{\left(\dfrac{1}{100} + \dfrac{1}{10}\right)(10^{-3}) v_S}{(5\times 10^3)(4.557\times 10^{-6})} = 4.828 \times 10^{-3} v_S \tag{3}$$

and

$$v_L = v_{ce} = \frac{\Delta_2}{\Delta} = \frac{\left(\dfrac{1}{R_F} - \dfrac{h_{fe}}{h_{ie}}\right) v_S}{R_S \Delta} = \frac{\left(\dfrac{1}{100} - \dfrac{50}{1.1}\right)(10^{-3}) v_S}{(5\times 10^3)(4.557\times 10^{-6})} = -1.995 v_S \tag{4}$$

So,
$$A_i = \frac{i_L}{i_S} = \frac{v_L/R_L}{(v_S - v_{be})/R_S} = \frac{R_S v_L}{R_L(v_S - v_{be})} = \frac{(5\times 10^3)(-1.995 v_S)}{(20\times 10^3)(v_S - 4.828\times 10^{-3} v_S)} = -0.501$$

7.46 Find and evaluate an expression for the current gain $A'_i = i_L/i_b$ of the amplifier described by Problem 7.45.

▌ Using the results of Problem 7.45,

$$A'_i = \frac{i_L}{i_b} = \frac{v_L/R_L}{v_{be}/h_{ie}} = \frac{h_{ie}v_L}{R_L v_{be}} = \frac{(1.1 \times 10^3)(-1.995 v_s)}{(20 \times 10^3)(4.828 \times 10^{-3} v_S)} = -22.73$$

7.47 Determine the voltage gain $A_v = v_L/v_S$ for the amplifier of Problem 7.45.

▌ Based on (4) of Problem 7.45,

$$A_v = \frac{v_L}{v_S} = \frac{-1.995 v_S}{v_S} = -1.995$$

7.48 Calculate the voltage gain $A'_v = v_L/v_{be}$ for the amplifier of Problem 7.45.

▌ Based on (3) and (4) of Problem 7.45,

$$A'_v = \frac{v_L}{v_{be}} = \frac{-1.995 v_S}{4.828 \times 10^{-3} v_S} = -413.2$$

7.49 For the CE amplifier described by Problem 7.45, find the value of input resistance R_i as indicated on Fig. 7-13.

▌ Referring to Fig. 7-13 and applying Ohm's law,

$$R_i = \frac{v_{be}}{i_S} = \frac{v_{be}}{(v_S - v_{be})/R_S} = \frac{R_S v_{be}}{v_S - v_{be}} \tag{1}$$

Substituting (3) of Problem 7.45 into (1), we have

$$R_i = \frac{(5 \times 10^3)(4.828 \times 10^{-3} v_S)}{v_S - 4.828 \times 10^{-3} v_S} = 24.26 \ \Omega$$

7.50 Form an expression for the output resistance R_o as indicated on Fig. 7-13.

▌ Deactive (short) v_S and replace R_L by a driving-point source such that $v_{dp} = v_L$ and $i_{dp} = -i_L$. By voltage division,

$$v_{be} = \frac{R_S \| h_{ie}}{R_S \| h_{ie} + R_F} v_{dp} \tag{1}$$

Using Ohm's law and (1),

$$i_b = \frac{v_{be}}{h_{ie}} = \frac{R_S \| h_{ie} v_{dp}}{h_{ie}(R_S \| h_{ie} + R_F)} = \frac{R_S v_{dp}}{R_S h_{ie} + R_F(R_S + h_{ie})} \tag{2}$$

By KCL at node C,

$$i_{dp} = \frac{v_{dp}}{R_C} + h_{fe} i_b + \frac{v_{dp}}{R_F + R_S \| h_{ie}} \tag{3}$$

Substitute (2) into (3) and rearrange to yield

$$R_o = \frac{v_{dp}}{i_{dp}} = \frac{R_C[R_S h_{ie} + R_F(R_S + h_{ie})]}{R_S h_{ie} + R_F(R_S + h_{ie}) + R_C[h_{ie} + (h_{fe} + 1)R_S]} \tag{4}$$

7.51 Calculate the value of output resistance R_o for the amplifier of Problem 7.45.

▌ From (4) of Problem 7.50,

$$R_o = \frac{(20 \times 10^3)[(5 \times 10^3)(1.1 \times 10^3) + (100 \times 10^3)(5 \times 10^3 + 1.1 \times 10^3)]}{(5 \times 10^3)(1.1 \times 10^3) + (100 \times 10^3)(5 \times 10^3 + 1.1 \times 10^3) + (20 \times 10^3)[1.1 \times 10^3 + (50 + 1)(5 \times 10^3)]}$$
$$= 2.154 \text{ k}\Omega$$

7.52 Draw a small-signal h-parameter equivalent circuit for the CE amplifier of Fig. 7-14 if h_{oe} and h_{re} are negligible.

▌ See Fig. 7-15.

SMALL-SIGNAL FREQUENCY-INDEPENDENT BJT AMPLIFIERS   207

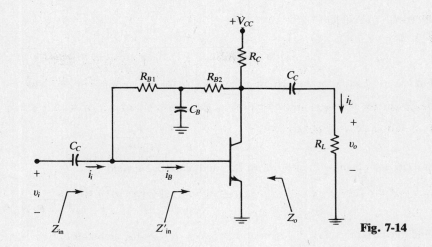

Fig. 7-14

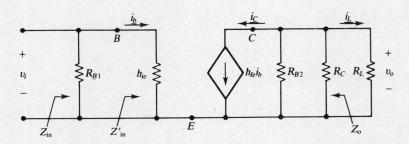

Fig. 7-15

7.53 For the CE amplifier of Fig. 7-14 that is modeled by Fig. 7-15, find an expression for the voltage gain ratio $A_v = v_o/v_i$.

▮ Referring to Fig. 7-15,

$$v_o = -h_{fe}i_b(R_{B2}\|R_C\|R_L) = -\frac{h_{fe}i_b R_{B2} R_C R_L}{R_{B2}R_C + R_{B2}R_L + R_C R_L} \tag{1}$$

But, by Ohm's law,

$$i_b = v_i/h_{ie} \tag{2}$$

Substitute (2) into (1) and rearrange to find

$$A_v = \frac{v_o}{v_i} = -\frac{h_{fe} R_{B2} R_C R_L}{h_{ie}(R_{B2}R_C + R_{B2}R_L + R_C R_L)} \tag{3}$$

7.54 Evaluate the voltage gain for the amplifier of Fig. 7-14 if $h_{fe} = 75$, $h_{ie} = 200\,\Omega$, $h_{oe} = h_{re} = 0$, $R_L = R_C = 2\,\text{k}\Omega$, and $R_{B1} = R_{B2} = 100\,\text{k}\Omega$.

▮ By (3) of Problem 7.53,

$$A_v = -\frac{75(100 \times 10^3)(2 \times 10^3)^2}{200[(100 \times 10^3)(2 \times 10^3) + (100 \times 10^3)(2 \times 10^3) + (2 \times 10^3)^2]} = -371.3$$

7.55 Find an expression for the current gain $A_i' = i_L/i_b$ for the amplifier of Fig. 7-14.

▮ By current division at the output of the applicable equivalent circuit of Fig. 7-15,

$$i_L = -\frac{R_{B2}\|R_C}{R_{B2}\|R_C + R_L} h_{fe}i_b \tag{1}$$

Whence,

$$A_i' = \frac{i_L}{i_b} = -\frac{h_{fe} R_{B2} R_C}{R_{B2}R_C + R_{B2}R_L + R_C R_L} \tag{2}$$

7.56 Evaluate the current gain $A_i' = i_L/i_b$ for the amplifier of Fig. 7-14 if it is described by the parameters of Problem 7.54.

208 □ CHAPTER 7

▎ By (2) of Problem 7.55,

$$A_i' = -\frac{75(100 \times 10^3)(2 \times 10^3)}{(100 \times 10^3)(2 \times 10^3) + (100 \times 10^3)(2 \times 10^3) + (2 \times 10^3)^2} = -37.13$$

7.57 Find an expression for the current gain $A_i = i_L/i_i$ for the amplifier of Fig. 7-14.

▎ By current division at the base node of the applicable equivalent circuit of Fig. 7-15,

$$i_b = \frac{R_{B1}}{R_{B1} + h_{ie}} i_i \tag{1}$$

Substituting (1) into (2) of Problem 7.55 and rearranging yield

$$A_i = \frac{i_L}{i_i} = -\frac{h_{fe} R_{B1} R_{B2} R_C}{(R_{B1} + h_{ie})(R_{B2} R_C + R_{B2} R_L + R_C R_L)} \tag{2}$$

7.58 Find the current gain $A_i = i_L/i_i$ in terms of $A_i' = i_L/i_i$ for the amplifier of Fig. 7-14 if it is described by the parameters of Problem 7.54. Evaluate A_i.

▎ Using (2) of Problem 7.55 and (1) of Problem 7.57,

$$i_L = A_i' i_b = A_i' \frac{R_{B1}}{R_{B1} + h_{ie}} i_i \tag{1}$$

Rearrange (1) and use the result of Problem 7.56 to find

$$A_i = \frac{i_L}{i_i} = \frac{A_i' R_{B1}}{R_{B1} + h_{ie}} = \frac{-37.13(100 \times 10^3)}{100 \times 10^3 + 200} = -37.06$$

7.59 Calculate the input impedance Z_{in}' for the equivalent circuit of Fig. 7-15.

▎ Referring to the equivalent circuit of Fig. 7-15, it is apparent that

$$Z_{in}' = v_i/i_b = h_{ie} = 200 \, \Omega$$

7.60 Find the input impedance Z_{in} for the amplifier of Fig. 7-14 if h_{oe} and h_{re} are negligible.

▎ From the equivalent circuit of Fig. 7-15, it is seen that

$$Z_{in} = R_{B1} \| h_{ie} = \frac{R_{B1} h_{ie}}{R_{B1} + h_{ie}} = \frac{(100 \times 10^3)(200)}{100 \times 10^3 + 200} = 199.6 \, \Omega$$

7.61 Determine the output impedance Z_o for the amplifier of Fig. 7-14 if it is modeled by Fig. 7-15.

▎ With $v_i = 0$ (shorted), $i_b = h_{fe} i_b = 0$. Hence, the driving-point impedance as seen with R_L removed is

$$Z_{dp} = Z_o = R_{B2} \| R_C = \frac{R_{B2} R_C}{R_{B2} + R_C} = \frac{(100 \times 10^3)(2 \times 10^3)}{100 \times 10^3 + 2 \times 10^3} = 1.96 \, k\Omega$$

7.62 Find the power gain for the amplifier of Fig. 7-14 if it is described by the parameters of Problem 7.54.

▎ Based on the results of Problems 7.54 and 7.58,

$$A_p = A_v A_i = -371.3(-37.06) = 13.76 \times 10^3$$

COMMON-BASE AMPLIFIERS

7.63 Draw a small-signal h-parameter equivalent circuit for the simplified (bias network omitted) CB amplifier of Fig. 7-16(a).

▎ See Fig. 7-16(b).

7.64 In the CB amplifier of Fig. 7-16(b), let $h_{ib} = 30 \, \Omega$, $h_{rb} = 4 \times 10^{-6}$, $h_{fb} = -0.99$, $h_{ob} = 8 \times 10^{-7}$ S, and $R_L = 20 \, k\Omega$. (These are typical CB amplifier values.) Find an expression for the current-gain ratio A_i.

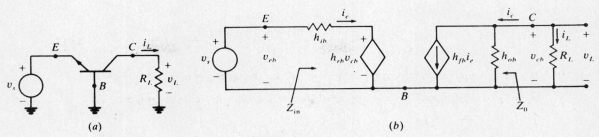

Fig. 7-16

By direct analogy with Fig. 7-10(b) and (2) of Problem 7.22,

$$A_i = -\frac{h_{fb}}{1 + h_{ob}R_L} = -\frac{-0.99}{1 + (8 \times 10^{-7})(20 \times 10^3)} = 0.974 \quad (1)$$

Note that $A_i \approx -h_{fb} < 1$, and that the input and output currents are in phase because $h_{fb} < 0$.

7.65 Derive an expression for the voltage-gain ratio A_v of the CB amplifier in Fig. 7-16(b) if described by the parameters of Problem 7.64.

By direct analogy with Fig. 7-10(b) and (3) of Problem 7.23,

$$A_v = -\frac{h_{fb}R_L}{h_{ib} + R_L(h_{ib}h_{oc} - h_{fb}h_{rb})} = -\frac{-0.99(20 \times 10^3)}{30 + (20 \times 10^3)[(30)(8 \times 10^{-7}) - (-0.99)(4 \times 10^{-6})]} = 647.9 \quad (1)$$

Observe that $A_v \approx -h_{fb}R_L/h_{ib}$, and the output and input voltages are in phase because $h_{fb} < 0$.

7.66 Find an expression for the input impedance Z_{in} of the CB amplifier of Fig. 7-16(b) if described by the parameters of Problem 7.64.

By direct analogy with Fig. 7-10(b) and (1) of Problem 7.24,

$$Z_{in} = h_{ib} - \frac{h_{rb}h_{fb}R_L}{1 + h_{ob}R_L} = 30 - \frac{(4 \times 10^{-6})(-0.99)(20 \times 10^3)}{1 + (8 \times 10^{-7})(20 \times 10^3)} = 30.08 \, \Omega \quad (1)$$

It is apparent that $Z_{in} \approx h_{ib}$.

7.67 Calculate the output impedance Z_o for the CB amplifier of Fig. 7-16(b) if described by the parameters of Problem 7.64.

By analogy with Fig. 7-10(b) and (3) of Problem 7.25,

$$Z_o = \frac{1}{h_{ob} - h_{fb}h_{rb}/h_{ib}} = \frac{1}{8 \times 10^{-7} - (-0.99)(4 \times 10^{-6})/30} = 1.07 \, M\Omega \quad (1)$$

Note that Z_o is decreased because of the feedback from the output mesh to the input mesh through $h_{rb}v_{cb}$.

7.68 Evaluate a typical CB amplifier from the results of Problems 7.64–7.67.

Based on the typical values of these problems, the characteristics of the CB amplifier can be summarized as follows:
1. Current gain of less than 1
2. High voltage gain
3. Power gain approximately equal to voltage gain
4. No phase shift for current or voltage
5. Small input impedance
6. Large output impedance

7.69 Draw a small-signal r-parameter equivalent circuit suitable for analysis of the CB amplifier of Fig. 4-21.

See Fig. 7-17 where the r-parameter model of Fig. 7-6 is used.

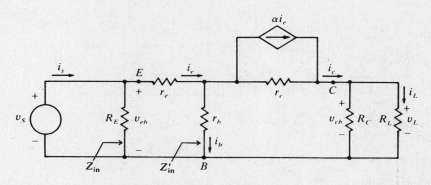

Fig. 7-17

7.70 For the CB amplifier of Fig. 4-21, find the voltage-gain ratio $A_v = v_L/v_S$.

▎ The small-signal circuit for the amplifier is given by Fig. 7-17. By Ohm's law,

$$i_c = \frac{v_{cb}}{R_C \| R_L} = \frac{(R_C + R_L)}{R_C R_L} \quad (1)$$

Substituting (1) into (5) and (6) of Problem 7.10 gives, respectively,

$$v_S = v_{eb} = (r_e + r_b)i_e - r_b \frac{(R_C + R_L)v_L}{R_C R_L} \quad (2)$$

$$v_L = v_{cb} = (\alpha r_c + r_b)i_e - (r_b + r_c)\frac{(R_C + R_L)v_L}{R_C R_L} \quad (3)$$

where we also made use of (7) of Problem 7.10. Solving (2) for i_e and substituting the result into (3) yield

$$v_L = (\alpha r_c + r_b)\frac{v_S + \dfrac{r_b(R_C + R_L)}{R_C + R_L}v_L}{r_e + r_b} - (r_b + r_c)\frac{R_C + R_L}{R_C R_L}v_L \quad (4)$$

The voltage-gain ratio follows directly from (4) as

$$A_v = \frac{v_L}{v_S} = \frac{(\alpha r_c + r_b)R_C R_L}{R_C R_L(r_e + r_b) + (R_C + R_L)[(1 - \alpha)r_c r_b + r_e(r_b + r_c)]} \quad (5)$$

7.71 For the CB amplifier of Fig. 4-21, let $r_b = 300\,\Omega$, $r_e = 30\,\Omega$, $r_c = 1\,\text{M}\Omega$, $R_E = 5\,\text{k}\Omega$, $R_C = R_L = 4\,\text{k}\Omega$, and $\alpha = 0.99$. Find the value of the voltage gain A_v.

▎ By (5) of Problem 7.70,

$$A_v = \frac{[0.99(1 \times 10^6) + 300](4 \times 10^3)^2}{(4 \times 10^3)^2(30 + 300) + (8 \times 10^3)[(1 - 0.99)(1 \times 10^6)(300) + 30(300 + 1 \times 10^6)]} = 58.83$$

7.72 Repeat Problem 7.71 if $\alpha = 0.995$ and all else is unchanged.

▎ From (5) of Problem 7.70,

$$A_v = \frac{[0.995(1 \times 10^6) + 300](4 \times 10^3)^2}{(4 \times 10^3)^2(30 + 300) + (8 \times 10^3)[(1 - 0.995)(1 \times 10^6)(300) + 30(300 + 1 \times 10^6)]} = 61.87$$

7.73 Assume that r_c is large enough so that $i_c \approx \alpha i_e$ for the CB amplifier of Fig. 4-21, whose small-signal circuit is given by Fig. 7-17. Find an expression for the input impedance Z_{in}.

▎ Letting $i_c \approx \alpha i_e$ in (5) of Problem 7.10 allows us to determine the input resistance R_{in}:

$$v_{eb} = (r_e + r_b)i_e - r_b(\alpha i_e)$$

from which

$$Z'_{in} = \frac{v_{eb}}{i_e} = r_e + (1 - \alpha)r_b \quad (1)$$

Therefore,

$$Z_{in} = R_E \| Z'_{in} = \frac{R_E[r_e + (1 - \alpha)r_b]}{R_E + r_e + (1 - \alpha)r_b} \quad (2)$$

7.74 Determine the value of input impedance Z_{in} for the CB amplifier of Fig. 4-21 if it is described by the parameters of Problem 7.71.

By (2) of Problem 7.73,
$$Z_{in} = \frac{(5 \times 10^3)[30 + (1 - 0.99)(300)]}{5 \times 10^3 + 30 + (1 - 0.99)(300)} = 32.78 \, \Omega$$

7.75 Determine the value of input impedance Z_{in} for the CB amplifier of Fig. 4-21 if it is described by the parameters of Problem 7.71 except $\alpha = 0.995$.

By (2) of Problem 7.73,
$$Z_{in} = \frac{(5 \times 10^3)[30 + (1 - 0.995)30]}{5 \times 10^3 + 30 + (1 - 0.995)(30)} = 31.3 \, \Omega$$

7.76 Find an expression for the current-gain ratio $A_i = i_L/i_s$ for the CB amplifier modeled by Fig. 7-17 if $i_c \approx \alpha i_e$.

By current division at node E,
$$i_e = \frac{R_E}{R_E + Z'_{in}} i_s$$

Solving for i_s and using (1) of Problem 7.73 give
$$i_s = \frac{R_E + R_{in}}{R_E} i_e = \frac{R_E + r_e + (1 - \alpha)r_b}{R_E} i_e \quad (1)$$

Current division at node C, again with $i_c \approx \alpha i_e$, yields
$$i_L = \frac{R_C}{R_C + R_L} i_c = \frac{R_C \alpha i_e}{R_C + R_L} \quad (2)$$

The current gain is now the ratio of (2) to (1):
$$A_i = \frac{i_L}{i_s} = \frac{\alpha R_C/(R_C + R_L)}{[R_E + r_e + (1 - \alpha)r_b]/R_E} = \frac{\alpha R_C R_E}{(R_C + R_L)[R_E + r_e + (1 - \alpha)r_b]} \quad (3)$$

7.77 Determine the value of current gain A_i for the CB amplifier of Fig. 4-21 if it is described by the parameters of Problem 7.71.

By (3) of Problem 7.76,
$$A_i = \frac{0.99(4 \times 10^3)(5 \times 10^3)}{(4 \times 10^3 + 4 \times 10^3)[5 \times 10^3 + 30 + (1 - 0.99)(300)]} = 0.492$$

7.78 Determine the value of current gain A_i for the CB amplifier of Fig. 4-21 if it is described by the parameters of Problem 7.71 except $\alpha = 0.995$.

By (3) of Problem 7.76,
$$A_i = \frac{0.995(4 \times 10^3)(5 \times 10^3)}{(4 \times 10^3 + 4 \times 10^3)[5 \times 10^3 + 30 + (1 - 0.995)(300)]} = 0.494$$

7.79 Based on Problems 7.70 through 7.78, evaluate the CB amplifier with regard to sensitivity to changes in α.

For these problems, A_v, A_i and Z_{in} were evaluated for $\alpha = 0.99$ and 0.995. This small change in α (approximately $\frac{1}{2}$ percent) had little effect on the amplifier performance. In contrast, such a change in α leads to approximately a 100-percent change in $h_{fe} \approx \beta = \alpha/(1 - \alpha)$; consequently, the performance of a CE amplifier is significantly sensitive to such small changes in α.

7.80 For the CB amplifier of Problem 4.70, determine graphically (a) h_{fb} and (b) h_{ob}.

(a) The Q point was established in Problem 4.70 and is indicated in Fig. 4-13. By (5) of Problem 7.3,
$$h_{fb} \approx \left.\frac{\Delta i_C}{\Delta i_E}\right|_{v_{CBQ} = -6.1 \, V} = \frac{(3.97 - 2.0) \times 10^{-3}}{(4 - 2) \times 10^{-3}} = 0.985$$

(b) By (6) of Problem 7.3,

$$h_{ob} \approx \left.\frac{\Delta i_C}{\Delta v_{CB}}\right|_{I_{EQ}=3\text{ mA}} = \frac{(3.05 - 2.95) \times 10^{-3}}{-10 - (-2)} = 12.5\ \mu\text{S}$$

7.81 Draw a small-signal, h-parameter equivalent circuit for the CB amplifier of Fig. 7-18(a).

▌ Using the CB h-parameter model of Fig. 7-2 leads to the equivalent circuit of Fig. 7-18(b).

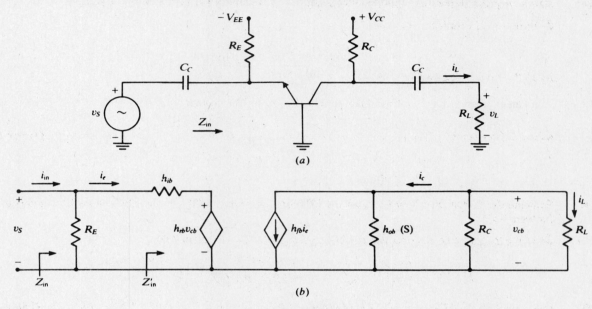

Fig. 7-18

7.82 Find an expression for the input impedance Z_{in} for the CB amplifier of Fig. 7-18(a).

▌ The h-parameter equivalent circuit is given in Fig. 7-18(b). By Ohm's law,

$$v_{cb} = -\frac{h_{fb} i_e}{h_{ob} + 1/R_C + 1/R_L} \equiv -\frac{h_{fb} i_e}{G} \qquad (1)$$

Application of KVL at the input gives

$$v_S = h_{rb} v_{cb} + h_{ib} i_e \qquad (2)$$

Now (1) may be substituted into (2) and the result solved for $Z'_{\text{in}} \equiv v_S/i_e$. Finally, Z_{in} may be found as the parallel combination of Z'_{in} and R_E:

$$Z_{\text{in}} = \frac{R_E(h_{ib} G - h_{rb} h_{fb})}{R_E G + h_{ib} G - h_{rb} h_{fb}} \qquad (3)$$

7.83 Evaluate the input impedance Z_{in} for the CB amplifier of Fig. 7-18(a) if $h_{ib} = 2\ \Omega$, $h_{rb} = 1 \times 10^{-4}$, $h_{fb} = -0.99$, $h_{ob} = 10^{-6}\ \mu\text{S}$, $R_E = 3\ \text{k}\Omega$, and $R_L = R_C = 8\ \text{k}\Omega$.

▌ Based on (1) and (3) of Problem 7.82,

$$G = h_{ob} + 1/R_C + 1/R_L = 1 \times 10^{-6} + 1/(8 \times 10^3) + 1/(8 \times 10^3) = 0.251\ \text{mS}$$

$$Z_{\text{in}} = \frac{(3 \times 10^3)[(2)(0.251 \times 10^{-3}) - (1 \times 10^{-4})(-0.99)]}{(3 \times 10^3)(0.251 \times 10^{-3}) + 2(0.251 \times 10^{-3}) - (1 \times 10^{-4})(-0.99)} = 1.89\ \text{k}\Omega$$

7.84 Find the expression for the voltage-gain ratio $A_v = v_{cb}/v_S$ for the CB amplifier of Fig. 7-18(a).

▌ By elimination of i_e between (1) and (2) of Problem 7.82 followed by rearrangement,

$$A_v = \frac{v_{cb}}{v_S} = -\frac{h_{fb}}{h_{ib} G - h_{rb} h_{fb}} \qquad (1)$$

7.85 Evaluate the voltage gain $A_v = v_{cb}/v_S = v_L/v_S$ for the CB amplifier of Fig. 7-18(a) if described by the parameters of Problem 7.83.

▌ Using G as determined in Problem 7.83 and (1) of Problem 7.84,

$$A_v = -\frac{h_{fb}}{h_{ib}G - h_{rb}h_{fb}} = -\frac{-0.99}{2(0.251 \times 10^{-3}) - (1 \times 10^{-4})(-0.99)} = 1647.3$$

7.86 Find an expression for the current-gain ratio $A_i = i_L/i_{in}$ for the CB amplifier of Fig. 7-18(a).

▌ From (1) of Problem 7.82,

$$i_L = \frac{v_{cb}}{R_L} = -\frac{h_{fb}i_e}{R_L G} \tag{1}$$

By KCL at the emitter node,

$$i_e = i_{in} - \frac{v_S}{R_E} = i_{in} - \frac{i_{in}Z_{in}}{R_E} = i_{in}\left(1 - \frac{Z_{in}}{R_E}\right) \tag{2}$$

where Z_{in} is given by (3) of Problem 7.82. Now elimination of i_e between (1) and (2) and rearrangement give

$$A_i = \frac{i_L}{i_{in}} = -\frac{h_{fb}}{R_L G}\left(1 - \frac{Z_{in}}{R_E}\right) \tag{3}$$

7.87 Evaluate the current gain $A_i = i_L/i_{in}$ for the CB amplifier of Fig. 7-18(a) if described by the parameters of Problem 7.83.

▌ Using Z_{in} and G as determined in Problem 7.83 and (3) of Problem 7.86,

$$A_i = -\frac{h_{fb}}{R_L G}\left(1 - \frac{Z_{in}}{R_e}\right) = -\frac{-0.99}{(8 \times 10^3)(0.251 \times 10^{-3})}\left(1 - \frac{1.89 \times 10^3}{3 \times 10^3}\right) = 0.812$$

7.88 Draw a small-signal h-parameter equivalent circuit for the CB amplifier of Fig. 7-19(a) if h_{rb} is negligibly small.

▌ With $h_{rb} = 0$, the controlled voltage source $h_{rb}v_{cb}$ is replaced by a short circuit leading to the equivalent circuit of Fig. 7-19(b).

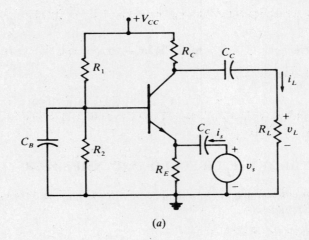

(a)

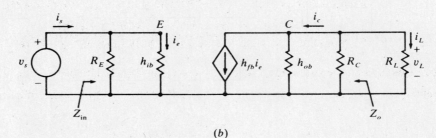

(b)

Fig. 7-19

7.89 For the CB amplifier of Fig. 7-19(a) as modeled by Fig. 7-19(b), find an expression for the voltage-gain ratio $A_v = v_L/v_s$.

▮ By Ohm's law at the input mesh of Fig. 7-19(b),
$$i_e = v_s/h_{ib} \tag{1}$$

Ohm's law at the output mesh requires that
$$v_L = \left(\frac{1}{h_{ob}} \| R_C \| R_L\right)(-h_{fb}i_e) = -\frac{R_C R_L h_{fb} i_e}{R_C + R_L + h_{ob} R_C R_L} \tag{2}$$

Substitution of (1) into (2) allows the formation of A_v:
$$A_v = \frac{v_L}{v_s} = -\frac{R_C R_L h_{fb}}{h_{ib}(R_C + R_L + h_{ob} R_C R_L)} \tag{3}$$

7.90 Evaluate the voltage gain $A_v = v_L/v_s$ for the amplifier of Fig. 7-19(b) if $R_1 = R_2 = 50\,\text{k}\Omega$, $R_C = 2.2\,\text{k}\Omega$, $R_E = 3.3\,\text{k}\Omega$, $R_L = 1.1\,\text{k}\Omega$, $h_{rb} = 0$, $h_{ib} = 25\,\Omega$, $h_{ob} = 10^{-6}\,\text{S}$, and $h_{fb} = -0.99$.

▮ By (3) of Problem 7.89,
$$A_v = -\frac{(2.2 \times 10^3)(1.1 \times 10^3)(-0.99)}{25[2.2 \times 10^3 + 1.1 \times 10^3 + (10^{-6})(2.2 \times 10^3)(1.1 \times 10^3)]} = 29.02$$

7.91 Find an expression for the current-gain ratio $A_i = i_L/i_s$ for a CB amplifier modeled by Fig. 7-19(b).

▮ By current division at node E,
$$i_e = \frac{R_E}{R_E + h_{ib}} i_s \tag{1}$$

Current division at node C gives
$$i_L = \frac{(1/h_{ob}) \| R_C}{(1/h_{ob}) \| R_C + R_L}(-h_{fb}i_e) = -\frac{R_C h_{fb} i_e}{R_C + R_L + h_{ob} R_L R_C} \tag{2}$$

Now substitution of (1) and (2) allows a direct calculation of A_i:
$$A_i = \frac{i_L}{i_s} = -\frac{R_E R_C h_{fb}}{(R_E + h_{ib})(R_C + R_L + h_{ob} R_L R_C)} \tag{3}$$

7.92 Evaluate the current gain $A_i = i_L/i_s$ for a CB amplifier modeled by Fig. 7-19(b) with the parameters of Problem 7.90.

▮ By (3) of Problem 7.91,
$$A_i = \frac{(3.3 \times 10^3)(2.2 \times 10^3)(-0.99)}{(3.3 \times 10^3 + 25)[2.2 \times 10^3 + 1.1 \times 10^3 + (10^{-6})(1.1 \times 10^3)(2.2 \times 10^3)]} = 0.655$$

COMMON-COLLECTOR (EMITTER-FOLLOWER) AMPLIFIERS

7.93 Draw a small-signal h-parameter equivalent circuit for the simplified (bias network omitted) CC (or EF) amplifier of Fig. 7-20(a).

▮ See Fig. 7-20(b).

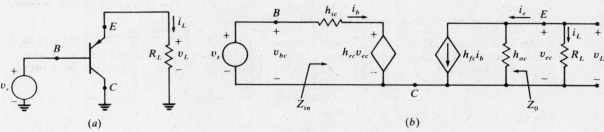

Fig. 7-20

SMALL-SIGNAL FREQUENCY-INDEPENDENT BJT AMPLIFIERS ▫ 215

7.94 In the CC amplifier of Fig. 7-20(b), let $h_{ic} = 1\ k\Omega$, $h_{rc} = 1$, $h_{fc} = -101$, $h_{oc} = 12\ \mu S$, and $R_L = 2\ k\Omega$. Drawing direct analogy with the CE amplifier of Problem 7.22, find an expression for the current-gain ratio A_i.

▮ In parallel with (2) of Problem 7.22,

$$A_i = \frac{h_{fc}}{1 + h_{oc}R_L} = -\frac{-101}{1 + (12 + 10^{-6})(2 \times 10^3)} = 98.6 \qquad (1)$$

Note that $A_i \approx -h_{fc}$, and that the input and output currents are in phase because $h_{fc} < 0$.

7.95 By direct analogy with the CE amplifier of Problem 7.23, find the voltage gain for the CB amplifier of Fig. 7-20(b) if it is characterized by the parameters of Problem 7.94.

▮ In parallel with (3) of Problem 7.23,

$$A_v = -\frac{h_{fc}R_L}{h_{ic} + R_L(h_{ic}h_{oc} - h_{fc}h_{rc})} = -\frac{-101(2 \times 10^3)}{1 \times 10^3 + (2 \times 10^3)[(1 \times 10^3)(12 \times 10^{-6}) - (-101)(1)]} = 0.995 \qquad (1)$$

Observe that $A_v \approx 1/(1 - h_{ic}h_{oc}/h_{fc}) \approx 1$. Since the gain is approximately 1 and the output voltage is in phase with the input voltage, this amplifier is commonly called a *unity follower*.

7.96 By direct analogy with the CE amplifier of Problem 7.24, find the input impedance for the CB amplifier of Fig. 7-20(b) if described by the parameters of Problem 7.94.

▮ In parallel with (1) of Problem 7.24,

$$Z_{in} = h_{ic} - \frac{h_{rc}h_{fc}R_L}{1 + h_{oc}R_L} = 1 \times 10^3 - \frac{1(-101)(2 \times 10^3)}{1 + (12 \times 10^{-6})(2 \times 10^3)} = 8.41\ M\Omega \qquad (1)$$

Note that $Z_{in} \approx -h_{fc}/h_{oc}$.

7.97 By direct analogy with the CE amplifier of Problem 7.25, find the output impedance for the CB amplifier of Fig. 7-20(b) if described by the parameters of Problem 7.94.

▮ In parallel with (3) of Problem 7.25,

$$Z_o = \frac{1}{h_{oc} - h_{fc}h_{rc}/h_{ic}} = \frac{1}{12 \times 10^{-6} - (-101)(1)/(1 \times 10^3)} = 9.9\ \Omega$$

Note that $Z_o \approx -h_{ic}/h_{fc}$.

7.98 Evaluate a typical CB amplifier from the results of Problems 7.94–7.97.

▮ Based on the typical values of these problems, the characteristics of the CB amplifier can be summarized as follows:
 1. High current gain
 2. Voltage gain of approximately unity
 3. Power gain approximately equal to current gain
 4. No current or voltage phase shift
 5. Large input impedance
 6. Small output impedance

7.99 Use the CC transistor model of Fig. 7-5 to find the Thévenin equivalent for the circuit to the right of terminals B, C in Fig. 7-3(b), assuming $h_{rc} \approx 1$ and $h_{oc} \approx 0$. Compare the results with (2) of Problem 7.6 to determine relationships between h_{ie} and h_{ic}, and between h_{fe} and h_{fc}.

▮ The circuit to be analyzed is Fig. 7-5 with a resistor R_E connected from E to C. With terminal pair B, C open, the voltage across terminals C, E is zero; thus, the Thévenin equivalent circuit consists only of $Z_{Th} = R_{Th}$. Now consider v_{bc} as a driving-point source, and apply KVL around the B, C loop to obtain

$$v_{dp} = v_{bc} = h_{ic}i_b + h_{rc}v_{ec} \approx h_{ic}i_b + v_{ec} \qquad (1)$$

Use KCL at node E to obtain

$$v_{ec} = -i_e R_E = -(h_{fc}i_b + h_{oc}v_{ce})R_E \approx -h_{fc}R_E i_b \qquad (2)$$

Substitute (2) into (1), and solve for the driving-point impedance:
$$R_{Th} = v_{bc}/i_b = h_{ic} - h_{fc}R_E \qquad (3)$$

Now when (3) is compared with (2) of Problem 7.6, it becomes apparent that $h_{ic} = h_{ie}$ and $h_{fc} = -(h_{fe} + 1)$, as is given in (3) of Problem 7.14.

7.100 Use the CC h-parameter model of Fig. 7-5 to draw a small-signal equivalent circuit suitable for analysis of the amplifier shown by Fig. 7-3(a).

▌ See Fig. 7-21.

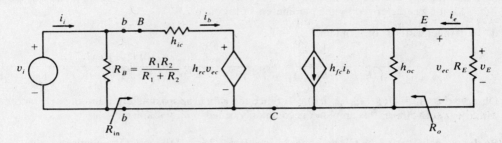

Fig. 7-21

7.101 For a CC amplifier modeled by the equivalent circuit of Fig. 7-21, find an expression for the current-gain ratio $A'_i = i_e/i_b$.

▌ At the output port of Fig. 7-21,

$$-i_e R_E = v_{ec} = -h_{fc}i_b\left(\frac{1}{h_{oe}} \parallel R_E\right) = -\frac{h_{fc}R_E}{h_{oc}R_E + 1}i_b \qquad (1)$$

and A'_i is obtained directly from (1) as

$$A'_i = \frac{i_e}{i_b} = \frac{h_{fc}}{h_{oc}R_E + 1}$$

7.102 Determine an expression for the overall current-gain ratio $A_i = i_e/i_i$ for the CB amplifier modeled by Fig. 7-21.

▌ With $R_{Th} = R_{in} = h_{ic} - h_{rc}h_{fc}R_E/(h_{oc}R_E + 1)$ and A'_i of Problem 7.100, current division at node B gives

$$\frac{i_b}{i_e} = \frac{1}{A'_i} = \frac{R_B}{R_B + R_{in}}\frac{i_i}{i_e} = \frac{R_B}{R_B + R_{in}}\frac{1}{A_i}$$

so

$$A_i = \frac{R_B}{R_B + R_{in}}A'_i = \frac{R_B}{R_B + h_{ic} + h_{rc}h_{fc}R_E/(h_{oc}R_E + 1)}\frac{h_{fc}}{h_{oc}R_E + 1} = \frac{h_{fc}R_B}{(R_B + h_{ic})(h_{oc}R_E + 1) + h_{rc}h_{fc}R_E}$$

7.103 The CE tee-equivalent circuit of Fig. 7-7(b) is suitable for use in the analysis of an EF amplifier if the collector and emitter branches are interchanged. Use this technique to draw a small-signal equivalent circuit suitable for analysis of the amplifier of Fig. 4-15(a).

▌ See Fig. 7-22.

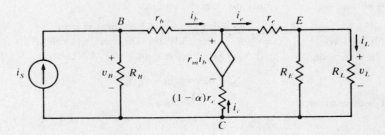

Fig. 7-22

7.104 Find the voltage-gain ratio $A_v = v_L/v_B$ for the EF amplifier of Fig. 4-15(a) as modeled by Fig. 7-22.

▮ By KVL around the B, C loop of Fig. 7-22 with $r_m = \alpha r_c$ (from Problem 7.12),

$$v_B = r_b i_b + r_m i_b + (1 - \alpha) r_c (i_b - i_e) = (r_b + r_c) i_b - (1 - \alpha) r_c i_e \qquad (1)$$

Application of KVL around the C, E loop, again with $r_m = \alpha r_c$, gives

$$0 = r_e i_e - r_m i_b - (1 - \alpha) r_c (i_b - i_e) + \frac{R_E R_L}{R_E + R_L} i_e = -r_c i_b + \left[r_e + (1 - \alpha) r_c + \frac{R_E R_L}{R_E + R_L} \right] i_e \qquad (2)$$

By Cramer's rule applied to the system consisting of (1) and (2), $i_e = \Delta_2/\Delta$, where

$$\Delta = r_b \left[r_e + (1 - \alpha) r_c + \frac{R_E R_L}{R_E + R_L} \right] + r_c \left(r_e + \frac{R_E R_L}{R_E + R_L} \right)$$

$$\Delta_2 = r_c v_B$$

Now, by Ohm's law,

$$v_L = (R_E \| R_L) i_e = \frac{R_E R_L}{R_E + R_L} \frac{\Delta_2}{\Delta}$$

Then, $\displaystyle A_v = \frac{v_L}{v_B} = \frac{R_E R_L r_c / (R_E + R_L)}{r_b [r_e + (1 - \alpha) r_c + R_E R_L/(R_E + R_L)] + r_c [r_e + R_E R_L/(R_E + R_L)]}$

7.105 Determine the input impedance Z_{in} for the EF amplifier of Fig. 4-15(a) as modeled by Fig. 7-22.

▮ The input impedance can be found as $Z_{in} = R_B \| (v_B/i_b)$. Now, in the system consisting of (1) and (2) of Problem 7.104, by Cramer's rule, $i_b = \Delta_1/\Delta$, where

$$\Delta_1 = \left[r_e + (1 - \alpha) r_c + \frac{R_E R_L}{R_E + R_L} \right] v_B$$

Hence,

$$Z_{in} = R_B \left\| \left(\frac{\Delta}{\Delta_1} v_B \right) \right. = \frac{R_B r_b \left[r_e + (1 - \alpha) r_c + \dfrac{R_E R_L}{R_E + R_L} \right] + R_B r_c \left(r_e + \dfrac{R_E R_L}{R_E + R_L} \right)}{(R_B + r_b) \left[r_e + (1 - \alpha) r_c + \dfrac{R_E R_L}{R_E + R_L} \right] + r_c \left(r_e + \dfrac{R_E R_L}{R_E + R_L} \right)}$$

7.106 For the EF amplifier of Fig. 7-23, let $h_{ic} = 500\ \Omega$, $h_{fc} = -100$, $h_{rc} = 1$, $h_{oc} = 0$, $R_B = 700\ \text{k}\Omega$, and $R_E = R_L = 2\ \text{k}\Omega$. Draw an h-parameter equivalent circuit to model the amplifier.

▮ See Fig. 7-24.

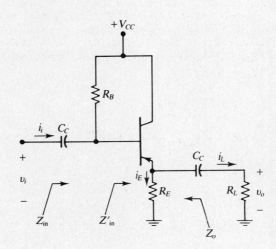

Fig. 7-23

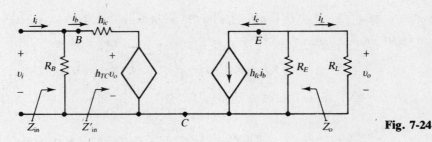

Fig. 7-24

7.107 For the amplifier of Problem 7.106, find an expression for the voltage-gain ratio $A_v = v_o/v_i$.

▎ Referring to Fig. 7-24,

$$v_o = -h_{fc}i_b(R_E \| R_L) = -\frac{h_{fe}R_ER_Li_b}{R_E + R_L} \tag{1}$$

But,

$$i_b = \frac{v_i - h_{rc}v_o}{h_{ic}} \tag{2}$$

Substitute (2) into (1) and rearrange to find

$$A_v = \frac{v_o}{v_i} = -\frac{h_{fc}R_FR_L}{(R_E + R_L)h_{ic} - h_{rc}h_{fc}R_ER_L} \tag{3}$$

7.108 Evaluate the voltage gain for the amplifier of Problem 7.106.

▎ By (3) of Problem 7.107 with values of Problem 7.106,

$$A_v = -\frac{-100(2 \times 10^3)(2 \times 10^3)}{(2 \times 10^3 + 2 \times 10^3)(500) - 1(-100)(2 \times 10^3)(2 \times 10^3)} = 0.995$$

7.109 Find an expression for the current gain $A_i' = i_L/i_b$ of the amplifier modeled by Fig. 7-24.

▎ By current division at the output of Fig. 7-24,

$$i_L = -\frac{R_E}{R_E + R_L}h_{fc}i_b \tag{1}$$

Whence,

$$A_i' = \frac{i_L}{i_b} = -\frac{h_{fc}R_E}{R_E + R_L} \tag{2}$$

7.110 Evaluate the current gain $A_i' = i_L/i_b$ for the amplifier of Problem 7.106.

▎ By (2) of Problem 7.109 and the values of Problem 7.106,

$$A_i' = -\frac{-100(2 \times 10^3)}{2 \times 10^3 + 2 \times 10^3} = 50$$

7.111 Find an expression for the current gain $A_i = i_L/i_i$ of the amplifier modeled by Fig. 7-24.

▎ Apply KVL around the base-collector mesh of Fig. 7-24 and use $v_o = R_Li_L$ to find

$$v_i = h_{ic}i_b + h_{rc}v_o = h_{ic}i_b + h_{rc}R_Li_L \tag{1}$$

Using Ohm's law and (3) of Problem 7.107,

$$i_L = \frac{v_o}{R_L} = \frac{A_vv_i}{R_L} \tag{2}$$

Solve (2) for v_i, back substitute into (1), and rearrange to yield

$$A_i = \frac{i_L}{i_i} = \frac{h_{ic}/R_L}{1/A_v - h_{rc}} \tag{3}$$

where A_v is given by (3) of Problem 7.107.

7.112 Determine the value of current gain $A_i = i_L/i_i$ for the EF amplifier of Problem 7.106.

▮ Based on (3) of Problem 7.111, the result of Problem 7.108, and values of Problem 7.106,

$$A_i = \frac{500/(2 \times 10^3)}{1/0.995 - 1} = 49.75$$

7.113 Find an expression for the input impedance Z'_{in} of the amplifier modeled by Fig. 7-24.

▮ By application of KVL around the base-collector loop of Fig. 7-24,

$$v_i = h_{ic}i_b + h_{rc}v_o \qquad (1)$$

Substitute v_o as given by (1) of Problem 7.107 into (1) and rearrange to give

$$Z'_{in} = \frac{v_i}{i_b} = h_{ic} - \frac{h_{rc}h_{fc}R_E R_L}{R_E + R_L} \qquad (2)$$

7.114 Evaluate Z'_{in} for the EF amplifier of Problem 7.106.

▮ Based on (2) of Problem 7.113 and values of Problem 7.106,

$$Z'_{in} = 500 - \frac{1(-100)(2 \times 10^3)(2 \times 10^3)}{2 \times 10^3 + 2 \times 10^3} = 100.5 \text{ k}\Omega$$

7.115 Determine the value of Z_{in} for the amplifier of Problem 7.106.

▮ Referring to Fig. 7-24,

$$Z_{in} = R_B \parallel Z'_{in} = \frac{R_B Z'_{in}}{R_B + Z'_{in}}$$

Using the result of Problem 7.114,

$$Z_{in} = \frac{(700 \times 10^3)(100.5 \times 10^3)}{700 \times 10^3 + 100.5 \times 10^3} = 87.88 \text{ k}\Omega$$

7.116 Find an expression for the output impedance Z_o of an amplifier modeled by Fig. 7-24.

▮ In Fig. 7-24, replace R_L by a driving-point source oriented so that $v_{dp} = v_o$ and deactivate (short) v_i; then

$$i_b = -\frac{h_{rc}v_{dp}}{h_{ic}} \qquad (1)$$

By KCL at the E node,

$$i_{dp} = \frac{v_{dp}}{R_E} + h_{fc}i_b \qquad (2)$$

Substitute (1) into (2) and rearrange to yield

$$Z_o = \frac{v_{dp}}{i_{dp}} = \frac{1}{1/R_E - h_{fc}h_{rc}/h_{ic}} \qquad (3)$$

7.117 Evaluate the output impedance Z_o for the EF amplifier of Problem 7.106.

▮ Based on (3) of Problem 7.116 and values of Problem 7.106,

$$Z_o = \frac{1}{1/(2 \times 10^3) - (-100)(1)/500} = 4.99 \text{ }\Omega$$

7.118 Find the power gain for the amplifier of Problem 7.106.

▮ Using the results of Problems 7.108 and 7.112,

$$A_p = A_v A_i = 0.995(49.75) = 49.5$$

7.119 For the EF amplifier of Fig. 4-15(a), use an appropriate r-parameter model of the transistor to calculate the current-gain ratio $A_i = i_L/i_s$.

220 ☐ CHAPTER 7

▮ The equivalent circuit of Fig. 7-22 is applicable. By current division and results of Problems 7.104 and 7.105,

$$A_i = \frac{i_L}{i_s} = \frac{\dfrac{R_E}{R_E + R_L} i_e}{v_B/R_B + i_b} = \frac{\dfrac{R_E}{R_E + R_L}\dfrac{\Delta_2}{\Delta}}{v_B/R_B + \Delta_1/\Delta} \quad (1)$$

Substituting for Δ, Δ_1 and Δ_2 from Problems 7.104 and 7.105 yields, after rearrangement,

$$A_i = \frac{R_E R_B r_c}{(R_E + R_L)(R_B + r_b)[r_e + (1-\alpha)r_c + R_E \parallel R_L] + (R_E + R_L)r_c(r_e + R_E \parallel R_L)}$$

7.120 The exact small-signal equivalent circuit for the CC amplifier of Fig. 7-3(a) is given by Fig. 7-21. Find the Thévenin equivalent for the circuit to the right of terminals b, b, assume that $h_{re} = h_{oe} \approx 0$, and show that the circuit of Fig. 7-4 results.

▮ Upon opening terminals b, b, the open circuit voltage (v_{Th}) is zero. Taking v_i as a driving-point source (R_B removed),

$$i_b = \frac{v_i - h_{rc}v_{ec}}{h_{ic}} = \frac{v_i}{h_{ic}} - \frac{h_{rc}}{h_{ic}}v_{ec} \quad (1)$$

But, Ohm's law at the output requires that

$$V_{ec} = -h_{fc}i_b\left(\frac{1}{h_{oc}} \parallel R_E\right) = -\frac{h_{fc}R_E}{h_{oc}R_E + 1}i_b \quad (2)$$

Substitute (2) into (1) and solve for the driving-point impedance given as

$$R_{Th} = \frac{v_i}{i_b} = h_{ic} - \frac{h_{rc}h_{fc}R_E}{h_{oc}R_E + 1} \quad (3)$$

Based on Problem 7.14 with $h_{re} = h_{oe} = 0$, $h_{rc} = 1$, $h_{fc} = -(h_{fe} + 1)$, and $h_{ic} = h_{ie}$. Hence, (3) can be rewritten as

$$R_{Th} = h_{ie} + (h_{fe} + 1)R_E$$

Thus, the Thévenin equivalent yields the identical circuit of Fig. 7-4.

7.121 Apply the CC h-parameter model of Fig. 7-5 to the amplifier of Fig. 7-3(a) to find an expression for the voltage-gain ratio $A_v = v_E/v_i$.

▮ The equivalent circuit of Fig. 7-21 is applicable. At the output network, Ohm's law requires that

$$v_{ec} = v_E = -h_{fc}i_b\left(\frac{1}{h_{ob}} \parallel R_E\right) = -\frac{h_{fc}R_E}{h_{oc}R_E + 1}i_b \quad (1)$$

At the input,
$$i_b = \frac{v_i - h_{rc}v_{ec}}{h_{ic}} = \frac{v_i}{h_{ic}} - \frac{h_{rc}}{h_{ic}}v_E \quad (2)$$

Substitute (2) into (1) and rearrange to find

$$A_v = \frac{v_E}{v_i} = -\frac{h_{fc}R_E}{h_{ic}(h_{oc}R_E + 1) - h_{rc}h_{fc}R_E} \quad (3)$$

7.122 For the CC amplifier of Fig. 7-3(a), let $h_{ic} = 100\,\Omega$, $h_{rc} = 1$, $h_{fc} = -100$, $h_{oc} = 10^{-5}$ S, and $R_E = 1$ kΩ. Find the value of voltage gain $A_v = v_E/v_i$.

▮ By (3) of Problem 7.121,

$$A_v = -\frac{-100(1 \times 10^3)}{100[(1 \times 10^{-5})(1 \times 10^3) + 1] - 1(-100)(1 \times 10^3)} = 0.999$$

7.123 Find an expression for R_o in the CC amplifier of Fig. 7-21; use the common approximations $h_{rc} \approx 1$ and $h_{oc} \approx 0$ to simplify the expression; and then evaluate it if $R_1 = 1$ kΩ, $R_2 = 10$ kΩ, $h_{fc} = -100$, and $h_{ic} = 100\,\Omega$.

▮ Deactivate (short) v_i and replace R_E by a driving-point source oriented such that $v_{dp} = v_{ec}$. Then, KCL applied at node E requires that

$$i_{dp} = h_{oc}v_{dp} + h_{fc}i_b \quad (1)$$

Current i_b of the input network is found by KVL as

$$i_b = -\frac{h_{rc}v_{ec}}{h_{ic}} = -\frac{h_{rc}v_{dp}}{h_{ic}} \qquad (2)$$

Substitute (2) into (1) to find

$$i_{dp} = h_{oc}v_{dp} - \frac{h_{fc}h_{rc}}{h_{ic}}v_{dp} \qquad (3)$$

Rearrangement of (3) and use of approximations give the driving-point impedance R_o:

$$R_o = \frac{v_{dp}}{i_{dp}} = \frac{h_{ic}}{h_{oc}h_{ic} - h_{fc}h_{re}} \approx -\frac{h_{ic}}{h_{fc}}$$

Evaluation yields

$$R_o \approx -\frac{100}{-100} = 1\,\Omega$$

Review Problems

7.124 Suppose the emitter-base junction of a Ge transistor is modeled as a forward-biased diode. Express h_{ie} in terms of the emitter current.

▌ The use of transistor notation in (1) of Problem 2.18 gives

$$i_B = I_{CBO}(e^{v_{BE}/v_T} - 1) \qquad (1)$$

Then, by (5) of Problem 7.1,

$$\frac{1}{h_{ie}} = \left.\frac{\partial i_B}{\partial v_{BE}}\right|_Q = \frac{1}{V_T}I_{CBO}e^{v_{BEQ}/v_T} \qquad (2)$$

But, by (1) and Problem 2.28,

$$I_{BQ} = I_{CBO}(e^{v_{BEQ}/v_T} - 1) \approx I_{CBO}e^{v_{BEQ}/v_T} \qquad (3)$$

and

$$I_{BQ} = I_{EQ}/(\beta + 1) \qquad (4)$$

Equations (2), (3), and (4) imply

$$h_{ie} = \frac{V_T(\beta + 1)}{I_{EQ}}$$

7.125 Find the input impedance Z_{in} of the circuit of Fig. 4-14(a) in terms of the h parameters, all of which are nonzero.

▌ The small-signal circuit of Fig. 7-12, with $R_B = R_1R_2/(R_1 + R_2)$, is applicable if a dependent source $h_{re}v_{ce}$ is added in series with h_{ie}, as in Fig. 7-1. The admittance of the collector circuit is given by

$$G = h_{oe} + \frac{1}{R_L} + \frac{1}{R_C}$$

and, by Ohm's law,

$$v_{ce} = -h_{fe}i_b/G \qquad (1)$$

By KVL applied to the input circuit,

$$i_b = \frac{v_i - h_{re}v_{ce}}{h_{ie}} \qquad (2)$$

Now (1) may be substituted in (2) to eliminate v_{ce}, and the result rearranged into

$$Z'_{\text{in}} = \frac{v_i}{i_b} = h_{ie} - \frac{h_{re}h_{fe}}{G} \qquad (3)$$

Then

$$Z_{\text{in}} = \frac{R_B Z'_{\text{in}}}{R_B + Z'_{\text{in}}} = \frac{R_B(h_{ie} - h_{re}h_{fe}/G)}{R_B + h_{ie} - h_{re}h_{fe}/G} \qquad (4)$$

7.126 For the CB amplifier of Fig. 4-21, $h_{re} = 10^{-4}$, $h_{ie} = 200\ \Omega$, $h_{fe} = 100$, $h_{oe} = 100\ \mu S$, $R_E = 3.3\ k\Omega$, $R_C = 7.1\ \Omega$, and $R_L = 1\ k\Omega$. Find the value of the CB h parameters for the transistor.

▌ Based on (9)–(12) of Problem 7.20,

$$h_{ib} = \frac{h_{ie}}{h_{fe} + 1} = \frac{200}{100 + 1} = 1.98\ \Omega$$

$$h_{rb} = \frac{h_{ie}h_{oe}}{h_{fe} + 1} - h_{re} = \frac{200(100 \times 10^{-6})}{100 + 1} - 1 \times 10^{-4} = 0.98 \times 10^{-4}$$

$$h_{fb} = -\frac{h_{fe}}{h_{fe} + 1} = -\frac{100}{100 + 1} = -0.99$$

$$h_{ob} = \frac{h_{oe}}{h_{fe} + 1} = \frac{1 \times 10^{-4}}{100 + 1} = 0.99 \times 10^{-4}\ S$$

7.127 Find the values of the tee-equivalent circuit parameters for the transistor of Problem 7.126.

▌ Based on the results of Problem 7.126 and (1)–(3) and (7) of Problem 7.10,

$$r_b = \frac{h_{rb}}{h_{ob}} = \frac{0.98 \times 10^{-4}}{0.99 \times 10^{-6}} = 98.99\ \Omega$$

$$r_e = h_{ib} - \frac{h_{rb}}{h_{ob}}(1 + h_{fb}) = 1.98 - \frac{0.98 \times 10^{-4}}{0.99 \times 10^{-6}}(1 - 0.99) = 0.99\ \Omega$$

$$r_c = \frac{1 - h_{rb}}{h_{ob}} = \frac{1 - 0.98 \times 10^{-4}}{0.99 \times 10^{-6}} = 1.01\ M\Omega$$

$$\alpha = -h_{fb} = -(-0.99) = 0.99$$

7.128 Find the voltage gain $A_v = v_L/v_S$ for the CB amplifier of Problem 7.126.

▌ Using (4) of Problem 7.70 and the results of Problem 7.127,

$$X = (\alpha r_c + r_b)R_C R_L = [0.99(1.01 \times 10^6) + 98.99](7.1 \times 10^3)(1 \times 10^3) = 7.1 \times 10^{12}$$

$$Y = R_C R_L (r_e + r_b) = (7.1 \times 10^3)(1 \times 10^3)(0.99 + 98.99) = 7.099 \times 10^8$$

$$Z = (R_C + R_L)[(1 - \alpha)r_c r_b + r_e(r_b + r_c)]$$

$$= (7.1 \times 10^3 + 1 \times 10^3)[(1 - 0.99)(1.01 \times 10^6)(98.99) + 0.99(98.99 + 1.01 \times 10^6)] = 1.62 \times 10^{10}$$

$$A_v = \frac{X}{Y + Z} = \frac{7.1 \times 10^{12}}{7.099 \times 10^8 + 1.62 \times 10^{10}} = 419.9$$

7.129 Find the current gain (i_S/i_L) for the CB amplifier of Problem 7.126.

▌ Using (3) of Problem 7.76 and results of Problem 7.127,

$$A_i = \frac{\alpha R_C R_E}{(R_C + R_L)[R_E + r_e + (1 - \alpha)r_b]} = \frac{0.99(7.1 \times 10^3)(3.3 \times 10^3)}{(7.1 \times 10^3 + 1 \times 10^3)[3.3 \times 10^3 + 0.99 + (1 - 0.99)(98.99)]} = 0.867$$

7.130 Determine an expression for the current-gain ratio $A_i' = i_L/i_b$ of the amplifier described by Problem 7.41.

▌ Directly from the expression for i_L of Problem 7.40,

$$A_i' = \frac{i_L}{i_b} = -\frac{h_{fe}R_C}{R_C + R_L + h_{oe}R_L R_C} \tag{1}$$

7.131 Evaluate the current gain $A_i' = i_L/i_b$ for the CB amplifier described by Problem 7.126.

▌ By (1) of Problem 7.130,

$$A_i' = \frac{h_{fe}R_C}{R_C + R_L + h_{oe}R_L R_C} = -\frac{90(800)}{800 + 800 + (100 \times 10^{-6})(800)(800)} = -43.27$$

7.132 Comparing A_i' of Problem 7.131 with A_i of Problem 7.42, it is seen that R_B is small enough so that it has reduced the current gain. Increase R_B over the value of Problem 7.42 by letting $R_1 = 3.6\ \text{k}\Omega$ and $R_2 = 8.1\ \Omega$. Find the new current gain $A_i = i_L/i_i$ for the amplifier of Problem 7.43 if all else is unchanged.

▮ $$R_B = R_1R_2/(R_1 + R_2) = (3.6 \times 10^3)(8.1 \times 10^3)/(3.6 \times 10^3 + 8.1 \times 10^3) = 2.49\ \text{k}\Omega$$

By (1) of Problem 7.42 and A_v of Problem 7.41,

$$A_i = \frac{R_B h_{ie}}{R_L(R_B + h_{ie})} A_v = \frac{(2.49 \times 10^3)(200)}{800(2.49 \times 10^3 + 200)}(-173.08) = -40.05$$

7.133 Repeat Problem 7.40 if $R_i = 50\ \Omega$ and all else is unchanged.

▮ The small-signal circuit of Fig. 7-12 is applicable if R_i is added to the input branch. By KVL at the input branch,

$$i_i = \frac{v_i}{R_i + R_B \| h_{ie}} = \frac{(R_B + h_{ie})v_i}{R_i R_B + R_i h_{ie} + R_B h_{ie}} \quad (1)$$

Apply current division at the base node and use (1) to find

$$i_b = \frac{R_B}{R_B + h_{ie}} i_i = \frac{R_B v_i}{R_i R_B + R_i h_{ie} + R_B h_{ie}} \quad (2)$$

Since $v_L = i_L R_L$, use of (1) of Problem 7.130 and (2) leads to

$$v_L = A_i' i_b R_L = -\frac{h_{fe} R_C}{R_C + R_L + h_{oe} R_L R_C} \frac{R_B v_i}{R_i R_B + R_i h_{ie} + R_B h_{ie}} R_L$$

Whence, $$A_v = \frac{v_L}{v_i} = \frac{-h_{fe} R_C R_L R_B}{(R_C + R_L + h_{oe} R_L R_C)(R_i R_B + R_i h_{ie} + R_B h_{ie})} \quad (3)$$

7.134 Evaluate the voltage gain expression of Problem 7.133 for the amplifier of Problem 7.41 where R_i is now 50 Ω.

▮ $$A_v = \frac{-90(800)(800)(831)}{[800 + 800 + (100 \times 10^{-6})(800)(800)][50(831) + 50(200) + 831(200)]} = -132.1$$

Comparing this value of voltage gain with A_v of Problem 7.41, it is seen that $R_i \neq 0$ results in a reduction in voltage gain.

7.135 For the CE amplifier of Fig. 7-25, $h_{fe} = 100$, $h_{ie} = 100\ \Omega$, $h_{oe} = h_{re} = 0$, $R_F = 150\ \text{k}\Omega$, and $R_C = R_L = 2\ \text{k}\Omega$. Draw a small-signal equivalent circuit suitable for analysis.

▮ The circuit of Fig. 7-13 is applicable if $R_S = 0$.

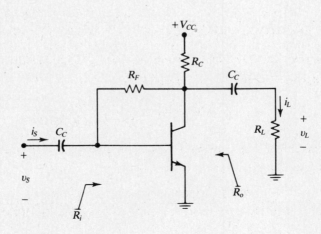

Fig. 7-25

7.136 For the amplifier of Problem 7.135, find an expression for the voltage-gain ratio $A_v = v_o/v_S$.

■ By Ohm's law applied to the circuit output network of Fig. 7-13,

$$v_L = -h_{fe}i_b(R_C \| R_L) = \frac{-h_{fe}R_C R_L i_b}{R_C + R_L} \tag{1}$$

But with $R_S = 0$,
$$i_b = v_S/h_{ie} \tag{2}$$

Use of (2) in (1) leads to

$$A_v = \frac{v_L}{v_S} = -\frac{h_{fe}R_C R_L}{h_{ie}(R_C + R_L)} \tag{3}$$

7.137 Evaluate the voltage gain for the amplifier of Problem 7.135.

■ By (3) of Problem 7.136 with values of Problem 7.135,

$$A_v = -\frac{100(2 \times 10^3)(2 \times 10^3)}{100(2 \times 10^3 + 2 \times 10^3)} = -1000$$

7.138 Find an expression for the current gain $A_i = i_L/i_S$ of the amplifier described in Problem 7.135.

■ By current division at the output network of Fig. 7-13,

$$i_L = -\frac{R_C}{R_C + R_L}h_{fe}i_b \tag{1}$$

By KCL at node B of Fig. 7-13 with use of $R_S = 0$ and Ohm's law,

$$i_b = i_S + \frac{v_L - v_S}{R_F} = i_S + \frac{i_L R_L - h_{ie}i_b}{R_F} \tag{2}$$

Solve (2) for i_b, substitute the result into (1), and rearrange to give

$$A_i = \frac{i_L}{i_S} = -\frac{h_{fe}R_F R_L}{(R_C + R_L)(R_F + h_{ie}) + h_{fe}R_C R_L} \tag{3}$$

7.139 Evaluate the current gain $A_i = i_L/i_S$ for the CE amplifier of Problem 7.135.

■ By (3) of Problem 7.138 with values from Problem 7.135,

$$A_i = -\frac{100(150 \times 10^3)(2 \times 10^3)}{(2 \times 10^3 + 2 \times 10^3)(150 \times 10^3 + 100) + 100(2 \times 10^3)(2 \times 10^3)} = -49.63$$

7.140 Derive an expression for the input resistance R_i of the CE amplifier of Fig. 7-25.

■ Consider,

$$\frac{A_i}{A_v} = \frac{i_L}{i_S}\frac{v_S}{v_L} = \frac{i_L}{i_S}\frac{v_S}{i_L R_L} = \frac{1}{R_L}\frac{v_S}{i_S} = \frac{1}{R_L}R_i$$

Whence,
$$R_i = R_L(A_i/A_v) \tag{1}$$

Substitute (3) of Problem 7.136 and (3) of Problem 7.138 into (1) and rearrange to yield

$$R_i = \frac{h_{ie}R_L R_F(R_C + R_L)/R_C}{(R_C + R_L)(R_F + h_{ie}) + h_{fe}R_C R_L} \tag{2}$$

7.141 Determine the value of input resistance for the amplifier of Problem 7.135.

■ From (2) of Problem 7.140 with values of Problem 7.135,

$$R_i = \frac{100(2 \times 10^3)(150 \times 10^3)(2 \times 10^3 + 2 \times 10^3)/(2 \times 10^3)}{(2 \times 10^3 + 2 \times 10^3)(150 \times 10^3 + 100) + 100(2 \times 10^3)(2 \times 10^3)} = 59.98 \, \Omega$$

7.142 Find an expression for the output resistance R_o of the CE amplifier shown in Fig. 7-25.

▮ In the equivalent circuit of Fig. 7-13, replace R_L by a driving-point source such that $v_{dp} = v_L$ and $i_{dp} = -i_L$. Deactivate (short) v_S; then, with $R_S = 0$,

$$i_b = v_{dp}/(R_F + h_{ie}) \tag{1}$$

By KCL at node C,
$$i_{dp} = (h_{fe} + 1)i_b + v_{dp}/R_C \tag{2}$$

Substitute (1) into (2) and rearrange to find

$$R_o = \frac{v_{dp}}{i_{dp}} = \frac{R_C(R_F + h_{ie})}{(h_{fe} + 1)R_C + R_F + h_{ie}} \tag{3}$$

7.143 Determine the value of output resistance R_o for the CE amplifier of Problem 7.135.

▮ By (3) of Problem 7.142 with values of Problem 7.135,

$$R_o = \frac{(2 \times 10^3)(150 \times 10^3 + 100)}{(100 + 1)(2 \times 10^3) + 150 \times 10^3 + 100} = 852.6 \, \Omega$$

7.144 Find the power gain for the CE amplifier of Problem 7.135.

▮ Using the results of Problems 7.137 and 7.139,

$$A_p = A_v A_i = -1000(-49.63) = 49.63 \times 10^3$$

7.145 For the CB amplifier of Fig. 4-21, find the voltage-gain ratio $A_v = v_L/v_{eb}$ using the tee-equivalent circuit of Fig. 7-6 if r_c is large enough that $i_c \approx \alpha i_e$.

▮ The equivalent circuit of Fig. 7-7 is applicable. Using the approximation $i_c \approx \alpha i_e$, KVL around the B, E mesh leads to

$$v_S = v_{eb} = (r_e + r_b)i_e - r_b(\alpha i_e) = [r_e + (1 - \alpha)r_b]i_e \tag{1}$$

Application of Ohm's law at the output and implementation of the approximation $i_c \approx \alpha i_e$ allows writing

$$v_L = i_c(R_C \| R_L) \approx \alpha i_e R_C R_L/(R_C + R_L) \tag{2}$$

The ratio of (2) to (1) gives the voltage gain:

$$A_v = \frac{v_L}{v_S} \approx \frac{\alpha R_C R_L}{(R_C + R_L)[r_e + (1 - \alpha)r_b]} \tag{3}$$

7.146 For the CB amplifier of Fig. 4-21 and Problem 7.145, $R_C = R_L = 4 \, k\Omega$, $r_e = 30 \, \Omega$, $r_b = 300 \, \Omega$, $r_c = 1 \, M\Omega$, and $\alpha = 0.99$. Determine the percentage error in the approximate voltage gain of Problem 7.145 (in which we assumed $i_c \approx \alpha i_e$), relative to the exact gain as determined in Problem 7.70.

▮ By (3) of Problem 7.145, the approximate voltage gain is

$$A_v \approx \frac{0.99(4 \times 10^3)(4 \times 10^3)}{(4 \times 10^3 + 4 \times 10^3)[30 + (1 - 0.99)(300)]} = 60$$

The exact voltage gain is found in Problem 7.71 as 58.83. Hence, the percentage error due to approximation is

$$\epsilon = \frac{60 - 58.83}{58.83}(100\%) = 1.99\%$$

7.147 Apply the definitions of the h parameters, given by (3) through (6) of Problem 1.76, to the r-parameter circuit of Fig. 7-6 to find the CB h parameters in terms of the r parameters.

▮ The method of mesh currents applied to the circuit of Fig. 7-6 yields the following linear system of equations:

$$\begin{bmatrix} r_e + r_b & r_b \\ r_b + \alpha r_c & r_b + r_c \end{bmatrix} \begin{bmatrix} i_e \\ i_c \end{bmatrix} = \begin{bmatrix} v_{eb} \\ v_{cb} \end{bmatrix} \tag{1}$$

By (3) of Problem 1.76,

$$h_{ib} = \left. \frac{v_{eb}}{i_e} \right|_{v_{cb}=0} \tag{2}$$

Simultaneous solution of (1) by Cramer's rule with $v_{cb} = 0$ results in

$$i_e = \frac{\Delta_1}{\Delta} = \frac{(r_b + r_c)v_{eb}}{r_e(r_b + r_c) + (1 - \alpha)r_b r_c} \tag{3}$$

Rearrangement of (3) and application of (2) leads to

$$h_{ib} = r_e + \frac{(1 - \alpha)r_b r_c}{r_b + r_c} \tag{4}$$

By (5) of Problem 1.76,

$$h_{fb} = \frac{i_c}{i_e}\bigg|_{v_{cb}=0} \tag{5}$$

Solution of (1) for i_c with $v_{cb} = 0$ yields

$$i_c = \frac{\Delta_2}{\Delta} = \frac{-(r_b + \alpha r_c)v_{eb}}{r_e(r_b + r_c) + (1 - \alpha)r_b r_c} \tag{6}$$

A ratio of (6) to (3) satisfies (5) giving

$$h_{fb} = -\frac{(r_b + \alpha r_c)}{r_b + r_c} \tag{7}$$

From (4) of Problem 1.76,

$$h_{rb} = \frac{v_{eb}}{v_{cb}}\bigg|_{i_e=0} \tag{8}$$

Setting $i_e = 0$ in (1), solving each equation of the set for i_c, and equating the results, we find

$$\frac{v_{eb}}{r_b} = \frac{v_{cb}}{r_b + r_c} \tag{9}$$

Whence, by (8),

$$h_{rb} = \frac{r_b}{r_b + r_c} \tag{10}$$

By (6) of Problem 1.76,

$$h_{ob} = \frac{i_c}{v_{cb}}\bigg|_{i_e=0} \tag{11}$$

Setting $i_e = 0$ in the second equation of (1) leads directly to h_{ob} of (11) as

$$h_{ob} = \frac{1}{r_b + r_c} \tag{12}$$

7.148 Apply the definitions of the z parameters, given by (3) through (6) of Problem 1.67, to the circuit of Fig. 7-6 to find values for the z parameters in the equivalent circuit of Fig. 7-26, which contains two dependent voltage sources.

▐ The linear system of equations given by (1) of Problem 7.147 describes the two-port network of Fig. 7-26. By (3) of Problem 1.67,

$$z_{11} = \frac{v_{eb}}{i_e}\bigg|_{i_c=0} \tag{1}$$

Set $i_c = 0$ in the first equation from (1) of Problem 7.147 to find

$$v_{eb} = (r_e + r_b)i_e \tag{2}$$

Whence by (1),

$$z_{11} = r_e + r_b \tag{3}$$

By (4) of Problem 1.67,

$$z_{12} = \frac{v_{eb}}{i_c}\bigg|_{i_e=0} \tag{4}$$

With $i_e = 0$ in the first equation from (1) of Problem 7.147, $v_{eb} = r_b i_c$. Hence, by (4),

$$z_{12} = r_b \tag{5}$$

By (5) of Problem 1.67,

$$z_{21} = \left.\frac{v_{cb}}{i_e}\right|_{i_c=0} \tag{6}$$

If $i_c = 0$ in the second equation from (1) of Problem 7.147, $v_{cb} = (r_b + \alpha r_c)i_e$. Thus, by (6),

$$z_{21} = r_b + \alpha r_c \tag{7}$$

By (6) of Problem 1.67,

$$z_{22} = \left.\frac{v_{cb}}{i_c}\right|_{i_e=0} \tag{8}$$

If $i_e = 0$ in the second equation from (1) of Problem 7.147, $v_{cb} = (r_b + r_c)i_c$. Therefore, by (8),

$$z_{22} = r_b + r_c \tag{9}$$

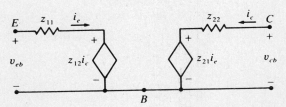

Fig. 7-26

7.149 Apply the definitions of the z parameters, given by (3) through (6) of Problem 1.67, to the CE h-parameter circuit of Fig. 7-1 to find values for the z parameters in the equivalent circuit of Fig. 7-27 in terms of the CE h parameters.

▌ Based on (3)–(8) of Problem 7.1, the linear system of equations describing the two-port CE h parameter circuit of Fig. 7-1 is as follows:

$$\begin{bmatrix} h_{ie} & h_{re} \\ h_{fe} & h_{oe} \end{bmatrix} \begin{bmatrix} i_b \\ v_{ce} \end{bmatrix} = \begin{bmatrix} v_{be} \\ i_c \end{bmatrix} \tag{1}$$

By (3) of Problem 1.67,

$$z_{11} = \left.\frac{v_{be}}{i_b}\right|_{i_c=0} \tag{2}$$

Set $i_c = 0$ in (1), solve the equation set simultaneously for i_b, and rearrange the result according to (2) yielding

$$z_{11} = h_{ie} - \frac{h_{fe}h_{re}}{h_{oe}} \tag{3}$$

By (5) of Problem 1.67,

$$z_{21} = \left.\frac{v_{ce}}{i_b}\right|_{i_c=0} \tag{4}$$

Letting $i_c = 0$ in the second equation of (1) allows direct calculation of z_{21} by (4) as

$$z_{21} = -\frac{h_{fe}}{h_{oe}} \tag{5}$$

By (4) of Problem 1.67,

$$z_{12} = \left.\frac{v_{be}}{i_c}\right|_{i_b=0} \tag{6}$$

Set $i_b = 0$ in (1), solve each equation for v_{ce}, and equate results to find

$$\frac{v_{be}}{h_{re}} = \frac{i_c}{h_{oe}} \tag{7}$$

Whence by (6),

$$z_{12} = \frac{h_{re}}{h_{oe}} \tag{8}$$

By (6) of Problem 1.67,

$$z_{22} = \left.\frac{v_{ce}}{i_c}\right|_{i_b=0} \tag{9}$$

Setting $i_b = 0$ in the second equation of (1) and applying (9) yield z_{22} directly as

$$z_{22} = \frac{1}{h_{oe}} \tag{10}$$

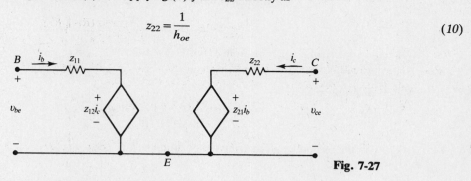

Fig. 7-27

7.150 Draw a small-signal z-parameter model suitable for analysis of the CE amplifier of Fig. 4-14(b).

▮ Application of the z-parameter model of Fig. 7-27 leads to the small-signal equivalent circuit of Fig. 7-28.

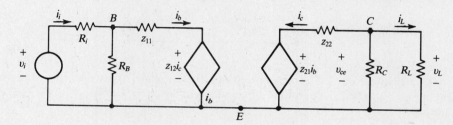

Fig. 7-28

7.151 Find an expression for the current-gain ratio of the CE amplifier modeled by Fig. 7-28.

▮ Application of the method of mesh currents leads to the following linear system of equations in i_i, i_b, and i_c.

$$\begin{bmatrix} R_i + R_B & -R_B & 0 \\ -R_B & R_B + z_{11} & z_{12} \\ 0 & z_{11} & z_{22} + R_C \parallel R_L \end{bmatrix} \begin{bmatrix} i_i \\ i_b \\ i_c \end{bmatrix} = \begin{bmatrix} v_i \\ 0 \\ 0 \end{bmatrix} \tag{1}$$

The determinant of coefficients is evaluated as

$$\Delta = (z_{22} + R_C \parallel R_L)[R_B R_i + z_{11}(R_B + R_i)] \tag{2}$$

Currents i_i and i_c are found by Cramer's rule giving

$$i_i = \frac{[(R_B + z_{11})(z_{22} + R_C \parallel R_L) - z_{12}z_{21}]v_i}{\Delta} \tag{3}$$

$$i_c = \frac{-R_B z_{21} v_i}{\Delta} \tag{4}$$

Now, by current division at node C and use of (4),

$$i_L = -\frac{R_C}{R_C + R_L} i_c = \frac{R_C R_B z_{21} v_i}{(R_C + R_L)\Delta} \tag{5}$$

The current-gain ratio follows from ratio of (5) to (3):

$$A_i = \frac{i_L}{i_i} = \frac{R_C R_B z_{21}}{(R_C + R_B)[(R_B + z_{11})(z_{22} + R_C \parallel R_L) - z_{12}z_{21}]} \tag{6}$$

7.152 Derive an expression for the voltage-gain ratio of the CE amplifier of Fig. 4-14(b) modeled by the circuit of Fig. 7-28.

▮ By Ohm's law and use of (5) in Problem 7.151,

$$v_L = R_L i_L = R_L \frac{R_C R_B z_{21} v_i}{(R_C + R_L)\Delta} = (R_C \parallel R_L) \frac{R_B z_{21} v_i}{\Delta} \tag{1}$$

Substitute Δ as given by (2) of Problem 7.151 into (1) and rearrange to find A_v.

$$A_v = \frac{v_L}{v_i} = \frac{z_{21}R_B(R_C \| R_L)}{(z_{22} + R_C \| R_L)[R_B R_i + z_{11}(R_B + R_i)]} \qquad (2)$$

7.153 In the circuit of Fig. 7-29, $h_{re} = 10^{-4}$, $h_{ie} = 200\ \Omega$, $h_{fe} = 100$, and $h_{oe} = 100\ \mu S$. Find the power gain as $A_p = |A_i A_v|$, the product of the current and voltage gains.

▌ Using (2) of Problem 7.22 and (3) of Problem 7.23,

$$A_p = |A_i A_v| = \left|-\frac{h_{fe}}{1 + h_{oe}R_L}\right| \left|-\frac{h_{fe}R_L}{h_{ie} + R_L(h_{ie}h_{oe} - h_{fe}h_{re})}\right| = \frac{h_{fe}^2 R_L}{(1 + h_{oe}R_L)|h_{ie} + R_L(h_{ie}h_{oe} - h_{fe}h_{re})|} \qquad (1)$$

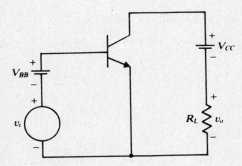

Fig. 7-29

7.154 Determine the numerical value of R_L that maximizes the power gain for the amplifier of Problem 7.153.

▌ Using (1) of Problem 7.153,

$$\frac{\partial A_p}{\partial R_L} = 0 = \frac{\partial}{\partial R_L}\left\{\frac{h_{fe}^2 R_L}{(1 + h_{oe}R_L)|h_{ie} + R_L(h_{ie}h_{oe} - h_{fe}h_{re})|}\right\} \qquad (1)$$

Carrying out the indicated partial differentiation of (1) leads to the result

$$R_L = \left[\frac{h_{ie}}{h_{oe}|h_{ie}h_{oe} - h_{fe}h_{re}|}\right]^{1/2} \qquad (2)$$

With the given parameters of Problem 7.153,

$$R_L = \left[\frac{200}{(100 \times 10^{-6})|200(100 \times 10^{-6}) - (100)(1 \times 10^{-4})|}\right]^{1/2} = 14.14\ k\Omega$$

7.155 The amplifier of Fig. 7-30 has an adjustable emitter resistor R_E, as indicated, with $0 \le \lambda \le 1$. Assume $h_{re} = h_{oe} = 0$ and draw a small-signal h-parameter equivalent circuit suitable for analysis of the amplifier.

▌ Using the CE model of Fig. 7-1 with $h_{re} = h_{oe} = 0$ leads to Fig. 7-31.

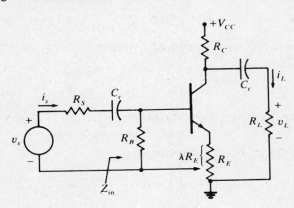

Fig. 7-30

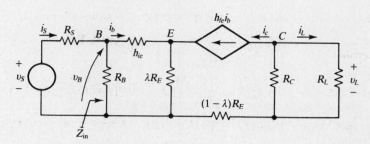

Fig. 7-31

7.156 Find an expression for the current-gain ratio $A_i = i_L/i_s$ for the amplifier modeled by Fig. 7-31.

▌ By current division, and using $i_c = h_{fe}i_b$,

$$i_L = -\frac{R_C}{R_C + R_L}i_c = -\frac{h_{fe}R_C}{R_C + R_L}i_b \tag{1}$$

By KVL around the B, E loop,

$$v_B = h_{ie}i_b + (h_{fe} + 1)\lambda R_E i_b = [h_{ie} + (h_{fe} + 1)\lambda R_E]i_b \tag{2}$$

KCL at node B requires

$$i_s = i_b + \frac{v_B}{R_B} \tag{3}$$

Substitute (2) into (3) to find

$$i_s = \left[\frac{R_B + h_{ie} + (h_{fe} + 1)\lambda R_E}{R_B}\right]i_b \tag{4}$$

Solve (1) for i_b, substitute the result into (4), and rearrange to yield the current-gain ratio:

$$A_i = \frac{i_L}{i_s} = -\frac{h_{fe}R_B R_L}{(R_C + R_L)[R_B + h_{ie} + (h_{fe} + 1)\lambda R_E]} \tag{5}$$

7.157 Derive an expression for the voltage-gain ratio $A_v = v_L/v_s$ for the amplifier of Fig. 7-30 as modeled by Fig. 7-31.

▌ By KVL around the input loop,

$$v_s = R_S i_s + v_B \tag{1}$$

Substituting (2) and (4) of Problem 7.156 into (1) leads to

$$v_s = R_S\left[\frac{R_B + h_{ie} + (h_{fe} + 1)\lambda R_E}{R_B}\right]i_b + [h_{ie} + (h_{fe} + 1)\lambda R_E]i_b \tag{2}$$

Now, by definition of A_v and using (1) of Problem 7.156,

$$A_v = \frac{v_L}{v_s} = \frac{R_L i_L}{v_s} = -\frac{h_{fe}R_C R_L}{v_s}i_b \tag{3}$$

Solve (2) for i_b and substitute the result into (3) to determine the voltage-gain ratio:

$$A_v = -\frac{h_{fe}R_C R_B R_L}{(R_C + R_L)\{R_S R_B + (R_S + R_B)[h_{ie} + (h_{fe} + 1)\lambda R_E]\}} \tag{4}$$

7.158 Find an expression for the input impedance Z_{in} for the amplifier modeled by Fig. 7-31.

▌ By the ratio of (2) to (4) of Problem 7.156,

$$Z_{in} = \frac{v_B}{i_s} = \frac{R_B[h_{ie} + (h_{fe} + 1)\lambda R_E]}{R_B + h_{ie} + (h_{fe} + 1)\lambda R_E}$$

7.159 For the CB amplifier of Problem 7.90, find the input impedance Z_{in}.

▌ With $h_{re} = 0$, it is apparent from Fig. 7-19(b) that

$$Z_{in} = \frac{v_s}{i_s} = R_E \parallel h_{ib} = \frac{R_E h_{ib}}{R_E + h_{ib}} = \frac{(3.3 \times 10^3)(25)}{3.3 \times 10^3 + 25} = 24.8\ \Omega$$

7.160 Determine the output impedance Z_o for the CB amplifier of Problem 7.90.

▌ Referring to Fig. 7-19(b), if v_s is deactivated (shorted), $i_e = 0$. Hence, $h_{fb}i_e = 0$ and Z_o is simply the Thévenin impedance

$$Z_o = Z_{Th} = \frac{1}{h_{ob}} \bigg\| R_C = \frac{R_C}{1 + h_{ob}R_C} = \frac{2.2 \times 10^3}{1 + (1 \times 10^{-6})(2.2 \times 10^3)} = 2.195 \text{ k}\Omega$$

7.161 Draw a small-signal h-parameter equivalent circuit suitable for analysis of the amplifier of Fig. 7-32 if $h_{re} = h_{oe} = 0$.

▌ See Fig. 7-33.

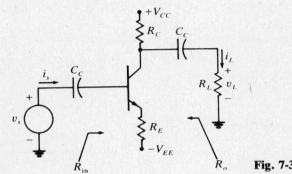

Fig. 7-32

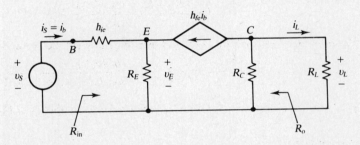

Fig. 7-33

7.162 Find an expression for the voltage-gain ratio $A_v = v_L/v_s$ for the amplifier of Fig. 7-32 when modeled as shown by Fig. 7-33.

▌ Using v_E and v_s as node voltages,

$$\frac{v_E - v_s}{h_{ie}} + \frac{v_E}{R_E} - h_{fe}\left(\frac{v_s - v_E}{h_{ie}}\right) = 0$$

or, after rearrangement,

$$v_E = \frac{(h_{fe} + 1)R_E}{(h_{fe} + 1)R_E + h_{ie}} \quad (1)$$

Applying KCL at node C,

$$h_{fe}i_b + v_L/(R_C \| R_L) = h_{fe}\left(\frac{v_s - v_E}{h_{ie}}\right) + \frac{v_L(R_C + R_L)}{R_C R_L}$$

or, rearranging,

$$v_L = \frac{R_C R_L}{R_C + R_L} \frac{h_{fe}}{h_{ie}} (v_E - v_s) \quad (2)$$

Substitute (1) into (2) and solve for the ratio v_L/v_s to give

$$A_v = \frac{v_L}{v_s} = -\frac{h_{fe}R_C R_L}{(R_C + R_L)[(h_{fe} + 1)R_E + h_{ie}]} \quad (3)$$

7.163 Derive an expression for the current-gain ratio $A_i = i_L/i_s$ for the amplifier modeled by Fig. 7-33.

▌
$$A_i = \frac{i_L}{i_s} = \frac{v_L/R_L}{(v_s - v_E)/h_{ie}} = \frac{h_{ie}}{R_L}\left(\frac{v_L}{v_s - v_E}\right) \quad (1)$$

Solve (2) of Problem 7.162 for $v_s - v_E$ and substitute the result into (1) to determine the expression for A_i:

$$A_i = -\frac{h_{fe}R_C}{R_C + R_L} \tag{2}$$

7.164 Find an expression for the input resistance R_{in} of the amplifier of Fig. 7-32 as modeled by Fig. 7-33.

▌ By KVL around the input mesh,

$$v_s = h_{ie}i_b + v_E = h_{ie}i_b + (h_{fe} + 1)i_b R_E$$

Whence,
$$R_{in} = \frac{v_s}{i_b} = h_{ie} + (h_{fe} + 1)R_E$$

7.165 Determine an expression for the output impedance R_o of the amplifier modeled by Fig. 7-33.

▌ Deactivate (short) v_s and replace R_L by a driving-point source oriented such that $v_{dp} = v_L$ and $i_{dp} = -i_L$; then KCL at node C requires that

$$i_{dp} = \frac{v_{dp}}{R_C} + h_{fe}i_b \tag{1}$$

But by current division at node E,

$$i_b = -h_{fe}i_b\left(\frac{R_E}{h_{ie} + R_E}\right) \tag{2}$$

Equation (2) can only hold for $i_b = 0$; thus, by (1),

$$R_o = \frac{v_{dp}}{i_{dp}} = R_C$$

CHAPTER 8
Small-Signal Frequency-Independent FET Amplifiers

FET EQUIVALENT CIRCUIT MODELS

8.1 For sufficiently small source-drain voltage and current excursions about the quiescent point, the FET may be replaced with a small-signal equivalent-circuit model. Of particular interest is the midfrequency range over which the amplifier can be considered independent of capacitive reactances. For a FET forming a two-port connection with source terminal common (CS), assume that the total drain current i_D goes through only small excursions about the Q point so that $\Delta i_D = i_d$ and determine a linear equation to relate the small-signal quantities i_d, v_{gs}, and v_{ds}.

▌ From the FET drain characteristics of Fig. 5-2, it is seen that if i_D is taken as the dependent variable, then
$$i_D = f(v_{GS}, v_{DS}) \tag{1}$$
For small excursions (ac signals) about the Q point, $\Delta i_D = i_d$; thus, application of the chain rule to (1) leads to
$$i_d = \Delta i_D \approx di_D = g_m v_{gs} + \frac{1}{r_{ds}} v_{ds} \tag{2}$$
where g_m and r_{ds} are defined as follows:

$$\text{Transconductance} \qquad g_m \equiv \left.\frac{\partial i_D}{\partial v_{GS}}\right|_Q \approx \left.\frac{\Delta i_D}{\Delta v_{GS}}\right|_Q \tag{3}$$

$$\text{Source-drain resistance} \qquad r_{ds} \equiv \left.\frac{\partial v_{DS}}{\partial i_D}\right|_Q \approx \left.\frac{\Delta v_{DS}}{\Delta i_D}\right|_Q \tag{4}$$

8.2 Draw a current-source equivalent circuit based on (2) of Problem 8.1 valid for use with a CS-connected FET whose signal excursions about the Q point are sufficiently small for g_m and r_{ds} to be treated as constants.

▌ As long as the JFET is operated in the pinchoff region, $i_G = i_g = 0$, so that the gate acts as an open circuit. This, along with (2) of Problem 8.1, leads to the current-source equivalent circuit of Fig. 8-1(a).

8.3 Derive the small-signal voltage-source model of Fig. 8-1(b) from the current-source model of Fig. 8-1(a).

▌ We find the Thévenin equivalent for the network to the left of the output terminals of Fig. 8-1(a). If all independent sources are deactivated, $v_{gs} = 0$; thus, $g_m v_{gs} = 0$, so that the dependent source also is deactivated (open circuit for a current source), and the Thévenin resistance is $R_{Th} = r_{ds}$. The open-circuit voltage appearing at the output terminals is $v_{Th} = v_{ds} = -g_m v_{gs} r_{ds} = -\mu v_{gs}$, where we have defined a new equivalent-circuit constant,

$$\text{Amplification factor} \qquad \mu \equiv g_m r_{ds}$$

Proper series arrangement of v_{Th} and R_{Th} leads to Fig. 8-1(b).

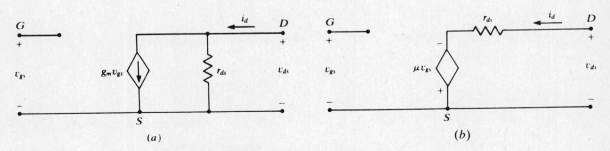

Fig. 8-1 Small-signal models for the CS FET.

8.4 For the JFET drain characteristics of Fig. 5-2, take v_{DS} as the dependent variable [so that $v_{DS} = f(v_{GS}, i_D)$] and derive the voltage-source small-signal model.

▌ For small variations about a Q point, the chain rule gives

$$v_{ds} = \Delta v_{DS} \approx dv_{DS} = \left.\frac{\partial v_{DS}}{\partial v_{GS}}\right|_Q v_{gs} + \left.\frac{\partial v_{DS}}{\partial i_D}\right|_Q i_d \qquad (1)$$

Now we may define

$$\left.\frac{\partial v_{DS}}{\partial v_{GS}}\right|_Q = \mu \quad \text{and} \quad \left.\frac{\partial v_{DS}}{\partial i_D}\right|_Q = r_{ds}$$

If the JFET operates in the pinchoff region, then the gate current is negligible and (1) is satisfied by the equivalent circuit of Fig. 8-1(b).

8.5 Find a small-signal equivalent circuit of two parallel-connected JFETs if the devices are not identical. Assume the connection assures gate to gate, source to source, and drain to drain.

▌ By KCL,

$$i_D = i_{D1} + i_{D2} \qquad (1)$$

Since the parallel connection assures that the gate-source and drain-source voltages are the same for both devices, (1) can be written as

$$i_D = f_1(v_{GS}, v_{DS}) + f_2(v_{GS}, v_{DS}) \qquad (2)$$

Application of the chain rule to (2) yields

$$i_d = \Delta i_D \approx di_D = (g_{m1} + g_{m2})v_{gs} + \left(\frac{1}{r_{ds1}} + \frac{1}{r_{ds2}}\right)v_{ds} \qquad (3)$$

where $\quad g_{m1} = \left.\frac{\partial i_{D1}}{\partial v_{GS}}\right|_Q \quad g_{m2} = \left.\frac{\partial i_{D2}}{\partial v_{GS}}\right|_Q \quad r_{ds1} = \left.\frac{\partial v_{DS}}{\partial i_{D1}}\right|_Q \quad r_{ds2} = \left.\frac{\partial v_{DS}}{\partial i_{D2}}\right|_Q$

Equation (3) is satisfied by the current-source circuit of Fig. 8-1(a) if $g_m = g_{m1} + g_{m2}$ and $r_{ds} = r_{ds1} \parallel r_{ds2}$.

8.6 For the JFET amplifier of Problem 5.26, use the drain characteristics of Fig. 5-12(b) to determine the small-signal equivalent-circuit constants g_m and r_{ds}.

▌ Let v_{gs} change by ± 1 V about the Q point of Fig. 5-12(b); then, by (3) of Problem 8.1,

$$g_m \approx \left.\frac{\Delta i_D}{\Delta v_{GS}}\right|_Q = \frac{(3.3 - 0.3) \times 10^{-3}}{2} = 1.5 \text{ mS}$$

At the Q point of Fig. 5-12(b), while v_{DS} changes from 5 to 20 V, i_D changes from 1.4 to 1.6 mA; thus, by (4) of Problem 8.1,

$$r_{ds} \approx \left.\frac{\Delta v_{DS}}{\Delta i_D}\right|_Q = \frac{20 - 5}{(1.6 - 1.4) \times 10^{-3}} = 75 \text{ k}\Omega$$

8.7 Carry out an alternative evaluation of g_m from that of Problem 8.6 through use of the transfer characteristic of Fig. 5-12(a).

▌ At the Q point of Fig. 5-12(a), while i_D changes from 1 to 2 mA, v_{GS} changes from -2.4 to -1.75 V; by (3) of Problem 8.1,

$$g_m \approx \left.\frac{\Delta i_D}{\Delta v_{GS}}\right|_Q = \frac{(2-1) \times 10^{-3}}{-1.75 - (-2.4)} = 1.54 \text{ mS}$$

8.8 Show that the transconductance of a JFET varies as the square root of the quiescent drain current.

▌ Apply (3) of Problem 8.1 to (1) of Problem 5.6 to find

$$g_m = \left.\frac{\partial i_D}{\partial v_{GS}}\right|_Q = \frac{2I_{DSS}}{V_{p0}}\left(1 + \frac{V_{GSQ}}{V_{p0}}\right) \qquad (1)$$

Solving (1) of Problem 5.6 $(1 + V_{GSQ}/V_{p0})$ and substituting into (1) lead to the desired result.

$$g_m = \frac{2\sqrt{I_{DSS}}}{V_{p0}}\sqrt{I_{DQ}} \qquad (2)$$

COMMON-SOURCE AMPLIFIERS

8.9 Draw a small-signal equivalent circuit for the CS FET amplifier of Fig. 8-2(a) where source resistor R_S is used to set the Q point but is bypassed by C_S for midfrequency operation.

▐ Use of the voltage-source model of Fig. 8-1(b) leads to the circuit of Fig. 8-2(b).

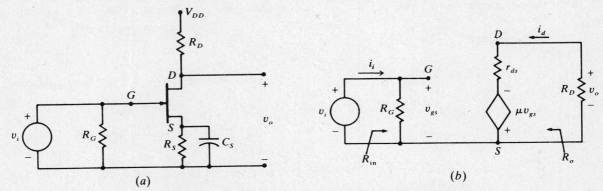

Fig. 8-2 (a) CS amplifier, (b) small-signal equivalent circuit.

8.10 In the CS amplifier of Fig. 8-2(a), let $R_D = 3\ \text{k}\Omega$, $R_G = 500\ \text{k}\Omega$, $\mu = 60$, and $r_{ds} = 30\ \text{k}\Omega$. Find an expression for the voltage-gain ratio $A_v = v_o/v_i$.

▐ By voltage division at the output network of the applicable equivalent circuit of Fig. 8-2(b),

$$v_o = -\frac{R_D}{R_D + r_{ds}} \mu v_{gs}$$

Substitution of $v_{gs} = v_i$ and rearrangement give

$$A_v = \frac{v_o}{v_i} = -\frac{\mu R_D}{R_D + r_{ds}} \tag{1}$$

8.11 Evaluate the voltage gain $A_v = v_o/v_i$ for the CS amplifier of Problem 8.10.

▐ Based on (1) of Problem 8.10,

$$A_v = -\frac{(60)(3 \times 10^3)}{3 \times 10^3 + 30 \times 10^3} = -5.45$$

where the minus sign indicates a 180°-phase shift between v_i and v_o.

8.12 Find an expression for the current-gain ratio $A_i = i_i/i_d$ for the CS amplifier of Fig. 8-2(b).

▐ KVL around the output network leads to

$$i_d = \frac{\mu v_{gs}}{r_{ds} + R_D} \tag{1}$$

But, Ohm's law requires that $v_{gs} = i_i R_G$ which, when substituted into (1) and the result rearranged, yields

$$A_i = \frac{i_d}{i_i} = \frac{\mu R_G}{r_{ds} + R_D} \tag{2}$$

8.13 Evaluate the current gain $A_i = i_d/i_i$ for the CS amplifier of Problem 8.10.

▐ By (2) of Problem 8.12,

$$A_i = \frac{(60)(500 \times 10^3)}{30 \times 10^3 + 3 \times 10^3} = 909.1$$

8.14 Find the input resistance R_{in} for the CS amplifier of Problem 8.10.

▐ By inspection of the applicable equivalent circuit of Fig. 8-2(b), it is apparent that

$$R_{in} = R_G = 500\ \text{k}\Omega$$

8.15 Find the output resistance R_o for the CS amplifier of Problem 8.10.

■ In the applicable equivalent circuit of Fig. 8-2(b), replace R_D with a driving-point source oriented such that $v_{dp} = v_o$ and $i_{dp} = i_d$. Deactivate (short) v_i from which it is seen that $v_{gs} = 0$. Hence, by KVL,

$$v_{dp} = i_{dp}r_{ds} + 0 \quad \text{or} \quad R_o = v_{dp}/i_{dp} = r_{ds} = 30 \text{ k}\Omega$$

8.16 Use the current-source model of Fig. 8-1(a) to draw a small-signal equivalent circuit suitable for analysis of the CS amplifier of Fig. 8-3.

■ See Fig. 8-4.

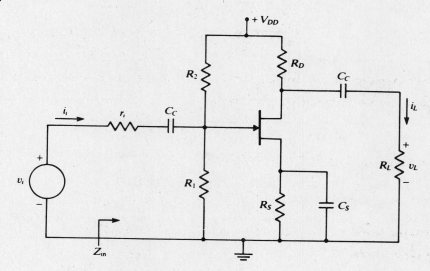

Fig. 8-3

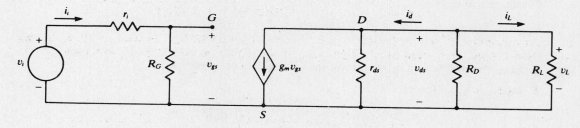

Fig. 8-4

8.17 For the JFET amplifier of Fig. 8-3, $g_m = 2$ mS, $r_{ds} = 30$ kΩ, $R_S = 3$ kΩ, $R_D = R_L = 2$ kΩ, $R_1 = 200$ kΩ, $R_2 = 800$ kΩ, and $r_i = 5$ kΩ. If C_C and C_S are large and the amplifier is biased in the pinchoff region, find Z_{in}.

■ The current-source small-signal equivalent circuit of Fig. 8-4 is applicable. Since the gate draws negligible current,

$$Z_{in} = R_G = \frac{R_1 R_2}{R_1 + R_2} = \frac{(200 \times 10^3)(800 \times 10^3)}{1000 \times 10^3} = 160 \text{ k}\Omega$$

8.18 Find an expression for the voltage-gain ratio $A_v = v_L/v_i$ for the CS FET amplifier of Fig. 8-3.

■ By voltage division at the input loop of the applicable circuit of Fig. 8-4,

$$v_{gs} = \frac{R_G}{R_G + r_i} v_i \quad (1)$$

The dependent source drives into R_{eq}, where

$$\frac{1}{R_{eq}} = \frac{1}{r_{ds}} + \frac{1}{R_D} + \frac{1}{R_L} \quad (2)$$

and, thus,

$$v_L = -g_m v_{gs} R_{eq} \quad (3)$$

Eliminating v_{gs} between (1) and (3) and using (2) yields

$$A_v = \frac{v_L}{v_i} = -\frac{g_m R_G R_{eq}}{R_G + r_i} = -\frac{g_m R_G R_D R_L r_{ds}}{(R_G + r_i)(R_D r_{ds} + R_L r_{ds} + R_D R_L)} \quad (4)$$

8.19 Evaluate the voltage gain $A_v = v_L/v_i$ for the amplifier of Problem 8.17.

By (2) and (4) of Problem 8.18,

$$\frac{1}{R_{eq}} = \frac{1}{30 \times 10^3} + \frac{1}{2 \times 10^3} + \frac{1}{2 \times 10^3} = \frac{1}{967.74} \text{ S}$$

$$A_v = -\frac{g_m R_G R_{eq}}{R_G + r_i} = -\frac{(2 \times 10^{-3})(160 \times 10^3)(967.74)}{160 \times 10^3 + 5 \times 10^3} = -1.88$$

8.20 Derive an expression for the current-gain ratio $A_i = i_L/i_i$ for the FET amplifier of Fig. 8-3.

Referring to the equivalent circuit of Fig. 8-4,

$$A_i = \frac{i_L}{i_i} = \frac{v_L/R_L}{v_i/(R_G + r_i)} = \frac{A_v(R_G + r_i)}{R_L} \quad (1)$$

Substituting (4) of Problem 8.18 into (1) yields

$$A_i = -\frac{g_m R_G R_D r_{ds}}{r_{ds}(R_D + R_L) + R_D R_L} \quad (2)$$

8.21 Evaluate the current gain $A_i = i_L/i_i$ for the amplifier of Problem 8.17.

By (2) of Problem 8.20,

$$A_i = -\frac{(2 \times 10^{-3})(160 \times 10^3)(2 \times 10^3)(30 \times 10^3)}{(30 \times 10^3)(2 \times 10^3 + 2 \times 10^3) + (2 \times 10^3)^2} = -154.84$$

8.22 In the drain-feedback-biased amplifier of Fig. 5-13(a), $R_F = 5$ MΩ, $R_L = 14$ kΩ, $r_{ds} = 40$ kΩ, and $g_m = 1$ mS. Draw an equivalent circuit based on the voltage-source FET model of Fig. 8-1(b) suitable for small-signal analysis.

See Fig. 8-5.

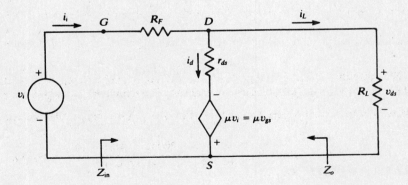

Fig. 8-5

8.23 Determine an expression giving the voltage-gain ratio $A_v = v_{ds}/v_i$ for the CS amplifier of Fig. 5-13(a) as modeled by Fig. 8-5.

With v_{ds} as a node voltage in Fig. 8-5,

$$\frac{v_i - v_{ds}}{R_F} = \frac{v_{ds}}{R_L} + \frac{v_{ds} + \mu v_i}{r_{ds}}$$

Substituting for $\mu = g_m r_{ds}$ and rearranging yield

$$A_v = \frac{v_{ds}}{v_i} = \frac{R_L r_{ds}(1 - R_F g_m)}{R_F r_{ds} + R_L r_{ds} + R_L R_F} \quad (1)$$

238 □ CHAPTER 8

8.24 Evaluate the voltage gain $A_v = v_{ds}/v_i$ for the amplifier of Problem 8.22.

▌ By (1) of Problem 8.23 with the parameter values of Problem 8.23, the voltage gain is

$$A_v = \frac{(14 \times 10^3)(40 \times 10^3)[1 - (5 \times 10^6)(1 \times 10^{-3})]}{(5 \times 10^6)(40 \times 10^3) + (14 \times 10^3)(40 \times 10^3) + (14 \times 10^3)(5 \times 10^6)} = -10.35$$

8.25 Find an expression for the input impedance Z_{in} of the CS amplifier in Fig. 5-13(a) as modeled by Fig. 8-5.

▌ KVL around the outer loop of Fig. 8-5 gives $v_i = i_i R_F + v_{ds} = i_i R_F + A_v v_i$, from which

$$Z_{\text{in}} = \frac{v_i}{i_i} = \frac{R_F}{1 - A_v} \tag{1}$$

where A_v is given by (1) of Problem 8.23.

8.26 Evaluate the input impedance Z_{in} for the amplifier of Problem 8.22.

▌ Based on (1) of Problem 8.25 and the result of Problem 8.24,

$$Z_{\text{in}} = \frac{5 \times 10^6}{1 - (-10.35)} = 440 \text{ k}\Omega$$

8.27 Determine an expression for the output impedance Z_o looking back through the drain-source terminals of the CS amplifier in Fig. 5-13(a) as modeled by Fig. 8-5.

▌ The driving-point impedance Z_o is found after deactivating the independent source v_i in Fig. 8-5. With $v_i = 0$, $\mu v_{gs} = \mu v_i = 0$ and

$$Z_o = \frac{r_{ds} R_F}{r_{ds} + R_F} \tag{1}$$

8.28 Evaluate the output impedance Z_o for the amplifier of Problem 8.22.

▌ By (1) of Problem 8.27,

$$Z_o = \frac{(40 \times 10^3)(5000 \times 10^3)}{5040 \times 10^3} = 39.68 \text{ k}\Omega$$

8.29 Derive an expression for current-gain ratio $A_i = i_i/i_L$ for the CS amplifier of Fig. 5-13(a) as modeled by Fig. 8-5.

▌
$$A_i = \frac{i_L}{i_i} = \frac{v_{ds}/R_L}{v_i/Z_{\text{in}}} = \frac{A_v Z_{\text{in}}}{R_L} \tag{1}$$

where A_v and Z_{in} are given by (1) of Problem 8.23 and (1) of Problem 8.25, respectively.

8.30 Evaluate the current gain $A_i = i_L/i_i$ for the amplifier of Problem 8.22.

▌ Based on (1) of Problem 8.29 and using the results of Problems 8.24 and 8.26,

$$A_i = \frac{-10.35(440 \times 10^3)}{14 \times 10^3} = -325.3$$

COMMON-DRAIN AMPLIFIERS

8.31 Use the voltage-source FET model of Fig. 8-1(b) to draw a small-signal equivalent circuit model for the common-drain (CD) or source-follower (SF) amplifier of Fig. 8-6.

▌ See Fig. 8-7(a).

8.32 Show that the small-signal equivalent circuit of Fig. 8-7(a) can be redrawn as shown in Fig. 8-7(b).

▌ Voltage v_{gd}, which is more easily determined than v_{gs}, has been labeled in Fig. 8-7(b). With terminals a, b opened in Fig. 8-7(a), KVL around the S, G, D loop yields

$$v_{gs} = \frac{v_{gd}}{\mu + 1}$$

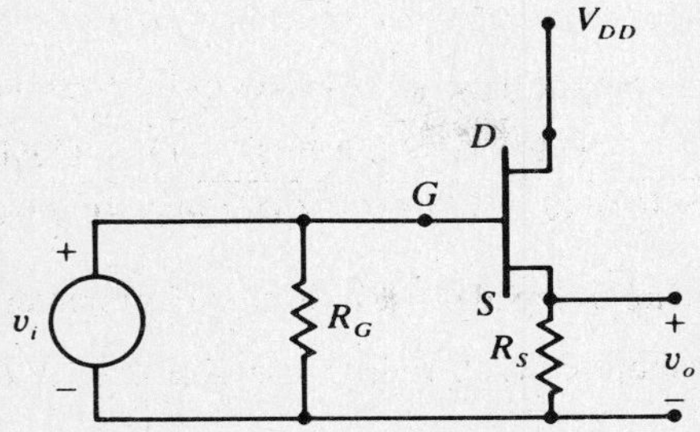

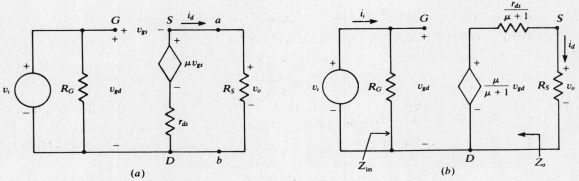

Fig. 8-6 CD or SF amplifier.

Fig. 8-7

Then the Thévenin voltage at the open-circuited terminals a, b is

$$v_{Th} = \mu v_{gs} = \frac{\mu}{\mu + 1} v_{gd} \tag{1}$$

The Thévenin impedance is found as the driving-point impedance to the left through a, b (with v_i deactivated or shorted), as seen by a source v_{ab} driving current i_a into terminal a. Since $v_{gs} = -v_{ab}$, KVL around the output loop of Fig. 8-7(a) gives

$$v_{ab} = \mu v_{gs} + i_a r_{ds} = -\mu v_{ab} + i_a r_{ds}$$

from which

$$R_{Th} = \frac{v_{ab}}{i_a} = \frac{r_{ds}}{\mu + 1} \tag{2}$$

Expressions (1) and (2) lead directly to the circuit of Fig. 8-7(b).

8.33 In the CD amplifier of Fig. 8-6 as modeled by the circuit of Fig. 8-7(b), let $R_G = 500\ \text{k}\Omega$, $R_S = 5\ \text{k}\Omega$, $\mu = 60$, and $r_{ds} = 30\ \text{k}\Omega$. Find an expression for the voltage-gain ratio $A_v = v_o/v_i$.

By voltage division in Fig. 8-7(b),

$$v_o = \frac{R_S}{R_S + r_{ds}/(\mu + 1)} \frac{\mu}{\mu + 1} v_{gd} = \frac{\mu R_S v_{gd}}{(\mu + 1)R_S + r_{ds}}$$

Replacement of v_{gd} by v_i and rearrangement give

$$A_v = \frac{v_o}{v_i} = \frac{\mu R_S}{(\mu + 1)R_S + r_{ds}} \tag{1}$$

8.34 Evaluate the voltage gain $A_v = v_o/v_i$ for the CD amplifier of Problem 8.33.

Substitution of the given values into (1) of Problem 8.33 leads to

$$A_v = \frac{60(5 \times 10^3)}{61(5 \times 10^3) + (30 \times 10^3)} = 0.895$$

Note that the gain is less than unity; its positive value indicates that v_o and v_i are in phase.

8.35 Determine an expression for the current-gain ratio $A_i = i_d/i_i$ for the CD amplifier modeled by Fig. 8-7(b).

▮ KVL applied around the output network leads to

$$i_d = \frac{\mu v_{gd}}{r_{ds} + (\mu + 1)R_S} \qquad (1)$$

But Ohm's law requires that $v_{gd} = i_i R_G$ which, when substituted into (1) and the result rearranged, yields

$$A_i = \frac{i_d}{i_i} = \frac{\mu R_G}{r_{ds} + (\mu + 1)R_S} \qquad (2)$$

8.36 Evaluate the current gain $A_i = i_d/i_i$ for the CD amplifier of Problem 8.33.

▮ By (2) of Problem 8.35,

$$A_i = \frac{i_d}{i_i} = \frac{60(500 \times 10^3)}{30 \times 10^3 + (60 + 1)(5 \times 10^3)} = 89.55$$

8.37 Find the input impedance Z_{in} for the CD amplifier of Problem 8.33.

▮ By inspection of the applicable equivalent circuit of Fig. 8-7(b), it is apparent that

$$Z_{in} = R_G = 500 \text{ k}\Omega$$

8.38 Derive an expression for the output impedance of the CD amplifier of Problem 8.33.

▮ Deactivate v_i in the applicable equivalent circuit of Fig. 8-7(b). With $v_i = 0$, $v_{ds} = 0 = \mu v_{ds}/(\mu + 1)$. Hence, when R_S is replaced by a driving-point source, the output impedance is

$$Z_o = \frac{v_{dp}}{i_{dp}} = \frac{r_{ds}}{\mu + 1} \qquad (1)$$

8.39 Evaluate the output impedance Z_o for the amplifier of Problem 8.33.

▮ By (1) of Problem 8.38,

$$Z_o = \frac{30 \times 10^3}{60 + 1} = 491.8 \text{ }\Omega$$

8.40 Find a current-source small-signal equivalent circuit for the CD FET amplifier modeled by Fig. 8-7(b).

▮ Norton's theorem can be applied to the voltage-source model of Fig. 8-7(b). The open-circuit voltage at terminals S, D (with R_S removed) is

$$v_{oc} = \frac{\mu}{\mu + 1} v_{gd} \qquad (1)$$

The short-circuit current at terminals S, D is

$$i_{SC} = \frac{\frac{\mu}{\mu + 1} v_{gd}}{r_{ds}/(\mu + 1)} = \frac{\mu}{r_{ds}} v_{gd} = g_m v_{gd} \qquad (2)$$

The Norton impedance is found as the ratio of (1) to (2):

$$R_N = \frac{v_{oc}}{i_{SC}} = \frac{\frac{\mu}{\mu + 1} v_{gd}}{g_m v_{gd}} = \frac{\mu}{(\mu + 1)g_m}$$

The equivalent circuit is given in Fig. 8-8. Usually, $\mu \gg 1$ and, thus, $R_N \approx 1/g_m$.

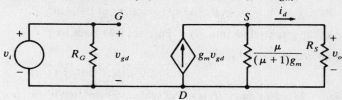

Fig. 8-8

8.41 Draw a small-signal equivalent circuit suitable for analysis of the CD FET amplifier of Fig. 8-9.

▌ Using the current-source small-signal model for the FET of Fig. 8-1(a), the circuit of Fig. 8-10 results.

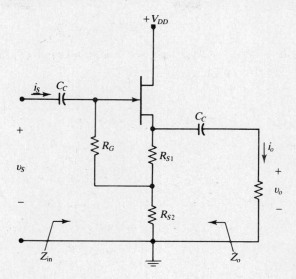

Fig. 8-9

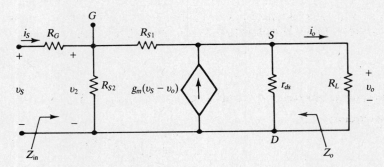

Fig. 8-10

8.42 Find an expression for the voltage-gain ratio $A_v = v_o/v_s$ for the CD FET amplifier of Fig. 8-9.

▌ Apply the method of node voltages to the applicable equivalent circuit of Fig. 8-10 to find
At node G:
$$\frac{v_2 - v_s}{R_G} + \frac{v_s}{R_{S2}} + \frac{v_2 - v_o}{R_{S1}} = 0 \tag{1}$$

At node S:
$$\frac{v_o - v_s}{R_{S1}} - g_m(v_s - v_o) + \frac{v_o}{r_{ds}} + \frac{v_o}{R_L} = 0 \tag{2}$$

Treating v_s as known and defining the conductances ($G_G = 1/R_G$, $G_L = 1/R_L$, $G_{S1} = 1/R_{S1}$, $G_{S2} = 1/R_{S2}$, $g_{ds} = 1/r_{ds}$), the linear system of equations above can be arranged as

$$\begin{bmatrix} G_G + G_{S1} + G_{S2} & -G_{S1} \\ -G_{S1} & G_{S1} + G_L + g_m + g_{ds} \end{bmatrix} \begin{bmatrix} v_2 \\ v_o \end{bmatrix} = \begin{bmatrix} G_G v_s \\ g_m v_s \end{bmatrix} \tag{3}$$

Solution of (3) for v_o by Cramer's rule gives

$$\Delta = (G_G + G_{S1} + G_{S2})(G_{S1} + G_L + g_m + g_{ds}) - G_{S1}^2 \tag{4}$$

$$v_o = \frac{\Delta_2}{\Delta} = \frac{[g_m(G_G + G_{S1} + G_{S2}) + G_G G_{S1}]v_s}{\Delta} \tag{5}$$

The voltage-gain ratio follows from (5) as

$$A_v = \frac{v_o}{v_s} = \frac{g_m(G_G + G_{S1} + G_{S2}) + G_G G_{S1}}{(G_G + G_{S1} + G_{S2})(G_{S1} + G_L + g_m + g_{ds}) - G_{S1}^2} \tag{6}$$

8.43 For the amplifier of Fig. 8-9, let $R_G = 1$ MΩ, $R_{S1} = 800$ Ω, $R_{S2} = 1.2$ kΩ, $R_L = 10$ kΩ, $g_m = 2 \times 10^{-3}$ S, and $r_{ds} = 20$ kΩ. Evaluate the voltage gain $A_v = v_o/v_s$.

▮ Based on (5) and (6) of Problem 8.42,

$$\Delta = \left(\frac{1}{1 \times 10^6} + \frac{1}{800} + \frac{1}{1.2 \times 10^3}\right)\left(\frac{1}{800} + \frac{1}{10 \times 10^3} + 2 \times 10^{-3} + \frac{1}{20 \times 10^3}\right) = 5.524 \times 10^{-6}$$

$$A_v = \frac{v_o}{v_s} = \frac{(2 \times 10^{-3})\left(\frac{1}{1 \times 10^6} + \frac{1}{800} + \frac{1}{1.2 \times 10^3}\right) + \left(\frac{1}{1 \times 10^6}\right)\left(\frac{1}{800}\right)}{5.524 \times 10^{-6}} = 0.755$$

8.44 Determine an expression for the input impedance Z_{in} of the CD FET amplifier in Fig. 8-9.

▮ Solving (3) of Problem 8.42 for v_2 by Cramer's rule yields

$$v_2 = \frac{\Delta_1}{\Delta} = \frac{[G_G(G_{S1} + G_L + g_m + g_{ds}) + G_{S1}g_m]v_s}{(G_G + G_{S1} + G_{S2})(G_{S1} + G_L + g_m + g_{ds}) - G_{S1}^2} \quad (1)$$

However, by the method of node voltages,

$$i_s = G_G(v_s - v_2) \quad (2)$$

Substitution of (1) into (2) and rearrangement lead to

$$Z_{in} = \frac{v_s}{i_s} = \frac{(G_G + G_{S1} + G_{S2})(G_{S1} + G_L + g_m + g_{ds}) - G_{S1}^2}{G_G[G_{S1}(g_{ds} + G_L) + G_{S2}(G_{S1} + G_L + g_m + g_{ds})]} \quad (3)$$

8.45 Evaluate Z_{in} for the FET amplifier of Fig. 8-9 if it is characterized by the parameters of Problem 8.43.

▮ Using the result of Problem 8.43 and (3) of Problem 8.44,

$$Z_{in} = \frac{5.524 \times 10^{-6}}{\frac{1}{1 \times 10^6}\left[\left(\frac{1}{800}\right)\left(\frac{1}{20 \times 10^3} + \frac{1}{10 \times 10^3}\right) + \left(\frac{1}{1.2 \times 10^3}\right)\left(\frac{1}{800} + \frac{1}{10 \times 10^3} + 2 \times 10^{-3} + \frac{1}{20 \times 10^3}\right)\right]}$$

$$= 1.829 \text{ M}\Omega$$

8.46 Derive an expression for the current-gain ratio $A_i = i_o/i_s$ of the CD FET amplifier in Fig. 8-9.

▮

$$A_i = \frac{i_o}{i_s} = \frac{v_o/R_L}{v_s/Z_{in}} = \frac{v_o}{v_s}\frac{Z_{in}}{R_L} = A_v\frac{Z_{in}}{R_L} \quad (1)$$

Using (6) of Problem 8.42 and (3) of Problem 8.44 in (1) leads to

$$A_i = \frac{i_o}{i_s} = \frac{g_m(G_G + G_{S1} + G_{S2}) + G_G G_{S1}}{R_L G_G[G_{S1}(g_{ds} + G_L) + G_{S2}(G_{S1} + G_L + g_m + g_{ds})]} \quad (2)$$

8.47 Evaluate the current gain $A_i = i_o/i_s$ for the CD FET amplifier of Fig. 8-9 if characterized by the parameters of Problem 8.43.

▮ By (2) of Problem 8.46,

$$A_i = \frac{(2 \times 10^{-3})\left(\frac{1}{1 \times 10^6} + \frac{1}{800} + \frac{1}{1.2 \times 10^3}\right) + \left(\frac{1}{1 \times 10^6}\right)\left(\frac{1}{800}\right)}{(10 \times 10^3)\left(\frac{1}{1 \times 10^6}\right)\left[\left(\frac{1}{800}\right)\left(\frac{1}{20 \times 10^3} + \frac{1}{10 \times 10^3}\right) + \left(\frac{1}{1.2 \times 10^3}\right)\left(\frac{1}{800} + \frac{1}{10 \times 10^3} + 2 \times 10^{-3} + \frac{1}{20 \times 10^3}\right)\right]}$$

$$= 138.04$$

8.48 Find an expression for the output impedance Z_o of the FET amplifier in Fig. 8-9.

▮ Deactivate (short) v_s in Fig. 8-10 and replace R_L by a driving-point source oriented so that $v_{dp} = v_o$ and $i_{dp} = -i_o$. Then KCL at node S requires that

$$i_{dp} = \frac{v_{dp}}{r_{ds}} + g_m v_{dp} + \frac{v_{dp}}{R_{S1} + R_G \parallel R_{S2}} \tag{1}$$

Whence,
$$Z_o = \frac{v_{dp}}{i_{dp}} = \frac{r_{ds}(R_{S1}R_G + R_{S1}R_{S2} + R_G R_{S2})}{(R_{S1}R_G + R_{S1}R_{S2} + R_G R_{S2})(1 + r_{ds}g_m) + r_{ds}(R_G + R_{S2})} \tag{2}$$

8.49 Evaluate the output impedance Z_o for the CD FET amplifier of Fig. 8-9 if the amplifier is characterized by the parameters of Problem 8.43.

▮ By (2) of Problem 8.48,

$$Z_o = \frac{(20 \times 10^3)[(800)(1 \times 10^6) + 800(1.2 \times 10^3) + (1 \times 10^6)(1.2 \times 10^3)]}{[800(1 \times 10^6) + 800(1.2 \times 10^3) + (1 \times 10^6)(1.2 \times 10^3)][1 + (20 \times 10^3)(2 \times 10^{-3})] + (20 \times 10^3)(1 \times 10^6 + 1.2 \times 10^3)}$$

$$= 264.6\ \Omega$$

8.50 Move capacitor C_S from its parallel connection across R_{S2} to a position across R_{S1} in Fig. 5-27. Let $R_G = 1$ MΩ, $R_{S1} = 800\ \Omega$, $R_{S2} = 1.2$ kΩ, and $R_L = 1$ kΩ. The JFET is characterized by $g_m = 0.002$ S and $r_{ds} = 30$ kΩ. Draw a small-signal equivalent circuit suitable for analysis of the resulting amplifier.

▮ See Fig. 8-11 for an equivalent circuit based on the current-source model of the JFET in Fig. 8-1(a).

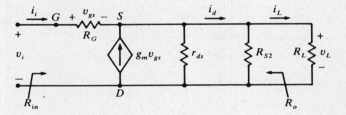

Fig. 8-11

8.51 Derive an expression for the voltage-gain ratio $A_v = v_L/v_i$ of the CD amplifier of Problem 8.50 as modeled by Fig. 8-11.

▮ By KVL,
$$v_{gs} = v_i - v_L \tag{1}$$

Using v_i and v_L as node voltages, we have
$$i_i = \frac{v_i - v_L}{R_G} \tag{2}$$

Now let
$$\frac{1}{R_{eq}} = \frac{1}{r_{ds}} + \frac{1}{R_{S2}} + \frac{1}{R_L} \tag{3}$$

By KCL and Ohm's law,
$$v_L = (i_i + g_m v_{gs})R_{eq} \tag{4}$$

Substitution of (1) and (2) into (4) and rearrangement lead to

$$A_v = \frac{v_L}{v_i} = \frac{(g_m R_G + 1)R_{eq}}{R_G + (g_m R_G + 1)R_{eq}} \tag{5}$$

8.52 Evaluate the voltage gain $A_v = v_L/v_i$ for the amplifier described by Problem 8.50.

▮ Based on (3) and (5) of Problem 8.51,

$$\frac{1}{R_{eq}} = \frac{1}{30 \times 10^3} + \frac{1}{1.2 \times 10^3} + \frac{1}{1 \times 10^3} = \frac{1}{536}$$

$$A_v = \frac{[(0.002)(1 \times 10^6) + 1](536)}{1 \times 10^6 + [(0.002)(1 \times 10^6) + 1](536)} = 0.517$$

8.53 Find an expression for the current-gain ratio $A_i = i_L/i_i$ of the CD amplifier described by Problem 8.50.

▌
$$A_i = \frac{i_L}{i_i} = \frac{v_L/R_L}{(v_i - v_L)/R_G} = \frac{A_v R_G}{(1 - A_v)R_L} \tag{1}$$

where A_v is given by (5) of Problem 8.51.

8.54 Evaluate the current gain $A_i = i_L/i_i$ for the amplifier described by Problem 8.50.

▌ Using the result of Problem 8.52 and (1) of Problem 8.53,

$$A_i = \frac{(0.517)(1 \times 10^6)}{(1 - 0.517)(1 \times 10^3)} = 1070.4$$

8.55 Determine an expression for the input resistance R_{in} of the amplifier described by Problem 8.50.

▌ From (2) of Problem 8.51,

$$i_i = \frac{v_i - v_L}{R_G} = \frac{v_i(1 - A_v)}{R_G} \tag{1}$$

R_{in} is found directly from (1) as

$$R_{\text{in}} = \frac{v_i}{i_i} = \frac{R_G}{1 - A_v} \tag{2}$$

where A_v is given by (5) of Problem 8.51.

8.56 Evaluate the input resistance R_{in} for the amplifier of Problem 8.50.

▌ Using the result of Problem 8.52 and (2) of Problem 8.55,

$$R_{\text{in}} = \frac{1 \times 10^6}{1 - 0.517} = 2.07 \text{ M}\Omega$$

8.57 Find an expression for the output resistance R_o of the amplifier described by Problem 8.50.

▌ We remove R_L and connect a driving-point source oriented such that $v_{dp} = v_L$ in the applicable equivalent circuit of Fig. 8-11. With v_i deactivated (shorted), $v_{gs} = -v_{dp}$. Then, by KCL,

$$i_{dp} = v_{dp}\left(\frac{1}{R_{S2}} + \frac{1}{r_{ds}} + \frac{1}{R_G}\right) - g_m v_{gs} = v_{dp}\left(\frac{1}{R_{S2}} + \frac{1}{r_{ds}} + \frac{1}{R_G} + g_m\right)$$

and
$$R_o = \frac{v_{dp}}{i_{dp}} = \frac{1}{\dfrac{1}{R_{S2}} + \dfrac{1}{r_{ds}} + \dfrac{1}{R_G} + g_m} \tag{1}$$

8.58 Evaluate the output resistance R_o for the amplifier of Problem 8.50.

▌ By (1) of Problem 8.57,

$$R_o = \frac{1}{\dfrac{1}{1.2 \times 10^3} + \dfrac{1}{30 \times 10^3} + \dfrac{1}{1 \times 10^6} + 0.002} = 348.7 \text{ }\Omega$$

COMMON-GATE AMPLIFIERS

8.59 Use the current-source FET model of Fig. 8-1(a) to draw a small-signal equivalent circuit model for the CG amplifier of Fig. 5-25.

▌ See Fig. 8-12.

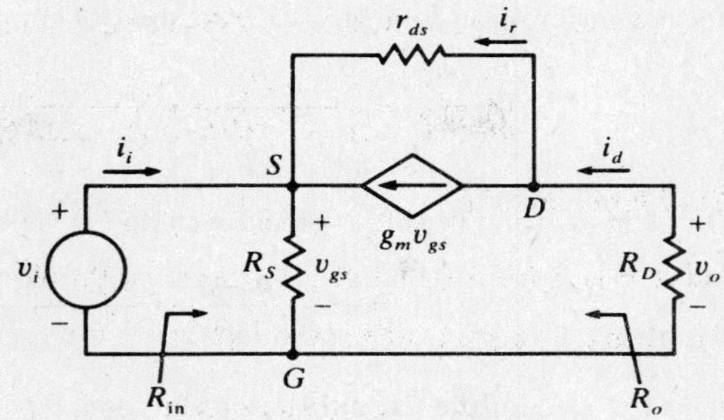

Fig. 8-12 CG small-signal equivalent circuit.

8.60 For the CG amplifier of Fig. 5-25 as modeled by Fig. 8-12, $R_D = R_S = 1\,\text{k}\Omega$, $g_m = 2 \times 10^{-3}\,\text{S}$, and $r_{ds} = 30\,\text{k}\Omega$. Find an expression for the voltage-gain ratio $A_v = v_o/v_i$.

▌ By KCL in Fig. 8-12, $i_r = i_d - g_m v_{gs}$. Applying KVL around the outer loop gives

$$v_o = (i_d - g_m v_{gs})r_{ds} - v_{gs}$$

But $v_{gs} = -v_i$ and $i_d = -v_o/R_D$; thus,

$$v_o = \left(-\frac{v_o}{R_D} + g_m v_i\right)r_{ds} + v_i$$

and
$$A_v = \frac{v_o}{v_i} = \frac{(g_m r_{ds} + 1)R_D}{R_D + r_{ds}} \qquad (1)$$

8.61 Evaluate the voltage gain $A_v = v_o/v_i$ for the CG amplifier of Problem 8.60.

▌ By (1) of Problem 8.60,

$$A_v = \frac{61(1 \times 10^3)}{1 \times 10^3 + 30 \times 10^3} = 1.97$$

8.62 For the CG amplifier of Problem 8.60, determine an expression for the input resistance R_{in}.

▌ Source v_i can be considered a driving-point source in Fig. 8-12. At the input mesh,

$$v_i = (i_i + i_d)R_S \qquad (1)$$

Application of KVL around the outer loop yields

$$v_i = -i_d R_D - (i_d - g_m v_{gs})r_{ds} \qquad (2)$$

Using $v_{gs} = -v_i$, recognizing that $g_m r_{ds} = \mu$, and rearranging (2) lead to

$$i_d = -\frac{(\mu + 1)}{R_D + r_{ds}} v_i \qquad (3)$$

Substitution of (3) into (1) allows solution for R_{in} as

$$R_{in} = \frac{v_i}{i_i} = \frac{R_S(R_D + r_{ds})}{(\mu + 1)R_S + R_D + r_{ds}} \qquad (4)$$

8.63 Evaluate the input resistance R_{in} for the CG amplifier of Problem 8.60.

▌ By (4) of Problem 8.62, where $\mu = g_m r_{ds} = (2 \times 10^{-3})(30 \times 10^3) = 60$, we have

$$R_{in} = \frac{(1 \times 10^3)(1 \times 10^3 + 30 \times 10^3)}{(60 + 1)(1 \times 10^3) + 1 \times 10^3 + 30 \times 10^3} = 337\,\Omega$$

8.64 Determine the output resistance for the CG amplifier of Problem 8.60 as modeled by Fig. 8-12.

▌ Deactivate (short) v_i. With $v_i = 0 = -v_{gs}$, $g_m v_{gs} = 0$, or the dependent current source is inactive. Hence,

$$R_o = r_{ds} = 30\,\text{k}\Omega$$

8.65 Derive an expression for the current-gain ratio $A_i = i_d/i_i$ for the CG amplifier of Fig. 5-25 as modeled by Fig. 8-12.

$$A_i = \frac{i_d}{i_i} = \frac{-v_o/R_D}{v_i/R_{\text{in}}} = -A_v \frac{R_{\text{in}}}{R_D} \quad (1)$$

Substitution of (1) of Problem 8.60 and (4) of Problem 8.62 into (1) yields

$$A_i = -\frac{(g_m r_{ds} + 1)R_S}{(\mu + 1)R_S + R_D + r_{ds}} \quad (2)$$

8.66 Evaluate the current gain $A_i = i_d/i_i$ for the CG amplifier of Problem 8.60.

Use of the results of Problems 8.61 and 8.63 in (1) of Problem 8.65 leads to

$$A_i = -(1.97)\frac{337}{1 \times 10^3} = -0.664$$

8.67 Figure 8-13(a) is a small-signal equivalent circuit (voltage-source model) of a common-gate JFET amplifier. Use the circuit to verify two *rules of impedance and voltage reflection* for FET amplifiers:
  (a) Voltages and impedances in the drain circuit are reflected to the source circuit divided by $\mu + 1$. [Verify this rule by finding the Thévenin equivalent for the circuit to the right of a, a' in Fig. 8-13(a) and showing that Fig. 8-13(b) results.]
  (b) Voltages and impedances in the source circuit are reflected to the drain circuit multiplied by $\mu + 1$. [Verify this rule by finding the Thévenin equivalent for the circuit to the left of b, b' in Fig. 8-13(a) and showing that Fig. 8-13(c) results.]

(a) With a, a' open, $i_d = 0$; hence, $v_{gs} = 0$ and $v_{Th} = 0$. After a driving-point source $v_{aa'}$ is connected to terminals a, a' to drive current i_a into terminal a, KVL gives

$$v_{aa'} = \mu v_{gs} + i_a(r_{ds} + R_D) \quad (1)$$

But $v_{gs} = -v_{aa'}$, which can be substituted into (1) to give

$$R_{Th} = \frac{v_{aa'}}{i_a} = \frac{r_{ds}}{\mu + 1} + \frac{R_D}{\mu + 1} \quad (2)$$

With $v_{Th} = 0$, insertion of R_{Th} in place of the network to the right of a, a' in Fig. 8-13(a) leads directly to Fig. 8-13(b).

(b) Applying KVL to the left of b, b' in Fig. 8-13(a) with b, b' open, while noting that $v_i = -v_{gs}$, yields

$$v_{Th} = v_i - \mu v_{gs} = (\mu + 1)v_i \quad (3)$$

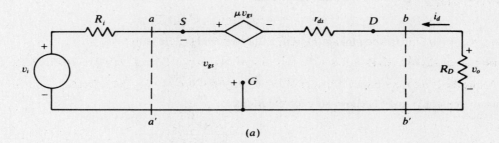

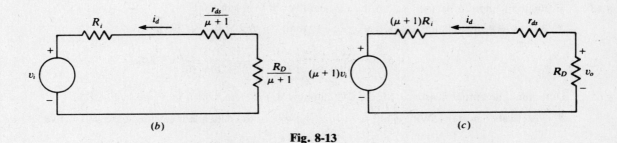

Fig. 8-13

Deactivating (shorting) v_i, connecting a driving-point source $v_{bb'}$ to terminals b, b' to drive current i_b into terminal b, noting that $v_{gs} = -i_b R_S$, and applying KVL around the outer loop of Fig. 8-13(a) yield

$$v_{bb'} = i_b(r_{ds} + R_S) - \mu v_{gs} = i_b[r_{ds} + (\mu+1)R_S] \quad (4)$$

The Thévenin impedance follows from (4) as

$$R_{Th} = \frac{v_{bb'}}{i_b} = r_{ds} + (\mu+1)R_S \quad (5)$$

When the Thévenin source of (3) and impedance of (5) are used to replace the network to the left of b, b', the circuit of Fig. 8-13(c) results.

8.68 Find the voltage-gain ratio $A_v = v_o/v_i$ for a CG amplifier modeled by the circuit of Fig. 8-13(a).

▌ The circuit of Fig. 8-13(c) is equivalent as shown in Problem 8.67. By voltage division in Fig. 8-13(c),

$$v_o = \frac{R_D}{R_D + r_{ds} + (\mu+1)R_i}(\mu+1)v_i$$

Whence,
$$A_v = \frac{v_o}{v_i} = \frac{(\mu+1)R_D}{R_D + r_{ds} + (\mu+1)R_i} \quad (1)$$

Review Problems

8.69 For the SF amplifier of Fig. 8-6, verify that the output resistance $R_o \approx 1/g_m$ if $\mu \gg 1$.

▌ Based on the result of Problem 8.38 with $\mu \gg 1$ and the definition of μ from Problem 8.3,

$$R_o = \frac{r_{ds}}{\mu+1} \approx \frac{r_{ds}}{\mu} = \frac{1}{g_m}$$

8.70 An FET is used as a load resistance with terminals D, S after connecting the gate to the drain. Find the resulting value of the load resistance R_L.

▌ Shorting G to D in Fig. 8-1(a), it is seen that $v_{gs} = v_{ds}$. Connect a driving-point source between the terminals D, S oriented such that $v_{dp} = v_{ds}$ and $i_{dp} = i_d$. Then, application of KCL at node D yields

$$i_{dp} = \frac{v_{dp}}{r_{ds}} + g_m v_{dp} = \left(\frac{1 + g_m r_{ds}}{r_{ds}}\right) v_{dp} \quad (1)$$

From (1) with $\mu = g_m r_{ds}$,

$$R_L = \frac{v_{dp}}{i_{dp}} = \frac{r_{ds}}{\mu+1}$$

8.71 Draw a small-signal equivalent circuit suitable for analysis of the CS FET amplifier in Fig. 8-14.

▌ Using the current-source FET model of Fig. 8-1(a), the circuit of Fig. 8-15 results.

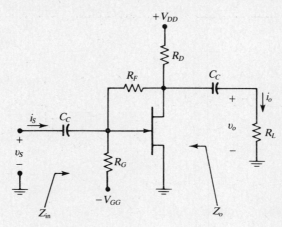

Fig. 8-14

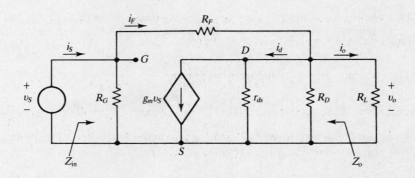

Fig. 8-15

8.72 Find an expression for the voltage-gain ratio $A_v = v_o/v_s$ for the FET amplifier of Fig. 8-14.

▌ KCL at node D of Fig. 8-15 requires that

$$i_F = i_o + i_d + v_o/R_D \tag{1}$$

Using the method of node voltages,

$$i_F = \frac{v_s - v_o}{R_F} \tag{2}$$

By Ohm's law,

$$i_o = v_o/R_L \tag{3}$$

Also,

$$i_d = v_o/r_{ds} + g_m v_s \tag{4}$$

Substitute (2)–(4) into (1) to find

$$\frac{v_s - v_o}{R_F} = \frac{v_o}{R_L} + \frac{v_o}{r_{ds}} + g_m v_s + \frac{v_o}{R_D} \tag{5}$$

After rearrangement, (5) yields

$$A_v = \frac{v_o}{v_s} = \frac{(1 - g_m R_F)/R_F}{1/R_F + 1/R_L + 1/r_{ds} + 1/R_D} \tag{6}$$

8.73 For the amplifier of Fig. 8-14, let $R_D = R_L = 10\text{ k}\Omega$, $R_F = 60\text{ k}\Omega$, $R_G = 50\text{ k}\Omega$, $\mu = 50$, and $r_{ds} = 25\text{ k}\Omega$. Evaluate the voltage gain $A_v = v_o/v_s$.

▌
$$g_m = \mu/r_{ds} = 50/(25 \times 10^3) = 2\text{ mS}$$

By (6) of Problem 8.72,

$$A_v = \frac{v_o}{v_s} = \frac{[1 - (2 \times 10^{-3})(60 \times 10^3)]/(60 \times 10^3)}{\dfrac{1}{60 \times 10^3} + \dfrac{1}{10 \times 10^3} + \dfrac{1}{25 \times 10^3} + \dfrac{1}{10 \times 10^3}} = -7.73$$

8.74 Determine an expression for the input impedance Z_{in} of the FET amplifier in Fig. 8-13.

▌ By KCL at node G in Fig. 8-15,

$$i_s = \frac{v_s}{R_G} + i_F \tag{1}$$

Using (2) of Problem 8.72 and $v_o = A_v v_s$ in (1) leads to

$$i_s = \frac{v_s}{R_G} + \frac{v_s - v_o}{R_F} = \frac{v_s}{R_G} + \frac{v_s}{R_F} - \frac{A_v v_s}{R_F} \tag{2}$$

After rearrangement, (2) yields

$$Z_{in} = \frac{v_s}{i_s} = \frac{R_F R_G}{R_F + (1 - A_v) R_G} \tag{3}$$

where A_v is given by (6) of Problem 8.72.

SMALL-SIGNAL FREQUENCY-INDEPENDENT FET AMPLIFIERS □ 249

8.75 Evaluate Z_{in} for the amplifier of Fig. 8-14 having the parameters of Problem 8.73.

▌ Using the result of Problem 8.73 and (3) of Problem 8.74,

$$Z_{in} = \frac{(60 \times 10^3)(50 \times 10^3)}{60 \times 10^3 + [1 - (-7.73)](50 \times 10^3)} = 6.042 \text{ k}\Omega$$

8.76 Find an expression for the current-gain ratio $A_i = i_o/i_s$ of the FET amplifier in Fig. 8-14.

▌

$$A_i = \frac{i_o}{i_s} = \frac{v_o/R_L}{v_s/R_{in}} = \frac{v_o}{v_s}\frac{Z_{in}}{R_L} = A_v \frac{Z_{in}}{R_L} \qquad (1)$$

Substitute (3) of Problem 8.74 into (1) to find

$$A_i = \frac{A_v R_F R_G}{R_L[R_F + (1 - A_v)R_G]} \qquad (2)$$

where A_v is given by (6) of Problem 8.72.

8.77 Evaluate the current gain $A_i = i_o/i_L$ for the amplifier of Fig. 8-13 if it is described by the parameters of Problem 8.73.

▌ Using Problem 8.76 and the result of Problem 8.73,

$$A_i = \frac{-7.73(60 \times 10^3)(50 \times 10^3)}{(10 \times 10^3)\{60 \times 10^3 + [1 - (-7.73)](50 \times 10^3)\}} = -4.67$$

8.78 Find an expression for the output impedance Z_o of the FET amplifier in Fig. 8-14.

▌ The small-signal equivalent circuit of Fig. 8-15 is applicable. With $v_s = 0$ (deactivated), the controlled source is inactive ($g_m v_s = 0$) and R_F is effectively connected to S. Hence,

$$Z_o = R_F \parallel R_D \parallel r_{ds} = \frac{R_F R_D r_{ds}}{R_F R_D + R_F r_{ds} + R_D r_{ds}} \qquad (1)$$

8.79 Evaluate the output impedance Z_o for the FET amplifier of Fig. 8-14 if the amplifier is described by the parameters of Problem 8.73.

▌ By (1) of Problem 8.78,

$$Z_o = \frac{(60 \times 10^3)(50 \times 10^3)(25 \times 10^3)}{(60 \times 10^3)(50 \times 10^3) + (60 \times 10^3)(25 \times 10^3) + (50 \times 10^3)(25 \times 10^3)} = 13.04 \text{ k}\Omega$$

8.80 Draw a small-signal equivalent circuit suitable for analysis of the CD FET amplifier of Fig. 8-16.

▌ The circuit of Fig. 8-17 results from use of the current-source FET model of Fig. 8-1(a).

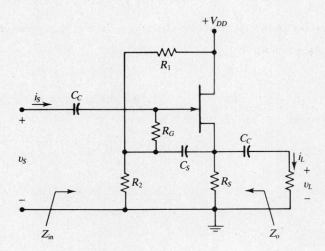

Fig. 8-16

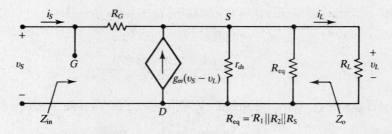

Fig. 8-17

8.81 Determine an expression for the voltage-gain ratio $A_v = v_L/v_s$ for the CD FET amplifier of Fig. 8-16.

▮ The equivalent circuit of Fig. 8-17 is applicable. KCL applied at node S requires that

$$i_s = \frac{v_L}{R_L} + \frac{v_L}{R_{eq}} + \frac{v_L}{r_{ds}} - g_m(v_s - v_L) \tag{1}$$

By method of node voltages,

$$i_s = \frac{v_s - v_L}{R_G} \tag{2}$$

Equate (1) and (2) and rearrange to find A_v.

$$A_v = \frac{v_L}{v_s} = \frac{1/R_G + g_m}{1/R_G + 1/R_L + 1/R_{eq} + 1/r_{ds} + g_m} \tag{3}$$

8.82 For the CD amplifier of Fig. 8-16, let $R_G = R_1 = 10$ MΩ, $R_2 = 2$ MΩ, $R_S = R_L = 4$ kΩ, $r_{ds} = 20$ kΩ, and $g_m = 2 \times 10^{-3}$ S. Evaluate the voltage gain $A_v = v_L/v_s$.

▮ $$R_{eq} = \frac{R_1 R_2 R_S}{R_1 R_2 + R_1 R_S + R_2 R_s} = \frac{(10 \times 10^6)(2 \times 10^6)(4 \times 10^3)}{(10 \times 10^6)(2 \times 10^6) + (10 \times 10^6)(4 \times 10^3) + (2 \times 10^6)(4 \times 10^3)} = 3.99 \text{ kΩ}$$

By (3) of Problem 8.81,

$$A_v = \frac{1/(10 \times 10^6) + 2 \times 10^{-3}}{1/(10 \times 10^6) + 1/(4 \times 10^3) + 1/(3.99 \times 10^3) + 1/(20 \times 10^3) + 2 \times 10^{-3}} = 0.784$$

8.83 Find an expression for the input impedance Z_{in} of the CD amplifier in Fig. 8-16.

▮ Substituting (2) into (3) of Problem 8.81 leads to

$$i_s = \frac{v_s - v_L}{R_G} = \frac{(1 - A_v)v_s}{R_G} \tag{1}$$

Whence,

$$Z_{in} = \frac{v_s}{i_s} = \frac{R_G}{1 - A_v} \tag{2}$$

Defining appropriate conductances, $G_G = 1/R_G$ and so forth, (2) can be written as

$$Z_{in} = \frac{v_s}{i_s} = \frac{G_G + G_L + G_{eq} + g_{ds} + g_m}{G_G(G_L + G_{eq} + g_{ds})} \tag{3}$$

8.84 Evaluate Z_{in} for the FET amplifier of Fig. 8-16 if it is characterized by the parameters of Problem 8.82.

▮ By (2) of Problem 8.83 and the result of Problem 8.82,

$$Z_{in} = \frac{R_G}{1 - A_v} = \frac{10 \times 10^6}{1 - 0.784} = 46.3 \text{ MΩ}$$

8.85 Derive an expression for the current-gain ratio $A_i = i_L/i_s$ of the CD amplifier of Fig. 8-16.

▮ $$A_i = \frac{i_L}{i_s} = \frac{v_L/R_L}{v_s/Z_{in}} = A_v \frac{Z_{in}}{R_L} \tag{1}$$

SMALL-SIGNAL FREQUENCY-INDEPENDENT FET AMPLIFIERS 251

Substituting (3) of Problem 8.81 and (3) of Problem 8.83 into (1) leads to

$$A_i = \frac{G_L(G_G + g_m)}{G_G(G_L + G_{eq} + g_{ds})} \qquad (2)$$

8.86 Evaluate the current gain $A_i = i_L/i_s$ for the FET amplifier of Fig. 8-16 if characterized by the parameters of Problem 8.82.

▎ By (1) of Problem 8.85 with the results of Problems 8.82 and 8.84,

$$A_i = (0.784)\frac{46.3 \times 10^6}{4 \times 10^3} = 9075$$

8.87 Find an expression for the output impedance Z_o of the CD amplifier in Fig. 8-16.

▎ Deactivate (short) v_s in Fig. 8-17 and replace R_L by a driving-point source oriented such that $v_{dp} = v_L$ and $i_{dp} = -i_L$. Then, KCL at node S requires that

$$i_{dp} = \frac{v_{dp}}{R_{eq}} + \frac{v_{dp}}{r_{ds}} + \frac{v_{dp}}{R_G} + g_m v_{dp} \qquad (1)$$

Whence,

$$Z_o = \frac{v_{dp}}{i_{dp}} = \frac{1}{G_{eq} + g_{ds} + G_G + g_m} \qquad (2)$$

where appropriate conductances have been defined.

8.88 Evaluate the output impedance Z_o for the FET amplifier of Fig. 8-16 if the amplifier is characterized by the parameters of Problem 8.82.

▎ By (2) of Problem 8.87 and using $R_{eq} = 1/G_{eq}$ as determined in Problem 8.82,

$$Z_o = \frac{1}{\frac{1}{3.99 \times 10^3} + \frac{1}{20 \times 10^3} + \frac{1}{10 \times 10^6} + 2 \times 10^{-3}} = 416.6\ \Omega$$

8.89 Based on the current-source FET model of Fig. 8-1(a), draw a small-signal equivalent circuit suitable for analysis of the CS amplifier of Fig. 8-18.

▎ See Fig. 8-19.

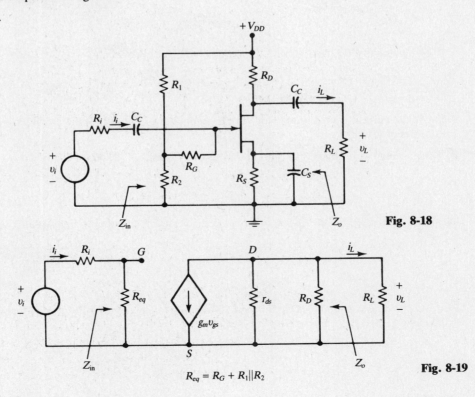

Fig. 8-18

Fig. 8-19

$R_{eq} = R_G + R_1 \| R_2$

8.90 Derive an expression for the voltage-gain ratio $A_v = v_L/v_i$ for the CS FET amplifier of Fig. 8-18.

■ Referring to the applicable equivalent circuit of Fig. 8-19, voltage division gives

$$v_{gs} = \frac{R_{eq}}{R_{eq} + R_i} v_i \quad (1)$$

Ohm's law applied at the output network leads to

$$v_L = -g_m v_{gs}(r_{ds} \| R_D \| R_L) = -\frac{g_m r_{ds} R_D R_L}{r_{ds} R_D + r_{ds} R_L + R_D R_L} v_{gs} \quad (2)$$

Substitute (1) into (2) and rearrange to find A_v as

$$A_v = \frac{v_L}{v_i} = -\frac{g_m r_{ds} R_D R_L R_{eq}}{(R_{eq} + R_i)(r_{ds} R_D + r_{ds} R_L + R_D R_L)} \quad (3)$$

8.91 For the CS amplifier of Fig. 8-18, let $R_i = 1\,\text{k}\Omega$, $R_G = 1\,\text{M}\Omega$, $R_1 = 180\,\text{k}\Omega$, $R_2 = 20\,\text{k}\Omega$, $R_D = R_L = 2\,\text{k}\Omega$, $r_{ds} = 15\,\text{k}\Omega$, and $g_m = 2 \times 10^{-3}$ S. Evaluate the voltage gain $A_v = v_L/v_i$.

■
$$R_{eq} = R_G + \frac{R_1 R_2}{R_1 + R_2} = 1 \times 10^6 + \frac{(180 \times 10^3)(20 \times 10^3)}{180 \times 10^3 + 20 \times 10^3} = 1.018\,\text{M}\Omega$$

Based on (3) of Problem 8.90,

$$A_v = -\frac{(2 \times 10^{-3})(15 \times 10^3)(2 \times 10^3)(2 \times 10^3)(1.018 \times 10^6)}{(1.018 \times 10^6 + 1 \times 10^3)[(15 \times 10^3)(2 \times 10^3) + (15 \times 10^3)(2 \times 10^3) + (2 \times 10^3)(2 \times 10^3)]} = -1.873$$

8.92 Evaluate the input impedance Z_{in} for the amplifier of Fig. 8-18 if it is characterized by the parameters of Problem 8.91.

■ From the equivalent circuit of Fig. 8-19, it is apparent that $Z_{in} = R_{eq}$. From the result of Problem 8.91,

$$Z_{in} = R_{eq} = R_G + \frac{R_1 R_2}{R_1 + R_2} = 1.018\,\text{M}\Omega$$

8.93 Find an expression for the current-gain ratio $A_i = i_L/i_i$ for the CS amplifier of Fig. 8-18.

■
$$A_i = \frac{i_L}{i_i} = \frac{v_L/R_L}{v_i/Z_{in}} = A_v \frac{Z_{in}}{R_L} \quad (1)$$

Substituting (3) of Problem 8.90 and Problem 8.92 into (1) yields

$$A_i = \frac{i_L}{i_i} = -\frac{g_m r_{ds} R_D R_L R_{eq}^2}{(R_{eq} + R_i)(r_{ds} R_D + r_{ds} R_L + R_D R_L) R_L} \quad (2)$$

8.94 Evaluate the current gain $A_i = i_L/i_i$ for the FET amplifier of Fig. 8-18 if the amplifier is characterized by the parameters of Problem 8.91.

■ Based on (1) of Problem 8.93 and the results of Problems 8.91 and 8.92,

$$A_i = (-1.873)\frac{1.018 \times 10^6}{2 \times 10^3} = -953.4$$

8.95 Derive an expression for the output impedance Z_o of the CS amplifier in Fig 8-18.

■ With v_i deactivated (shorted) in Fig. 8-19, $v_{gs} = 0$ and the controlled source is inactive ($g_m v_{gs} = 0$). Hence,

$$Z_o = r_{ds} \| R_D = \frac{r_{ds} R_D}{r_{ds} + R_D} \quad (1)$$

8.96 Evaluate the output impedance Z_o for the FET amplifier of Fig. 8-18 if the amplifier parameters are those of Problem 8.91.

■ By (1) of Problem 8.95,

$$Z_o = \frac{(15 \times 10^3)(2 \times 10^3)}{15 \times 10^3 + 2 \times 10^3} = 1.76\,\text{k}\Omega$$

SMALL-SIGNAL FREQUENCY-INDEPENDENT FET AMPLIFIERS 253

8.97 Suppose capacitor C_S is removed from the circuit of Problem 8.17 (Fig. 8-3), and all else remains unchanged. Based on the voltage-source FET model of Fig. 8-1(b), draw a small-signal equivalent circuit suitable for analysis of the amplifier.

▌ See Fig. 8-20.

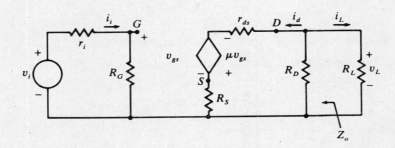

Fig. 8-20

8.98 Derive an expression for the voltage-gain ratio $A_v = v_L/v_i$ of the amplifier described by Problem 8.97.

▌ Referring to the equivalent circuit of Fig. 8-20, voltage division and KVL give

$$v_{gs} = \frac{R_G}{R_G + r_i} v_i - i_d R_S \tag{1}$$

But by Ohm's law,

$$i_d = \frac{\mu v_{gs}}{r_{ds} + R_S + R_D \| R_L} \tag{2}$$

Substituting (2) into (1) and solving for v_{gs} yield

$$v_{gs} = \frac{R_G(r_{ds} + R_S + R_D \| R_L) v_i}{(R_G + r_i)[r_{ds} + (\mu + 1)R_S + R_D \| R_L]} \tag{3}$$

Now voltage division gives

$$v_L = -\frac{R_D \| R_L}{r_{ds} + R_S + R_D \| R_L} \mu v_{gs} \tag{4}$$

and substitution of (3) into (4) and rearrangement give

$$A_v = \frac{v_L}{v_i} = \frac{-\mu R_G R_D R_L}{(R_G + r_i)\{(R_D + R_L)[r_{ds} + (\mu + 1)R_S] + R_D R_L\}} \tag{5}$$

8.99 Evaluate the voltage gain $A_v = v_L/v_i$ for the amplifier described by Problem 8.97.

▌ By (5) of Problem 8.98 and using $\mu = g_m r_{ds}$ along with the given parameter values of Problem 8.17, the voltage gain is found as

$$A_v = \frac{-(2 \times 10^{-3})(30 \times 10^3)(160)(2)(2)}{(160 + 5)\{(2 + 2)[30 + (60 + 1)3] + 2(2)\}} = -0.272$$

8.100 Find an expression for the current-gain ratio $A_i = i_L/i_i$ of the FET amplifier described by Problem 8.97 and modeled by Fig. 8-20.

▌

$$A_i = \frac{i_L}{i_i} = \frac{v_L/R_L}{v_i/(R_G + r_i)} = A_v \frac{R_G + r_i}{R_L} \tag{1}$$

where A_v is given by (5) of Problem 8.98.

8.101 Evaluate the current gain ratio $A_i = i_L/i_i$ for the FET amplifier of Problem 8.97.

▌ By (1) of Problem 8.100 along with the given values of Problem 8.97 and the result of Problem 8.99, the current gain is

$$A_i = \frac{A_v(R_G + r_i)}{R_L} = \frac{-0.272(160 + 5)}{2} = -22.4$$

8.102 Determine an expression for the output impedance Z_o of the FET amplifier described by Problem 8.97.

▮ R_L is disconnected, and a driving-point source is added such that $v_{dp} = v_L$. With v_i deactivated (short-circuited), $v_{gs} = 0$ and

$$Z_o = R_D \parallel (r_{ds} + R_S) = \frac{R_D(r_{ds} + R_S)}{R_D + r_{ds} + R_S} \quad (1)$$

8.103 Evaluate the output impedance Z_o for the FET amplifier of Problem 8.97.

▮ By (1) of Problem 8.102 with given values of Problem 8.97,

$$Z_o = \frac{(2 \times 10^3)(30 \times 10^3 + 3 \times 10^3)}{2 \times 10^3 + 30 \times 10^3 + 3 \times 10^3} = 1.89 \text{ k}\Omega$$

8.104 Use the small-signal equivalent circuit to predict the peak values of i_d and v_{ds} in Problem 5.29. Compare your result with Problem 5.29, and comment on any differences.

▮ The values of g_m and r_{ds} for operation near the Q point of Fig. 5-12 were determined in Problem 8.6. We may use the current-source model of Fig. 8-1(a) to form the equivalent circuit for Fig. 5-9. In that circuit, with $v_{gs} = \sin t$ V, Ohm's law requires that

$$v_{ds} = -g_m v_{gs}(r_{ds} \parallel R_D) = \frac{-g_m r_{ds} R_D v_{gs}}{r_{ds} + R_D} = \frac{-(1.5 \times 10^{-3})(75 \times 10^3)(3 \times 10^3)v_{gs}}{75 \times 10^3 + 3 \times 10^3} = -4.33 v_{gs}$$

Thus, $\qquad V_{dsm} = 4.33 V_{gsm} = 4.33(1) = 4.3$ V

Also, from Fig. 8-1(a),

$$i_d = g_m v_{gs} + \frac{v_{ds}}{r_{ds}}$$

so $\qquad I_{dm} = g_m V_{gsm} + \frac{V_{dsm}}{r_{ds}} = (1.5 \times 10^{-3})(1) + \frac{1}{75 \times 10^3} = 1.513$ mA

The ± 1 V excursion of v_{gs} leads to operation over a large portion of the nonlinear drain characteristics. Consequently, the small-signal equivalent circuit predicts greater positive peaks and smaller negative peaks of i_d and v_{ds} than the graphical solution of Problem 5.29, which inherently accounts for the nonlinearities.

8.105 In the amplifier of Fig. 5-14, $R_1 = 20$ kΩ, $R_2 = 100$ kΩ, $R_3 = 1$ MΩ, $r_{ds} = 30$ kΩ, $\mu = 150$, and $R_S = 1$ kΩ. Find the voltage gain $A_v = v_o/v_i$.

▮ The amplifier can be modeled by the circuit of Fig. 8-7(b) where $R_G = R_3 + R_1 \parallel R_2$. Hence, by (1) of Problem 8.33,

$$A_v = \frac{\mu R_S}{(\mu + 1)R_S + r_{ds}} = \frac{150(1 \times 10^3)}{(150 + 1)(1 \times 10^3) + 30 \times 10^3} = 0.829$$

8.106 Evaluate the current gain $A_i = i_d/i_i$ for the amplifier of Problem 8.105.

▮

$$R_G = R_3 + \frac{R_1 R_2}{R_1 + R_2} = 1 \times 10^6 + \frac{(20 \times 10^3)(100 \times 10^3)}{20 \times 10^3 + 100 \times 10^3} = 1.017 \text{ M}\Omega$$

By (2) of Problem 8.35,

$$A_i = \frac{\mu R_G}{r_{ds} + (\mu + 1)R_S} = \frac{150(1.017 \times 10^6)}{30 \times 10^3 + (150 + 1)(1 \times 10^3)} = 843$$

8.107 Evaluate the output impedance Z_o for the amplifier of Problem 8.105.

▮ By (1) of Problem 8.38.

$$Z_o = \frac{r_{ds}}{\mu + 1} = \frac{30 \times 10^3}{150 + 1} = 198.7 \text{ }\Omega$$

8.108 Draw a small-signal equivalent circuit suitable for determination of v_2 in the MOSFET amplifier of Fig. 8-21(a).

▮ See Fig. 8-21(b) where impedance reflection (see Problem 8.67) has been used for simplification.

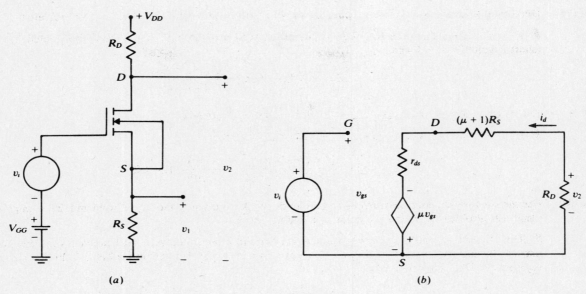

Fig. 8-21

8.109 Find an expression for the voltage-gain ratio $A_{v2} = v_2/v_i$ for the amplifier of Fig. 8-21(a).

■ By voltage division applied to Fig. 8-21(b).

$$v_2 = \frac{R_D}{R_D + r_{ds} + (\mu + 1)R_S} \mu v_{gs}$$

Whence, since $v_{gs} = v_i$,

$$A_{v2} = \frac{v_2}{v_i} = -\frac{\mu R_D}{R_D + r_{ds} + (\mu + 1)R_S} \tag{1}$$

8.110 Determine an expression for the voltage-gain ratio $A_{v1} = v_1/v_i$ for the amplifier of Fig. 8-21(a).

■ The equivalent circuit of Fig. 8-20 is applicable if $r_i = 0$ and $R_L, R_G \to \infty$. Hence, by (5) of Problem 8.98,

$$A_{v1} = \frac{v_1}{v_i} = \frac{\mu R_S}{r_{ds} + (\mu + 1)R_S + R_D} \tag{1}$$

8.111 If $R_D = R_S$ in the amplifier of Fig. 8-21(a), the circuit is commonly called a *phase splitter*, since $v_2 = -v_1$ (the outputs are equal in magnitude but 180° out of phase). Verify that the circuit actually is a phase splitter.

■ Comparing (1) of Problem 8.109 with (1) of Problem 8.110, it is seen that if $R_D = R_S$,

$$\frac{A_{v1}}{A_{v2}} = \frac{v_1}{v_2} = -1 \quad \text{or} \quad v_2 = -v_1$$

8.112 The JFET amplifier of Fig. 5-27 has $R_G = 1\,\text{M}\Omega$, $R_{S1} = 800\,\Omega$, $R_{S2} = 1.2\,\text{k}\Omega$, and $R_L = 1\,\text{k}\Omega$. The JFET obeys (1) of Problem 5.6 and is characterized by $I_{DSS} = 10\,\text{mA}$, $V_{p0} = 4\,\text{V}$, $V_{GSQ} = -2\,\text{V}$, and $\mu = 60$. Determine (a) g_m by use of (3) of Problem 8.1 and (b) r_{ds}.

■ (a) By (3) of Problem 8.1,

$$g_m = \left.\frac{\partial i_D}{\partial v_{GS}}\right|_Q = \frac{\partial}{\partial v_{GS}}\left[I_{DSS}\left(1 + \frac{v_{GS}}{V_{p0}}\right)^2\right]\bigg|_Q = \frac{2I_{DSS}}{V_{p0}}\left(1 + \frac{V_{GSQ}}{V_{p0}}\right) = \frac{2(10 \times 10^{-3})}{4}\left(1 + \frac{-2}{4}\right) = 2.5 \times 10^{-3}\,\text{S}$$

(b)

$$r_{ds} = \frac{\mu}{g_m} = \frac{60}{2.5 \times 10^{-3}} = 24\,\text{k}\Omega$$

8.113 Evaluate the voltage gain $A_v = v_L/v_i$ for the amplifier of Problem 8.112.

▎ The small-signal circuit of Fig. 8-7(b) is applicable if R_S is replaced by $R_{S1} \| R_L$. Hence, by the result of Problem 8.105,

$$A_v = \frac{v_L}{v_i} = \frac{\mu(R_{S1} \| R_L)}{(\mu + 1)(R_{S1} \| R_L) + r_{ds}} = \frac{\mu R_{S1} R_L}{(\mu + 1)R_{S1}R_L + r_{ds}(R_{S1} + R_L)}$$

$$= \frac{60(800)(1 \times 10^3)}{(60 + 1)(800)(1 \times 10^3) + (24 \times 10^3)(800 + 1 \times 10^3)} = 0.52$$

CHAPTER 9
Multiple-Transistor Amplifiers

MULTIPLE-BJT AMPLIFIERS

9.1 Draw a small-signal equivalent circuit suitable for analysis of the two-stage amplifier of Fig. 9-1 if $h_{re1} = h_{re2} \approx 0$.

▌ Based on the CE h-parameter BJT model of Fig. 7-1 if $h_{re} = 0$, the circuit of Fig. 9-2 results where $R_{B1} = R_{11} \parallel R_{12}$ and $R_{B2} = R_{22} \parallel R_{21}$.

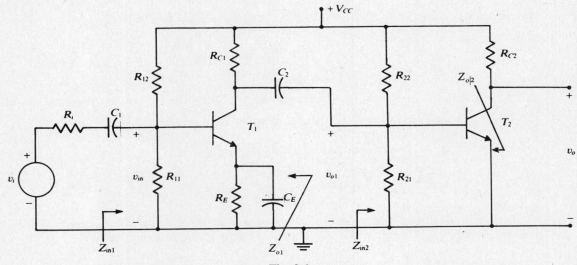

Fig. 9-1

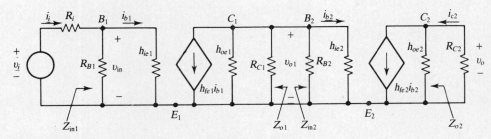

Fig. 9-2

9.2 In the two-stage amplifier of Fig. 9-1, the transistors are identical, having $h_{ie} = 1500\ \Omega$, $h_{fe} = 40$, $h_{re} \approx 0$, and $h_{oe} = 30\ \mu S$. Also, $R_i = 1\ k\Omega$, $R_{C2} = 20\ k\Omega$, $R_{C1} = 10\ k\Omega$,

$$R_{B1} \equiv \frac{R_{11}R_{12}}{R_{11}+R_{12}} = 5\ k\Omega \quad \text{and} \quad R_{B2} \equiv \frac{R_{21}R_{22}}{R_{21}+R_{22}} = 5\ k\Omega$$

Find the final-stage voltage gain $A_{v2} \equiv v_o/v_{o1}$.

▌ Comparing Fig. 7-12 with the applicable equivalent circuit of Fig. 9-2, it is apparent that the final-stage voltage gain is given by the result of Problem 7.40 if the parallel combination of R_L and R_C is replaced with R_{C2}:

$$A_{v2} = \frac{v_o}{v_{o1}} = -\frac{h_{fe}R_{C2}}{h_{ie}(1+h_{oe}R_{C2})} = -\frac{40(20 \times 10^3)}{1500[1+(30 \times 10^{-6})(20 \times 10^3)]} = -333.3 \quad (1)$$

9.3 Determine the final-stage input impedance Z_{in2} for the two-stage amplifier of Problem 9.2 as modeled by Fig. 9-2.

258 □ CHAPTER 9

▮ From (4) of Problem 7.125 with $h_{re} \approx 0$ and R_B replaced by R_{B2},

$$Z_{in2} = \frac{R_{B2}h_{ie}}{R_{B2} + h_{ie}} = \frac{(5 \times 10^3)(1500)}{5 \times 10^3 + 1500} = 1.154 \text{ k}\Omega$$

9.4 Find the value of the initial-stage voltage gain $A_{v1} = v_{o1}/v_{in}$ for the two-stage BJT amplifier of Problem 9.2 as modeled by Fig. 9-2.

▮ The initial-stage voltage gain is given by the result of Problem 7.40 if R_C and R_L are replaced with R_{C1} and Z_{in2}, respectively:

$$A_{v1} = \frac{v_{o1}}{v_{in}} = -\frac{h_{fe}Z_{in2}R_{C1}}{h_{ie}(R_{C1} + Z_{in2} + h_{oe}Z_{in2}R_{C1})} = -\frac{40(1154)(10^4)}{1500(10^4 + 1154 + 346.2)} = -26.8 \quad (1)$$

9.5 Evaluate the input impedance Z_{in1} for the amplifier of Problem 9.2 as modeled by Fig. 9-2.

▮ As in Problem 9.3,

$$Z_{in1} = \frac{R_{B1}h_{ie}}{R_{B1} + h_{ie}} = 1.154 \text{ k}\Omega \quad (1)$$

9.6 Find the overall voltage gain $A_v = v_o/v_i$ for the two-stage BJT amplifier of Problem 9.2 as modeled by Fig. 9-2.

▮ By voltage division and use of the results of Problems 9.2–9.4,

$$\frac{v_{in}}{v_i} = \frac{Z_{in1}}{Z_{in1} + R_i} = \frac{1154}{1154 + 1000} = 0.5357 \quad (1)$$

and

$$A_v \equiv \frac{v_o}{v_i} = \frac{v_{in}}{v_i} A_{v1} A_{v2} = 0.5357(-26.8)(-333.3) = 4786 \quad (2)$$

Note that, in this problem, we made use of the labor-saving technique of applying results determined for single-stage amplifiers to the individual stages of a cascaded (multistage) amplifier.

9.7 Derive an expression for the current-gain ratio $A_i = -i_{C2}/i_i$ for the two-stage BJT amplifier of Fig. 9-1.

▮
$$A_i = \frac{-i_{C2}}{i_i} = \frac{v_o/R_{C2}}{v_i/(R_i + Z_{in1})} = A_v \frac{R_i + Z_{in1}}{R_{C2}} \quad (1)$$

Use of (1) and (2) of Problem 9.6 allows (1) to be written as

$$A_i = \frac{Z_{in1}}{Z_{in1} + R_i} A_{v1} A_{v2} \frac{R_i + Z_{in1}}{R_{C2}} = A_{v1} A_{v2} \frac{Z_{in1}}{R_{C2}} \quad (2)$$

Based on the expressions for A_{v2}, A_{v1}, and Z_{in1} of Problems 9.2, 9.4, and 9.5, respectively, (2) becomes

$$A_i = \frac{h_{fe}^2 h_{ie} R_{B1} R_{B2} R_{C1}}{h_{ie}(1 + h_{oe}R_{C2})(R_{B1} + h_{ie})[R_{C1}(R_{B2} + h_{ie}) + R_{B2}h_{ie} + h_{oe}h_{ie}R_{B2}R_{C1}]}$$

9.8 Evaluate the current gain $A_i = -i_{C2}/i_i$ for the amplifier of Problem 9.2.

▮ By (1) of Problem 9.7 with the results of Problems 9.5 and 9.6,

$$A_i = A_v \frac{R_i + Z_{in1}}{R_{C2}} = (4786) \frac{1 \times 10^3 + 1.154 \times 10^3}{20 \times 10^3} = 515.4$$

9.9 Find the values of output impedances Z_{o1} and Z_{o2} as indicated in Fig. 9-2 if the amplifier is described by Problem 9.2.

▮ Deactivate (short) v_i in Fig. 9-2. Whence, $i_{b1} = 0$ and $h_{fe1}i_{b1} = 0$. Consequently, $i_{b2} = 0 = h_{fe}i_{b2}$. With both controlled current sources inactive, it is apparent that

$$Z_{o1} = R_{C1} \parallel \left(\frac{1}{h_{oe1}}\right) = \frac{R_{C1}}{1 + h_{oe1}R_{C1}} = \frac{10 \times 10^3}{1 + (30 \times 10^{-6})(10 \times 10^3)} = 7.692 \text{ k}\Omega$$

$$Z_{o2} = \frac{1}{h_{oe2}} = \frac{1}{30 \times 10^{-6}} = 33.33 \text{ k}\Omega$$

9.10 In the amplifier of Fig. 9-3(a), the transistors are identical and have $h_{re} = h_{oe} \approx 0$. Use the CE h-parameter model to draw an equivalent circuit.

▌ See Fig. 9-3(b).

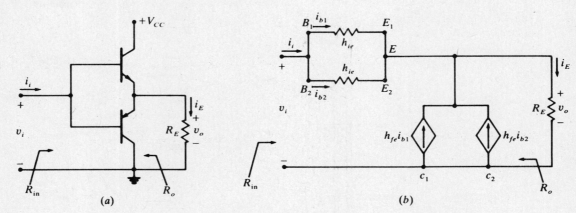

Fig. 9-3

9.11 Find an expression for the current-gain ratio $A_i = i_E/i_i$ for the amplifier of Fig. 9-3.

▌ KCL at node E of Fig. 9-3(b) gives

$$i_E = h_{fe}i_{b1} + h_{fe}i_{b2} + i_{b1} + i_{b2} = (h_{fe} + 1)(i_{b1} + i_{b2}) \tag{1}$$

Since $i_i = i_{b1} + i_{b2}$, the current-gain ratio follows directly from (1) and is $A_i = h_{fe} + 1$.

9.12 Determine an expression for the input resistance R_{in} for the amplifier of Fig. 9-3.

▌ KVL applied around the outer loop of Fig. 9-3(b) gives

$$v_i = (h_{ie} \| h_{ie})i_i + R_E(h_{fe} + 1)i_i$$

so that

$$R_{in} = \frac{v_i}{i_i} = \frac{1}{2}h_{ie} + (h_{fe} + 1)R_E \tag{1}$$

9.13 Derive an expression for the voltage-gain ratio $A_v = v_o/v_i$ for the BJT amplifier of Fig. 9-3.

▌ By KVL applied in Fig. 9-3(b),

$$v_o = v_i - (h_{ie} \| h_{ie})i_i = v_i - \tfrac{1}{2}h_{ie}i_i \tag{1}$$

But

$$i_i = v_i/R_{in} \tag{2}$$

Substitution of (2) and then (1) of Problem 9.12 into (1) allows solution for the voltage-gain ratio as

$$A_v = \frac{v_o}{v_i} = 1 - \frac{1}{2}\frac{h_{ie}}{R_{in}} = 1 - \frac{\tfrac{1}{2}h_{ie}}{\tfrac{1}{2}h_{ie} + (h_{fe} + 1)R_E} = \frac{(h_{fe} + 1)R_E}{\tfrac{1}{2}h_{ie} + (h_{fe} + 1)R_E}$$

9.14 Find an expression for the output resistance R_o for the amplifier of Fig. 9-3.

▌ If R_E is replaced by a driving-point source with v_i shorted in Fig. 9-3(b), KCL requires that

$$i_{dp} = h_{fe}(i_{b1} + i_{b2}) + \frac{v_{dp}}{h_{ie} \| h_{ie}} \tag{1}$$

But

$$i_{b1} + i_{b2} = i_i = -\frac{v_{dp}}{h_{ie} \| h_{ie}} = -\frac{v_{dp}}{\tfrac{1}{2}h_{ie}} \tag{2}$$

Substituting (2) into (1) leads to

$$R_o = \frac{v_{dp}}{i_{dp}} = \frac{1}{h_{fe}/\tfrac{1}{2}h_{ie} + 1/\tfrac{1}{2}h_{ie}} = \frac{h_{ie}}{2(h_{fe} + 1)}$$

9.15 The cascaded amplifier of Fig. 9-4(a) uses a CC first stage followed by a CE second stage. If $h_{re} = h_{oe} \approx 0$ for both transistors, draw a small-signal equivalent circuit suitable for analysis of the amplifier.

▌ From (3) of Problem 7.14 with $h_{re} = h_{oe} = 0$, it is seen that $h_{rc} = 1$ and $h_{oc} = 0$. Based on the CC and CE h-parameter models of Figs. 7-5 and 7-1, respectively, the equivalent circuit of Fig. 9-4(b) results where $R_{B1} = R_{11} \| R_{12}$ and $R_{B2} = R_{22} \| R_{21}$.

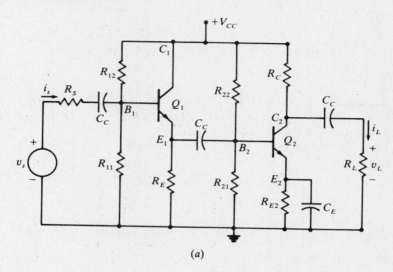

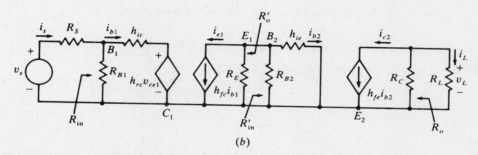

Fig. 9-4

9.16 For the cascaded amplifier of Fig. 9-4(a), let $R_S = 0$, $R_{11} = 100$ kΩ, $R_{12} = 90$ kΩ, $R_{21} = 10$ kΩ, $R_{22} = 90$ kΩ, $R_L = R_C = 5$ kΩ, and $R_E = 9$ kΩ. For transistor Q_1, $h_{oc} \approx 0$, $h_{ic} = 1$ kΩ, $h_{rc} \approx 1$, and $h_{fc} = -100$. For Q_2, $h_{re} = h_{oe} \approx 0$, $h_{fe} = 100$, and $h_{ie} = 1$ kΩ. Find the overall voltage ratio $A_v = v_L/v_s$.

▌ The small-signal equivalent circuit of Fig. 9-4(b) is applicable, where

$$R_{B1} = R_{11} \| R_{12} = \frac{(90 \times 10^3)(100 \times 10^3)}{90 \times 10^3 + 100 \times 10^3} = 47.37 \text{ k}\Omega$$

and

$$R_{B2} = R_{22} \| R_{21} = \frac{(90 \times 10^3)(10 \times 10^3)}{90 \times 10^3 + 10 \times 10^3} = 4.5 \text{ k}\Omega$$

From the results of Problem 7.121 where R_E is replaced by $R_E \| R_{B2} \| h_{ie}$,

$$A_{v1} = -\frac{h_{fc}(R_E \| R_{B2} \| h_{ie})}{h_{ic} - h_{rc}h_{fc}(R_E \| R_{B2} \| h_{ie})} = -\frac{-100(818.2)}{1 \times 10^3 - 1(-100)(818.2)} = 0.9879$$

and from the results of Problem 7.40,

$$A_{v2} = -\frac{h_{fe}R_L R_C}{h_{ie}(R_L + R_C)} = -\frac{100(5 \times 10^3)(5 \times 10^3)}{(1 \times 10^3)(5 \times 10^3 + 5 \times 10^3)} = -100$$

Then

$$A_v = A_{v1}A_{v2} = 0.9879(-100) = -98.79$$

9.17 Determine the overall current gain $A_i = i_L/i_s$ for the cascaded BJT amplifier of Problem 9.16.

❙ From the results of Problem 7.102 with $h_{oc} = 0$, R_B replaced by R_{B1}, and R_E replaced by $R_E \| R_{B2} \| h_{ie}$,

$$A_{i1} = \frac{-i_{e1}}{i_s} = \frac{h_{fc}R_{B1}}{R_{B1} + h_{ic} + h_{rc}h_{fc}(R_E \| R_{B2} \| h_{ie})} = \frac{-100(47.37 \times 10^3)}{47.37 \times 10^3 + 1 \times 10^3 + 1(-100)(818.2)} = 36.38$$

and again from Problem 7.40,

$$A_{i2} = \frac{(R_E \| R_{B2})h_{ie}}{R_L(R_E \| R_{B2} + h_{ie})} A_{v2} = \frac{(4.5 \times 10^3)(91 \times 10^3)}{(5 \times 10^3)(4.5 \times 10^3 + 1 \times 10^3)}(-100) = -16.36$$

Then $\quad A_i = A_{i1}A_{i2} = 36.38(-16.36) = -595.2$

9.18 The cascaded amplifier of Fig. 9-5(a) is built up with identical transistors for which $h_{re} = h_{oe} \approx 0$, $h_{fe} = 100$, and $h_{ie} = 1\,k\Omega$. Let $R_{E1} = 1\,k\Omega$, $R_{C1} = 10\,k\Omega$, $R_{E2} = 100\,\Omega$, and $R_{C2} = R_L = 3\,k\Omega$. Draw a small-signal equivalent circuit suitable for analysis of the amplifier.

❙ See Fig. 9-5(b).

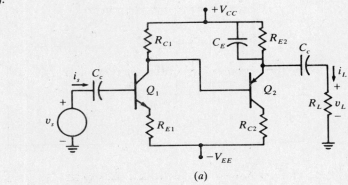

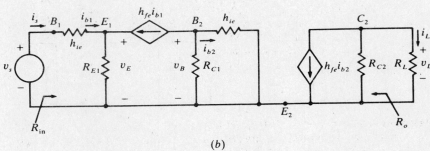

Fig. 9-5

9.19 Find the overall voltage gain $A_v = v_L/v_s$ for the cascaded BJT amplifier of Problem 9.18.

❙ The small-signal equivalent circuit of Fig. 9-5(b) is applicable. From the results of Problem 7.40 with $h_{oe} = 0$ and R_C replaced with R_{C2},

$$A_{v2} = -\frac{h_{fe}R_L R_{C2}}{h_{ie}(R_L + R_{C2})} = -\frac{100(3 \times 10^3)(3 \times 10^3)}{(1 \times 10^3)(3 \times 10^3 + 3 \times 10^3)} = -150$$

From the results of Problem 7.162, in which R_C, R_L, and R_E are replaced with R_{C1}, h_{ie}, and R_{E1}, respectively,

$$A_{v1} = -\frac{h_{fe}R_{C1}h_{ie}}{(RC_1 + h_{ie})[(h_{fe} + 1)R_{E1} + h_{ie}]} = -\frac{100(10 \times 10^3)(1 \times 10^3)}{(11 \times 10^3)[(100 + 1)(1 \times 10^3) + 1 \times 10^3]} = -0.891$$

Thus, $\quad A_v = A_{v1}A_{v2} = (-0.891)(-150) = 133.6$

9.20 Determine the overall current gain $A_i = i_L/i_s$ for the amplifier of Problem 9.18 as modeled by Fig. 9-5(b).

❙ From the results of Problem 7.163 with R_C and R_L replaced with R_{C1} and h_{ie}, respectively,

$$A_{i1} = -\frac{h_{fe}R_{C1}}{R_{C1} + h_{ie}} = -\frac{100(10 \times 10^3)}{10 \times 10^3 + 1 \times 10^3} = -90.91$$

Now, by current division at the output network of Fig. 9-5(b),

$$i_L = -h_{fe}i_{b2}\frac{R_{C2}}{R_{C2}+R_L}$$

Hence,

$$A_{i2} = \frac{i_L}{i_{b2}} = -\frac{h_{fe}R_{C2}}{R_{C2}+R_L} = -\frac{100(3\times 10^3)}{3\times 10^3 + 3\times 10^3} = -50$$

and

$$A_i = A_{i1}A_{i2} = -90.91(-50) = 4545.4$$

9.21 In the cascaded CB-CC amplifier of Fig. 9-6(a), transistor Q_1 is characterized by $h_{rb1} = h_{ob1} \approx 0$, $h_{ib1} = 50\,\Omega$, and $h_{fb1} = -0.99$. The h parameters of transistor Q_2 are $h_{oc2} \approx 0$, $h_{rc2} = 1$, $h_{ic2} = 500\,\Omega$, and $h_{fc2} = -100$. Let $R_L = R_{E2} = 2\,\text{k}\Omega$, $R_{B1} = 30\,\text{k}\Omega$, $R_{B2} = 60\,\text{k}\Omega$, $R_1 = 50\,\text{k}\Omega$, $R_2 = 100\,\text{k}\Omega$, and $R_{E1} = 5\,\text{k}\Omega$. Draw a small-signal equivalent circuit suitable for analysis of the amplifier.

▮ See Fig. 9-6(b).

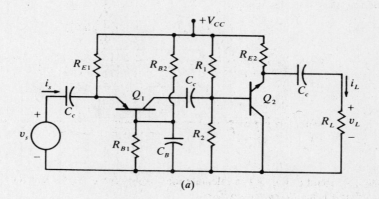

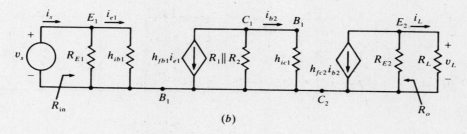

Fig. 9-6

9.22 Find the overall voltage gain $A_v = v_L/v_s$ for the cascaded BJT amplifier of Problem 9.21.

▮ The small-signal equivalent circuit of Fig. 9-6(b) is applicable. From the results of Problem 7.89 with $R_B = R_1 \| R_2$,

$$A_{v1} = -\frac{h_{fb1}R_B h_{ic2}}{h_{ib1}(R_B + h_{ic2})} = -\frac{-0.99(33.3 \times 10^3)(500)}{50(33.3\times 10^3 + 500)} = 9.75$$

By the results of Problem 7.121,

$$A_{v2} = -\frac{h_{fc2}(R_{E2}\|R_L)}{h_{ic2} - h_{rc2}h_{fc2}(R_{E2}\|R_L)} = -\frac{-100(1\times 10^3)}{500 - 1(-100)(1\times 10^3)} = 0.995$$

Thus,

$$A_v = A_{v1}A_{v2} = 9.75(0.995) = 9.70$$

9.23 Determine the overall current gain $A_i = i_L/i_s$ for the amplifier described by Problem 9.21.

▮ Based on the results of Problem 7.89,

$$A_{i1} = -\frac{h_{fb2}R_{E1}R_B}{(R_{E1}+h_{ib1})(R_B+h_{ic2})} = -\frac{-0.99(5\times 10^3)(33.3\times 10^3)}{(5\times 10^3 + 50)(33.3\times 10^3 + 500)} = 0.966$$

By current division at node E_2 in Fig. 9-6(b),

$$\frac{i_L}{i_{b2}} = A_{i2} = -\frac{h_{fc2}R_{E2}}{R_{E2} + R_L} = -\frac{-100(2 \times 10^3)}{(2 \times 10^3) + (2 \times 10^3)} = 50$$

Then,
$$A_i = A_{i1}A_{i2} = 0.966(50) = 48.3$$

9.24 Use the CE h-parameter model of Fig. 7-1 to draw a small-signal equivalent circuit of the amplifier of Fig. 4-17 if the two transistors are identical and $h_{re} = h_{oe} \approx 0$.

▌ See Fig. 9-7.

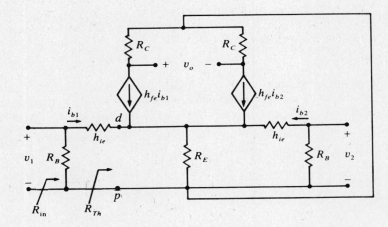

Fig. 9-7

9.25 Use the equivalent circuit of Fig. 9-7 to calculate the output voltage v_o for the amplifier of Fig. 4-17, thus demonstrating that it is a *difference amplifier*.

▌ Let $a = h_{ie} + (h_{fe} + 1)R_E$ and $b = (h_{fe} + 1)R_E$; then, by KVL,

$$v_1 = ai_{b1} + bi_{b2} \tag{1}$$

$$v_2 = b_1 i_{b1} + a i_{b2} \tag{2}$$

$$v_o = h_{fe}R_C(i_{b1} - i_{b2}) \tag{3}$$

Solving (1) and (2) simultaneously using Cramer's rule gives

$$\Delta = a^2 - b^2$$

and
$$i_{b1} = \frac{\Delta_1}{\Delta} = \frac{av_1 - bv_2}{a^2 - b^2} \tag{4}$$

$$i_{b2} = \frac{\Delta_2}{\Delta} = \frac{av_2 - bv_1}{a^2 - b^2} \tag{5}$$

Substituting (4) and (5) into (3) gives, finally,

$$v_o = \frac{h_{fe}R_C}{a^2 - b^2}(av_1 - bv_2 - av_2 + bv_1) = \frac{h_{fe}R_C}{a - b}(v_1 - v_2) = \frac{h_{fe}}{h_{ie}}R_C(v_1 - v_2)$$

which clearly shows that the circuit amplifies the *difference* between signals v_1 and v_2.

9.26 Suppose v_2 is replaced with a short circuit in the differential amplifier of Fig. 4-17. Find the input impedance R_{in1} looking into the terminal across which v_1 appears if $R_B = 20$ kΩ, $R_E = 1$ kΩ, $h_{ie} = 25$ Ω, $h_{fe} = 100$, and $h_{re} = h_{oe} \approx 0$.

▌ Short the terminal pair across which v_2 appears in Fig. 9-7. Remove the circuit to the left of terminal pair d, p and connect a driving-point source v_{dp} to feed the balance of the network to the right of terminal pair d, p so that $i_{dp} = i_{b1}$. Hence,

$$i_{b1} = \frac{v_{dp}}{R_E} - h_{fe}(i_{b1} + i_{b2}) - i_{b2} \tag{1}$$

But, by Ohm's law,
$$i_{b2} = -\frac{v_{dp}}{R_B + h_{ie}} \tag{2}$$

Substitution of (2) into (1) and rearrangement lead to solution for the Thévenin's resistance to the right of terminal d, p as

$$R_{Th} = \frac{v_{dp}}{i_{b1}} = \frac{(h_{fe}+1)R_E(h_{ie}+R_B)}{h_{ie}+R_B+(h_{fe}+1)R_E}$$

$$= \frac{(100+1)(1\times 10^3)(25+20\times 10^3)}{25+20\times 10^3 + (100+1)(1\times 10^3)} = 16.71 \text{ k}\Omega \tag{3}$$

Referring to Fig. 9-7, it is apparent that

$$R_{in} = R_B \parallel (h_{ie}+R_{Th}) = \frac{R_B(h_{ie}+R_{Th})}{R_B+h_{ie}+R_{Th}} = \frac{(20\times 10^3)(25+16.71\times 10^3)}{20\times 10^3 + 25 + 16.71\times 10^3} = 9.11 \text{ k}\Omega$$

9.27 The cascaded amplifier circuit of Fig. 9-4(a) matches a high-input-impedance CC first stage with a high-output-impedance CE second stage to produce an amplifier with high input and output impedances. Refer to Fig. 9-4(b) and determine the value for $Z_{in} = R_{in}$ if circuit values are as given in Problem 9.16.

❚ Equation (3) of Problem 7.120 gives the Thévenin resistance to the right of R_{B1} in Fig. 9-4(b) if R_E is replaced by $R_E \parallel R_{B2} \parallel h_{ie}$. Hence, with $h_{oc} = 0$,

$$R_{Th} = h_{ic} - \frac{h_{rc}h_{fc}h_{ie}R_E R_{B2}}{h_{ie}(R_E+R_{B2})+R_E R_{B2}} = 1\times 10^3 - \frac{1(-100)(1\times 10^3)(9\times 10^3)(4.5\times 10^3)}{(1\times 10^3)(9\times 10^3 + 4.5\times 10^3)+(9\times 10^3)(4.5\times 10^3)}$$

$$= 76 \text{ k}\Omega$$

And
$$R_{in} = R_{B1} \parallel R_{Th} = \frac{R_{B1}R_{Th}}{R_{B1}+R_{Th}} = \frac{(47.37\times 10^3)(76\times 10^3)}{47.37\times 10^3 + 76\times 10^3} = 29.18 \text{ k}\Omega$$

9.28 Find the output impedance R_o for the cascaded amplifier of Fig. 9-4 if the parameter values are given by Problem 9.16.

❚ With v_s deactivated (shorted) in Fig. 9-4(b), $i_{b1} = i_{b2} = 0 = h_{fe}i_{b2}$. Since the right most controlled current source is inactive, it is apparent that $R_o = R_C = 5 \text{ k}\Omega$.

9.29 To illustrate the effect of signal-source internal impedance, calculate the voltage-gain ratio $A_v = v_L/v_s$ for the cascaded amplifier of Fig. 9-4(a) if $R_S = 20 \text{ k}\Omega$ and all other values are as given in Problem 9.16; then compare your result with the value of A_v found in Problem 9.16.

❚ Based on the result of Problem 9.16 and letting v_{RB1} be the voltage across R_{B1},

$$A_v = v_L/v_{RB1} = -98.79 \tag{1}$$

The voltage gain with $R_S \neq 0$ is simply

$$A_{vs} = \frac{R_{in}}{R_{in}+R_S}A_v = \frac{29.18\times 10^3}{29.18\times 10^3 + 20\times 10^3}(-98.79) = -58.61$$

where the value of R_{in} was determined in Problem 9.27. The voltage gain has been reduced by approximately 40 percent due to R_S.

9.30 For the Darlington-pair emitter-follower of Fig. 9-8, $h_{re1} = h_{re2} = h_{oe1} = h_{oe2} = 0$. Draw a small-signal h-parameter equivalent circuit suitable for analysis of the amplifier.

❚ Extending the work of Problem 7.6 and the resulting circuit of Fig. 7-4 leads to the circuit of Fig. 9-9.

9.31 Find an expression for the voltage-gain ratio $A_v = v_E/v_S$ for the Darlington-pair amplifier of Fig. 9-8 as modeled by Fig. 9-9.

❚ The voltage-gain ratio follows directly from application of voltage division in Fig. 9-9 as

$$A_v = \frac{v_E}{v_S} = \frac{(h_{fe1}+1)(h_{fe2}+1)R_E}{h_{ie1}+(h_{fe1}+1)[h_{ie2}+(h_{fe2}+1)R_E]}$$

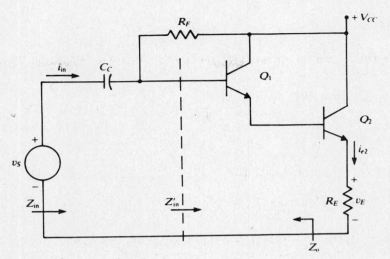

Fig. 9-8

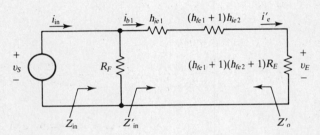

Fig. 9-9

9.32 Determine an expression for the current gain ratio $A_i = i_{e2}/i_{in}$ for the amplifier of Fig. 9-8 as modeled by Fig. 9-9.

▌ Applying current division in Fig. 9-9,

$$i_{b1} = \frac{R_F i_{in}}{R_F + h_{ie} + (h_{fe1}+1)[h_{ie2} + (h_{fe2}+1)R_E]} \quad (1)$$

Based on the work of Problem 7.6,

$$i_{e2} = (h_{fe1}+1)(h_{fe2}+1)i_{b1} \quad (2)$$

Substitution of i_{b1} as determined from (2) into (1) and rearrangement give

$$A_i = \frac{i_{e2}}{i_{in}} = \frac{(h_{fe1}+1)(h_{fe2}+1)R_F}{R_F + h_{ie1} + (h_{fe1}+1)[h_{ie2} + (h_{fe2}+1)R_E]}$$

9.33 Derive an expression for the input impedance Z_{in} for the Darlington-pair emitter follower of Fig. 9-8 as modeled by Fig. 9-9.

▌ From Fig. 9-9, it is apparent that

$$Z'_{in} = \frac{v_S}{i_{b1}} = h_{ie1} + (h_{fe1}+1)[h_{ie2} + (h_{fe2}+1)R_E]$$

Hence, $$Z_{in} = R_F \parallel Z'_{in} = \frac{R_F\{h_{ie1} + (h_{fe1}+1)[h_{ie2} + (h_{fe2}+1)R_E]\}}{R_F + h_{ie1} + (h_{fe1}+1)[h_{ie2} + (h_{fe2}+1)R_E]}$$

9.34 Find an expression for the output Z_o for the amplifier of Fig. 9-8 as modeled by Fig. 9-9.

▌ Impedance Z'_o of Fig. 9-9 is the output impedance as seen from base B_1 point of reference; specifically,

$$i'_e = i_{e2}/[(h_{fe1}+1)(h_{fe2}+1)] \quad (1)$$

With v_S deactivated (shorted) and $(h_{fe1}+1)(h_{fe2}+1)R_E$ replaced by a driving-point source, KVL yields

$$v_{dp} = -i'_e[h_{ie1} + (h_{fe1}+1)h_{ie2}] \quad (2)$$

By use of (1) in (2),

$$Z_o = \frac{v_{dp}}{-i_{e2}} = \frac{h_{ie1} + (h_{fe1}+1)h_{ie2}}{(h_{fe1}+1)(h_{fe2}+1)}$$

MULTIPLE-FET AMPLIFIERS

9.35 In the circuit of Fig. 9-10, $R_S = 3$ kΩ, $R_D = R_L = 2$ kΩ, $r_i = 5$ kΩ, and $R_G = 100$ kΩ. Assume that the two JFETs are identical with $r_{ds} = 25$ kΩ and $g_m = 0.0025$ S. Draw a small-signal equivalent circuit suitable for analysis of the amplifier.

▌ Based on the results of Problem 8.5 for modeling parallel-connected JFETs, the equivalent circuit of Fig. 9-11 results.

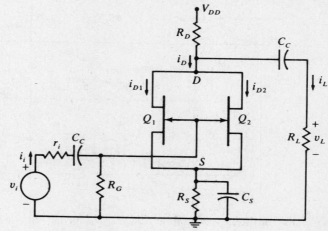

Fig. 9-10

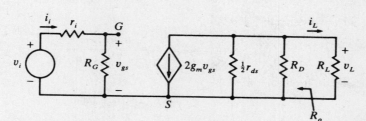

Fig. 9-11

9.36 Find an expression for the voltage-gain ratio $A_v = v_L/v_i$ for the amplifier of Fig. 9-10 as modeled by Fig. 9-11.

▌ By voltage division in Fig. 9-11,

$$v_{gs} = \frac{R_G}{R_G + r_i} v_i \tag{1}$$

Now let

$$R_{eq} = \tfrac{1}{2} r_{ds} \parallel R_D \parallel R_L = \frac{r_{ds} R_D R_L}{2 R_L R_D + r_{ds}(R_L + R_D)} \tag{2}$$

Then, by Ohm's law, $v_L = -2 g_m v_{gs} R_{eq}$; with (1) and (2), this gives

$$A_v = \frac{v_L}{v_i} = -2 g_m \frac{R_G}{R_G + r_i} R_{eq} \tag{3}$$

9.37 Evaluate the voltage gain $A_v = v_L/v_i$ for the amplifier of Problem 9.35.

▌ By (2) and (3) of Problem 9.36,

$$R_{eq} = \frac{(25 \times 10^3)(2 \times 10^3)(2 \times 10^3)}{2(2 \times 10^3)(2 \times 10^3) + (25 \times 10^3)(2 \times 10^3 + 2 \times 10^3)} = 962 \;\Omega$$

$$A_v = -2(0.0025) \frac{100 \times 10^3}{100 \times 10^3 + 5 \times 10^3} (962) = -4.58$$

9.38 Derive an expression for the current-gain ratio $A_i = i_L/i_i$ for the amplifier of Fig. 9-10 as modeled by Fig. 9-11.

▌ The current-gain ratio is

$$A_i = \frac{i_L}{i_i} = \frac{v_L/R_L}{v_i/(R_G + r_i)} = \frac{A_v(R_G + r_i)}{R_L} \tag{1}$$

where A_v is given by (3) of Problem 9.36.

9.39 Evaluate the current gain $A_i = i_L/i_i$ for the amplifier of Problem 9.35.

▌ By (1) of Problem 9.38 and using the result of Problem 9.37,

$$A_i = \frac{A_v(R_G + r_i)}{R_L} = \frac{-4.58(100 \times 10^3 + 5 \times 10^3)}{2 \times 10^3} = -240.4$$

9.40 Determine the value of output resistance R_o for the multiple-FET amplifier of Problem 9.35.

▌ We replace R_L with a driving-point source oriented such that $v_{dp} = v_L$ in Fig. 9-11. With v_i deactivated (short-circuited), $v_{gs} = 0$; thus,

$$R_o = R_D \parallel (\tfrac{1}{2}r_{ds}) = \frac{R_D r_{ds}}{2R_D + r_{ds}} = \frac{2(25) \times 10^3}{2(2) + 25} = 1.72 \text{ k}\Omega$$

9.41 Draw a small-signal equivalent circuit suitable for analysis of the cascaded MOSFET amplifier of Fig. 9-12.

▌ Application of the current source FET model of Fig. 8-1(a) leads to the equivalent circuit of Fig. 9-13.

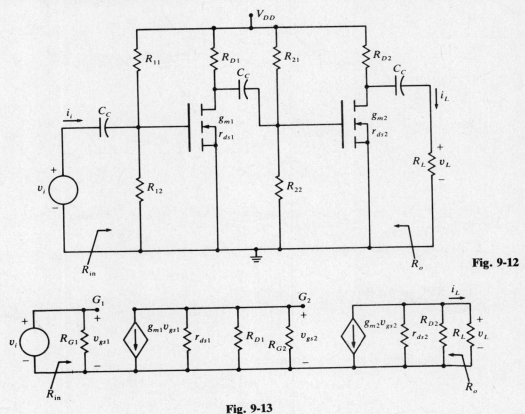

Fig. 9-12

Fig. 9-13

9.42 Determine an expression for the voltage-gain ratio $A_v = v_L/v_i$ for the cascaded MOSFET amplifier of Fig. 9-12 as modeled by Fig. 9-13.

▌ Using the result of Problem 8.10, but replacing μ with $g_m r_{ds}$ and R_D with $R_{D1} \parallel R_{G2}$ where $R_{G2} = R_{21} \parallel R_{22}$, we have

$$A_{v1} = \frac{-g_{m1} r_{ds1}(R_{D1} \parallel R_{G2})}{r_{ds1} + (R_{D1} \parallel R_{G2})} \tag{1}$$

Similarly,
$$A_{v2} = \frac{-g_{m2}r_{ds2}(R_{D2} \| R_L)}{r_{ds2} + (R_{D2} \| R_L)} \quad (2)$$

Then,
$$A_v = A_{v1}A_{v2} = \frac{g_{m1}g_{m2}r_{ds1}r_{ds2}(R_{D1} \| R_{G2})(R_{D2} \| R_L)}{[r_{ds1} + (R_{D1} \| R_{G2})][r_{ds2} + (R_{D2} \| R_L)]} \quad (3)$$

9.43 Derive an expression for the current-gain ratio $A_i = i_L/i_i$ for the cascaded MOSFET amplifier of Fig. 9-12.

▮ Realizing that $R_{G1} = R_{11} \| R_{12}$ in the applicable equivalent circuit of Fig. 9-13, we have

$$A_i = \frac{i_L}{i_i} = \frac{v_o/R_L}{v_i/R_{G1}} = A_v \frac{R_{G1}}{R_L}$$

where A_v is given by (3) of Problem 9.42.

9.44 The series-connected JFETs of Fig. 5-22 are identical, with $\mu = 70$, $r_{ds} = 30$ kΩ, $R_G = 100$ kΩ, and $R_D = R_L = 4$ kΩ. Draw a small-signal equivalent circuit suitable for analysis of the amplifier.

▮ See Fig. 9-14.

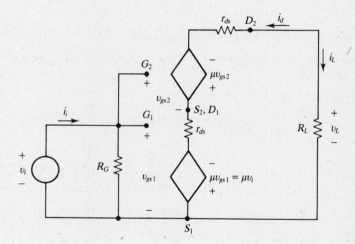

Fig. 9-14

9.45 Determine an expression for the voltage-gain ratio $A_v = v_L/v_i$ for the amplifier of Fig. 5-22 as modeled by Fig. 9-14.

▮ Application of KVL around a loop that includes the signal source v_i and nodes G_2, S_2, and S_1 yields

$$v_{gs2} = v_i + \mu v_i - i_d r_{ds} \quad (1)$$

Use of KVL around the output loop leads to

$$i_d = \frac{\mu(v_i + v_{gs2})}{2r_{ds} + R_L} \quad (2)$$

▮ Substitution of (2) into (1) and rearrangement give

$$v_{gs2} = \left[\frac{(\mu+2)r_{ds} + (\mu+1)R_L}{(\mu+2)r_{ds} + R_L}\right] v_i \quad (3)$$

By voltage division,

$$v_L = -\frac{R_L}{R_L + 2r_{ds}} \mu(v_i + v_{gs2}) \quad (4)$$

Substitute (3) into (4) and rearrange to find

$$A_v = \frac{v_L}{v_i} = -\left[\frac{\mu R_L}{R_L + 2r_{ds}}\right]\left[1 + \frac{(\mu+2)r_{ds} + (\mu+1)R_L}{(\mu+2)R_{ds} + R_L}\right] \quad (5)$$

9.46 Evaluate the voltage gain $A_v = v_L/v_i$ for the amplifier of Problem 9.44.

By (5) of Problem 9.45,

$$A_v = -\left[\frac{70(4 \times 10^3)}{4 \times 10^3 + 2(30 \times 10^3)}\right]\left[1 + \frac{(70+2)(30 \times 10^3) + (70+1)(4 \times 10^3)}{(70+2)(30 \times 10^3 + 4 \times 10^3)}\right] = -9.32$$

9.47 Find the value of the current gain $A_i = i_L/i_i$ for the amplifier of Problem 9.44.

$$A_i = \frac{i_L}{i_i} = \frac{v_L/R_L}{v_i/R_G} = A_v \frac{R_G}{R_L} = (-9.32)\frac{100 \times 10^3}{4 \times 10^3} = -233$$

9.48 Evaluate the large output resistance R_o displayed by the amplifier of Problem 9.44.

With v_i deactivated (shorted) in Fig. 9-14, $v_i = v_{g1} = 0 = \mu v_i$. Replace R_L by a driving-point source oriented such that $v_{dp} = v_L$ and $i_{dp} = i_d$. Then by KVL,

$$v_{dp} = 2r_{ds}i_{dp} - \mu v_{gs2} \tag{1}$$

But

$$v_{gs2} = -i_{dp}r_{ds} \tag{2}$$

Substitute (2) into (1) and rearrange to find

$$R_o = v_{dp}/i_{dp} = (\mu + 2)r_{ds} = (70+2)(30 \times 10^3) = 2.16 \text{ M}\Omega$$

9.49 In the circuit of Fig. 9-15, the two JFETs are identical. Draw a small-signal equivalent circuit suitable for analysis of the amplifier.

See Fig. 9-16.

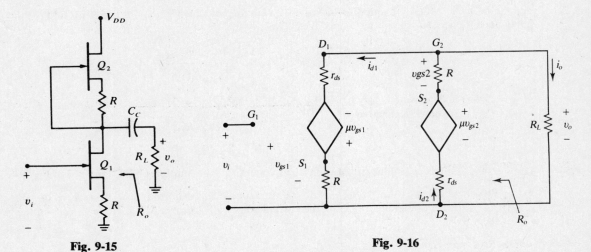

Fig. 9-15   Fig. 9-16

9.50 Derive an expression for the voltage-gain ratio $A_v = v_o/v_i$ for the multiple-JFET amplifier of Fig. 9-15 as modeled by Fig. 9-16.

Referring to Fig. 9-16 and applying the method of node voltages give

$$\frac{v_o + \mu v_{gs1}}{R + r_{ds}} + \frac{v_o - \mu v_{gs2}}{R + r_{ds}} + \frac{v_o}{R_L} = 0 \tag{1}$$

By KVL,

$$v_{gs1} = v_i - \frac{v_o + \mu v_{gs1}}{R + r_{ds}} R \tag{2}$$

or

$$v_{gs1} = \frac{(R + r_{ds})v_i - v_o R}{(\mu + 1)R + r_{ds}} \tag{3}$$

Again applying KVL,

$$v_{gs2} = \frac{v_o - \mu v_{gs2}}{R + r_{ds}} R \tag{4}$$

or

$$v_{gs2} = \frac{v_o R}{(\mu + 1)R + r_{ds}} \quad (5)$$

Substitution of (2) and (5) into (1) and rearrangement lead to

$$A_v = \frac{v_o}{v_i} = -\frac{\mu R_L}{2R_L + (\mu + 1)R + r_{ds}}$$

9.51 Find an expression for the output resistance R_o for the amplifier of Fig. 9-15 as modeled by Fig. 9-16.

▌ With v_i deactivated (shorted), $v_{gs1} = -i_{d1}R$. Also, $v_{gs2} = -i_{d2}R$. Replace R_L with a driving-point source oriented such that $v_{dp} = v_o$ and $i_{dp} = -i_o$; then

$$i_{d1} = \frac{v_{dp}}{(\mu + 1)R + r_{ds}} \quad (1)$$

and

$$i_{d2} = -\frac{v_{dp}}{(\mu + 1)R + r_{ds}} \quad (2)$$

By KCL and use of (1) and (2),

$$i_{dp} = i_{d1} - i_{d2} = \frac{2v_{dp}}{(\mu + 1)R + r_{ds}} \quad (3)$$

R_o follows directly from (3) as

$$R_o = \frac{v_{dp}}{i_{dp}} = \frac{(\mu + 1)R + r_{ds}}{2}$$

9.52 For the cascaded MOSFET amplifier of Fig. 9-12 with the equivalent circuit in Fig. 9-13, find (a) the input impedance R_{in} and (b) the output impedance R_o.

▌ (a)

$$R_{in} = R_{G1} = R_{11} \parallel R_{12} = \frac{R_{11}R_{12}}{R_{11} + R_{12}}$$

▌ (b) With v_i deactivated, $v_i = v_{gs1} = 0 = g_{m1}v_{gs1}$. Hence, $v_{gs2} = 0 = g_{m2}v_{gs2}$. Since the right-hand controlled current source is inactive,

$$R_o = r_{ds2} \parallel R_{D2} = \frac{r_{ds2}R_{D2}}{r_{ds2} + R_{D2}}$$

9.53 Draw a small-signal equivalent circuit suitable for analysis of the multiple-JFET amplifier of Fig. 9-17.

▌ By Problem 8.70, we know that Q_2, with the gate-source leads shorted, acts as a resistor of value $r_{ds2}/(\mu_2 + 1)$. Hence, modeling Q_1 using the current-source FET model of Fig. 8-1(a) results in the circuit of Fig. 9-18.

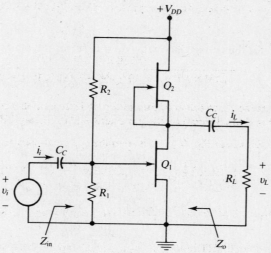

Fig. 9-17

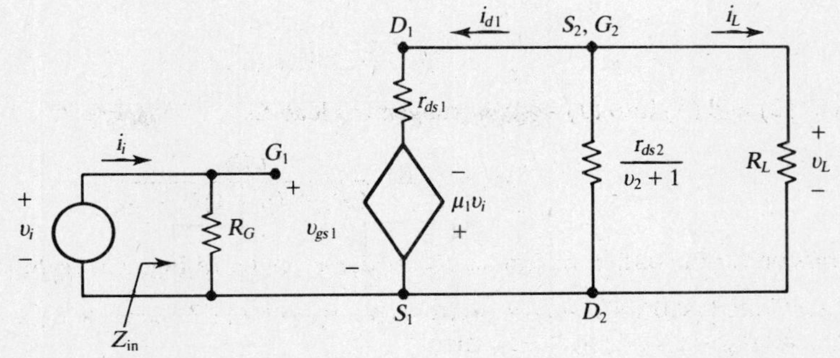

MULTIPLE-TRANSISTOR AMPLIFIERS ◊ 271

Fig. 9-18

9.54 Find an expression for the voltage-gain ratio $A_v = v_L/v_i$ for the amplifier of Fig. 9-17.

▌ Voltage division applied to the pertinent equivalent circuit of Fig. 9-18 yields

$$v_L = -\frac{R_L \left\|\left(\frac{r_{ds2}}{\mu_2+1}\right)\right.}{r_{ds1} + R_L \left\|\left(\frac{r_{ds2}}{\mu_2+1}\right)\right.} \mu_1 v_i$$

After rearrangement,

$$A_v = \frac{v_L}{v_i} = -\frac{\mu_1 R_L r_{ds2}}{(\mu_2+1)R_L r_{ds1} + r_{ds2}(r_{ds1} + R_L)} \tag{1}$$

9.55 For the amplifier of Fig. 9-17, $R_1 = 250\,\Omega$, $R_2 = 1\,\text{M}\,k\Omega$, and $R_L = 10\,k\Omega$. The JFETs are identical and characterized by $\mu = 60$ and $r_{ds} = 20\,k\Omega$. Evaluate the voltage gain $A_v = v_L/v_i$.

▌ By (1) of Problem 9.54 considering identical JFETs,

$$A_v = \frac{v_L}{v_i} = -\frac{\mu R_L}{(\mu+2)R_L + r_{ds}} = -\frac{60(10 \times 10^3)}{(60+2)(10 \times 10^3) + 20 \times 10^3} = -0.9375$$

9.56 Derive an expression for the current-gain ratio $A_i = i_L/i_i$ of the amplifier in Fig. 9-17 as modeled by Fig. 9-18 if the JFETs are identical.

▌

$$A_i = \frac{i_L}{i_i} = \frac{v_L/R_L}{v_i/Z_{in}} = A_v \frac{Z_{in}}{R_L} \tag{1}$$

Realizing that $Z_{in} = R_G = R_1 R_2/(R_1 + R_2)$ and using (1) of Problem 9.54 for the identical JFETs, (1) becomes

$$A_i = \frac{i_L}{i_i} = -\frac{\mu R_1 R_2}{(R_1 + R_2)[(\mu+2)R_L + r_{ds}]} \tag{2}$$

9.57 Evaluate the current gain $A_i = i_L/i_i$ for the amplifier of Fig. 9-17 if it is described by the parameters of Problem 9.55.

▌ By (2) of Problem 9.56,

$$A_i = -\frac{60(250 \times 10^3)(1 \times 10^6)}{(250 \times 10^3 + 1 \times 10^6)[(60+2)(10 \times 10^3) + 20 \times 10^3]} = -18.75$$

9.58 Find the output impedance Z_o of the amplifier of Fig. 9-17 as modeled by the equivalent circuit of Fig. 9-18. Assume the amplifier is characterized by the parameters of Problem 9.55.

▌ Deactivating (shorting) v_i in Fig. 9-18, it is seen that $\mu_i v_i = 0$. Hence,

$$Z_o = \left(\frac{r_{ds2}}{\mu_2+1}\right) \| r_{ds1} = \frac{r_{ds1} r_{ds2}}{(\mu_2+1)r_{ds1} + r_{ds2}} = \frac{r_{ds}}{\mu+2} = \frac{20 \times 10^3}{60+2} = 322.6\,\Omega$$

9.59 Draw a small-signal equivalent circuit suitable for analysis of the cascaded, three-stage JFET amplifier of Fig. 9-19.

▌ Use of the voltage-source FET models of Figs. 8-1(b) and 8-7(b) leads to the circuit of Fig. 9-20.

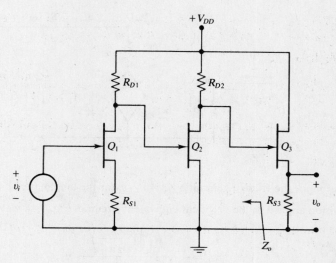

Fig. 9-19

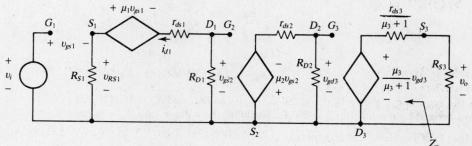

Fig. 9-20

9.60 Derive an expression for the overall voltage-gain ratio $A_v = v_o/v_i$ for the cascaded JFET amplifier of Fig. 9-19.

▌ By KVL around the input loop of the applicable equivalent circuit of Fig. 9-20,

$$v_{gs1} = v_i - v_{RS1} \tag{1}$$

Voltage division gives

$$v_{RS1} = \frac{R_{S1}}{R_{S1} + R_{D1} + r_{ds1}} \mu_1 v_{gs1} \tag{2}$$

Substitution of (2) into (1) yields, after rearrangement,

$$v_{gs1} = \frac{R_{S1} + R_{D1} + r_{ds1}}{(\mu_1 + 1)R_{S1} + R_{D1} + r_{ds1}} v_i \tag{3}$$

By voltage division and then use of (3),

$$v_{gs2} = -\frac{R_{D1}}{R_{D1} + R_{S1} + r_{ds1}} \mu_1 v_{gs1} = -\frac{\mu_1 R_{D1}}{(\mu_1 + 1)R_{S1} + R_{D1} + r_{ds1}} v_i \tag{4}$$

By voltage division and then use of (4),

$$v_{gd3} = -\frac{R_{D2}}{R_{D2} + r_{ds2}} \mu_2 v_{gs2} = \frac{\mu_1 \mu_2 R_{D1} R_{D2}}{(R_{D2} + r_{ds2})[(\mu_1 + 1)R_{S1} + R_{D1} + r_{ds1}]} v_i \tag{5}$$

Finally, application of voltage division at output mesh yields

$$v_o = \frac{R_{S3}}{R_{S3} + \dfrac{r_{ds3}}{\mu_3 + 1}} \left(\frac{\mu_3}{\mu_3 + 1}\right) v_{gd3} = \frac{\mu_3 R_{S3}}{(\mu_3 + 1)R_{S3} + r_{ds3}} v_{gd3} \tag{6}$$

Substituting (5) into (6) and rearranging lead to A_v:

$$A_v = \frac{v_o}{v_i} = \frac{\mu_1 \mu_2 \mu_3 R_{D1} R_{D2} R_{S3}}{(R_{D2} + r_{ds2})[(\mu_1 + 1)R_{S1} + R_{D1} + r_{ds1}][(\mu_3 + 1)R_{S3} + r_{ds3}]} \tag{7}$$

9.61 The three JFETs of Fig. 9-19 are identical with $\mu = 60$ and $r_{ds} = 30$ kΩ. If $R_{S1} = R_{S3} = 2$ kΩ, $R_{D1} = 30$ kΩ, and $R_{D2} = 10$ kΩ, evaluate the voltage gain of the first stage $A_{v1} = v_{gs2}/v_i$.

▌ By (4) of Problem 9.60,
$$A_{v1} = \frac{v_{gs2}}{v_i} = -\frac{\mu_1 R_{D1}}{(\mu_1 + 1)R_{S1} + R_{D1} + r_{ds1}} = -\frac{60(30 \times 10^3)}{(60+1)(2 \times 10^3) + 30 \times 10^3 + 30 \times 10^3} = -9.89$$

9.62 Evaluate the voltage gain of the second stage, $A_{v2} = v_{gd3}/v_{gs2}$, for the amplifier of Fig. 9-19 if the amplifier is described by the parameters of Problem 9.61.

▌ By (5) of Problem 9.60,
$$A_{v2} = \frac{v_{gd3}}{v_{gs2}} = -\frac{\mu_2 R_{D2}}{R_{D2} + r_{ds2}} = -\frac{60(10 \times 10^3)}{10 \times 10^3 + 30 \times 10^3} = -15$$

9.63 Find the value of voltage gain of the third stage, $A_{v3} = v_o/v_{gd3}$, for the amplifier of Fig. 9-19 if the parameters of Problem 9.61 are pertinent.

▌ By (6) of Problem 9.61,
$$A_{v3} = \frac{v_o}{v_{gd3}} = \frac{\mu_3 R_{S3}}{(\mu_3 + 1)R_{S3} + r_{ds3}} = \frac{60(2 \times 10^3)}{(60+1)(2 \times 10^3) + 30 \times 10^3} = 0.789$$

9.64 Evaluate the overall voltage gain $A_v = v_o/v_i$ for the amplifier of Fig. 9-19 if the amplifier is characterized by the parameters of Problem 9.61.

▌ Applying the results of Problems 9.61–9.63,
$$A_v = A_{v1}A_{v2}A_{v3} = \left(\frac{v_{gs2}}{v_i}\right)\left(\frac{v_{gd3}}{v_{gs2}}\right)\left(\frac{v_o}{v_{gd3}}\right) = \frac{v_o}{v_i} = -9.89(-15)(0.789) = 117.05$$

9.65 Determine the output impedance Z_o for the amplifier of Problem 9.61.

▌ With v_i deactivated (shorted) in Fig. 9-20, $v_{gs1} = v_{gs2} = v_{gd3} = 0$. Hence,
$$Z_o = \frac{r_{ds3}}{\mu_3 + 1} = \frac{30 \times 10^3}{60 + 1} = 491.8 \ \Omega$$

Review Problems

9.66 Draw an h-parameter equivalent circuit suitable for analysis of the cascaded, three-stage BJT amplifier of Fig. 9-21. Assume $h_{re} = h_{oe} \approx 0$ for all transistors.

▌ See Fig. 9-22.

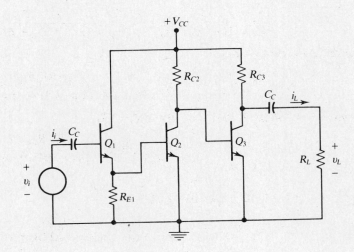

Fig. 9-21

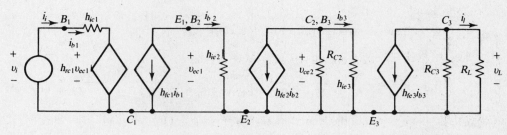

Fig. 9-22

9.67 For the CC-CE-CE cascaded amplifier of Fig. 9-21, the three transistors are identical with $h_{ie} = 100\ \Omega$ and $h_{fe} = 50$. Let $R_{E1} = 200\ \Omega$, $R_L = 2\ k\Omega$, and $R_{C2} = R_{C3} = 5\ k\Omega$. Determine the voltage gain $A_{v1} = v_{ec1}/v_i$ for the first stage.

■ By (3) of Problem 7.14,

$$h_{ic1} = h_{ie1} = 200\ \Omega$$

$$h_{rc1} = 1 - h_{re1} = 1 - 0 = 1$$

$$h_{fc1} = -(h_{fe1} + 1) = -(50 + 1) = -51$$

The voltage gain of this CC stage is given by (1) of Problem 7.95 where R_L is replaced by h_{ie2}.

$$A_{v1} = \frac{v_{ec1}}{v_i} = -\frac{h_{fc1}h_{ie2}}{h_{ic1} - h_{ie2}h_{fc1}h_{rc1}} = -\frac{-51(200)}{200 - 200(-51)(1)} = 0.981$$

9.68 Find the voltage gain $A_{v2} = v_{ce2}/v_{ec1}$ for the second stage of the amplifier in Fig. 9-21 as modeled by Fig. 9-22. Assume that the amplifier is described by the parameters of Problem 9.67.

■ The voltage gain of this CE stage is given by (3) of Problem 7.23 if R_L is replaced by $R_{C2} \parallel h_{ie3}$.

$$A_{v2} = \frac{v_{ce2}}{v_{ec1}} = -\frac{h_{fe2}(R_{C2} \parallel h_{ie3})}{h_{ie2}} = -\frac{h_{fe2}R_{C2}h_{ie3}}{h_{ie2}(R_{C2} + h_{ie3})} = -\frac{50(5 \times 10^3)(200)}{200(5 \times 10^3 + 200)} = -48.08$$

9.69 Evaluate the voltage gain $A_{v3} = v_L/v_{ce2}$ for the third stage of the amplifier of Fig. 9-21 if the amplifier is modeled by Fig. 9-22 and characterized by the parameters of Problem 9.67.

■ By (3) of Problem 7.23 where R_L is replaced by $R_L \parallel R_{C3}$,

$$A_{v3} = \frac{v_L}{v_{ce2}} = -\frac{h_{fe3}(R_L \parallel R_{C3})}{h_{ie3}} = -\frac{h_{fe3}R_L R_{C3}}{h_{ie3}(R_L + R_{C3})} = -\frac{50(2 \times 10^3)(5 \times 10^3)}{200(10 \times 10^3 + 5 \times 10^3)} = -166.7$$

9.70 Determine the overall voltage gain $A_v = v_L/v_i$ for the amplifier of Problem 9.67.

■ Based on the results of Problems 9.67–9.69,

$$A_v = A_{v1}A_{v2}A_{v3} = \left(\frac{v_{ec1}}{v_i}\right)\left(\frac{v_{ce2}}{v_{ec1}}\right)\left(\frac{v_L}{v_{ce2}}\right) = \frac{v_L}{v_i} = 0.981(-48.08)(-166.7) = 7862.6$$

9.71 Find the overall current gain $A_i = i_L/i_i$ for the cascaded BJT amplifier of Problem 9.67.

■ The input impedance Z_{in} of the amplifier is found by use of (1) from Problem 7.96 where R_L is replaced by h_{ie2}. The CC h parameters determined in Problem 9.67 are pertinent.

$$Z_{in} = h_{ic1} - h_{rc1}h_{fc1}h_{ie2} = 200 - 1(-51)(200) = 10.4\ k\Omega$$

The overall current gain follows from use of A_v determined in Problem 9.70 as

$$A_i = \frac{i_L}{i_i} = \frac{v_L/R_L}{v_i/Z_{in}} = A_v \frac{Z_{in}}{R_L} = (7862.6)\frac{10.4 \times 10^3}{2 \times 10^3} = 40{,}888.5$$

9.72 The Darlington-pair amplifier of Fig. 9-23 with the added resistor R_B allows setting the bias point Q_2 independent of the bias point Q_1. Establish a procedure to select R_B so that $I_{CQ1} = I_{CQ2}$ for Si devices.

At node B_2, KCL requires that

$$I_{EQ1} = I_{BQ2} + V_{BEQ2}/R_B \tag{1}$$

Using $I_{EQ1} = (\beta_1 + 1)I_{CQ1}/\beta_1$ and $I_{BQ2} = I_{CQ2}/\beta_2$ in (1) leads to

$$\frac{\beta_1 + 1}{\beta_1} I_{CQ1} - \frac{1}{\beta_2} I_{CQ2} = \frac{V_{BEQ2}}{R_B} \tag{2}$$

Letting $I_{CQ1} = I_{CQ2}$ and $V_{BEQ2} = 0.7$ in (2) and rearranging yield

$$R_B = \frac{0.7\beta_1\beta_2}{(\beta_1\beta_2 + \beta_2 - \beta_1)I_{CQ1}} \tag{3}$$

Once a desired bias current I_{CQ1} is selected, then R_B is sized according to (3).

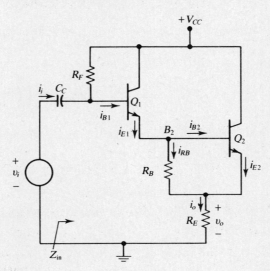

Fig. 9-23

9.73 Draw an h-parameter equivalent circuit suitable for analysis of the Darlington-pair amplifier of Fig. 9-23 if both transistors are identical and $h_{oe} = h_{re} = 0$.

See Fig. 9-24.

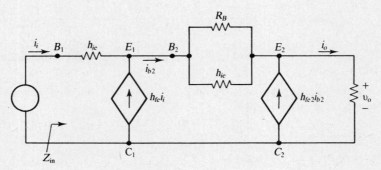

Fig. 9-24

9.74 Find an expression for the overall current gain $A_i = i_o/i_i$ for the Darlington-pair amplifier of Fig. 9-23.

Referring to the applicable equivalent circuit of Fig. 9-24, KCL at node E_1 requires that

$$i_{b2} = i_i + h_{fe}i_i = (h_{fe} + 1)i_i \tag{1}$$

KCL at node E_2 gives

$$i_o = h_{fe}i_{b2} + i_{b2} = (h_{fe} + 1)i_{b2} \tag{2}$$

Substitute (1) into (2) and solve directly for

$$A_i = i_o/i_i = (h_{fe} + 1)^2 \tag{3}$$

9.75 If the identical transistors of the Darlington-pair amplifier in Fig. 9-23 are described by $h_{ie} = 150\,\Omega$ and $h_{fe} = 50$, find the overall current gain $A_i = i_o/i_i$. Let $R_B = 70\,\Omega$ and $R_E = 1\,\text{k}\Omega$.

▮ By (3) of Problem 9.74,
$$A_i = i_o/i_i = (50 + 1)^2 = 2601$$

9.76 Derive an expression for the input impedance Z_{in} of the Darlington-pair amplifier of Fig. 9-23.

▮ KVL applied around the outer loop of the applicable equivalent circuit in Fig. 9-24 yields
$$v_i = i_i h_{ie} + i_{b2}(R_B \parallel h_{ie}) + i_o R_E \qquad (1)$$

Substitute i_{b2} and i_o as determined in Problem 9.74 into (1) allowing direct solution for Z_{in} as
$$Z_{\text{in}} = v_i/i_i = h_{ie} + (h_{fe} + 1)(R_B \parallel h_{ie}) + (h_{fe} + 1)^2 R_E \qquad (2)$$

9.77 Evaluate the input impedance Z_{in} for the amplifier of Fig. 9-23 if the parameter values of Problem 9.75 are pertinent.

▮ By (2) of Problem 9.76,
$$Z_{\text{in}} = 150 + (50 + 1)\frac{70(150)}{70 + 150} + (50 + 1)^2(1 \times 10^3) = 2.604\,\text{M}\Omega$$

9.78 Derive an expression for the overall voltage gain of the Darlington amplifier of Fig. 9-23 if Q_1 and Q_2 are identical.

▮
$$A_v = \frac{v_o}{v_i} = \frac{i_o R_E}{i_i Z_{\text{in}}} = A_i \frac{R_E}{Z_{\text{in}}} \qquad (1)$$

Substituting (3) of Problem 9.74 and (2) of Problem 9.76 in (1) finds A_v as
$$A_v = \frac{(h_{fe} + 1)^2 R_E}{h_{ie} + (h_{fe} + 1)(R_B \parallel h_{ie}) + (h_{fe} + 1)^2 R_E} \qquad (2)$$

9.79 Evaluate the overal voltage gain $A_v = v_o/v_i$ for the amplifier of Fig. 9-23 if the parameters of Problem 9.75 describe the amplifier.

▮ By (2) of Problem 9.78,
$$A_v = \frac{(50 + 1)^2(1 \times 10^3)}{150 + (50 + 1)(150)(70)/(150 + 70) + (50 + 1)^2(1 \times 10^3)} = 0.999$$

9.80 Determine an expression for the output impedance Z_o of the Darlington amplifier in Fig. 9-23 as described by the equivalent circuit of Fig. 9-24.

▮ Deactivate (short) v_i and replace R_E by a driving-point source oriented such that $v_{dp} = v_o$ and $i_{dp} = -i_o$ in Fig. 9-24. Then, KCL at node E_2 gives
$$i_{dp} = -(h_{fe} + 1)i_{b2} \quad \text{or} \quad i_{b2} = -i_{dp}/(h_{fe} + 1) \qquad (1)$$

Use of (1) in (1) of Problem 9.74 leads to
$$i_i = \frac{i_{b2}}{h_{fe} + 1} = -\frac{i_{dp}}{(h_{fe} + 1)^2} \qquad (2)$$

Application of KVL around the outer loop of Fig. 9-24 yields
$$v_{dp} = -i_{b2}(R_B \parallel h_{ie}) - i_i h_{ie} \qquad (3)$$

Substitute (1) and (2) into (3) and rearrange to find
$$Z_o = \frac{v_{dp}}{i_{dp}} = \frac{R_B \parallel h_{ie}}{h_{fe} + 1} + \frac{h_{ie}}{(h_{fe} + 1)^2} = \frac{R_B h_{ie}}{(h_{fe} + 1)(R_B + h_{ie})} + \frac{h_{ie}}{(h_{fe} + 1)^2} \qquad (4)$$

9.81 Evaluate the output impedance Z_o for the amplifier of Problem 9.75.

▌ By (4) of Problem 9.80,
$$Z_o = \frac{70(150)}{(50+1)(70+150)} + \frac{150}{(50+1)^2} = 0.993 \, \Omega$$

9.82 Draw a small-signal equivalent circuit suitable for analysis of the JFET amplifier of Fig. 9-25.

▌ See Fig. 9-26.

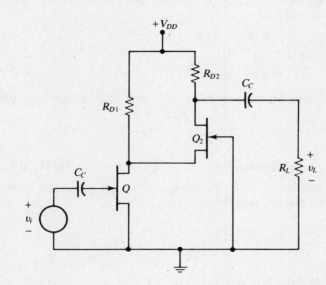

Fig. 9-25

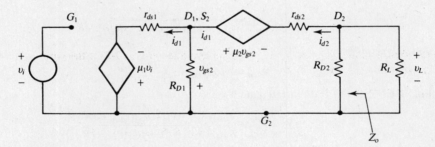

Fig. 9-26

9.83 Find an expression for the voltage-gain ratio $A_v = v_L/v_i$ of the amplifier in Fig. 9-25.

▌ By KCL at node D_1,
$$i_{D1} = i_{d2} - i_{d1} \tag{1}$$

Application of KVL and use of (1) lead to the following set of equations:
$$i_{d1}r_{ds1} - i_{D1}R_{D1} = (r_{ds1} + R_{D1})i_{d1} - R_{D1}i_{d2} = \mu_1 v_i \tag{2}$$
$$i_{D1}R_{D1} + i_{d2}(r_{ds2} + R_{D2} \parallel R_L) = -R_{D1}i_{d1} + (R_{D1} + r_{ds2} + R_{D2} \parallel R_L)i_{d2} = \mu_2 v_{gs2} \tag{3}$$

But
$$v_{gs2} = i_{D1}R_{D1} = -(i_{d2} - i_{d1})R_{D1} \tag{4}$$

Substitution of (4) into (3) leads to
$$-(1+\mu_2)R_{D1}i_{d1} + [(1+\mu_2)R_{D1} + r_{ds2} + R_{D2} \parallel R_L]i_{d2} = 0 \tag{5}$$

Simultaneous solution of (2) and (5) for i_{d2} yields
$$\Delta = (r_{ds1} + R_{D1})(r_{ds2} + R_{D2} \parallel R_L) + (1+\mu_2)r_{ds1}R_{D1} \tag{6}$$
$$i_{d2} = \frac{\Delta_2}{\Delta} = \frac{\mu_1(1+\mu_2)R_{D1}v_i}{\Delta} \tag{7}$$

By Ohm's law,
$$v_L = -(R_{D2} \| R_L)i_{d2} \tag{8}$$

Substituting (7) into (8) and rearranging give A_v as

$$A_v = \frac{v_L}{v_i} = -\frac{\mu_1(1+\mu_2)R_{D1}(R_{D2} \| R_L)}{\Delta} \tag{9}$$

9.84 Evaluate the voltage gain $A_v = v_L/v_i$ for the amplifier of Fig. 9-25 if $R_L = R_{D1} = R_{D2} = 20$ kΩ. The JFETs are identical with $\mu = 40$ and $r_{ds} = 20$ kΩ.

Based on (6) and (9) of Problem 9.83,

$$R_{D2} \| R_L = \frac{(20 \times 10^3)(20 \times 10^3)}{20 \times 10^3 + 20 \times 10^3} = 10 \text{ k}\Omega$$

$$\Delta = (20 \times 10^3 + 20 \times 10^3)(20 \times 10^3 + 10 \times 10^3) + (1+40)(20 \times 10^3)(20 \times 10^3) = 1.76 \times 10^{10}$$

$$A_v = \frac{40(1+40)(20 \times 10^3)(10 \times 10^3)}{1.76 \times 10^{10}} = 18.64$$

9.85 Derive an expression for the output impedance Z_o of the amplifier in Fig. 9-25.

Deactivating (shorting) v_i, $\mu v_i = 0$ in the applicable equivalent circuit of Fig. 9-26. Replace R_L by a driving-point source oriented such that $v_{dp} = v_L$. Then,

$$i_{d2} = \frac{v_{dp} + \mu_2 v_{gs2}}{r_{ds2} + r_{ds1} \| R_{D1}} \tag{1}$$

But
$$v_{gs2} = -i_{d2}(r_{ds1} \| R_{D1}) \tag{2}$$

Substitute (2) into (1) and rearrange to obtain

$$i_{d2} = \frac{v_{dp}}{r_{ds2} + (1+\mu_2)(r_{ds1} + R_{D1})} \tag{3}$$

KCL at node D_2 requires that

$$i_{dp} = v_{dp}/R_{D2} + i_{d2} \tag{4}$$

Substitution of (3) into (4) and rearrangement yield

$$Z_o = \frac{v_{dp}}{i_{dp}} = \frac{R_{D2}[r_{ds2} + (1+\mu_2)(r_{ds1} + R_{D1})]}{R_{D2} + r_{ds2} + (1+\mu_2)(r_{ds1} + R_{D1})} \tag{5}$$

9.86 Evaluate the output impedance Z_o for the amplifier of Fig. 9-25 if the amplifier is characterized by the parameters of Problem 9.84.

By (5) of Problem 9.85,

$$Z_o = \frac{(20 \times 10^3)[20 \times 10^3 + (1+40)(20 \times 10^3 + 20 \times 10^3)]}{20 \times 10^3 + 20 \times 10^3 + (1+40)(20 \times 10^3 + 20 \times 10^3)} = 19.762 \text{ k}\Omega$$

9.87 Draw a small-signal equivalent circuit suitable for analysis of the amplifier of Fig. 9-27(a). Assume $h_{re} = h_{oe} = 0$ and that $R_G \gg R_1, R_2$.

See Fig. 9-27(b), where the CD model of the JFET (Problem 8.32) has been used.

9.88 For the JFET-BJT Darlington amplifier of Fig. 9-27(a), find the voltage-gain ratio $A_v = v_e/v_i$.

Since $i_b = i_d$ and $v_{gd} = v_i$ in Fig. 9-27(b), KVL yields

$$\frac{\mu}{\mu+1} v_i = i_d \left(\frac{r_{ds}}{\mu+1} + h_{ie} \right) + (h_{fe}+1)i_d(R_1 + R_2) \tag{1}$$

By Ohm's law,

$$v_e = (h_{fe}+1)i_d(R_1 + R_2) \tag{2}$$

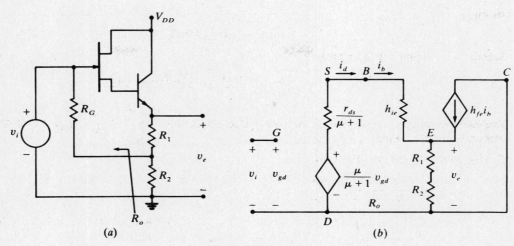

Fig. 9-27

Solving (1) for i_d, substituting the result into (2), and rearranging give

$$A_v = \frac{v_e}{v_i} = \frac{\mu(h_{fe}+1)(R_1+R_2)}{r_{ds}+(\mu+1)[h_{ie}+(h_{fe}+1)(R_1+R_2)]}$$

9.89 Determine an expression for the output resistance R_o of the Darlington amplifier of Fig. 9-27.

▌ We replace $R_1 + R_2$ in Fig. 9-27(b) with a driving-point source oriented such that $v_{dp} = v_e$. With v_i deactivated (short-circuited), $v_{gd} = 0$. Then, by Ohm's law,

$$i_b = -\frac{v_{dp}}{h_{ie}+r_{ds}/(\mu+1)} \quad (1)$$

and, by KCL,
$$i_{dp} = -(h_{fe}+1)i_b \quad (2)$$

Substituting (1) into (2) and rearranging give

$$R_o = \frac{v_{dp}}{i_{dp}} = \frac{r_{ds}+(\mu+1)h_{ie}}{(\mu+1)(h_{fe}+1)}$$

9.90 In the cascaded FET-BJT amplifier of Fig. 9-28, assume $h_{re} = h_{oe} = 0$ and $h_{ie} \ll R_D$ so that $h_{ie} \| R_D \approx h_{ie}$. Draw a small-signal equivalent circuit suitable for analysis of the amplifier.

▌ See Fig. 9-29.

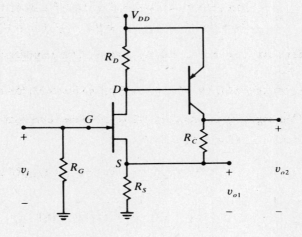

Fig. 9-28

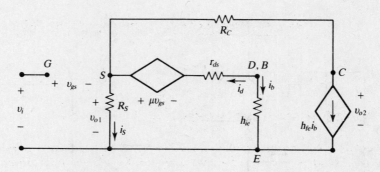

Fig. 9-29

9.91 Find an expression for the voltage-gain ratio $A_{v1} = v_{o1}/v_i$ of the FET/BJT amplifier of Fig. 9-28 as modeled by Fig. 9-29.

▌ By KCL at node S,

$$i_s = -i_b - h_{fe}i_b = -(h_{fe} + 1)i_b \tag{1}$$

Applying KVL around the input loop and using (1) lead to

$$v_{gs} = v_i - i_s R_S = v_i + (h_{fe} + 1)i_b R_S \tag{2}$$

By KVL around the loop including S-D-E and using (1),

$$\mu v_{gs} = i_s R_S - (h_{ie} + r_{ds})i_b = -[(h_{fe} + 1)R_S + h_{ie} + r_{ds}]i_b \tag{3}$$

Substitute (2) into (3) and rearrange to find

$$i_b = -\frac{\mu v_i}{(\mu + 1)(h_{fe} + 1)R_S + h_{ie} + r_{ds}} \tag{4}$$

By Ohm's law and use of (1),

$$v_{o1} = i_s R_S = -(h_{fe} + 1)R_S i_b \tag{5}$$

Use of (4) in (5) leads to

$$A_{v1} = \frac{v_{o1}}{v_i} = \frac{\mu(h_{fe} + 1)R_S}{(\mu + 1)(h_{fe} + 1)R_S + h_{ie} + r_{ds}} \tag{6}$$

9.92 Determine an expression for the voltage-gain ratio $A_{v2} = v_{o2}/v_i$ of the amplifier of Fig. 9-28 when modeled by Fig. 9-29.

▌ By KVL and use of $A_{v1} = v_{o1}/v_i$,

$$v_{o2} = -h_{fe}i_b R_C + v_{o1} = -h_{fe}R_C i_b + A_{v1}v_i \tag{1}$$

Substitution of (4) and (6) of Problem 9.91 into (1) yields

$$v_{o2} = \frac{\mu h_{fe} R_C v_i}{(\mu + 1)(h_{fe} + 1)R_S + h_{ie} + r_{ds}} + \frac{\mu(h_{fe} + 1)R_S v_i}{(\mu + 1)(h_{fe} + 1)R_S + h_{ie} + r_{ds}} \tag{2}$$

A_{v2} follows directly from (2):

$$A_{v2} = \frac{v_{o2}}{v_i} = \frac{\mu[h_{fe}R_C + (h_{fe} + 1)R_S]}{(\mu + 1)(h_{fe} + 1)R_S + h_{ie} + r_{ds}} \tag{3}$$

CHAPTER 10
Frequency Effects in Amplifiers

BODE PLOTS AND FREQUENCY RESPONSE

10.1 For operation of an amplifier with signals of frequency outside the midfrequency range illustrated by Fig. 10-1(a), the gain A (voltage or current gain) is attenuated. Low-frequency range attenuation is attributable to the presence of coupling and bypass capacitors whereas high-frequency range gain reduction is due to inherent capacitive reactances associated with active devices (transistors). Consequently, small-signal amplifier analysis for other than midfrequency operation is conveniently carried out by frequency-domain techniques that apply operation mathematics as suggested by the Laplace-domain two-port network of Fig. 10-1(b), where *transfer function* $T(s)$ may be either a voltage-gain or current-gain ratio. For that network, express (a) the voltage-gain ratio and (b) the current-gain ratio in frequency-domain formulation.

(a) The Laplace-domain voltage-gain ratio is

$$A_v(s) = T(s) = \frac{V_2(s)}{V_1(s)} \tag{1}$$

(b) Similarly,

$$A_i(s) = T(S) = \frac{I_2(s)}{I_1(s)} \tag{2}$$

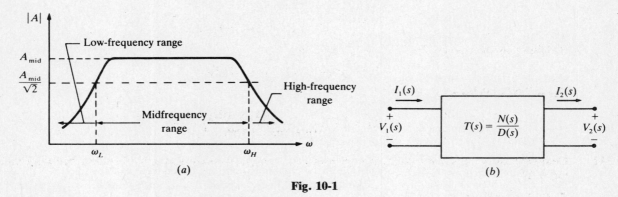

Fig. 10-1

10.2 For a sinusoidal input voltage signal feeding a linear two-port amplifier network such as that of Fig. 10-1(b), the Laplace transform pair

$$v_1(t) = V_{1m} \sin \omega t \quad \leftrightarrow \quad V_1(s) = \frac{V_{1m}\omega}{s^2 + \omega^2}$$

is applicable, and the network frequency-domain response is given by

$$V_2(s) = A_v(s) V_1(s) = \frac{A_v(s) V_{1m} \omega}{s^2 + \omega^2} \tag{1}$$

Determine the time-domain response of the network.

Without loss of generality, we may assume that the polynomial $D(s) = 0$ has n distinct roots. Then the partial-fraction expansion of (1) yields

$$V_2(s) = \frac{k_1}{s - j\omega} + \frac{k_2}{s + j\omega} + \frac{k_3}{s + p_1} + \frac{k_4}{s + p_2} + \cdots + \frac{k_{n+2}}{s + p_n} \tag{2}$$

where the first two terms on the right-hand side are forced-response terms (called the *frequency response*), and the balance of the terms constitute the transient response. The transient response diminishes to zero with time, provided the roots of $D(s) = 0$ are located in the left-half plane of complex numbers (the condition for a *stable* system).

The coefficients k_1 and k_2 are evaluated by the method of residues, and the results are used in an inverse transformation to the time-domain steady-state sinusoidal response given by

$$v_2(t) = V_{1m}\,|A_v(j\omega)|\sin(\omega t + \phi) = V_{2m}\sin(\omega t + \phi) \tag{3}$$

The *network phase angle* ϕ is defined as

$$\phi = \tan^{-1}\frac{\operatorname{Im}\{A_v(j\omega)\}}{\operatorname{Re}\{A_v(j\omega)\}} \tag{4}$$

From (3), it is apparent that a sinusoidal input to a stable, linear, two-port network results in a steady-state output that is also sinusoidal; the input and output waveforms differ only in amplitude and phase angle.

10.3 Show that (3) of Problem 10.2 follows from the evaluation of k_1 and k_2 in (2) of Problem 10.2.

▌ Our interest is in the steady-state response, that, is the response after all transient terms have decayed to zero. Hence, only constants k_1 and k_2 are of interest, and their value follows from residue theory applied to (1) of Problem 10.2.

$$k_1 = (s - j\omega)V_2(s)\big|_{s=j\omega} = \frac{A_v(j\omega)V_{1m}\omega}{j\omega + j\omega} = \frac{A_v(j\omega)V_{1m}}{j2} \tag{1}$$

$$k_2 = (s + j\omega)V_2(s)\big|_{s=-j\omega} = \frac{A_v(-j\omega)V_{1m}\omega}{-j\omega - j\omega} = \frac{A_v(-j\omega)V_{1m}}{-j2} = k_1^* \tag{2}$$

Taking the inverse transform of (2) from Problem 10.2, using (1) and (2), and evaluating for t large enough that transient terms have decayed to zero leads to

$$v_2(t) = k_1 e^{j\omega t} + k_2 e^{-j\omega t} = k_1 e^{j\omega t} + k_1^* e^{-j\omega t} = 2\operatorname{Re}\{k_1 e^{j\omega t}\}$$

$$= 2\operatorname{Re}\left\{\frac{A_v(j\omega)V_{1m}}{j2}e^{j\omega t}\right\} \tag{3}$$

$A_v(j\omega)$ can be expressed in polar form as

$$A_v(j\omega) = |A_v(j\omega)|\,e^{j\phi}$$

where ϕ is given by (4) of Problem 10.2. Whence, (3) can be rewritten as

$$v_2(t) = \operatorname{Re}\{|A_v(j\omega)|V_{1m}e^{j(\omega t + \phi - \pi/2)}\} = V_{1m}|A_v(j\omega)|\sin(\omega t + \phi) \tag{4}$$

10.4 For convenience, we make the following definitions:
1. Call $A(j\omega)$ the *frequency transfer function*.
2. Define $M \equiv |A(j\omega)|$, the *gain ratio*.
3. Define $M_{db} \equiv 20\log M = 20\log|A(j\omega)|$, the *amplitude ratio*, measured in *decibels* (db).

The subscript v or i may be added to any of these quantities to specifically denote reference to voltage or current, respectively. The graph of M_{db} (simultaneously with ϕ if desired) versus the logarithm of the input signal frequency (positive values only) is called a *Bode plot*.

A simple first-order network has Laplace-domain transfer function and frequency transfer function

$$A(s) = \frac{1}{\tau s + 1} \quad \text{and} \quad A(j\omega) = \frac{1}{1 + j\omega\tau}$$

where τ is the system time constant. Determine the network phase angle ϕ and the amplitude ratio M_{db}.

▌ In polar form, the given frequency transfer function is

$$A(j\omega) = \frac{1}{\sqrt{1 + (\omega\tau)^2}\,\underline{/\tan^{-1}(\omega\tau/1)}} = \frac{1}{\sqrt{1 + (\omega\tau)^2}}\,\underline{/-\tan^{-1}\omega\tau}$$

Hence,
$$\phi = -\tan^{-1}\omega\tau \tag{1}$$

and
$$M_{db} = 20\log|A(j\omega)| = 20\log\frac{1}{\sqrt{1 + (\omega\tau)^2}} = -10\log[1 + (\omega\tau)^2] \tag{2}$$

10.5 Construct the Bode plot for the first-order network of Problem 10.4.

▌ If values of (1) and (2) of Problem 10.4 are calculated and plotted for various values of ω, then a Bode plot is generated. This is done in Fig. 10-2, where ω is given in terms of time constants τ rather than, say, hertz. This particular system is called a *lag network* because its phase angle ϕ is negative for all ω.

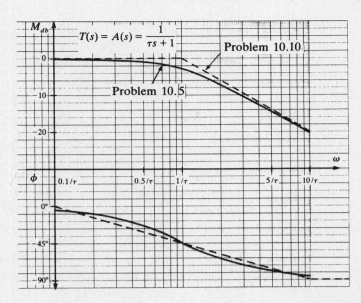

Fig. 10-2

10.6 A simple first-order network has Laplace-domain transfer function and frequency transfer function

$$A(s) = \tau s + 1 \quad \text{and} \quad A(j\omega) = 1 + j\omega\tau$$

Determine the network phase angle ϕ and the amplitude ratio M_{db}.

▌ After $A(j\omega)$ is converted to polar form, it becomes apparent that

$$\phi = \tan^{-1} \omega\tau \tag{1}$$

and $$M_{db} = 20 \log |A(j\omega)| = 20 \log \sqrt{1 + (\omega\tau)^2} = 10 \log [1 + (\omega\tau)^2] \tag{2}$$

10.7 Discuss the nature of the Bode plot for the first-order network of Problem 10.6.

▌ Comparison of (1) of Problem 10.4 and (1) of Problem 10.6 reveals that the network phase angle is the mirror image of the phase angle for the network of Problem 10.4. (As ω increases, ϕ ranges from 0° to 90°.) Further, (2) of Problem 10.6 shows that the amplitude ratio is the mirror image of the amplitude ratio of Problem 10.4. (As ω increases, M_{db} ranges from 0 to positive values.) Thus, the complete Bode plot consists of the mirror images about zero of M_{db} and ϕ of Fig. 10-2. Since here the phase angle ϕ is everywhere positive, this network is called a *lead network*.

10.8 A *break frequency* or *corner frequency* is the frequency $1/\tau$. Corner frequencies serve as key points in the construction of Bode plots. For a simple lag or lead network, it is the frequency at which $M^2 = |A(j\omega)|^2$ has changed by 50 percent from its value at $\omega = 0$. Determine the change in M_{db} at the corner frequency of a simple lag (or lead) network from its value at $\omega = 0$.

▌ Based on (2) of Problem 10.4, the change in M_{db} for a simple lag network is

$$\Delta M_{db} = M_{db}|_{\omega=0} - M_{db}|_{\omega=1/\tau} \tag{1}$$
$$= -10 \log [1 + 0] - (-10 \log [1 + (1)^2]) = 3.01 \approx 3 \text{ db}$$

For a simple lead network, use of (2) of Problem 10.6 in (1) finds $\Delta M_{db} = -3$ db.

10.9 Describe the Bode plot of a network whose output is the time derivative of its input.

▌ The network has Laplace-domain transfer function $A(s) = s$ and frequency transfer function $A(j\omega) = j\omega$. Converting $A(j\omega)$ to polar form shows that

$$\phi = \tan^{-1} \frac{\omega}{0} = 90° \tag{1}$$

and $$M_{db} = 20 \log \omega \tag{2}$$

Obviously, the network phase angle is a constant 90°. By (2), $M_{db} = 0$ when $\omega = 1$; further, M_{db} increases by 20 db for each order-of-magnitude (*decade*) change in ω. A graph of M_{db} versus the logarithm of ω would thus have a slope of 20 db per decade of frequency. A complete Bode plot is shown in Fig. 10-3.

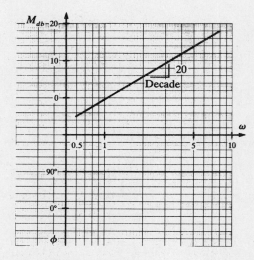

Fig. 10-3

10.10 The exact Bode plot of a network frequency transfer function is tedious to construct. Frequently, sufficiently accurate information can be obtained from an *asymptotic* Bode plot. The exact Bode plot for the first-order system of Problem 10.4 is given in Fig. 10-2. Add the asymptotic Bode plot to that figure.

▌ Asymptotic Bode plots are piecewise-linear approximations. The asymptotic plot of M_{db} for a simple lag network has value zero out to the single break frequency $\omega = 1/\tau$ and then *decreases* at 20 db per decade. The asymptotic plot of ϕ has the value zero out to $\omega = 0.1/\tau$, decreases linearly to $-90°$ at $\omega = 10/\tau$, and then is constant at $-90°$. Both asymptotic plots are shown dashed in Fig. 10-2.

10.11 Describe the asymptotic Bode plot for the system of Problems 10.6 and 10.7.

▌ The asymptotic Bode plot for a simple lead network is the mirror image of that for a simple lag network. Thus, the asymptotic plot of M_{db} in Problem 10.7 is zero out to $\omega = 1/\tau$ and then *increases* at 20 db per decade; the plot of ϕ is zero out to $\omega = 0.1/\tau$, increases to 90° at $\omega = 10/\tau$, and then remains constant.

10.12 Calculate and tabulate the difference between the asymptotic and exact plots of Fig. 10-2, for use in correcting asymptotic plots to exact plots.

The difference ϵ may be found by subtraction. For the M_{db} plot,

For $0 \leq \omega \leq \dfrac{1}{\tau}$:

$$\epsilon_{Mdb} = 0 - \{-10 \log [1 + (\omega\tau)^2]\} = 10 \log [1 + (\omega\tau)^2] \quad (1)$$

For $\omega > \dfrac{1}{\tau}$:

$$\epsilon_{Mdb} = -10 \log (\omega\tau)^2 - (-10 \log [1 + (\omega\tau)^2]) = 10 \log [1 + 1/(\omega\tau)^2] \quad (2)$$

and for the ϕ plot,

For $0 \leq \omega \leq \dfrac{0.1}{\tau}$:

$$\epsilon_\phi = 0 - (-\tan^{-1} \omega\tau) = \tan^{-1} \omega\tau \quad (3)$$

For $\dfrac{0.1}{\tau} < \omega < \dfrac{10}{\tau}$:

$$\epsilon_\phi = -45° \log 10\, \omega\tau + \tan^{-1} \omega\tau \quad (4)$$

For $\omega \geq \dfrac{10}{\tau}$:

$$\epsilon_\phi = -90° - (-\tan^{-1} \omega\tau) = \tan^{-1} \omega t - 90° \quad (5)$$

Application of (*1*) to (*5*) yields Table 10-1.

FREQUENCY EFFECTS IN AMPLIFIERS □ 285

TABLE 10-1 Bode-Plot Corrections

ω	ϵ_{Mdb}	ϵ_ϕ
$0.1/\tau$	0.04	5.7°
$0.5/\tau$	1	−4.9°
$0.76/\tau$	2	−2.4°
$1/\tau$	3	0°
$1.32/\tau$	2	2.4°
$2/\tau$	1	4.9°
$10/\tau$	0.04	−5.7°

10.13 Show that if two linear networks are connected in cascade to form a new network such that

$$T(j\omega) = T_1(j\omega)T_2(j\omega) \tag{1}$$

then the composite Bode plot is obtained by adding the individual amplitude ratios M_{db1} and M_{db2} and phase angles (ϕ_1 and ϕ_2) associated with $T_1(j\omega)$ and $T_2(j\omega)$ at each frequency.

▌ Express (1) in polar form to see immediately that the composite phase angle is the sum of the individual phase angles.

$$|T(j\omega)| e^{j\phi} = |T_1(j\omega)| e^{j\phi_1} |T_2(j\omega)| e^{j\phi_2} = |T_1(j\omega)| \| T_2(j\omega)| e^{j(\phi_1+\phi_2)} \tag{2}$$

Application of the amplitude ratio definition of Problem 10.4 shows that the composite amplitude ratio is simply the sum of the individual amplitude ratios:

$$M_{db} = \log |T(j\omega)| = \log [|T_1(j\omega)| \| T_2(j\omega)|] = \log |T_1(j\omega)| + \log |T_2(j\omega)| \tag{3}$$

Whence,
$$M_{db} = M_{db1} + M_{db2} \tag{4}$$

10.14 The s-domain transfer function for a system can be written in the form

$$T(s) = \frac{K_b(\tau_{z1}s + 1)(\tau_{z2}s + 1)\cdots}{s^n(\tau_{p1}s + 1)(\tau_{p2}s + 1)\cdots} \tag{1}$$

where n may be positive, negative, or zero. Show that the Bode plot (for M_{db} only) may be generated as a composite of individual Bode plots for three basic types of terms.

The frequency transfer function corresponding to (1) is

$$T(j\omega) = \frac{K_b(1 + j\omega\tau_{z1})(1 + j\omega\tau_{z2})\cdots}{(j\omega)^n(1 + j\omega\tau_{p1})(1 + j\omega\tau_{p2})\cdots} \tag{2}$$

From Definition 3 of Problem 10.4,

$$M_{db} = 20 \log |T(j\omega)| = 20 \log \left[\frac{K_b |1 + j\omega\tau_{z1}| |1 + j\omega\tau_{z2}|\cdots}{|(j\omega)^n| |1 + j\omega\tau_{p1}| |1 + j\omega\tau_{p2}|\cdots} \right] \tag{3}$$

which may be written as

$$M_{db} = 20 \log K_b + 20 \log |1 + j\omega\tau_{z1}| + 20 \log |1 + j\omega\tau_{z2}| + \cdots$$
$$- 20n \log |j\omega| - 20 \log |1 + j\omega\tau_{p1}| - 20 \log |1 + j\omega\tau_{p2}| - \cdots \tag{4}$$

It is apparent from (4) that the Bode plot of $T(j\omega)$ can be formed by point-by-point addition of the plots of three types of terms:
  1. A frequency-invariant or gain-constant term K_b whose Bode plot is a horizontal line at $M_{db} = 20 \log K_b$.
  2. Poles or zeroes of multiplicity n, $(j\omega)^{\pm n}$, whose amplitude ratio is $M_{db} = \pm 20n \log \omega$, where the plus sign corresponds to zeroes and the minus sign to poles of the transfer function. (See Problem 10.9.)
  3. First-order lead and lag factors, $(1 + j\omega\tau)^{\pm n}$, as discussed in Problems 10.4 and 10.6. They are usually approximated with asymptotic Bode plots; if greater accuracy is needed, the asymptotic plots are corrected using Table 10-1.

10.15 An amplifier has a Laplace-domain transfer function (voltage-gain ratio) given by

$$A_v(s) = \frac{V_o}{V_i} = \frac{Ks}{(s + 100)(s + 10^5)} \tag{1}$$

If an asymptotic Bode plot of $A_v(j\omega)$ is made, over what values of frequency (in the midfrequency range) is the gain constant in amplitude?

▌ The frequency transfer function follows from (1) as

$$A(j\omega) = \frac{jK\omega(10^{-7})}{(1+j0.01\omega)(1+j10^{-5}\omega)} \quad (2)$$

Since the value for K is of no consequence in determination of midfrequency range, select $K = 10^7$ for convenience. The resulting Bode plot of (2) is displayed by Fig. 10-4 from which it is seen that the amplitude ratio, thus the gain of the amplifier, is constant for $100 \leq \omega \leq 10^5$ rad/s.

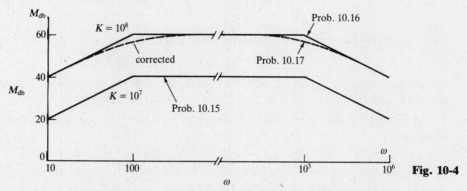

Fig. 10-4

10.16 Determine the midfrequency gain in decibels for the amplifier of Problem 10.15 if $K = 10^8$.

▌ For $K = 10^8$ in (2) of Problem 10.15, the Bode gain K_b is 10 (see Problem 10.14) resulting in an upward shift of the asymptotic Bode plot of Fig. 10-4 by $20 \log(K_b) = 20$ db. Consequently, the midfrequency gain is 60 db.

10.17 Within 2 percent accuracy, over what range of frequencies is the exact midfrequency gain constant for the amplifier of Problem 10.15?

▌ The asymptotic amplitude ratio curve with $K = 10^8$ of Fig. 10-4 is corrected to give the exact gain in accordance with Table 10.1. For no more than 2 percent error, $M_{db} \geq 0.98(60) = 58.8$ db. Inspection of the corrected M_{db} plot of Fig. 10-4 shows the 2 percent accuracy midfrequency range to be $500 \leq \omega \leq 5 \times 10^4$ rad/s.

10.18 The circuit of Fig. 10-5(a) is driven by a sinusoidal source v_S. Sketch the asymptotic Bode plot (M_{db} only) associated with the Laplace-domain transfer function $T(s) = V_o/V_S$.

▌ By voltage division in Fig. 10-5(a)

$$V_o = \frac{R_L}{R_L + R_S + 1/sC} V_S \quad (1)$$

so that

$$\frac{V_o}{V_S} = \frac{sR_L C}{1 + sC(R_L + R_S)} = \frac{K_b s}{1 + \tau s} \quad (2)$$

Using the result of Problem 10.15, we recognize (2) as the combination of a first-order lag, a constant gain, and a zero of multiplicity 1. The components of the asymptotic Bode plot are shown dashed in Fig. 10-5(b), and the composite is solid. For purposes of illustration, it was assumed that $1/[C(R_L + R_S)] > 1$, which is true in most cases.

10.19 Use Table 10-1 to correct the asymptotic Bode plot of Problem 10.18, so as to show the exact Bode plot.

▌ The correction factors of Table 10-1 lead to the exact Bode plot as drawn in Fig. 10-5(b).

10.20 Determine the Laplace-domain transfer function $T(s) = V_2/V_S$ for the RC circuit of Fig. 10-6(a).

▌ By voltage division,

$$V_2 = \frac{R_2 \parallel (1/sC_2)}{R_1 + R_2 + 1/sC_2} V_S = \frac{\dfrac{R_2}{sR_2C_2 + 1}}{R_1 + \dfrac{R_2}{sR_2C_2 + 1}} V_S$$

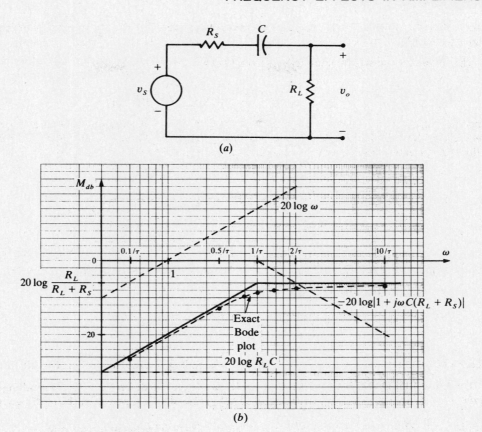

Fig. 10-5

and the Laplace-domain transfer function is

$$T(s) = \frac{V_2}{V_S} = \frac{R_2/(R_1 + R_2)}{s\left(\dfrac{R_1 R_2}{R_1 + R_2}\right)C + 1} = \frac{K_b}{sR_{eq}C_2 + 1} \qquad (1)$$

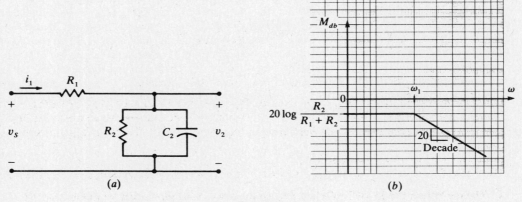

Fig. 10-6

10.21 Sketch the asymptotic Bode plot (M_{db} only) associated with the output-to-input voltage ratio of the circuit in Fig. 10-6(a).

▌ From $T(s)$ given by (1) of Problem 10.20, it is apparent that the circuit forms a low-pass filter with low-frequency gain $T(0) = R_2/(R_1 + R_2)$ and a corner frequency at $\omega_1 = 1/\tau_1 = 1/R_{eq}C_2$. Its Bode plot is sketched in Fig. 10-6(b).

LOW-FREQUENCY EFFECT OF BYPASS AND COUPLING CAPACITORS

10.22 For the amplifier of Fig. 4-14, assume that $C_C \to \infty$ but that the bypass capacitor C_E cannot be neglected. If $h_{re} = h_{oe} = R_i = 0$, draw a small-signal equivalent circuit suitable for analysis of the amplifier.

■ See Fig. 10-7.

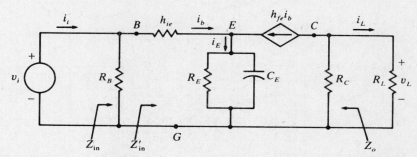

Fig. 10-7

10.23 For the amplifier of Problem 10.22, find an expression for the voltage-gain ratio $A_v(s)$ at any frequency.

■ In the Laplace domain, we have from Fig. 10-7,

$$Z_E = R_E \left\| \frac{1}{sC_E} = \frac{(R_E)(1/sC_E)}{R_E + 1/sC_E} = \frac{R_E}{sR_EC_E + 1} \right. \tag{1}$$

We next note that

$$I_E = I_b + h_{fe}I_b = (h_{fe} + 1)I_b \tag{2}$$

Then KVL and (2) yield

$$V_i = h_{ie}I_b + Z_E I_E = [h_{ie} + (h_{fe} + 1)Z_E]I_b \tag{3}$$

But, by Ohm's law,

$$V_L = -(h_{fe}I_b)(R_C \| R_L) = -\frac{h_{fe}R_CR_L}{R_C + R_L} I_b \tag{4}$$

Solving (3) for I_b, substituting the result into (4), using (1), and rearranging give the desired voltage-gain ratio:

$$A_v(s) = \frac{V_L}{V_i} = -\frac{h_{fe}R_CR_L}{R_C + R_L} \frac{sR_EC_E + 1}{sR_EC_Eh_{ie} + h_{ie} + (h_{fe} + 1)R_E} \tag{5}$$

10.24 As the frequency of the input signal to an amplifier decreases below the midfrequency range, the voltage (or current) gain ratio decreases in magnitude. The *low-frequency cutoff point* ω_L is the frequency at which the gain ratio equals $1/\sqrt{2}$ ($= 0.707$) times its midfrequency value [Fig. 10-1(a)], or at which M_{db} has decreased by exactly 3 db from its midfrequency value. The range of frequencies below ω_L is called the *low-frequency region*. Low-frequency amplifier performance (attenuation, really) is a consequence of the use of bypass and coupling capacitors to fashion the dc bias characteristics. When viewed from the low-frequency region, such amplifier response is analogous to that of a *high-pass filter* (signals for which $\omega < \omega_L$ are appreciably attenuated, whereas higher-frequency signals with $\omega \geq \omega_L$ are unattenuated). Determine the voltage-gain ratio for the amplifier of Problem 10.22 for low-frequency operation.

■ The low-frequency voltage-gain ratio is obtained by letting $s \to 0$ in (5) of Problem 10.23:

$$A_v(0) = \lim_{s \to 0} \frac{V_L}{V_i} = \frac{-h_{fe}R_CR_L}{(R_C + R_L)[h_{ie} + (h_{fe} + 1)R_E]} \tag{1}$$

Comparison of (1) with (1) of Problem 7.40 (but with $h_{oe} = 0$) shows that inclusion of the bypass capacitor in the analysis can significantly change the expression obtained for the voltage-gain ratio.

10.25 Find an expression for the higher-frequency (midfrequency) voltage-gain ratio of the amplifier described by Problem 10.22.

■ The higher-frequency (midfrequency) voltage-gain ratio is obtained by letting $s \to \infty$ in (5) of Problem 10.23:

$$A_v(\infty) = \lim_{s \to \infty} \frac{V_L}{V_i} = \lim_{s \to \infty} \left\{ -\frac{h_{fe}R_CR_L}{R_C + R_L} \frac{R_EC_E + 1/s}{R_EC_Eh_{ie} + [h_{ie} + (h_{fe} + 1)R_E]/s} \right\} = \frac{-h_{fe}R_CR_L}{h_{ie}(R_C + R_L)} \tag{1}$$

10.26 Determine the low-frequency cutoff point ω_L for the amplifier of Problem 10.22.

▌ Equation (5) of Problem 10.23 can be rearranged to give

$$A_v(s) = \frac{-h_{fe}R_C R_L}{(R_C + R_L)[h_{ie} + (h_{fe} + 1)R_E]} \frac{sR_E C_E + 1}{s\dfrac{R_E C_E h_{ie}}{h_{ie} + (h_{fe} + 1)R_E} + 1} \tag{1}$$

which clearly is of the form

$$A_v(s) = k_v \frac{\tau_1 s + 1}{\tau_2 s + 1}$$

Thus, we may use (1) to write

$$\omega_1 = \frac{1}{\tau_1} = \frac{1}{C_E R_E} \tag{2}$$

and

$$\omega_2 = \frac{1}{\tau_2} = \frac{h_{ie} + (h_{fe} + 1)R_E}{R_E C_E h_{ie}} \tag{3}$$

Typically, $h_{fe} \gg 1$ and $h_{fe}R_E \gg h_{ie}$, so a reasonable approximation of ω_2 is

$$\omega_2 \approx \frac{1}{C_E h_{ie}/h_{fe}} \tag{4}$$

Since h_{ie}/h_{fe} is typically an order of magnitude smaller than R_E, ω_2 is an order of magnitude greater than ω_1, and $\omega_L = \omega_2$.

10.27 Sketch the asymptotic Bode plot (amplitude ratio only) for the amplifier of Problem 10.22.

▌ The low- and midfrequency asymptotic Bode plot is depicted in Fig. 10-8, where ω_1 and ω_2 are given by (2) and (4) of Problem 10.26, respectively. From (1) of Problem 10.24 and (1) of Problem 10.25,

$$M_{dbL} = 20 \log \frac{h_{fe}R_C R_L}{(R_C + R_L)[h_{ie} + (h_{fe} + 1)R_E]} \tag{1}$$

and

$$M_{dbM} = 20 \log \frac{h_{fe}R_C R_L}{h_{ie}(R_C + R_L)} \tag{2}$$

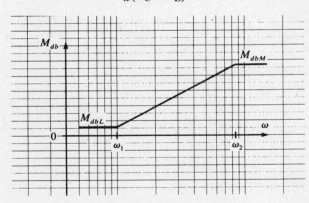

Fig. 10-8

10.28 In the circuit of Fig. 4-18, the battery V_S is replaced with a sinusoidal source v_S. The impedance of the coupling capacitor is not negligibly small. Draw a small-signal equivalent circuit suitable for analysis of the amplifier if $h_{re} = h_{oe} = 0$.

▌ See Fig. 10-9.

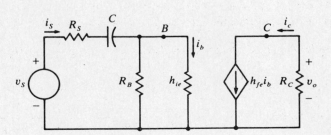

Fig. 10-9

10.29 Derive an expression for the voltage-gain ratio $M = |A_v(j\omega)| = |v_o/v_S|$ of the amplifier of Problem 10.28.

▌ By Ohm's law applied in the Laplace-domain to the pertinent equivalent circuit of Fig. 10-9,

$$I_S = \frac{V_S}{R_S + h_{ie} \parallel R_B + 1/sC} \qquad (1)$$

Then current division gives

$$I_b = \frac{R_B}{R_B + h_{ie}} I_S = \frac{R_B V_S}{(R_B + h_{ie})(R_S + h_{ie} \parallel R_B + 1/sC)} \qquad (2)$$

But Ohm's law requires that

$$V_o = -h_{fe} R_C I_b \qquad (3)$$

Substituting (2) into (3) and rearranging give

$$A_v(s) = \frac{V_o}{V_S} = \frac{-h_{fe} R_C R_B C s}{(R_B + h_{ie})[1 + sC(R_S + h_{ie} \parallel R_B)]} \qquad (4)$$

Now, with $s = j\omega$ in (4), its magnitude is

$$M = |A_v(j\omega)| = \frac{h_{fe} R_C R_B C \omega}{(R_B + h_{ie})\sqrt{1 + (\omega C)^2(R_S + h_{ie} \parallel R_B)^2}} \qquad (5)$$

10.30 Determine the midfrequency voltage-gain ratio for the amplifier of Problem 10.28.

▌ The midfrequency gain follows from letting $s = j\omega \to \infty$ in (4) of Problem 10.29. We may do so because reactances associated with inherent capacitances have been assumed infinitely large (neglected) in the equivalent circuit. We have, then,

$$A_{\text{mid}} = \frac{-h_{fe} R_C R_B}{(R_B + h_{ie})(R_S + h_{ie} \parallel R_B)} \qquad (1)$$

10.31 Determine the low-frequency cutoff point ω_L and sketch the asymptotic Bode plot for the amplifier of Problem 10.28.

▌ From (4) of Problem 10.29,

$$\omega_L = 1/\tau = \frac{1}{C(R_S + h_{ie} \parallel R_B)} = \frac{R_B + h_{ie}}{C[R_S(h_{ie} + R_B) + h_{ie} R_B]} \qquad (1)$$

The asymptotic Bode plot is sketched in Fig. 10-10.

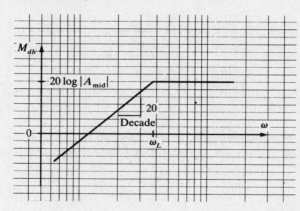

Fig. 10-10

10.32 For the amplifier of Fig. 4-14, assume that $C_C \to \infty$, $h_{re} = h_{oe} = 0$, and $R_i = 0$. The bypass capacitor C_E cannot be neglected. Find an expression for the current-gain ratio $A_i(s)$.

▌ The small-signal low-frequency equivalent circuit is given in Fig. 10-7. By current division for Laplace-domain quantities,

$$I_b = \frac{R_B}{R_B + h_{ie} + Z_E} I_i \qquad (1)$$

FREQUENCY EFFECTS IN AMPLIFIERS □ 291

where
$$Z_E = R_E \left\| \frac{1}{sC_E} = \frac{R_E}{sR_EC_E + 1} \right. \quad (2)$$

Also,
$$I_L = \frac{-R_C}{R_C + R_L} h_{fe} I_b \quad (3)$$

Substitution of (1) into (3) gives the current-gain ratio as

$$\frac{I_L}{I_i} = \frac{-R_C}{R_C + R_L} \frac{h_{fe} R_B}{R_B + h_{ie} + Z_E} \quad (4)$$

Using (2) in (4) and rearranging lead to the desired current-gain ratio:

$$A_i(s) = \frac{I_L}{I_i} = \frac{\dfrac{-h_{fe}R_C R_B}{(R_C + R_L)(R_E + R_B + h_{ie})}(sR_E C_E + 1)}{s\dfrac{R_E C_E(R_B + h_{ie})}{R_E + R_B + h_{ie}} + 1} \quad (5)$$

10.33 Find an expression for the current-gain ratio of the amplifier described in Problem 10.32.

▎ The low-frequency current-gain ratio follows from letting $s \to 0$ in (5) of Problem 10.32:

$$A_i(0) = \lim_{s \to 0} \frac{I_L}{I_i} = \frac{-h_{fe}R_C R_B}{(R_C + R_L)(R_E + R_B + h_{ie})} \quad (1)$$

10.34 Determine an expression for the current-gain ratio of the amplifier described in Problem 10.32.

▎ The midfrequency current-gain ratio is obtained by letting $s \to \infty$ in (5) of Problem 10.32:

$$A_i(\infty) = \lim_{s \to \infty} \frac{I_L}{I_i} = \frac{-h_{fe} R_C R_B}{(R_C + R_L)(R_B + h_{ie})} \quad (1)$$

10.35 Determine the low-frequency cutoff point and sketch the asymptotic Bode plot for the amplifier of Problem 10.32.

▎ Inspection of (5) of Problem 10.32 shows that the Laplace-domain transfer function is of the form

$$A_i(s) = K_b \frac{\tau_1 s + 1}{\tau_2 s + 1}$$

where
$$\omega_1 = \frac{1}{\tau_1} = \frac{1}{R_E C_E} \quad \text{and} \quad \omega_2 = \frac{1}{\tau_2} = \frac{R_E + R_B + h_{ie}}{R_E C_E (R_B + h_{ie})} \quad (1)$$

With ω_1 and ω_2 as given by (1) and with

$$M_{dbL} = 20 \log A_i(0) \quad \text{and} \quad M_{dbM} = 20 \log A_i(\infty)$$

the Bode plot is identical to that of Fig. 10-8. Since $\omega_2 > \omega_1$, ω_2 is closer to the midfrequency region and thus is the low-frequency cutoff point, provided the parameter values are such that $\omega_2 \gg \omega_1$.

10.36 In the amplifier of Fig. 4-14, $C_C \to \infty$, $R_i = 0$, $R_E = 1 \text{ k}\Omega$, $R_1 = 3.2 \text{ k}\Omega$, $R_2 = 17 \text{ k}\Omega$, $R_L = 10 \text{ k}\Omega$, and $h_{oe} = h_{re} = 0$. The transistors used are characterized by $75 \le h_{fe} \le 100$ and $300 \le h_{ie} \le 1000 \, \Omega$. By proper selection of R_C and C_E, design an amplifier with low-frequency cutoff $f_L \le 200 \text{ Hz}$ and high-frequency voltage gain $|A_v| \ge 50$.

▎ According to (1) of Problem 10.25, the worst-case transistor parameters for high $A_v(\infty)$ are minimum h_{fe} and maximum h_{ie}. Using those parameter values allows us to determine a value for the parallel combination of R_C and R_L:

$$R_{eq} = R_C \| R_L \ge |A_v(\infty)| \frac{h_{ie}}{h_{fe}} = (50) \frac{1000}{75} = 666.7 \, \Omega$$

Then
$$R_C = \frac{R_{eq} R_L}{R_L - R_{eq}} \ge \frac{666.7(10{,}000)}{9333.3} = 714.3$$

Now, from (3) of Problem 10.26, for $f_L \leq 200$ Hz,

$$C_E \geq \frac{h_{ie} + (h_{fe} + 1)R_E}{\omega_L R_E h_{ie}} = \frac{300 + 101(1000)}{2\pi(200)(1000)(300)} = 268.7\ \mu\text{F}$$

10.37 For the amplifier design of Problem 10.36, determine the low-frequency voltage-gain ratio if h_{ie} and h_{fe} have median values.

▌ By (1) of Problem 10.24,

$$A_v(0) = -\frac{h_{fe}R_{eq}}{h_{ie} + (h_{fe} + 1)R_E} = \frac{\frac{1}{2}(100 + 75)(666.7)}{\frac{1}{2}(300 + 1000) + [\frac{1}{2}(100 + 75) + 1](1000)} = -0.654$$

10.38 Let $C_1, C_E \to \infty$ in the capacitor-coupled amplifier of Fig. 9-1. Assume $h_{oe1} = h_{re1} = h_{oe2} = h_{re2} = 0$. Find an expression for the voltage-gain ratio $A_v(s)$.

▌ The first-stage amplifier can be replaced with a Thévenin equivalent, and the second stage represented by its input impedance, as shown in Fig. 10-11. A'_v follows from voltage division and (1) of Problem 7.23 if R_L, h_{fe}, and h_{ie} are replaced with R_{C1}, h_{fe1}, and h_{ie1}, respectively:

$$A'_v = \frac{R_{eq}}{R_{eq} + R_i}\frac{-h_{fe1}R_{C1}}{h_{ie1}} = -\frac{h_{fe1}R_{C1}R_{eq}}{h_{ie1}(R_{eq} + R_i)} \tag{1}$$

where

$$R_{eq} = h_{ie1} \| R_{B1} = h_{ie1} \| R_{11} \| R_{12} = \frac{h_{ie1}R_{11}R_{12}}{h_{ie1}(R_{11} + R_{12}) + R_{11}R_{12}} \tag{2}$$

Z_{o1} is given by (3) of Problem 7.25 with h_{oe} replaced with R_{C1} (and with $h_{re1} = h_{oe1} = 0$):

$$Z_{o1} = R_{C1} \tag{3}$$

The second-stage input impedance is given by (1) of Problem 7.24 if h_{ie} is replaced with $h_{ie2} \| R_{B2} = h_{ie2} \| R_{21} \| R_{22}$:

$$Z_{in2} = \frac{h_{ie2}R_{B2}}{h_{ie2} + R_{B2}} = \frac{h_{ie2}R_{21}R_{22}}{h_{ie2}(R_{21} + R_{22}) + R_{21}R_{22}} \tag{4}$$

Now, from (2) of Problem 10.18,

$$\frac{V_{o1}}{A'_v V_i} = \frac{sZ_{in2}C_2}{sC_2(Z_{in2} + Z_{o1}) + 1} \tag{5}$$

and rearranging yields the first-stage gain as

$$A_{v1} = \frac{V_{o1}}{V_i} = A'_v \frac{sZ_{in2}C_2}{sC_2(Z_{in2} + Z_{o1}) + 1} \tag{6}$$

The second-stage gain follows directly from (1) of Problem 7.23 if R_L is replaced with R_{C2}:

$$A_{v2} = -\frac{h_{fe2}R_{C2}}{h_{ie2}}$$

Consequently, the overall gain is

$$A_v = A_{v1}A_{v2} = -A'_v \frac{sZ_{in}C_2}{sC_2(Z_{in2} + Z_{o1}) + 1}\frac{h_{fe2}R_{C2}}{h_{ie2}} \tag{7}$$

Substituting (1) into (7) and simplifying yield the desired gain:

$$A_v(s) = \frac{h_{fe1}h_{fe2}R_{C1}R_{C2}R_{eq}}{h_{ie1}h_{ie2}(R_{eq} + R_i)}\frac{sZ_{in2}C_2}{sC_2(Z_{in2} + Z_{o1}) + 1} \tag{8}$$

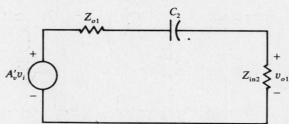

Fig. 10-11

10.39 In the cascaded amplifier of Problem 10.38 (Fig. 9-1 with C_1 and $C_E \to \infty$), let $h_{ie1} = h_{ie2} = 1500\,\Omega$, $h_{fe1} = h_{fe2} = 40$, $C_2 = 1\,\mu\text{F}$, $R_i = 1\,\text{k}\Omega$, $R_{C1} = 10\,\text{k}\Omega$, $R_{C2} = 20\,\text{k}\Omega$, and $R_{B1} = R_{B2} = 5\,\text{k}\Omega$. Determine the low-frequency voltage gain.

❙ Letting $s \to 0$ in (8) of Problem 10.38 makes apparent the fact that the low-frequency gain $A_v(0) = 0$.

10.40 Evaluate the midfrequency voltage gain for the cascaded BJT amplifier of Problem 10.39.

❙ The midfrequency gain is determined by letting $s \to \infty$ in (8) of Problem 10.38. From (2), (3), and (4) of Problem 10.38:

$$A_v(\infty) = \lim_{s \to \infty} A_v(s) = \frac{h_{fe1}h_{fe2}R_{C1}R_{C2}R_{eq}}{h_{ie1}h_{ie2}(R_{eq} + R_i)} \frac{Z_{in2}}{Z_{in2} + Z_{o1}}$$

$$R_{eq} = \frac{h_{ie1}R_{B1}}{h_{ie1} + R_{B1}} = \frac{1500(5000)}{6500} = 1153.8\,\Omega$$

$$Z_{o1} = R_{C1} = 10\,\text{k}\Omega$$

and

$$Z_{in2} = \frac{h_{ie2}R_{B2}}{h_{ie2} + R_{B2}} = \frac{1500(5000)}{6500} = 1153.8\,\Omega$$

Then

$$A_v(\infty) = \frac{40(40)(10 \times 10^3)(20 \times 10^3)(1153.8)}{1500(1500)(2153.8)} \frac{1153.8}{1153.8 + 10 \times 10^3} = 7881.3$$

10.41 Find the value of the low-frequency cutoff point for the amplifier of Problem 10.39.

❙ The low-frequency cutoff point is computed from the lag term in (8) of Problem 10.38:

$$f_L = \frac{\omega_L}{2\pi} = \frac{1}{2\pi C_2(Z_{in2} + Z_{o1})} = \frac{1}{2\pi(1 \times 10^{-6})(1153.8 + 10 \times 10^3)} = 14.3\,\text{Hz}$$

10.42 The two coupling capacitors in the CB amplifier of Fig. 7-18 are identical and cannot be neglected. Assume $h_{rb} = h_{ob} = 0$. Find an expression for the voltage-gain ratio V_L/V_S.

❙ The small-signal low-frequency equivalent circuit is given in Fig. 10-12. Applying Ohm's law in the Laplace domain, we obtain

$$I_S = \frac{V_S}{1/sC_C + R_E h_{ib}/(R_E + h_{ib})} = \frac{sC_C V_S}{sC_C R_E h_{ib}/(R_E + h_{ib}) + 1}$$

Voltage division then gives

$$I_e = \frac{R_E}{h_{ib} + R_E} I_S = \frac{R_E}{h_{ib} + R_E} \frac{sC_C}{sC_C R_E h_{ib}/(R_E + h_{ib}) + 1} V_S \tag{1}$$

By current division at the output,

$$V_L = R_L I_L = -R_L \frac{R_C}{R_C + 1/sC_C + R_L} h_{fb} I_e = -\frac{s h_{fb} R_L R_C C_C I_e}{sC_C(R_L + R_C) + 1} \tag{2}$$

Substituting (1) into (2) and rearranging lead to the desired voltage-gain ratio:

$$A_v(s) = \frac{V_L}{V_S} = -\frac{R_E R_L R_C h_{fb} C_C^2 s^2}{(h_{ib} + R_E)[sC_C R_E h_{ib}/(R_E + h_{ib}) + 1][sC_C(R_L + R_C) + 1]} \tag{3}$$

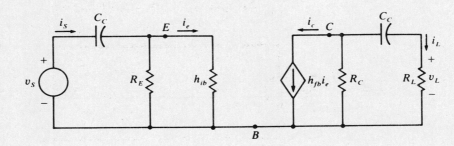

Fig. 10-12

10.43 Determine an expression for the midfrequency voltage-gain ratio of the amplifier of Problem 10.42.

▌ Letting $s \to \infty$ in (3) of Problem 10.42 leads to the midfrequency gain:

$$A_v(\infty) = -\frac{R_L R_C h_{fb}}{h_{ib}(R_L + R_C)} \tag{1}$$

10.44 The two coupling capacitors in the CB amplifier of Fig. 7-18 are identical. Also, $h_{rb} = h_{ob} = 0$. Find an expression for the current-gain ratio $A_i(s)$ that is valid at any frequency.

▌ The low-frequency equivalent circuit is displayed in Fig. 10-12. By current division,

$$I_e = \frac{R_E}{h_{ib} + R_E} I_S \tag{1}$$

and

$$I_L = -\frac{R_C}{R_C + 1/sC_C + R_L} h_{fb} I_e = -\frac{s h_{fb} R_L R_C C_C I_e}{sC_C(R_L + R_C) + 1} \tag{2}$$

Substituting (1) into (2) and dividing by I_S give the desired current-gain ratio:

$$A_i(s) = \frac{I_L}{I_i} = -\frac{s h_{fb} R_L R_C R_E C_C}{(h_{ib} + R_E)[sC_C(R_L + R_C) + 1]} \tag{3}$$

10.45 Determine an expression for the midfrequency current-gain ratio of the amplifier of Problem 10.44:

▌ The midfrequency current-gain ratio is found by letting $s \to \infty$ in (3) of Problem 10.44:

$$A_i(\infty) = -\frac{h_{fb} R_L R_C R_E}{(h_{ib} + R_E)(R_L + R_C)} \tag{1}$$

10.46 On a common set of axes, sketch the asymptotic Bode plots (M_{db} only) for the voltage- and current-gain ratios of the CB amplifier of Fig. 7-18, and then correct them to exact plots. Assume that the coupling capacitors are identical and that, for typical values, $1 \ll C_C(R_E \parallel h_{ib}) \ll C_C(R_L + R_E)$.

▌ The Laplace-domain transfer functions that serve as bases for Bode plots of the voltage- and current-gain ratios are, respectively, (3) of Problem 10.42 and (3) of Problem 10.44. Under the given assumptions, inspection shows that the two transfer functions share a break frequency at $\omega = 1/[C_C(R_L + R_C)]$ and the voltage-gain transfer function has another at a higher frequency. Moreover, the voltage plot rises at 40 db per decade to its first break point, and the current plot at 20 db per decade. With

$$\omega_{1v} = \omega_{1i} = \omega_{Li} = \frac{1}{C_C(R_L + R_C)} \quad \text{and} \quad \omega_{2v} = \omega_{Lv} = \frac{R_E + h_{ib}}{C_C R_E h_{ib}}$$

the low-frequency asymptotic Bode plots of voltage and current gain are sketched in Fig. 10-13. The given assumption assures a separation of at least a decade between ω_{1v} and ω_{2v} and between $\omega = 1$ and ω_{1v}.

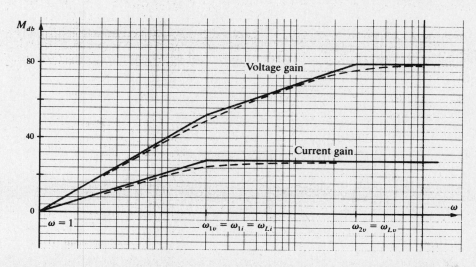

Fig. 10-13

Since the parameter values are not known, the sketches were made under the assumption that $K_b = 1$ in both plots. When values become known, the Bode plots must be shifted upward by

$$20 \log K_{bv} = 20 \log \frac{R_E R_L R_C h_{fb} C_C^2}{h_{ib} + R_E} \quad \text{for the voltage plot}$$

and

$$20 \log K_{bi} = 20 \log \frac{h_{fb} R_L R_C R_E C_C}{h_{ib} + R_E} \quad \text{for the current plot}$$

Correction of the asymptotic plot requires only the application of Table 10-1. The exact plots are shown dashed.

10.47 For the CE amplifier of Fig. 4-14, determine Z'_{in} if $C_C \to \infty$ but C_E cannot be neglected.

▮ The small-signal low-frequency equivalent circuit is given in Fig. 10-7. Using (1) and (3) of Problem 10.23, we have

$$Z'_{in} = \frac{V_i}{I_b} = h_{ie} + (h_{fe} + 1) Z_E = \frac{s h_{ie} R_E C_E + h_{ie} + (h_{fe} + 1) R_E}{s R_E C_E + 1} \tag{1}$$

10.48 Find an expression for Z_{in} of the amplifier of Fig. 4-14 as modeled by Fig. 10-7 if $C_C \to \infty$ but C_E cannot be neglected.

▮ By inspection of Fig. 10-7,

$$Z_{in} = R_B \parallel Z'_{in} = \frac{R_B Z'_{in}}{R_B + Z'_{in}} \tag{1}$$

Substituting (1) of Problem 10.47 into (1) and rearranging give

$$Z_{in} = \frac{R_B [s h_{ie} R_E C_E + h_{ie} + (h_{fe} + 1) R_E]}{s R_E C_E (R_B + h_{ie}) + R_B + h_{ie} + (h_{fe} + 1) R_E} \tag{2}$$

10.49 Determine the output impedance Z_o for the amplifier of Fig. 4-14 as modeled by Fig. 10-7.

▮ With voltage source v_i deactivated (shorted), KVL requires that

$$I_b = \frac{-h_{fe} I_b (Z_E \parallel h_{ie})}{h_{ie}}$$

so that

$$\left[1 + \frac{h_{fe} Z_E h_{ie}}{h_{ie}(Z_E + h_{ie})} \right] I_b = 0 \tag{1}$$

Since (1) can be satisfied in general only by $I_b = 0$, the output impedance is simply

$$Z_o = R_C \tag{2}$$

In this particular case, (2) of Problem 10.48 shows that the input impedance is frequency-dependent, while (2) shows that the output impedance is independent of frequency. In general, however, the output impedance does depend on frequency, through a finite-valued coupling capacitor C_C. (See Problem 10.52.)

10.50 To examine the combined effects of coupling and bypass capacitors, let the input coupling capacitor be infinitely large while the output coupling capacitor and the bypass capacitor have practical values in the CE amplifier of Fig. 4-14. For simplicity, assume $h_{re} = h_{oe} = 0$ and $R_B \gg Z'_{in}$. Draw a small-signal equivalent circuit suitable for analysis of the amplifier.

▮ See Fig. 10-14.

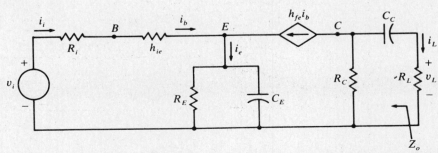

Fig. 10-14

10.51 Find the voltage-gain ratio $A_v(s) = V_L/V_i$ for the amplifier of Problem 10.50.

▌ From the pertinent small-signal equivalent circuit of Fig. 10-14, we first define

$$Z_E = R_E \left\| \frac{1}{sC_E} = \frac{R_E}{sR_EC_E + 1} \right. \tag{1}$$

Then, by KCL,

$$I_e = I_b + h_{fe}I_b = (h_{fe} + 1)I_b \tag{2}$$

KVL around the input mesh requires that

$$V_i = (R_i + h_{ie})I_b + Z_EI_e \tag{3}$$

Substituting (2) into (3) and solving for I_b yields

$$I_b = \frac{V_i}{R_i + h_{ie} + (h_{fe} + 1)Z_E} \tag{4}$$

Current division at the collector node gives

$$I_L = -\frac{R_C}{R_C + R_L + 1/sC_C} h_{fe}I_b \tag{5}$$

and Ohm's law and (5) yield

$$V_L = R_L I_L = -\frac{R_L R_C}{R_C + R_L + 1/sC_C} h_{fe}I_b \tag{6}$$

Substituting (4) and (1) into (6) and rearranging now lead to the desired voltage-gain ratio:

$$A_v(s) = \frac{V_L}{V_i} = -\frac{\dfrac{h_{fe}R_LR_CC_C}{(h_{fe}+1)R_E + h_{ie} + R_i} s(sR_EC_E + 1)}{[sC_C(R_C + R_L) + 1]\left[s\dfrac{C_ER_E(R_i + h_{ie})}{(h_{fe}+1)R_E + R_i + h_{ie}} + 1\right]} \tag{7}$$

10.52 For the amplifier of Problem 10.50, let $C_E = 200\ \mu F$, $C_C = 10\ \mu F$, $R_i = R_E = 100\ \Omega$, $R_C = R_L = 2\ k\Omega$, $h_{ie} = 1\ k\Omega$, and $h_{fe} = 100$. Determine what parameters control the low-frequency cutoff point and whether it is below 100 Hz.

▌ The pertinent Laplace-domain transfer function given by (7) from Problem 10.50 is of the form

$$T(s) = \frac{-K_b s(\tau_2 s + 1)}{(\tau_1 s + 1)(\tau_3 s + 1)}$$

where

$$\omega_1 = \frac{1}{\tau_1} = \frac{1}{C_C(R_C + R_L)} = \frac{1}{(10 \times 10^{-6})(4000)} = 25\ \text{rad/s}$$

$$\omega_2 = \frac{1}{\tau_2} = \frac{1}{R_EC_E} = \frac{1}{100(200 \times 10^{-6})} = 50\ \text{rad/s}$$

$$\omega_3 = \frac{1}{\tau_3} = \frac{(h_{fe}+1)R_E + R_i + h_{ie}}{C_ER_E(R_i + h_{ie})} = \frac{101(100) + 100 + 1000}{(200 \times 10^{-6})(100)(1100)} = 509.1\ \text{rad/s}$$

Since there is at least a decade of frequency (in which the gain can attenuate from its midfrequency value) between ω_3 and the other (lower) break frequencies, ω_3 must be the low-frequency cutoff ω_L. Then

$$f_L = \frac{\omega_3}{2\pi} = \frac{509.1}{2\pi} = 81.02\ \text{Hz} < 100\ \text{Hz}$$

10.53 For the amplifier of Problem 10.50, determine an expression for the output impedance Z_o at any frequency.

▌ As in Problem 10.49, $I_b = 0$ if v_i is deactivated; a driving-point source replacing R_L would then see a frequency-dependent output impedance given by

$$Z_o = Z_{dp} = R_C + \frac{1}{sC_C} \tag{1}$$

10.54 Assume that the coupling capacitors in the CS MOSFET amplifier of Fig. 5-24 are identical. Draw a small-signal equivalent circuit suitable for analysis of the amplifier if the capacitors cannot be neglected.

▮ See Fig. 10-15.

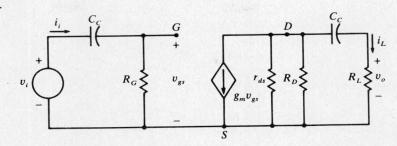

Fig. 10-15

10.55 For the MOSFET amplifier of Fig. 5-24 as modeled by Fig. 10-15, find an expression for the voltage-gain ratio $A_v(s) = V_o/V_i$ valid for any frequency.

▮ By voltage division in Fig. 10-15,

$$V_{gs} = \frac{R_G}{R_G + 1/sC_C} V_i = \frac{sR_G C_C}{sR_G C_C + 1} V_1 \quad \text{where} \quad R_G = R_1 \| R_2 = \frac{R_1 R_2}{R_1 + R_2} \quad (1)$$

Current division at the drain node yields

$$I_L = -\frac{R_D \| r_{ds}}{R_D \| r_{ds} + 1/sC_C + R_L} g_m V_{gs} = -\frac{sC_C[R_D r_{ds}/(R_D + r_{ds})]g_m V_{gs}}{sC_C[R_D r_{ds}/(R_D + r_{ds}) + R_L] + 1} \quad (2)$$

from which

$$V_o = R_L I_L = -\frac{sg_m R_D R_L r_{ds} C_C/(R_D + r_{ds})}{sC_C[R_D r_{ds}/(R_D + r_{ds}) + R_L] + 1} V_{gs} \quad (3)$$

Substitution of (1) into (3) and rearrangement then give

$$A_v(s) = \frac{V_o}{V_i} = -\frac{s^2 g_m R_G R_D R_L r_{ds} C_C^2/(R_D + r_{ds})}{\left[sC_C\left(\frac{R_D r_{ds}}{R_D + r_{ds}} + R_L\right) + 1\right][sC_C R_G + 1]} \quad (4)$$

10.56 Determine an expression for the midfrequency voltage-gain ratio V_L/V_i for the amplifier of Fig. 5-24 that is modeled by Fig. 10-15.

▮ Since high-frequency capacitances have not been modeled, the midfrequency gain follows from letting $s \to \infty$ in (4) of Problem 10.55:

$$A_{\text{mid}} = A_v(\infty) = -\frac{g_m R_D R_L r_{ds}}{R_D r_{ds} + R_L(R_D + r_{ds})}$$

10.57 Draw a small-signal equivalent circuit suitable for analysis of the CS JFET amplifier of Fig. 8-2 if the source bypass capacitor C_S is not negligible.

▮ See Fig. 10-16.

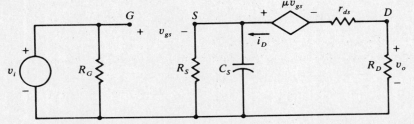

Fig. 10-16

10.58 For the CS JFET amplifier of Fig. 8-2, find an expression for the voltage-gain ratio $A_v(s) = V_o/V_i$.

▮ The low-frequency equivalent circuit of Fig. 10-16 is applicable. By KVL,

$$I_d = \frac{\mu V_{gs}}{(R_S \| 1/sC_S) + r_{ds} + R_D} = \frac{\mu(sR_S C_S + 1)V_{gs}}{sC_S R_S(R_D + r_{ds}) + R_S + R_D + r_{ds}} \quad (1)$$

But KVL requires that

$$V_{gs} = V_i - I_D\left(R_S \Big\| \frac{1}{sC_S}\right) = V_i - \frac{R_S I_D}{SR_S C_S + 1} \quad (2)$$

Substituting (1) into (2) and solving for V_{gs} give

$$V_{gs} = \frac{sC_S R_S(R_D + r_{ds}) + R_S + R_D + r_{ds}}{sC_S R_S(R_D + r_{ds}) + R_D + r_{ds} + (\mu + 1)R_S} V_i \quad (3)$$

Now, by Ohm's law and (1),

$$V_o = -R_D I_D = -\frac{\mu R_D(sR_S C_S + 1)V_{gs}}{sC_S R_S(R_D + r_{ds}) + R_S + R_D + r_{ds}} \quad (4)$$

Substituting V_{gs} as given by (3) into (4) and rearranging yield, finally,

$$A_v(s) = \frac{V_o}{V_i} = -\frac{\mu R_D}{R_D + r_{ds} + (\mu + 1)R_S} \frac{sR_S C_S + 1}{s\dfrac{C_S R_S(R_D + r_{ds})}{R_D + r_{ds} + (\mu + 1)R_S} + 1} \quad (5)$$

10.59 Determine an expression for the low-frequency cutoff point of the amplifier described by Problem 10.58.

❙ It is apparent that the low-frequency cutoff is the larger of the two break frequencies; from (5) of Problem 10.58; it is

$$\omega_L = \frac{R_D + r_{ds} + (\mu + 1)R_S}{C_S R_S(R_D + r_{ds})}$$

HIGH-FREQUENCY BJT AND FET MODELS

10.60 Because of capacitance that is inherent within the transistor, amplifier current- and voltage-gain ratios decrease in magnitude as the frequency of the input signal increases beyond the midfrequency range. The *high-frequency cutoff point* ω_H is the frequency at which the gain ratio equals $1/\sqrt{2}$ times its midfrequency value [see Fig. 10-1(a)], or at which M_{db} has decreased by 3 db from its midfrequency value. The range of frequencies above ω_H is called the *high-frequency region*. Like ω_L, ω_H is a break frequency.

The most useful high-frequency model for the BJT is called the *hybrid-π* equivalent circuit. In this model, the reverse voltage ratio h_{re} and output admittance h_{oe} are assumed negligible. The *base ohmic resistance* $r_{bb'}$, assumed to be located between the base terminal B and the base junction B', has a constant value (typically 10 to 50 Ω) that depends directly on the base width. The *base-emitter-junction resistance* $r_{b'e}$ is usually much larger than $r_{bb'}$ and can be calculated as

$$r_{b'e} = \frac{V_T(\beta + 1)}{I_{EQ}} = \frac{V_T \beta}{I_{CQ}} \quad (1)$$

(see Problem 7.124). Capacitance C_μ is the depletion capacitance associated with the reverse-biased collector-base junction; its value is a function of V_{BCQ}. Capacitance $C_\pi(\gg C_\mu)$ is the diffusion capacitance associated with the forward-biased base-emitter junction; its value is a function of I_{EQ}. Draw the hybrid-π model for the BJT as suggested by the above discussion.

❙ See Fig. 10-17.

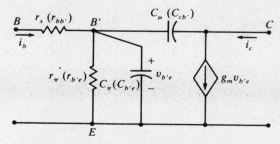

Fig. 10-17 Hybrid-π model for the BJT.

10.61 Apply the hybrid-π model of Fig. 10-17 to the amplifier of Fig. 4-14 to find a model valid for high-frequency analysis if $R_i = 0$. Then, show how a Thévenin's equivalent can be used to produce a more simple circuit for analysis.

┃ Since our interest is in high-frequency analysis, reactances of capacitors C_C and C_E of Fig. 4-14 are considered negligibly small. Direct application of the hybrid-π model of Fig. 10-17 leads to the small-signal high-frequency circuit of Fig. 10-18(a). To simplify the analysis, a Thévenin equivalent circuit may be found for the network to the left of terminal pair B', E, with

$$V_{Th} = \frac{r_\pi}{r_\pi + r_x} V_S \tag{1}$$

and
$$R_{Th} = r_\pi \parallel r_x = \frac{r_\pi r_x}{r_\pi + r_x} \tag{2}$$

Figure 10-18(b) shows the circuit with the Thévenin's equivalent in position.

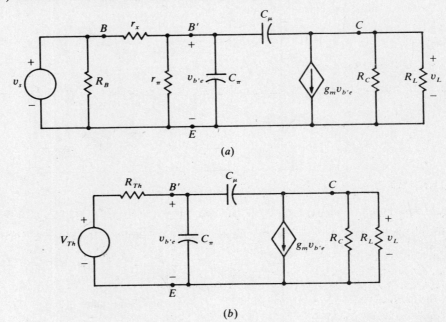

Fig. 10-18

10.62 Derive an expression for the high-frequency voltage-gain ratio V_L/V_S of the amplifier of Fig. 4-14 as modeled by Fig. 10-18(b).

┃ Using $v_{b'e}$ and v_L as node voltages in Fig. 10-18(b) and working in the Laplace domain, we may write the following two equations:

$$\frac{V_{b'e} - V_{Th}}{R_{Th}} + \frac{V_{b'e}}{1/sC_\pi} + \frac{V_{b'e} - V_L}{1/sC_\mu} = 0 \tag{1}$$

$$\frac{V_L}{R_C \parallel R_L} + g_m V_{b'e} + \frac{V_L - V_{b'e}}{1/sC_\mu} = 0 \tag{2}$$

The latter equation can be solved for $V_{b'e}$, then substituted into (1), and the result rearranged to give the voltage ratio V_{Th}/V_L:

$$\frac{V_{Th}}{V_L} = \frac{s^2 C_\mu C_\pi R_{Th}(R_C \parallel R_L) + s[(1 - g_m)C_\mu(R_C \parallel R_L)] + 1}{(R_C \parallel R_L)(sC_\mu - g_m)} \tag{3}$$

For typical values, the coefficient of s^2 on the right side of (3) is several orders of magnitude smaller than the other terms; by approximating this coefficient to zero (i.e., neglecting the s^2 term), we neglect a breakpoint at a frequency much greater than ω_H. Doing so and using (1) of Problem 10.61, we obtain the desired high-frequency voltage-gain ratio:

$$A_v(s) = \frac{V_L}{V_S} = \frac{r_\pi}{r_\pi + r_x} \frac{R_C \parallel R_L(sC_\mu - g_m)}{s(1 - g_m)C_\mu(R_C \parallel R_L) + 1} \tag{4}$$

10.63 The hybrid-π equivalent circuit for the CE amplifier of Fig. 4-14 with the output shorted is shown in Fig. 10-19. Find an expression for the so-called *β cutoff frequency f_β*, which is simply the high-frequency current-gain cutoff point of the transistor with the collector and emitter terminals shorted.

▌ Ohm's law gives

$$V_{b'e} = \frac{I_b}{g_\pi + s(C_\pi + C_\mu)} \quad \text{where} \quad g_\pi = \frac{1}{r_\pi} \tag{1}$$

But with the collector and emitter terminals shorted,

$$I_L = -g_m V_{b'e} \tag{2}$$

Substituting (1) into (2) and rearranging give the current-gain ratio

$$\frac{i_L}{I_b} = -\frac{g_m}{g_\pi + s(C_\pi + C_\mu)} = -\frac{g_m r_\pi}{sr_\pi(C_\pi + C_\mu) + 1} \tag{3}$$

From (3), the β cutoff frequency is seen to be

$$f_\beta = \frac{\omega_\beta}{2\pi} = \frac{1}{2\pi r_\pi (C_\pi + C_\mu)} \tag{4}$$

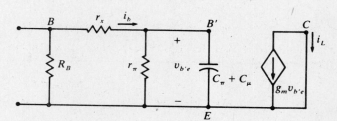

Fig. 10-19

10.64 Evaluate the β cutoff frequency for a BJT characterized by $r_x = 100\,\Omega$, $r_\pi = 1\,\text{k}\Omega$, $C_\mu = 3\,\text{pF}$, and $C_\pi = 100\,\text{pF}$.

▌ Substituting the given high-frequency parameters in (4) of Problem 10.63 yields

$$f_\beta = \frac{1}{2\pi(1000)(103 \times 10^{-12})} = 1.545\,\text{MHz}$$

10.65 Apply the hybrid-π high-frequency model of Fig. 10-17 to the CB amplifier of Fig. 7-18(b).

▌ See Fig. 10-20 for the resulting circuit where the coupling capacitors are assumed to act as short circuits at high frequency.

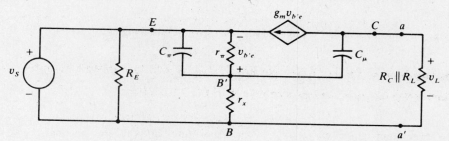

Fig. 10-20

10.66 Find an expression for the high-frequency voltage-gain ratio V_L/V_S of the CB amplifier of Fig. 7-18 as modeled by Fig. 10-20.

▌ For typical values, $r_x \ll 1/sC_\pi$, r_π, $1/sC_\mu$ for frequencies near the break frequencies; thus, letting $r_x = 0$ introduces little error (but considerable simplicity).

A Thévenin equivalent can be found for the network to the left of terminal pair a,a'. With $r_x = 0$, current from the dependent source flows only through C, so

$$V_{Th} = -\frac{1}{sC_\mu} g_m V_{b'e} \tag{1}$$

By the method of node voltages,

$$\frac{V_S + V_{b'e}}{R_S} + g_m V_{b'e} + V_{b'e}(sC_\pi + G_E + g_\pi) = 0 \tag{2}$$

Solving (2) for $V_{b'e}$ and substituting the result into (1) yield

$$V_{Th} = \frac{g_m V_S}{sC_\mu[1 + R_S g_m + R_S(sC_\pi + G_E + g_\pi)]} \tag{3}$$

Deactivating (shorting) V_S also shorts E to B'. Consequently, $V_{b'e} = 0$, the dependent current source is open-circuited, and $Z_{Th} = 1/sC_\mu$.

Now, the Thévenin equivalent and voltage division lead to

$$V_L = \frac{R_C \| R_L}{R_C \| R_L + Z_{Th}} V_{Th} \tag{4}$$

Substitution of (3) into (4) and rearrangement give the desired voltage-gain ratio

$$A_v = \frac{V_L}{V_S} = \frac{g_m R_C \| R_L}{[sC_\mu(R_C \| R_L) + 1][sC_\pi R_S + R_S(g_m + g_\pi + G_E) + 1]} \tag{5}$$

10.67 Describe the high-frequency behavior of the CB amplifier of Problem 10.66.

▌ Since (5) of Problem 10.66 involves the upper-frequency range, it describes the amplifier as a low-pass (midfrequency) filter with break frequencies at

$$\omega_1 = \frac{1}{C_\mu(R_C \| R_L)} \quad \text{and} \quad \omega_2 = \frac{R_S(g_m + g_\pi + G_E) + 1}{C_\pi R_S} \tag{1}$$

10.68 Draw a small-signal equivalent circuit suitable for high-frequency analysis of the EF amplifier of Fig. 7-3(a).

▌ See Fig. 10-21.

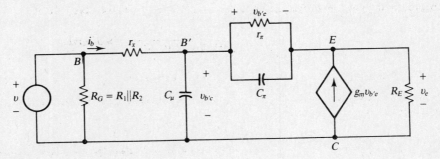

Fig. 10-21

10.69 Derive an expression for the high-frequency voltage-gain ratio V_e/V of the EF amplifier of Fig. 7-3(a).

▌ Referring to the pertinent equivalent circuit of Fig. 10-21, define

$$Z_\pi = r_\pi \left\| \left(\frac{1}{sC_\pi}\right) = \frac{r_\pi}{sr_\pi C_\pi + 1} \tag{1}$$

By the method of node voltages at node B',

$$\frac{V_{b'c} - V}{r_x} + \frac{V_{b'c}}{1/sC_\mu} + \frac{V_{b'c} - V_e}{Z_\pi} = 0 \tag{2}$$

Use of (1) in (2) and rearrangement give

$$-r_\pi V + [r_\pi(1 + sr_x C_\mu) + r_x(1 + sr_\pi C_\pi)]V_{b'c} = r_x(1 + sr_\pi C_\pi)V_e \tag{3}$$

At node E,

$$\frac{V_e - V_{b'c}}{Z_\pi} - g_m V_{b'e} + \frac{V_e}{R_E} = 0 \tag{4}$$

Substitute (1) into (4), use $V_{b'e} = V_{b'c} - V_e$, and rearrange to find

$$V_{b'c} = 1 + \frac{r_\pi}{R_E(1 + g_m r_\pi + sr_\pi C_\pi)} \approx 1 + \frac{1}{g_m R_E(1 + sC_\pi/g_m)} \tag{5}$$

where use has been made of the fact that typically $g_m r_\pi \gg 1$. Substitution of (5) into (3) and rearrangement lead to

$$g_m r_\pi R_E V = \left[\frac{r_\pi(g_m R_E + 1 + sR_E C_\pi)(1 + sr_x C_\mu) + r_x(1 + sr_\pi C_\pi)}{1 + sC_\pi/g_m} \right] V_e \tag{6}$$

For typical values, $g_m R_E \gg 1$. Use of this approximation in (6) and solution for the voltage-gain ratio yield

$$A_v(s) = \frac{V_e}{V} = \frac{1 + sC_\pi/g_m}{(1 + sC_\pi/g_m)(1 + sr_x C_\mu) + \dfrac{r_x}{g_m r_\pi R_E}(1 + sr_\pi C_\pi)} \tag{7}$$

10.70 Find an expression for the midfrequency voltage-gain ratio of the EF amplifier of Fig. 7-3(a) in terms of the hybrid-π parameters.

⬛ The midfrequency gain follows from (7) of Problem 10.69 by setting $s = 0$:

$$A_{v\,\text{mid}} = A_v(0) = \frac{1}{1 + \dfrac{r_x}{g_m r_\pi R_E}} = \frac{g_m r_\pi R_E}{g_m r_\pi R_E + r_x} \tag{1}$$

10.71 If the EF amplifier of Fig. 7-3(a) is described by $R_E = 10\,\text{k}\Omega$, $g_m = 0.03\,\text{S}$, $r_\pi = 5\,\text{k}\Omega$, $r_x = 25\,\Omega$, $C_\pi = 25\,\text{pF}$, and $C_\mu = 4\,\text{pF}$, evaluate the midfrequency gain.

⬛ By (1) of Problem 10.70,

$$A_{v\,\text{mid}} = \frac{0.03(5 \times 10^3)(10 \times 10^3)}{0.03(5 \times 10^3)(10 \times 10^3) + 25} = 0.99998 \approx 1$$

10.72 Determine the high-frequency break point for the EF amplifier of Fig. 7-3(a) if the parameters of Problem 10.71 apply.

⬛ Equation (7) of Problem 10.69 can be written as

$$A_v(s) = \frac{1 + \tau_2 s}{(1 + \tau_2 s)(1 + \tau_3 s) + K(1 + \tau_1 s)} \tag{1}$$

where

$$\tau_1 = r_\pi C_\pi = (5 \times 10^3)(25 \times 10^{-12}) = 1.25 \times 10^{-7} \quad \text{or} \quad \omega_1 = \frac{1}{\tau_1} = 8 \times 10^6 \text{ rad/s}$$

$$\tau_2 = \frac{C_\pi}{g_m} = \frac{25 \times 10^{-12}}{0.03} = 8.333 \times 10^{-10} \quad \text{or} \quad \omega_2 = \frac{1}{\tau_2} = 1.2 \times 10^9 \text{ rad/s}$$

$$\tau_3 = r_x C_\mu = 25(4 \times 10^{-12}) = 1 \times 10^{-10} \quad \text{or} \quad \omega_3 = \frac{1}{\tau_3} = 1 \times 10^{10} \text{ rad/s}$$

$$K = \frac{r_x}{g_m r_\pi R_E} = \frac{25}{0.03(5 \times 10^3)(10 \times 10^3)} = 1.6667 \times 10^{-5}$$

Since K is small, (1) can be approximated as

$$A_v(s) = \frac{1}{1 + \tau_3 s} \tag{2}$$

Whence, the high-frequency break point is

$$f_H = \frac{\omega_H}{2\pi} = \frac{\omega_3}{2\pi} = \frac{1 \times 10^{10}}{2\pi} = 1.59 \times 10^9 \text{ Hz}$$

This value of f_H may be the cutoff frequency of the transistor which is commonly the case for an EF amplifier.

FREQUENCY EFFECTS IN AMPLIFIERS □ 303

10.73 Determine the gain ratio for the EF amplifier of Fig. 7-3(a) at a frequency of 2×10^9 Hz if the parameters of Problem 10.71 are pertinent.

▌ Replace s by $j\omega = j2\pi(2 \times 10^9) = j12.566 \times 10^9$ rad/s in (7) of Problem 10.69.

$$M = |A(j\omega)| \approx \frac{1}{|1 + j\omega\tau_3|} = \frac{1}{|1 + j(12.566 \times 10^9)(1 \times 10^{-10})|} = \frac{1}{|1 + j1.2566|} = 0.623$$

10.74 The small-signal high-frequency model for the FET is an extension of the midfrequency model of Fig. 8-1(a). Three capacitors are added: C_{gs} between gate and source, C_{gd} between gate and drain, and C_{ds} between drain and source. They are all of the same order of magnitude—typically 1 to 10 pF. Draw a circuit of the described model.

▌ Figure 10-22 shows the small-signal high-frequency model based on the current-source model of Fig. 8-1(a). Another model, based on the voltage-source model of Fig. 8-1(b), can also be drawn.

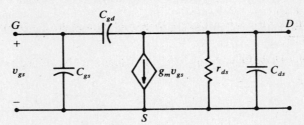

Fig. 10-22 High-frequency small-signal current-source FET model.

10.75 Draw a high-frequency small-signal equivalent circuit for the CS JFET amplifier of Fig. 5-10(b).

▌ Based on the high-frequency FET model of Fig. 10-22 and replacing bypass and coupling capacitors by short circuits lead to the network of Fig. 10-23.

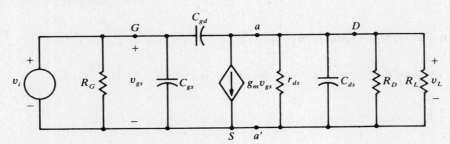

Fig. 10-23

10.76 For the JFET amplifier of Fig. 5-10(b), find an expression for the high-frequency voltage-gain ratio $A_v(s) = V_L/V_i$.

▌ We first find a Thévenin equivalent for the network to the left of terminal pair a, a' in Fig. 10-23. Noting that $v_{gs} = v_i$, we see that the open-circuit voltage is given by

$$V_{Th} = V_i - \frac{g_m}{sC_{gd}} V_i = \frac{sC_{gd} - g_m}{sC_{gd}} V_i \tag{1}$$

If V_i is deactivated, $V_i = V_{gs} = 0$ and the dependent current source is zero (open-circuited). A driving-point source connected to a, a' sees only

$$Z_{Th} = \frac{V_{dp}}{I_{dp}} = \frac{1}{sC_{gd}} \tag{2}$$

Now, with the Thévenin equivalent in place, voltage division leads to

$$V_L = \frac{Z_{eq}}{Z_{eq} + Z_{Th}} V_{Th} = \frac{1}{1 + Z_{Th}/Z_{eq}} \frac{sC_{gd} - g_m}{sC_{gd}} V_i \tag{3}$$

where

$$\frac{1}{Z_{eq}} = Y_{eq} = sC_{ds} + \frac{1}{r_{ds}} + \frac{1}{R_D} + \frac{1}{R_L} = sC_{ds} + g_{ds} + G_D + G_L \tag{4}$$

Rearranging (3) and using (4), we get

$$A_v(s) = \frac{V_L}{V_i} = \frac{sC_{gd} - g_m}{s(C_{ds} + C_{gd}) + g_{ds} + G_D + G_L} \quad (5)$$

10.77 Determine the high-frequency cutoff point for the JFET amplifier of Problem 10.76.

▮ From (5) of Problem 10.76, the high-frequency cutoff point is obviously

$$\omega_H = \frac{g_{ds} + G_D + G_L}{C_{ds} + C_{gd}} \quad (1)$$

Note that the high-frequency cutoff point is independent of C_{gs} as long as the source internal impedance is negligible.

10.78 High-frequency models of transistors characteristically include a capacitor path from input to output, modeled as admittance Y_F in the two-port network of Fig. 10-24(a). This added conduction path generally increases the difficulty of analysis. Establish a procedure to replace the input-to-output connected admittance Y_F with equivalent shunt elements.

▮ Referring to Fig. 10-24(a) and using KCL, we have

$$Y_{in} = \frac{I_i}{V_1} = \frac{I_1 + I_F}{V_1} \quad (1)$$

But
$$I_F = (V_1 - V_2)Y_F \quad (2)$$

Substitution of (2) into (1) gives

$$Y_{in} = \frac{I_1}{V_1} + \frac{(V_1 - V_2)Y_F}{V_1} = Y_1 + (1 - K_F)Y_F \quad (3)$$

where $K_F = V_2/V_1$ is obviously the forward voltage-gain ratio of the amplifier.

In a similar manner,

$$Y_o = \frac{-I_o}{V_2} = \frac{-(I_2 + I_F)}{V_2} \quad (4)$$

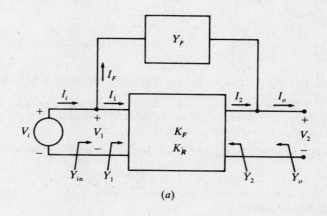

(a)

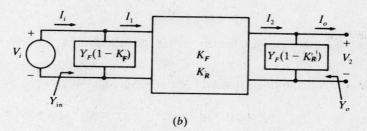

(b)

Fig. 10-24

FREQUENCY EFFECTS IN AMPLIFIERS □ 305

and the use of (2) in (4) gives us

$$Y_o = -\left(\frac{I_2}{V_2} + \frac{V_1 - V_2}{V_2} Y_F\right) = -[-Y_2 + (K_R - 1)Y_F] = Y_2 + (1 - K_R)Y_F \quad (5)$$

where $K_R = V_1/V_2$ is the reverse voltage-gain ratio of the amplifier.

Equations (3) and (5) suggest that the feedback admittance Y_F can be replaced with two shunt-connected admittances as shown in Fig. 10-24(b). When this two-port network is used to model an amplifier, the voltage gain K_F frequently turns out to have a large negative value, so that $(1 - K_F)Y_F \approx |K_F| Y_F$. Hence, a small feedback capacitance appears as a large shunt capacitance (called the *Miller capacitance*). On the other hand, K_R is typically small, so that $(1 - K_R)Y_F \approx Y_F$.

10.79 Apply the results of Problem 10.78 to the small-signal equivalent circuit of Fig. 10-18(a) to determine the Miller capacitance.

▌ First, the gain K_F must be found with capacitor C_μ and load resistor R_L removed. Since

$$V_L = -g_m V_{b'e} R_C$$

the desired gain is

$$K = \frac{V_L}{V_{b'e}} = -g_m R_C \quad (1)$$

The Miller capacitance C_M is the input shunt capacitance suggested by (3) of Problem 10.78:

$$C_M = (1 - K)\frac{Y_F}{s} = (1 + g_m R_C)C_\mu \quad (2)$$

since comparison of Figs. 10-18(a) and 10-24(a) shows that C_μ forms a feedback path analogous to Y_F.

10.80 Using the Miller capacitance determined in Problem 10.79, draw the equivalent circuit of Fig. 10-18(a) associated with Fig. 10-24(b).

▌ The output shunt capacitance, as suggested by (5) of Problem 10.78, must also be determined. Since $h_{re} = 0$ underlies the hybrid-π model, the reverse voltage-gain ratio $K_R = 0$; hence,

$$Y_o = Y_2 + (1 - K_R)Y_F = Y_2 + Y_F = Y_2 + sC_\mu \quad (1)$$

Comparison of Fig. 10-18(a) with Fig. 10-24(b) and the use of (2) of Problem 10.79 and (1) lead to the equivalent circuit of Fig. 10-25.

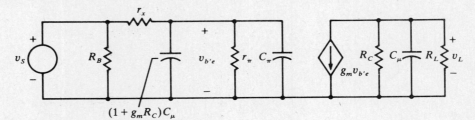

Fig. 10-25

10.81 Determine the high-frequency voltage-gain ratio for the CE amplifier modeled by Fig. 10-25.

▌ Let

$$C_{eq} = C_M + C_\pi = (1 + g_m R_C)C_\mu + C_\pi$$

Then, by voltage division,

$$V_{b'e} = \frac{r_\pi/(sr_\pi C_{eq} + 1)}{r_x + r_\pi/(sr_\pi C_{eq} + 1)} V_s = \frac{r_\pi/(r_x + r_\pi)}{s(r_x \| r_\pi)C_{eq} + 1} V_s \quad (1)$$

and by Ohm's law,

$$V_L = -\frac{R_C \| R_L}{s(R_C \| R_L)C_\mu + 1} g_m V_{b'e} \quad (2)$$

Substitution of (1) into (2) and rearrangement yield the desired voltage-gain ratio:

$$A_v(s) = \frac{V_L}{V_s} = -\frac{g_m(R_C \| R_L)r_\pi/(r_x + r_\pi)}{[s(R_C \| R_L)C_\mu + 1][s(r_x \| r_\pi)C_{eq} + 1]} \quad (3)$$

10.82 Apply the results of Problem 10.78 to the small-signal equivalent circuit of Fig. 10-23 to determine the Miller admittance.

▌ With load resistor R_L and feedback capacitor C_{gd} removed from the circuit of Fig. 8-23, the forward gain K_F follows from an application of Ohm's law:

$$K_F = \frac{V_L}{V_{gs}} = -\frac{g_m(r_{ds} \| R_D)}{s(r_{ds} \| R_D)C_{ds} + 1} \tag{1}$$

The Miller admittance suggested by (3) of Problem 10.78 is

$$Y_M = (1 - K)Y_F = \left[1 + \frac{g_m(r_{ds} \| R_D)}{s(r_{ds} \| R_D)C_{ds} + 1}\right]sC_{gd} \tag{2}$$

In the frequency range of interest and for typical values of r_{ds}, R_D, and C_{ds}, generally $|s(r_{ds} \| R_D)C_{ds}| \ll 1$; thus, the Miller admittance can be synthesized as a capacitor with value

$$C_M = \frac{Y_M}{s} = [1 + g_m(r_{ds} \| R_D)]C_{gd} \tag{3}$$

10.83 Using the Miller admittance determined by Problem 10.82, draw the equivalent circuit of Fig. 10-23 associated with Fig. 10-24(b).

▌ Since there is no feedback of output voltage to the input network of Fig. 10-23, $K_R = 0$. Hence, the output shunt admittance, as suggested by (5) of Problem 10.78, is simply

$$(1 - K_R)Y_F = Y_F = sC_{gd} \tag{1}$$

Comparison of Fig. 10-23 with Fig. 10-24(b) and the use of (3) from Problem 10.82 and (1) lead to the equivalent circuit of Fig. 10-26.

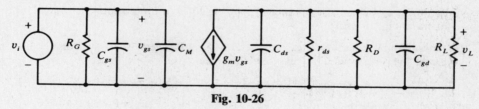

Fig. 10-26

10.84 Derive an expression for the high-frequency voltage-gain ratio of the JFET amplifier modeled by Fig. 10-26.

▌ By Ohm's law,

$$V_L = -\frac{g_m V_{gs}}{s(C_{ds} + C_{gd}) + g_{ds} + G_D + G_L} \tag{1}$$

Since $V_{gs} = V_i$, the required voltage-gain ratio follows as

$$A_v(s) = \frac{V_L}{V_i} = -\frac{g_m}{s(C_{ds} + C_{gd}) + g_{ds} + G_D + G_L} \tag{2}$$

As long as the source resistance is negligible, A_v is independent of C_M. (See Problem 10.85.)

Review Problems

10.85 In Problem 10.84, the gain of the FET amplifier does not depend on the Miller capacitance C_M; however, the situation changes if the source resistance is nonzero. Add a source resistance R_i to Fig. 10-26, and find an expression for the voltage-gain ratio V_L/V_i.

▌ With R_i added to Fig. 10-26, voltage division at the input network gives

$$V_{gs} = \frac{R_G \left\|\left[\dfrac{1}{s(C_{gs} + C_M)}\right]\right.}{R_i + R_G \left\|\left[\dfrac{1}{s(C_{gs} + C_M)}\right]\right.} V_i = \frac{R_G V_i}{R_i[sR_G(C_{gs} + C_M) + 1] + R_G} \tag{1}$$

FREQUENCY EFFECTS IN AMPLIFIERS □ 307

Ohm's law at the output network requires that

$$V_L = -\frac{g_m V_{gs}}{s(C_{ds} + C_{gd}) + g_{ds} + G_D + G_L} \qquad (2)$$

Substitute (1) into (2) and rearrange to find

$$A_v(s) = \frac{V_L}{V_i} = \frac{-g_m R_G}{[s(C_{ds} + C_{gd}) + g_{ds} + G_D + G_L][sR_i R_G (C_{gs} + C_M) + R_i + R_G]} \qquad (3)$$

10.86 Evaluate the gain or frequency transfer function of the CS JFET amplifier modeled by Fig. 10-26 for $R_i = 0$ and for $R_i = 100\ \Omega$ if $C_{gs} = 3$ pF, $C_{ds} = 1$ pF, $C_{gd} = 2.7$ pF, $r_{ds} = 50$ kΩ, $g_m = 0.016$ S, $R_L = R_D = 2$ kΩ, $R_G = 1$ MΩ, and $f = 50$ MHz.

▎ $s = j\omega = j2\pi f = j2\pi(50 \times 10^6) = j\pi \times 10^8$ rad/s

If $R_i = 0$, (3) of Problem 10.85 becomes

$$A_v(s) = \frac{-g_m}{s(C_{ds} + C_{gd}) + g_{ds} + G_D + G_L} \qquad (1)$$

Whence,

$$A_v(j\omega) = \frac{-0.016}{(j\pi \times 10^8)(1 \times 10^{-12} + 2.7 \times 10^{-12}) + 1/(50 \times 10^3) + 1/(2 \times 10^3) + 1/(2 \times 10^3)} = 10.348\ \underline{/131.53°}$$

If $R_i \neq 0$, by (3) of Problem 10.82,

$$C_M = \left[1 + \frac{g_m r_{ds} R_D}{r_{ds} + R_D}\right] C_{gd} = \left[1 + \frac{0.016(50 \times 10^3)(2 \times 10^3)}{50 \times 10^3 + 2 \times 10^3}\right](2.7 \times 10^{-12}) = 85.77\ \text{pF}$$

From (3) of Problem 10.85 and using the result above for $R_i = 0$,

$$A_v(j\omega) = \frac{A_v(j\omega)|_{R_i=0}(R_G)}{sR_i R_G (C_{gs} + C_M) + R_i + R_G}$$

$$= \frac{10.348\ \underline{/131.53°}(1 \times 10^6)}{(j\pi \times 10^8)(100)(1 \times 10^6)(3 \times 10^{-12} + 85.77 \times 10^{-12}) + 100 + 1 \times 10^6} = 3.49\ \underline{/61.26°}$$

10.87 Consider the high-pass filter circuit of Fig. 10-5(a). Show that as ω becomes large, the amplitude ratio M_{db} actually approaches $20 \log [R_L/(R_L + R_S)]$ as indicated in Fig. 10-5(b).

▎ By (2) of Problem 10.18 and Definition 2 of Problem 10.4,

$$M(j\omega) = \left|\frac{j\omega R_L C}{1 + j\omega C(R_L + R_S)}\right| \qquad (1)$$

Let $\omega \to \infty$ in (1) to find

$$M(j\infty) = \frac{R_L}{R_L + R_S} \qquad (2)$$

Apply Definition 3 of Problem 10.4 to (2) to yield

$$M_{db} = 20 \log M(j\infty) = 20 \log [R_L/(R_L + R_S)] \qquad (3)$$

10.88 For the high-pass filter circuit of Fig. 10-5(a), show that $M^2(j\omega_L)$, where $\omega_L = 1/C(R_L + R_S)$, has the value $\tfrac{1}{2}M^2(j\infty) = \tfrac{1}{2}[R_L/(R_L + R_S)]^2$.

▎ From (1) of Problem 10.87,

$$M^2(j\omega) = \left|\frac{j\omega R_L C}{1 + j\omega C(R_L + R_S)}\right|^2 = \frac{\omega^2 (R_L C)^2}{1 + \omega^2 C^2 (R_L + R_S)^2} \qquad (1)$$

Setting $\omega = \omega_L = 1/C(R_L + R_S)$ in (1) yields, after simplification,

$$M^2(j\omega_L) = \frac{1}{2}\left[\frac{R_L}{R_L + R_S}\right]^2 \qquad (2)$$

Comparing (2) with (2) of Problem 10.87 leads to the conclusion that $M^2(j\omega_L) = \tfrac{1}{2} M^2(j\infty)$.

10.89 In the high-pass filter circuit of Fig. 10-5(a), the source impedance $R_S = 5$ kΩ. If the circuit is to have a high-frequency gain of 0.75 and a break or cutoff frequency of 100 rad/s, size R_L and C.

▌ For a high-frequency gain of 0.75, (2) of Problem 10.87 yields

$$M(j\infty) = 0.75 = \frac{R_L}{R_L + R_S} = \frac{R_L}{R_L + 5 \times 10^3} \quad (1)$$

Whence, $R_L = 15$ kΩ. It is evident from Problem 10.88 that ω_L is the break frequency; hence,

$$C = \frac{1}{\omega_L(R_L + R_S)} = \frac{1}{(100)(15 \times 10^3 \times 5 \times 10^3)} = 0.5 \, \mu\text{F}$$

10.90 In the circuit of Fig. 4-18, replace V_S with a sinusoidal source to give the small-signal circuit of Fig. 10-9. If the impedance of the coupling capacitor is not negligible, find the current-gain ratio $A_i(s) = I_c/I_s$.

▌ By current division,

$$I_b = \frac{R_B}{R_B + h_{ie}} \quad (1)$$

But
$$I_c = h_{fe} I_b \quad (2)$$

Hence, substitution of (1) into (2) yields

$$A_i(s) = \frac{I_c}{I_b} = \frac{h_{fe} R_B}{R_B + h_{ie}} \quad (3)$$

10.91 Determine the low-frequency cutoff point for the current gain of the amplifier of Problem 10.90.

▌ Since (3) of Problem 10.90 shows the current gain $A_i(s)$ to be independent of s (or frequency), the gain is independent of frequency down to $f = 0$.

10.92 Show that the RC network of Fig. 10-27 is a high-pass filter. Determine its low-frequency cutoff point.

▌ By voltage division,

$$V_o = \frac{R_2 + R_3 \left\| \left(\frac{1}{sC_3}\right)\right.}{R_1 + R_2 + R_3 \left\| \left(\frac{1}{sC_3}\right)\right.} V_S \quad (1)$$

Rearranging (1) leads to

$$\frac{V_o}{V_S} = \frac{R_2 + R_3}{R_1 + R_2 + R_3} \frac{s\left[\frac{R_2 R_3 C_3}{R_2 + R_3}\right] + 1}{s\left[\frac{(R_1 + R_2)R_3 C_3}{R_1 + R_2 + R_3}\right] + 1} \quad (2)$$

It can be seen that the transfer function given by (2) has a zero at $\omega_1 = (R_2 + R_3)/R_2 R_3 C_3$ and a pole at $\omega_2 = (R_1 + R_2 + R_3)/(R_1 + R_2)R_3 C_3$. Since $\omega_2 > \omega_1$, its low-frequency cutoff point is $\omega_L = \omega_2$ if the parameters are chosen so that $\omega_2 \geq 10\omega_1$.

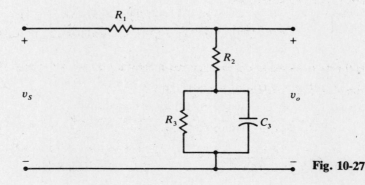

Fig. 10-27

10.93 The amplifier of Fig. 4-14 is modeled for small-signal operation by Fig. 10-7 where C_C is assumed to be infinitely large. Let $C_E = 100\ \mu\text{F}$, $R_E = 100\ \Omega$, $R_C = R_L = 2\ \text{k}\Omega$, $h_{ie} = 200\ \Omega$, and $h_{fe} = 75$. Determine the low-frequency voltage gain.

▌ By (1) of Problem 10.24,

$$A_v(0) = \frac{V_L}{V_i} = \frac{-(75)(2 \times 10^3)(2 \times 10^3)}{(2 \times 10^3 + 2 \times 10^3)[200 + (75 + 1)(100)]} = -9.62$$

10.94 Evaluate the midfrequency voltage gain for the amplifier of Problem 10.93.

▌ Based on (1) of Problem 10.25,

$$A_{v\ \text{mid}} = A_v(\infty) = \frac{V_L}{V_i} = \frac{-(75)(2 \times 10^3)(2 \times 10^3)}{(200)(2 \times 10^3 + 2 \times 10^3)} = -375$$

10.95 Determine the voltage-gain low-frequency cutoff point for the amplifier of Problem 10.93.

▌ Equations (2) and (4) of Problem 10.26 are applicable since $h_{fe}R_E \gg h_{ie}$. Hence,

$$\omega_1 = \frac{1}{C_E R_E} = \frac{1}{(100 \times 10^{-6})(100)} = 100\ \text{rad/s}$$

$$\omega_2 = \frac{h_{ie} + (h_{fe} + 1)R_E}{R_E C_E h_{ie}} = \frac{200 + (75 + 1)(100)}{100(200 \times 10^{-6})(200)} = 3750\ \text{rad/s}$$

Since $\omega_2 > 10\omega_1$,

$$\omega_L = \omega_2 = 3750\ \text{rad/s}$$

10.96 For the amplifier of Fig. 4-14, let $C_C \to \infty$. Show that if the source internal impedance R_i is not negligible, but $R_i \ll R_B = R_1 \| R_2$, then the low-frequency cutoff point is given by

$$\omega_L = \frac{h_{ie} + R_i + (h_{fe} + 1)R_E}{R_E C_E (h_{ie} + R_i)}$$

▌ With R_i placed in the applicable small-signal equivalent circuit of Fig. 10-7 and since $R_i \ll R_B$, a Thévenin equivalent circuit to the left of terminals A, G yields $v_{Th} = v_i R_B/(R_B + R_i) \approx v_i$ and $R_{Th} = R_B \| R_i \approx R_i$. Hence, the results of Problems 10.24–10.26 are valid if h_{ie} were replaced by $h_{ie} + R_i$. The desired result follows from such replacement of h_{ie} in (3) of Problem 10.26.

10.97 In the amplifier of Fig. 4-14, let $C_C \to \infty$, $C_E = 100\ \mu\text{F}$, $R_E = 20\ \text{k}\Omega$, $h_{ie} = 100\ \Omega$, $h_{fe} = 75$, $R_C = R_L = 2\ \text{k}\Omega$, $R_1 = 2\ \text{k}\Omega$, and $R_2 = 20\ \text{k}\Omega$. Determine the low-frequency current gain.

▌ Now

$$R_B = R_1 \| R_2 = \frac{R_1 R_2}{R_1 + R_2} = \frac{(2 \times 10^3)(20 \times 10^3)}{2 \times 10^3 + 20 \times 10^3} = 1.818\ \text{k}\Omega$$

Then, by (1) of Problem 10.33,

$$A_i(0) = \frac{I_L}{I_i} = \frac{-75(2 \times 10^3)(1.818 \times 10^3)}{(2 \times 10^3 + 2 \times 10^3)(20 \times 10^3 + 1.818 \times 10^3 + 100)} = -3.11$$

10.98 Evaluate the midfrequency current gain for the amplifier of Problem 10.97.

▌ Based on (1) of Problem 10.34,

$$A_{i\ \text{mid}} = A_i(\infty) = \frac{I_L}{I_i} = \frac{-75(2 \times 10^3)(1.818 \times 10^3)}{(2 \times 10^3 + 2 \times 10^3)(1.818 \times 10^3 + 100)} = -35.54$$

10.99 Determine the current-gain low-frequency cutoff point for the amplifier of Problem 10.97.

▌ Equation set (1) of Problem 10.35 is applicable.

$$\omega_1 = \frac{1}{R_E C_E} = \frac{1}{(20 \times 10^3)(100 \times 10^{-6})} = 0.5\ \text{rad/s}$$

$$\omega_2 = \frac{R_E + R_B + h_{ie}}{R_E C_E (R_B + h_{ie})} = \frac{20 \times 10^3 + 1.818 \times 10^3 + 100}{(20 \times 10^3)(100 \times 10^{-6})(1.818 \times 10^3 + 100)} = 5.71\ \text{rad/s}$$

Since $\omega_2 > 10\omega_1$,
$$\omega_L = \omega_2 = 5.71 \text{ rad/s}$$

10.100 In the amplifier of Problem 10.36, let $R_i = 500\ \Omega$ and all else remain unchanged. Determine the value of the emitter bypass capacitor required to ensure that $f_L \leq 200$ Hz. Compare your result with that of Problem 10.36 to see that consideration of the source internal impedance allows the use of a smaller bypass capacitor.

▌ Applying the result of Problem 10.96,

$$C_E \geq \frac{h_{ie} + R_i + (h_{fe} + 1)R_E}{\omega_L R_E(h_{ie} + R_i)} = \frac{300 + 500 + (100 + 1)(1 \times 10^3)}{2\pi(200)(1 \times 10^3)(300 + 500)} = 101.3\ \mu\text{F}$$

Obviously, the above value of C_E is less than 40% of the value from Problem 10.36 wherein $R_i = 0$.

10.101 In the amplifier of Fig. 4-14, $C_C \to \infty$, $R_i = 500\ \Omega$, $R_E = 30\ \text{k}\Omega$, $R_1 = 3.2\ \text{k}\Omega$, $R_2 = 17\ \text{k}\Omega$, $R_L = 10\ \text{k}\Omega$, $h_{oe} = h_{re} = 0$, $h_{fe} = 100$, and $h_{ie} = 100\ \Omega$. Determine R_C and C_E so that the amplifier has a midfrequency current-gain ratio $|A_i| \geq 30$ with low-frequency cutoff $f_L \geq 20$ Hz.

▌
$$R_B = R_1 \parallel R_2 = \frac{(3.2 \times 10^3)(17 \times 10^3)}{3.2 \times 10^3 + 17 \times 10^3} = 2.68\ \text{k}\Omega$$

Based on (1) of Problem 10.34,

$$R_C \geq \frac{A_{i\ \text{mid}}R_L(h_{ie} + R_B)}{h_{fe}R_B - A_{i\ \text{mid}}(h_{ie} + R_B)} = \frac{30(10 \times 10^3)(100 + 2.68 \times 10^3)}{100(2.68 \times 10^3) - (30)(100 + 2.68 \times 10^3)}$$

$$R_C \geq 4.52\ \text{k}\Omega$$

By (1) of Problem 10.35,

$$C_E \geq \frac{R_E + R_B + h_{ie}}{\omega_L R_E(R_B + h_{ie})} = \frac{30 \times 10^3 + 2.68 \times 10^3 + 100}{2\pi(20)(30 \times 10^3)(2.68 \times 10^3 + 100)} = 3.13\ \mu\text{F}$$

Since $\omega_2 = 2\pi(20) = 125.7$ rad/s $> 10\omega_1 = 10/R_E C_E = 10/(30 \times 10^3)(3.13 \times 10^{-6}) = 106.5$ rad/s, $w_2 = \omega_L$ and the above analysis for C_E holds.

10.102 In the CE amplifier of Fig. 4-14, let $C_C \to \infty$, $C_E = 100\ \mu\text{F}$, $R_E = 100\ \Omega$, $R_i = 0$, $R_B = 5\ \text{k}\Omega$, $R_C = R_L = 2\ \text{k}\Omega$, $h_{oe} = h_{re} = 0$, $h_{fe} = 75$, and $h_{ie} = 1\ \text{k}\Omega$. The small-signal ac equivalent circuit is given by Fig. 10-7. If a sinusoidal signal $v_i = V_M \sin \omega t$ is impressed (with $\omega = 400$ rad/s), determine the phase angle between v_i and i_i.

▌ The phase angle between v_i and i_i is given by the angle of Z_{in}. From (2) of Problem 10.48 with $s = j\omega = j400$,

$$\frac{Z_{\text{in}}}{R_B} = \frac{j400(1 \times 10^3)(100)(100 \times 10^{-6}) + 1 \times 10^3 + (75 + 1)(100)}{j400(100)(100 \times 10^{-6})(5 \times 10^3 + 1 \times 10^3) + 5 \times 10^3 + 1 \times 10^3 + (75 + 1)(100)}$$

$$\frac{Z_{\text{in}}}{5 \times 10^3} = \frac{8600 + j4000}{13600 + j24000} = \frac{9.485 \times 10^3 \underline{/24.94°}}{27.586 \times 10^3 \underline{/60.46°}} = 0.344\ \underline{/-35.52°}$$

$$Z_{\text{in}} = 1.72 \times 10^3\ \underline{/-35.52°}$$

Whence, current i_i leads voltage v_i by 35.52°.

10.103 Find the phase shift between input and output voltages for the amplifier of Problem 10.102.

▌ The phase angle between v_L and v_i is determined by the phase angle of the voltage-gain ratio of (5) from Problem 10.23 with $s = j\omega = j400$.

$$\frac{V_L}{V_i} \frac{(R_C + R_L)}{h_{fe}R_C R_L} = \frac{-[j400(100)(100 \times 10^{-6}) + 1]}{j400(100)(100 \times 10^{-6})(1 \times 10^3) + 1 \times 10^3 + (75 + 1)(100)}$$

$$= \frac{-1 - j4}{8600 + j4000} = \frac{4.123\ \underline{/-104.04°}}{9.485 \times 10^3\ \underline{/24.94°}} = 0.435 \times 10^{-3}\ \underline{/-128.98°}$$

Whence, output voltage v_L lags input voltage v_i by 128.98°.

FREQUENCY EFFECTS IN AMPLIFIERS □ 311

10.104 Determine the phase shift between the input and output currents for the amplifier of Problem 10.102.

▮ The phase angle between i_L and i_i is identically the phase angle of the current-gain ratio given by (5) of Problem 10.32 for $s = j\omega = j400$. Define $k_i = (R_C + R_L)(R_E + R_B + h_{ie})/(h_{fe}R_C R_B)$. Then, by (5) of Problem 10.32,

$$k_i \frac{I_L}{I_i} = \frac{-[j400(100)(100 \times 10^{-6}) + 1]}{j400 \dfrac{100(100 \times 10^{-6})(5 \times 10^3 + 1 \times 10^3)}{100 + 5 \times 10^3 + 1 \times 10^3} + 1}$$

$$= \frac{-1 - j4}{1 + j3.93} = \frac{4.123 \,\underline{/-104.04°}}{4.055 \,\underline{/75.72°}} = 1.017 \,\underline{/-179.76°}$$

Thus, output current i_L lags input current i_i by $179.76° \approx 180°$.

10.105 In the amplifier of Problem 10.50, let $C_E = 200\,\mu\text{F}$, $C_C = 10\,\mu\text{F}$, $R_E = 50\,\Omega$, $R_C = R_L = 2\,\text{k}\Omega$, $R_i = 100\,\Omega$, $h_{re} = h_{oe} = 0$, $h_{ie} = 1\,\text{k}\Omega$, and $h_{fe} = 50$. Sketch the asymptotic Bode plot (M_{db} only) for the voltage-gain ratio.

▮ Substitution of given values into (7) of Problem 10.50 yields

$$\frac{V_L}{V_i} = \frac{-\dfrac{50(2 \times 10^3)(2 \times 10^3)(10 \times 10^{-6})}{(50+1)(50) + 1 \times 10^3 + 100}s[s(50)(200 \times 10^{-6}) + 1]}{[s(10 \times 10^{-6})(2 \times 10^3 + 2 \times 10^3) + 1]\left[s\dfrac{(200 \times 10^{-6})(50)(100 + 1 \times 10^3)}{(50+1)(50) + 100 + 1 \times 10^3} + 1\right]}$$

$$= \frac{-0.548s(s/100 + 1)}{(s/25 + 1)(s/332 + 1)} \qquad (1)$$

The associated Bode plot is given by Fig. 10-28.

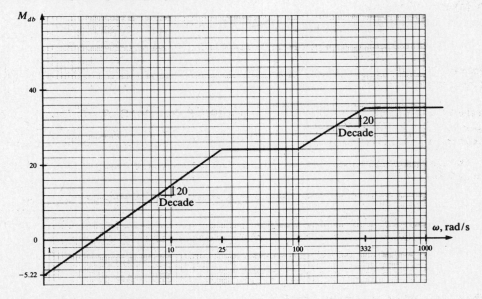

Fig. 10-28

10.106 Is the 3-db attenuation point for the amplifier of Problem 10.105 below 40 Hz?

▮ The midfrequency amplitude ratio follows from Definition 3 of Problem 10.4 and by use of (7) from Problem 10.50 as $s \to j\infty$. The result is

$$M_{db}(j\infty) = 20 \log \left|\frac{h_{fe}R_L R_C}{(R_C + R_L)(R_i + h_{ie})}\right| = 20 \log \left|\frac{50(2 \times 10^3)(2 \times 10^3)}{(2 \times 10^3 + 2 \times 10^3)(100 + 1 \times 10^3)}\right| = 33.15\,\text{db}$$

For the amplitude ratio at 40 Hz, let $s = j\omega = j2\pi(40) = j80\pi$ in (1) of Problem 10.105 giving

$$M_{db}(j80\pi) = 20 \log \left|\frac{-0.548(j80\pi)(j80\pi/100 + 1)}{(j80\pi/25 + 1)(j80\pi/332 + 1)}\right| = 29.36\,\text{db}$$

Now, $\qquad M_{db}(j\infty) - M_{db}(j80\pi) = 33.15 - 29.36 = 3.79\,\text{db}$

Since the amplitude ratio is decreased by more than 3 db over the midfrequency value for $f = 40$ Hz, the 3-db attenuation point lies at a frequency greater than 40 Hz.

10.107 In the CE amplifier of Problem 10.62, let $g_m = 0.035$ S, $r_\pi = 8$ kΩ, $r_x = 30$ Ω, $R_C = R_L = 10$ kΩ, $C_\pi = 10$ pF, and $C_\mu = 2$ pF. Determine the high-frequency cutoff point.

▎ From the voltage-gain ratio of (4) from Problem 10.62, it is apparent that the high-frequency break point is given by

$$f_H = \frac{R_C + R_L}{2\pi(1 - g_m)C_\mu R_C R_L} = \frac{10 \times 10^3 + 10 \times 10^3}{2\pi(1 - 0.035)(2 \times 10^{-12})(10 \times 10^3)(10 \times 10^3)} = 16.49 \text{ MHz}$$

10.108 Evaluate the midfrequency voltage gain for the amplifier of Problem 10.107.

▎ The midfrequency voltage gain follows from setting $s = 0$ in (4) of Problem 10.62:

$$A_{v \text{ mid}} = \frac{r_\pi}{r_\pi + r_x} \frac{R_C R_L(-g_m)}{R_C + R_L} = \frac{8 \times 10^3}{8 \times 10^3 + 30} \frac{(10 \times 10^3)(10 \times 10^3)(-0.035)}{10 \times 10^3 + 10 \times 10^3} = -174.3$$

10.109 In the CB amplifier of Problem 10.66, let $R_S = 100$ Ω, $R_E = 1$ kΩ, $R_C = R_L = 10$ kΩ, $C_\mu = 2$ pF, $C_\pi = 40$ pF, $g_m = 0.035$ S, and $r_\pi = 5$ kΩ. Determine the midfrequency gain.

▎ Setting $s = 0$ in (5) of Problem 10.66 yields

$$A_{v \text{ mid}} = \frac{g_m R_C R_L}{(R_C + R_L)[R_S(g_m + g_\pi + G_E) + 1]}$$

$$= \frac{0.035(10 \times 10^3)(10 \times 10^3)}{(10 \times 10^3 + 10 \times 10^3)[100(0.035 + 1/5 \times 10^3 + 1/1 \times 10^3) + 1]} = 37.88$$

10.110 Evaluate the high-frequency cutoff point for the CB amplifier described by Problem 10.109.

▎ By (1) of Problem 10.67,

$$f_1 = \frac{R_C + R_L}{2\pi C_\mu R_C R_L} = \frac{10 \times 10^3 + 10 \times 10^3}{2\pi(2 \times 10^{-12})(10 \times 10^3)(10 \times 10^3)} = 15.91 \text{ MHz}$$

CHAPTER 11
Audio-Frequency Power Amplifiers

DIRECT-COUPLED AMPLIFIERS

11.1 The analysis of amplifiers with regard to power centers on three quantities: the power supplied by the bias sources, the power delivered to the load, and the power dissipated by the transistor. Find general expressions (independent of particular amplifier configuration) for (a) the *supplied power* P_{CC}, (b) the *load power* P_L, and (c) the *collector power dissipation* P_C.

(a) The supplied power includes all dc power flowing from bias supplies to the amplifier circuit; however, the power dissipated by the emitter-base network is practically negligible (see Problem 11.93). Hence, supplied power is calculated as

$$P_{CC} = V_{CC} \frac{1}{T} \int_0^T i_C \, dt = V_{CC} i_{C\,\text{av}} = V_{CC} I_{CQ} \qquad (1)$$

where T is the period of the collector current i_C. The right-hand equality holds only if the presence of a signal does not alter the average value of i_C (that is, there is no distortion).

(b) Load power, the average value of the ac power delivered to the amplifier load, is given by

$$P_L = \frac{1}{T} \int_0^T i_L^2 R_L \, dt = \frac{I_{Lm}^2}{2} R_L \qquad (2)$$

The right-hand equality holds only if the load current has zero average value.

(c) The power dissipated by the transistor is the sum of the power delivered to the base-emitter port and that delivered to the collector-emitter port, but the former is negligible compared to the latter. Thus, practical analysis requires only calculation of the collector power dissipation, which is found as the difference

$$P_C = P_{CC} - (I^2 R \text{ losses}) - P_L \qquad (3)$$

11.2 The *efficiency* of an amplifier is defined as the percentage ratio of load power to supplied power. Based on the results of Problem 11.1, determine an expression for the efficiency of an amplifier.

$$\eta = \frac{\text{load power}}{\text{supplied power}} (100\%) = \frac{\frac{1}{T} \int_0^T i_L^2 R \, dt}{V_{CC} i_{C\,\text{av}}} (100\%) \qquad (1)$$

11.3 Although its collector characteristics involve considerably larger values of current, the class A direct-coupled power amplifier of Fig. 11-1(a) does not differ in operating principle from the small-signal, low-power amplifiers of Chapter 4. Maximum symmetrical swing (Problem 4.90) is universally implemented to yield maximum power output. Find an expression for the supplied power P_{CC} of the amplifier.

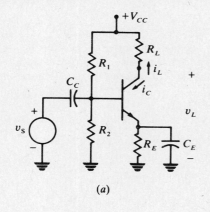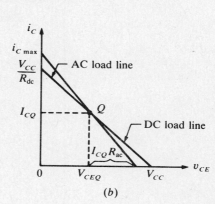

(a)   (b)   **Fig. 11-1**

We must find the Q point of Fig. 11-1(b) for maximum symmetrical swing. We first note that

$$R_{dc} = R_L + R_E \quad \text{and} \quad R_{ac} = R_L$$

Then, by (3) of Problem 4.93,

$$I_{CQ} = \frac{V_{CC}}{R_{ac} + R_{dc}} = \frac{V_{CC}}{2R_L + R_E} \tag{1}$$

The delivered power is then determined by (1) of Problem 11.1 as

$$P_{CC} = V_{CC} i_{C \text{ av}} = V_{CC} I_{CQ} = \frac{V_{CC}^2}{2R_L + R_E} \tag{2}$$

11.4 For the class A, direct-coupled power amplifier of Fig. 11-1, let $R_E = 0.5\ \Omega$, $R_L = 8\ \Omega$, $R_1 = 20\ \text{k}\Omega$, $R_2 = 1\ \text{k}\Omega$, and $V_{CC} = 24\ \text{V}$. Find the values of quiescent collector current and supplied power.

Based on (1) and (2) of Problem 11.3,

$$I_{CQ} = \frac{24}{2(8) + 0.5} = 1.454\ \text{mA}$$

$$P_{CC} = \frac{(24)^2}{2(8) + 0.5} = 34.896\ \text{W}$$

11.5 Determine an expression for the maximum value of load power.

The maximum power is delivered to the load by the maximum signal; hence, by (2) of Problem 11.1 and (1) of Problem 11.3,

$$P_{L\ \text{max}} = \frac{I_{CQ}^2}{2} R_L = \frac{V_{CC}^2 R_L}{(2R_L + R_E)^2} \tag{1}$$

11.6 Evaluate the maximum value of load power for the amplifier described by Problem 11.4.

By (1) of Problem 11.5,

$$P_{L\ \text{max}} = \frac{(24)^2 (8)}{[2(8) + 0.5]^2} = 8.456\ \text{W}$$

11.7 Neglecting the small losses of the bias network and emitter resistor, determine the range of collector power dissipation for the power amplifier of Problem 11.4.

The collector power dissipation is found by use of (3) of Problem 11.1. Maximum collector power occurs for $P_L = 0$, so max $(P_C) = P_{CC} = 34.896\ \text{W}$. Minimum collector power coincides with $P_L = P_{L\ \text{max}}$, so min $(P_C) = P_{CC} - P_{L\ \text{max}} = 34.896 - 8.456 = 26.44\ \text{W}$.

11.8 Find an expression for the maximum efficiency of the power amplifier in Fig. 11-1.

Based on (1) of Problem 11.2, (2) of Problem 11.1, and (1) of Problem 11.3,

$$\eta_{\text{max}} = \frac{\frac{1}{2} I_{CQ}^2 R_L}{V_{CC} I_{CQ}} (100\%) = \frac{I_{CQ} R_L}{2 V_{CC}} (100\%) = \frac{[V_{CC}/(2R_L + R_E)] R_L}{2 V_{CC}} (100\%)$$

$$= \frac{R_L}{4R_L + 2R_E} (100\%) \tag{1}$$

Equation (1) shows that if bias stability is sacrificed by allowing R_E to go to zero, then η_{max} has 25 percent as an upper limit. In power amplifier applications, such low efficiency means ineffective use of the transistor and presents significant cooling problems; it is the impetus for seeking alternative circuit designs for power amplifiers.

11.9 Evaluate the maximum value of efficiency for the amplifier of Problem 11.4.

The amplifier efficiency is at a maximum when maximum power is being delivered to the load. By (1) of Problem 11.8,

$$\eta_{\text{max}} = \frac{R_L}{4R_L + 2R_E} (100\%) = \frac{8}{4(8) + 2(0.5)} (100\%) = 24.2\%$$

11.10 In the amplifier of Fig. 4-14(a), assume that the power to R_1 and R_2 is negligible, but that the values of these resistances can be selected to place the Q point at any position on the ac load line. Find an expression for the maximum possible efficiency for this amplifier with the given values.

▌ Maximum efficiency occurs with maximum ac load current, so we assume R_1 and R_2 are selected for maximum symmetrical swing. Then, by (3) of Problem 4.93,

$$I_{CQ} = \frac{V_{CC}}{R_{ac} + R_{dc}} = \frac{V_{CC}}{R_C \| R_L + R_E + R_C} \tag{1}$$

The delivered power is found with (1):

$$P_{CC} = V_{CC} I_{CQ} = \frac{V_{CC}^2}{R_C \| R_L + R_E + R_C}$$

Typically $V_{CE\,sat} \approx 0$, and we assume $I_{CBO} = 0$, so that

$$I_{cm} = I_{CQ} = \frac{V_{CC}}{R_C \| R_L + R_E + R_C}$$

$$I_{Lm} = \frac{R_C}{R_L + R_C} I_{cm} = \frac{R_C V_{CC}}{(R_L + R_C)(R_C \| R_L + R_E + R_C)}$$

The maximum efficiency follows from (2) of Problem 11.1 and (1) of Problem 11.2:

$$\eta_{max} = \frac{P_{L\,max}}{P_{CC}}(100\%) = \frac{\frac{1}{2}\left[\frac{R_C V_{CC}}{(R_L + R_C)(R_C \| R_L + R_E + R_C)}\right]^2 R_L}{\frac{V_{CC}^2}{R_C \| R_L + R_E + R_C}}(100\%)$$

$$= \frac{R_C^2 R_L (100\%)}{2(R_L + R_C)^2 (R_C \| R_L + R_E + R_C)} \tag{2}$$

11.11 For the amplifier of Fig. 4-14, let $R_E = 470\,\Omega$, $R_C = 800\,\Omega$, and $R_L = 1.2\,k\Omega$. Determine the value of maximum efficiency possible from this amplifier.

▌
$$R_C \| R_L = \frac{800(1.2 \times 10^3)}{800 + 1.2 \times 10^3} = 480\,\Omega$$

By (2) of Problem 11.10,

$$\eta_{max} = \frac{(800)^2 (1.2 \times 10^3)(100\%)}{2(1.2 \times 10^3 + 800)^2 (480 + 470 + 800)} = 5.49\%$$

11.12 Add a large emitter bypass capacitor to the circuit of Fig. 4-5(a), and let $V_{CC} = 24$ V, $R_C = 5$ kΩ, and $R_E = 200\,\Omega$. Assume that R_1 and R_2 are selected for maximum symmetrical swing but dissipate negligible power. Also, let $V_{CE\,sat} \approx 0$ and $I_{CBO} = 0$. Find the greatest possible amplitude for the undistorted collector current. The output signal is taken from the common to the collector terminal.

▌
$$I_{cm} = I_{CQ} = \frac{1}{2}\frac{V_{CC}}{R_C + R_E} = \frac{1}{2}\frac{24}{5000 + 200} = 2.308\text{ mA} \tag{1}$$

11.13 For values of 0, 50, and 100 percent of the maximum value of undistorted collector current, determine the values of supplied power P_{CC} to the amplifier of Problem 11.12.

▌ The supplied power is independent of I_{cm} and is given by

$$P_{CC} = V_{CQ} I_{CQ} = 24(2.308 \times 10^{-3}) = 55.38\text{ mW}$$

11.14 For values of 0, 50, and 100 percent of the maximum value of undistorted collector current, determine the values of load power P_L for the amplifier of Problem 11.12.

▌ Based on (1) of Problem 11.12, calculations are to made for $I_{cm} = 0$, 1.154 and 2.308 mA. The average value of the ac power delivered to the load R_C is

$$P_L = \tfrac{1}{2} I_{cm}^2 R_C = 2500 I_{cm}^2\text{ W}$$

which, when evaluated at the three values of I_{cm}, gives 0, 3.33, and 13.32 mW.

11.15 For values of 0, 50, and 100 percent of the maximum value of undistorted collector current, determine the values of collector power dissipation for the amplifier of Problem 11.12.

▮ From (3) of Problem 11.1, the collector power dissipation is

$$P_C = P_{CC} - I_{CQ}^2(R_C + R_E) - P_L = 27.7 \times 10^{-3} - P_L \text{ W} \tag{1}$$

Based on the results for I_{cm} and P_L of Problem 11.14, (1) gives P_C values of 27.7, 24.37, and 14.38 mW.

11.16 For values of 0, 50, and 100 percent of the maximum value of undistorted collector current, calculate the resulting values of efficiency for the amplifier of Problem 11.12.

▮ Since $\eta = (P_L/P_{CC})(100\%)$, use of the results of Problems 11.13 and 11.14 gives the three efficiencies as 0, 6.01, and 24.04 percent.

11.17 In order to illustrate the variation of power and efficiency with collector (load) current, sketch P_{CC}, P_C, P_L and η for the amplifier of Problem 11.12.

▮ The sketch of Fig. 11-2 follows from values calculated in Problems 11.13–11.16.

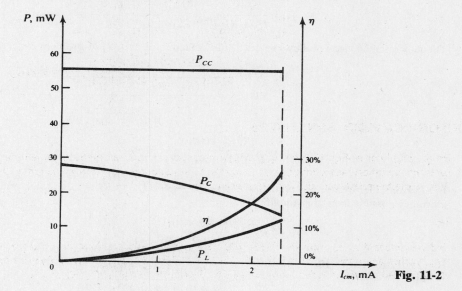

Fig. 11-2

11.18 Transistors are rated according to the power-dissipating capability $P_{C\,max}$ of the collector. A plot of $P_{C\,max} = v_{CE}i_C$ on the collector characteristics (i_C vs. v_{CE}) is called the *maximum-dissipation hyperbola*. Show that if a transistor is biased for maximum symmetrical swing, then the ac load line is tangent to the maximum-dissipation hyperbola at the Q point.

▮ From the maximum-dissipation equation,

$$i_C = \frac{P_{C\,max}}{v_{CE}} \tag{1}$$

We may differentiate (1) with respect to v_{CE} and evaluate the result at the Q point to find the slope of the maximum-dissipation hyperbola at that point:

$$\left.\frac{\partial i_C}{\partial v_{CE}}\right|_Q = -\left.\frac{P_{C\,max}}{v_{CE}^2}\right|_Q = -\frac{V_{CEQ}I_{CQ}}{V_{CEQ}^2} = -\frac{I_{CQ}}{V_{CEQ}} \tag{2}$$

But for maximum symmetrical swing, $I_{CQ} = V_{CEQ}/R_{ac}$, so (2) becomes

$$\left.\frac{\partial i_C}{\partial v_{CE}}\right|_Q = -\frac{V_{CEQ}/R_{ac}}{V_{CEQ}} = -\frac{1}{R_{ac}} \tag{3}$$

From (3) it is apparent that the slope of the maximum-dissipation hyperbola is the same as that of the ac load line. Since the line and the curve have identical slopes at a common point (the Q point), they are tangent at that point.

11.19 Find an expression for R_L to maximize the efficiency of the amplifier of Fig. 4-14.

▌ Differentiate the expression for η_{max} as given by (2) of Problem 11.10 with respect to R_L and equate the result to zero to find

$$\frac{d\eta_{max}}{dR_L} = \frac{d}{dR_L}\left[\frac{R_C^2 R_L(100\%)}{2(R_L+R_C)(2R_C R_L + R_C R_E + R_L R_E + R_C^2)}\right] = 0$$

Carry out the indicated differentiation and rearrange the result to find

$$R_L = R_C\left[\frac{R_C + R_E}{2R_C + R_E}\right]^{1/2} \tag{1}$$

11.20 For the amplifier of Fig. 4-14, let $R_E = 2\,\Omega$ and $R_C = 100\,\Omega$. Find the value of R_L to give maximum efficiency and the resulting value of efficiency.

▌ By (1) of Problem 11.19 and (2) of Problem 11.10,

$$R_L = (100)\left[\frac{100+2}{2(100)+2}\right]^{1/2} = 71.06\,\Omega$$

$$R_C \| R_L = \frac{100(71.06)}{100+71.06} = 41.54\,\Omega$$

$$\eta_{max} = \frac{(100)^2(71.06)(100\%)}{2(71.06+100)^2(41.54+2+100)} = 8.46\%$$

INDUCTOR-COUPLED AMPLIFIERS

11.21 In the amplifier of Fig. 11-3(a), I_{CQ} (dc bias current) can flow through the large inductor, but the inductor presents an infinite impedance to an ac signal. Consequently, the collector is not an ac ground. If the inductor has negligible resistance, the dc load-line equation is

$$i_C = -\frac{1}{R_E}v_{CE} + \frac{V_{CC}}{R_E} \tag{1}$$

If R_E is selected to be quite small (just large enough to allow stabilized bias design), then the dc load line approaches a vertical line as is illustrated in Fig. 11-3(b). For the class A inductor-coupled amplifier of Fig. 11-3(a), find expressions for I_{CQ} and V_{CEQ} if $(R_E/R_L)^2 \ll 1$ and if $R_E/R_L \ll 1$.

▌ By (3) of Problem 4.92,

$$I_{CQ} = \frac{V_{CC}}{R_{ac}+R_{dc}} = \frac{V_{CC}}{R_L+R_E} = \frac{V_{CC}/R_L}{1+R_E/R_L}$$

$$= \frac{V_{CC}/R_L}{1+R_E/R_L}\frac{1-R_E/R_L}{1-R_E/R_L} = \frac{(V_{CC}/R_L)(1-R_E/R_L)}{1-(R_E/R_L)^2} \tag{1}$$

If $(R_E/R_L)^2 \ll 1$, (1) becomes

$$I_{CQ} \approx \frac{V_{CC}}{R_L}\left(1-\frac{R_E}{R_L}\right) \tag{2}$$

which, for $R_E/R_L \ll 1$, can be further approximated as

$$I_{CQ} \approx V_{CC}/R_L \tag{3}$$

Application of KVL around the collector circuit of Fig. 11-3(a) and the use of (1) yield

$$V_{CEQ} = V_{CC} - I_{CQ}R_E = V_{CC} - \frac{(V_{CC}/R_L)(1-R_E/R_L)}{1-(R_E/R_L)^2}R_E \tag{4}$$

If $(R_E/R_L)^2 \ll 1$, (4) reduces to

$$V_{CEQ} \approx V_{CC}\left(1-\frac{R_E}{R_L}\right) \tag{5}$$

318 □ CHAPTER 11

For $R_E/R_L \ll 1$, (5) can be further simplified to

$$V_{CEQ} \approx V_{CC} \qquad (6)$$

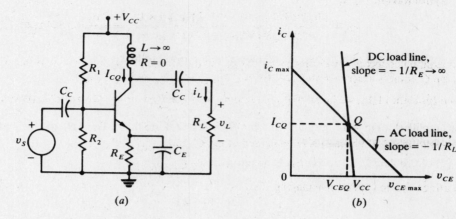

Fig. 11-3

11.22 Find an expression for the power delivered to the inductor-coupled amplifier of Fig. 11-3(a) if $R_E/R_L \ll 1$.

▮ Based on (1) of Problem 11.1 and using (3) and (6) of Problem 11.21,

$$P_{CC} = V_{CEQ}I_{CQ} = V_{CC}\left(\frac{V_{CC}}{R_L}\right) = \frac{V_{CC}^2}{R_L} \qquad (1)$$

11.23 Determine the maximum value of load power for the amplifier of Fig. 11-3(a) if $R_E/R_L \ll 1$ and the amplifier is biased for maximum symmetrical swing.

▮ For maximum symmetrical swing, it is apparent from Fig. 11-3(b) that $I_{cm} = I_{CQ} = V_{CC}/R_L$. Since no ac current flows through $L \to \infty$, $I_{Lm} = I_{cm} = I_{CQ}$, and

$$P_{L\,max} = \frac{I_{Lm}^2}{2}R_L = \frac{I_{CQ}^2}{2}R_L = \frac{(V_{CC}/R_L)^2 R_L}{2} = \frac{V_{CC}^2}{2R_L} \qquad (1)$$

11.24 Determine an expression for η_{max} of the inductor-coupled amplifier in Fig. 11-3(a) if $R_E/R_L \ll 1$ and the amplifier is biased for maximum symmetrical swing.

▮ Based on the results of Problems 11.22 and 11.23,

$$\eta_{max} = \frac{P_{L\,max}}{P_{CC}}(100\%) = \frac{V_{CC}^2/2R_L}{V_{CC}^2/R_L}(100\%) = 50\%$$

11.25 For the inductor-coupled amplifier of Fig. 11-3(a) with $R_E/R_L \ll 1$, find an expression for the value of R_L to place the Q point at the midpoint of the ac load line, thus assuring maximum symmetrical swing.

▮ Equating I_{CQ} as determined by (1) of Problem 4.40 and (3) of Problem 11.21,

$$\frac{V_{BB} - V_{BEQ}}{R_B/(\beta+1) + R_E} = \frac{V_{CC}}{R_L} \qquad (1)$$

Substitute for V_{BB} and R_B as given by (1) of Problem 4.38 into (1) and solve for R_L to find

$$R_L = \frac{V_{CC}[R_1R_2 + (\beta+1)R_E(R_1 + R_2)]}{(\beta+1)[V_{CC}R_2 - V_{BEQ}(R_1 + R_2)]} \qquad (2)$$

11.26 Let $R_1 = 4\text{ k}\Omega$, $R_2 = 500\ \Omega$, $R_E = 1\ \Omega$, $V_{CC} = 15\text{ V}$, and $\beta = 20$ for the inductor-coupled amplifier of Fig. 11-3(a). Assume that the transistor is a Si device and size R_L for maximum symmetrical swing.

▮ By (2) of Problem 11.25,

$$R_L = \frac{15[(4 \times 10^3)(500) + (20+1)(1)(4 \times 10^3 + 500)]}{(20+1)[15(500) - 0.7(4 \times 10^3 + 500)]} = 343.9\ \Omega$$

11.27 If the transistor for the amplifier of Problem 11.26 were a Ge device and all else remains unchanged, determine the value of R_L for maximum symmetrical swing.

By (2) of Problem 11.25,
$$R_L = \frac{15[(4 \times 10^3)(500) + (20 + 1)(1)(4 \times 10^3 + 500)]}{(20 + 1)[15(500) - 0.3(4 \times 10^3 + 500)]} = 2.554 \text{ k}\Omega$$

11.28 If the maximum collector power dissipation $P_{C\,\text{max}}$ is known, show that, for the case of maximum symmetrical swing, the Q point is described by $I_{CQ} = \sqrt{P_{C\,\text{max}}/R_{\text{ac}}}$ and $V_{CEQ} = \sqrt{R_{\text{ac}}P_{C\,\text{max}}}$.

From Problem 11.18, we know that the Q point lies on the maximum-dissipation hyperbola; thus,
$$I_{CQ} = P_{C\,\text{max}}/V_{CEQ} \tag{1}$$

Moreover, from (1) of Problem 4.71 evaluated at the Q point,
$$V_{CEQ} = I_{CQ}R_{\text{ac}} \tag{2}$$

Substituting (2) into (1) and rearranging yield
$$I_{CQ} = \frac{P_{C\,\text{max}}}{I_{CQ}R_{\text{ac}}} \quad \text{so that} \quad I_{CQ} = \sqrt{P_{C\,\text{max}}/R_{\text{ac}}} \tag{3}$$

and then
$$V_{CEQ} = \frac{P_{C\,\text{max}}}{I_{CQ}} = \frac{P_{C\,\text{max}}}{\sqrt{P_{C\,\text{max}}/R_{\text{ac}}}} = \sqrt{R_{\text{ac}}P_{C\,\text{max}}} \tag{4}$$

11.29 A Si transistor with ratings $P_{C\,\text{max}} = 6$ W, $BV_{ECO} = 40$ V, and $i_{C\,\text{max}} = 2$ A is to be used in the amplifier of Fig. 11-3(a), where $\beta = 20$, $R_L = 24\,\Omega$, and $R_E \approx 0$. Select V_{CC}, R_1, and R_2 to give maximum symmetrical swing.

Equations (3) and (4) of Problem 11.28 are equally applicable to an induction-coupled amplifier. Thus,
$$I_{CQ} = \sqrt{P_{C\,\text{max}}/R_{\text{ac}}} = \sqrt{6/24} = 0.5 \text{ A} \quad \text{and} \quad V_{CEQ} = \sqrt{R_L P_{C\,\text{max}}} = \sqrt{24(6)} = 12 \text{ V}$$

For maximum symmetrical swing, we would select $V_{CC} = V_{CEQ} = 12$ V.

Applying KVL around the base-emitter loop gives
$$V_{BB} = \frac{I_{CQ}}{\beta}R_B + V_{BEQ} = \frac{0.5}{20}R_B + 0.7 \tag{1}$$

We arbitrarily select $R_B = 400\,\Omega$; then, by (1), $V_{BB} = 10.7$ V. Next we simultaneously solve equation set (1) of Problem 4.38 for R_1 and R_2 and use the results to find
$$R_1 = \frac{R_B}{1 - V_{BB}/V_{CC}} = \frac{400}{1 - 10.7/12} = 3.69 \text{ k}\Omega$$

and
$$R_2 = R_B \frac{V_{CC}}{V_{BB}} = 400 \frac{12}{10.7} = 448.6\,\Omega$$

11.30 Determine the maximum value of load power delivered by the amplifier of Problem 11.29.

By (1) of Problem 11.23,
$$P_{L\,\text{max}} = \frac{V_{CC}^2}{2R_L} = \frac{(12)^2}{2(24)} = 3 \text{ W}$$

TRANSFORMER-COUPLED AMPLIFIERS

11.31 For the class A transformer-coupled amplifier of Fig. 11-4(a), find expressions for I_{CQ} and V_{CEQ} if $(R_E/R_L')^2 \ll 1$ and if $(R_E/R_L') \ll 1$, where for the ideal transformer
$$v_L' = av_L \tag{1}$$
$$i_C = \frac{1}{a}i_L \tag{2}$$
$$R_L' = \frac{v_L'}{i_C} = \frac{av_L}{(1/a)i_L} = a^2 R_L \tag{3}$$

▌ Note that $R_{ac} = a^2 R_L = R'_L$. Further, since the resistance of the windings of an ideal transformer is negligible, $R_{dc} = R_E$. Hence, by symmetry, the results of Problem 11.21 hold if R_L is replaced with $R'_L = a^2 R_L$. The load lines are shown on the collector characteristic of Fig. 11-4(b).

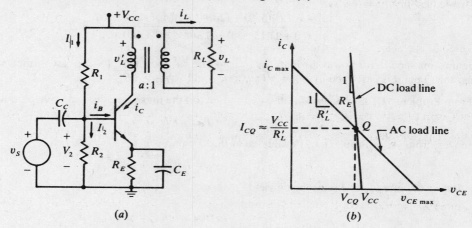

Fig. 11-4

11.32 Find the maximum power that can be delivered to the load R_L of a class A transformer-coupled amplifier.

▌ An ideal transformer is a lossless device; hence, the power delivered to the primary is identically the power delivered to R_L. Using (2) and (3) of Problem 11.31, we have

$$P_L = \frac{I_{Lm}^2}{2} R_L = \frac{(aI_{cm})^2}{2} R_L = \frac{I_{cm}^2}{2} R'_L \tag{1}$$

From Fig. 11-4(b), with $R_E \ll R'_L$, for maximum symmetrical swing $I_{cm} \approx I_{CQ} \approx V_{CC}/R'_L$. Therefore,

$$P_{L\,max} \approx \frac{I_{CQ}^2}{2} R'_L \approx \frac{V_{CC}^2}{2R'_L} = \frac{V_{CC}^2}{2a^2 R_L} \tag{2}$$

11.33 Explain how the maximum power that can be delivered to a particular load R_L of a class A transformer-coupled amplifier can be changed.

▌ From (1) of Problem 11.32, it is obvious that proper selection of the turns ratio a allows adjustment of the power delivered to R_L.

11.34 The transformer-coupled amplifier of Fig. 11-4(a) is biased for maximum symmetrical swing. The ideal transformer is coupled to an 8-Ω speaker load R_L which requires an input of 9 W. Let $V_{CC} = 24$ V and $R_E = 0$. Select the transformer turns ratio a.

▌ The average power to R_L must be 9 W; thus,

$$\frac{V_{Lm}^2}{2R_L} = 9 \quad \text{so} \quad V_{Lm} = \sqrt{9(2R_L)} = \sqrt{9(2)(8)} = 12 \text{ V}$$

For maximum symmetrical swing with $R_E = 0$, $V'_{Lm} = V_{CEQ} = V_{CC}$. Since the transformer is ideal,

$$a = \frac{V'_{Lm}}{V_{Lm}} = \frac{V_{CC}}{V_{Lm}} = \frac{24}{12} = 2$$

11.35 Determine the minimum power rating for the transistor in the amplifier of Problem 11.34.

▌ For maximum symmetrical swing,

$$I_{CQ} = \frac{V_{CC}}{R_{ac} + R_{dc}} = \frac{V_{CC}}{R'_L + 0} = \frac{V_{CC}}{a^2 R_L} = \frac{24}{(2)^2(8)} = 0.75 \text{ A}$$

and the power supplied is

$$P_{CC} = V_{CC} I_{CQ} = 24(0.75) = 18 \text{ W}$$

Since there are no $I^2 R$ losses, by (3) of Problem 11.1,

$$P_C = P_{CC} - P_L = 18 - 9 = 9 \text{ W}$$

11.36 For the transistor in the transformer-coupled amplifier of Fig. 11-4(a), assume that $V_{CE\,sat} = 0$ and $I_{CBO} = 0$. Derive a formula for easily determining V_{CEQ} for maximum symmetrical swing.

▌ The dc load-line equation is

$$V_{CC} = v_{CE} + i_E R_E \approx v_{CE} + i_C R_{dc} \qquad (1)$$

Using (3) of Problem 4.92 in (1) and evaluating at the Q point give

$$V_{CC} \approx V_{CEQ} + \frac{V_{CC}}{R_{ac} + R_{dc}} R_{dc} \quad \text{so} \quad V_{CEQ} \approx \frac{V_{CC}}{1 + R_{dc}/R_{ac}} \qquad (2)$$

11.37 Find an expression for the supplied power to the transformer-coupled amplifier of Fig. 11-4(a).

▌ Using (3) of Problem 4.92 in (1) of Problem 11.1 and the result of Problem 11.31, we find

$$P_{CC} = V_{CC} I_{CQ} = V_{CC} \frac{V_{CC}}{R_{ac} + R_{dc}} = \frac{V_{CC}^2}{a^2 R_L + R_E} \qquad (1)$$

11.38 Determine an expression for the maximum load power if R_E is not negligible, but maximum symmetrical swing exists.

▌ Realizing that $I_{Lm} = I_{CQ}$, we substitute (3) of Problem 4.92 into (1) of Problem 11.32 and use the result of Problem 11.31 to find

$$P_{L\,max} = \frac{I_{Lm}^2}{2} R_L = \frac{I_{CQ}^2}{2} a^2 R_L = \frac{1}{2}\left(\frac{V_{CC}}{R_{ac} + R_{dc}}\right)^2 a^2 R_L = \frac{a^2 V_{CC}^2 R_L}{2(a^2 R_L + R_E)^2} \qquad (1)$$

11.39 Derive an expression for the maximum efficiency possible for the transformer-coupled amplifier of Fig. 11-4(a) with maximum symmetrical swing if the power dissipated in R_1 and R_2 may be neglected.

▌ Since there are no $I^2 R$ losses, the maximum efficiency is given by use of results of Problems 11.38 and 11.39 as

$$\eta_{max} = \frac{P_{L\,max}}{P_{CC}} (100\%) = \frac{a^2 V_{CC}^2 R_L / [2(a^2 R_L + R_E)^2]}{V_{CC}^2 / (a^2 R_L + R_E)} (100\%) = \frac{a^2 R_L}{2(a^2 R_L + R_E)} (100\%)$$

$$= \frac{1}{2(1 + R_{dc}/R_{ac})} (100\%) = \frac{50}{1 + R_{dc}/R_{ac}} \% \qquad (1)$$

If $R_{dc} \to 0$, then $\eta_{max} \to 50$ percent as an upper limit.

11.40 In the transformer-coupled class A amplifier of Fig. 11-4(a), $V_{CC} = 15$ V, $R_E = 0.5\,\Omega$, and $R_L = 1.5\,\Omega$. Neglect power losses in R_1 and R_2, but assume their values were selected for maximum symmetrical swing. The resistance R_p of the primary coil of the transformer is $2.5\,\Omega$, and that of the secondary (R_s) is $0.1\,\Omega$. The transformer turns ratio is 5:1. Find I_{CQ} for the transistor used in this amplifier. Assume that the transformer leakage reactance is negligible at the frequencies of interest.

▌ The secondary winding resistance can be reflected, along with R_L, to the primary side to give

$$R_{ac} = R_p + a^2(R_s + R_L) = 2.5 + (5)^2(0.1 + 1.5) = 42.5\,\Omega$$

$$R_{dc} = R_p + R_E = 2.5 + 0.5 = 3\,\Omega$$

By (3) of Problem 4.92,

$$I_{CQ} = \frac{V_{CC}}{R_{ac} + R_{dc}} = \frac{15}{42.5 + 3} = 0.33\,\text{A}$$

11.41 Find the supplied power for the amplifier of Problem 11.40.

▌ The supplied power follows from the result of Problem 11.40 and (1) of Problem 11.1:

$$P_{CC} = V_{CC} I_{CQ} = 15(0.33) = 4.95\,\text{W}$$

11.42 Specify the minimum values of $i_{C\,max}$, BV_{CEO}, and $P_{C\,max}$ for the transistor used in the amplifier of Problem 11.40.

▌ Since $R_{dc}/R_{ac} = 3/42.5 \approx 0.07 \ll 1$, the conclusions of Problem 11.31 allow us to write $V_{CEQ} \approx V_{CC} = 15$ V. Thus, we specify

$$i_{C\,max} \geq 2I_{CQ} = 2(0.33) = 0.66 \text{ A}$$

and
$$BV_{CEO} \geq 2V_{CC} = 2(15) = 30 \text{ V}$$

The maximum collector dissipation occurs for $P_L = 0$, so we specify

$$P_{C\,max} \geq P_{CC} - I_{CQ}^2(R_E + R_p) = 4.95 - (0.33)^2(0.5 + 2.5) = 4.623 \text{ W}$$

11.43 In the transformer-coupled class A amplifier of Fig. 11-4(a), $R_L = 1.5\ \Omega$, $V_{CC} = 15$ V, and $\beta = 20$. The maximum delivered load power is 3 W. Neglect R_E, and assume R_1 and R_2 are sized for maximum symmetrical swing but dissipate negligible power. The transformer is ideal. Find the quiescent collector current.

▌ For maximum symmetrical swing, the dc load line should bisect the ac load line; thus, $V_{CEQ} = V_{CC} = 15$ V. For the given resistive load,

$$P_{L\,max} = \frac{V_{cem}I_{cm}}{2} = \frac{V_{CEQ}I_{CQ}}{2}$$

so that
$$I_{CQ} = \frac{2P_{L\,max}}{V_{CEQ}} = \frac{2(3)}{15} = 0.4 \text{ A}$$

11.44 Determine the supplied power for the transformer-coupled amplifier of Problem 11.43.

▌ Based on (1) of Problem 11.1 and the results of Problem 11.43,

$$P_{CC} = V_{CC}I_{CQ} = 15(0.4) = 6 \text{ W}$$

11.45 Choose a transistor for the amplifier of Problem 11.43 by specifying $i_{C\,max}$, BV_{CEO}, and $P_{C\,max}$.

▌ Maximum v_{CE} occurs at cutoff; consequently, we specify

$$BV_{CEO} \geq 2V_{CC} = 2(15) = 30 \text{ V}$$

With $R_{dc}/R_{ac} \ll 1$, (3) of Problem 11.21 leads to

$$R_{ac} = \frac{V_{CC}}{I_{CQ}} = \frac{15}{0.4} = 37.5\ \Omega$$

and since collector current is a maximum at saturation, we specify

$$i_{C\,max} \geq \frac{2V_{CC}}{R_{ac}} = \frac{2(15)}{37.5} = 0.8 \text{ A}$$

With R_E neglected and for an ideal transformer, we specify

$$P_{C\,max} \geq V_{CEQ}I_{CQ} = 15(0.4) = 6 \text{ W}$$

11.46 Determine the transformer turns ratio for the amplifier of Problem 11.43.

▌ Using the result of Problem 11.45 and the impedance-reflection property of the ideal transformer,

$$a^2 = \frac{R_{ac}}{R_L} = \frac{37.5}{1.5} = 25 \quad \text{so} \quad a = 5$$

11.47 For the Si power transistor of Fig. 11-4(a), $h_{FE} = \beta = 40$ and $h_{ie} = 25\ \Omega$; the transformer is ideal. Let $R_E = 2\ \Omega$, $R_L = 2.5\ k\Omega$, $V_{CC} = 12$ V, and $a = 1/6$. Select values of R_1 and R_2 to produce β-independent bias while giving maximum symmetrical swing.

▌ Using the usual order of magnitude to satisfy the margin of inequality in (2) of Problem 4.40,

$$R_B = \frac{(\beta + 1)R_E}{10} = \frac{(40 + 1)(2)}{10} = 8.2\ \Omega$$

Based on results of Problem 11.31,

$$R_{ac} = a^2 R_L = (\tfrac{1}{6})^2(2.5 \times 10^3) = 69.4\ \Omega \quad \text{and} \quad R_{dc} = R_E = 2\ \Omega$$

Then, by (3) of Problem 4.92,
$$I_{CQ} = \frac{V_{CC}}{R_{ac} + R_{dc}} = \frac{12}{69.4 + 2} = 168.07 \text{ mA}$$

Since $I_{CQ} = \beta I_{EQ}/(\beta + 1)$, (1) of Problem 4.39 leads to
$$V_{BB} = \frac{I_{CQ}}{\beta} R_B + V_{BEQ} + \frac{\beta+1}{\beta} I_{CQ} R_E = \frac{168.07 \times 10^{-3}}{40}(8.2) + 0.7 + \frac{(40+1)}{40}(168.07 \times 10^{-3})(2) = 1.079 \text{ V}$$

By use of (1) of Problem 4.38,
$$R_1 = \frac{R_B}{1 - V_{BB}/V_{CC}} = \frac{8.2}{1 - 1.079/12} = 9.01 \; \Omega \quad \text{and} \quad R_2 = R_B \frac{V_{CC}}{V_{BB}} = (8.2)\frac{12}{1.079} = 91.19 \; \Omega$$

11.48 For ideal collector characteristics, determine the maximum load voltage and current for class A operation of the amplifier of Problem 11.47.

▌ Now i_C can swing ± 168.07 mA about $I_{CQ} = 168.07$ mA, and
$$I_{Lm} = aI_{cm} = aI_{CQ} = \tfrac{1}{6}(168.07 \times 10^{-3}) = 28.01 \text{ mA}$$
$$V_{CEQ} = \frac{V_{CC}}{1 + R_{dc}/R_{ac}} = \frac{12}{1 + 2/69.4} = 11.66 \text{ V}$$

Since $V_{CEQ} < V_{CC}$, saturation limits symmetrical swing, and
$$V_{Lm} = \frac{1}{a} V'_{Lm} = \frac{1}{a} V_{CEQ} = 6(11.66) = 69.96 \text{ V}$$

11.49 Find the maximum possible efficiency of the transformer-coupled amplifier of Problem 11.47.

▌ By (1) of Problem 11.39,
$$\eta_{max} = \frac{50\%}{1 + R_{dc}/R_{ac}} = \frac{50\%}{1 + 2/69.4} = 48.6\%$$

11.50 In the class A transformer-coupled amplifier of Fig. 11-4(a), $\beta = 30$, $R_E = 2 \; \Omega$, $R_1 = 200 \; \Omega$, $R_2 = 25 \; \Omega$, $R_L = 8 \; \Omega$, and $V_{CC} = 15$ V. The transistor is a Si device, and the transformer is ideal. Specify the transformer turns ratio for maximum symmetrical swing.

▌ Based on (1) of Problem 4.38 and (1) of Problem 4.40,
$$V_{BB} = \frac{R_2}{R_1 + R_2} V_{CC} = \frac{25}{200 + 25}(15) = 1.70 \text{ V}$$
$$R_B = \frac{R_1 R_2}{R_1 + R_2} = \frac{200(25)}{200 + 25} = 22.22 \; \Omega$$
$$I_{CQ} = \frac{\beta}{\beta + 1} I_{EQ} = \frac{\beta(V_{BB} - V_{BEQ})}{R_B + (\beta + 1)R_E} = \frac{30(1.7 - 0.7)}{22.22 + (30 + 1)(2)} = 0.356 \text{ A}$$

For maximum symmetrical swing, (3) of Problem 4.92 leads to
$$R_{ac} = \frac{V_{CC} - I_{CQ} R_{dc}}{I_{CQ}} = \frac{15 - 0.356(2)}{0.356} = 40.13 \; \Omega$$

For the ideal transformer, $R_{ac} = a^2 R_L$, or
$$a = \sqrt{R_{ac}/R_L} = \sqrt{40.13/8} = 2.24$$

11.51 Specify the minimum ratings for the transistor of the amplifier in Problem 11.50.

▌
$$V_{CEQ} \approx V_{CC} - I_{CQ} R_E = 15 - 0.356(2) = 14.29 \text{ V}$$
$$BV_{CEQ} \geq 2V_{CC} = 2(14.29) = 28.58 \text{ V}$$
$$i_{C\,max} \geq 2I_{CQ} = 2(0.356) = 0.712 \text{ A}$$
$$P_{C\,max} \geq V_{CEQ} I_{CQ} = 14.29(0.356) = 5.09 \text{ W}$$

PUSH-PULL AMPLIFIERS

11.52 In a class B amplifier, the dc collector current I_{CQ} is lower than the peak value of ac current. Consequently, there is less collector dissipation and the efficiency can be greater than for a class A amplifier. The common-emitter push-pull amplifier of Fig. 11-5 has one active transistor and one transistor in cutoff at any instant of time. Owing to the symmetry, we need only study the operation of the active transistor, as diagrammed in Fig. 11-6(a). Determine the load lines for the half class B push-pull amplifier of Fig. 11-6(a).

▌ Since the resistance of the windings of an ideal transformer is negligible, the dc load-line equation is simply $V_{CEQ} = V_{CC}$; the dc load line is then a vertical line, as shown in Fig. 11-6(b). When no ac signal is present (quiescent conditions), $i_{B2} = 0$, so that the transistor is cut off. Hence, $I_{CQ} = 0$, and the ac load-line equation follows from (2) of Problem 4.71 as

$$i_{C2} = \frac{V_{CEQ2}}{R'_L} - \frac{v_{CE2}}{R'_L} = \frac{V_{CC}}{R'_L} - \frac{v_{CE2}}{R'_L} \tag{1}$$

When Q_2 is active, (1) describes a line [on the collector characteristics of Fig. 11-6(b)] with vertical intercept $i_{C2} = V_{CC}/R'_L$ and horizontal intercept $v_{CE2} = V_{CC}$; when Q_2 is cut off, $i_{C2} = 0$ and $v_{CE2} = V_{CC}$, giving the horizontal line segment from $v_{CE2} = V_{CC}$ to $v_{CE2} = 2V_{CC}$.

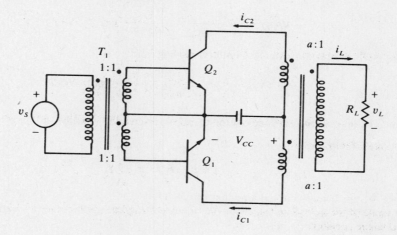

Fig. 11-5

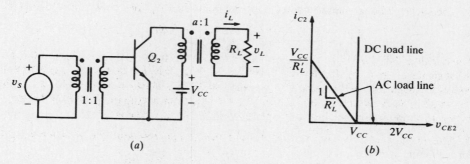

Fig. 11-6

11.53 For the class B push-pull amplifier of Fig. 11-5, find the maximum possible efficiency (which occurs at maximum signal swing) for a sinusoidal input signal.

▌ For a sinusoidal signal $v_s = V_m \sin \omega t$, the current through each transistor is a positive half sine wave such that

$$i_{C1} = I_{cm} \sin \omega t \qquad 0 \le \omega t < \pi$$
$$i_{C2} = I_{cm} \sin \omega t \qquad \pi \le \omega t < 2\pi$$

Thus, the average power supplied by V_{CC} is found as

$$P_{CC} = \frac{1}{\pi} \int_0^\pi V_{CC} I_{cm} \sin \omega t \, d(\omega t) = \frac{V_{CC}}{\pi} I_{cm} \cos \omega t \Big|_\pi^0 = \frac{2}{\pi} V_{CC} I_{cm} \tag{1}$$

For maximum swing, $V_{Lm} = V_{CC}/a$, and for any signal, $I_{Lm} = aI_{cm}$. The power delivered to the load at maximum swing is then

$$P_{L\,\max} = \frac{V_{Lm}}{\sqrt{2}} \frac{I_{Lm}}{\sqrt{2}} = \frac{V_{CC}}{a\sqrt{2}} \frac{aI_{cm}}{\sqrt{2}} = \frac{V_{CC}I_{cm}}{2} \quad (2)$$

Hence, the maximum possible efficiency is

$$\eta_{\max} = \frac{P_{L\,\max}}{P_{CC}}(100\%) = \frac{V_{CC}I_{cm}/2}{(2/\pi)V_{CC}I_{cm}}(100\%) = \frac{\pi}{4}(100\%) = 78.54\%$$

11.54 A loudspeaker with an 8-Ω input resistance requiring a power of 0.5 W is to be driven by the push-pull amplifier of Fig. 11-5. V_{CC} is a 9-V transistor battery, and the identical transistors have $V_{CE\,\text{sat}} = 0.5$ V and $I_{BEO} = 0$. Select a suitable turns ratio for the output transformer.

▌ We design for maximum symmetrical swing; then

$$I_{cm} = \frac{V_{CC} - V_{CE\,\text{sat}}}{a^2 R_L}$$

and the power delivered to the load is

$$P_{L\,\max} = \frac{I_{cm}^2}{2} a^2 R_L = \frac{1}{2}\left(\frac{V_{CC} - V_{CE\,\text{sat}}}{a^2 R_L}\right)^2 a^2 R_L = \frac{(V_{CC} - V_{CE\,\text{sat}})^2}{2a^2 R_L}$$

Solving for a and evaluating give the desired turns ratio:

$$a = \frac{V_{CC} - V_{CE\,\text{sat}}}{\sqrt{2R_L P_{L\,\max}}} = \frac{9 - 0.5}{\sqrt{2(8)(0.5)}} \approx 3 \text{ turns}$$

11.55 Determine the value of supplied power for the push-pull amplifier of Problem 11.54.

▌ From (1) of Problem 11.53,

$$P_{CC} = \frac{2}{\pi} V_{CC} I_{cm} = \frac{2}{\pi} V_{CC} \frac{2P_{L\,\max}}{V_{CC}} = \frac{4}{\pi} P_{L\,\max} = \frac{4}{\pi} 0.5 = 0.636 \text{ W}$$

11.56 Find the value of the collector dissipated power for the amplifier of Problem 11.54 when 500 mW is being delivered to the speaker.

▌ The collector power dissipated per transistor is found using the result of Problem 11.55 as

$$P_C = \tfrac{1}{2}(P_{CC} - P_L) = \tfrac{1}{2}(0.636 - 0.5) = 68 \text{ mW}$$

11.57 The circuit of Fig. 11-7 is a class A push-pull amplifier. Both Q_1 and Q_2 are biased for class A operation. The input signal is "split" by transformer T_1 (the phase inverter of Problem 11.60 could be substituted). Since collector currents i_{C1} and i_{C2} are 180° out of phase, they establish additive fluxes in transformer T_2. Assume that the value of R_B is selected so that Q_1 and Q_2 are biased for maximum symmetrical swing. If $R_L = 5$ Ω and $V_{CC} = 15$ V, determine the transformer turns ratio for $P_{L\,\max} = 10$ W.

▌ The output voltage can be determined from the load power requirement. Since $V_{Lm}^2/2R_L = P_{L\,\max}$, we have

$$V_{Lm} = \sqrt{2P_{L\,\max} R_L} = \sqrt{2(10)(5)} = 10 \text{ V}$$

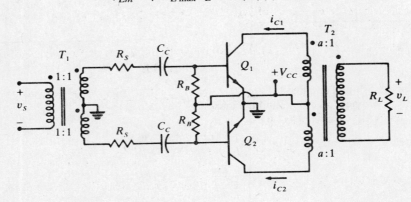

Fig. 11-7

If we assume that the transformer primary resistance is small, the dc load line is nearly vertical; thus,

$$v_{CE\,max} \approx 2V_{CC} = 2(15) = 30\text{ V} \quad \text{and} \quad V_{cem} = \tfrac{1}{2}v_{CE\,max} = V_{CC} = 15\text{ V}$$

The transformer turns ratio is then

$$a = \frac{V_{cem}}{V_{Lm}} = \frac{15}{10} = 1.5$$

11.58 Specify the transistor ratings BV_{CEO}, $i_{C\,max}$, and $P_{C\,max}$ for the push-pull amplifier of Problem 11.57.

▮ The transistor avalanche breakdown voltage must be greater than or equal to $v_{CE\,max}$ of Problem 11.57, so $BV_{CEO} \geq 30$ V. One-half the reflected load current must be supplied by each transistor, so

$$I_{cm} = \frac{1}{2}\left(\frac{1}{a}\right)\frac{V_{Lm}}{R_L} = \frac{1}{2}\left(\frac{1}{1.5}\right)\frac{10}{5} = 0.667\text{ A}$$

Since the circuit is designed for maximum symmetrical swing, $I_{CQ} = I_{cm}$; consequently, specify $i_{C\,max} = 2I_{CQ} = 2(0.667) = 1.334$ A. The collector power dissipation for a class A amplifier is largest when no signal is present; hence, by (3) of Problem 11.1,

$$P_{C\,max} = P_{CC} = V_{CC}I_{CQ} = 15(0.667) = 10\text{ W}$$

11.59 Determine the maximum possible efficiency of the amplifier of Problem 11.57.

▮ The power supplied to the entire amplifier is $2P_{CC}$, so the maximum amplifier efficiency is

$$\eta = \frac{P_{L\,max}}{2P_{CC}}(100\%) = \frac{10}{20}(100\%) = 50\%$$

11.60 Rather than use a bulky center-tapped transformer to provide input signals that are 180° out of phase to a push-pull amplifier, a *phase-inverter* circuit such as that of Fig. 11-8(a) can be used. In that circuit, assume that $R_3 = R_4$ and that $h_{oe} = h_{re} = 0$. Use small-signal analysis to show that $v_1/v_2 = -1$, proving that the signals are equal in magnitude and 180° out of phase.

▮ The small-signal equivalent circuit is drawn in Fig. 11-8(b). By Ohm's law,

$$v_1 = h_{fe}i_bR_4 \quad \text{and} \quad v_2 = -h_{fe}i_bR_3$$

so that

$$\frac{v_1}{v_2} = \frac{h_{fe}i_bR_4}{-h_{fe}i_bR_3} = -1 \tag{1}$$

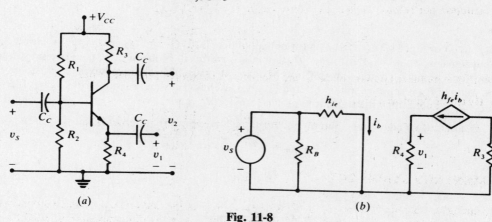

Fig. 11-8

11.61 In the class A push-pull amplifier of Fig. 11-7, let $V_{CC} = 15$ V, $R_L = 20\ \Omega$, and $a = 1/2$. If Q_1 and Q_2 are biased for maximum symmetrical swing, find the maximum possible efficiency.

▮

$$V_{Lm} = \frac{1}{2}V_{cem} = 2(15) = 30\text{ V}$$

$$P_{L\,max} = \frac{V_{Lm}^2}{2R_L} = \frac{(30)^2}{2(20)} = 22.5\text{ W}$$

$$I_{CQ} = I_{cm} = \frac{1}{2}\left(\frac{1}{a}\right)\frac{V_{Lm}}{R_L} = \frac{1}{2}(2)\frac{30}{20} = 1.5 \text{ A}$$

$$P_{CC} = 2V_{CEQ}I_{CQ} = 2V_{CC}I_{CQ} = 2(15)(1.5) = 45 \text{ W}$$

$$\eta_{max} = \frac{P_{L\,max}}{P_{CC}}(100\%) = \frac{22.5}{45}(100\%) = 50\%$$

11.62 In the class B push-pull amplifier of Fig. 11-5, let $R_L = 5\,\Omega$, $a = 2$, and $V_{CC} = 15$ V. If, under operating conditions, the power delivered to the load is 4 W, find the amplifier efficiency.

$$P_L = \frac{I_{Lm}^2}{2}R_L = \frac{a^2 I_{cm}^2}{2}R_L \quad \text{or} \quad I_{cm} = \frac{1}{a}\sqrt{2P_L/R_L} \quad (1)$$

Substitute (1) into (1) of Problem 11.53 to find

$$P_{CC} = \frac{2V_{CC}}{\pi}\left(\frac{1}{a}\right)\sqrt{2P_L/R_L} = \frac{2(15)}{\pi}\left(\frac{1}{2}\right)\sqrt{2(4/5)} = 6.04 \text{ W}$$

Hence,
$$\eta = \frac{P_L}{P_{CC}}(100\%) = \frac{4}{6.04}(100\%) = 66.2\%$$

11.63 For a class B amplifier with negligible I^2R losses (such as that in Fig. 11-5 or 11-9), derive expressions for the value of collector current at which maximum collector dissipation occurs and for the maximum collector power dissipation.

For each transistor, the collector dissipation is one-half the total, or

$$P_C = \frac{1}{2}(P_{CC} - P_L) = \frac{1}{2}\left(\frac{2}{\pi}V_{CC}I_{cm} - \frac{I_{cm}^2}{2}a^2 R_L\right) \quad (1)$$

We differentiate (1) with respect to I_{cm} and equate the result to zero to find the collector current at $P_{C\,max}$:

$$\frac{\partial P_C}{\partial I_{cm}} = 0 = \frac{2}{\pi}V_{CC} - I_{cm}a^2 R_L$$

from which
$$I_{cm} = \frac{2V_{CC}}{\pi a^2 R_L} \quad \text{at} \quad P_{C\,max} \quad (2)$$

Substitution of (2) into (1) gives the desired expression for $P_{C\,max}$:

$$P_{C\,max} = \frac{1}{2}\left(\left[\frac{2}{\pi}V_{CC}\frac{2V_{CC}}{\pi a^2 R_L} - \frac{1}{2}\left(\frac{2V_{CC}}{\pi a^2 R_L}\right)^2 a^2 R_L\right]\right) = \frac{1}{R_L}\left(\frac{V_{CC}}{\pi a}\right)^2 \quad (3)$$

11.64 Specify the minimum power rating for the identical transistors of Problem 11.62.

By (3) of Problem 11.63,

$$P_{C\,max} = \frac{1}{R_L}\left(\frac{V_{CC}}{\pi a}\right)^2 = \frac{1}{5}\left[\frac{15}{\pi(2)}\right]^2 = 1.14 \text{ W}$$

COMPLEMENTARY-SYMMETRY AMPLIFIERS

11.65 By taking advantage of the two possible polarities of BJTs (*npn* and *pnp*), a complementary-symmetry amplifier can be devised to use direct coupling of both the source and the load, eliminating the need for transformers. Each transistor is essentially a class B emitter follower. The load line and power relations are the same as for the class B push-pull amplifier. (See Problems 11.52 and 11.53.) In the complementary-symmetry amplifier of Fig. 11-9, let $R_L = 5\,\Omega$ and $i_{C\,max} = 2$ A (for both Q_1 and Q_2). Find the minimum value of V_{CC} for maximum power to R_L.

Since the amplifier is biased for class B operation, each transistor can be considered an independent class B emitter follower. Assume $V_{CE\,sat} \approx 0$ and $I_{CEO} = 0$. Then

$$\text{Minimum } V_{CC} = i_{C\,max}R_L = 2(5) = 10 \text{ V}$$

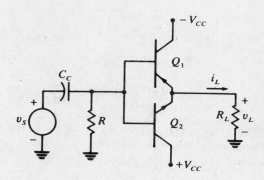

Fig. 11-9

11.66 Find the maximum value of load power that the amplifier of Problem 11.65 can deliver.

$$P_{L\,max} = \tfrac{1}{2}i_{C\,max}^2 R_L = \tfrac{1}{2}(2)^2(5) = 10\text{ W}$$

11.67 Determine the efficiency of the complementary-symmetry amplifier of Problem 11.65 when $P_{L\,max}$ as determined by Problem 11.66 is being delivered.

From (1) of Problem 11.53,

$$P_{CC} = \frac{2}{\pi}V_{CC}I_{cm} = \frac{2}{\pi}(10)(2) = 12.73\text{ W}$$

and

$$\eta_{max} = \frac{P_{L\,max}}{P_{CC}}(100\%) = \frac{10}{12.73}(100\%) = 78.54\%$$

11.68 In the complementary-symmetry amplifier of Fig. 11-10, $R_L = 4\,\Omega$, $R_C = 1\,\Omega$, and $V_{CC} = 10$ V. Find the maximum power delivered to the load R_L.

When $v_s > 0$, Q_2 conducts and Q_1 is cut off; when $v_s < 0$, Q_1 conducts and Q_2 is cut off. If $V_{CE\,sat} \approx 0$, then $I_{cm1} = I_{cm2} = V_{CC}/(R_L + R_C)$, and

$$P_{L\,max} = \frac{1}{2}I_{cm}^2 R_L = \frac{1}{2}\left(\frac{V_{CC}}{R_L + R_C}\right)^2 R_L = \frac{1}{2}\left(\frac{10}{4+1}\right)^2(4) = 8\text{ W}$$

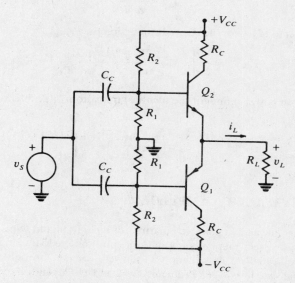

Fig. 11-10

11.69 Find the power supplied by each dc bias source for the amplifier of Problem 11.68.

By (1) of Problem 11.53, since each source must supply half the total power P_{CC}, we have

$$\frac{1}{2}P_{CC} = \frac{1}{2}\left(\frac{2}{\pi}V_{CC}I_{cm}\right) = \frac{1}{\pi}V_{CC}\frac{V_{CC}}{R_L + R_C} = \frac{V_{CC}^2}{\pi(R_L + R_C)} = \frac{(10)^2}{\pi(4+1)} = 6.37\text{ W}$$

11.70 Determine the maximum power dissipated by each transistor of the amplifier in Problem 11.68.

▌ The maximum collector power dissipation per transistor follows from (3) of Problem 11.63 with $a = 1$ and R_L replaced with $R_L + R_C$:

$$P_{C\,\text{max}} = \frac{1}{R_L + R_C}\left(\frac{V_{CC}}{\pi}\right)^2 = \frac{1}{4+1}\left(\frac{10}{\pi}\right)^2 = 2.026 \text{ W} \tag{1}$$

11.71 The complementary-symmetry common-collector amplifier of Fig. 11-11 offers the advantage of using a single-ended (one-polarity, here positive) power supply. Let $V_{CC} = 15$ V, $R_E = 0.25\ \Omega$, $R_L = 10\ \Omega$, $V_{CE\,\text{sat}} \approx 0$, and $I_{CBO} = 0$. The transistors are a *matched-complement pair* (with identical parameters, but with characteristic curves described by complementary voltages and currents). Find the peak values of output voltage and current.

▌ From Fig. 11-11, it is apparent that $I_{EQ1} = I_{EQ2}$. Further, $V_{CEQ2} = -V_{CEQ1}$ since the transistors are a matched-complement pair; hence, $V_{AQ} = V_{CC}/2 = v_C$, and the output voltage will exhibit symmetrical swing. Thus, determination of the magnitude of the positive half-cycle of v_L is sufficient. At the point of maximum positive swing of v_s (at which $V_{CE2\,\text{sat}} = 0$ and Q_1 is cut off), voltage division gives

$$V_{Am} = \frac{R_L}{R_E + R_L} V_{CC} \tag{1}$$

Then KVL and (1) give

$$V_{Lm} = V_{Am} - v_C = \left(\frac{R_L}{R_E + R_L} - \frac{1}{2}\right)V_{CC} = \left(\frac{10}{10.25} - 0.5\right)(15) = 7.134 \text{ V}$$

and

$$I_{Lm} = \frac{V_{Lm}}{R_L} = \frac{7.134}{10} = 713.4 \text{ mA}$$

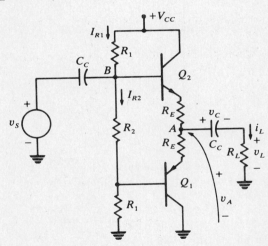

Fig. 11-11

11.72 Determine the maximum value of average power delivered to the load by the complementary-symmetry amplifier of Problem 11.71.

▌ The maximum load power is

$$P_{L\,\text{max}} = \tfrac{1}{2} I_{Lm}^2 R_L = \tfrac{1}{2}(0.7134)^2(10) = 2.54 \text{ W}$$

11.73 In the CC complementary-symmetry amplifier of Fig. 11-11, let $\beta_1 = \beta_2 = 30$, $R_E = 0.25\ \Omega$, and $V_{CC} = 15$ V. The transistors are Si devices. As a matter of design choice, select values of R_1 and R_2 so that $I_{R1} = 10 I_{BQ2}$ (and, thus, $I_{R1} \approx I_{R2}$) and so that $I_{CQ1} = I_{CQ2} = 10$ mA.

▌ Since $I_{R1} \approx I_{R2}$, KVL around a mesh that includes both transistor base-emitter junctions yields

$$I_{R1} R_2 \approx I_{R2} R_2 = V_{BEQ2} + I_{EQ2} R_E + I_{EQ1} R_E - V_{BEQ1} \tag{1}$$

Using $I_{EQ1} = I_{EQ2} = (\beta_2 + 1)I_{CQ2}/\beta_2$, $I_{R1} = 10 I_{BQ2} = 10 I_{CQ2}/\beta_2$, and $V_{BEQ2} = -V_{BEQ1} = 0.7$ V in (1) leads to

$$R_2 = \frac{2\beta_2 V_{BEQ2}}{10 I_{CQ2}} + \frac{2R_E(\beta_2 + 1)}{10} = \frac{2(30)(0.7)}{10(10 \times 10^{-3})} + \frac{2(0.25)(30+1)}{10} = 421.5\ \Omega$$

Now, $$I_{R1} = 10I_{BQ2} = \frac{10I_{CQ2}}{\beta_2 + 1} = \frac{10(10 \times 10^{-3})}{30 + 1} = 3.226 \text{ mA}$$

By KVL, using $I_{R1} = I_{R2}$,

$$I_{R1} = R_1 + R_2 + R_1 = V_{CC}$$

or $$R_1 = \frac{1}{2}\left[\frac{V_{CC}}{I_{R1}} - R_2\right] = \frac{1}{2}\left[\frac{15}{3.226 \times 10^{-3}} - 421.5\right] = 2.114 \text{ k}\Omega$$

11.74 Find the value of $V_{CEQ2} = -V_{CEQ1}$ for the amplifier of Problem 11.73.

▌ By KCL,

$$V_{CEQ2} - V_{CEQ2} = 2V_{CEQ2} = V_{CC} - 2R_E I_{EQ2} = V_{CC} - 2R_E\left(\frac{\beta_2 + 1}{\beta_2}\right)I_{CQ2}$$

Whence,

$$V_{CEQ2} = \frac{V_{CC}}{2} - \left(\frac{\beta_2 + 1}{\beta_2}\right)R_E I_{CQ2} = \frac{15}{2} - \left(\frac{30 + 1}{30}\right)(0.25)(10 \times 10^{-3}) = 7.474 \text{ V}$$

11.75 Determine the value of power supplied to the amplifier of Problem 11.73.

▌ Using the results of Problems 11.71 and 11.73, and (*1*) of Problem 11.53 but realizing that i_{C2} only flows for one-half of the source cycle (negative half-cycle load current is supplied by the large coupling capacitor C_C),

$$P_{CC} = V_{CC}I_{R1} + \frac{V_{CC}I_{cm}}{\pi} = V_{CC}\left(I_{R1} + \frac{I_{Lm}}{\pi}\right) = 15\left(3.226 \times 10^{-3} + \frac{713.4 \times 10^{-3}}{\pi}\right) = 3.455 \text{ W}$$

11.76 Find the value of maximum efficiency possible for the amplifier of Problem 11.73.

▌ Based on the results of Problems 11.72 and 11.75,

$$\eta_{\max} = \frac{P_{L\max}}{P_{CC}}(100\%) = \frac{2.54}{3.455}(100\%) = 73.5\%$$

DISTORTION CONSIDERATIONS

11.77 For small-signal operation, the base and collector currents of a BJT are related by $i_c \approx h_{fe}i_b$. However, for large signal excursions in a power amplifier, the relationship between i_c and i_b becomes nonlinear, and the following parabolic model often is used:

$$i_c = h_1 i_b + h_2 i_b^2 \tag{1}$$

Show that if the base current signal is given by

$$i_b = I_{bm} \cos \omega t \tag{2}$$

then the collector current contains both a dc component and a second-harmonic component, as well as the fundamental.

▌ Directly substituting (*2*) into (*1*) gives

$$i_c = h_1 I_{bm} \cos \omega t + h_2 I_{bm}^2 \cos^2 \omega t \tag{3}$$

After application of the trigonometric identity $\cos^2 x = \frac{1}{2} + \frac{1}{2}\cos 2x$, the total collector current can be written as

$$i_C = I_{CQ} + i_c = I_{CQ} + \tfrac{1}{2}h_2 I_{bm}^2 + h_1 I_{bm} \cos \omega t + \tfrac{1}{2}h_2 I_{bm}^2 \cos 2\omega t \tag{4}$$

from which the added dc and second-harmonic components are obvious.

11.78 The collector current of a BJT is experimentally observed to swing about a quiescent point I_{CQ} between extremes $I_{\max}$ and $I_{\min}$. Find expressions for h_1 and h_2 in (*4*) of Problem 11.77 in terms of the experimentally measured values of I_{CQ}, $I_{\min}$, and $I_{\max}$ if $i_b = I_{bm} \cos \omega t$.

▌ When $\omega t = 0$, $i_c = I_{max}$, and when $\omega t = \pi$, $i_c = I_{min}$. Substituting these values into (4) of Problem 11.77 gives the set of equations

$$I_{max} = I_{CQ} + \tfrac{1}{2}h_2 I_{bm}^2 + h_1 I_{bm} + \tfrac{1}{2}h_2 I_{bm}^2$$

$$I_{min} = I_{CQ} + \tfrac{1}{2}h_2 I_{bm}^2 - h_1 I_{bm} + \tfrac{1}{2}h_2 I_{bm}^2$$

whose simultaneous solution yields

$$h_1 = \frac{I_{max} - I_{min}}{2 I_{bm}} \qquad h_2 = \frac{I_{max} + I_{min} - 2 I_{CQ}}{2 I_{bm}^2} \qquad (1)$$

11.79 The *second-harmonic distortion factor* D_2 is defined as the ratio of the magnitude of the second harmonic to that of the fundamental component. Find an expression for the second-harmonic distortion factor of the collector current of a BJT in terms of experimentally measured values I_{max}, I_{min}, and I_{CQ} for an amplifier driven by a base current $i_b = I_{bm} \cos \omega t$.

▌ By the definition of D_2 and (4) of Problem 11.77,

$$D_2 = \left| \frac{\tfrac{1}{2} h_2 I_{bm}^2}{h_1 I_{bm}} \right| = \left| \frac{h_2 I_{bm}}{2 h_1} \right| \qquad (1)$$

Substituting the expressions for h_1 and h_2 from Problem 11.78 into (1) leads to

$$D_2 = \left| \frac{\dfrac{I_{max} + I_{min} - 2 I_{CQ}}{2 I_{bm}^2} I_{bm}}{\dfrac{2(I_{max} - I_{min})}{2 I_{bm}}} \right| = \left| \frac{I_{max} + I_{min} - 2 I_{CQ}}{2(I_{max} - I_{min})} \right| \qquad (2)$$

11.80 In the EF amplifier of Problem 4.76, let $i_s = i_b = 50 \cos \omega t$ μA. Find the parabolic gain coefficients h_1 and h_2.

▌ The Q point is shown in Fig. 4-15(b), where by inspection, $I_{CQ} \approx 6.3$ mA. Also, if $i_b = 50 \cos \omega t$ μA, it is determined that $I_{max} = 11.5$ mA and $I_{min} = 0.1$ mA. Hence, by (1) of Problem 11.78,

$$h_1 = \frac{I_{max} - I_{min} - 2 I_{CQ}}{2 I_{bm}^2} = \frac{11.5 \times 10^{-3} - 0.1 \times 10^{-3}}{2(50 \times 10^{-6})} = 114$$

$$h_2 = \frac{I_{max} + I_{min}}{2 I_{bm}^2} = \frac{11.5 \times 10^{-3} + 0.1 \times 10^{-3} - 2(6.3 \times 10^{-3})}{2(50 \times 10^{-6})^2} = -2 \times 10^5 \text{ A}^{-1}$$

11.81 Determine the second-harmonic distortion factor for the amplifier of Problem 4.76 if $i_s = i_b = 50 \cos \omega t$ μA.

▌ Based on (2) of Problem 11.79,

$$D_2 = \left| \frac{I_{max} + I_{min} - 2 I_{CQ}}{2(I_{max} - I_{min})} \right| = \left| \frac{11.5 + 0.1 - 2(6.3)}{2(11.5 - 0.1)} \right| = 0.0439$$

11.82 Distortion is greater in FET amplifiers than BJT amplifiers, owing to the obviously nonlinear separation of the drain characteristics. For FETs, the parabolic-gain model is written in terms of voltage as $v_{ds} = \mu_1 v_{gs} + \mu_2 v_{gs}^2$. As is done for BJTs in Problem 11.78, determine the gain constants μ_1 and μ_2 for FET amplifiers, in terms of the quiescent drain-source voltage V_{DSQ}, the peak value of the gate-source voltage V_{gsm}, and the maximum and minimum values of drain-source voltage excursion V_{max} and V_{min}.

▌ If $v_{gs} = V_{gsm} \cos \omega t$, then

$$v_{ds} = \mu_1 V_{gsm} \cos \omega t + \mu_2 V_{gsm}^2 \cos^2 \omega t \qquad (1)$$

The total drain-source voltage is given by

$$v_{DS} = V_{DSQ} + \tfrac{1}{2} \mu_2 V_{gsm}^2 + \mu_1 V_{gsm} \cos \omega t + \tfrac{1}{2} \mu_2 V_{gsm}^2 \cos 2\omega t \qquad (2)$$

When $\omega t = 0$, $v_{ds} = V_{max}$, and when $\omega t = \pi$, $v_{ds} = V_{min}$. Substituting these values into (2) gives the set of equations

$$V_{max} = V_{DSQ} + \tfrac{1}{2} \mu_2 V_{gsm}^2 + \mu_1 V_{gsm} + \tfrac{1}{2} \mu_2 V_{gsm}^2$$

$$V_{min} = V_{DSQ} + \tfrac{1}{2} \mu_2 V_{gsm}^2 - \mu_1 V_{gsm} + \tfrac{1}{2} \mu_2 V_{gsm}^2$$

Simultaneous solution of the above equations yields

$$\mu_1 = \frac{V_{\max} - V_{\min}}{2V_{gsm}} \qquad \mu_2 = \frac{V_{\max} + V_{\min} - 2V_{DSQ}}{2V_{gsm}^2} \qquad (3)$$

11.83 In an analogous manner to that of Problem 11.79, determine the second-harmonic distortion factor associated with a FET amplifier drain-source voltage if $v_{gs} = V_{gsm} \cos \omega t$.

▌ Based on the definition of D_2 given by Problem 11.79 and (2) of Problem 11.82,

$$D_2 = \left| \frac{\frac{1}{2}\mu_2 V_{gsm}^2}{\mu_1 V_{gsm}} \right| = \left| \frac{\mu_2 V_{gsm}}{2\mu_1} \right| \qquad (1)$$

Substituting the expressions for μ_1 and μ_2 from Problem 11.82 into (1) leads to

$$D_2 = \left| \frac{\frac{V_{\max} + V_{\min} - 2V_{DSQ}}{2V_{gsm}^2} V_{gsm}}{\frac{2(V_{\max} - V_{\min})}{2V_{gsm}}} \right| = \left| \frac{V_{\max} + V_{\min} - 2V_{DSQ}}{2(V_{\max} - V_{\min})} \right| \qquad (2)$$

11.84 For the FET amplifier of Problem 5.29, find the parabolic gain coefficients μ_1 and μ_2.

▌ From Fig. 5-12(b), $V_{DSQ} = 12.5$ V, $V_{\max} = 16$ V, $V_{\min} = 7.3$ V, and $V_{gsm} = 1$ V. By (3) of Problem 11.83,

$$\mu_1 = \frac{V_{\max} - V_{\min}}{2V_{gsm}} = \frac{16 - 7.3}{2(1)} = 4.35$$

$$\mu_2 = \frac{V_{\max} + V_{\min} - 2V_{DSQ}}{2V_{gsm}^2} = \frac{16 + 7.3 - 2(12.5)}{2(1)^2} = -0.85 \text{ V}^{-1}$$

11.85 Compute the second-harmonic distortion factor for the FET amplifier of Problem 5.29.

▌ Using the graphically determined values of Problem 11.84 and (2) of Problem 11.83,

$$D_2 = \left| \frac{V_{\max} + V_{\min} - 2V_{DSQ}}{2(V_{\max} - V_{\min})} \right| = \left| \frac{16 + 7.3 - 2(12.5)}{2(16 - 7.3)} \right| = 0.098$$

11.86 The second-harmonic distortion factor for the amplifier of Fig. 11-3(a) is 5 percent ($D_2 = 0.05$). Determine the error in load power calculation if only the power delivered at the fundamental frequency is considered.

▌ Based on the result of Problem 1.107,

$$P_L = \frac{I_{Lm}^2}{2} R_L = (I_1^2 + I_2^2) \frac{R_L}{2} \qquad (1)$$

From (4) of Problem 11.77, (1) can be written as

$$P_L = \left[(h_1 I_{bm})^2 + \left(\frac{1}{2} h_2 I_{bm} \right)^2 \right] \frac{R_L}{2} = \left[1 + \left(\frac{h_2 I_{bm}}{2h_1} \right)^2 \right] h_1 I_{bm}^2 \frac{R_L}{2} \qquad (2)$$

Recognize the post-factored term of (2) as the fundamental frequency power P_1 and use (1) of Problem 11.79 to rewrite (2) as

$$P_L = (1 + D_2^2) P_1 \qquad (3)$$

Substitution for D_2 in (3) gives

$$P_L = [1 + (0.05)^2] P_1 = 1.0025 P_1$$

Hence, the error if second-harmonic power were neglected is 0.25%.

11.87 *Turn-on distortion* occurs at the beginning of an output-signal half-cycle, and *turn-off distortion* at the end; in both cases, distortion exists if actual $v_{ce} = 0$ for $0 < v_{be} < V_{BEQ}$. The amplifier of Fig. 11-4(a) is to be operated as a class B amplifier with maximum swing and no *turn-on distortion*. If $R_L = 16 \, \Omega$, $R_E = 0$, $V_{CC} = 24$ V, $a = 0.5$, and $R_2 = 2 \, \text{k}\Omega$, select R_1 to establish class B operation for this Si transistor.

▮ For class B operation, the transistor must be biased so that $V_{BEQ} = 0.7$ V, or

$$V_{BEQ} = 0.7 = \frac{R_2}{R_1 + R_2} V_{CC} \tag{1}$$

Solving (1) for R_1 yields

$$R_1 = \left(\frac{V_{CC}}{V_{BEQ}} - 1\right) R_1 = \left(\frac{24}{0.7} - 1\right)(2 \times 10^3) = 66.57 \text{ k}\Omega$$

11.88 For maximum signal swing with $v_s = V_m \sin \omega t$, neglect power dissipated by the R_1-R_2 bias network and find an expression for the power supplied to amplifier of Fig. 11-4(a) for class B operation.

▮ Since the class B amplifier is biased at cutoff, $I_{CQ} = 0$ and V_{CC} only supplies power during the 180° conduction interval. Hence,

$$P_{CC} = \frac{1}{T}\int_0^T V_{CC} i_C \, dt = \frac{V_{CC}}{2\pi}\int_0^\pi I'_{Lm} \sin \omega t \, d(\omega t) = \frac{V_{CC} I'_{Lm}}{\pi} = \frac{V_{CC}}{\pi}\frac{V'_{Lm}}{R'_L} = \frac{V_{CC}}{\pi}\frac{V_{CC}}{a^2 R_L} = \frac{V_{CC}^2}{\pi a^2 R_L} \tag{1}$$

11.89 Neglecting the power dissipated by the bias network, find the value of power supplied to the amplifier of Problem 11.87 for maximum signal swing.

▮ By (1) of Problem 11.88,

$$P_{CC} = \frac{V_{CC}^2}{\pi a^2 R_L} = \frac{(24)^2}{\pi (0.5)^2 (16)} = 45.84 \text{ W}$$

11.90 For maximum signal swing with $v_s = V_m \sin \omega t$, find an expression for the power delivered to the load R_L by the amplifier of Fig. 11-4(a) for class B operation.

▮ The maximum power delivered to R_L for the half-cycle conduction is

$$P_{L\,\max} = \frac{1}{T}\int_0^T i_L^2 R_L \, dt = \frac{1}{2\pi}\int_0^\pi \frac{(V'_{Lm})^2}{R'_L}\sin^2 \omega t \, d(\omega t) = \frac{V_{CC}^2}{2\pi a^2 R_L}\int_0^\pi \sin^2 \omega t \, d(\omega t)$$

$$= \frac{V_{CC}^2}{2\pi a^2 R_L}\int_0^\pi \frac{1}{2}(1 - \cos 2\omega t)\, d(\omega t) = \frac{V_{CC}^2}{4 a^2 R_L} \tag{1}$$

11.91 Find the maximum value of load power delivered by the amplifier of Problem 11.87.

▮ By (1) of Problem 11.90,

$$P_{L\,\max} = \frac{V_{CC}^2}{4 a^2 R_L} = \frac{(24)^2}{4(0.5)^2(16)} = 36 \text{ W}$$

11.92 Find the maximum possible efficiency for the class B amplifier of Problem 11.87.

▮ Based on the results of Problems 11.89 and 11.91,

$$\eta_{\max} = \frac{P_{L\,\max}}{P_{CC}}(100\%) = \frac{36}{45.84}(100\%) = 78.54\%$$

THERMAL CONSIDERATIONS

11.93 Effective and safe utilization of power transistors requires primary attention to voltage, current, power, and temperature limitations, which are of only secondary concern in small-signal amplifiers. Manufacturer-furnished transistor specifications include

$i_{C\,\max} \equiv$ maximum continuous collector current
$BV_{CEO} \equiv$ avalanche breakdown voltage ($v_{CE} > 0$) with $i_B = 0$
$P_{C\,\max} \equiv$ maximum average power that the transistors can dissipate with case temperature $\leq T_{CO}$
$T_{CO} \equiv$ maximum case temperature without derating from $P_{C\,\max}$ (typical value is 25°C)
$T_{J\,\max} \equiv$ maximum allowable junction (collector-base) temperature (80 to 100°C for Ge, 125 to 200°C for Si)

In active-region operation, the reverse-biased collector junction is the greatest source of heat; thus, the power dissipation is associated with the collector junction. A Si power transistor is biased so that $V_{CEQ} = 14$ V and

$I_{CQ} = 100$ mA. At quiescent conditions, determine the power dissipated by the base-emitter junction and the power dissipated by the collector-base junction.

▌ For active-region operation, $V_{BEQ} \approx 0.7$ V and

$$P_B = V_{BEQ}I_{CQ} = 0.7(0.1) = 0.07 \text{ W}$$

By KVL,
$$V_{CBQ} = V_{CEQ} - V_{BEQ} = 14 - 0.7 = 13.3 \text{ V}$$

so that
$$P_C = V_{CBQ}I_{CQ} = 13.7(0.1) = 1.37 \text{ W}$$

In this typical example, 97.8 percent of the power dissipated by the transistor is associated with the collector junction.

11.94 The manufacturer's specifications of Problem 11.93 can be used to sketch the *dissipation derating curve* of Fig. 11-12, which shows the allowable collector power dissipation P_C as a function of transistor case temperature T_C. The dissipation derating curve follows from the average-value heat-flow model for the BJT in Fig. 11-13, wherein the average value of the collector power dissipation, temperature, and *thermal resistance* are treated, respectively, as electrical analogs of current, voltage, and resistance. The value of the junction-to-case thermal resistance θ_{jc} (measured in degrees Celsius per watt) is given by the manufacturer. A typical value for the case-to-heat-sink thermal resistance θ_{cs} is usually suggested by the transistor manufacturer, but the user must ensure the integrity of the case-to-heat-sink junction. The heat-sink-to-ambient thermal resistance θ_{sa} is specified by the heat-sink manufacturer. The case-to-ambient thermal impedance θ_{ca} is formed by series combination as $\theta_{ca} = \theta_{cs} + \theta_{sa}$. A power transistor is to operate with $I_{CQ} = 1.25$ A and $V_{CEQ} = 7.5$ V. If $\theta_{jc} = 3°$C/W and $\theta_{ca} = 10°$C/W, determine the junction temperature if the transistor is operating in a 40°C ambient.

▌ The average power dissipated by the transistor collector is

$$P_C = V_{CEQ}I_{CQ} = 7.5(1.25) = 9.375 \text{ W}$$

Applying KVL to the analog circuit of Fig. 11-13 yields

$$T_J = P_C(\theta_{jc} + \theta_{ca}) + T_A = 9.375(3 + 10) + 40 = 161.9°C \qquad (1)$$

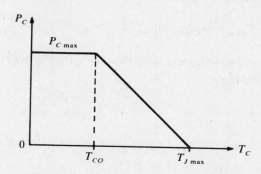

Fig. 11-12 Dissipation derating curve.

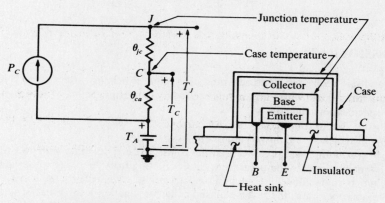

Fig. 11-13 Average-value heat-flow model for BJT.

11.95 A power transistor is to dissipate 10 W of power. Let $\theta_{jc} = 2.5°C/W$ and $\theta_{cs} = 0.5°C/W$. The maximum allowable junction temperature is $T_{J\,max} = 150°C$. The device must operate in a maximum ambient of 50°C. Three different heat sinks are available, with thermal resistances of 3, 6, and 9°C/W. Select the best heat sink.

▮ Letting $\theta_{ca} = \theta_{cs} + \theta_{sa}$ in (1) of Problem 11.94 and rearranging yield

$$\theta_{sa} \leq \frac{T_{J\,max} - T_A}{P_C} - \theta_{jc} - \theta_{cs} = \frac{150 - 50}{10} - 2.5 - 0.5 = 7°C/W$$

We choose the heat sink with $\theta_{sa} = 6°C/W$, as it is the smaller of the heat sinks that will be adequate.

11.96 A 5-W Si power transistor is to dissipate 5 W in an ambient temperature of 40°C; the device is mounted on a heat sink for which $\theta_{sa} = 10°C/W$. Find the junction temperature T_J if the thermal resistance is $\theta_{jc} = 2.4°C/W$.

▮ Application of KVL to the heat-flow model of Fig. 11-13 gives

$$T_J = T_A + P_C(\theta_{jc} + \theta_{ca}) = 40 + 5(2.4 + 10) = 102°C$$

11.97 Determine the maximum ambient temperature in which the transistor of Problem 11.96 can operate if the maximum allowable junction temperature is 125°C.

▮ Applying the heat-flow model of Fig. 11-13 leads to

$$T_A = T_J - P_C(\theta_{jc} + \theta_{ca}) = 125 - 5(2.4 + 10) = 63°C$$

Review Problems

11.98 For the Si power transistor of Fig. 11-4(a), $h_{FE} = \beta = 40$ and $h_{ie} = 25\,\Omega$. Assume that the transformer is ideal and that $V_{CE\,sat} \approx 0$ and $I_{CBO} = 0$. Let $V_{CC} = 12$ V, $a = 1/6$, $R_E = 0$, and $R_L = 2.5\,k\Omega$. Find the values of R_1 and R_2 that are needed to set $I_{CQ} = 100$ mA.

▮ By KVL around base-emitter loop,

$$V_{BB} = V_{BEQ} + I_{BQ}R_B = V_{BEQ} + \frac{I_{CQ}}{\beta}R_B = 0.7 + \frac{100 \times 10^{-3}}{40}R_B = 0.7 + 0.0025R_B \quad (1)$$

Since $R_E = 0$, there is no emitter circuit constraint, and thus, any value of R_B that satisfies (1) giving $V_{BB} \leq V_{CC}$ is acceptable. Arbitrarily select $R_B = 2\,k\Omega$, then by (1)

$$V_{BB} = 0.7 + 0.0025(2 \times 10^3) = 5.7\,V$$

By (1) of Problem 4.38,

$$R_1 = \frac{R_B}{1 - V_{BB}/V_{CC}} = \frac{2 \times 10^3}{1 - 5.7/12} = 3.809\,k\Omega$$

$$R_2 = R_B \frac{V_{CC}}{V_{BB}} = (2 \times 10^3)\frac{12}{5.7} = 4.21\,k\Omega$$

11.99 Determine the peak values of output voltage and current for class A operation of the amplifier of Problem 11.98.

▮ $$R_{ac} = a^2 R_L = (\tfrac{1}{6})^2(2.5 \times 10^3) = 69.4\,\Omega$$

Now assume that i_C can swing ± 100 mA about $I_{CQ} = 100$ mA and (2) of Problem 4.72:

$$v_{CE\,max} = V_{CEQ} + I_{CQ}R_{ac} = 12 + (100 \times 10^{-3})(69.4) = 18.94\,V$$

Since $v_{CE\,max} < 2V_{CC}$, cutoff occurs before saturation and the assumption on i_C is valid. Therefore,

$$V_{Lm} = \frac{1}{a}V'_{Lm} = \frac{1}{a}I_{CQ}R_{ac} = 6(100 \times 10^{-3})(69.4) = 41.64\,V$$

$$I_{Lm} = aI_{cm} = aI_{CQ} = \tfrac{1}{6}(100 \times 10^{-3}) = 16.7\,mA$$

11.100 In the inductor-coupled class A power amplifier of Fig. 11-3(a), $V_{CC} = 15$ V, $R_2 = 20$ Ω, $R_1 = 180$ Ω, $R_E = 1$ Ω, $R_L = 75$ Ω, $R = 2$ Ω, $L \to \infty$, and $\beta = 20$. Assume a Si transistor. Find equations for the instantaneous values of maximum undistorted v_{CE}, i_C, and i_L if $v_s = V_m \sin \omega t$.

By (1) of Problem 4.38,

$$V_{BB} = \frac{R_2}{R_2 + R_1} V_{CC} = \frac{20}{20 + 180}(15) = 1.5 \text{ V}$$

$$R_B = \frac{R_1 R_2}{R_1 + R_2} = \frac{180(20)}{180 + 20} = 18 \text{ Ω}$$

From (4) of Problem 6.4 with $I_{CBO} = 0$,

$$I_{CQ} = \frac{V_{BB} - V_{BEQ}}{R_B/\beta + R_E} = \frac{1.5 - 0.7}{18/20 + 1} = 0.421 \text{ A}$$

From Problem 4.41,

$$V_{CEQ} = V_{CC} - I_{CQ} R_{dc} = V_{CC} - I_{CQ}(R_E + R) = 15 - 0.421(1 + 2) = 13.737 \text{ V}$$

$$v_{CE\,max} = V_{CEQ} + I_{CQ} R_{ac} = 13.737 + 0.421(75) = 45.312 \text{ V}$$

$$i_{C\,max} = \frac{v_{CE\,max}}{R_L} = \frac{45.312}{75} = 0.604 \text{ A}$$

Thus,

$$i_C = I_{CQ} + i_c = I_{CQ} + (i_{C\,max} - I_{CQ}) \sin \omega t = 0.421 + (0.604 - 0.421) \sin \omega t = 0.421 + 0.183 \sin \omega t \text{ A}$$

$$v_{CE} = V_{CEQ} - V_{CEQ} \sin \omega t = 13.737(1 - \sin \omega t)$$

$$i_L = \frac{v_{ce}}{R_L} = \frac{-13.737 \sin \omega t}{75} = -0.183 \sin \omega t \text{ A}$$

11.101 Specify the minimum ratings for the transistor of the inductor-coupled amplifier in Problem 11.100.

$$i_{C\,max} \geq \frac{v_{CE\,max}}{R_L} = \frac{45.312}{75} = 0.604 \text{ A}$$

$$v_{CE\,max} \geq 45.312 \text{ V}$$

$$P_{C\,max} \geq V_{CEQ} I_{CQ} = 13.737(0.421) = 5.78 \text{ W}$$

11.102 Determine the maximum value of efficiency for the amplifier of Problem 11.100 considering the power dissipated in the R_1-R_2 bias network.

$$V_{R2} \approx I_{CQ} R_E + V_{BEQ} = 0.421(1) + 0.7 = 1.121 \text{ V}$$

$$I_{R1} = \frac{V_{CC} - V_{R2}}{R_1} = \frac{15 - 1.121}{180} = 0.077 \text{ A}$$

$$P_{CC} = V_{CC}(I_{CQ} + I_{R1}) = 15(0.421 + 0.077) = 7.47 \text{ W}$$

$$P_{L\,max} = \frac{I_{Lm}^2}{2} R_L = \frac{(0.183)^2}{2}(75) = 1.256 \text{ W}$$

Hence,

$$\eta_{max} = \frac{P_{L\,max}}{P_{CC}}(100\%) = \frac{1.256}{7.47}(100\%) = 16.81\%$$

11.103 The phase-inverter circuit of Problem 11.60 was shown to yield output voltages v_1 and v_2 that are equal in magnitude and 180° out of phase with no load present. Show that if this circuit is used to replace the input transformer T_1 of Fig. 11-5, then $v_1/v_2 = -1$ if Q_1 and Q_2 are identical, and the amplifier gives class B operation without turn-on distortion.

If both Q_1 and Q_2 are biased for class B operation without distortion (see Problem 11.87), then the input impedance (Z_{in}) looking into the base-emitter port of each transistor is identical. Hence, R_3 and R_4 in (1) of

Problem 11.60 are replaced, respectively, by $R_3 \| Z_{in}$ and $R_4 \| Z_{in}$. With $R_3 = R_4$,

$$\frac{v_1}{v_2} = \frac{h_{fe}i_b R_4 \| Z_{in}}{-h_{fe}i_b R_3 \| Z_{in}} = -1$$

11.104 The complementary-symmetry push-pull amplifier of Fig. 11-9 is required to deliver a maximum power of 20 W into a 10-Ω load resistor R_L. Find the necessary value of V_{CC}.

▮ The maximum load voltage is equal to V_{CC}; whence,

$$V_{CC} = V_{Lm} = \sqrt{2 P_{L\,max} R_L} = \sqrt{2(20)(10)} = 20 \text{ V}$$

11.105 Determine the minimum ratings for the transistors of the amplifier described by Problem 11.104.

▮ Since the maximum efficiency of a class B amplifier is 78.54 percent,

$$0.7854 = \frac{P_{L\,max}}{P_{CC}} = \frac{P_{L\,max}}{(2/\pi)V_{CC}I_{cm}}$$

or

$$i_{C\,max} \geq I_{cm} = \frac{\pi P_{L\,max}}{2V_{CC}(0.7854)} = \frac{\pi(20)}{2(20)(0.7854)} = 2 \text{ A}$$

By (1) of Problem 11.70 with $R_C = 0$,

$$P_{C\,max} \geq \frac{V_{CC}^2}{\pi^2 R_L} = \frac{(20)^2}{\pi^2(20)} = 2.026 \text{ W}$$

$$v_{CE\,max} \geq 2V_{CC} = 2(20) = 40 \text{ V}$$

11.106 To eliminate *crossover distortion* (combined turn-on and turn-off distortion in the transistors) in push-pull amplifiers, the base-emitter junctions are forward-biased so that transistor operation is linear for small values of the input signal. In the complementary-symmetry push-pull amplifier of Fig. 11-10, let $R_1 = 15$ kΩ, and select R_2 so that $V_{BEQ2} = -V_{BEQ1} = 0.7$ V if $V_{CC} = 10$ V.

▮ Each transistor must be biased so that $V_{BEQ} = 0.7$ V. Hence, we want

$$\frac{R_1}{R_1 + R_2} V_{CC} = V_{BEQ} = 0.7$$

or

$$R_2 = R_1\left(\frac{V_{CC}}{0.7} - 1\right) = (15 \times 10^3)\left(\frac{10}{0.7} - 1\right) = 199.3 \text{ k}\Omega$$

11.107 A 5-W Si power transistor is to dissipate 3 W in a particular circuit. The maximum allowable junction temperature is 125°C, the ambient temperature is 40°C, and $\theta_{jc} = 2.4$°C/W. Determine the maximum allowable thermal resistance θ_{ca} between the case and ambient.

▮ Based on the thermal model of Fig. 11-13,

$$T_J - T_A = P_C(\theta_{jc} + \theta_{ca})$$

or

$$\theta_{ca} = \frac{T_J - T_A}{P_C} - \theta_{jc} = \frac{125 - 40}{3} - 2.4 = 25.93 \text{°C/W}$$

11.108 Find the case temperature of the transistor in Problem 11.107.

▮
$$T_C = T_J - P_C \theta_{jc} = 125 - 3(2.4) = 117.8 \text{°C}$$

11.109 For the transformer-coupled amplifier of Fig. 11-4(a), let $R_1 = 1$ kΩ, $R_2 = 250$ Ω, $V_{CC} = 15$ V, $R_L = 4$ Ω, $R_E = 10$ Ω, $a = 4$, transformer primary resistance $R_p = 5$ Ω, transformer secondary resistance $R_s = 1$ Ω, and $\beta = 20$ for the Si transistor. Find the largest value of power that can be delivered to the load with symmetrical swing if the transistor has ideal collector characteristics.

▮ By KCL at the base node,

$$I_1 - I_2 = \frac{15 - V_2}{1 \times 10^3} - \frac{V_2}{100} = I_{BQ} = \frac{I_{CQ}}{\beta} = \frac{I_{CQ}}{20} \qquad (1)$$

Apply KVL around the base-emitter loop to find

$$V_2 = \frac{\beta+1}{\beta} I_{CQ} R_E + V_{BEQ} = \frac{20+1}{20} I_{CQ}(10) + 0.7 \qquad (2)$$

Simultaneous solution of (1) and (2) gives $I_{CQ} = 94.07$ mA. The transistor is not necessarily biased for maximum symmetrical swing, so we must check to see if undistorted swing is limited by saturation or cutoff. By KVL,

$$V_{CEQ} = V_{CC} - I_{CQ} R_{dc} = V_{CC} - I_{CQ}(R_E + R_p) = 15 - (94.07 \times 10^{-3})(10+5) = 13.59 \text{ V}$$

By (1) of Problem 4.72,

$$i_{C\max} = \frac{V_{CEQ}}{R_{ac}} + I_{CQ} = \frac{V_{CEQ}}{R_p + a^2(R_s + R_L)} + I_{CQ} = \frac{13.59}{5 + (4)^2(1+4)} + 94.07 \times 10^{-3} = 253.9 \text{ mA}$$

Since $i_{C\max} > 2I_{CQ}$, undistorted operation is limited by cutoff. Consequently, $I_{cm} = I_{CQ}$, and

$$P_{L\max} = \frac{(aI_{cm})^2}{2} R_L = \frac{[4(94.07 \times 10^{-3})]^2}{2}(4) = 283 \text{ mW}$$

11.110 Determine the power supplied to the amplifier of Problem 11.109.

▎ By (2) of Problem 11.109,

$$V_2 = 10.5 I_{CQ} + 0.7 = 10.5(94.07 \times 10^{-3}) + 0.7 = 11.86 \text{ V}$$

From Fig. 11-4(a),

$$I_1 = \frac{V_{CC} - V_2}{R_1} = \frac{15 - 11.86}{1 \times 10^3} = 3.14 \text{ mA}$$

Hence,
$$P_{CC} = V_{CC}(I_1 + I_{CQ}) = 15(3.14 \times 10^{-3} + 94.07 \times 10^{-3}) = 1.458 \text{ W}$$

11.111 Find the maximum efficiency of the amplifier in Problem 11.109.

▎ Based on the results of Problems 11.109 and 11.110,

$$\eta_{\max} = \frac{P_{L\max}}{P_{CC}}(100\%) = \frac{0.283}{1.458}(100\%) = 19.41\%$$

CHAPTER 12
Feedback Amplifiers

FEEDBACK CONCEPTS

12.1 Any single- or multiple-stage amplifier can be modeled as a two-port network with a forward transfer ratio A (which may be a current-gain ratio $A_i = i_L/i_S$; a voltage-gain ratio $A_v = v_L/v_S$; a *transresistance ratio* $A_r = v_L/i_S$; or a *transconductance ratio* $A_g = i_L/v_S$). If an output port variable of the basic amplifier is sampled, adjusted by a feedback transfer ratio β (which may be a voltage-gain ratio $\beta_{vv} = v_f/v_a$; a current-gain ratio $\beta_{ii} = i_f/i_L$; a transresistance ratio $\beta_{iv} = v_f/i_L$; or a transconductance ratio $\beta_{vi} = i_L/v_f$) and compared (mixed) with an external signal (from a signal source) to form the driving input signal for the basic amplifier, a *feedback amplifier* is formed. This result is illustrated by the *functional block diagram* of Fig. 12-1(a). Convert the functional block diagram of Fig. 12-1(a) to a *mathematical block diagram* (or simply *block diagram*).

▌ See Figure 12-1(b).

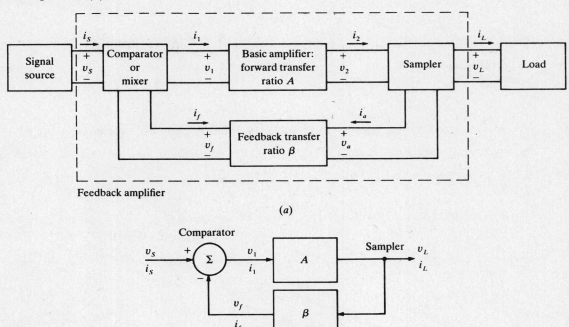

Fig. 12-1 Feedback amplifier.

12.2 Let the forward transfer ratio A and the feedback transfer ratio β for an amplifier modeled by the block diagram of Fig. 12-1(b) both be voltage-gain ratios. Find an expression for the overall voltage gain A_{vf} for the feedback amplifier.

▌ The forward path through gain block A gives

$$v_L = A_v v_1 \tag{1}$$

Using the comparator relationship and the feedback path gain,

$$v_1 = v_S - v_f = v_S - \beta_{vv} v_L \tag{2}$$

Substituting (2) into (1) and rearranging give the overall voltage gain of the feedback amplifier as

$$A_f = A_{vf} \equiv \frac{v_L}{v_S} = \frac{A_v}{1 + \beta_{vv} A_v} \tag{3}$$

To ensure *negative feedback* (so that the signal v_f will be subtracted from the signal v_S by the comparator to form v_1), the signs of A_v and β_{vv} must be identical; thus, from (3) it is apparent that $|A_{vf}| < |A|$.

12.3 Let an amplifier have the upper cutoff frequency ω_H without feedback; then its gain A for high-frequency operation can usually be written in terms of the midfrequency gain $A_o (= A_{\text{mid}})$ as

$$A = \frac{A_o}{1 + j\omega/\omega_H} \qquad (1)$$

Show that negative feedback extends the range of high-frequency response for this amplifier.

▌ From (3) of Problem 12.2 and (1), the gain with negative feedback is then

$$A_f = \frac{A}{1 + \beta A} = \frac{\dfrac{A_o}{1 + j\omega/\omega_H}}{1 + \dfrac{\beta A_o}{1 + j\omega/\omega_H}} = \frac{A_o}{1 + \beta A_o + j\omega/\omega_H} \qquad (2)$$

The upper cutoff frequency with negative feedback is the frequency at which the real and imaginary parts of the denominator of (2) are equal; thus,

$$\omega_{Hf} \equiv \omega_H(1 + \beta A_o) \qquad (3)$$

from which we see that negative feedback acts to increase the upper cutoff frequency.

12.4 The low-frequency gain of an amplifier without feedback can usually be written in terms of the lower cutoff frequency ω_L as

$$A = \frac{A_o}{1 + j\omega_L/\omega} \qquad (1)$$

Show that negative feedback extends the range of low-frequency response for this amplifier.

▌ Paralleling the process by which we obtained (3) of Problem 12.3, we find that the lower cutoff frequency with feedback is

$$\omega_{Lf} \equiv \frac{\omega_L}{1 + \beta A_o} \qquad (2)$$

indicating that it is reduced by the negative feedback. The combined results of this problem and Problem 12.3 show that negative feedback increases the *bandwidth* (or midfrequency response range) of an amplifier.

12.5 An amplifier has a midfrequency gain $A_o = 100$. For a feedback transfer ratio $\beta = 0.1$, find the overall gain at midfrequency.

▌ From the general form of (3) of Problem 12.2,

$$A_{fo} = \frac{A_o}{1 + \beta A_o} = \frac{100}{1 + 0.1(100)} = 9.09$$

12.6 Given an upper cutoff frequency $f_H = 40$ kHz for the amplifier of Problem 12.5, determine the upper cutoff frequency with negative feedback implemented.

▌ From (3) of Problem 12.3,

$$f_{Hf} = f_H(1 + \beta A_o) = (40 \times 10^3)[1 + 0.1(100)] = 440 \text{ kHz}$$

12.7 Given a lower cutoff frequency $f_L = 20$ Hz for the amplifier of Problem 12.5, evaluate the lower cutoff frequency with negative feed-back implemented.

▌ From (2) of Problem 12.4,

$$f_{Lf} = \frac{f_L}{1 + \beta A_o} = \frac{20}{1 + 0.1(100)} = 1.82 \text{ Hz}$$

12.8 The addition of negative feedback to an amplifier reduces the midfrequency gain while increasing the bandwidth. If $f_L \ll f_H$ and $f_{Lf} \ll f_{Hf}$, show that the *gain-bandwidth product* $\text{GBW} \equiv A_o(f_H - f_L) \approx A_o f_H$ remains approximately constant; that is, show that $\text{GBW} \approx \text{GBW}_f$.

■ From (2) of Problem 12.3, the midfrequency gain ($\omega \ll \omega_H$) is given by

$$A_f = \frac{A_o}{1 + \beta A_o} \quad (1)$$

Use (2) of Problem 12.3 and (1) to form the midfrequency gain-bandwidth product with feedback product as

$$\text{GBW}_f = A_f f_{Hf} = \left[\frac{A_o}{1 + \beta A_o}\right] f_H (1 + \beta A_o) = A_o f_H \quad (2)$$

Clearly, $\text{GBW} = \text{GBW}_f$.

12.9 If distortion is introduced into the output voltage of a high-gain amplifier or, as commonly occurs, into the output (power) stage of a cascaded amplifier, then feedback can be used to reduce the amplitude of the distortion. A model for a voltage-series feedback amplifier, with distortion introduced as voltage v_D, is shown in Fig. 12-2, where the gain A_v may be the product of the gains of cascaded stages as long as v_1 and v_2 are in phase. Find the output v_L in terms of v_s and v_D, to show that the amplitude of the distortion voltage is reduced if $\beta_{vv} A_v \gg 1$.

■ KVL around the input loop requires that

$$v_1 = v_s - v_f = v_s - \beta_{vv} v_L \quad (1)$$

The gain relationship for the basic amplifier and an application of KVL around the output loop then lead to

$$v_1 = \frac{v_2}{A_v} = \frac{v_L - v_D}{A_v} \quad (2)$$

Equating (1) and (2) and solving for v_L give

$$v_L = \frac{A_v}{1 + \beta_{vv} A_v} v_S + \frac{v_D}{1 + \beta_{vv} A_v} \quad (3)$$

Equation (3) shows that the magnitude of the distortion voltage is reduced significantly if $\beta_{vv} A_v \gg 1$.

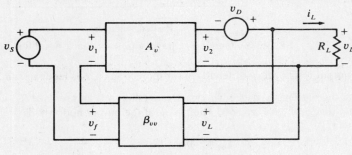

Fig. 12-2

VOLTAGE-SERIES FEEDBACK

12.10 The general feedback amplifier model of Fig. 12-3, in which output voltage v_L is the sampled variable and feedback voltage v_f is the mixed variable, is called a *voltage-series* (or *parallel-output, series-input*; or *node-sampling, loop-comparison*) feedback amplifier. For the voltage-series feedback amplifier of Fig. 12-3, find the input resistance with feedback.

■ The method of node voltages (with $G_i = 1/R_i$) gives

$$(v_f - v_S)G_{11} + v_f G_1 + (v_f - v_L)G_2 = 0 \quad (1)$$
$$(v_L - v_f)G_2 - \beta i_1 + v_L G_{22} + v_L G_L = 0 \quad (2)$$

But
$$i_1 = (v_S - v_f)G_{11} \quad (3)$$

Substituting (3) into (2) and rearranging yield the following set of linear equations:

$$\begin{bmatrix} G_{11} + G_1 + G_2 & -G_2 \\ -(G_2 - \beta G_{11}) & G_2 + G_{22} + G_L \end{bmatrix} \begin{bmatrix} v_f \\ v_L \end{bmatrix} = \begin{bmatrix} G_{11} v_s \\ \beta G_{11} v_S \end{bmatrix} \quad (4)$$

Simultaneous solution of (4) by means of Cramer's rule gives the feedback voltage v_f as

$$v_f = \frac{\Delta_1}{\Delta} = \frac{\begin{vmatrix} G_{11}v_S & -G_2 \\ \beta G_{11}v_S & G_2 + G_{22} + G_L \end{vmatrix}}{\begin{vmatrix} G_{11} + G_1 + G_2 & -G_2 \\ -(G_2 - \beta G_{11}) & G_2 + G_{22} + G_L \end{vmatrix}} = \frac{v_S G_{11} G}{\Delta} \qquad (5)$$

where
$$G = (1 + \beta)G_2 + G_{22} + G_L \qquad (6)$$

and
$$\Delta = G_{11}G + G_1 G_2 + (G_1 + G_2)(G_{22} + G_L) \qquad (7)$$

From Fig. 12-3 and (3) and (5), we see that

$$R_{\text{in}f} = R_{11} + \frac{v_f}{i_1} = R_{11} + \frac{v_f}{(v_S - v_f)G_{11}} = R_{11} + \frac{G}{\Delta - G_{11}G} \qquad (8)$$

Substituting (6) and (7) into (8) and rearranging then lead to

$$R_{\text{in}f} = R_{11} + \frac{(1 + \beta)G_2 + G_{22} + G_L}{G_1 G_2 + (G_1 + G_2)(G_{22} + G_L)} \qquad (9)$$

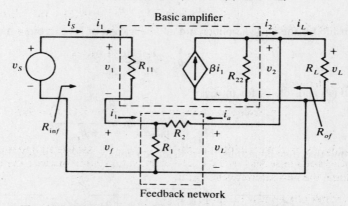

Fig. 12-3 General voltage-series feedback amplifier model.

12.11 Determine the output resistance with feedback for the voltage-series feedback amplifier of Fig. 12-3.

▌ To determine the output resistance, we deactivate (short) v_S and replace R_L with a driving-point source with polarity matching that of v_L in Fig. 12-3. Then KCL at the output node yields

$$i_{dp} = -\beta i_1 + G_{22}v_{dp} + \frac{v_{dp}}{R_2 + R_{11} \| R_1} \qquad (1)$$

and current division at the node within the feedback network gives

$$i_1 = \frac{-R_1}{R_{11} + R_1} \quad i_a = \frac{-R_1}{R_{11} + R_1} \frac{v_{dp}}{R_2 + R_{11} \| R_1} \qquad (2)$$

Substituting (2) into (1) leads to an expression for the output conductance:

$$G_{of} = \frac{1}{R_{of}} = \frac{i_{dp}}{v_{dp}} = \frac{1}{R_{22}} + \frac{(1 + \beta R_1)/(R_{11} + R_1)}{R_2 + R_{11} \| R_1} = G_{22} + \frac{(1 + \beta)G_{11} + G_1}{(G_{11} + G_1)R_2 + 1} \qquad (3)$$

12.12 Find the input and output resistances for the amplifier of Fig. 12-3 if the feedback is inactive.

▌ The feedback is deactivated if $R_1 = 1/G_1 = 0$ and $R_2 = 1/G_2 = \infty$; thus, from (9) of Problem 12.10 and (3) of Problem 12.11,

$$R_{\text{in}} = \lim_{\substack{G_1 \to \infty \\ G_2 \to 0}} R_{\text{in}f} = R_{11} \qquad (1)$$

$$R_o = \lim_{\substack{G_1 \to \infty \\ G_2 \to 0}} \frac{1}{G_{of}} = R_{22} \qquad (2)$$

12.13 In the voltage-series feedback amplifier of Fig. 12-3, let $G_1(=1/R_1) = 1 \times 10^{-3}$ S, $G_2 = 5 \times 10^{-5}$ S, $G_{11} = 0.01$ S, $G_{22} = 1 \times 10^{-6}$ S, $G_L = 0.5 \times 10^{-3}$ S, and $\beta = 75$. Determine the percentage change in input resistance R_{in} due to the addition of feedback.

▮ Without feedback, by (1) of Problem 12.12, $R_{in} = R_{11} = 1/G_{11} = 100\ \Omega$. With feedback, (9) of Problem 12.10 gives

$$R_{in\,f} = R_{11} + \frac{(1+\beta)G_2 + G_{22} + G_L}{G_1 G_2 + (G_1 + G_2)(G_{22} + G_L)} = 100 + \frac{76(5 \times 10^{-5}) + 10^{-6} + 0.5 \times 10^{-3}}{(10^{-3})(5 \times 10^{-5}) + (1.05 \times 10^{-3})(501 \times 10^{-6})} = 7.566\ \text{k}\Omega$$

The percentage change is then

$$\text{percentage increase} = \frac{7466}{100}(100\%) = 7466\%$$

12.14 Find the percentage change in output resistance R_o due to the addition of feedback to the amplifier of Fig. 12-3 with the parameters of Problem 12.13.

▮ By (2) of Problem 12.12, $R_o = R_{22} = 1/G_{22} = 1\ \text{M}\Omega$. With feedback, from (3) of Problem 12.11,

$$G_{of} = G_{22} + \frac{(1+\beta)G_{11} + G_1}{(G_{11} + G_1)R_2 + 1} = 10^{-6} + \frac{76(0.01) + 10^{-3}}{0.011[1/(5 \times 10^{-5})] + 1} = 0.003444\ \text{S}$$

so
$$R_{of} = 1/G_{of} = 290.36\ \Omega$$

and the percentage change is

$$\text{percentage decrease} = \frac{99.971 \times 10^4}{1 \times 10^6}(100\%) = 99.97\%$$

12.15 A one-pole amplifier ($f_L = 0$) with midfrequency gain $A_{vo} = -1000$ has an upper cutoff frequency (3-db point) of 20 kHz. Negative voltage-series feedback is to be implemented to increase the upper cutoff frequency to 100 kHz. Determine the new midfrequency gain.

▮ Based on the result of Problem 12.8,

$$A_{vfo} = \frac{\text{GBW}}{f_{Hf}} = \frac{A_{vo} f_H}{f_{Hf}} = \frac{-1000(20 \times 10^3)}{100 \times 10^3} = -200$$

12.16 Determine the necessary feedback transfer ratio for the amplifier of Problem 12.15.

▮ For $\omega \ll \omega_H$, (2) of Problem 12.3 applied to voltage-series feedback gives

$$A_{vfo} = \frac{A_{vo}}{1 + \beta_{vv} A_{vo}} \tag{1}$$

from which
$$\beta_{vv} = \frac{1}{A_{vfo}} - \frac{1}{A_{vo}} = \frac{1}{-200} - \frac{1}{-1000} = -0.004 \tag{2}$$

12.17 To see that voltage-series feedback is characterized by an increase in input resistance for all types of amplifiers, replace the current-source model within the basic amplifier of Fig. 12-3 with a voltage-source model, to obtain the voltage-series feedback amplifier of Fig. 12-4. Find the input resistance with feedback and without feedback.

▮ Application of the method of node voltages to Fig. 12-4 gives

$$(v_f - v_S)G_{11} + v_f G_1 + (v_f - v_L)G_2 = 0 \tag{1}$$
$$(v_L - v_f)G_2 + (v_L - \gamma v_1)G_{22} + v_L G_L = 0 \tag{2}$$

But
$$v_1 = v_S - v_f \tag{3}$$

Substitution of (3) into (2) and rearrangement result in the linear system of equations

$$\begin{bmatrix} G_{11} + G_1 + G_2 & -G_2 \\ -(G_2 - \gamma G_{22}) & G_2 + G_{22} + G_L \end{bmatrix} \begin{bmatrix} v_f \\ v_L \end{bmatrix} = \begin{bmatrix} G_{11} v_S \\ \gamma G_{22} v_S \end{bmatrix} \tag{4}$$

Simultaneous solution of (4) by Cramer's rule gives

$$v_f = \frac{\Delta_1}{\Delta} = \frac{\begin{vmatrix} G_{11}v_S & -G_2 \\ \gamma G_{22}v_S & G_2 + G_{22} + G_L \end{vmatrix}}{\begin{vmatrix} G_{11} + G_1 + G_2 & -G_2 \\ -(G_2 - \gamma G_{22}) & G_2 + G_{22} + G_L \end{vmatrix}} = \frac{G_{11}(G_2 + G_{22} + G_L) + \gamma G_2 G_{22}}{\Delta} v_S$$

where
$$\Delta = (G_{11} + G_1)(G_2 + G_{22} + G_L) + G_2[(1 + \gamma)G_{22} + G_L] \tag{5}$$

Now,
$$i_1 = (v_S - v_f)G_{11}$$

so that
$$R_{\text{in}f} = R_{11} + \frac{v_f}{i_1} = R_{11} + \frac{v_f}{(v_S - v_f)G_{11}} \tag{6}$$

Substitution of (5) into (6) and simplification give

$$R_{\text{in}f} = R_{11} + \frac{G_{11}(G_2 + G_{22} + G_L) + \gamma G_2 G_{22}/G_{11}}{G_1(G_2 + G_{22} + G_L) + G_{22}(G_2 + G_L)} \tag{7}$$

To deactivate the feedback network, let $R_1(= 1/G_1) = 0$ and $R_2(= 1/G_2) = \infty$. Then the input resistance without feedback is found from (7) as

$$R_{\text{in}} = \lim_{\substack{G_1 \to 0 \\ G_2 \to \infty}} R_{\text{in}f} = R_{11} \tag{8}$$

Comparing (7) and (8), it is apparent that the second right-hand term of (7) represents the increase in input resistance due to feedback.

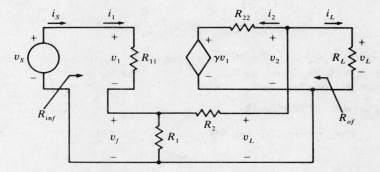

Fig. 12-4

12.18 By use of the voltage-source model of Fig. 12-4, show that voltage-series feedback is characterized by a decrease in output impedance for all types of amplifiers.

▎ To find the output resistance (Thévenin resistance), we deactivate (short) v_S and replace R_L with a driving-point source in Fig. 12-4. Then, by KCL,

$$i_{dp} = \frac{v_{dp}}{R_2 + R_1 \| R_{11}} + \frac{v_{dp} - \gamma v_1}{R_{22}} \tag{1}$$

By voltage division,

$$v_1 = -\frac{R_1 \| R_{11}}{R_1 \| R_{11} + R_2} v_{dp} \tag{2}$$

Substituting (2) into (1), converting to conductances, and solving for the output conductance give

$$G_{of} = \frac{1}{R_{of}} = \frac{i_{dp}}{v_{dp}} = G_{22} + \frac{(G_1 + G_{11} + \gamma G_{22})G_2}{(G_1 + G_{11} + G_{22})G_{22}} \tag{3}$$

Deactivate the feedback network ($G_1 = \infty$, $G_2 = 0$) to find the output resistance without feedback from (3) as

$$R_o = \lim_{\substack{G_1 \to \infty \\ G_2 \to 0}} \frac{1}{G_{of}} = \frac{1}{G_{22}} = R_{22} \tag{4}$$

Comparing (3) and (4), it is apparent that the second right-hand term of (3) represents the increase in output conductance (or decrease in output resistance) due to feedback.

12.19 The ideal voltage-series feedback amplifier model, shown in block-diagram form in Fig. 12-5, is based on the assumption that the feedback network does not *load* the basic amplifier (i.e., does not alter the forward transfer ratio A). Determine the overall voltage gain of this ideal voltage-series feedback amplifier.

▮ Since the basic amplifier gain $A_v = v_L/v_1$, we have

$$v_L = A_v v_1 = \frac{\beta G_{11}}{G_L + G_{22}} v_1 \tag{1}$$

The comparator output is

$$v_1 = v_S - v_f = v_S - \beta_{vv} v_L \tag{2}$$

Substituting (2) into (1) and rearranging give the desired gain:

$$A_{vf} = \frac{v_L}{v_S} = \frac{A_v}{1 + \beta_{vv} A_v} = \frac{\beta G_{11}}{G_L + G_{22} + \beta_{vv} \beta G_{11}} \tag{3}$$

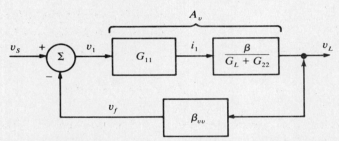

Fig. 12-5 Ideal voltage-series feedback model.

12.20 For the voltage-series feedback amplifier of Fig. 12-3, determine the feedback transfer ratio β_{vv}.

▮ Simultaneous solution of (4) of Problem 12.10 for v_L gives

$$v_L = \frac{\Delta_2}{\Delta} = \frac{\begin{vmatrix} G_{11} + G_1 + G_2 & G_{11} v_S \\ -(G_2 - \beta G_{11}) & \beta G_{11} v_S \end{vmatrix}}{\Delta} = \frac{G_{11}[(1 + \beta)G_2 + \beta G_1] v_S}{\Delta} \tag{1}$$

Then dividing (5) of Problem 12.10 by (1) gives

$$\beta_{vv} = \frac{v_f}{v_L} = \frac{(1 + \beta)G_2 + G_{22} + G_L}{(1 + \beta)G_2 + \beta G_1} \tag{2}$$

12.21 Find an expression for the basic amplifier gain A_v of the voltage-series feedback amplifier in Fig. 12-3.

▮ With the feedback network deactivated ($R_1 = 0$, $R_2 = \infty$),

$$i_1 = G_{11} v_1 \tag{1}$$

and

$$v_2 = \beta i_1 (R_{22} \parallel R_L) = \frac{\beta i_1}{G_{22} + G_L} \tag{2}$$

Substitution of (1) into (2) then allows solution for the basic amplifier voltage gain:

$$A_v = \frac{v_2}{v_1} = \frac{\beta G_{11}}{G_{22} + G_L} \tag{3}$$

12.22 Use β_{vv} and A_v as determined in Problems 12.20 and 12.21, respectively, to find the overall voltage gain for the voltage-series feedback amplifier of Fig. 12-3. Assume that the feedback network does not load or alter A_v.

▮ Since the feedback network does not alter A_v, (3) of Problem 12.19 yields

$$A_{vf} = \frac{A_v}{1 + \beta_{vv} A_v} = \frac{\beta G_{11}/(G_{22} + G_L)}{1 + \frac{(1 + \beta)G_2 + G_{22} + G_L}{(1 + \beta)G_2 + \beta G_1} \frac{\beta G_{11}}{G_{22} + G_L}}$$

$$= \frac{\beta G_{11}[(1 + \beta)G_2 + \beta G_1]}{(G_{22} + G_L)[(1 + \beta)G_2 + \beta G_1] + \beta G_{11}[(1 + \beta)G_2 + G_{22} + G_L]} \tag{1}$$

12.23 Using the voltage-series feedback amplifier of Fig. 12-4, show that if loading by the feedback network is negligible, then $R_{\text{in}f} = R_{\text{in}}(1 + \beta_{vv}A_v)$ where, for the specific circuit at hand, $R_{\text{in}} = R_{11}$, $A_v = \gamma R_L/(R_L + R_{22})$, and $\beta_{vv} = R_1/(R_1 + R_2)$.

▌ By KVL around the input loop of Fig. 12-4,

$$v_S = i_1 R_{11} + v_f = i_1 R_{11} + \frac{R_1}{R_1 + R_2} v_L = i_1 R_{11} + \beta_{vv} v_L \tag{1}$$

Neglecting feedback-network loading, voltage division applied to the output loop gives

$$v_L = \frac{R_L}{R_L + R_{22}} \gamma v_1 = \frac{R_L}{R_L + R_{22}} \gamma (R_{11} i_1) = A_v R_{11} i_1 \tag{2}$$

where $A_v = \gamma R_L/(R_L + R_{22})$ is the gain of the basic amplifier with feedback deactivated. Substituting (2) into (1) and rearranging give

$$v_S = i_1 R_{11} + \beta_{vv}(A_v R_{11} i_1) = i_1 R_{11}(1 + \beta_{vv} A_v)$$

Hence,
$$R_{\text{in}f} = \frac{v_S}{i_1} = \frac{v_S}{i_S} = R_{11}(1 + \beta_{vv}A_v) = R_{\text{in}}(1 + \beta_{vv}A_v) \tag{3}$$

where $R_{\text{in}} = R_{11}$ is the input resistance with feedback deactivated.

12.24 Neglecting load by the feedback network in the voltage-series feedback amplifier of Fig. 12-4, show that $R_{of} = R_o/(1 + \beta_{vv}A'_v)$ where for this specific circuit, $R_o = R_{22}$, $A'_v = \gamma$, and $\beta_{vv} = R_1/(R_1 + R_2)$.

▌ With v_S deactivated (shorted) and R_L replaced with a driving-point source in Fig. 12-4, KCL applied at the node of the output network requires that

$$i_{dp} = \frac{v_{dp} - \gamma v_1}{R_{22}} \tag{1}$$

But with $v_S = 0$,

$$v_1 = -v_F = -\beta_{vv} v_{dp} \tag{2}$$

Substituting (2) into (1) and rearranging yield

$$R_{of} = \frac{v_{dp}}{i_{dp}} = \frac{R_{22}}{1 + \beta_{vv}\gamma} = \frac{R_o}{1 + \beta_{vv}A'_v} \tag{3}$$

12.25 Voltage-series feedback is used in the cascaded FET amplifiers of Fig. 12-6(a) and the FETs are identical. The ideal transformer is introduced to establish the proper polarity for v_f for negative feedback at the mixing point. Draw a small-signal equivalent circuit suitable for analysis of the amplifier.

▌ See Fig. 12-6(b).

12.26 For the cascaded amplifiers of Fig. 12-6(a) using identical FETs voltage-series feedback, let $R_1 = 10$ kΩ, $R_2 = 190$ kΩ, $R_D = R_L = 2$ kΩ, $R_G = 1$ MΩ, $r_{ds} = 30$ kΩ, and $\mu = 60$. Neglect loading by the feedback network and find the overall voltage gain.

▌ The small-signal equivalent circuit of Fig. 12-6(b) is applicable. Since $R_G \| R_D \approx R_D$ and $R_G \| R_L \| R_D \approx R_L \| R_D$, the voltage gain without feedback is

$$A_v = A_{v1}A_{v2} \approx \frac{-\mu R_D}{R_D + r_{ds}} \frac{-\mu R_d \| R_L}{R_D \| R_L + r_{ds}} = \frac{\mu^2 R_D^2 R_L}{(R_D + r_{ds})[R_D R_L + r_{ds}(R_D + R_L)]} \tag{1}$$

$$= \frac{(60)^2(2000)^2(2000)}{(32 \times 10^3)[2000(2000) + (30 \times 10^3)(4000)]} = 7.26$$

Voltage division gives the feedback transfer ratio as

$$\beta_{vv} = \frac{R_1}{R_1 + R_2} = \frac{10 \times 10^3}{(10 + 190) \times 10^3} = 0.05$$

With feedback loading neglected, (3) of Problem 12.19 gives the overall voltage gain as

$$A_{vf} = \frac{A_v}{1 + \beta_{vv}A_v} = \frac{7.26}{1 + 0.05(7.26)} = 5.33$$

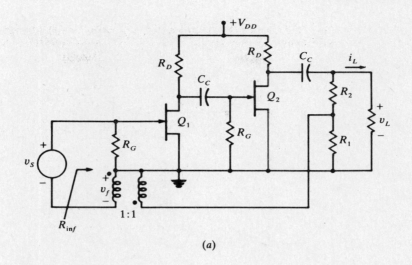

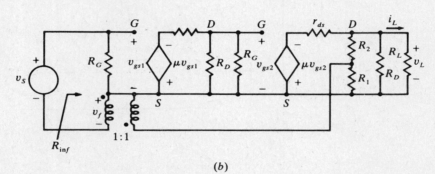

Fig. 12-6

12.27 Determine the input resistance for the amplifier of Problem 12.26.

> Without feedback ($R_1 = 0$, $R_2 = \infty$), $R_{in} = R_G = 1$ MΩ. Then by (3) of Problem 12.23 and using the results of Problem 12.26,
>
> $$R_{inf} = R_{in}(1 + \beta_{vv}A_v) = (1 \times 10^6)[1 + 0.05(7.26)] = 1.363 \text{ M}\Omega$$

12.28 Evaluate the output resistance for the amplifier of Problem 12.26.

> Without feedback, the output resistance is
>
> $$R_o = r_{ds} \| R_D = \frac{r_{ds}R_D}{r_{ds} + R_D} = \frac{(30 \times 10^3)(2 \times 10^3)}{(30 + 2) \times 10^3} = 1.875 \text{ k}\Omega$$

The output resistance with feedback active is given by (3) of Problem 12.24, where A'_v is the voltage gain with $i_L = 0$. That is,

$$V_{Th} = -\mu(A_{v1}v_1)\frac{R_D}{R_D + r_{ds}} = -\mu\frac{-\mu R_D}{R_D + r_{ds}}\frac{R_D}{R_D + r_{ds}}v_1$$

and

$$A'_v = \frac{V_{Th}}{v_1} = \left(\frac{\mu R_D}{R_D + r_{ds}}\right)^2 = \left[\frac{60(2000)}{(2+30) \times 10^3}\right]^2 = 14.06$$

Then,

$$R_{of} = \frac{R_o}{1 + \beta_{vv}A'_v} = \frac{1875}{1 + 0.05(14.06)} = 1101 \text{ }\Omega$$

12.29 Voltage-series feedback is used in the cascaded BJT amplifier of Fig. 12-7(a). Draw a small-signal equivalent circuit suitable for analysis of the amplifier if $h_{re1} = h_{re2} = h_{oe1} = h_{oe2} = 0$.

> See Fig. 12-7(b).

348 □ CHAPTER 12

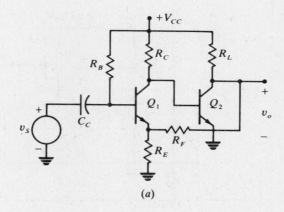

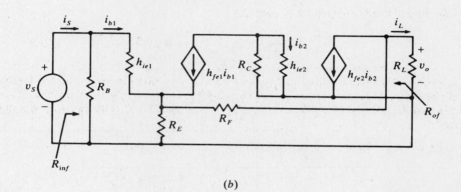

Fig. 12-7

12.30 For the voltage-series feedback amplifier of Fig. 12-7(a), let $R_E = 100\,\Omega$, $R_B = 500\,\text{k}\Omega$, $R_C = 2\,\text{k}\Omega$, $R_L = 3\,\text{k}\Omega$, $R_F = 15\,\text{k}\Omega$, $h_{ie1} = 100\,\Omega$, $h_{ie2} = 250\,\Omega$, $h_{fe1} = 75$, $h_{fe2} = 90$, and $h_{re1} = h_{re2} = h_{oe1} = h_{oe2} = 0$. Find the input resistance. Neglect any loading by the feedback network.

▌ The small-signal equivalent circuit of Fig. 12-7(b) is applicable, from which, without feedback, $R_{in} = R_{11} = h_{ie1}$ and $R_o = 1/h_{oe2} = \infty$. With no feedback and with R_L open-circuited, the controlled current source βi_1 of Fig. 12-3 becomes

$$\beta i_1 = \beta i_{b1} = -h_{fe2} i_{b2} = -h_{fe2}(-h_{fe1} i_{b1})\frac{R_C}{R_C + h_{ie2}}$$

so that

$$\beta = \frac{h_{fe1} h_{fe2} R_C}{R_C + h_{ie2}} = \frac{75(90)(2000)}{2100} = 6428.6$$

Then according to (9) of Problem 12.10,

$$R_{inf} = R_{11} + \frac{(1+\beta)G_2 + G_{22} + G_L}{G_1 G_2 + (G_1 + G_2)(G_{22} + G_L)}$$

$$= 100 + \frac{6429.6(0.0667 \times 10^{-3}) + 0 + 3.333 \times 10^{-4}}{0.01(0.0667 \times 10^{-3}) + 0.010067(3.333 \times 10^{-4})} = 106.79\,\text{k}\Omega$$

12.31 Determine the output resistance for the amplifier of Problem 12.30.

▌ From (3) of Problem 12.11,

$$G_{of} = G_{22} + \frac{(1+\beta)G_{11} + G_1}{(G_{11} + G_1)R_2 + 1} = 0 + \frac{6429.6(0.01) + 0.01}{0.02(15 \times 10^3) + 1} = 0.2136$$

so

$$R_{of} = \frac{1}{G_{of}} = \frac{1}{0.2136} = 4.68\,\Omega$$

12.32 Evaluate the overall voltage gain for the amplifier of Problem 12.30.

▌ By voltage division applied to the feedback network of Fig. 12-7(b),

$$\beta_{vv} = \frac{R_E}{R_E + R_F} = \frac{100}{15,100} = 0.00662$$

and from (3) of Problem 12.21,

$$A_v = \frac{\beta G_{11}}{G_{22} + G_L} = \frac{6428.6(0.01)}{0 + 3.333 \times 10^{-4}} = 192,877$$

Then, (3) of Problem 12.19 gives

$$A_{vf} = \frac{A_v}{1 + \beta_{vv} A_v} = \frac{192,877}{1 + 0.00662(192,877)} = 150.94$$

12.33 Find the overall current gain for the amplifier of Problem 12.30.

▌ Using the results of Problems 12.30 and 12.32,

$$A_{if} = \frac{i_L}{i_s} = \frac{v_o/R_L}{v_S/R_{\text{in}f}} = A_{vf} \frac{R_{\text{in}f}}{R_L} = (150.94) \frac{106.79 \times 10^3}{3 \times 10^3} = 5372.9$$

12.34 A mass-produced amplifier with voltage-series feedback has a nominal open-loop (no-feedback) voltage gain A_v of 2000; however, its gain can range from 1200 to 2800 as component parameters vary. If negative feedback is added so that the nominal overall (closed-loop) gain is reduced to 50, what range of overall gain is to be expected?

▌ Solving (3) of Problem 12.19 for the feedback transfer ratio gives

$$\beta_{vv} = \frac{1}{A_{vf}} - \frac{1}{A_v} = \frac{1}{50} - \frac{1}{2000} = 0.0195$$

And the range to be expected is, again from (3) of Problem 12.19,

$$\frac{1200}{1 + 0.0195(1200)} \leq A_{vf} \leq \frac{2800}{1 + 0.0195(2800)} \quad \text{or} \quad 49.18 \leq A_{vf} \leq 50.36$$

12.35 The CE amplifier of Fig. 4-14, as analyzed in Problems 10.50 and 10.51, has a low-frequency cutoff f_L of 81.02 Hz. The amplifier is to be used for frequencies down to the low end of the audio range (20 Hz), and a transformer is to be added, as in Problem 12.25, to introduce voltage-series feedback. Specify a feedback gain that will reduce the low-frequency cutoff point to 20 Hz. Assume the transformer turns ratio is 1:1.

▌ The midfrequency gain of the amplifier without feedback follows from (7) of Problem 10.51, if we let C_C and then S approach infinity as

$$A_{vo} = -\frac{h_{fe} R_L R_C}{(R_C + R_L)(R_i + h_{ie})} = -\frac{100(2 \times 10^3)(2 \times 10^3)}{(2 \times 10^3 + 2 \times 10^3)(100 + 1 \times 10^3)} = -90.91$$

Rearranging (2) of Problem 12.4, the required feedback gain is

$$\beta_{vv} = \frac{1}{A_{vo}} \left(\frac{f_L}{f_{Lf}} - 1 \right) = \frac{1}{-90.91} \left(\frac{81.02}{20} - 1 \right) = -0.03356$$

12.36 Determine the resulting midfrequency voltage gain for the amplifier of Problem 12.35 after the addition of the specified feedback.

▌ Based on (2) of Problem 12.3,

$$A_{vf} = \frac{A_{vo}}{1 + \beta_{vv} A_{vo}} = \frac{-90.91}{1 + (-0.03356)(-90.91)} = -22.44$$

12.37 For the voltage-source amplifier with voltage-series feedback of Fig. 12-4, determine the overall voltage gain with feedback.

Simultaneous solution of (4) from Problem 12.17 by Cramer's rule gives

$$v_L = \frac{\Delta_2}{\Delta} = \frac{\begin{vmatrix} G_{11} + G_1 + G_2 & G_{11}v_S \\ -(G_2 - \gamma G_{22}) & \gamma G_{22}v_S \end{vmatrix}}{\Delta} = \frac{\gamma G_{22}(G_{11} + G_1 + G_2) + G_{11}(G_2 - \gamma G_{22})}{\Delta} v_S$$

Whence,
$$A_{vf} = \frac{v_L}{v_S} = \frac{G_{11}G_2 + \gamma G_{22}(G_1 + G_2)}{(G_{11} + G_1)(G_2 + G_{22} + G_L) + G_2[(1 + \gamma)G_{22} + G_L]} \quad (1)$$

12.38 Find an expression for the overall voltage gain of the amplifier of Fig. 12-4 if the feedback is inactive.

▌ To deactivate the feedback let $G_2 = 0$, then $G_1 = \infty$ in (1) of Problem 12.37 to find

$$A_v = \frac{\gamma G_{22}}{G_{22} + G_L} \quad (1)$$

12.39 For the general voltage-series feedback amplifier of Fig. 12-3, let $G_1 = 1 \times 10^{-3}$ S, $G_2 = 0.05 \times 10^{-3}$ S, $G_{11} = 0.01$ S, $G_{22} \approx 0$, $G_L = 0.5 \times 10^{-3}$ S, and $\beta = 75$. Determine A_v.

▌ By (3) of Problem 12.21,

$$A_v = \frac{\beta G_{11}}{G_{22} + G_L} = \frac{75(0.01)}{0 + 0.5 \times 10^{-3}} = 1500$$

12.40 Find the overall voltage gain with feedback active for the general voltage-series feedback amplifier described by Problem 12.39.

▌ By (2) of Problem 12.20,

$$\beta_{vv} = \frac{(1 + \beta)G_2 + G_{22} + G_L}{(1 + \beta)G_2 + \beta G_1} = \frac{(1 + 75)(0.05 \times 10^{-3}) + 0 + 0.5 \times 10^{-3}}{(1 + 75)(0.05 \times 10^{-3}) + 75(1 \times 10^{-3})} = 0.0546$$

Hence, (3) of Problem 12.2 and result of Problem 12.39 lead to

$$A_{vf} = \frac{A_v}{1 + \beta_{vv}A_v} = \frac{1500}{1 + 0.0546(1500)} = 18.094$$

12.41 Evaluate the current gain for the general voltage-series feedback amplifier of Problem 12.39 if the feedback network is inactive.

▌ Deactivating the feedback network ($R_1 = 0$, $R_2 = \infty$) and using current division at the output network,

$$i_L = \frac{R_{22}}{R_{22} + R_L}\beta i_1 = \frac{\beta G_L}{G_L + G_{22}} i_1$$

Whence,
$$A_i = \frac{\beta G_L}{G_L + G_{22}} = \frac{75(0.5 \times 10^{-3})}{0.5 \times 10^{-3} + 0} = 75$$

12.42 Find the current gain for the general voltage-series feedback amplifier of Problem 12.39 with feedback active.

▌
$$A_{if} = \frac{i_L}{i_S} = \frac{v_L/R_L}{v_S/R_{\inf}} = A_{vf}R_{\inf}G_L$$

Using the results of Problems 12.13 and 12.40,

$$A_{if} = 18.094(7.566 \times 10^3)(0.5 \times 10^{-3}) = 68.45$$

12.43 Determine the error in overall voltage gain that results from neglecting the effects of feedback loading in the voltage-series feedback amplifier of Problem 12.40 by treating the feedback network as a true voltage divider.

▌ If the feedback network were a true voltage divider,

$$v_f = \frac{R_1}{R_1 + R_2}v_L = \frac{G_2}{G_1 + G_2}v_L$$

Hence,
$$\beta'_{vv} = \frac{v_f}{v_L} = \frac{G_2}{G_1 + G_2} = \frac{0.05 \times 10^{-3}}{1 \times 10^{-3} + 0.05 \times 10^{-3}} = 0.0476$$

Based on (3) of Problem 12.3,
$$A'_{vf} = \frac{A_v}{1 + \beta'_{vv}A_v} = \frac{1500}{1 + 0.0476(1500)} = 20.72$$

Thus,
$$\text{percent error} = \frac{20.72 - 18.094}{18.094}(100\%) = 14.51\%$$

12.44 Show that for the ideal voltage-series feedback amplifier (no loading by feedback network), $R_{inf} = R_{in}(1 + \beta_{vv}A'_i)$, where $A'_i = A_i R_L / R_{11} = \beta R_{22} R_L / R_{11}(R_{22} + R_L)$.

▌ Refer to Fig. 12-3, apply KVL around the input network, and apply the feedback relationship to find
$$v_S = i_1 R_{11} + v_f = i_1 R_{11} + \beta_{vv} v_L \tag{1}$$

Neglecting loading by the R_1-R_2 voltage divider network,
$$v_L = \beta i_1 \frac{R_{22} R_L}{R_{22} + R_L} \tag{2}$$

Substitute (2) into (1) and rearrange to give
$$R_{inf} = \frac{v_S}{i_1} = R_{11}\left[1 + \beta_{vv}\frac{\beta R_{22} R_L}{R_{11}(R_{22} + R_L)}\right] \tag{3}$$

By (1) of Problem 12.2 and result of Problem 12.41, $R_{11} = R_{in}$ and $\beta R_{22} R_L / R_{11}(R_{22} + R_L) = A_i R_L / R_{11} \equiv A'_i$. Thus, (3) can be written as
$$R_{inf} = R_{in}(1 + \beta_{vv} A'_i) \tag{4}$$

12.45 Show that for the ideal voltage-series feedback amplifier, $R_{of} = R_o/(1 + \beta_{vv} A''_i)$, where $A''_i = \beta R_{22}/R_{11}$.

▌ Deactivate (short) v_S and replace R_L with a driving-point source oriented such that $v_{dp} = v_L$ and $i_{dp} = -i_L$ in Fig. 12-3. Then,
$$i_1 = \frac{v_f}{R_{11}} = \frac{\beta_{vv} v_{dp}}{R_{11}} \tag{1}$$

and by KCL,
$$i_{dp} = \frac{v_{dp}}{R_{22}} - \beta i_1 \tag{2}$$

Substitute (1) into (2), rearrange the result, and recall $R_{22} = R_o$ from Problem 12.12 to find
$$R_{of} = \frac{v_{dp}}{i_{dp}} = \frac{R_{22}}{1 + \beta_{vv}\beta(R_{22}/R_{11})} = \frac{R_o}{1 + \beta_{vv} A''_i} \tag{3}$$

12.46 For the cascaded FET amplifier of Fig. 12-6, find the parameters R_{11}, R_{22}, and γ of the general voltage-source amplifier of Fig. 12-4 with voltage-series feedback.

▌ With the feedback deactivated ($v_f = 0$), $R_{11} = R_{in} = R_G$. Based on the results of Problems 12.2 and 12.28, $R_{22} = R_o = r_{ds} R_D / (r_{ds} + R_D)$. Further, inspection of Fig. 12-4 shows that γ is the gain with feedback inactive and R_L is open-circuited ($R_L = \infty$). Hence, letting $R_L \to \infty$ in (1) of Problem 12.26,
$$A_v(R_L = \infty) = \gamma = \left(\frac{\mu R_D}{R_D + r_{ds}}\right)^2$$

12.47 Use the parameters determined in Problem 12.46 to evaluate the overall voltage gain of the cascaded FET amplifier of Fig. 12-6 if the amplifier values are described by Problem 12.26. Assume negligible loading by feedback network.

▌
$$R_{11} = R_G = 1 \text{ M}\Omega$$
$$R_{22} = \frac{r_{ds} R_D}{r_{ds} + R_D} = \frac{(30 \times 10^3)(2 \times 10^3)}{30 \times 10^3 + 2 \times 10^3} = 1.875 \text{ k}\Omega$$

352 ☐ CHAPTER 12

$$\beta_{vv} = \frac{R_1}{R_1 + R_2} = \frac{10 \times 10^3}{10 \times 10^3 + 190 \times 10^3} = 0.05$$

$$\gamma = \left[\frac{\mu R_D}{R_D + r_{ds}}\right]^2 = \left[\frac{60(2 \times 10^3)}{2 \times 10^3 + 30 \times 10^3}\right]^2 = 14.063$$

With no feedback, A_v follows from voltage division applied to the output network of Fig. 12-4:

$$A_v = \frac{v_L}{v_1} = \frac{\gamma R_L}{R_L + R_{22}} = \frac{14.063(2 \times 10^3)}{2 \times 10^3 + 1.875 \times 10^3} = 7.26$$

Based on (3) of Problem 12.2,

$$A_{vf} = \frac{A_v}{1 + \beta_{vv} A_v} = \frac{7.26}{1 + 0.05(7.26)} = 5.33$$

The result agrees with that of Problem 12.26.

12.48 For the general voltage-series feedback amplifier of Fig. 12-3, model the feedback network as a two-port network using h parameters.

▌ In terms of the feedback network variables of Fig. 12-3, (1) to (6) of Problem 1.76 lead to

$$v_f = h_{11} i_1 + h_{12} v_L \tag{1}$$

$$i_a = h_{21} i_1 + h_{22} v_L \tag{2}$$

where

$$h_{11} = \left.\frac{v_f}{i_1}\right|_{v_L=0} \quad h_{12} = \left.\frac{v_f}{v_L}\right|_{i_1=0} \quad h_{21} = \left.\frac{i_a}{i_1}\right|_{v_L=0} \quad h_{22} = \left.\frac{i_a}{v_L}\right|_{i_1=0} \tag{3}$$

For the particular feedback network at hand,

$$h_{11} = R_1 \parallel R_1 = \frac{R_1 R_2}{R_1 + R_2} \quad h_{12} = \frac{R_1}{R_1 + R_2} \quad h_{21} = -\frac{R_1}{R_1 + R_2} \quad h_{22} = \frac{1}{R_1 + R_2}$$

12.49 Using the h parameters determined in Problem 12.48 for the feedback network of Fig. 12-3, modify the basic amplifier parameters β, R_{11}, and R_{22} to account for any loading introduced by the feedback network, thus leaving it ideal.

▌ The series combination of R_{11} and h_{11} can be added to give

$$R'_{11} = R_{11} + h_{11} \tag{1}$$

The parallel combination of R_{22} and h_{22} can be replaced with

$$R'_{22} = R_{22} \parallel \frac{1}{h_{22}} = \frac{R_{22}}{h_{22} R_{22} + 1} \tag{2}$$

And an equivalent current source $\beta' i_1$ can be formed as

$$\beta' = \beta - h_{21} \tag{3}$$

12.50 Use the h parameters of Problem 12.48 and the modified basic amplifier parameters of Problem 12.49 to draw an equivalent circuit model of the general voltage-series feedback amplifier, where the feedback network is ideal.

▌ See Fig. 12-8.

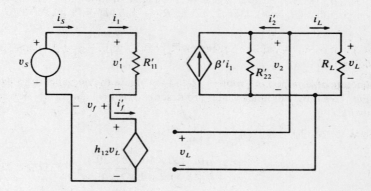

Fig. 12-8

12.51 For the cascaded voltage-series feedback amplifier of Fig. 12-7(a), transfer the loading effects of the feedback network to the basic amplifier (determine R'_{11}, R'_{22}, and β') so that the model of Fig. 12-8 is applicable. Use the parameter values of Problem 12.30.

By (3) of Problem 12.48,

$$h_{11} = \frac{v_f}{i_b}\bigg|_{v_L=0} = R_E \parallel R_F = \frac{R_E R_F}{R_E + R_F} = \frac{100(15 \times 10^3)}{100 + 15 \times 10^3} = 99.34\ \Omega$$

$$h_{12} = \frac{v_f}{v_L}\bigg|_{i_b=0} = \frac{R_E}{R_E + R_F} = \frac{100}{100 + 15 \times 10^3} = 0.00662$$

$$h_{21} = \frac{i_a}{i_b}\bigg|_{v_L=0} = -\frac{G_F v_F}{(G_E + G_F)v_F} = -\frac{R_E}{R_E + R_F} = -\frac{100}{100 + 15 \times 10^3} = -0.00662$$

$$h_{22} = \frac{i_a}{v_L}\bigg|_{i_b=0} = \frac{1}{R_E + R_F} = \frac{1}{100 + 15 \times 10^3} = 6.62 \times 10^{-5}\ \text{S}$$

Hence, by (1)–(3) of Problem 12.49,

$$R'_{11} = R_{11} + h_{11} = h_{ie1} + h_{11} = 100 + 99.34 = 199.34\ \Omega$$

$$R'_{22} = \frac{R_{22}}{h_{22}R_{22}+1} = \frac{1}{h_{22} + G_{22}} = \frac{1}{6.62 \times 10^{-5} + 0} = 15.1\ \text{k}\Omega$$

$$\beta' = \beta - h_{21} = \frac{h_{fe1} h_{fe2} R_C}{R_C + h_{ie2}} - h_{21} = 6428.6 - (-0.00662) = 6428.59$$

12.52 Find the overall voltage gain for the cascaded voltage-series feedback amplifier of Fig. 12-7 with parameter values of Problem 12.30.

The ideal voltage-series feedback model of Fig. 12-5 is applicable if G_{11}, G_{22}, and β are replaced, respectively, with G'_{11} ($=1/R'_{11}$), G'_{22} ($=1/R'_{22}$), and β' as determined in Problem 12.51. Hence,

$$G'_{11} = \frac{1}{R'_{11}} = \frac{1}{199.34} = 0.00502\ \text{S}$$

$$G'_{22} = \frac{1}{R'_{22}} = \frac{1}{15.1 \times 10^3} = 6.62 \times 10^{-5}\ \text{S}$$

From Fig. 12-5,

$$A_v = \frac{G'_{11}\beta'}{G_L + G'_{22}} = \frac{0.00502(6428.59)}{1/(3 \times 10^3) + 6.62 \times 10^{-5}} = 80.773 \times 10^3$$

From Problem 12.51 we see that

$$\beta_{vv} = h_{12} = 0.00662$$

Whence by (3) of Problem 12.2,

$$A_{vf} = \frac{A_v}{1 + \beta_{vv} A_v} = \frac{80.773 \times 10^3}{1 + 0.00662(80.773 \times 10^3)} = 150.78$$

12.53 Determine the value of R_{inf} for the voltage-series feedback amplifier of Problem 12.30 through use of the ideal feedback model of Fig. 12-8.

Apply KVL to the input network of Fig. 12-8 and use the gain of Problem 12.52 to find

$$v_S = i_1 R'_{11} + h_{12} v_L = i_1 R_{11} + h_{12} A_{vf} v_S$$

Whence, along with the results of Problem 12.51,

$$R_{inf} = \frac{v_S}{i_1} = \frac{R'_{11}}{1 - h_{12} A_{vf}} = \frac{199.34}{1 - 0.00662(150.77)} = 106.79\ \text{k}\Omega$$

12.54 Find the value of R_{of} for the amplifier of Problem 12.30 by application of the ideal feedback model of Fig. 12-8.

▎ Deactivate (short) v_S and replace R_L in Fig. 12-8 by a driving-point source oriented such that $v_{dp} = v_L$ and $i_{dp} = -i_L$. KVL around the input loop yields

$$i_1 = -\frac{h_{12}v_L}{R'_{11}} = -\frac{h_{21}v_{dp}}{R'_{11}} \tag{1}$$

Apply KCL to the output network and use (1) to find

$$i_{dp} = \frac{v_{dp}}{R'_{22}} - \beta' i_1 = \frac{v_{dp}}{R'_{22}} - \beta'\left(\frac{-h_{21}v_{dp}}{R'_{11}}\right) \tag{2}$$

Rearrangement of (2) and use of values from Problems 12.51 and 12.52 lead to

$$R_{of} = \frac{v_{dp}}{i_{dp}} = \frac{1}{G'_{22} + \beta' h_{21} G'_{11}} = \frac{1}{6.62 \times 10^{-5} + 6428.59(0.00662)(0.00502)} = 4.68\ \Omega$$

12.55 The h-parameter small-signal equivalent circuit model of the BJT includes ideal voltage-series feedback through controlled source $h_{re}v_{ce}$. For the simple CE amplifier of Fig. 7-10(a), find expressions for the basic amplifier voltage gain A_v and feedback transfer ratio β_{vv} of Fig. 12-5. Write the overall voltage gain, showing that positive feedback exists.

▎ Comparing Fig. 7-10(b) with Fig. 12-5, the following correspondences are seen: $v_1 = v_{be}$, $i_1 = i_b$, $G_{11} = 1/h_{ie}$, $\beta = -h_{fe}$, $G_{22} = h_{oe}$, $v_f = v_{ce}$, $v_L = v_{ce}$, and $\beta_{vv} = h_{re}$. Whence,

$$A_v = \frac{\beta G_{11}}{G_L + G_{22}} = \frac{-h_{fe}/h_{ie}}{G_L + h_{oe}}$$

By (3) of Problem 12.2,

$$A_{vf} = \frac{v_{ce}}{v_S} = \frac{A_v}{1 + h_{re}\left[\dfrac{-h_{fe}/h_{ie}}{G_L + h_{oe}}\right]} = \frac{A_v}{1 - h_{re}h_{fe}/h_{ie}(G_L + h_{oe})}$$

The minus sign in the denominator of A_{vf} indicates positive feedback.

CURRENT-SERIES FEEDBACK

12.56 The general feedback amplifier model of Fig. 12-9, in which output current i_L is the sampled variable and feedback voltage v_f is the mixed variable, is called a *current-series* (or *series-input, series-output*; or *loop-sampling, loop-comparison*) feedback amplifier. For the current-series feedback amplifier of Fig. 12-9, find the overall voltage gain.

▎ The method of node voltages requires that

$$(v_f - v_S)G_{11} + v_f G_F - \beta i_1 + (v_f - v_L)G_{22} = 0 \tag{1}$$

and
$$\beta i_1 + (v_L - v_f)G_{22} + v_L G_L = 0 \tag{2}$$

But
$$i_1 = (v_S - v_f)G_{11} \tag{3}$$

Substituting (3) into (1) and (2) and rearranging lead to the linear system of equations

$$\begin{bmatrix} (1+\beta)G_{11} + G_F + G_{22} & -G_{22} \\ -(\beta G_{11} + G_{22}) & G_{22} + G_L \end{bmatrix} \begin{bmatrix} v_f \\ v_L \end{bmatrix} = \begin{bmatrix} (1+\beta)G_{11}v_S \\ -\beta G_{11}v_s \end{bmatrix} \tag{4}$$

Simultaneous solution of (4) for the load voltage v_L by Cramer's rule leads to

$$v_L = \frac{\Delta_2}{\Delta} = \frac{\begin{vmatrix} (1+\beta)G_{11} + G_F + G_{22} & (1+\beta)G_{11}v_S \\ -(\beta G_{11} + G_{22}) & -\beta G_{11}v_S \end{vmatrix}}{\begin{vmatrix} (1+\beta)G_{11} + G_F + G_{22} & -G_{22} \\ -(\beta G_{11} + G_{22}) & G_{22} + G_L \end{vmatrix}} = \frac{G_{11}(G_{22} - \beta G_F)v_S}{\Delta} \tag{5}$$

where
$$\Delta = G_{22}(G_{11} + G_F) + G_L[(1+\beta)G_{11} + G_{22} + G_F] \tag{6}$$

Using (6) in (5) and rearranging yield

$$A_{vf} = \frac{v_L}{v_s} = \frac{G_{11}(G_{22} - \beta G_F)}{G_{22}(G_{11} + G_F) + G_L[(1+\beta)G_{11} + G_{22}] + G_L G_F} \tag{7}$$

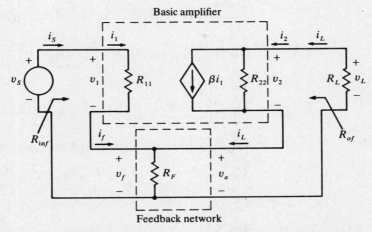

Fig. 12-9 General current-series feedback amplifier model.

12.57 For the current-series feedback amplifier model of Fig. 12-9, find the feedback transfer ratio $\beta_{iv} = v_f/i_L$.

▎ Simultaneous solution of (4) of Problem 12.26 for v_f yields

$$v_f = \frac{\begin{vmatrix} (1+\beta)G_{11}v_S & -G_{22} \\ -\beta G_{11}v_S & G_{22} + G_L \end{vmatrix}}{\Delta} = \frac{G_{11}[(1+\beta)G_L + G_{22}]v_S}{\Delta} \tag{1}$$

But

$$\beta_{iv} = \frac{v_f}{i_L} = \frac{v_f}{-v_L/R_L} = -G_L \frac{v_f}{v_L} \tag{2}$$

Substituting (5) of Problem 12.56 and (1) in (2) and simplifying thus give

$$\beta_{iv} = \frac{(1+\beta)G_L + G_{22}}{G_L(\beta G_F - G_{22})} \tag{3}$$

12.58 The ideal current-series feedback amplifier model, shown in block-diagram form in Fig. 12-10, is based on the assumption that the feedback network does not load the basic amplifier. The basic amplifier is, in this case, a transconductance amplifier. For the ideal current-series feedback amplifier of Fig. 12-10, find the overall transconductance gain.

▎ The basic amplifier gain $A_g = i_L/v_1$ requires that

$$i_L = A_g v_1 = G_{11} A_i v_1 = \frac{\beta G_L G_{11}}{G_L + G_{22}} v_1 \tag{1}$$

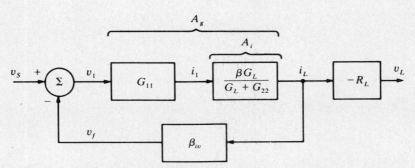

Fig. 12-10 Ideal current-series feedback model.

From the comparator output,
$$v_1 = v_S - v_f = v_S - \beta_{iv}v_L \tag{2}$$

Substitution of (2) into (1) allows solution for the overall transconductance gain as
$$A_{gf} = \frac{i_L}{v_S} = \frac{A_g}{1 + \beta_{iv}A_g} = \frac{G_{11}A_i}{1 + \beta_{iv}G_{11}A_i} = \frac{\beta G_L G_{11}}{G_L + G_{22} + \beta_{iv}\beta G_L G_{11}} \tag{3}$$

12.59 Find the overall voltage gain for the ideal current-series feedback amplifier of Fig. 12-10.

∎ Since $v_L = -i_L R_L$,
$$A_{vf} = \frac{v_L}{v_S} = -R_L \frac{i_L}{v_S} = -R_L A_{gf} = \frac{-R_L A_g}{1 + \beta_{iv}A_g} \tag{1}$$

12.60 For the general current-series feedback amplifier of Fig. 12-9, find the input resistance.

∎ From Fig. 12-9, with $i_1 = G_{11}(v_S - v_f)$,
$$R_{inf} = R_{11} + \frac{v_f}{i_1} = R_{11} + \frac{v_f}{G_{11}(v_S - v_f)} \tag{1}$$

Substituting for v_f from (1) of Problem 12.57 and simplifying then give
$$R_{inf} = R_{11} + \frac{(1+\beta)G_L + G_{22}}{G_{22}G_F + G_L G_{22} + G_L G_F} \tag{2}$$

12.61 Find an expression for the output resistance of the general current-series feedback amplifier in Fig. 12-9.

∎ To find R_{of}, we deactivate (short) v_S and replace R_L with a driving-point source in Fig. 12-9. Then the method of node voltages gives
$$-\beta i_1 + (v_f - v_{dp})G_{22} + v_f G_F + v_f G_{11} = 0 \tag{1}$$

But now $i_1 = -v_f G_{11}$ so that, from (1),
$$v_f = \frac{G_{22}v_{dp}}{(1+\beta)G_{11} + G_{22} + G_F} \tag{2}$$

KCL applied at the node of the output network requires
$$i_{dp} = (v_{dp} - v_f)G_{22} + \beta i_1 = (v_{dp} - v_f)G_{22} - \beta v_f G_{11} \tag{3}$$

The use of (2) in (3) and rearrangement yield
$$R_{of} = \frac{v_{dp}}{i_{dp}} = \frac{(1+\beta)G_{11} + G_{22} + G_F}{G_{22}(G_{11} + G_F)} \tag{4}$$

12.62 Model the feedback network of the current-series feedback amplifier of Fig. 12-9 as a two-port network using z parameters.

∎ In terms of the feedback network variables of Fig. 12-9, (1) to (6) of Problem 1.67 lead to
$$v_f = z_{11}i_1 + z_{12}i_L \equiv z_{11}i_1 + v_f' \tag{1}$$
$$v_a = z_{21}i_1 + z_{22}i_L \tag{2}$$

where
$$z_{11} = \left.\frac{v_f}{i_1}\right|_{i_L=0} \quad z_{12} = \left.\frac{v_f}{i_L}\right|_{i_1=0} \quad z_{21} = \left.\frac{v_a}{i_1}\right|_{i_L=0} \quad z_{22} = \left.\frac{v_a}{i_L}\right|_{i_1=0} \tag{3}$$

For the particular feedback network at hand,
$$z_{11} = z_{12} = z_{21} = z_{22} = R_F$$

12.63 Modify the basic amplifier parameters β, R_{11}, and R_{22} of Fig. 12-9 to account for any loading introduced by the feedback circuitry.

▌ Impedance (resistance) z_{11} can immediately be combined with R_{11} of Fig. 12-9 to form

$$R'_{11} = R_{11} + z_{11} \qquad (1)$$

leaving only the current-controlled voltage source $v'_f = z_{12}i_L$ in the feedback network.

A Norton equivalent circuit can be formed for the network to the left of R_L. With R_L replaced with a driving-point source and v_S deactivated (shorted),

$$i_1 = \frac{-z_{12}i_L}{R'_{11}} \qquad (2)$$

Applying KCL at the upper node of the output network gives, with (2) of Problem 12.62,

$$i_{dp} = \beta i_1 + \frac{v_{dp} - v_a}{R_{22}} = \beta i_1 + \frac{1}{R_{22}}(v_{dp} - z_{21}i_i - z_{22}i_{dp}) \qquad (3)$$

Substituting (2) into (3) and solving for the driving-point impedance give

$$R_{Th} = \frac{1}{Y_N} = R_{22}\left(1 + \frac{\beta z_{12}}{R'_{11}}\right) + z_{22} - \frac{z_{12}z_{21}}{R'_{11}} = R'_{22} \qquad (4)$$

If the output were open-circuited (R_L removed), we would have $i_L = 0$ and $i_1 = v_S/R'_{11}$. Hence, the Thévenin voltage (with polarity opposite that of v_L) is found from (2) of Problem 12.62 as

$$V_{Th} = \beta i_1 R_{22} - v_a = \beta i_1 R_{22} - z_{12}i_1 = (\beta R_{22} - z_{12})\frac{v_S}{R'_{11}} \qquad (5)$$

It is convenient to determine the input resistance. By KVL applied at the input and output networks,

$$v_S = R'_{11}i_1 + z_{12}i_L \qquad (6)$$
$$V_{Th} = (R_{Th} + R_L)i_L \qquad (7)$$

Substituting (5) into (7), solving for i_L, and substituting the result in (6) give the input impedance as

$$R_{\text{in} f} = \frac{v_S}{i_1} = \frac{R'_{11}(R_{Th} + R_L) + z_{12}(\beta R_{22} - z_{12})}{R_{Th} + R_L} \qquad (8)$$

The Norton source follows as

$$I_N = \beta'i_1 = \frac{V_{Th}}{R_{Th}} = \frac{(\beta R_{22} - z_{21})v_S}{R_{Th}R'_{11}} = \frac{(\beta R_{22} - z_{21})R_{\text{in} f}i_1}{R'_{11}R'_{22}} \qquad (9)$$

so

$$\beta' = \frac{(\beta R_{22} - z_{21})R_{\text{in} f}}{R'_{11}R'_{22}} \qquad (10)$$

12.64 Use the results of Problem 12.63 to redraw the current-series amplifier as ideal in that loading effects of the feedback network have been accounted for by adjustment of basic amplifier parameters.

▌ The current-series feedback amplifier of Fig. 12-9 can be redrawn as shown in Fig. 12-11. The ideal feedback network simply multiplies load current i_L by the transresistance gain and feeds the result (v'_f) to the input loop for mixing with the input signal v_S.

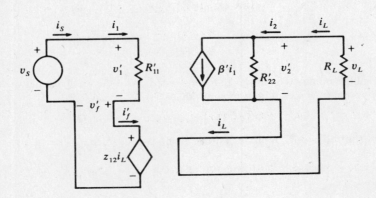

Fig. 12-11

12.65 The cascaded BJT amplifier of Fig. 12-12(a) uses current-series feedback. Draw a small-signal equivalent circuit suitable for use in analysis of the amplifier. Assume for all transistors that $h_{re} = h_{oe} = 0$.

▌ See Fig. 12-12(b).

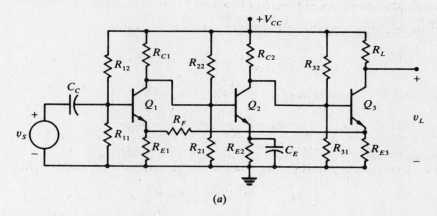

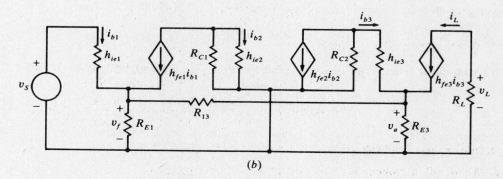

Fig. 12-12

12.66 Figure 12-12(a) displays a current-series feedback amplifier. Let $h_{fe1} = h_{fe2} = h_{fe3} = 50\,\Omega$, $h_{ie1} = h_{ie2} = h_{ie3} = 200\,\Omega$, $R_{C1} = R_{C2} = R_L = 2\,\text{k}\Omega$, $R_{E1} = 50\,\Omega$, $R_{E3} = 150\,\Omega$, and $R_{13} = 100\,\Omega$. Assume $h_{re1} = h_{re2} = h_{re3} = h_{oe1} = h_{oe2} = h_{oe3} = 0$. For this amplifier, find R_{of}.

▌ The small-signal equivalent circuit of Fig. 12-12(b) is applicable. By (3) of Problem 12.62, the z parameters are

$$z_{11} = \left.\frac{v_f}{i_{b1}}\right|_{i_L=0} = R_{E1} \parallel (R_{13} + R_{E3}) = \frac{50(250)}{300} = 41.67\,\Omega$$

$$z_{12} = \left.\frac{v_f}{i_L}\right|_{i_{b1}=0} = \frac{R_{E1}R_{E3}}{R_{E1} + R_{13} + R_{E3}} = \frac{50(150)}{300} = 25\,\Omega$$

$$z_{21} = \left.\frac{v_a}{i_1}\right|_{i_L=0} = \frac{R_{E1}R_{E3}}{R_{E1} + R_{13} + R_{E3}} = z_{12} = 25\,\Omega$$

$$z_{22} = \left.\frac{v_a}{i_L}\right|_{i_1=0} = R_{E3} \parallel (R_{13} + R_{E1}) = \frac{150(50)}{200} = 75\,\Omega$$

The basic amplifier parameters of Fig. 12-9 are needed. With the feedback network of Fig. 12-12(b) deactivated, so that $v_f = v_a = 0$, it is apparent that

$$R_{11} = R_{in} = h_{ie1} = 200\,\Omega$$

Also, with v_S shorted and R_L replaced with a driving-point source,

$$R_{22} = R_o = \frac{1}{h_{oe3}} = \infty$$

FEEDBACK AMPLIFIERS □ 359

From (4) of Problem 12.63 with $R_{22} = \infty$,

$$R_{of} = R_{Th} = R'_{22} = \infty$$

12.67 Evaluate $R_{in\,f}$ for the amplifier of Problem 12.66.

▌ With the feedback network still deactivated, current division gives

$$i_{b2} = \frac{R_{C1}}{R_{C1} + h_{ie2}} f_{fe1} i_{b1} \quad \text{and} \quad i_{b3} = \frac{R_{C2}}{R_{C2} + h_{ie3}} h_{fe2} i_{b2} \tag{1}$$

so that

$$i_L = h_{fe3} i_{b3} = \frac{R_{C1} R_{C2} h_{fe1} h_{fe2} h_{fe3}}{(R_{C1} + h_{ie2})(R_{C2} + h_{ie3})} i_{b1}$$

Whence,

$$\beta = \frac{i_L}{i_{b1}} = \frac{R_{C1} R_{C2} h_{fe1} h_{fe2} h_{fe3}}{(R_{C1} + h_{ie2})(R_{C2} + h_{ie3})} = \frac{(2000)^2 (50)^3}{(2200)^2} = 103.3 \times 10^3$$

Now, substituting R_{Th} as given by (4) of Problem 12.63 into (8) of Problem 12.63 gives

$$R_{in\,f} = \frac{R'_{11}\left[R_{22}\left(1 + \dfrac{\beta z_{12}}{R'_{11}}\right) + z_{22} + \dfrac{z_{12} z_{21}}{R'_{11}} + R_L\right] + z_{12}(\beta R_{22} - z_{12})}{R_{22}\left(1 + \dfrac{\beta z_{12}}{R'_{11}}\right) + z_{22} + \dfrac{z_{12} z_{21}}{R'_{11}} + R_L}$$

which in the limit as $R_{22} \to \infty$ yields

$$R_{in\,f} = R'_{11} + \frac{\beta R'_{11} z_{12}}{R'_{11} + \beta z_{12}} = h_{ie1} + z_{11} + \frac{\beta (h_{ie1} + z_{11}) z_{12}}{h_{ie1} + z_{11} + \beta z_{12}}$$

$$= 200 + 41.67 + \frac{(103.3 \times 10^3)(200 + 41.67)(25)}{200 + 41.67 + (103.3 \times 10^3)(25)} = 483.34 \ \Omega$$

12.68 Determine the overall voltage gain for the amplifier of Problem 12.66.

▌ We use (4) of Problem 12.63 to rewrite (10) of Problem 12.63 as

$$\beta' = \frac{(\beta R_{22} - z_{21}) R_{in\,f}}{[R_{22}(1 + \beta z_{12}/R'_{11}) + z_{22} - z_{12} z_{21}/R'_{11}] R'_{11}}$$

which, in the limit as $R_{22} \to \infty$, becomes

$$\beta' = \frac{\beta R_{in\,f}}{R'_{11} + \beta z_{12}} = \frac{\beta R_{in\,f}}{h_{ie1} + z_{11} + \beta z_{12}} = \frac{(103.3 \times 10^3)(483.34)}{200 + 41.67 + (103.3 \times 10^3)(25)} = 19.33 \tag{1}$$

The transconductance gain is found from Fig. 12-10 by replacing G_{11}, G_{22}, and β with $G'_{11} = 1/R'_{11}$, $G'_{22} = 0$, and β' respectively. The result is

$$A'_g = \frac{\beta' G'_{11} G_L}{G_L + G'_{22}} = \beta' G'_{11} = \frac{\beta'}{h_{ie1} + z_{11}} = \frac{19.33}{241.67} = 0.0452 \tag{2}$$

Then (1) of Problem 12.59 with A_g replaced by A'_g yields

$$A_{vf} = \frac{-R_L A'_g}{1 + \beta_{iv} A'_g} = \frac{-R_L A'_g}{1 + z_{12} A'_g} = \frac{-(2000)(0.0452)}{1 + 25(0.0452)} = -42.44 \tag{3}$$

12.69 Find the overall current gain for the amplifier of Problem 12.66.

▌ Using the results of Problems 12.67 and 12.68, the overall current gain follows as

$$A_{if} = \frac{i_L}{i_S} = \frac{-v_L/R_L}{v_S/R_{in\,f}} = -A_{vf} \frac{R_{in\,f}}{R_L} = -\frac{-42.44(483.34)}{2 \times 10^3} = 10.26$$

12.70 Show that if $\beta \gg 1$ and $G_{22} \to 0$ for the current-source amplifier with current-series feedback of Fig. 12-9, then

$$A_{vf} = \frac{A_v}{1 + \beta_{vv} A_v} = \frac{A_v}{1 - G_L \beta_{iv} A_v}$$

Based on (2) of Problem 12.56 and (3) of Problem 12.57 with $G_{22} = 0$ and $\beta \gg 1$,

$$A_{vf} = \frac{-\beta G_F G_{11}}{G_L(1+\beta)G_{11} + G_L G_F} \approx \frac{-\beta G_{11}/G_L}{1 + \beta G_{11}/G_F} \quad (1)$$

$$\beta_{iv} = \frac{1+\beta}{\beta G_F} \approx 1/G_F \quad (2)$$

Now **if** feedback is deactivated ($G_F = \infty$), then from (1)

$$A_v = -\beta G_{11}/G_L \quad (3)$$

Equation (1) can be rewritten by use of (2) and (3) as

$$A_{vf} = \frac{-\beta G_{11}/G_L}{1 - G_L \left(\frac{1}{G_F}\right)\left(-\frac{\beta G_{11}}{G_L}\right)} = \frac{A_v}{1 - G_L \beta_{iv} A_v} \quad (4)$$

Further,

$$\beta_{vv} \equiv \frac{v_f}{v_L} = \frac{v_f}{i_L}\frac{i_L}{v_L} = \beta_{iv}(-G_L) = -G_L \beta_{iv} \quad (5)$$

Substitute (5) into (4) to yield

$$A_{vf} = \frac{A_v}{1 + \beta_{vv} A_v} \quad (6)$$

12.71 Show that if $\beta \gg 1$ and $G_{22} \to 0$ for the current-source amplifier with current-series feedback of Fig. 12-9, then $R_{inf} = R_{in}(1 + \beta_{vv}A_v)$.

Based on (2) of Problem 12.60 with $G_{22} = 0$ and $\beta \gg 1$,

$$R_{inf} = R_{11} + \frac{1+\beta}{G_F} \approx R_{11} + \beta R_F \quad (1)$$

Comparison of the denominators of (3) of Problem 12.2 and (1) of Problem 12.70,

$$\beta_{vv} A_v = \frac{\beta G_{11}}{G_F} = \frac{\beta R_F}{R_{11}} \quad (2)$$

Whence,

$$\beta R_F = R_{11}\beta_{vv}A_v \quad (3)$$

Substitute (3) into (1) and realize from (2) of Problem 12.60 that, without feedback ($G_F = \infty$), $R_{in} = R_{11}$ to find

$$R_{inf} = R_{11}(1 + \beta_{vv}A_v) = R_{in}(1 + \beta_{vv}A_v) \quad (4)$$

12.72 Show that if $\beta \gg 1$, $\beta_{vv}A_v/\beta \ll 1$, and $G_{22}(=1/R_o) \to 0$ for the amplifier of Fig. 12-9 with current-series feedback, then $R_{of} = R_o(1 + \beta_{vv}A_v) \to \infty$.

From (4) of Problem 12.61 with $\beta \gg 1$ and using $G_{22} = 1/R_o$,

$$R_{of} \approx \frac{R_{22}(\beta G_{11} + G_{22} + G_F)}{G_{11} + G_F} = \frac{R_{22}(\beta G_{11}/G_F + G_{22}/G_F + 1)}{G_{11}/G_F + 1} \quad (1)$$

Using (1) of Problem 12.71 and $\beta_{vv}A_v/\beta \ll 1$ in (2) yields

$$R_{of} = \frac{R_{22}(G_{22}/G_F + 1 + \beta_{vv}A_v)}{1 + \beta_{vv}A_v/\beta} \approx R_{22}(1 + G_{22}/G_F + \beta_{vv}A_v) \quad (2)$$

Clearly, as $G_{22} \to 0$ ($R_o = R_{22} \to \infty$), (2) shows that

$$R_{of} = R_o(1 + \beta_{vv}A_v) \to \infty$$

12.73 Let $R_B = 200\text{ k}\Omega$, $R_L = 2\text{ k}\Omega$, $R_E = 100\text{ }\Omega$, $h_{ie} = 100\text{ }\Omega$, $h_{fe} = 60$, $h_{re} = 0$, and $h_{oe} = 5 \times 10^{-5}$ S for the BJT amplifier of Fig. 12-13. Find the exact feedback transfer ratio β_{iv}.

Based on (3) of Problem 12.57 and using parameters of the amplifier,

$$\beta_{iv} = \frac{(1+\beta)G_L + G_{22}}{G_L(\beta G_F - G_{22})} = \frac{(1+h_{fe})/R_L + h_{oe}}{1/R_L(h_{fe}/R_E - h_{oe})} = \frac{(1+60)/(2\times 10^3) + 5\times 10^{-5}}{1/(2\times 10^3)(60/100 - 5\times 10^{-5})} = 101.84$$

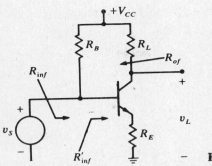

Fig. 12-13

12.74 For the amplifier of Problem 12.73, determine the overall voltage gain.

▮ Based on (7) of Problem 12.56 using the parameters of the amplifier,

$$A_{vf} = \frac{\frac{1}{h_{ie}}(h_{oe} - h_{fe}/R_E)}{h_{oe}(1/h_{ie} + 1/R_E) + \frac{1}{R_L}[(1+h_{fe})/h_{ie} + h_{oe}] + 1/R_L R_E}$$

$$= \frac{(1/100)(5 \times 10^{-5} - 60/100)}{(5 \times 10^{-5})(1/100 + 1/100) + (1/2 \times 10^3)[(1+60)/100 + 5 \times 10^{-5}] + (1/2 \times 10^3)(1/100)} = -19.289$$

12.75 Evaluate the input resistance for the amplifier of Problem 12.73.

▮ Based on (2) of Problem 12.60 using the parameters of the amplifier,

$$R'_{inf} = h_{ie} + \frac{(1+h_{fe})/R_L + h_{oe}}{h_{oe}/R_E + h_{oe}/R_L + 1/R_L R_E}$$

$$= 100 + \frac{(1+60)/(2 \times 10^3) + 5 \times 10^{-5}}{(5 \times 10^{-5})/100 + (5 \times 10^{-5})/(2 \times 10^3) + (1/2 \times 10^3)(1/100)} = 5629.4\,\Omega$$

and $\quad R_{inf} = R'_{inf} \| R_B = \dfrac{R'_{inf} R_B}{R'_{inf} + R_B} = \dfrac{5629.4(200 \times 10^3)}{5629.4 + 200 \times 10^3} = 5.47\,\text{k}\Omega$

12.76 Determine the output resistance for the amplifier of Problem 12.73.

▮ Based on (4) of Problem 12.61 with the parameters of the amplifier,

$$R_{of} = \frac{(1+h_{fe})/h_{ie} + h_{oe} + 1/R_E}{h_{oe}(1/h_{ie} + 1/R_E)} = \frac{(1+60)/100 + 5 \times 10^{-5} + 1/100}{(5 \times 10^{-5})(1/100 + 1/100)} = 620.05\,\text{k}\Omega$$

12.77 For the BJT amplifier of Problem 12.73, transfer the loading effects of the feedback network to the basic amplifier so that the circuit of Fig. 12-11 is applicable. Specifically, find R'_{11}, R'_{22}, and β'.

▮ Based on (3) of Problem 12.62, $z_{11} = z_{12} = z_{21} = z_{22} = R_E = 100\,\Omega$. From (1) of Problem 12.63,

$$R'_{11} = R_{11} + z_{11} = h_{ie} + z_{11} = 100 + 100 = 200\,\Omega$$

From (4) of Problem 12.63,

$$R'_{22} = \frac{1}{h_{oe}}(1 + h_{fe} z_{12}/R'_{11}) + z_{22} - z_{12} z_{21}/R'_{11}$$

$$= \frac{1}{5 \times 10^{-5}}[1 + 60(100)/200] + 100 - 100(100)/200 = 620.05\,\text{k}\Omega$$

Based on (10) of Problem 12.63 and result of Problem 12.75,

$$\beta' = \frac{(h_{fe}/h_{oe} - z_{21})R_{inf}}{R'_{11} R'_{22}} = \frac{(60/5 \times 10^{-5} - 100)(5.47 \times 10^3)}{200(620.05 \times 10^3)} = 52.93$$

12.78 Use the results of Problem 12.77 to find the overall voltage gain of the amplifier of Problem 12.73.

▌ Replacing G_{11}, G_{22}, and β of Fig. 12-10 by G'_{11}, G'_{22}, and β', respectively,

$$A'_g = \frac{\beta' G'_{11} G_L}{G_L + G'_{22}} = \frac{52.93(1/200)(1/2 \times 10^3)}{1/2 \times 10^3 + 1/620.05 \times 10^3} = 0.2638$$

Noting that $\beta_{iv} = z_{12}$ and using (3) of Problem 12.68,

$$A_{vf} = \frac{-R_L A'_g}{1 + \beta_{iv} A'_g} = \frac{-(2 \times 10^3)(0.2638)}{1 + 100(0.2638)} = -19.27$$

12.79 Determine the overall current gain for the amplifier of Problem 12.73.

▌

$$A_{if} = \frac{i_L}{i_1} = \frac{-v_L/R_L}{v_S/R_{\inf}} = -A_{vf} \frac{R_{\inf}}{R_L} = -(-19.27) \frac{5.47 \times 10^3}{2 \times 10^3} = 52.7$$

12.80 Rework Problem 12.66 with $h_{oe3} = 5 \times 10^{-5}$ S and all else unchanged.

▌ Only the calculation of $R_{of} = R_{Th}$ need be recomputed. Using β as determined in Problem 12.67 and (4) of Problem 12.63,

$$R_{Th} = R_{of} = \frac{1}{h_{oe3}} \left(1 + \frac{\beta z_{12}}{h_{ie1} + z_{11}}\right) + z_{22} - \frac{z_{12} z_{21}}{h_{ie1} + z_{11}}$$

$$= \frac{1}{5 \times 10^{-5}} \left[1 + \frac{(103.3 \times 10^3)(25)}{200 + 41.67}\right] + 75 - \frac{25(25)}{200 + 41.67} = 213.7 \text{ M}\Omega$$

12.81 Rework Problem 12.67 with $h_{oe3} = 5 \times 10^{-5}$ S and all else unchanged.

▌ Based on (8) of Problem 12.63,

$$R_{\inf} = \frac{(h_{ie1} + z_{11})(R_{Th} + R_L) + z_{12}(\beta/h_{oe3} - z_{12})}{R_{Th} + R_L}$$

$$= \frac{(200 + 41.67)(213.7 \times 10^6 + 2 \times 10^3) + 25(103.3 \times 10^3/5 \times 10^{-5} - 25)}{213.7 \times 10^6 + 2 \times 10^3} = 483.3 \, \Omega$$

12.82 Rework Problem 12.68 with $h_{oe3} = 5 \times 10^{-5}$ S and all else unchanged.

▌ Based on (10) of Problem 12.63,

$$\beta' = \frac{(\beta/h_{oe3} - z_{21})R_{\inf}}{(h_{ie1} + z_{11})R_{Th}} = \frac{(103.3 \times 10^3/5 \times 10^{-5} - 25)(483.3)}{(200 + 41.67)(213.7 \times 10^6)} = 19.33$$

From Fig. 12-10,

$$A'_g = \frac{\beta' G'_{11} G_L}{G_L + G'_{22}} = \frac{\beta'/(h_{ie1} + z_{11})}{1 + R_L h_{oe3}} = \frac{19.33/(200 + 41.67)}{1 + (2 \times 10^3)(5 \times 10^{-5})} = 0.0727$$

$$A_{vf} = \frac{-R_L A'_g}{1 + \beta_{iv} A'_g} = \frac{-(2 \times 10^3)(0.0727)}{1 + 25(0.0727)} = -51.61$$

12.83 Rework Problem 12.69 with $h_{oe3} = 5 \times 10^{-5}$ S and all else unchanged.

▌

$$A_{if} = -A_{vf} \frac{R_{\inf}}{R_L} = -(-51.61) \frac{483.3}{2 \times 10^3} = 12.47$$

VOLTAGE-SHUNT FEEDBACK

12.84 The general feedback amplifier model of Fig. 12-14, with output voltage v_L as the sampled variable and feedback current i_f as the mixed variable, is called a *voltage-shunt* (or *parallel-output, series input*; or *node-sampling, node-comparison*) feedback amplifier. For the voltage-shunt feedback amplifier of Fig. 12-14, find the input resistance.

■ By the method of node voltages for v_1 ($= v_f$) and v_L,

$$(v_1 - v_S)G_S + v_1 G_{11} + (v_1 - v_L)G_F = 0 \tag{1}$$

and

$$\beta i_1 + v_L G_{22} + (v_L - v_1)G_F + v_L G_L = 0 \tag{2}$$

But

$$i_1 = G_{11} v_1 \tag{3}$$

We substitute (3) into (2) and rearrange to get the following system of linear equations:

$$\begin{bmatrix} G_F + G_S + G_{11} & -G_F \\ -(G_F - \beta G_{11}) & G_{22} + G_F + G_L \end{bmatrix} \begin{bmatrix} v_1 \\ v_L \end{bmatrix} = \begin{bmatrix} G_S v_S \\ 0 \end{bmatrix} \tag{4}$$

Simultaneous solution of (4) by Cramer's rule gives v_1 as

$$v_1 = \frac{\Delta_1}{\Delta} = \frac{\begin{vmatrix} G_S v_S & -G_F \\ 0 & G_{22} + G_F + G_L \end{vmatrix}}{\begin{vmatrix} G_F + G_S + G_{11} & -G_F \\ -(G - \beta G_{11}) & G_{22} + G_F + G_L \end{vmatrix}} = \frac{G_S(G_{22} + G_F + G_L)}{\Delta} \tag{5}$$

where

$$\Delta = G_{11}[(1 + \beta)G_F + G_{22} + G_L] + G_S(G_{22} + G_F) + G_L(G_S + G_F) \tag{6}$$

Utilizing $i_S = G_S(v_S - v_1)$ and (5), we find the input resistance as

$$R_{\inf} = \frac{v_i}{i_S} = \frac{v_1}{G_S(v_S - v_1)} = \frac{G_{22} + G_F + G_L}{\Delta - G_S(G_{22} + G_F + G_L)} \tag{7}$$

Substituting (6) into (7) and rearranging, we obtain

$$R_{\inf} = \frac{G_{22} + G_F + G_L}{G_{11}[(1 + \beta)G_F + G_{22} + G_L] + G_L G_F} \tag{8}$$

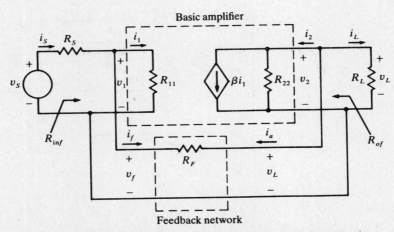

Fig. 12-14 General voltage-shunt feedback amplifier model.

12.85 For the voltage-shunt amplifier of Fig. 12-14, determine an expression for output resistance.

■ If we replace R_L with a driving-point source and short v_S in Fig. 12-14, then we have

$$i_{dp} = \beta i_1 + \frac{v_{dp}}{R_{22}} + \frac{v_{dp}}{R_F + (R_S \parallel R_{11})} = \beta i_1 + G_{22} v_{dp} + \frac{(G_{11} + G_S) v_{dp}}{1 + R_F(G_{11} + G_S)} \tag{1}$$

But

$$i_1 = \frac{v_{dp}}{R_F + (R_S \parallel R_{11})} \frac{R_S}{R_S + R_{11}} = \frac{G_{11} v_{dp}}{1 + R_F(G_{11} + G_S)} \tag{2}$$

Using (2) in (1) and solving for the output conductance now give

$$G_{of} = \frac{1}{R_{of}} = \frac{i_{dp}}{v_{dp}} = G_{22} + \frac{(1 + \beta)G_{11} + G_S}{1 + R_F(G_{11} + G_S)} \tag{3}$$

12.86 Using the results of Problems 11.84 and 11.85, determine the input and output resistances for the amplifier of Fig. 12-14 if the feedback is inactive.

■ To deactivate the feedback network, we let $G_F = 1/R_F = 0$; then (8) of Problem 12.84 and (3) of Problem 12.85 yield

$$R_{in} = \lim_{G_F \to 0} R_{inf} = \frac{1}{G_{11}} = R_{11} \qquad (1)$$

$$\frac{1}{R_o} = G_o = \lim_{G_F \to 0} G_{of} = G_{22} \qquad (2)$$

12.87 The ideal voltage-shunt feedback amplifier, shown in block-diagram form in Fig. 12-15, is based on the assumption that the feedback network does not load the basic amplifier. In this case, the basic amplifier is a transresistance amplifier. For the ideal voltage-shunt amplifier of Fig. 12-15, find the overall voltage gain.

■ The basic amplifier gain provides that

$$v_L = A_r i_1 = R_L A_i i_1 = -\frac{\beta}{G_L + G_{22}} i_1 \qquad (1)$$

and from the comparator output,

$$i_1 = i_S - i_f = \frac{v_S}{R_{inf} + R_S} - \beta_{vi} v_L \qquad (2)$$

Substituting (2) into (1) and rearranging give

$$A_{vf} = \frac{v_L}{v_S} = \frac{A_r/(R_{inf} + R_S)}{1 + \beta_{vi} A_r} = \frac{R_L A_i/(R_{inf} + R_S)}{1 + \beta_{vi} R_L A_i} = \frac{-\beta/(R_{inf} + R_S)}{G_L + G_{22} - \beta_{vi}\beta} \qquad (3)$$

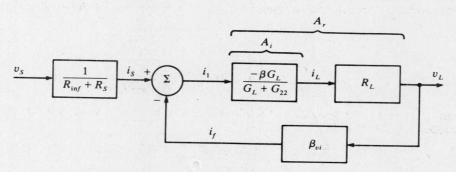

Fig. 12-15 Ideal-voltage-shunt feedback model.

12.88 Find an expression for the overall current gain of the ideal voltage-shunt feedback amplifier.

■ Using the results of Problem 12.87, the overall current gain follows as

$$A_{if} = \frac{i_L}{i_S} = \frac{v_L/R_L}{v_S/(R_{inf} + R_S)} = A_{vf} \frac{R_{inf} + R_S}{R_L} = \frac{A_r/R_L}{1 + \beta_{vi} A_r} \qquad (1)$$

12.89 For the general voltage-shunt feedback amplifier of Fig. 12-14, find the overall voltage gain.

■ Building on the work of Problem 12.84, we have

$$v_L = \frac{\Delta_2}{\Delta} = \frac{\begin{vmatrix} G_F + G_S + G_{11} & G_S v_S \\ -(G_F - \beta G_{11}) & 0 \end{vmatrix}}{\Delta} = \frac{G_S(G_F - \beta G_{11})}{\Delta} v_S \qquad (1)$$

so that, with Δ as in (6) of Problem 12.84, the overall voltage gain is

$$A_{vf} = \frac{v_L}{v_S} = \frac{G_S(G_F - \beta G_{11})}{G_{11}[(1+\beta)G_F + G_{22} + G_L] + G_S(G_{22} + G_F) + G_L(G_S + G_F)} \qquad (2)$$

12.90 Determine an expression for the overall current gain of the voltage-shunt feedback amplifier in Fig. 12-14.

▌ The overall current gain is found as

$$A_{if} = \frac{i_L}{i_1} = \frac{v_L/R_L}{v_S/(R_{inf}+R_S)} = A_{vf}\frac{R_{inf}+R_S}{R_L} \tag{1}$$

Substituting (2) of Problem 12.89 and (8) of Problem 12.84 into (1) then yields

$$A_{if} = \frac{G_L G_S (G_F - \beta G_{11})(G_{22}+G_F+G_L+R_S G)}{G[G+G_S(G_{22}+G_F+G_L)]}$$

where
$$G = G_{11}[(1+\beta)G_F + G_{22} + G_L] + G_L G_F \tag{2}$$

12.91 The loading effects of the feedback network in the voltage-shunt feedback amplifier can be accounted for by modifying the basic amplifier through the use of *short-circuit admittance parameters* or *y parameters* defined in terms of the variables of Fig. 12-14 to model the feedback network as a two-port network:

$$i_a = y_{11}v_L + y_{12}v_f \tag{1}$$

$$i_f = y_{21}v_L + y_{22}v_f \tag{2}$$

where
$$y_{11} = \frac{i_a}{v_L}\bigg|_{v_f=0} \quad y_{12} = \frac{i_a}{v_f}\bigg|_{v_L=0} \quad y_{21} = \frac{i_f}{v_L}\bigg|_{v_f=0} \quad y_{22} = \frac{i_f}{v_f}\bigg|_{v_L=0} \tag{3}$$

Draw the modified voltage-shunt feedback amplifier that results when the loading of the feedback network of Fig. 12-14 is transferred to the basic amplifier through use of the y parameters, leaving an ideal feedback network.

▌ Equations (1) and (2) each suggest a current source in parallel with an admittance. Admittance y_{11} of (1) can be combined with R_{11}, and admittance y_{22} of (2) can be combined with R_{22}, to give

$$R'_{11} = R_{11} \left\| \frac{1}{y_{11}} = \frac{R_{11}}{1+y_{11}R_{11}} \right. \tag{4}$$

$$R'_{22} = R_{22} \left\| \frac{1}{y_{22}} = \frac{R_{22}}{1+y_{22}R_{22}} \right. \tag{5}$$

Further, the current-controlled current source of Fig. 12-14 can be combined with the suggested voltage-controlled current source of (2) to obtain

$$\beta' i_1 = \beta i_1 + y_{12}v_f = \beta i_1 + y_{12}R_{11}i_1$$

so that
$$\beta' = \beta + y_{12}R_{11} \tag{6}$$

The network of Fig. 12-16 results from the use of (4) to (6).

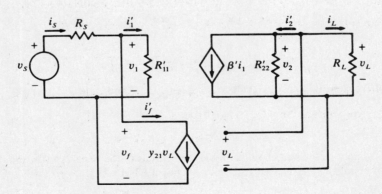

Fig. 12-16

12.92 Voltage-shunt feedback is used in the BJT amplifier of Fig. 12-17. Let $R_C = 4\text{ k}\Omega$, $R_F = 40\text{ k}\Omega$, $R_S = 10\text{ k}\Omega$, $h_{ie} = 1100\text{ }\Omega$, $h_{fe} = 50$, and $h_{re} = h_{oe} = 0$. Using the exact model of Fig. 12-14, find R_{inf}.

By (8) of Problem 12.84 with $G_{22} = h_{oe} = 0$, $G_L = G_C$, and $G_{11} = 1/h_{ie}$,

$$R_{inf} = \frac{h_{oe} + G_F + G_C}{\frac{1}{h_{ie}}[(1+h_{fe})G_F + h_{oe} + G_C] + G_C G_F}$$

$$= \frac{\frac{1}{40 \times 10^3} + \frac{1}{4 \times 10^3}}{\frac{1}{1100}\left(\frac{51}{40 \times 10^3} + \frac{1}{4 \times 10^3}\right) + \frac{1}{(4 \times 10^3)(40 \times 10^3)}} = 197.47\ \Omega$$

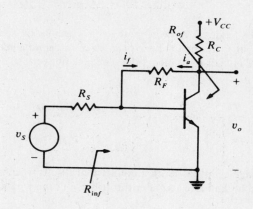

Fig. 12-17

12.93 Evaluate the input resistance for the amplifier of Problem 12.92.

From (3) of Problem 12.85,

$$G_{of} = h_{oe} + \frac{(1+h_{fe})/h_{ie} + 1/R_S}{1 + R_F(1/h_{ie} + 1/R_S)} = \frac{51/1100 + 1/10^4}{1 + 40 \times 10^3(1/1100 + 1/10^4)} = 0.001123$$

and $\quad R_{of} = \dfrac{1}{G_{of}} = \dfrac{1}{0.001123} = 890.2\ \Omega$

12.94 Determine the value of overall voltage gain for the amplifier of Problem 12.92.

From (2) of Problem 12.89,

$$A_{vf} = \frac{G_S(G_F - h_{fe}/h_{ie})}{(1/h_{ie})[(1+h_{fe})G_F + h_{oe} + G_C] + G_S(h_{oe} + G_F) + G_C(G_S + G_F)}$$

$$= \frac{(1 \times 10^{-4})(2.5 \times 10^{-5} - 50/1100)}{9.09 \times 10^{-4}[51(2.5 \times 10^{-5}) + 2.5 \times 10^{-4}] + (1 \times 10^{-4})(2.5 \times 10^{-5}) + (2.5 \times 10^{-4})(1 \times 10^{-4} + 2.5 \times 10^{-5})}$$

$$= -3.2$$

12.95 Find the overall current gain for the amplifier of Problem 12.92.

By (1) of Problem 12.90 using results of Problems 12.92 and 12.94,

$$A_{if} = A_{vf} \frac{R_{inf} + R_S}{R_C} = \frac{-3.2(197.47 + 10 \times 10^3)}{4 \times 10^3} = -8.16$$

12.96 Draw a small-signal equivalent circuit suitable for analysis of the cascaded BJT amplifier of Fig. 12-18(a) that makes use of voltage-shunt feedback. Assume identical BJTs with $h_{re} = h_{oe} = 0$.

See Fig. 12-18(b).

12.97 Voltage shunt feedback is used in the cascaded BJT amplifier of Fig. 12-18(a). The transistors are identical, with $h_{re} = h_{oe} = 0$, $h_{fe} = 100$, and $h_{ie} = 2\ k\Omega$. Let $R_S = 100\ \Omega$, $R_E = 4\ k\Omega$, $R_F = 3\ k\Omega$, and $R_{C2} = 6\ k\Omega$. Find R_{inf}.

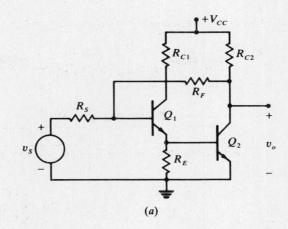

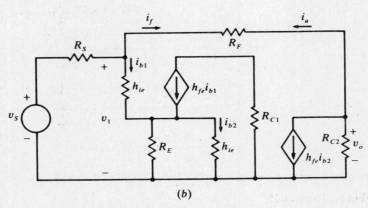

Fig. 12-18

▌ The small-signal equivalent circuit of Fig. 12-18(b) is applicable. With the feedback network deactivated (R_F removed), $\beta i_{b1} = h_{fe} i_{b2}$. But by current division,

$$i_{b2} = \frac{R_E}{R_E + h_{ie}} (h_{fe} + 1) i_{b1}$$

so

$$\beta = \frac{h_{fe}(h_{fe} + 1) R_E}{R_E + h_{ie}} = \frac{100(101)(4000)}{6000} = 6733.3$$

We deactivate (short) v_S and replace R_{C2} with a driving-point source to see that $R_o = R_{22} = \infty$.

The input impedance (resistance) with the feedback network deactivated can be found by applying KVL around the input loop:

$$v_S = i_{b1} h_{ie} + (h_{fe} + 1) i_{b1} (R_E \parallel h_{ie})$$

so

$$R_{11} = R_{in} = \frac{v_S}{i_{b1}} = h_{ie} + \frac{(h_{fe} + 1) R_E h_{ie}}{R_E + h_{ie}} = 2000 + \frac{101(4000)(2000)}{6000} = 136.67 \text{ k}\Omega$$

Then, from (8) of Problem 12.84,

$$R_{inf} = \frac{G_{22} + G_F + G_{C2}}{G_{11}[(1 + \beta) G_F + G_{22} + G_{C2}] + G_{C2} G_F}$$

$$= \frac{1/3000 + 1/6000}{\dfrac{1}{136{,}670} \left(\dfrac{6734.3}{3000} + \dfrac{1}{6000} \right) + \dfrac{1}{(6000)(3000)}} = 30.34 \ \Omega$$

12.98 Determine the output resistance for the amplifier of Problem 12.97.

From (3) of Problem 12.85,

$$G_{of} = G_{22} + \frac{(1+\beta)G_{11} + G_S}{1 + R_F(G_{11} + G_S)} = \frac{6734.3/136{,}670 + 1/100}{1 + 3000(1/136{,}670 + 1/100)} = 0.001911 \text{ S}$$

and

$$R_{of} = \frac{1}{G_{of}} = \frac{1}{0.001911} = 523.3 \ \Omega$$

12.99 Find the overall voltage gain for the amplifier of Problem 12.97.

From (3) of Problem 12.91,

$$y_{11} = \left.\frac{i_a}{v_o}\right|_{v_1=0} = \frac{1}{R_F} = \frac{1}{3000} = 3.333 \times 10^{-4} \text{ S}$$

$$y_{12} = \left.\frac{i_a}{v_1}\right|_{v_o=0} = -\frac{1}{R_F} = -3.333 \times 10^{-4} \text{ S}$$

$$y_{21} = \left.\frac{i_f}{v_o}\right|_{v_1=0} = -\frac{1}{R_F} = -3.333 \times 10^{-4} \text{ S}$$

$$y_{22} = \left.\frac{i_f}{v_1}\right|_{v_o=0} = \frac{1}{R_F} = 3.333 \times 10^{-4} \text{ S}$$

Thus, by (5) and (6) of Problem 12.91,

$$R'_{22} = \lim_{R_{22}\to\infty} \frac{R_{22}}{1 + y_{22}R_{22}} = \frac{1}{y_{22}} = R_F = 3 \text{ k}\Omega$$

$$\beta' = \beta + y_{12}R_{11} = 6733.3 + (-3.333 \times 10^{-4})(136.67 \times 10^3) = 6687.7$$

From Fig. 12-16 with β and G_{22} replaced by β' and G'_{22}, respectively, and with $G_L = G_{C2}$, the transresistance gain is

$$A_r = \frac{-\beta'}{G_{C2} + G'_{22}} = \frac{-6687.7}{1/6000 + 1/3000} = -13.375 \times 10^6$$

The feedback transfer ratio is $\beta_{vi} = y_{21} = -1/R_F$; hence, by (3) of Problem 12.87,

$$A_{vf} = \frac{A_r/(R_{\text{in}f} + R_S)}{1 + \beta_{vi}A_r} = \frac{-13.375 \times 10^6/(130.09)}{1 + (-3.333 \times 10^{-4})(-13.375 \times 10^6)} = -23.05$$

CURRENT-SHUNT FEEDBACK

12.100 The general feedback amplifier of Fig. 12-19, in which output current i_L is the sampled variable and feedback current i_f is the mixed variable, is called a *current-shunt* (or *parallel-output, parallel-input*; or *node-sampling, node-comparison*) feedback amplifier. For the current-shunt feedback amplifier of Fig. 12-19, model the feedback network as a two-port network using h parameters.

Using the feedback-network variables of Fig. 12-19 in (1) to (6) of Problem 1.76 leads to

$$v_a = h_{11}i_L + h_{12}v_f \tag{1}$$

$$i_f = h_{21}i_L + h_{22}v_f \equiv i'_f + h_{22}v_f \tag{2}$$

where

$$h_{11} = \left.\frac{v_a}{i_L}\right|_{v_f=0} \quad h_{12} = \left.\frac{v_a}{v_f}\right|_{i_L=0} \quad h_{21} = \left.\frac{i_f}{i_L}\right|_{v_f=0} \quad h_{22} = \left.\frac{i_f}{v_f}\right|_{i_L=0} \tag{3}$$

For the particular feedback network at hand,

$$h_{11} = R_E \parallel R_F = \frac{R_E R_F}{R_E + R_F} \quad h_{12} = \frac{R_E}{R_E + R_F} \quad h_{21} = \frac{-R_E}{R_E + R_F} \quad h_{22} = \frac{1}{R_E + R_F}$$

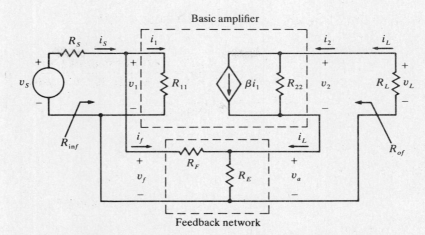

Fig. 12-19 General current-shunt feedback amplifier model.

12.101 Modify the basic amplifier parameters β, R_{11}, and R_{22} of Fig. 12-19 to account for any loading introduced by the feedback network, thus leaving the latter ideal.

▮ Since $v_1 = v_f$ in Fig. 12-19, admittance h_{22} in (2) of Problem 12.100 is connected in parallel with R_{11}; thus, an equivalent resistance,

$$R'_{11} = R_{11} \left\| \frac{1}{h_{22}} = \frac{R_{11}}{1 + h_{22}R_{11}} \right. \tag{1}$$

can be formed, leaving only a controlled current source within the left port of the feedback network.

A Thévenin equivalent circuit can be found for the network to the left of R_L in Fig. 12-19. With v_S deactivated (shorted) and a driving-point source replacing R_L, KCL yields

$$i_{dp} = \beta i_1 + G_{22}[v_{dp} - (i_{dp}h_{11} + h_{12}v_f)] \tag{2}$$

But, by current division at the node of the input network,

$$i_1 = \frac{-1/h_{22}}{1/h_{22} + R_{11}} h_{21}i_L = \frac{-h_{21}}{1 + h_{22}R_{11}} i_L = \frac{-h_{21}}{1 + h_{22}R_{11}} i_{dp} \tag{3}$$

Also,
$$v_f = R_{11} i_1 \tag{4}$$

Substituting (3) and (4) into (2) and rearranging yield the Thévenin resistance:

$$R_{Th} = \frac{1}{Y_N} = R'_{22} = \frac{v_{dp}}{i_{dp}} = R_{22}\left(1 + \frac{\beta h_{21}}{1 + h_{22}R_{11}}\right) + h_{11} - \frac{G_{22}h_{12}h_{21}R_{11}}{1 + h_{22}R_{11}} \tag{5}$$

The Thévenin voltage is found by allowing $R_L \to \infty$:

$$V_{Th} = -v_L = (\beta R_{22} + h_{12}R_{11})i_1 \tag{6}$$

The Norton current that would flow if R_L were zero is found by use of (5) and (6):

$$I_N = \frac{V_{Th}}{R_{Th}} = \frac{(\beta R_{22} + h_{12}R_{11})(1 + h_{22}R_{11})i_1}{R_{22}(1 + h_{22}R_{11} + \beta h_{21}) + h_{11}(1 + h_{22}R_{11}) - G_{22}h_{12}h_{21}R_{11}} = \beta' i_1 \tag{7}$$

The expression for β is apparent from (7).

12.102 Based on the results of Problem 12.101, replace the general current-shunt feedback amplifier model of Fig. 12-19 with a new model with ideal feedback wherein any loading by the original feedback network has been appropriately accounted for by adjustment of the basic amplifier parameters.

▮ With R'_{11} of (1) in Problem 12.101 and the elements of the Norton equivalent circuit as given by (5) and (7) of Problem 12.101, the output network of Fig. 12-19 can be redrawn as in Fig. 12-20. The ideal feedback network merely multiplies current i_L by h_{21} and feeds the result to the input node for mixing with the input signal i_S.

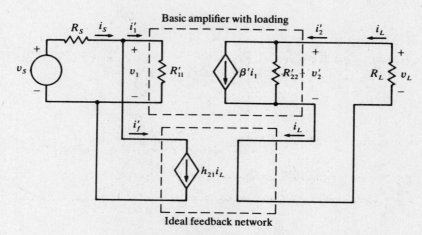

Fig. 12-20 Ideal feedback network.

12.103 The ideal current-shunt feedback amplifier model shown in block-diagram form in Fig. 12-21 is based on the assumption that the feedback network does not load the basic amplifier; however, if the loading were not negligible, the procedure of Problem 12.101 could be utilized to replace β with β' and G_{22} with $G'_{22} = 1/R'_{22}$. The basic amplifier in this case is a current amplifier. For the ideal current-shunt feedback amplifier of Fig. 12-21, find the overall current gain.

❙ The basic amplifier gain relationship is

$$i_L = A_i i_1 = \frac{\beta G_L}{G_L + G_{22}} i_1 \qquad (1)$$

and the comparator output gives

$$i_1 = i_S - i_f = i_S - \beta_{ii} i_L \qquad (2)$$

We substitute (1) into (2) and rearrange to find

$$A_{if} = \frac{i_L}{i_S} = \frac{A_i}{1 + \beta_{ii} A_i} = \frac{\beta G_L}{G_L + G_{22} + \beta_{ii} G_L} \qquad (3)$$

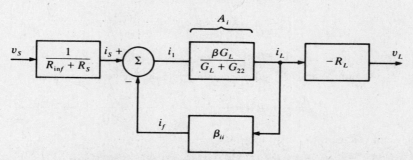

Fig. 12-21 Ideal current-shunt feedback model.

12.104 Based on the ideal current-shunt feedback amplifier model of Fig. 12-21, find an expression for the overall voltage gain.

❙ The overall voltage gain follows, with use of (3) of Problem 12.103, as

$$A_{vf} = \frac{v_L}{v_S} = \frac{-i_L R_L}{i_S(R_{\inf} + R_S)} = \frac{-A_{if} R_L}{R_{\inf} + R_S} = \frac{-\beta}{(G_L + G_{22} + \beta_{ii} G_L)(R_{\inf} + R_S)} \qquad (1)$$

12.105 For the general current-shunt feedback amplifier of Fig. 12-19, find $R_{\inf}$.

❙ For the model of Fig. 12-20, in which the feedback network loading is accounted for within the basic amplifier, KCL at the node of the input network yields

$$i_S = h_{21} i_L + \frac{v_S}{R'_{11}} \qquad (1)$$

But
$$i_L = \frac{V_{Th}}{R_{Th} + R_L} \quad (2)$$

where V_{Th} is given by (6) of Problem 12.101, and R_{Th} by (5) of Problem 12.101. Substituting (6) of Problem 12.101 into (2) with $i_1 = v_S/R_{11}$ and using the result in (1) give

$$i_S = h_{21} \frac{\beta R_{22} + h_{12} R_{11}}{R_{Th} + R_L} \frac{v_S}{R_{11}} + \frac{v_S}{R'_{11}} \quad (3)$$

from which
$$R_{inf} = \frac{v_S}{i_S} = \frac{R_{11} R'_{11} (R_{Th} + R_L)}{h_{21} R'_{11} (\beta R_{22} + h_{12} R_{11}) + R_{11} (R_{Th} + R_L)} \quad (4)$$

12.106 Determine R_{of} for the general current-shunt feedback amplifier of Fig. 12-19.

▮ If R_L were replaced with a driving-point source and v_S deactivated, we would have

$$R_{of} = R'_{22} = R_{Th}$$

as determined by (5) of Problem 12.101.

12.107 Current-shunt feedback is used in the cascaded BJT amplifier of Fig. 12-22(a). If $h_{oe1} = h_{oe2} = h_{re1} = h_{re2} = 0$, draw a small-signal equivalent circuit suitable for analysis of this amplifier.

▮ See Fig. 12-22(b).

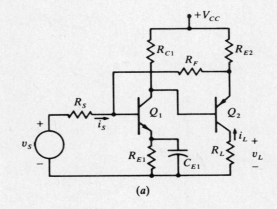

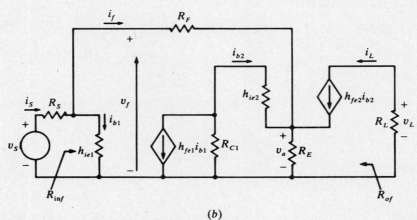

Fig. 12-22

12.108 For the current-shunt feedback amplifier of Fig. 12-22(a), let $R_{E2} = R_S = 1\,k\Omega$, $R_{C1} = 12\,k\Omega$, $R_L = 4\,k\Omega$, $R_F = 20\,k\Omega$, $h_{fe1} = h_{fe2} = 100$, $h_{ie1} = h_{ie2} = 2\,k\Omega$, and $h_{re1} = h_{re2} = h_{oe1} = h_{oe2} = 0$. Find R_{of}.

▮ The small-signal equivalent circuit of Fig. 12-22(b) is applicable. The parameters β, R_{11}, and R_{22} of the basic amplifier are needed. By inspection, with the feedback loop deactivated ($R_E = 0$ and $R_F = \infty$), $R_{in} = R_{11} = h_{ie1} = 2\,k\Omega$. Further, replacing R_L with a driving-point source leads to the conclusion that

$R_o = R_{22} = 1/h_{oe2} = \infty$. And current division gives

$$\beta i_1 = \beta i_{b1} = h_{fe2}i_{b2} = h_{fe2}\left(-h_{fe1}i_{b1}\frac{R_{C1}}{R_{C1}+h_{ie2}}\right)$$

so that
$$\beta = \frac{-h_{fe1}h_{fe2}R_{C1}}{R_{C1}+h_{ie2}} = -\frac{(100)^2(12,000)}{14,000} = -8571.4$$

The h parameters of the feedback network are determined by use of (3) of Problem 12.100. We neglect i_{b2} typically $(h_{fe2}+1)i_{b2} \approx h_{fe2}i_{b2}$; then,

$$h_{11} \approx \left.\frac{v_a}{i_L}\right|_{v_f=0} = R_E \| R_F = \frac{1000(20,000)}{21,000} = 952.4\,\Omega$$

$$h_{12} \approx \left.\frac{v_a}{v_L}\right|_{i_L=0} = \frac{R_E}{R_F+R_E} = \frac{1000}{21,000} = 0.04762$$

$$h_{21} \approx \left.\frac{i_f}{i_L}\right|_{v_f=0} = \frac{-R_E}{R_F+R_E} = \frac{-1000}{21,000} = -0.04762$$

$$h_{22} \approx \left.\frac{i_f}{v_f}\right|_{i_L=0} = \frac{1}{R_F+R_E} = \frac{1}{21,000} = 4.762\times 10^{-5}\,\text{S}$$

Since $R_{22} = \infty$, (5) of Problem 12.101 gives, regardless of the values of the h parameters,

$$R_{of} = R_{Th} = R'_{22} = \infty$$

12.109 Evaluate $R_{\text{in}f}$ for the amplifier of Problem 12.108.

▌ Direct use of (1) of Problem 12.101, along with the h parameters of Problem 12.108, yields

$$R'_{11} = \frac{R_{11}}{1+h_{22}R_{11}} = \frac{h_{ie1}}{1+h_{22}h_{ie1}} = \frac{2000}{1+(4.762\times 10^{-5})(2000)} = 1826.1\,\Omega$$

By (4) of Problem 12.105 and (5) of Problem 12.101, since $R_{22}=\infty$,

$$R_{\text{in}f} = \lim_{R_{22}\to\infty}\frac{R_{11}R'_{11}(R_{Th}/R_{22}+R_L/R_{22})}{h_{21}R'_{11}(\beta+h_{12}R_{11}/R_{22})+R_{11}(R_{Th}/R_{22}+R_L/R_{22})}$$

$$= \frac{R_{11}R'_{11}(1+R_{11}h_{22}+\beta h_{21})}{h_{21}R'_{11}\beta(1+R_{11}h_{22})+R_{11}(1+R_{11}h_{22}+\beta h_{21})}$$

$$= \frac{2000(1826.1)[1+2000(4.672\times 10^{-5})+(-8571.4)(-0.04762)]}{-0.04762(1826.1)(-8571.4)[1+2000(4.672\times 10^{-5})]}$$
$$+2000[1+2000(4.672\times 10^{-5})+(-8571.4)(-0.04762)]$$

$$= 27.03\,\Omega$$

12.110 Determine the overall current gain for the amplifier of Problem 12.108.

▌ From Fig. 12-21 with $G_{22}=1/R_{22}=0$,

$$A_i = \frac{\beta G_L}{G_L+G'_{22}} = \beta = -8571.4 \quad \text{and} \quad \beta_{ii} = h_{21} = -0.04762$$

Then from (3) of Problem 12.103,

$$A_{if} = \frac{A_i}{1+\beta_{ii}A_i} = \frac{-8571.4}{1+(-0.04762)(-8571.4)} = -20.95$$

12.111 Find the overall voltage gain for the amplifier of Problem 12.108.

▌ The overall voltage gain is found by direct use of (1) of Problem 12.104 and the results of Problems 12.109 and 12.110:

$$A_{vf} = \frac{-A_{if}R_L}{R_{\text{in}f}+R_S} = \frac{-(-20.95)(4000)}{1027.03} = 81.59$$

Review Problems

12.112 The gain-bandwidth product GBW of an amplifier remains essentially constant with the addition of feedback (see Problem 12.8). Illustrate that fact for the amplifier of Problem 12.5.

▌ Without feedback,

$$\text{GBW} = A_o(f_H - f_L) = 100(40 \times 10^3 - 20) = 3.998 \times 10^6 \text{ Hz}$$

With feedback, using the results of Problems 12.7 and 12.8,

$$\text{GBW}_f = A_{fo}(f_{Hf} - f_{Lf}) = 9.09(440 \times 10^3 - 1.82) = 3.999 \times 10^6 \text{ Hz}$$

12.113 Draw a small-signal equivalent circuit suitable for analysis of the FET amplifier in Fig. 12-23 and classify the feedback.

▌ The equivalent circuit is shown by Fig. 12-24. Comparison with Fig. 12-3 leads to the conclusion that voltage-series feedback exists.

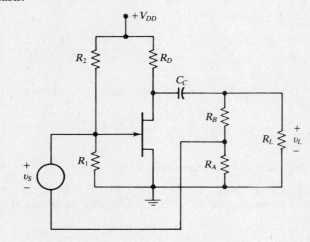

Fig. 12-23

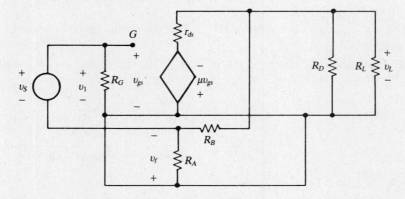

Fig. 12-24

12.114 Find expressions for the voltage-gain ratio of the amplifier in Fig. 12-23 with feedback and without feedback ($R_A = 0$, $R_B = \infty$). Assume that R_A and R_B are large enough to result in negligible loading by the feedback network.

▌ Consider the case without feedback ($R_A = 0$, $R_B = \infty$). By voltage division applied to the output network of Fig. 12-24,

$$v_L = \frac{R_D \parallel R_L}{R_D \parallel R_L + r_{ds}} \mu v_{gs} \tag{1}$$

Since $v_{gs} = v_S$, the voltage-gain ratio follows from (1) as

$$A_v = \frac{v_L}{v_S} = \frac{\mu R_D R_L}{R_D R_L + r_{ds}(R_D + R_L)} \tag{2}$$

R_G is typically large. Hence, for large R_A and R_B, β_{vv} is found by voltage division, giving

$$\beta_{vv} = \frac{v_f}{v_L} = \frac{R_A}{R_A + R_B} \tag{3}$$

By (3) of Problem 12.2 and using (2) and (3),

$$A_{vf} = \frac{A_v}{1 + \beta_{vv}A_v} = \frac{\mu R_D R_L / [R_D R_L + r_{ds}(R_D + R_L)]}{1 + \dfrac{R_A}{R_A + R_B}(\mu R_D R_L)/[R_D R_L + r_{ds}(R_D + R_L)]}$$

$$= \frac{\mu R_D R_L (R_A + R_B)}{(\mu + 1)R_A R_D R_L + R_B R_D R_L + r_{ds}(R_A + R_B)(R_D + R_L)} \tag{4}$$

12.115 For the FET amplifier of Fig. 12-23, let $R_G = 500\ \text{k}\Omega$, $R_D = R_L = 5\ \text{k}\Omega$, $R_A = 50\ \text{k}\Omega$, $R_B = 200\ \text{k}\Omega$, $r_{ds} = 20\ \text{k}\Omega$, and $\mu = 40$. Evaluate the voltage gain with and without feedback.

▎ By (2), (3), and (4) of Problem 12.114,

$$A_v = \frac{\mu R_D R_L}{R_D R_L + r_{ds}(R_D + R_L)} = \frac{40(5 \times 10^3)(5 \times 10^3)}{(5 \times 10^3)(5 \times 10^3) + (20 \times 10^3)(5 \times 10^3 + 5 \times 10^3)} = 4.44$$

$$\beta_{vv} = \frac{R_A}{R_A + R_B} = \frac{50 \times 10^3}{50 \times 10^3 + 200 \times 10^3} = 0.2$$

$$A_{vf} = \frac{A_v}{1 + \beta_{vv}A_v} = \frac{4.44}{1 + 0.2(4.44)} = 2.35$$

12.116 Find expressions for R_{in} and $R_{\text{in}f}$ for the amplifier of Fig. 12-23.

▎ With $R_A = 0$ and $R_B = \infty$ in Fig. 12-24, it is apparent that $R_{\text{in}} = r_G$. If the current-source rather than the voltage-source model were used for the FET in Fig. 12-24, then based on (9) of Problem 12.10,

$$R_{\text{in}f} = R_G + \frac{(1 - g_m)G_B + g_{ds} + G_D + G_L}{G_A G_B + (G_A + G_B)(g_{ds} + G_D + G_L)} \tag{1}$$

12.117 Evaluate R_{in} and $R_{\text{in}f}$ for the amplifier of Fig. 12-23 if the amplifier is characterized by the parameters of Problem 12.115.

▎ Based on the results of Problem 12.116, $R_{\text{in}} = R_G = 500\ \text{k}\Omega$. Also, $g_m = \mu / r_{ds} = 40/(20 \times 10^3) = 2 \times 10^{-3}$ S. By (1) of Problem 12.116,

$$R_{\text{in}f} = 500 \times 10^3 + \frac{\dfrac{1 - 2 \times 10^3}{200 \times 10^3} + \dfrac{1}{20 \times 10^3} + \dfrac{1}{5 \times 10^3} + \dfrac{1}{5 \times 10^3}}{\left(\dfrac{1}{50 \times 10^3}\right)\left(\dfrac{1}{200 \times 10^3}\right) + \left(\dfrac{1}{50 \times 10^3} + \dfrac{1}{200 \times 10^3}\right)\left(\dfrac{1}{20 \times 10^3} + \dfrac{1}{5 \times 10^3} + \dfrac{1}{5 \times 10^3}\right)}$$

$$= 540.09\ \text{k}\Omega$$

12.118 Find expressions for R_o and R_{of} for the FET amplifier of Fig. 12-23.

▎ For the case of no feedback, (1) of Problem 12.12 gives

$$R_o = R_{22} = r_{ds}\ \|\ R_D = \frac{r_{ds}R_D}{r_{ds} + R_D} \tag{1}$$

With the current-source model for the FET in Fig. 12-24, (3) of Problem 12.11 leads to

$$\frac{1}{R_{of}} = G_{of} = g_{ds} + G_D + \frac{(1 - g_m)G_G + G_A}{(G_G + G_A)R_B + 1} \tag{2}$$

12.119 Evaluate R_o and R_{of} for the FET amplifier of Fig. 12-23 if the pertinent parameters of the amplifier are given by Problem 12.115.

FEEDBACK AMPLIFIERS ▯ 375

▎ By (1) of Problem 12.118,

$$R_o = \frac{(20 \times 10^3)(5 \times 10^3)}{20 \times 10^3 + 5 \times 10^3} = 4 \text{ k}\Omega$$

By (2) of Problem 12.118,

$$G_{of} = \frac{1}{20 \times 10^3} + \frac{1}{5 \times 10^3} + \frac{\dfrac{1 - 2 \times 10^{-3}}{500 \times 10^3} + \dfrac{1}{50 \times 10^3}}{\left(\dfrac{1}{500 \times 10^3} + \dfrac{1}{50 \times 10^3}\right)(200 \times 10^3) + 1} = 2.5407 \times 10^{-4} \text{ S}$$

$$R_{of} = \frac{1}{G_{of}} = 3.94 \text{ k}\Omega$$

12.120 Draw a small-signal equivalent circuit for the BJT amplifier with the unbypassed emitter resistor of Fig. 12-25 and classify the feedback. Assume $h_{re} = h_{oe} = 0$.

▎ The equivalent circuit is drawn in Fig. 12-26 where

$$R'_{Th} = R_B \| R_S = \frac{R_B R_S}{R_B + R_S} \quad (1)$$

$$R_B = R_1 \| R_2$$

$$v'_{Th} = \frac{R_B}{R_B + R_S} v_S \quad (2)$$

Comparison of Fig. 12-26 with Fig. 12-9 leads to the conclusion that current-series feedback exists.

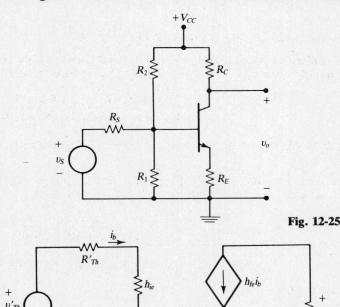

Fig. 12-25

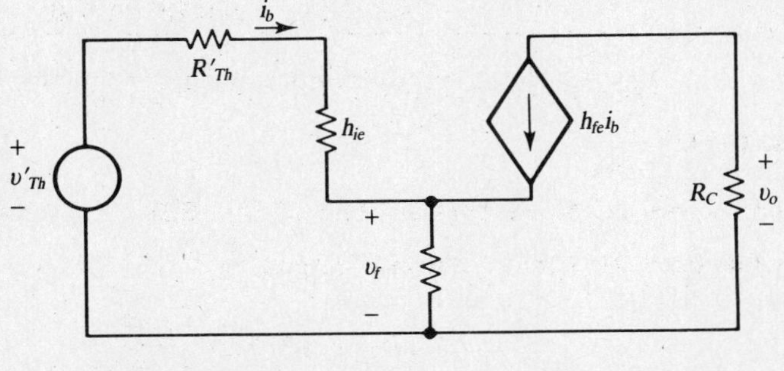

Fig. 12-26

12.121 Find expressions for the voltage-gain ratio of the amplifier of Fig. 12-25 with feedback and without feedback ($R_E = 0$).

▎ From Fig. 12-26 with $R_E = 0$ and using (1) and (2) of Problem 12.120,

$$i_b = \frac{v'_{Th}}{R'_{Th} + h_{ie}} = \frac{R_B v_S}{R_B R_S + h_{ie}(R_B + R_S)} \quad (1)$$

But

$$v_o = -h_{fe} i_b R_C \quad (2)$$

Substitute (1) into (2) and rearrange to find
$$A_v = \frac{v_o}{v_S} = \frac{-h_{fe}R_B R_C}{R_B R_S + h_{ie}(R_B + R_S)} \quad (3)$$

If $R_E \neq 0$, (7) of Problem 12.56 is applicable with $G_{22} = 0$. Hence,

$$A'_{vf} = \frac{v_o}{v'_{Th}} = \frac{v_o}{\dfrac{R_B v_S}{R_B + R_S}} = \frac{\dfrac{1}{R'_{Th} + h_{ie}} \dfrac{-h_{fe}}{R_E}}{\dfrac{1}{R_C}\left[(1 + h_{fe})\dfrac{1}{R'_{Th} + h_{ie}}\right] + \dfrac{1}{R_C R_E}}$$

Whence,
$$A_{vf} = \frac{v_o}{v_S} = \frac{-h_{fe} R_B R_C}{R_B R_S + (R_B + R_S)[(h_{fe} + 1)R_E + h_{ie}]} \quad (4)$$

12.122 For the amplifier of Fig. 12-25, let $R_C = R_E = 1\text{ k}\Omega$, $R_S = 100\text{ }\Omega$, $h_{ie} = 500\text{ }\Omega$, $h_{fe} = 60$, and $R_B = 4\text{ k}\Omega$. Evaluate the voltage gain with and without feedback.

▮ By (3) and (4) of Problem 12.121,

$$A_v = \frac{-60(4 \times 10^3)(1 \times 10^3)}{(4 \times 10^3)(100) + 500(4 \times 10^3 + 100)} = -97.96$$

$$A_{vf} = \frac{-60(4 \times 10^3)(1 \times 10^3)}{(4 \times 10^3)(100) + (4 \times 10^3 + 100)[(60 + 1)(1 \times 10^3) + 500]} = -0.95$$

12.123 Find the expressions for R_{in} and R_{inf} for the amplifier of Fig. 12-25.

▮ With $R_E = 0$ in Fig. 12-26, it is apparent that
$$R_{in} = R'_{Th} + h_{ie} = R_B \parallel R_S + h_{ie} \quad (1)$$

Using the results of Problems 12.62 and 12.63,

$$z_{11} = z_{12} = z_{21} = z_{22} = R_E$$
$$R_{11} = R_B \parallel R_S + h_{ie}$$
$$R'_{11} = R_B \parallel R_S + h_{ie} + R_E$$
$$R_{Th} = R_{22}\left(1 + \frac{h_{fe}R_E}{R'_{11}}\right) + R_E$$
$$R_{22} = \infty$$

Substitute the above values into (8) of Problem 12.63, take the limit as $R_{22} \to \infty$, and rearrange the result to find

$$R_{inf} = R_B \parallel R_S + h_{ie} + R_E\left[1 + \frac{h_{fe}(R_B \parallel R_S + h_{ie} + R_E)}{R_B \parallel R_S + h_{ie} + (h_{fe} + 1)R_E}\right] \quad (2)$$

12.124 Evaluate R_{in} and R_{inf} for the amplifier of Fig. 12-25 if the amplifier is described by the parameters of Problem 12.122.

▮ By (1) of Problem 12.123,
$$R_{in} = \frac{R_B R_S}{R_B + R_S} + h_{ie} = \frac{(4 \times 10^3)(100)}{4 \times 10^3 + 100} + 500 = 597.56\text{ }\Omega$$

Equation (2) of Problem 12.123 can be rewritten as

$$R_{inf} = R_{in} + R_E\left[1 + \frac{h_{fe}(R_{in} + R_E)}{R_{in} + (h_{fe} + 1)R_E}\right]$$
$$= 597.56 + (1 \times 10^3)\left[1 + \frac{60(597.56 + 1 \times 10^3)}{597.56 + (60 + 1)(1 \times 10^3)}\right] = 3.154\text{ k}\Omega$$

12.125 For the amplifier of Fig. 12-25, find the output resistance with and without feedback.

▮ With $R_E = 0$ in Fig. 12-26, if v_{Th} is deactivated (shorted) and R_C is replaced with a driving-point source, it is apparent that $R_o = \infty$.

With $R_E \neq 0$ in Fig. 12-26, short v_{Th} and replace R_C with a driving-point source oriented such that $v_{dp} = v_o$. Then, by Ohm's law,

$$i_b = -\frac{v_f}{R'_{Th} + h_{ie}} \tag{1}$$

However, the current through R_E is given by $(h_{fe} + 1)i_b$. Hence, by Ohm's law and (1),

$$v_f = (h_{fe} + 1)i_b R_E = \frac{-(h_{fe} + 1)R_E v_f}{R'_{Th} + h_{ie}} \tag{2}$$

Equation (2) can only be satisfied in general if $v_f = 0$. Whence by (1), $i_b = 0$. Consequently, $i_{dp} = h_{fe}i_b = 0$, or

$$R_{of} = \frac{v_{dp}}{i_{dp}} = \infty$$

12.126 For the BJT amplifier of Fig. 12-17, determine the parameters of the voltage-shunt feedback model as defined in Fig. 12-16 if $R_C = 4\,\text{k}\Omega$, $R_F = 40\,\text{k}\Omega$, $R_S = 10\,\text{k}\Omega$, $h_{ie} = 1100\,\Omega$, $h_{fe} = 50$, and $h_{re} = h_{oe} = 0$.

▮ Following the procedure established in Problem 12.91,

$$y_{11} = \frac{i_a}{v_o}\bigg|_{v_f=0} = \frac{1}{R_F}$$

$$y_{12} = \frac{i_a}{v_f}\bigg|_{v_o=0} = -\frac{1}{R_F}$$

$$y_{21} = \frac{i_f}{v_L}\bigg|_{v_f=0} = -\frac{1}{R_F}$$

$$y_{22} = \frac{i_f}{v_f}\bigg|_{v_o=0} = \frac{1}{R_F}$$

Since $\beta = h_{fe}$ and $R_{11} = h_{ie}$,

$$\beta' = \beta + y_{12}R_{11} = h_{fe} - h_{ie}/R_F$$

$$R'_{11} = \frac{R_{11}}{1 + y_{11}R_{11}} = \frac{h_{ie}}{1 + h_{ie}/R_F}$$

$$R'_{22} = \frac{R_{22}}{1 + y_{22}R_{22}} = \frac{1}{h_{oe} + y_{22}} = R_F$$

12.127 Determine the overall current gain for the amplifier of Fig. 12-16 through use of (1) from Problem 12.88.

▮ Based on Fig. 12-15 and results of Problem 12.126,

$$A_r = \frac{-\beta'}{G_L + G'_{22}} = \frac{-(h_{fe} - h_{ie}/R_F)}{1/R_C + 1/R_F} = \frac{-(50 - 1100/40 \times 10^3)}{1/4 \times 10^3 + 1/40 \times 10^3} = -1.817 \times 10^5$$

Now,
$$\beta_{vi} = y_{21} = -1/R_F = -2.5 \times 10^{-5}$$

Hence, by (1) of Problem 12.88,

$$A_{if} = \frac{A_r/R_C}{1 + \beta_{vi}A_r} = \frac{(-1.817 \times 10^5)/4 \times 10^3}{1 + (-2.5 \times 10^{-5})(-1.817 \times 10^5)} = -8.196$$

12.128 Show that for $\beta \to \infty$ (approximately realizable with a large odd number of cascaded BJT stages), the voltage-shunt feedback amplifier of Fig. 12-14 displays the characteristics of an operational amplifier in that $A_{vf} \approx -R_F/R_S$ and $R_{of} \approx 0$.

▮ Taking the limit as $\beta \to \infty$ in (2) of Problem 12.89,

$$A_{vf} \approx \frac{-\beta G_{11}G_S}{G_{11}\beta G_F} = -\frac{R_F}{R_S}$$

As $\beta \to \infty$ in (3) of Problem 12.85, it is seen that $G_{of} \to \infty$. Hence, $R_{of} = 1/G_{of} \to 0$.

12.129 The MOSFET amplifier of Fig. 5-13(a) includes voltage-shunt feedback if a gate-to-source resistor R_G is added. Draw a small-signal equivalent circuit suitable for analysis of this amplifier with R_G in place.

▌ See Fig. 12-27.

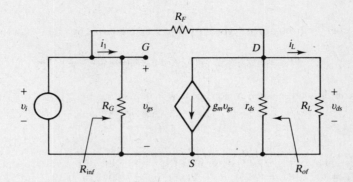

Fig. 12-27

12.130 For the MOSFET amplifier described by Problem 12.129, let $R_F = 5$ MΩ, $R_L = 14$ kΩ, $r_{ds} = 40$ kΩ, $R_G = 10$ MΩ, and $g_m = 1$ mS. Find R_{inf}.

▌ The circuit of Fig. 12-27 is applicable. Now,

$$g_m v_{gs} = g_m v_i = g_m R_G i_1$$

or $$\beta = g_m R_G = (1 \times 10^{-3})(10 \times 10^6) = 1 \times 10^4$$

By (8) of Problem 12.84,

$$R_{inf} = \frac{g_{ds} + G_F + G_L}{G_G[(1+\beta)G_F + g_{ds} + G_L] + G_L G_F}$$

$$= \frac{1/40 \times 10^3 + 1/5 \times 10^6 + 1/14 \times 10^3}{1/10 \times 10^6[(1 + 1 \times 10^4)/5 \times 10^6 + 1/40 \times 10^3 + 1/14 \times 10^3] + (1/14 \times 10^3)(1/5 \times 10^6)}$$

$$= 431.5 \text{ k}\Omega$$

12.131 For the MOSFET amplifier of Problem 12.130, find R_{of}.

▌ Based on (3) of Problem 12.85 with $R_S = \infty$ ($G_S = 0$),

$$G_{of} = \frac{1}{r_{ds}} + \frac{(1+\beta)/R_G}{1 + R_F/R_G} = \frac{1}{40 \times 10^3} + \frac{(1 + 1 \times 10^4)/10 \times 10^6}{1 + (5 \times 10^6)/10 \times 10^6} = 6.917 \times 10^{-4} \text{ S}$$

Thus, $$R_{of} = \frac{1}{G_{of}} = 1.446 \text{ k}\Omega$$

12.132 Determine the overall voltage gain for the amplifier of Problem 12.130.

▌ Based on (3) of Problem 12.87 with $R_S = 0$ and $\beta_{vi} = -1/R_F = -0.2 \times 10^{-6}$ S,

$$A_{vf} = \frac{-\beta/R_{inf}}{1/R_L + 1/r_{ds} - \beta_{vi}\beta} = \frac{-1 \times 10^4/431.5 \times 10^3}{1/14 \times 10^3 + 1/40 \times 10^3 - (-0.2 \times 10^{-6})(1 \times 10^4)} = -11.05$$

12.133 Find the overall current gain for the amplifier of Problem 12.130.

▌ Based on (1) of Problem 12.90 with $R_S = 0$,

$$A_{if} = A_{vf} \frac{R_{inf}}{R_L} = (-11.05) \frac{431.5 \times 10^3}{14 \times 10^3} = -340.6$$

CHAPTER 13
Operational Amplifiers

OP AMP FUNDAMENTALS

13.1 An op amp amplifies the difference $v_d \equiv v_1 - v_2$ between two input signals (see Fig. 13-1), exhibiting the open-loop voltage gain

$$A_{OL} \equiv \frac{v_o}{v_d} \tag{1}$$

In Fig. 13-1, terminal 1 is the *inverting input* (labeled with a minus sign on the actual amplifier); signal v_1 is amplified in magnitude and appears phase-inverted at the output. Terminal 2 is the *noninverting input* (labeled with a plus sign); output due to v_2 is phase-preserved. In magnitude, the open-loop voltage gain in op amps ranges from 10^4 to 10^7. The maximum magnitude of the output voltage from an op amp is called its *saturation voltage*; this voltage is approximately 2 V smaller than the power-supply voltage. In other words, the amplifier is linear over the range

$$-V_{o\,\text{sat}} = -(V_{CC} - 2) < v_o < V_{CC} - 2 = V_{o\,\text{sat}} \tag{2}$$

An op amp has saturation voltage $V_{o\,\text{sat}} = 10$ V, an open-loop voltage gain of -10^5, and input resistance 100 kΩ. Find the value of v_d that will just drive the amplifier to saturation.

❚ By (2),
$$v_d = \frac{\pm V_{o\,\text{sat}}}{A_{OL}} = \frac{\pm 10}{-10^5} = \pm 0.1 \text{ mV}$$

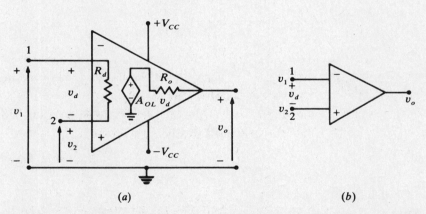

Fig. 13-1 Operational amplifier. (a) Complete representation, (b) simplified representation.

13.2 Determine the op amp input current at the onset of saturation for the device of Problem 13.1.

❚ Let i_{in} be the current into terminal 1 of Fig. 13-1(b); then
$$i_{\text{in}} = \frac{v_d}{R_d} = \frac{\pm 0.1 \times 10^{-3}}{100 \times 10^3} = \pm 1 \text{ nA}$$

13.3 The *ideal* op amp has three essential characteristics which serve as standards for assessing the goodness of a *practical* op amp:
 1. The open-loop voltage gain A_{OL} is negatively infinite.
 2. The input impedance R_d between terminals 1 and 2 is infinitely large; thus, the input current is zero.
 3. The output impedance R_o is zero; consequently, the output voltage is independent of the load.

Figure 13-1(a) models the practical characteristics.

The *inverting amplifier* of Fig. 13-2 has its noninverting input connected to ground or common. A signal is applied through input resistor R_1, and negative current feedback is implemented through *feedback resistor* R_F. Output v_o has polarity opposite that of input v_S. For this inverting amplifier, find the voltage gain v_o/v_S using only characteristic 1.

379

■ By the method of node voltages at the inverting input, the current balance is

$$\frac{v_S - v_d}{R_1} + \frac{v_o - v_d}{R_F} = i_{in} = \frac{v_d}{R_d} \tag{1}$$

where R_d is the differential input resistance. By (1) of Problem 13.1, $v_d = v_o/A_{OL}$ which, when substituted into (1), gives

$$\frac{v_S - v_o/A_{OL}}{R_1} + \frac{v_o - v_o/A_{OL}}{R_F} = \frac{v_o/R_d}{A_{OL}} \tag{2}$$

In the limit as $A_{OL} \to \infty$, (2) becomes

$$\frac{v_S}{R_1} + \frac{v_o}{R_F} = 0 \quad \text{so that} \quad A_v \equiv \frac{v_o}{v_S} = -\frac{R_F}{R_1} \tag{3}$$

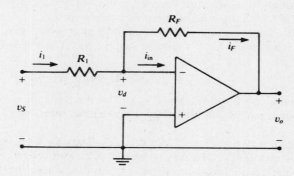

Fig. 13-2 Inverting amplifier.

13.4 Determine the voltage gain v_o/v_S for the op amp circuit of Fig. 13-2 using only characteristic 2 of the ideal op amp from Problem 13.3.

■ If $i_{in} = 0$, then $v_d = i_{in} R_d = 0$ and $i_1 = i_F \equiv i$. The input and feedback-loop equations are, respectively,

$$v_S = iR_1 \quad \text{and} \quad v_o = -iR_F$$

Whence
$$A_v \equiv \frac{v_o}{v_S} = -\frac{R_F}{R_1} \tag{1}$$

in agreement with (3) of Problem 13.3.

13.5 For the inverting amplifier of Fig. 13-2, show that as $A_{OL} \to -\infty$, $v_d \to 0$; thus, the inverting input remains nearly at ground potential (and is called a *virtual ground*).

■ By KVL around the outer loop,

$$v_S - v_o = i_1 R_1 + i_F R_F \tag{1}$$

Using (1) of Problem 13.1 in (1), rearranging, and taking the limit give

$$\lim_{A_{OL} \to -\infty} v_d = \lim_{A_{OL} \to -\infty} \frac{-i_1 R_1 - i_F R_F + v_S}{A_{OL}} = 0 \tag{2}$$

13.6 Show that the current feedback for the inverting amplifier of Fig. 13-2 is actually negative feedback.

■ The feedback is negative if i_F counteracts i_1; that is, the two currents must have the same algebraic sign. By two applications of KVL, with $v_d \approx 0$,

$$i_1 = \frac{v_S - v_d}{R_1} \approx \frac{v_S}{R_1} \quad \text{and} \quad i_F = \frac{-v_o + v_d}{R_F} \approx \frac{-v_o}{R_F}$$

But in an inverting amplifier, v_o and v_S have opposite signs; therefore, i_1 and i_F have like signs.

13.7 Use (2) of Problem 13.3 to derive an exact formula for the gain of a practical inverting op amp.

▎ Rearranging (2) of Problem 13.3 to obtain the voltage-gain ratio gives

$$A_v \equiv \frac{v_o}{v_S} = \frac{A_{OL}}{1 + (R_1/R_F)(1 - A_{OL}) + R_1/R_d} \quad (1)$$

13.8 If $R_1 = 1\,\text{k}\Omega$, $R_F = 10\,\text{k}\Omega$, $R_d = 1\,\text{k}\Omega$, and $A_{OL} = -10^4$ for a practical inverting op amp amplifier, evaluate the voltage gain.

▎ Substitution of the given values into (1) of Problem 13.7 yields

$$A_v = \frac{-10^4}{1 - (1/10)(1 + 10^4) + 1/1} = -9.979$$

13.9 Compare the result of Problem 13.8 with the ideal op amp approximation determined in Problem 13.3.

▎ From (3) of Problem 13.3,

$$A_{v\,\text{ideal}} = -\frac{R_F}{R_1} = -\frac{10 \times 10^3}{1 \times 10^3} = -10$$

so the error is

$$\frac{-9.979 - (-10)}{-9.979}(100\%) = -0.21\%$$

Note that R_d and A_{OL} are far removed from the ideal, yet the error is quite small.

13.10 The *noninverting amplifier* of Fig. 13-3 is realized by grounding R_1 of Fig. 13-2 and applying the input signal at the noninverting op amp terminal. When v_2 is positive, v_o is positive and current i is positive. Voltage $v_1 = iR_1$ then is applied to the inverting terminal as negative voltage feedback. For this noninverting amplifier, assume that the current into the inverting terminal of the op amp is zero, so that $v_d \approx 0$ and $v_1 \approx v_2$. Derive an expression for the voltage gain v_o/v_2.

▎ With zero input current to the basic op amp, the currents through R_2 and R_1 must be identical; thus,

$$\frac{v_o - v_1}{R_2} = \frac{v_1}{R_1} \quad \text{and} \quad A_v \equiv \frac{v_o}{v_2} \approx \frac{v_o}{v_1} = 1 + \frac{R_2}{R_1} \quad (1)$$

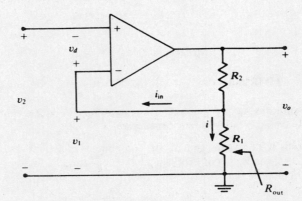

Fig. 13-3 Noninverting amplifier.

13.11 For the noninverting amplifier of Fig. 13-3, find an exact expression for the voltage-gain ratio.

▎ Referring to Fig. 13-3, the feedback voltage can be written as

$$v_1 = v_o - (i + i_{\text{in}})R_2 \quad (1)$$

But, $i = v_1/R_1$ and $i_{\text{in}} = v_d/R_d = v_o/A_{OL}R_d$ which, when substituted into (1), yield, after rearrangement,

$$v_1 = \left[\frac{1 - R_2/A_{OL}R_d}{1 + R_2/R_1}\right] v_o \quad (2)$$

By (1) of Problem 13.1,
$$v_d = v_1 - v_2 = v_o/A_{OL} \quad \text{or} \quad v_2 = v_1 - v_o/A_{OL} \tag{3}$$

Substitute (2) into (3) and rearrange to find
$$A_v = \frac{v_o}{v_2} = \frac{R_1 + R_2}{R_1 - \dfrac{R_1 R_2}{A_{OL} R_d} - \dfrac{R_1 + R_2}{A_{OL}}} \tag{4}$$

13.12 If $R_1 = 1\,\text{k}\Omega$, $R_2 = 10\,\text{k}\Omega$, $R_d = 1\,\text{k}\Omega$, and $A_{OL} = -10^4$, for a practical noninverting op amp amplifier, evaluate the voltage gain.

▌ Substitution of the given values into (4) of Problem 13.11 yields
$$A_v = \frac{1 \times 10^3 + 10 \times 10^3}{1 \times 10^3 - \dfrac{(1 \times 10^3)(10 \times 10^3)}{(-10^4)(1 \times 10^3)} - \dfrac{1 \times 10^3 + 10 \times 10^3}{-10^4}} = 10.997$$

13.13 Compare the result of Problem 13.12 with the ideal op amp approximation determined in Problem 13.10.

▌ From (1) of Problem 13.10,
$$A_{v\,\text{ideal}} = 1 + \frac{R_2}{R_1} = 1 + \frac{10 \times 10^3}{1 \times 10^3} = 11$$

Thus, the error is
$$\frac{11 - 10.997}{11}(100\%) = 0.21\%$$

13.14 The *common-mode gain* is defined (see Fig. 13-1) as
$$A_{cm} \equiv -\frac{v_o}{v_2} \tag{1}$$

where $v_1 = v_2$ by explicit connection. Usually, A_{cm} is much less than unity ($A_{cm} = -0.01$ being typical). Common-mode gain sensitivity is frequently quantized via the *common-mode rejection ratio* (CMMR), defined as
$$\text{CMMR} = \frac{A_{OL}}{A_{cm}} \tag{2}$$

and expressed in decibels as
$$\text{CMMR}_{db} = 20 \log \frac{A_{OL}}{A_{cm}} = 20 \log \text{CMMR} \tag{3}$$

Typical values for the CMMR range from 100 to 10,000, with corresponding CMMR_{db} values of from 40 to 80 db.

Find the voltage-gain ratio A_v of the noninverting amplifier of Fig. 13-3 in terms of its CMMR. Assume $v_1 = v_2$ insofar as the common-mode gain is concerned.

▌ The amplifier output voltage is the sum of two components. The first results from amplification of the difference voltage v_d as given by (1) of Problem 13.1. The second, defined by (1), is a direct consequence of the common-mode gain. The total output voltage is, then,
$$v_o = A_{OL} v_d - A_{cm} v_2 \tag{4}$$

Voltage division (with $i_{\text{in}} = 0$) gives
$$v_d = v_1 - v_2 = \frac{R_1}{R_1 + R_2} v_o - v_2 \tag{5}$$

and substituting (5) into (4) and rearranging give
$$v_o \left(1 - \frac{A_{OL} R_1}{R_1 + R_2}\right) = -(A_{OL} + A_{cm}) v_2$$

Then
$$A_v = \frac{v_o}{v_2} = \frac{-(A_{OL} + A_{cm})}{1 - A_{OL} R_1/(R_1 + R_2)} = \frac{-A_{OL}}{1 - A_{OL} R_1/(R_1 + R_2)} - \frac{A_{OL}/\text{CMMR}}{1 - A_{OL} R_1/(R_1 + R_2)} \tag{6}$$

13.15 For the noninverting amplifier of Fig. 13-3, compare the expressions obtained for voltage gain with common-mode rejection and without (in the ideal amplifier of Problem 13.10), for $A_{OL} \to -\infty$.

▮ We let $A_{OL} \to -\infty$ in (6) of Problem 13.14 since that is implicit in Problem 13.10:

$$\lim_{A_{OL} \to -\infty} A_v = \lim_{A_{OL} \to -\infty} \left[\frac{-A_{OL}}{1 - A_{OL}R_1/(R_1+R_2)} + \frac{-A_{OL}/\text{CMMR}}{1 - A_{OL}R_1/(R_1+R_2)} \right]$$

$$= 1 + \frac{R_2}{R_1} + \frac{1}{\text{CMMR}}\left(1 + \frac{R_2}{R_1}\right) \qquad (1)$$

Now we can compare (1) above with (1) of Problem 13.10; the difference is the last term on the right-hand side of (1) above.

13.16 Show that if CMMR is very large, then it need not be considered in computing the gain of a noninverting amplifier.

▮ Let CMMR $\to \infty$ in (1) of Problem 13.15 to get

$$\lim_{\substack{\text{CMMR} \to \infty \\ A_{OL} \to -\infty}} A_v = 1 + \frac{R_2}{R_1}$$

which is identical to the ideal case of Problem 13.10.

13.17 The noninverting amplifier circuit of Fig. 13-3 has an infinite input impedance if the basic op amp is ideal. If the op amp is not ideal, but instead $R_d = 1\text{ M}\Omega$ and $A_{OL} = -10^6$, find the input impedance. Let $R_2 = 10\text{ k}\Omega$ and $R_1 = 1\text{ k}\Omega$.

▮ Consider v_2 a driving-point source and apply KVL to find

$$v_2 = v_1 - v_d = iR_1 - v_d \qquad (1)$$

By Ohm's law, KVL around the output network, and use of (1) from Problem 13.1,

$$v_o = i(R_1 + R_2) + i_{in}R_2 = A_{OL}v_d$$

Whence, $$i = \frac{A_{OL}v_d - i_{in}R_2}{R_1 + R_2} \qquad (2)$$

Substitute (2) into (1) and rearrange to find

$$Z_{in} = -\frac{v_2}{i_{in}} = R_2 + R_d\left[1 + \frac{R_2}{R_1} - A_{OL}\right] \qquad (3)$$

Use the given values in (3) to give

$$Z_{in} = 10 \times 10^3 + (1 \times 10^6)\left[1 + \frac{10 \times 10^3}{1 \times 10^3} - (-10^6)\right] \approx 10^{12}\,\Omega$$

LINEAR FUNCTION OPERATIONS

13.18 The *inverting summer amplifier* (or *inverting adder*) of Fig. 13-4 is formed by adding parallel inputs to the inverting amplifier of Fig. 13-2. Its output is a weighted sum of the inputs, but inverted in polarity. Find an expression for the output of the inverting summer amplifier of Fig. 13-4, assuming the basic op amp is ideal.

▮ We use the principle of superposition. With $v_{S2} = v_{S3} = 0$, the current in R_1 is not affected by the presence of R_2 and R_3, since the inverting node is a virtual ground (see Problem 13.5). Hence, the output voltage due to v_{S1} is, by (3) of Problem 13.3, $v_{o1} = -(R_F/R_1)v_{S1}$. Similarly, $v_{o2} = -(R_F/R_2)v_{S2}$ and $v_{o3} = -(R_F/R_3)v_{S3}$. Then, by superposition,

$$v_o = v_{o1} + v_{o2} + v_{o3} = -R_F\left(\frac{v_{S1}}{R_1} + \frac{v_{S2}}{R_2} + \frac{v_{S3}}{R_3}\right) \qquad (1)$$

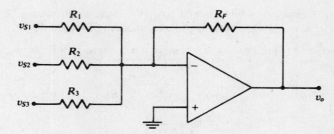

Fig. 13-4 Inverting summer amplifier.

13.19 Let $R_1 = R_2 = R_3 = 3R_F$ in the inverting summer amplifier of Fig. 13-4. What mathematical operation does this circuit perform?

▌ By (1) of Problem 13.18,

$$v_o = -R_F\left(\frac{v_{S1}}{3R_F} + \frac{v_{S2}}{3R_F} + \frac{v_{S3}}{3R_F}\right) = -\frac{1}{3}(v_{S1} + v_{S2} + v_{S3})$$

The circuit gives the negative of the instantaneous average value of the input signals.

13.20 An inverting summer (Fig. 13-4) has n inputs with $R_1 = R_2 = R_3 = \cdots = R_n = R$. Assume that the open-loop basic op amp gain A_{OL} is finite, but that the inverting-terminal input current is negligible. Derive a relationship that shows how gain magnitude is reduced in the presence of multiple inputs for a practical op amp.

▌ Apply KCL at the input node to find

$$\frac{v_{S1} - v_d}{R} + \frac{v_{S2} - v_d}{R} + \cdots + \frac{v_{Sn} - v_d}{R} + \frac{v_o - v_d}{R_F} = 0 \qquad (1)$$

Use (1) of Problem 13.1 in (1) and rearrange to yield

$$v_{S1} + v_{S2} + \cdots + v_{Sn} = \frac{nv_o}{A_{OL}} - \frac{R}{R_F}\left(1 - \frac{1}{A_{OL}}\right)v_o \qquad (2)$$

Rearrangement of (2) allows us to write

$$A_n \equiv \frac{v_o}{v_{S1} + v_{S2} + \cdots + v_{Sn}} = -\frac{R_F/R}{1 - \dfrac{nR_F}{(R+1)A_{OL}}}$$

For a single input v_{S1}, the gain is A_1. For the same input v_{S1} together with $n-1$ zero inputs $v_{S2} = \cdots = v_{Sn} = 0$, the gain is A_n. But since $A_{OL} < 0$, $|A_n| < |A_1|$ for $n > 1$.

13.21 A *differential amplifier* (sometimes called a *subtractor*) responds to the difference between two input signals, removing any identical portions (often a bias or noise) in a process called *common-mode rejection*. Find an expression for v_o in Fig. 13-5 that shows this circuit to be a differential amplifier. Assume an ideal op amp.

▌ Since the current into the ideal op amp is zero, a loop equation gives

$$v_1 = v_{S1} - Ri_1 = v_{S1} - R\frac{v_{S1} - v_o}{R + R_1}$$

By voltage division at the noninverting node,

$$v_2 = \frac{R_1}{R + R_1} v_{S2}$$

In the ideal op amp, $v_d = 0$, so that $v_1 = v_2$, which leads to

$$v_o = \frac{R_1}{R}(v_{S2} - v_{S1})$$

Thus, the output voltage is directly proportional to the difference between the input voltages.

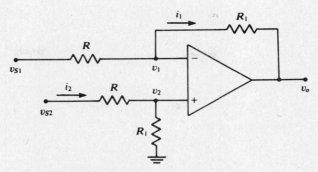

Fig. 13-5 Differential amplifier.

13.22 Find the input impedance Z_1 of the inverting amplifier of Fig. 13-2, assuming the basic op amp is ideal.

▌ Consider v_S a driving-point source. Since the op amp is ideal, the inverting terminal is a virtual ground, and a loop equation at the input leads to

$$v_S = i_1 R_1 + 0 \quad \text{so that} \quad Z_1 = \frac{v_S}{i_1} = R_1$$

13.23 The *unity-follower* amplifier of Fig. 13-6 has a voltage gain of 1, and the output is in phase with the input. It also has an extremely high input impedance, leading to its use as an intermediate-stage (*buffer*) amplifier to prevent a small load impedance from loading a source. Assume a practical op amp having $A_{OL} = -10^6$ (a typical value). Show that $v_o \approx v_S$.

▌ Writing a loop equation and using (*1*) of Problem 13.1,

$$v_S = v_o - v_d = v_o \left(1 - \frac{1}{A_{OL}}\right)$$

from which

$$v_o = \frac{v_S}{1 - 1/A_{OL}} = \frac{v_S}{1 + 10^{-6}} = 0.999999 v_S \approx v_S$$

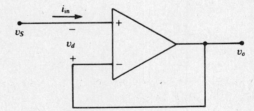

Fig. 13-6 Unity follower.

13.24 Find an expression for the input impedance of the unity follower amplifier of Fig. 13-6 and evaluate it if $R_d = 1\ M\Omega$ (a typical value).

▌ Considering v_S a driving-point source and using (*1*) of Problem 13.1, we have

$$v_S = i_{in} R_d + v_o = i_{in} R_d - A_{OL} v_d = i_{in} R_d (1 - A_{OL})$$

and

$$Z_{in} = \frac{v_S}{i_{in}} = R_d(1 - A_{OL}) \approx -A_{OL} R_d = -(-10^6)(10^6) = 1\ T\Omega$$

13.25 Find an expression for the output v_o of the amplifier circuit of Fig. 13-7. Assume an ideal op amp. What mathematical operation does the circuit perform?

▌ The principle of superposition is applicable to this linear circuit. With $v_{S2} = 0$ (shorted), the voltage appearing at the noninverting terminal is found by voltage division to be

$$v_2 = \frac{R}{R+R} v_{S1} = \frac{v_{S1}}{2} \tag{1}$$

Let v_{o1} be the value of v_o with $v_{S2} = 0$. By the result of Problem 13.10 and (1),

$$v_{o1} = \left(1 + \frac{R_2}{R_1}\right)v_2 = \left(1 + \frac{R_2}{R_1}\right)\frac{v_{S1}}{2}$$

Similarly, with $v_{S1} = 0$,

$$v_{o2} = \left(1 + \frac{R_2}{R_1}\right)\frac{v_{S2}}{2}$$

By superposition, the total output is then

$$v_o = v_{o1} + v_{o2} = \frac{1}{2}\left(1 + \frac{R_2}{R_1}\right)(v_{S1} + v_{S2}) \qquad (2)$$

The circuit is a noninverting adder.

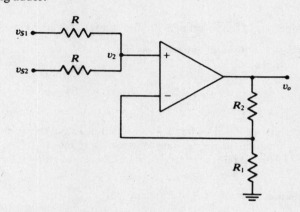

Fig. 13-7

13.26 Show that the transfer function for the op amp circuit of Fig. 13-8 is $v_o/v_i = 1$.

▌ Because the op amp draws negligible current, $i_2 = 0$. Hence, $v_2 = v_i$. However, since $v_d \approx 0$, $v_1 \approx v_2 = v_i$ and

$$i_1 = \frac{v_i - v_1}{R} \approx 0$$

Also, by the method of node voltages,

$$i_1 = \frac{v_i - v_o}{2R} = 0$$

Thus, $v_i = v_o$ and so $v_o/v_i = 1$.

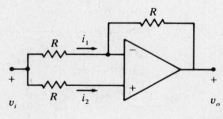

Fig. 13-8

13.27 The basic op amp in Fig. 13-9 is ideal. Find v_o and determine what mathematical operation is performed by the amplifier circuit.

▌ By (1) of Problem 13.10,

$$v_{o1} = \left(1 + \frac{R_1}{R_2}\right)v_{S1} \qquad (1)$$

Since $v_d \approx 0$, the voltage at the inverting terminal of the second op amp is v_{S2}. Application of KCL at the inverting node of the second op amp gives

$$\frac{v_{o1} - v_{S2}}{R_1} = \frac{v_{S2} - v_o}{R_2} \qquad (2)$$

Substitute (1) into (2) and rearrange to find

$$v_o = \left(1 + \frac{R_2}{R_1}\right)(v_{S2} - v_{S1}) \qquad (3)$$

The circuit is a subtractor.

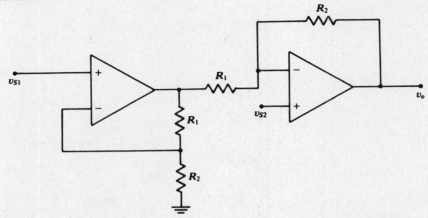

Fig. 13-9

13.28 If the nonideal op amp of the circuit of Fig. 13-10 has an open-loop gain $A_{OL} = -10^4$, find v_o.

▮ By KVL, $\qquad v_d = -E_b + v_o \qquad (1)$

Substitute (1) into (1) of Problem 13.10 and solve for v_o.

$$v_o = A_{OL} v_d = A_{OL}(-E_b + v_o)$$

or $\qquad v_o = \dfrac{-A_{OL}}{1 - A_{OL}} E_b = \dfrac{-(-10^4)}{1-(-10^4)} E_b = 0.9999 E_b$

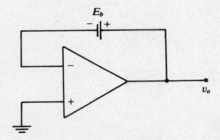

Fig. 13-10

13.29 Find an expression for the output of the *inverting differentiator* of Fig. 13-11, assuming the basic op amp is ideal.

▮ Since the op amp is ideal, $v_d \approx 0$, and the inverting terminal is a virtual ground. Consequently, v_S appears across capacitor C:

$$i_S = C \frac{dv_S}{dt}$$

But the capacitor current is also the current through R (since $i_{in} = 0$). Hence,

$$v_o = -I_F R = -i_S R = -RC \frac{dv_S}{dt}$$

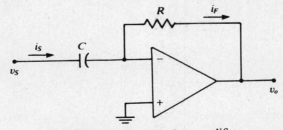

Fig. 13-11 Differentiating amplifier.

13.30 Show that the output of the *inverting integrator* of Fig. 13-12 actually is the time integral of the input signal, assuming the op amp is ideal.

▌ If the op amp is ideal, the inverting terminal is a virtual ground, and v_S appears across R. Thus, $i_S = v_S/R$. But, with negligible current into the op amp, the current through R must also flow through C. Then,

$$v_o = -\frac{1}{C}\int i_F\,dt = -\frac{1}{C}\int i_S\,dt = -\frac{1}{RC}\int v_S\,dt$$

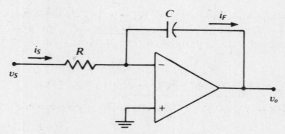

Fig. 13-12 Integrating amplifier.

13.31 The circuit of Fig. 13-13(a) (represented in the s domain) is a more practical differentiator than that of Fig. 13-11, because it will attenuate high-frequency noise. Find the s-domain transfer function relating V_o and V_S.

▌ In an ideal op amp the inverting terminal is a virtual ground, so $I_S(s) = -I_F(s)$. Let $Z_F(s)$ represent the parallel combination of R and C; then,

$$Z_F(s) = \frac{R}{sRC + 1}$$

and

$$I_F(s) = \frac{V_o(s)}{Z_F(s)} = \frac{sRC + 1}{R}V_o(s)$$

But

$$V_S(s) = I_S(s)Z_{in}(s) = -I_F(s)Z_{in}(s) = -\frac{sRC + 1}{R}V_o(s)\frac{sRC + 1}{sC}$$

Whence,

$$A(s) \equiv \frac{V_o(s)}{V_S(s)} = -\frac{sRC}{(sRC + 1)^2} \qquad (1)$$

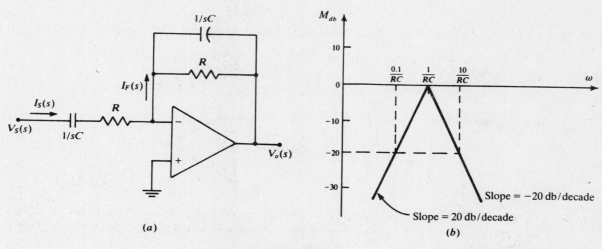

Fig. 13-13

13.32 Sketch the Bode plot (M_{db} only) for the practical integrator of Fig. 13-13(a) and explain how high-frequency noise effects are reduced.

OPERATIONAL AMPLIFIERS □ 389

▌ Letting $s = j\omega$ in (1) of Problem 13.31,

$$M_{db} \equiv 20\log|A(j\omega)| = 20\log\omega RC - 40\log|j\omega RC + 1| \approx \begin{cases} 20\log\omega RC & \text{for } \omega RC \leq 1 \\ -20\log\omega RC & \text{for } \omega RC \geq 1 \end{cases}$$

Figure 13-13(b) is a plot of this approximate (asymptotic) expression for M_{db}. For a true differentiator, we would have

$$v_o = K\frac{dv_S}{dt} \quad \text{or} \quad V_o = sKV_S$$

which would lead to $M_{db} = 20\log\omega K$. Thus, the practical circuit differentiates only components of the signal whose frequency is less than the break frequency $f_1 \equiv 1/2\pi RC$ Hz. Spectral components above the break frequency, including (and especially) noise, will be attenuated; the higher the frequency, the greater the attenuation.

13.33 Find the relationship between v_o and v_i in the circuit of Fig. 13-14.

▌ Since the inverting terminal is a virtual ground, the Laplace-domain input current is given by

$$I_i = \frac{V_i}{R + (R \parallel 1/sC)} = \frac{V_i(sRC + 1)}{sR^2C + 2R}$$

With zero current flowing into the op amp inverting terminal, current division yields

$$I_2 = I_1 = \frac{1/sC}{R + 1/sC}I_i = \frac{1}{sRC + 1}\frac{V_i(sRC + 1)}{R(sRC + 2)} = \frac{V_i}{R(sRC + 2)}$$

Again because the inverting terminal is a virtual ground,

$$I_3 = \frac{V_o}{\dfrac{1}{sC/2} + \dfrac{1}{sC/2} \parallel \dfrac{R}{2}} = \frac{sC(sRC + 4)}{4(sRC + 2)}V_o$$

and, by current division,

$$I_2 = \frac{-R/2}{2/sC + R/2}I_3 = \frac{-sRC}{sRC + 4}\frac{sC(sRC + 4)}{4(sRC + 2)}V_o = \frac{-s^2RC^2V_o}{4(sRC + 2)}$$

Equating the two expressions for I_2 yields a Laplace-domain expression relating V_o and V_i:

$$V_o = -\frac{4}{s^2R^2C^2}V_i \tag{1}$$

or, after inverse transformation,

$$v_o = -\frac{4}{(RC)^2}\int\left(\int v_i\,dt\right)dt$$

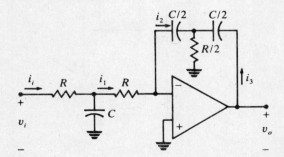

Fig. 13-14

13.34 The circuit of Fig. 13-15 is, in essence, a noninverting amplifier with a feedback impedance Z_N and is known as a *negative-impedance converter* (NIC). Find the Thévenin or driving-point impedance to the right of the input terminals, and explain why such a name is appropriate.

▌ At the inverting node, the phasor input current is given by

$$I_i = I_N = \frac{V_i - V_o}{Z_N}$$

so that
$$V_o = V_i = I_i Z_N \tag{1}$$

Since $V_d \approx 0$,
$$I_P = \frac{V_o - V_i}{Z_P} = I_Z = \frac{V_i}{Z}$$

so that
$$V_o = \frac{Z_P}{Z} V_i + V_i \tag{2}$$

If (1) and (2) are equated and rearranged, they result in
$$Z_{dp} = \frac{V_i}{I_i} = -\frac{Z_N}{Z_P} Z \tag{3}$$

Observe that if $Z_P = Z_N$, then the impedance Z appears to be converted to the negative of its value; hence the name.

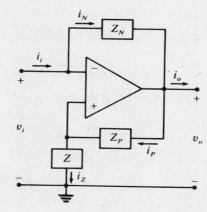

Fig. 13-15

13.35 Describe a circuit arrangement that makes use of the NIC of Problem 13.34 and Fig. 13-15, with only resistors and capacitors, to simulate a pure inductor.

▮ Consider the circuit of Fig. 13-16. According to (3) of Problem 13.34,

$$Z'_{IN} = \frac{Z_N}{Z_P} Z = -\frac{R}{R} R = -R$$

and
$$Z_{IN} = -\frac{Z_N}{Z_P} Z = -\frac{R}{1/sC}(-R) = sR^2 C \equiv sL_{eq} \tag{1}$$

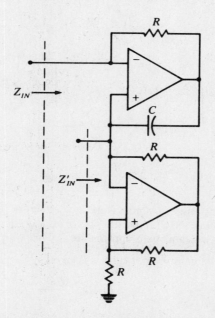

Fig. 13-16

13.36 If only four 10-kΩ resistors and a 0.01-μF capacitor are available for use in the circuit of Fig. 13-16, determine the value of L that can be simulated.

▌ By (1) of Problem 13.35, the value of L_{eq} is
$$L_{eq} = R^2 C = (10^4)^2 (0.01 \times 10^{-6}) = 1 \text{ H}$$

13.37 Show that the op amp circuit of Fig. 13-17 is actually a *current-to-voltage converter* by determining that $v_o = -R_F I_S$.

▌ Since the inverting terminal of the op amp is a virtual ground ($v_d = 0$), no component of I_S flows through R_S ($i_{RS} = 0$); otherwise, $v_d = i_{RS} R_S \neq 0$. Consequently, $i_F = I_S$, and
$$v_o = -i_F R_F = -R_F I_S \tag{1}$$

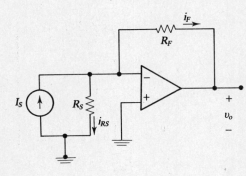

Fig. 13-17

13.38 For the op amp circuit of Fig. 13-18, determine that $i_L = v_S / R_L$, thereby showing that the circuit is a *voltage-to-current converter*.

▌ Based on (2) of Problem 13.33,
$$v_o = \frac{R}{R \| R_L} v_S + v_S = \frac{R + 2R_L}{R_L} v_S \tag{1}$$

Since the noninverting terminal of the op amp draws negligible current, voltage division leads to
$$v_L = \frac{R \| R_L}{R \| R_L + R} v_o = \frac{R_L}{R + 2R_L} v_o \tag{2}$$

Substitute (1) into (2) to find
$$v_L = \left(\frac{R_L}{R + 2R_L} \right) \left(\frac{R + 2R_L}{R_L} v_S \right) = v_S \tag{3}$$

By Ohm's law and use of (3),
$$i_L = \frac{v_L}{R_L} = \frac{v_S}{R_L} \tag{4}$$

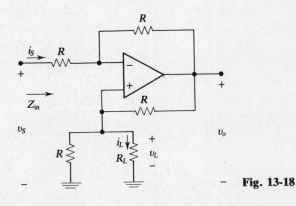

Fig. 13-18

13.39 Find the input impedance Z_{in} for the voltage-to-current converter of Fig. 13-18.

▎ Since $v_d = 0$,

$$i_S = \frac{v_S - v_L}{R} \tag{1}$$

Using (3) of Problem 13.38 in (1),

$$i_S = \frac{v_S - v_S}{R} = 0$$

Hence,

$$Z_{in} = \frac{v_S}{i_S} = \frac{v_S}{0} = \infty$$

13.40 Find the relationship between v_o and v_S for the circuit of Fig. 13-19.

▎ The Laplace-domain transfer function is

$$\frac{V_o}{V_S} = -\frac{Z_F}{Z_1} = -\frac{R}{sL} \tag{1}$$

After inverse transformation of (1),

$$v_o = -\frac{R}{L} \int v_S \, dt \tag{2}$$

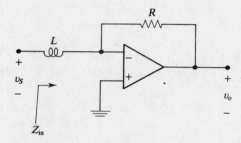

Fig. 13-19

13.41 Determine the input impedance $Z_{in}(s)$ for the amplifier of Fig. 13-19.

▎ Based on (3) of Problem 13.34 with $Z = 0$ and $Z_P = \infty$,

$$Z_{in} = sL + \lim_{\substack{Z \to 0 \\ Z_P \to \infty}} \left(-\frac{R}{Z_P} Z \right) = sL$$

13.42 If the op amp in the circuit of Fig. 13-20 is represented by the model of Fig. 13-1(a), the amplifier can be analyzed as a voltage-shunt feedback amplifier. Applying this method, find an expression for the overall voltage gain, and show that as $A_{OL} \to -\infty$, the resulting gain is consistent with (1) of Problem 13.4.

▎ By voltage division,

$$v_d = \frac{R_d}{R_d + R_2} v_1 \tag{1}$$

The controlled voltage source with series resistor R_o of Fig. 13-1(a) can be converted with use of (1) to a controlled current source of magnitude

$$\frac{A_{OL} v_d}{R_o} = \frac{A_{OL} R_d}{R_o} \frac{v_1}{R_d + R_2} = \frac{A_{OL} R_d}{R_o} i_1 = \beta i_1 \tag{2}$$

connected in parallel with resistance R_o. Since the output of the controlled current source of (2) is directed upward and the output of the controlled current source of Fig. 12-24 is directed downward,

$$\beta = -A_{OL} R_d / R_o \tag{3}$$

Based on (2) of Problem 12.89,

$$A_{vf} = \frac{G_1 [G_F - A_{OL} R_d (R_d + R_2)/R_o}{(R_d + R_2)[(1 - A_{OL} R_d/R_o) G_F + G_o + G_L] + G_S(G_o + G_F) + G_L(G_S + G_F)} \tag{4}$$

Taking the limit of (4) as $A_{OL} \to -\infty$,

$$A_{vf} = -\frac{G_1}{G_F} = -\frac{R_F}{R_1} \quad (5)$$

where (5) is identical to (1) of Problem 13.4.

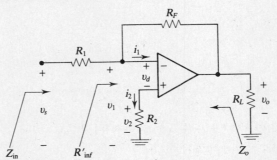

Fig. 13-20

13.43 Find the input impedance Z_{in} for the op amp circuit of Fig. 13-20 based on (8) of Problem 12.84 if $A_{OL} \to -\infty$.

▌ Using (3) of Problem 13.42 in (8) of Problem 12.84,

$$R'_{inf} = \frac{G_o + G_F + G_L}{1/(R_d + R_2)[(1 - A_{OL}R_d/R_o)G_F + G_o + G_L] + G_L G_F} \quad (1)$$

Taking the limit of (1) as $A_{OL} \to -\infty$, $R'_{inf} = 0$. Hence,

$$Z_{in} = R_1 + R'_{inf} = R_1 \quad (2)$$

13.44 Find the output impedance Z_o for the op amp circuit of Fig. 13-20 based on (3) of Problem 12.85 if $A_{OL} \to -\infty$.

▌ Using (3) of Problem 13.42 in (3) of Problem 12.85,

$$G_{of} = \frac{(1 - A_{OL}R_d/R_o)(R_d + R_2) + G_1}{1 + R_F[1/(R_d + R_2) + G_S]} \quad (1)$$

Taking the limit of (1) as $A_{OL} \to -\infty$ yields $G_{of} = \infty$; whence,

$$R_{of} = 1/G_{of} = 0 \quad (2)$$

13.45 Show that if $A_{OL} \to -\infty$ and $R_d \to \infty$ for the op amp circuit of Fig. 13-20, then the gain v_o/v_S is independent of the value for R_2.

▌ If $R_d \to \infty$, $i_2 = v_d/R_d = 0$. Thus, $v_2 = i_2 R_2 = 0$. Consequently, the balance of the analysis of Problem 13.4 is pertinent, and $A_v = -R_F/R_1$, which is clearly independent of R_2.

13.46 Is it necessary that $R_d \to \infty$ for the result of Problem 13.45 to be valid?

▌ No, the result is valid for finite R_d. See Problem 13.42 where identically the same value of gain v_o/v_S is obtained without any restriction on the value of R_d.

13.47 For the op amp circuit of Fig. 13-21, show that $A_v = v_o/v_S = k + 1$.

▌ Recognize the circuit as a noninverting amplifier. Then, by (1) of Problem 13.10 with R_2 replaced by kR and R_1 replaced by R,

$$A_v = \frac{v_o}{v_S} = 1 + \frac{kR}{R} = 1 + k \quad (1)$$

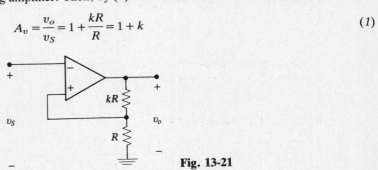

Fig. 13-21

13.48 Determine the input-output relationship for the op amp circuit of Fig. 13-22.

▌ Assume the inverting terminal to be virtual ground; then by KCL at the inverting node,

$$i_1 + i_2 = \frac{v_1}{R} + \frac{v_2}{R} = i_C = -C\frac{dv_o}{dt} \qquad (1)$$

Rearrange (1) and integrate to find

$$v_o(t) = \int_0^t dv_o = -\frac{1}{RC}\int_0^t (v_1 + v_2)\,d\tau + v_o(0) \qquad (2)$$

where $v_o(0)$ is the capacitor voltage at $t = 0$. Clearly, the circuit is a *summing integrator*.

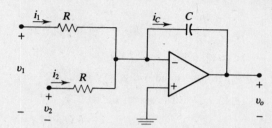

Fig. 13-22

13.49 For the op amp circuit of Fig. 13-23, find the gain v_o/v_1.

▌ By Ohm's law,

$$i_1 = \frac{v_1}{R_1} \qquad (1)$$

and

$$i_{11} = -\frac{v_{11}}{R_{11}} \qquad (2)$$

By voltage division,

$$v_{11} = \frac{R_{11} \parallel R_{12}}{R_{11} \parallel R_{12} + R_{22}} v_o \qquad (3)$$

Substitute (3) into (2) to find

$$i_{11} = -\frac{R_{12} v_o}{R_{11} R_{12} + R_{11} R_{22} + R_{12} R_{22}} \qquad (4)$$

Since $i_{in} = 0$,

$$i_1 = i_{11} \qquad (5)$$

Use (1) and (4) in (5) and rearrange to yield

$$A_v = \frac{v_o}{v_1} = -\frac{R_{11} + R_{22}}{R_1} - \frac{R_{11} R_{22}}{R_1 R_{12}} \qquad (6)$$

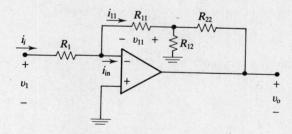

Fig. 13-23

13.50 To illustrate that a large gain is realizable from the amplifier of Fig. 13-23 without use of large resistor values, let $R_1 = 10\,\text{k}\Omega$, $R_{11} = R_{22} = 5\,\text{k}\Omega$, and $R_{12} = 10\,\text{k}\Omega$; find A_v.

▌ By (6) of Problem 13.49,

$$A_v = \frac{v_o}{v_i} = -\frac{5 \times 10^3 + 5 \times 10^3}{10 \times 10^3} - \frac{(5 \times 10^3)(5 \times 10^3)}{(10 \times 10^3)(10)} = -251$$

To have realized this gain with the simple inverting amplifier of Fig. 13-2, while maintaining $R_1 = R_{in} = 10\,k\Omega$, would have required $R_F = 2.5\,M\Omega$ according to (3) of Problem 13.3.

13.51 For the op amp circuit of Fig. 13-24, show that current i_L is independent of the value of R_L; hence, a voltage-controlled current source is formed.

▮ Since the op amp inverting terminal draws negligible current,

$$\frac{v_1}{R} = i_1 = i_F = -\frac{v_o}{100R} \tag{1}$$

or

$$v_o = -100 v_1 \tag{2}$$

Whence, by Ohm's law,

$$i_o = \frac{v_o}{R} = \frac{-100 v_1}{R} \tag{3}$$

Applying KCL at the output node, and using (1) and (3),

$$i_L = i_F - i_o = \frac{v_1}{R} - \frac{-100 v_1}{R} = \frac{101 v_1}{R} \tag{4}$$

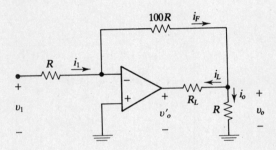

Fig. 13-24

13.52 Determine the op amp output voltage v_o' for the circuit of Fig. 13-24.

▮ By KVL and the result of Problem 13.51,

$$v_o' = v_o - i_L R_L = -100 v_1 - \frac{101 v_1}{R} R_L = -\left(100 + 101\frac{R_L}{R}\right) v_1 \tag{1}$$

13.53 Find the overall gain $A_v = v_{o2}/v_1$ for the op amp circuit of Fig. 13-25.

▮ Since the first stage is a unity follower (see Problem 13.23),

$$v_{o1} = v_1 \tag{1}$$

The second stage is an inverting amplifier (see Problem 13.4),

$$v_{o2} = -\frac{nR}{R} v_{o1} = -n v_{o1} \tag{2}$$

Substitute (1) into (2) to find

$$A_v = \frac{v_{o2}}{v_1} = -n \tag{3}$$

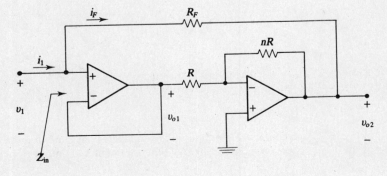

Fig. 13-25

13.54 Determine an expression for Z_{in} of the amplifier in Fig. 13-25.

■ By method of node voltages and use of (3) of Problem 13.53,

$$i_1 = \frac{v_1 - v_{o2}}{R_F} = \frac{v_1 - (-nv_1)}{R_F} = \frac{(n+1)v_1}{R_F}$$

Hence,
$$Z_{in} = \frac{v_1}{i_1} = \frac{v_1}{(n+1)v_1/R_F} = \frac{R_F}{n+1} \quad (1)$$

13.55 Determine the gain $A_v = v_o/v_S$ for the op amp circuit of Fig. 13-26.

■ Since the inverting terminal of the op amp draws negligible current and is a virtual ground,

$$\frac{v_A}{R_2} = -\frac{v_o}{R_3} \quad \text{or} \quad v_A = -\frac{R_2}{R_3}v_o \quad (1)$$

By method of node voltages,

$$\frac{v_A - v_S}{R_1} + \frac{v_A}{R_2} + \frac{v_A - v_o}{R_4} = 0 \quad (2)$$

Substitute (1) into (2) and rearrange to find

$$A_v = \frac{v_o}{v_S} = -\frac{R_3 R_4}{R_4(R_1 + R_2) + R_1(R_2 + R_3)} \quad (3)$$

13.56 Find the input impedance for the op amp circuit in Fig. 13-26.

■ By the method of node voltages,

$$i_S = \frac{v_S - v_A}{R_1} \quad (1)$$

Substitute (1) of Problem 13.55 into (1) to yield

$$i_S = \frac{v_S - \left(-\frac{R_2}{R_3}v_o\right)}{R_1} = \frac{v_S + \frac{R_2}{R_3}v_o}{R_1} \quad (2)$$

Use (3) of Problem 13.55 in (2) to give

$$i_S = \frac{\left(1 + \frac{R_2}{R_3}A_v\right)v_S}{R_1} \quad (3)$$

Whence,
$$Z_{in} = \frac{v_S}{i_S} = \frac{R_1}{1 + \frac{R_2}{R_3}A_v} \quad (4)$$

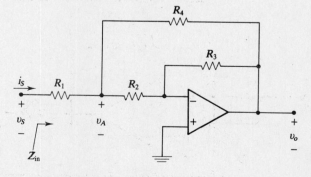

Fig. 13-26

13.57 The circuit of Fig. 13-27 is an *adjustable-output voltage regulator*. Assume that the basic op amp is ideal. Regulation of the Zener is preserved if $i_Z \geq 0.1 I_Z$. Find the regulated output v_o in terms of V_Z.

Since V_Z is the voltage at node a, (3) of Problem 13.3 gives

$$v_o = -\frac{R_2}{R_1}V_Z$$

So long as $i_Z \geq 0.1 I_Z$, a regulated value of v_o can be achieved by adjustment of R_2.

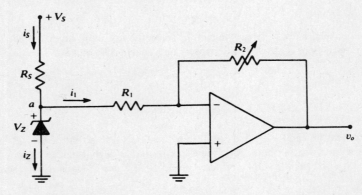

Fig. 13-27

13.58 Given a specific Zener diode and the values of R_S and R_1, over what range of V_S would there be no loss of regulation for the adjustable-output voltage regulator of Problem 13.57?

I Regulation is preserved and the diode current $i_Z = i_S - i_1$ does not exceed its rated value I_Z if

$$0.1I_Z \leq i_S - i_1 \leq i_Z \quad \text{or} \quad 0.1I_Z \leq \frac{V_S - V_Z}{R_S} - \frac{V_Z}{R_1} \leq I_Z$$

or

$$0.1I_Z R_S + \left(1 + \frac{R_S}{R_1}\right)V_Z \leq V_S \leq I_Z R_S + \left(1 + \frac{R_S}{R_1}\right)V_Z$$

13.59 Use an op amp to design a noninverting voltage source (see Problem 13.57). Determine the conditions under which regulation is maintained in your source.

I Simply replace the inverting amplifier of Fig. 13-27 with the noninverting amplifier of Fig. 13-3. Since the op amp draws negligible current, regulation is preserved if V_S and R_S are selected so that i_Z remains within the regulation range of the Zener diode. Specifically, regulation is maintained if $0.1I_Z \leq V_S/R_S \leq I_Z$.

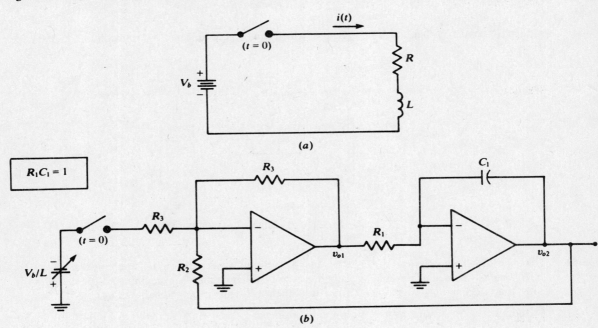

Fig. 13-28

13.60 The analog computer utilizes operational amplifiers to solve differential equations. Devise an analog solution for $i(t)$, $t \geq 0$, in the circuit of Fig. 13-28(a). Assume that you have available an inverting integrator with unity gain ($R_1 C_1 = 1$), inverting amplifiers, a variable dc source, and a switch.

▌ For $t > 0$, the governing differential equation for the circuit of Fig. 13-28(a) may be written as

$$-\frac{di}{dt} = -\frac{V_b}{L} + \frac{R}{L}i \tag{1}$$

The sum on the right side of (1) can be simulated by the left-hand inverting adder of Fig. 13-28(b), where $v_{o1} = -di/dt$ and where R_2 and R_3 are chosen such that $R_3/R_2 = R/L$. Then $v_{o2} = -\int v_{o1} \, dt$ will be an analog of $i(t)$ on a scale of $1 \, A/V$.

NONLINEAR FUNCTION OPERATIONS

13.61 Analog multiplication can be carried out with a basic circuit like that of Fig. 13-29. Essential to the operation of the logarithmic amplifier is the use of a feedback-loop device that has an exponential terminal characteristic curve; one such device is the semiconductor diode of Chapter 2, which is characterized by

$$i_D = I_o(e^{v_D/\eta V_T} - 1) \approx I_o e^{v_D/\eta V_T} \tag{1}$$

A grounded-base BJT can also be utilized, since its emitter current and base-to-emitter voltage are related by

$$i_E = I_S e^{v_{BE}/V_T} \tag{2}$$

Determine the condition under which the output voltage v_o is proportional to the logarithm of the input voltage v_i in the circuit of Fig. 13-29.

▌ Since the op amp draws negligible current,

$$i_i = \frac{v_i}{R} = i_D \tag{3}$$

Since $v_D = -v_o$, substitution of (3) into (1) yields

$$v_i = R I_o e^{-v_o/V_T} \tag{4}$$

Taking the logarithm of both sides of (4) leads to

$$\ln v_i = \ln R I_o - \frac{v_o}{V_T} \tag{5}$$

Under the condition that $\ln R I_o$ is negligible (which can be accomplished by controlling R so that $R I_o \approx 1$), (5) gives $v_o \approx -V_T \ln v_i$.

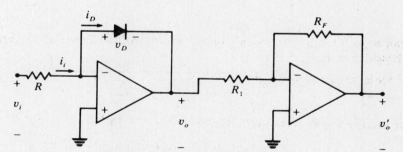

Fig. 13-29 Logarithmic amplifier.

13.62 The logarithmic amplifier of Fig. 13-29 has two undesirable aspects: V_T and I_o are temperature-dependent, and $\ln R I_o$ may not be negligibly small. A circuit that can overcome these shortcomings is presented in Fig. 13-30. Show that if Q_1 and Q_2 are matched transistors, then v_o is truly proportional to $\ln v_S$.

▌ In matched transistors, reverse saturation currents are equal. By KVL, with $v_1 \approx 0$,

$$v_2 = v_{BE2} - v_{BE1} \tag{1}$$

Taking the logarithm of both sides of (2) from Problem 13.61 leads to

$$v_{BE} = V_T \ln \frac{I_E}{I_S} \tag{2}$$

Now the use of (2) in (1), with $I_C \approx I_E$, gives

$$v_2 = V_T \ln \frac{I_{E2}}{I_S} - V_T \ln \frac{I_{E1}}{I_S} = -V_T \ln \frac{I_{C1}}{I_{C2}} \tag{3}$$

According to (1), v_2 is the difference between two small voltages. Thus, if V_R is several volts in magnitude, then $v_s \ll V_R$, and

$$I_{C2} \approx I_{E2} = \frac{V_R - v_2}{R_2} \approx \frac{V_R}{R_2} \tag{4}$$

Also, since $v_1 \approx 0$,

$$I_{C1} \approx I_{E1} = \frac{v_S - v_1}{R_1} \approx \frac{v_S}{R_1} \tag{5}$$

Thus, by (1) of Problem 13.10 along with (3) to (5),

$$v_o = \frac{R_3 + R_4}{R_3} v_2 = -V_T \frac{R_3 + R_4}{R_3} \ln \frac{I_{C1}}{I_{C2}} = -V_T \left(1 + \frac{R_4}{R_3}\right)\left[\ln v_S - \ln\left(\frac{R_1}{R_2} V_R\right)\right] \tag{6}$$

The selection of $(R_1/R_2)V_R = 1$ forces the last term on the right-hand side of (6) to zero. Also, R_3 can be selected with a temperature sensitivity similar to that of V_T, to offset changes in V_T. Further, it is simple to select $R_4/R_3 \gg 1$, so that (6) becomes

$$v_o \approx -V_T \frac{R_4}{R_3} \ln v_S$$

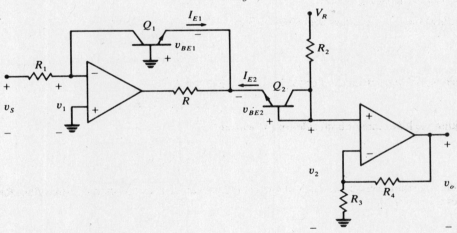

Fig. 13-30

13.63 The circuit of Fig. 13-31 is an *exponential* or *inverse log* amplifier. Show that the output v_o is proportional to the inverse logarithm of the input v_i.

▌ Since the input current to the op amp is negligible,

$$i_R \approx i_D = I_o e^{v_D/\eta V_T}$$

But since the inverting terminal is a virtual ground, $v_D = v_i$. Thus,

$$v_o = -i_R R \approx -R I_o e^{v_i/\eta V_T} = -R I_o \ln^{-1} \frac{v_i}{\eta V_T}$$

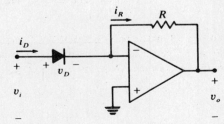

Fig. 13-31

13.64 Having now at your disposal a logarithmic amplifier (Problems 10.61 and 10.62) and an exponential (inverse log) amplifier (Problem 13.63), devise a circuit that will multiply two numbers together.

▌ Since $xy = e^{\ln x + \ln y}$, the circuit of Fig. 13-32 is a possible realization.

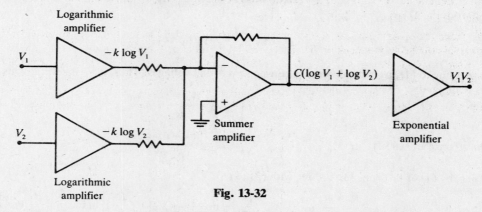

Fig. 13-32

13.65 The signal-conditioning amplifier of Fig. 13-33 changes gain depending upon the polarity of V_S. Find the circuit voltage gain for positive v_S and negative v_S if diode D_2 is ideal.

▌ If $v_S > 0$, then $v_o < 0$ and D_2 is forward-biased and appears as a short circuit. The equivalent feedback resistance is then
$$R_{F\,eq} = \frac{R_2 R_3}{R_2 + R_3}$$

and by (3) of Problem 13.3,
$$A_v = -\frac{R_{F\,eq}}{R_1} = -\frac{R_2 R_3}{R_1(R_2 + R_3)} \tag{1}$$

If $v_S < 0$, then $v_o > 0$ and D_2 is reverse-biased and appears as an open circuit. The equivalent feedback resistance is now $R_{F\,eq} = R_3$, and
$$A_v = -\frac{R_{F\,eq}}{R_1} = -\frac{R_3}{R_1} \tag{2}$$

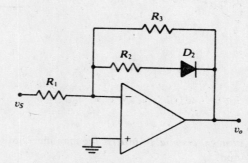

Fig. 13-33

13.66 Sketch the transfer characteristic for the amplifier of Problem 13.65.

▌ See Fig. 13-34.

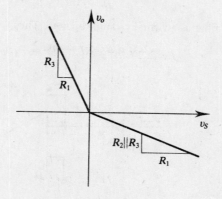

Fig. 13-34

13.67 In analog signal processing, the need often arises to introduce a *level clamp* (linear amplification to a desired output level or value, and then no further increase in output level as the input continues to increase). One level-clamp circuit, shown in Fig. 13-35(a), uses series Zener diodes in a negative feedback path. Assuming ideal Zeners and op amp, find the relationship between v_o and v_S. Sketch the results on a transfer characteristic.

▎ Since the op amp is ideal, the inverting terminal is a virtual ground, and v_o appears across the parallel-connected feedback paths. There are two distinct possibilities:

Case I: $v_S > 0$. For $v_o < 0$, Z_2 is forward-biased and Z_1 reverse-biased. The Zener feedback path is an open circuit until $v_o = -V_{Z1}$; then Z_1 will limit v_o at $-V_{Z1}$ so that no further negative excursion is possible.

Case II: $v_S < 0$. For $v_o > 0$, Z_1 is forward-biased and Z_2 reverse-biased. The Zener feedback path acts as an open circuit until v_o reaches V_{Z2}, at which point Z_2 limits v_o to that value. In summary, for both cases

$$v_o = \begin{cases} V_{Z2} & \text{for } v_S < -\dfrac{R_1}{R_2} V_{Z2} \\ -\dfrac{R_2}{R_1} v_S & \text{for } -\dfrac{R_1}{R_2} V_{Z2} \leq v_S \leq \dfrac{R_1}{R_2} V_{Z1} \\ -V_{Z1} & \text{for } v_S > \dfrac{R_1}{R_2} V_{Z1} \end{cases}$$

Figure 13-35(b) gives the transfer characteristic.

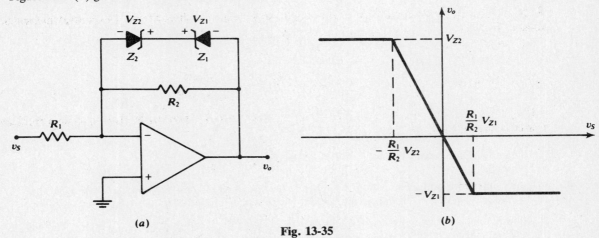

Fig. 13-35

13.68 The circuit of Fig. 13-36(a) is a *limiter*; it reduces the signal gain to some limiting level rather than imposing the abrupt clamping action of the circuit of Problem 13.67. Determine the limiting value V_ℓ of v_o at which the diode

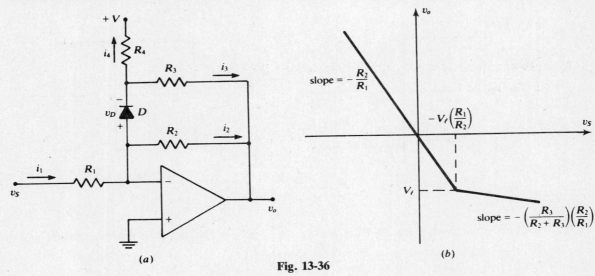

Fig. 13-36

D becomes forward-biased, thus establishing a second feedback path through R_3. Assume an ideal op amp and an ideal diode.

▌ The diode voltage v_D is found by writing a loop equation. Since the inverting input is a virtual ground, v_o appears across R_2 and

$$v_D = -v_o - i_3 R_3 = v_o - \frac{V - v_o}{R_3 + R_4} R_3 \tag{1}$$

When $v_D = 0$, $v_o = V_\ell$, and (1) gives

$$V_\ell = -\frac{R_3}{R_4} V \tag{2}$$

13.69 Determine the relationship between v_o and v_S for the limiter of Problem 13.68. Sketch the associated transfer characteristic.

▌ For $v_o > V_\ell$, as determined in Problem 13.68, the diode blocks and R_2 constitutes the only feedback path. Since $i_1 = i_2$,

$$\frac{v_S}{R_1} = -\frac{v_o}{R_2} \tag{1}$$

For $v_o \leq V_\ell$, the diode conducts and the parallel combination of R_2 and R_3 forms the feedback path. Since now $i_1 = i_2 + i_3 + i_4$,

$$\frac{v_S}{R_1} = -\left(\frac{v_o}{R_2} + \frac{v_o}{R_3} + \frac{V}{R_4}\right) \tag{2}$$

It follows from (2) of Problem 13.68, (1) and (2) that

$$v_o \begin{cases} -\dfrac{R_2}{R_1} v_S & \text{for } v_S < \dfrac{R_1 R_3}{R_2 R_4} V \\ -\dfrac{R_3}{R_2 + R_3} \dfrac{R_2}{R_1} v_S - \dfrac{R_2}{R_2 + R_3} \dfrac{R_3}{R_4} V & \text{for } v_S \geq \dfrac{R_1 R_3}{R_2 R_4} V \end{cases}$$

This transfer characteristic is plotted in Fig. 13-36(b).

13.70 What modifications and specifications will change the circuit of Fig. 13-35(a) into a 3-V square-wave generator, if $v_S = 0.02 \sin \omega t$ V? Sketch the circuit transfer characteristic and the input and output waveforms.

▌ Modify the circuit by removing R_2, and specify Zener diodes such that $V_{Z1} = V_{Z2} = 3$ V. The transfer characteristic of Fig. 13-35(b) will change to that of Fig. 13-37(a). The time relationship between v_S and v_o will be that displayed in Fig. 13-37(b).

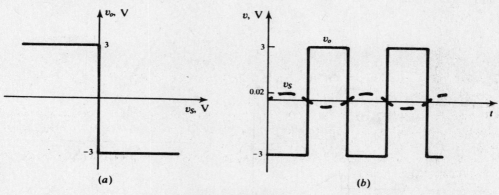

Fig. 13-37

13.71 Let $R_2 = \infty$ (removed) in Fig. 13-35(a). Describe the output v_o and sketch the resulting transfer characteristic.

▌ Proceeding in manner similar to the work of Problem 13.67, there are two possibilities:
Case I: $v_S > 0$ and $v_0 < 0$. Z_2 is forward-biased and Z_1 is reverse-biased; then Z_1 will limit v_o at $-V_{Z1}$.

Case II: $v_S < 0$ and $v_o > 0$. Z_1 is forward-biased and Z_2 is reverse-biased; then Z_2 limits v_o at V_{Z2}.

$$v_o = \begin{cases} V_{Z2} & \text{for } v_S < 0 \\ -V_{Z1} & \text{for } v_S > 0 \end{cases}$$

The resulting transfer characteristic is shown by Fig. 13-38.

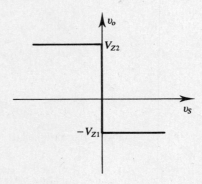

Fig. 13-38

13.72 Let $R_2 = \infty$ (removed) and replace Zener diode Z_2 by a short circuit in Fig. 13-35(a). Describe the output v_o and sketch the resulting transfer characteristic.

■ For $v_S > 0$, the analysis is identical to that of Problem 13.71 (Case I). However, for $v_S < 0$, Z_1 is forward-biased, ideally limiting v_o to zero. Hence,

$$v_o = \begin{cases} 0 & \text{for } v_S < 0 \\ -V_{Z1} & \text{for } v_S > 0 \end{cases}$$

The associated transfer characteristic is presented in Fig. 13-39.

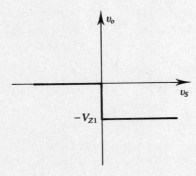

Fig. 13-39

13.73 Replace Z_2 with a rectifying diode D_2 of same orientation in Fig. 13-35(a). Describe the output v_o and sketch the resulting transfer characteristic.

■ v_o appears across the parallel-connected feedback paths with two distinct conduction possibilities:
Case I: $v_S > 0$ and $v_o < 0$. D_2 is forward-biased, but Z_1 is open circuit until $v_o = -V_{Z1}$, at which point Z_1 limits v_o so that a more negative value of v_o is possible.
Case II: $v_S < 0$ and $v_o > 0$. Z_2 is forward-biased, but D_2 is reverse-biased. Hence, the diode feedback path remains open circuit for all $v_o > 0$.

$$v_o = \begin{cases} -\dfrac{R_2}{R_1} v_S & \text{for } v_S \leq \dfrac{R_1}{R_2} V_{Z1} \\ -V_{Z1} & \text{for } v_S > \dfrac{R_1}{R_2} V_{Z1} \end{cases}$$

The resulting transfer characteristic is sketched in Fig. 13-40.

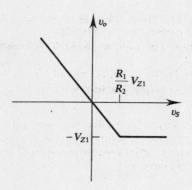

Fig. 13-40

13.74 Let $v_S = V_m \sin \omega t$ for the circuit of Fig. 13-35(a). Assume $V_m > (R_1/R_2)V_{Z2} > (R_1/R_2)V_{Z1}$ and sketch the resulting waveform for v_o.

▌ Signal v_o is a clipped sinusoidal waveform with 180° phase shift from v_S and of magnitude $(R_2/R_1)V_m$ over the linear region of operation.

$$v_o = \begin{cases} V_{Z2} & \text{for } v_S < -\dfrac{R_1}{R_2}V_{Z2} \\ -\dfrac{R_2}{R_1}V_m \sin \omega t & \text{for } -\dfrac{R_1}{R_2}V_{Z2} \leq v_S \leq \dfrac{R_1}{R_2}V_{Z1} \\ -V_{Z1} & \text{for } v_S > \dfrac{R_1}{R_2}V_{Z1} \end{cases}$$

The resulting transfer characteristic is sketched in Fig. 13-41.

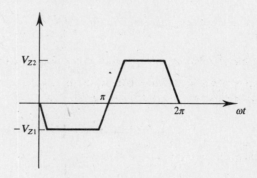

Fig. 13-41

13.75 Describe the output v_o and sketch the transfer characteristic for the *zero limiter* op amp circuit of Fig. 13-42 if (a) diode D is ideal, and (b) if diode D is nonideal in that the voltage drop in forward conduction is 0.7 V.

▌ (a) If $v_S < 0$, then $v_o > 0$, and diode D is OFF, leaving only R_F as a feedback path. Hence, $v_o = -(R_F/R_1)v_o$. On the other hand, if $v_S > 0$, then v_o attempts to become negative and diode D is ON, so that $v_o = 0$ for an ideal diode. The transfer characteristic is shown by Fig. 13-43.

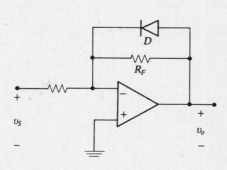

Fig. 13-42

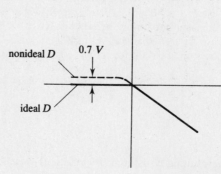

Fig. 13-43

(b) For a nonideal diode, the analysis of $v_S < 0$ is identical to that of part (a). However, when $v_S > 0$, $v_S = -0.7$ V. Thus, the output is not limited to the ideal value of zero. The resulting transfer characteristic is added to Fig. 13-43 as shown by the dashed line.

13.76 Explain how the *precision zero limiter* op amp circuit of Fig. 13-44 overcomes the objection of the circuit of Fig. 13-42 by limiting v_o to zero for $v_S < 0$ to give the ideal transfer characteristic of Fig. 13-43 while using nonideal diodes.

▌ For $v_S > 0$, thus $v_o < 0$, the diode D_1 conducts the current flowing into the op amp output acting only to effectively increase the output resistance of the op amp. Consequently, $v_o = -(R_F/R_1)v_o$. For $v_S < 0$, $v'_o = v_{D2} = 0.7$ V and the necessary current that must be supplied to the source through R_1 to assure $v_d \approx 0$ flows through D_2. D_1 blocks (OFF) with $v_{D1} = -0.7$ V, leaving $v_o = v_{D1} + v'_o = -0.7 + 0.7 = 0$.

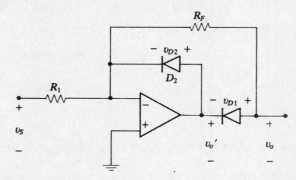

Fig. 13-44

13.77 If $v_S = V_m \sin \omega t$ is supplied to the precision zero limiter circuit of Fig. 13-44, determine the output voltage v_o for $0 \le \omega t \le 2\pi$ if $R_F = R_1$.

▌ The applicable transfer characteristic of Fig. 13-43 with $R_F = R_1$ describes negative half-wave rectification. Consequently,
$$v_o = \begin{cases} 0 & \text{for } 0 \le \omega t < \pi \\ V_m \sin \omega t & \text{for } \pi \le \omega t \le 2\pi \end{cases}$$

13.78 Describe the output and transfer characteristic of the Schmitt trigger circuit of Fig. 13-45.

▌ If $v_S < v_1$, the op amp is driven to positive saturation, or $v_o = V_{o\,\text{sat}}$. Since the noninverting input draws negligible current, voltage v_1 is found using KVL and voltage division as

$$v'_1 = V_b + v_{R2} = V_b + \frac{R_2}{R_1 + R_2}(V_{o\,\text{sat}} - V_b) \tag{1}$$

As v_S is increased, no change in $v_o = V_{o\,\text{sat}}$ occurs until v_S exceeds v'_1. At this upper trigger level, v_o switches to $-V_{o\,\text{sat}}$ and

$$v''_1 = V_b + v_{R2} = V_b + \frac{R_2}{R_1 + R_2}(-V_{o\,\text{sat}} - V_b) \tag{2}$$

Further increase in v_S does not change $v_o = -V_{o\,\text{sat}}$. However, if v_S is decreased so that $v_S < v''_1$, the op amp output v_o is again driven to $V_{o\,\text{sat}}$. Since $v''_1 < v'_1$, or the lower trigger level is less than the upper trigger level, hysteresis is present as illustrated by the transfer characteristic of Fig. 13-46 where the arrows indicate the manner by which the characteristic is traversed as v_S varies.

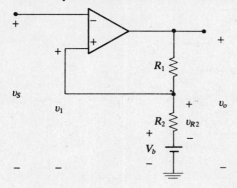

Fig. 13-45

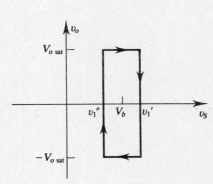

Fig. 13-46

13.79 Let $V_{o\,sat} = 10.5$ V, $V_b = 5$ V, $R_1 = 50$ kΩ, and $R_2 = 5$ kΩ for the Schmitt trigger circuit of Fig. 13-45. Determine the hysteresis voltage $v_1' - v_1''$.

▐ Based on (1) and (2) of Problem 13.78,

$$v_1' = 5 + \frac{5 \times 10^3}{50 \times 10^3 + 5 \times 10^3}(10.5 - 5) = 5.5 \text{ V}$$

$$v_1'' = 5 + \frac{5 \times 10^3}{50 \times 10^3 + 5 \times 10^3}(-10.5 - 5) = 3.59 \text{ V}$$

Hence, $\quad V_H = v_1' - v_1'' = 5.5 - 3.59 = 1.91$ V

13.80 Repeat Problem 13.79 if $V_b = 0$, $R_2 = 500$ Ω, and all else is unchanged.

▐ By (1) and (2) of Problem 13.78,

$$v_1' = 0 + \frac{500}{50 \times 10^3 + 500}(10.5 - 0) = 0.104 \text{ V}$$

$$v_1'' = 0 + \frac{500}{50 \times 10^3 + 500}(-10.5 - 0) = -0.104 \text{ V}$$

$$V_H = v_1' - v_1'' = 0.104 - (-0.104) = 0.208 \text{ V}$$

Since this circuit switches upon polarity reversal of v_S, it is useful as a zero crossing detector.

13.81 The *analog multiplier* of Fig. 13-47 has the characteristic $v_P = v_1 v_2$. Determine the output v_o for the op amp circuit.

▐ Since $i_S = i_P$ and $v_d = 0$,

$$\frac{v_S}{R} = -\frac{v_P}{R} = -\frac{v_1 v_2}{R} = -\frac{v_o^2}{R} \quad (1)$$

from which $\quad v_o = \sqrt{-v_S} \quad (2)$

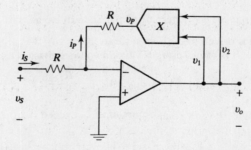

Fig. 13-47

13.82 Show that the op amp circuit of Fig. 13-48 uses the analog multiplier to perform analog division.

▐ Since $i_S = i_P$ and $v_d = 0$,

$$\frac{v_S}{R} = -\frac{v_P}{R} = -\frac{v_o v_{SS}}{R} \quad (1)$$

Hence, $\quad v_o = -\dfrac{v_S}{v_{SS}} \quad (2)$

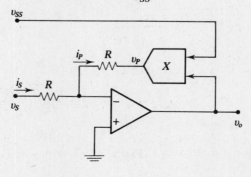

Fig. 13-48

13.83 Determine v_o for the circuit of Fig. 13-49.

▌ Since $v_d = 0$,

$$i_{S1} + i_{S2} = \frac{v'_{S1}}{R} + \frac{v'_{S2}}{R} = \frac{v_{S1}^2}{R} + \frac{v_{S2}^2}{R} \qquad (1)$$

Also,

$$i_P = -\frac{v_P}{R} = -\frac{v_o^2}{R} \qquad (2)$$

However, the noninverting terminal of the op amp draws negligible current, and thus, by KCL,

$$i_{S1} + i_{S2} = i_P \qquad (3)$$

Substitute (1) and (2) into (3) to find

$$v_o = \sqrt{-(v_{S1}^2 + v_{S2}^2)} \qquad (4)$$

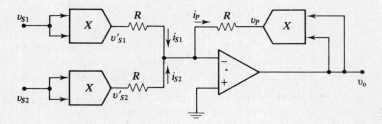

Fig. 13-49

FILTERS

13.84 Active filter realizations can eliminate the need for bulky inductors, which do not satisfactorily lend themselves to integrated circuitry. Further, active filters do not necessarily attenuate the signal over the band pass, as do their passive-element counterparts. A simple *inverting, first-order, low-pass filter* using an op amp as the active device is shown in Fig. 13-50(a). Find the transfer function (voltage-gain ratio) $A_v(s) = V_o(s)/V_S(s)$.

▌ The feedback impedance $Z_F(s)$ and the input impedance $Z_1(s)$ are

$$Z_F(s) = \frac{R(1/sC)}{R + (1/sC)} = \frac{R}{sRC + 1} \quad \text{and} \quad Z_1(s) = R_1 \qquad (1)$$

The resistive circuit analysis of Problem 13.4 extends directly to the s domain; thus,

$$A_v(s) = -\frac{Z_F(s)}{Z_1(s)} = -\frac{R/R_1}{sRC + 1} \qquad (2)$$

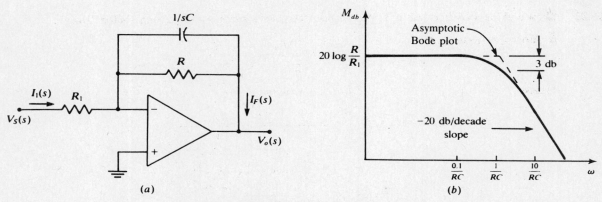

Fig. 13-50 First-order low-pass filter.

13.85 Draw the Bode plot (M_{db} only) associated with the transfer function given by (2) of Problem 13.84 to show that the filter passes low-frequency signals and attenuates high-frequency signals.

Letting $s = j\omega$ in (2) of Problem 13.84 gives

$$M_{db} \equiv 20 \log |A_v(j\omega)| = 20 \log \frac{R}{R_1} - 20 \log |j\omega RC + 1|$$

A plot of M_{db} is displayed in Fig. 13-50(b). The curve is essentially flat below $\omega = 0.1/RC$; thus, all frequencies below $0.1/RC$ are passed with the dc gain R/R_1. A 3-db reduction in gain is experienced at the corner frequency $1/\tau = 1/RC$, and the gain is attenuated by 20 db per decade of frequency change for frequencies greater than $10/RC$.

13.86 Design a first-order low-pass filter with dc gain of magnitude 2 and input impedance 5 kΩ. The gain should be flat to 100 Hz.

▌ The filter is shown in Fig. 13-50(a). For an ideal op amp, Problem 13.22 gives $Z_1 = R_1 = 5$ kΩ. The dc gain is given by (2) of Problem 13.84 as $A(0) = -R/R_1$, whence $R = 2R_1 = 10$ kΩ. Figure 13-50(b) shows that the magnitude of the gain is flat to $\omega = 0.1/RC$, so the capacitor must be sized such that

$$C = \frac{0.1}{2\pi f R} = \frac{0.1}{2\pi (100)(10 \times 10^3)} = 15.9 \text{ nF}$$

13.87 Two identical passive RC low-pass filter sections are to be connected in cascade so as to create a double-pole filter with corner frequency at $1/\tau = 1/RC$. Will simple cascade connection of these filters yield the desired transfer function $T(s) = (1/\tau)^2/(s + 1/\tau)^2$?

▌ With simple cascading, the overall transfer function would be

$$\frac{V_o}{V_i} = T' = \frac{(1/\tau)^2}{s^2 + 3(1/\tau)s + (1/\tau)^2}$$

which has two distinct negative roots. The desired result is not obtained because the impedance looking into the second stage is not infinite, and thus, the transfer function of the first stage is not simply $(1/\tau)/(s + 1/\tau)$.

13.88 How may the desired result of Problem 13.87 be realized?

▌ The desired result can be obtained by adding a unity follower between stages, as illustrated in Fig. 13-51.

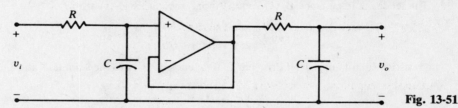

Fig. 13-51

13.89 Find the transfer function for the circuit of Fig. 13-52.

▌ By extension of (3) of Problem 13.3,

$$T(s) = \frac{V_o}{V_S} = -\frac{Z_2}{Z_1} = -\frac{\dfrac{R_2}{sR_2C_2 + 1}}{\dfrac{R_1}{sR_1C_1 + 1}} = -\frac{C_1 s + 1/\tau_1}{C_2 s + 1/\tau_2} \quad (1)$$

where $\tau_1 = R_1 C_1$ and $\tau_2 = R_2 C_2$.

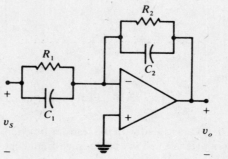

Fig. 13-52

13.90 In control theory, there is a compensation network whose transfer function is of the form $(s + 1/\tau_1)/(s + 1/\tau_2)$; it is called a *lead-lag network* if $1/\tau_1 < 1/\tau_2$, and a *lag-lead network* if $1/\tau_2 < 1/\tau_1$. Explain how the circuit of Fig. 13-52 may be used as such a compensation network.

▌ To obtain unity gain, set $C_1 = C_2$ in (1) of Problem 13.89. To obtain a positive transfer function, insert an inverter stage either before or after the circuit. Then, the selection of $R_1 > R_2$ yields $1/\tau_1 < 1/\tau_2$, giving the lead-lag network, and $R_1 < R_2$ results in $1/\tau_2 < 1/\tau_1$, giving the lag-lead network.

13.91 Replace R_4 with a capacitor C in Fig. 13-26. Find the Laplace-domain transfer function V_o/V_S and classify the circuit.

▌ Based on (3) of Problem 13.55,

$$\frac{V_o}{V_S} = -\frac{R_3(1/sC)}{(1/sC)(R_1+R_2) + R_1(R_2+R_3)} = -\frac{R_3/(R_1+R_2)}{sCR_1(R_2+R_3)/(R_1+R_2) + 1} \quad (1)$$

Inspection of (1) shows the circuit is an inverting low-pass filter.

13.92 Replace R_3 with an inductor L in Fig. 13-26. Find the Laplace-domain transfer function V_o/V_S and classify the circuit.

▌ Based on (3) of Problem 13.55,

$$\frac{V_o}{V_S} = -\frac{sLR_4}{R_4(R_1+R_2) + R_1(R_2+sL)} = -\frac{sLR_4/(R_1R_2 + R_1R_4 + R_2R_4)}{sLR_1/(R_1R_2 + R_1R_4 + R_2R_4) + 1} \quad (1)$$

From (1), it is seen that the circuit is a high-pass filter.

13.93 Find the Laplace-domain transfer function for the high-pass filter of Fig. 13-53.

▌ Recognize the circuit between v_A and v_o as a noninverting op amp and use (1) of Problem 13.10 to give

$$\frac{V_o}{V_A} = 1 + \frac{R}{R} = 2 \quad (1)$$

By application of the method of node voltages at node 1,

$$(V_1 - V_B)sC + (V_A - V_B)sC + (V_o - V_B)/R = 0 \quad (2)$$

Use $V_A = \tfrac{1}{2}V_o$ from (1) and (2) and rearrange to find

$$V_B = \frac{(sRC/2 + 1)V_o + sRCV_1}{s2RC} \quad (3)$$

Since the op amp input current is negligible, voltage division yields

$$V_A = \frac{1}{2}V_o = \frac{R}{R + 1/sC}V_B = \frac{sRC}{sRC + 1}V_B \quad (4)$$

Substitute (3) into (4) and rearrange to produce the transfer function

$$\frac{V_o}{V_1} = \frac{2s^2}{s^2 + \dfrac{1}{RC}s + (1/RC)^2} \quad (5)$$

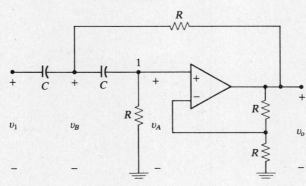

Fig. 13-53

13.94 Sketch the asymptotic Bode plot (magnitude only) for the high-pass filter of Problem 13.93.

▌ Replace s by $j\omega$ in (5) of Problem 13.93, let $\omega_n \equiv 1/RC$, and rearrange to find

$$\frac{V_o}{V_1} = T(j\omega) = \frac{-2(\omega/\omega_n)^2}{1 - (\omega/\omega_n)^2 + \omega/\omega_n} \quad (1)$$

The amplitude ratio follows from Definition 3 of Problem 10.4 as

$$M_{db} = 20 \log |T(j\omega)| = 20 \log (2) + 20 \log (\omega/\omega_n)^2 + 20 \log \left[\frac{1}{(1 - \omega^2/\omega_n^2)^2 + (\omega/\omega_n)^2}\right]^{1/2} \quad (2)$$

The three terms on the right-hand side of (2) yield, respectively, the following components of the Bode plot:
1. A constant value of $20 \log (2) \approx 6$ db.
2. A straight line segment with slope of 40 db/decade passing through the point $M_{db} = 0$ and $\omega/\omega_n = 1$.
3. A piecewise linear plot consisting of a 0 db magnitude for $\omega/\omega_n \leq 1$ and breaking to a slope of -40 db/decade for $\omega/\omega_n > 1$.

The three components (dashed lines) and the composite Bode plot (solid line) are shown in Fig. 13-54.

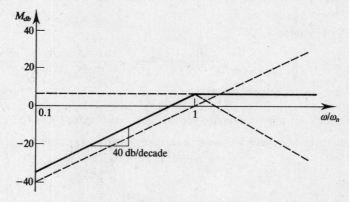

Fig. 13-54

13.95 For the op amp circuit of Fig. 13-55, (a) find the transfer function relating V_o and V_1, and (b) classify the circuit.

▌ (a) Recognize the op amp circuit relating v_A and v_o as a noninverting amplifier. Thus, by (1) of Problem 13.10,

$$V_o = \left(1 + \frac{R_1}{R_1}\right)V_A = 2V_A \quad (1)$$

Since the op amp draws negligible input current, voltage division gives

$$V_A = \frac{1/sC}{R + 1/sC} V_1 = \frac{1}{sRC + 1} V_1 \quad (2)$$

Solve (1) for V_A and substitute the result into (2) to yield

$$V_A = \frac{1}{2} V_o = \frac{1}{sRC + 1} V_1 \quad (3)$$

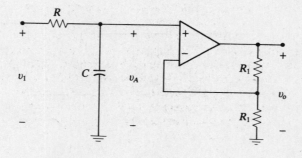

Fig. 13-55

Rearrangement of (3) leads to the transfer function

$$\frac{V_o}{V_1} = \frac{2}{sRC + 1} \qquad (4)$$

(b) From (4), it is apparent that the circuit is a low-pass filter with break frequency $\tau = 1/RC$.

13.96 Sketch the asymptotic Bode plot (magnitude only) for the low-pass filter of Problem 13.95.

▌ Based on Definition 3 of Problem 10.4 and using (4) of Problem 13.95,

$$M_{db} = 20 \log(2) - 20 \log|j\omega C + 1|$$

The plot is similar to that of Fig. 13-50(b) and is presented by Fig. 13-56.

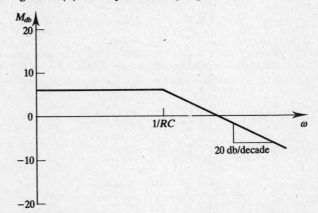

Fig. 13-56

13.97 Determine the transfer function relating V_o and V_1 for the circuit of Fig. 13-57 to show that the circuit is a second-order low-pass filter.

▌ Recognize the op amp circuit relating v_B and v_o as a noninverting amplifier; hence, by (1) of Problem 13.10,

$$V_o = \left(1 + \frac{R_1}{R_1}\right) V_B = 2V_B \qquad (1)$$

Apply the method of node voltages at node 1 to yield

$$\frac{V_1 - V_A}{R} + \frac{V_B - V_A}{R} + (V_o - V_A)sC = 0 \qquad (2)$$

Use $V_B = \tfrac{1}{2} V_o$ from (1) and (2) and rearrange to find

$$V_A = \frac{(sRC + \tfrac{1}{2})V_o + V_1}{sRC + 2} \qquad (3)$$

For negligible op amp input current, voltage division leads to

$$V_B = \frac{1}{2} V_o = \frac{1/sC}{R + 1/sC} V_A \qquad (4)$$

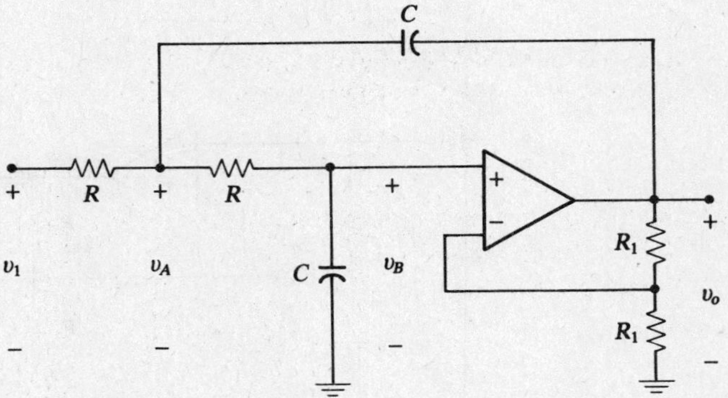

Fig. 13-57

Substitute (3) into (4) and solve for the transfer function

$$\frac{V_o}{V_1} = \frac{2/(RC)^2}{s^2 + (1/RC)s + (1/RC)^2} \tag{5}$$

13.98 Sketch the asymptotic Bode plot (magnitude only) for the low-pass filter of Problem 13.97.

▌ Let $\omega_n \equiv 1/RC$ and proceed in a manner similar to the work of Problem 13.94 to find

$$\frac{V_o}{V_1} = T(j\omega) = \frac{2}{1 - (\omega/\omega_n)^2 + \omega/\omega_n} \tag{1}$$

Based on Definition 3 of Problem 10.4, the amplitude ratio is found as

$$M_{db} = 20 \log |T(j\omega)| = 20 \log (2) + 20 \log \left[\frac{1}{(1 - \omega^2/\omega_n^2)^2 + (\omega/\omega_n)^2} \right]^{1/2} \tag{2}$$

The first term on the right-hand side of (2) represents a constant component of the Bode plot with value $2 \log (2) \approx 6$ db; the second term is a low-pass filter term with a value of 0 db for $\omega/\omega_n \leq 1$ and a slope of -40 db/decade for $\omega/\omega_n > 1$. See Fig. 13-58 for the resulting Bode plot.

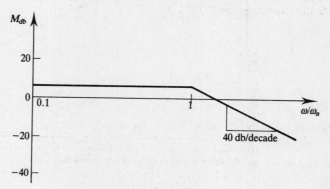

Fig. 13-58

13.99 For the op amp circuit of Fig. 13-59, find the transfer function relating v_1 and v_o and classify the circuit.

▌ Since the op amp circuit relating v_A and v_o is a noninverting amplifier, application of (1) from Problem 13.10 gives

$$V_o = \left(1 + \frac{R_1}{R_1}\right) V_A = 2V_A \tag{1}$$

By voltage division assuming negligible op amp input current,

$$V_A = \frac{R}{R + 1/sC} V_1 = \frac{sRC}{sRC + 1} V_1 \tag{2}$$

Solve (1) for V_A and substitute the result into (2) to find

$$V_A = \frac{1}{2} V_o = \frac{sRC}{sRC + 1} V_1 \tag{3}$$

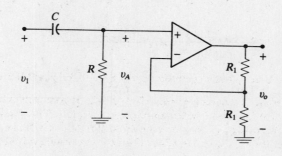

Fig. 13-59

Rearrangement of (3) leads to the transfer function

$$\frac{V_o}{V_1} = \frac{s2RC}{sRC + 1} \tag{4}$$

Equation (4) shows the amplifier to be a high-pass filter with $\omega_L = 1/RC$.

13.100 Sketch the asymptotic Bode plot (magnitude only) for the high-pass filter of Problem 13.99.

▎ Replace s by $j\omega$ in (4) of Problem 13.99 and apply Definition 3 of Problem 10.4 to find

$$M_{db} = 20 \log (2) + 20 \log |j\omega RC| + 20 \log \left|\frac{1}{j\omega RC + 1}\right| \tag{1}$$

The three terms on the right-hand side of (1) yield, respectively, the following components of the Bode plot:
1. Constant value of $20 \log (2) \approx 6$ db.
2. A straight line segment of 20 db/decade slope passing through $M_{db} = 0$ and $\omega = 1/RC$.
3. A low-pass filter term with a value of 0 db for $\omega \le 1/RC$ and a slope of -20 db/decade for $\omega > 1/RC$.

The resulting Bode plot is sketched in Fig. 13-60.

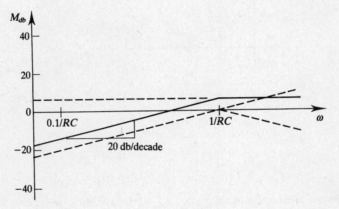

Fig. 13-60

13.101 Use the low-pass filter of Fig. 13-55 and the high-pass filter of Fig. 13-59 to design a band-pass filter where the pass band ranges from $f_1 = 1$ kHz to $f_2 = 50$ kHz.

▎ The two filters are cascade-connected as shown in Fig. 13-61 where the low-frequency cutoff for the high-pass filter is set for $f_2 = f_L = 1/2\pi R_L C_L$ and the high-frequency cutoff for the low-pass filter is set for $f_1 = f_H = 1/2\pi R_H C_H$.

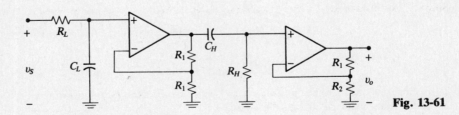

Fig. 13-61

13.102 Sketch the asymptotic Bode plot (magnitude only) for the band-pass filter of Problem 13.101.

▎ Since the amplifiers are cascaded (see Fig. 13-61), their overall gain is the product of individual amplifier gains in the time domain or the addition of their amplitude ratio plots. The resulting Bode plot is a composite of the low-pass filter Bode plot of Fig. 13-56 with $1/2\pi RC$ set for $f_2 = 50$ kHz, and the high-pass filter Bode plot of Fig. 13-60 with $1/2\pi RC$ set for $f_1 = 1$ kHz. See Fig. 13-62 for the resulting Bode plot.

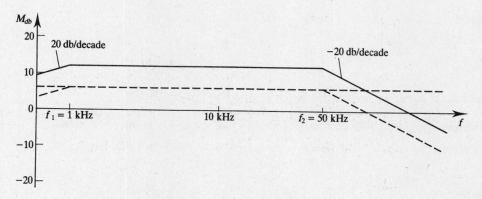

Fig. 13-62

13.103 Use the low-pass filter of Fig. 13-55 and the high-pass filter of Fig. 13-59 to design a notch or band-reject filter where the stop band ranges from $f_1 = 1$ kHz to $f_2 = 50$ kHz.

▌ The two filters are parallel-connected as shown in Fig. 13-63, where the high-frequency cutoff for the low-pass filter is set at $f_1 = f_H = 1/2\pi R_L C_L$ and the high-pass filter cutoff is set at $f_2 = f_L = 1/2\pi R_H C_H$. The final op amp stage acts as a summer.

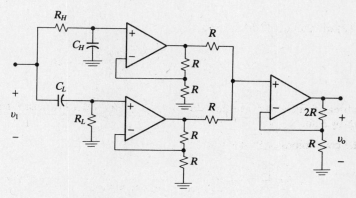

Fig. 13-63

13.104 Sketch the asymptotic Bode plot (magnitude only) for the band-reject filter of Problem 13.103.

▌ Using (2) of Problem 13.25, (4) of Problem 13.95, and (4) of Problem 13.99, the overall transfer function for the circuit of Fig. 13-63 is

$$\frac{V_o}{V_1} = \frac{2}{sR_L C_L + 1} + \frac{s2R_H C_H}{sR_H C_H + 1}$$

$$= \frac{2R_L C_L R_H C_H [s^2 + (1/R_L C_L + 1/R_H C_H)s + 1/R_L C_L R_H C_H]}{(sR_L C_L + 1)(sR_H C_H + 1)} \quad (1)$$

Replace s with $j\omega$ in (1), let $\omega_n \equiv \sqrt{1/R_L C_L R_H C_H}$, and apply Definition 3 of Problem 10.4 to find

$$M_{db} = 20 \log(2) + 20 \log[(1 - \omega^2/\omega_n^2)^2 + (2\zeta\omega/\omega_n)^2]^{1/2}$$

$$+ 20 \log \left| \frac{1}{j\omega R_L C_L + 1} \right| + 20 \log \left| \frac{1}{j\omega R_H C_H + 1} \right| \quad (2)$$

The four terms on the right-hand side of (2) yield, respectively, the following components of the Bode plot:
1. Constant value of $20 \log(2) \approx 6$ db.
2. A piecewise linear plot with constant value of 0 db for $\omega/\omega_n \leq 1$ and with increasing gain of 40 db/decade for $\omega/\omega_n > 1$.
3. A low-pass filter term with constant value of $M_{db} = 0$ to $\omega/\omega_n = \sqrt{R_H C_H / R_L C_L}$ and a slope of -20 db/decade for $\omega/\omega_n > \sqrt{R_H C_H / R_L C_L}$.
4. A low-pass filter term with constant value of $M_{dp} = 0$ to $\omega/\omega_n = \sqrt{R_L C_L / R_H C_H}$ and a slope of -20 db/decade for $\omega/\omega_n > \sqrt{R_L C_L / R_H C_H}$.

The composite sketch of the Bode plot is shown by Fig. 13-64.

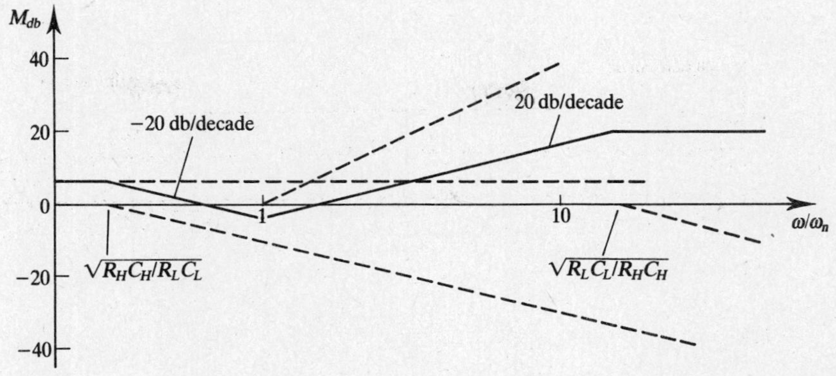

Fig. 13-64

Review Problems

13.105 In the first-order low-pass filter of Problem 13.84, $R = 10 \text{ k}\Omega$, $R_1 = 1 \text{ k}\Omega$, and $C = 0.1 \text{ }\mu\text{F}$. Find the gain for dc signals.

▌ Evaluating (2) of Problem 13.84 at $s = 0$,

$$A_v(0) = -\frac{R}{R_1} = -\frac{10 \times 10^3}{1 \times 10^3} = -10$$

13.106 Determine the break frequency f_1 at which the gain of the first-order low-pass filter with parameters of Problem 13.105 drops off by 3 db.

▌ Based on the conclusions of Problem 13.85,

$$f_1 = \frac{\omega_1}{2\pi} = \frac{1/\tau}{2\pi} = \frac{1}{2\pi RC} = \frac{1}{2\pi(10 \times 10^4)(0.1 \times 10^{-6})} = 159.2 \text{ Hz}$$

13.107 For the low-pass filter of Problem 13.105, determine the frequency f_u at which the gain has dropped to unity (called the *unity-gain bandwidth*).

▌ Set $s = j\omega_u$ in (2) of Problem 13.84 and solve for $f_u = \omega_u/2\pi$ at which $|A(j\omega_u)| = 1$.

$$|A(j\omega_u)| = 1 = \frac{|-R/R_1|}{|j\omega_u RC + 1|} = \frac{R/R_1}{[(\omega_u RC)^2 + 1]^{1/2}} \quad (1)$$

Solving (1) for f_u gives

$$f_u = \frac{\omega_u}{2\pi} = \left[\left(\frac{1}{R_1 C}\right)^2 + \left(\frac{1}{RC}\right)^2\right]^{1/2} = \left[\left(\frac{1}{(1 \times 10^3)(0.1 \times 10^{-6})}\right)^2 + \left(\frac{1}{(10 \times 10^3)(0.1 \times 10^{-6})}\right)^2\right]^{1/2} = 1583.6 \text{ Hz}$$

13.108 Describe the transfer characteristic of the level-clamp circuit of Fig. 13-35(a) if diode Z_2 is shorted.

▌ The analysis would differ from that of Problem 13.67 for $v_o > 0$. If v_o attempted a positive excursion, Z_1 conducts, limiting v_o to zero. Hence, the transfer characteristic is given by Fig. 13-35(b) with $V_{Z2} = 0$.

13.109 Find the gain of the inverting amplifier of Fig. 13-65 if the op amp and diodes are ideal.

▌ If $v_S > 0$, then $v_o < 0$, D_3 blocks, and D_2 conducts. If $v_S < 0$, then $v_o > 0$, D_2 blocks, and D_3 conducts. Consequently,

$$A_v = \begin{cases} -R_2/R_1 & \text{for } v_S > 0 \\ -R_3/R_1 & \text{for } v_S < 0 \end{cases}$$

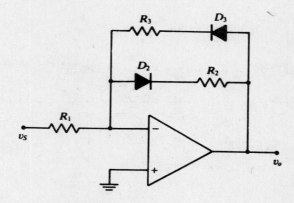

Fig. 13-65

13.110 Sketch the transfer characteristic for the amplifier of Problem 13.109. Assume $R_2 > R_3$.

❙ Since $R_2 > R_3$, the gain for $v_S > 0$ is greater than the gain for $v_S < 0$. The resulting transfer characteristic is presented by Fig. 13-66.

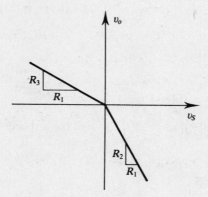

Fig. 13-66

13.111 The op amp in the circuit of Fig. 13-67 is ideal. Find an expression for v_o in terms of v_S, and determine the function of the circuit.

❙ Since $v_d \approx 0$ and the op amp input current is negligible, voltage division yields

$$v_2 = v_1 = \frac{R_2}{R_2 + R_2} v_o = \frac{v_o}{2} \tag{1}$$

Finding current i_F and using (1) leads to

$$i_F = \frac{v_2 - v_o}{R_1} = \frac{v_1 - v_o}{R_1} = \frac{v_o/2 - v_o}{R_1} = -\frac{v_o}{2R_1} \tag{2}$$

Writing an expression for i_S and using (1) give

$$i_S = \frac{v_S - v_2}{R_1} = \frac{v_S - v_o/2}{R_1} = \frac{v_S}{R_1} - \frac{v_o}{2R_1} \tag{3}$$

KCL at the input node requires that

$$i_C = i_S - i_F \tag{4}$$

Substitute (2) and (3) into (4) and simplify to get

$$i_S = \frac{v_S}{R_1} - \frac{v_o}{2R_1} - \left(-\frac{v_o}{2R_1}\right) = \frac{v_S}{R_1} \tag{5}$$

By use of (5),

$$v_2 = \frac{1}{C} \int i_C \, dt = \frac{1}{R_1 C} \int v_s \, dt \tag{6}$$

Based on (1) and (6),

$$v_o = 2v_2 = \frac{2}{R_1 C} \int v_S \, dt \tag{7}$$

From (7), it is apparent that the circuit is a noninverting integrator.

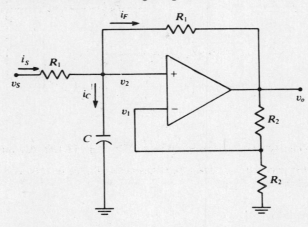

Fig. 13-67

13.112 How can the square-wave generator of Problem 13.70 be used to make a triangular-wave generator?

■ Since the integral of a square wave is a triangular wave, cascade the integrator of Fig. 13-12 to the output of the square-wave generator.

13.113 Describe an op amp circuit that will simulate the equation $3v_1 + 2v_2 + v_3 = v_o$.

■ One arrangement would be to use the summer of Fig. 13-4, with $R_F/R_1 = 3$, $R_F/R_2 = 2$, and $R_F/R_3 = 1$, cascaded into the inverting amplifier of Fig. 13-2, with $R_F/R_1 = 1$.

13.114 The circuit of Fig. 13-68 (called a *gyrator*) can be used to simulate an inductor in active RC filter design. Assuming ideal op amps, find the s-domain input impedance $Z(s)$ for the circuit.

■ The left-hand op amp circuit is a noninverting amplifier; hence,

$$V_{o1} = \left(1 + \frac{R_4}{R_3}\right) V_1 \tag{1}$$

Since $v_d = 0$ for the right-hand op amp,

$$V_i = V_n = V_1 \tag{2}$$

Since the noninverting input terminal of the right-hand op amp draws negligible current, KCL yields

$$\frac{V_{o1} - V_i}{R_1} + (V_o - V_i)sC = 0 \tag{3}$$

Substitute (1) and (2) into (3) and rearrange to find

$$V_1 = \frac{sR_1 R_3 C}{R_4}(V_1 - V_o) \tag{4}$$

Since the noninverting input of the right-hand op amp draws negligible current,

$$I_1 = I_2 = \frac{V_1 - V_o}{R_2} \tag{5}$$

The ratio (4) and (5) determines the input impedance as

$$Z(s) = \frac{V_1}{I_1} = \frac{sR_1 R_2 R_3 C}{R_4} \tag{6}$$

Clearly, from (6), the circuit simulates an inductance

$$L_{eq} = \frac{R_1 R_2 R_3 C}{R_4} \tag{7}$$

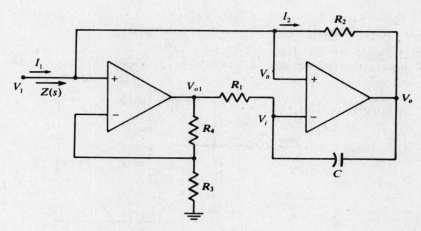

Fig. 13-68

13.115 Find the value of inductance simulated by the gyrator of Fig. 13-68 if $C = 10$ nF, $R_1 = 20$ kΩ, $R_2 = 100$ kΩ, and $R_3 = R_4 = 10$ kΩ.

▌ By (7) of Problem 13.114,

$$L_{eq} = \frac{R_1 R_2 R_3 C}{R_4} = \frac{(20 \times 10^3)(100 \times 10^3)(10 \times 10^3)(10 \times 10^{-9})}{10 \times 10^3} = 2 \text{ H}$$

13.116 For the double integrator circuit of Problem 13.33, if the output is connected to the input so that $v_i = v_o$, an oscillator is formed. Show that this claim is so, and that the frequency of oscillation is $f = 1/\pi RC$ Hz.

▌ Replace V_o by V_i in (1) of Problem 13.33 to find

$$V_i = -\frac{1}{s^2 (RC/2)^2} V_i \tag{1}$$

Rearranging (1),

$$\left[\frac{s^2 (RC/2)^2 + 1}{s^2 (RC/2)^2}\right] V_i = 0 \tag{2}$$

Since, in general, $V_i \neq 0$ the first term of (2) must be zero. Hence,

$$S_{1,2} = \pm j \frac{2}{RC} = \pm j\omega \tag{3}$$

Thus, the circuit must operate without damping at a frequency given by

$$f = \frac{\omega}{2\pi} = \frac{1}{\pi RC}$$

13.117 In the logarithmic amplifier of Fig. 13-30, let $v_S = 5$ V, $V_R = 10$ V, $R_1 = 1$ kΩ, $R_2 = 10$ kΩ, $R_3 = 1$ kΩ, and $R_4 = 50$ kΩ. The matched BJTs are operating at 25°C, with $V_T = 0.026$ V. Find v_2.

▌ Based on (4) and (5) of Problem 13.62,

$$I_{C2} \approx \frac{V_R}{R_2} = \frac{10}{10 \times 10^3} = 1 \text{ mA} \quad \text{and} \quad I_{C1} \approx \frac{v_S}{R_1} = \frac{5}{1 \times 10^3} = 5 \text{ mA}$$

By (3) of Problem 13.62,

$$v_2 = -V_T \ln\left(\frac{I_{C1}}{I_{C2}}\right) = -0.026 \ln\left(\frac{5 \times 10^{-3}}{1 \times 10^{-3}}\right) = -41.8 \text{ mV}$$

13.118 Find v_o for the logarithmic amplifier described by Problem 13.117.

▌ Based on (6) of Problem 13.62 and the result of Problem 13.117,

$$v_o = \frac{R_3 + R_4}{R_3} v_2 = \frac{1 \times 10^3 + 50 \times 10^3}{1 \times 10^3} (-41.8 \times 10^{-3}) = -2.13 \text{ V}$$

13.119 Having at your disposal a logarithmic amplifier and an exponential amplifier, devise a circuit that will produce the quotient of two numbers.

▮ Recall that $x/y = e^{\ln x - \ln y}$. Thus, if the divisor and the dividend are each fed to a logarithmic amplifier, the two resulting signals are fed to a difference amplifier, and the difference is fed to an exponential amplifier, the result is identical to division. See Fig. 13-69 for connection of the circuit.

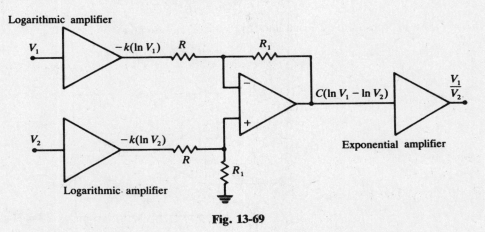

Fig. 13-69

13.120 The circuit of Fig. 13-52 is to be used as a high-pass filter having a gain of 0.1 at low frequencies, unity gain at high frequencies, and a gain of 0.707 at 1 k rad/s. Arbitrarily select $C_1 = C_2 = 0.1\ \mu F$, and size R_1 and R_2.

▮ The high-frequency gain follows from (1) of Problem 13.89 as $s \to \infty$, or

$$|T(\infty)| = C_1/C_2$$

Hence, with $C_1 = C_2$, the unity gain for high frequency is satisfied. The 3-db gain of 0.707 is attained if $1/\tau_2 = 1/R_2C_2 = 1$ k rad/s. Thus,

$$R_2 = \frac{1}{C_2(1 \times 10^3)} = \frac{1}{(0.1 \times 10^{-6})(1 \times 10^3)} = 10\ k\Omega$$

The low-frequency gain follows from (1) of Problem 13.89 with $s = 0$, or

$$|T(o)| = \frac{C_1 \tau_2}{C_2 \tau_1} = \frac{C_1(R_2C_2)}{C_2(R_1C_1)} = \frac{R_2}{R_1} = 0.1$$

Whence, $$R_1 = R_2/0.1 = 10 \times 10^3/0.1 = 100\ k\Omega$$

13.121 Find the transfer function for the circuit of Fig. 13-70, and explain the use of the circuit.

▮ Recognize the op amp circuit relating v_C and v_o is a unity follower, or $v_o = v_C$. Since the op amp noninverting input terminal draws negligible current, voltage division yields

$$V_o = V_C = \frac{1/sC}{R + 1/sC} V_i$$

or

$$T(s) = \frac{V_o}{V_i} = \frac{1}{sRC + 1} \quad (1)$$

From (1), it is apparent that the circuit is a low-pass filter.

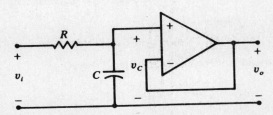

Fig. 13-70

13.122 For the circuit of Fig. 13-71, show that $I_o = -(1 + R_1/R_2)I_i$, so that the circuit is a true current amplifier.

▌ Since $v_d = 0$,
$$I_i = v_o/R_1 \tag{1}$$

By KCL applied at the output node,
$$I_o = -I_i + v_o/R_2 \tag{2}$$

Solve (1) for v_o, substitute the result into (2), and rearrange to find
$$I_o = -(1 + R_1/R_2)I_i \tag{3}$$

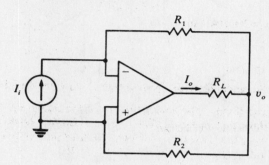

Fig. 13-71

13.123 If the noninverting terminal of the op amp in Fig. 13-8 is grounded, find the transfer function v_o/v_i.

▌ Due to virtual ground of inverting terminal when the noninverting terminal is grounded,
$$i_2 = i_1 = v_i/R \tag{1}$$

Since the inverting terminal of the op amp draws negligible current,
$$v_o = -i_1 R \tag{2}$$

Substitute (1) into (2) to find directly
$$v_o = -(v_i/R)R = -v_i \quad \text{or} \quad v_o/v_i = -1 \tag{3}$$

13.124 Devise a method for using the inverting op amp circuit of Fig. 13-2 as a current source.

▌ Let I_F be the output current; then $i_F = i_1 = v_S/R_1$ regardless of the value of R_F.

13.125 A noninverting amplifier with gain $A_v = 21$ is desired. Based on ideal op amp theory, values of $R_1 = 10$ kΩ and $R_2 = 200$ kΩ are selected for the circuit of Fig. 13-3. If the op amp is recognized as nonideal in that $A_{OL} = -10^4$ and $\text{CMMR}_{db} = 40$ db, find the actual gain $A_v = v_o/v_2$.

▌ By (2) and (3) of Problem 13.14,
$$\text{CMMR} = 10^{\text{CMMR}_{db}/20} = 10^{40/20} = 100$$
$$A_{cm} = \frac{A_{OL}}{\text{CMMR}} = \frac{-10^4}{100} = -100$$

Then, based on (6) of Problem 13.14,
$$A_v = \frac{v_o}{v_2} = \frac{-(A_{OL} + A_{cm})}{1 - A_{OL}R_1/(R_1 + R_2)} = \frac{-(-10^4 - 100)}{1 - (-10^4)(10 \times 10^3)/(10 \times 10^3 + 200 \times 10^3)} = 21.17$$

13.126 Show that if $1/RC \leq 2\omega/10$, the op amp circuit of Fig. 13-72 is a power factor detector.

▌ The analog multiplier output is
$$v_P = v_1 v_2 = A\cos(\omega t)B\cos(\omega t + \phi) \tag{1}$$

Apply the trigonometric identity $\cos A \cos B = \tfrac{1}{2}[\cos(A+B) + \cos(A-B)]$ to (1).
$$v_P = \frac{AB}{2}[\cos(2\omega t + \phi) + \cos\phi] \tag{2}$$

Since $1/RC \leq 2\omega/10$, the low-pass filter will effectively remove the $\cos(2\omega t + \phi)$ term from (2), and

$$v_o = \frac{AB}{2} \cos \phi \qquad (3)$$

If v_1 were the voltage across the input of a two-port network and v_2 were proportional to the current flowing into the network, then, based on (3), the output v_o is directly proportional to the power factor of the network.

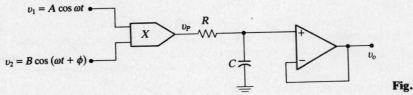

Fig. 13-72

13.127 Use an analog multiplier and a filter circuit to design a frequency doubler circuit if $v_S = V_m \cos \omega t$ is available.

▌ If v_S were fed to both inputs of an analog multiplier, the output $v_P = v_S^2 = V_m^2 \cos^2 \omega t = \frac{1}{2}V_m^2(1 + \cos 2\omega t)$. If v_P were then passed through a high-pass filter with $\omega_L < 2\omega/10$, the dc component of v_P is removed (filtered), leaving an output of twice the frequency of the input source v_S. A circuit to implement the described scheme is shown by Fig. 13-73; the high-pass filter is that of Fig. 13-59.

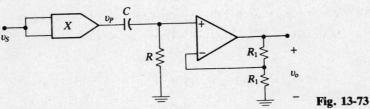

Fig. 13-73

CHAPTER 14
Switching Circuits for Digital Logic

BINARY FUNCTIONS

14.1 Since binary variables can take on only two levels or values (0 or 1), it is convenient to show the functional relationship between one independent binary variable and the dependent binary variable, formed by action of a bistate switching circuit, in tabular form (called a *truth table*). If A is a binary variable that is fed to a bistate switching circuit with output $x = b_1(A)$, discuss the possible results for x.

▌ There are only two possible cases for processing of the input signal by the bistate switching circuit:
  Case I. $x = b_{11}(A) = A$, or the signal remains in the state that it occupied when entering the circuit. In this case, the circuit is called a *buffer*.
  Case II. $x = b_{12}(A) = \bar{A}$ (read as A NOT or A complement), or the signal exits the bistate switching circuit in the opposite state from which it entered. Such a circuit is called an *inverter*.

14.2 Generate the truth tables for the two logic circuits of Problem 14.1.

▌ See Tables 14-1 and 14-2 for the truth tables of the buffer and the inverter, respectively.

TABLE 14-1
$x = b_{11}(A)$

A	x
0	0
1	1

TABLE 14-2
$x = b_{12}(A)$

A	x
0	1
1	0

14.3 Draw a general truth table for a circuit that processes two input binary variables (A and B) to produce a single binary output, $x = b_2(A, B)$.

▌ Since there are two inputs, each of which could be one of two states, there are $2^2 = 4$ possible combinations of input variables to form output x. The resulting truth table is shown by Table 14-3 where each x_i could be a 0 or 1.

TABLE 14-3
$x = b_2(A, B)$

A	B	x
0	0	x_1
0	1	x_2
1	0	x_3
1	1	x_4

14.4 How many different two-input, one-output binary logic circuits (*logic gates*) is it possible to construct?

▌ Referring to Table 14-3, there are four possible outputs from each two-input, one-output binary logic circuit. Since there are two possible states for each x_i ($i = 1, 2, 3, 4$), a total of $2^4 = 16$ possible output combinations are possible. Out of these sixteen possible logic circuits, only six find a significant use.

14.5 One of the common two-input, one-output binary logic gates (called an *AND gate*) gives an output $x = 1$ if and only if $A = B = 1$. Construct the truth table for a two-input, one-output AND gate.

▌ Based on the notation of Problem 14.3,

$$x_1 = b_{21}(0, 0) = 0 \tag{1}$$

$$x_2 = b_{21}(0, 1) = 0 \tag{2}$$

$$x_3 = b_{21}(1, 0) = 0 \tag{3}$$
$$x_4 = b_{21}(1, 0) = 1 \tag{4}$$

The truth table corresponding to (1)–(4) is given by Table 14-4.

TABLE 14-4 AND Gate

A	B	x
0	0	0
0	1	0
1	0	0
1	1	1

14.6 The AND gate described by Problem 14.5 uses *positive logic*. A *negative-logic* AND gate gives an output $x = 0$ if and only if $A = B = 0$. Describe the truth table for the negative-logic, two-input, one-output AND gate.

▌ Proceeding in a manner similar to Problem 14.5, but replacing each binary number by its *complement*, or opposite state,

$$x_1 = \bar{b}_{21}(1, 1) = 1 \tag{1}$$
$$x_2 = \bar{b}_{21}(1, 0) = 1 \tag{2}$$
$$x_3 = \bar{b}_{21}(0, 1) = 1 \tag{3}$$
$$x_4 = \bar{b}_{21}(0, 0) = 0 \tag{4}$$

Hence, the truth table, given by Table 14-5, is everywhere the complement of Table 14-4.

TABLE 14-5 Negative Logic AND Gate

A	B	x
1	1	1
1	0	1
0	1	1
0	0	0

Positive logic is more commonly used than negative logic and it is so assumed unless a statement to the contrary is made.

14.7 Another of the common two-input, one-output binary logic circuits (called an *OR gate*) gives an output $x = 0$ if and only if $A = B = 0$. Construct the truth table for the two-input OR gate.

▌ Based on the notation of Problem 14.3,

$$x_1 = b_{22}(0, 0) = 0 \tag{1}$$
$$x_2 = b_{22}(0, 1) = 1 \tag{2}$$
$$x_3 = b_{22}(1, 0) = 1 \tag{3}$$
$$x_4 = b_{22}(1, 1) = 1 \tag{4}$$

The truth table corresponding to (1)–(4) is given by Table 14-6.

TABLE 14-6 OR Gate

A	B	x
0	0	0
0	1	1
1	0	1
1	1	1

14.8 A common two-input, one-output binary logic circuit (called a *NAND gate*) gives an output $x = 0$ if and only if $A = B = 1$. Construct the truth table for this NAND gate.

▎ Based on the notation of Problem 14.3,

$$x_1 = b_{23}(0, 0) = 1 \qquad (1)$$

$$x_2 = b_{23}(0, 1) = 1 \qquad (2)$$

$$x_3 = b_{23}(1, 0) = 1 \qquad (3)$$

$$x_4 = b_{23}(1, 1) = 0 \qquad (4)$$

The truth table for (1)–(4) is presented by Table 14-7.

TABLE 14-7
NAND Gate

A	B	x
0	0	1
0	1	1
1	0	1
1	1	0

14.9 The two-input, one-output binary logic gate that gives an output $x = 1$ if and only if $A = B = 0$ is called a *NOR gate*. Construct the truth table.

▎ Using the notation of Problem 14.3,

$$x_1 = b_{24}(0, 0) = 1 \qquad (1)$$

$$x_2 = b_{24}(0, 1) = 0 \qquad (2)$$

$$x_3 = b_{24}(1, 0) = 0 \qquad (3)$$

$$x_4 = b_{24}(1, 1) = 0 \qquad (4)$$

See Table 14-8 for the truth table corresponding to (1)–(4).

TABLE 14-8   NOR Gate

A	B	x
0	0	1
0	1	0
1	0	0
1	1	0

14.10 The *exclusive-OR* (XOR) gate is a two-input, one-output binary logic circuit that gives an output $x = 0$ if and only if $A = B$. Construct the truth table.

▎ By the notation of Problem 14.3,

$$x_1 = b_{25}(0, 0) = 0 \qquad (1)$$

$$x_2 = b_{25}(0, 1) = 1 \qquad (2)$$

$$x_3 = b_{25}(1, 0) = 1 \qquad (3)$$

$$x_4 = b_{25}(1, 1) = 0 \qquad (4)$$

The truth table corresponding to (1)–(4) is given by Table 14-9.

TABLE 14-9   XOR Gate

A	B	x
0	0	0
0	1	1
1	0	1
1	1	0

14.11 The *exclusive-NOR* (XNOR) gate is a two-input, one output binary logic circuit that gives an output $x = 1$ if and only if $A = B$. Construct the truth table for the XNOR gate.

| Based on the notation of Problem 14.3,

$$x_1 = b_{26}(0, 0) = 1 \tag{1}$$
$$x_2 = b_{26}(0, 1) = 0 \tag{2}$$
$$x_3 = b_{26}(1, 0) = 0 \tag{3}$$
$$x_4 = b_{26}(1, 1) = 1 \tag{4}$$

See Table 14-10 for the associated truth table.

TABLE 14-10
XNOR Gate

A	B	x
0	0	1
0	1	0
1	0	0
1	1	1

14.12 Use two switches (A and B), a battery, and a light bulb to illustrate the AND gate truth table.

| Let an open (closed) switch represent $A = 0$ ($= 1$) and a completed circuit so that the light bulb when illuminated represents $x = 1$; then Table 14-4 is satisfied by the circuit of Fig. 14-1.

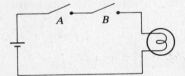

Fig. 14-1

14.13 Connect two switches (A and B), a battery, and a light bulb to illustrate the OR gate truth table.

| Let an open (closed) switch represent $A = 0$ ($= 1$) and a completed circuit so that the light bulb when illuminated represents $x = 1$; then Table 14-6 is satisfied by the parallel-connected switch arrangement of Fig. 14-2.

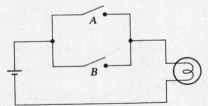

Fig. 14-2

14.14 Connect two switches (A and B), a battery, and a light bulb to illustrate the NAND gate truth table.

| Let an open (closed) switch represent $A = 0$ ($= 1$) and an illuminated light bulb represent $x = 1$. Then Table 14-7 is satisfied by the series-connected switches in parallel with the light bulb as shown by Fig. 14-3. The small resistor R is added to limit the current supplied by the battery when both switches are closed.

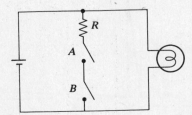

Fig. 14-3

14.15 Use two switches (A and B), a battery, and a light bulb to illustrate the NOR gate truth table.

▮ Let an open (closed) switch represent $A = 0$ ($= 1$) and an illuminated light bulb represent $x = 1$. Then, Table 14-8 is satisfied by the parallel switch and light bulb arrangement of Fig. 14-4. The small resistor R is added to limit the battery current when either or both switches A and B are closed.

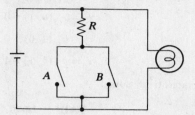

Fig. 14-4

TRANSISTOR SWITCHING CONCEPTS

14.16 Practical switching models for the BJT are derived by considering the boundaries between the saturation, active, and cutoff regions of Fig. 4-4(c) to be straight line segments, as indicated in Fig. 14-5. The typical manufacturer's specification sheet gives the value of $V_{CE\,sat}$ along the corresponding values of $i_C = I_{C\,sat}$ and $i_B = I_{B\,sat}$. The value of I_{CEO} and the associated value of $v_{CE} = V_{CEO} = V_{CE\,cutoff}$ are also given. For the case of cutoff (the OFF state), $i_B = 0$, and the BJT is modeled as a collector-to-emitter cutoff resistance

$$R_{CO} = \frac{V_{CEO}}{I_{CEO}} \tag{1}$$

as shown in Fig. 14-6(a). Saturation operation (the ON state) can be modeled by describing the base-to-emitter junction as a piecewise-linear diode and the collector-to-emitter path as a saturation resistor

$$R_{sat} = \frac{V_{CE\,sat}}{I_{C\,sat}} \tag{2}$$

as shown in Fig. 14-6(b). (The ideal diodes of Fig. 14-6 are reversed for a *pnp* device.)

The manufacturer's specification sheet for a particular Si *npn* BJT shows $I_{CBO} = 20\,\mu\text{A}$ at test condition $V_{CE} = 25$ V, $V_{CE\,sat} = 0.2$ V at test conditions $I_C = 15$ mA and $I_B = 200\,\mu\text{A}$, and $V_{BE} = 0.7$ V at test conditions $I_C = 15$ mA and $I_B = 200\,\mu\text{A}$. Determine the switching-model parameters of Fig. 14-6.

▮ Using given data,

$$\beta = \frac{I_C}{I_B} = \frac{15 \times 10^{-3}}{200 \times 10^{-6}} = 75$$

The cutoff resistance follows directly from (1):

$$R_{CO} = \frac{V_{CEO}}{I_{CEO}} = \frac{V_{CEO}}{(\beta + 1)I_{CBO}} = \frac{25}{(75 + 1)(20 \times 10^{-6})} = 16.45\,\text{k}\Omega$$

For the typical value $V_F = 0.65$ V, KVL applied to Fig. 14-6(b) requires that

$$R_F = \frac{V_{BE} - V_F}{I_{CBO}} = \frac{0.7 - 0.65}{20 \times 10^{-6}} = 2.5\,\text{k}\Omega$$

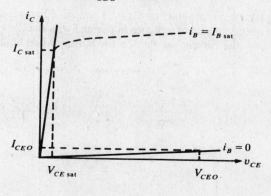

Fig. 14-5

Then, by (2)

$$R_{sat} = \frac{V_{CE\,sat}}{I_{C\,sat}} = \frac{0.2}{15 \times 10^{-3}} = 13.33\ \Omega$$

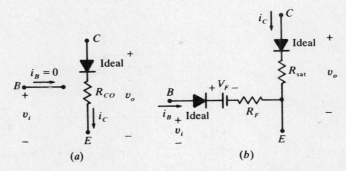

Fig. 14-6 (a) Cutoff model of *npn* BJT, (b) Saturation model of *npn* BJT.

14.17 Frequently, the error of analysis is acceptably small if the practical BJT model of Fig. 14-6 is simplified to the ideal BJT model of Fig. 14-7. Determine the implications of this ideal model with regard to the BJT characteristic curves.

▮ For the practical cutoff model of Fig. 14-6(a) to approach the ideal cutoff model of Fig. 14-7(a), R_{CO} must become infinitely large. By (1) of Problem 14.16, it is seen that I_{CEO} must be zero; hence, the $i_B = 0$ curve on the collector characteristic of Fig. 14-5 must lie along the abscissa axis.

Comparing the ideal saturation model of Fig. 14-7(b) with the nonideal model of Fig. 14-6(b), it is seen that we must require $V_F = R_F = R_{sat} = 0$. By (2) of Problem 14.16, $R_{sat} = 0$ if $V_{CE\,sat} = 0$. Thus, the saturation line of the collector characteristic in Fig. 14-5 must lie along the ordinate axis. In order for $V_F = R_F = 0$, the emitter characteristic curves [see Fig. 4-4(b)] must lie coincidentally along the ordinate axis.

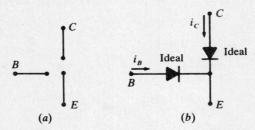

Fig. 14-7 (a) Ideal cutoff model of *npn* BJT, (b) ideal saturation model of *npn* BJT.

14.18 As a BJT approaches saturation, the collector-base junction (reverse-biased for active-region operation) becomes forward-biased, reducing the width of the collector-base depletion region. Consequently, a lower percentage of majority carriers are swept into the collector, and both α and $\beta = h_{FE} = \alpha/(1-\alpha)$ are decreased. The manufacturer's specification sheet for switching transistors gives a value $h_{FE\,min}$, measured at or near saturation, that is valid for relating collector and base currents for saturation operation. For the circuit of Fig. 14-8, $V_{CC} = 15$ V and the transistor is to be modeled with Fig. 14-7(b). Determine the minimum value of $i_B = I_{B\,sat}$ if $h_{FE\,min} = 30$ and $R_L = 1$ kΩ.

▮ The model of Fig. 14-7(b) is based on the assumption that $V_{CE\,sat} = 0$; hence, for the saturation condition,

$$I_{C\,sat} = \frac{V_{CC}}{R_L} = \frac{15}{1 \times 10^3} = 15\ mA$$

and the minimum base current is

$$I_{B\,sat} = \frac{I_{C\,sat}}{h_{FE\,min}} = \frac{15 \times 10^{-3}}{30} = 500\ \mu A$$

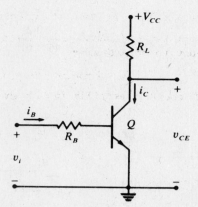

Fig. 14-8

14.19 For the circuit of Fig. 14-8, $V_{CC} = 15$ V, $h_{FE\,min} = 30$, and $i_B = 100\,\mu$A. Find the maximum value of R_L that will ensure saturation operation.

▌ The collector current at saturation is
$$I_{C\,sat} = h_{FE\,min}I_{B\,sat} = 30(100 \times 10^{-6}) = 3 \text{ mA}$$
whence
$$R_L \leq \frac{V_{CC}}{I_{C\,sat}} = \frac{15}{3 \times 10^{-3}} = 5 \text{ k}\Omega$$

14.20 The practical and ideal models of Figs. 14-6 and 14-7 describe the transistor at the initial and final conditions of a switching operation. However, the transistor must traverse the active region as it switches from cutoff to saturation, or vice versa. Since the switching operation spans more than zero time, study of the time variation of a switching transistor circuit is facilitated by construction of a (*static*) *voltage transfer characteristic* (a plot of output voltage v_o as a function of the input voltage v_i for the switching circuit). If the BJT circuit of Fig. 14-9 is to be operated in the switching mode, draw the equivalent circuits that must be used to construct the voltage transfer characteristic based on the practical BJT switching models of Fig. 14-6.

▌ See Fig. 14-10 where parts (*a*), (*b*), and (*c*) represent, respectively, the models for cutoff, active region, and saturation operation.

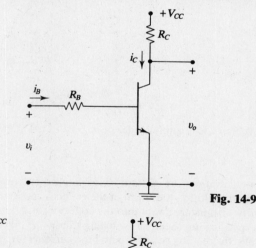

Fig. 14-9

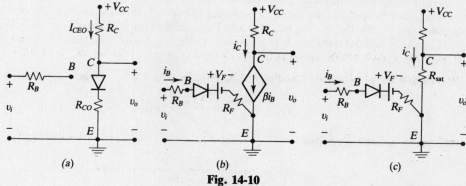

Fig. 14-10

14.21 Based on the equivalent circuits of Fig. 14-10, construct the asymptotic voltage transfer characteristic.

▎ As long as v_i is small enough that $i_B \approx 0$, the transistor remains in cutoff and the output voltage of Fig. 14-10(a) remains approximately constant with value

$$v_o = I_{CEO}R_{CO} = V_{CC} - I_{CEO}R_C \tag{1}$$

Over the active region of BJT operation, use of Fig. 14-10(b) gives

$$i_B = \frac{v_i - V_F}{R_B + R_F} \tag{2}$$

and

$$v_o = V_{CC} - \beta i_B R_C \tag{3}$$

Substitute (2) into (3) and rearrange to find

$$v_o = V_{CC} + \frac{\beta R_C}{R_B + R_F} V_F - \frac{\beta R_C}{R_B + R_F} v_i \tag{4}$$

As v_i is increased until saturation region operation prevails, the output voltage becomes independent of v_i and follows from Fig. 14-10(c), by voltage division, as

$$v_o = \frac{R_{\text{sat}}}{R_{\text{sat}} + R_C} V_{CC} \approx \frac{R_{\text{sat}}}{R_C} V_{CC} \tag{5}$$

where the approximation is based on typical values for $R_{\text{sat}} \ll R_C$.

The asymptotic voltage transfer characteristic is sketched in Fig. 14-11 where (1) applies for $v_i < v_{iC}$, (2) governs for $v_{iC} \leq v_i \leq v_{iS}$, and (3) describes v_o for $v_i > v_{iS}$. The actual curve in the neighborhood of the breakpoints is indicated, although not theoretically justified.

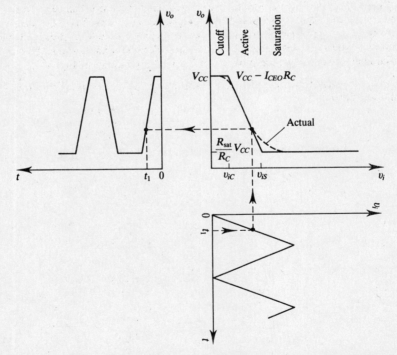

Fig. 14-11

14.22 Determine the breakpoints v_{iC} and v_{iS} for the asymptotic voltage transfer characteristic of Fig. 14-11.

▎ Rearrangement of (4) from Problem 14.21 gives

$$v_i = V_F + \frac{R_B + R_F}{\beta R_C}(V_{CC} - v_o) \tag{1}$$

Breakpoint v_{iC} is found by substituting (1) of Problem 14.21 into (1) and rearranging the result:

$$v_{iC} = V_F + \frac{R_B + R_F}{\beta} I_{CEO} \approx V_F + (R_B + R_F)I_{CBO} \tag{2}$$

The breakpoint v_{iS} follows from substitution of (5) from Problem 14.21 into (1) as

$$v_{iS} = V_F + \frac{R_B + R_F}{\beta R_C}\left(V_{CC} - \frac{R_{sat}}{R_{sat} + R_C}V_{CC}\right) = V_F + \frac{R_B + R_F}{\beta(R_C + R_{sat})}V_{CC} \qquad (3)$$

14.23 A triangular input voltage waveform such that $v_i \geq 0$ is applied to the BJT switching circuit of Fig. 14-9. The amplitude of the triangular waveform is large enough to drive the transistor into saturation. Sketch the output voltage v_o if the voltage transfer characteristic of Fig. 14-11 is applicable.

❚ Auxiliary time axes for both v_o and v_i are constructed on Fig. 14-11. Point $v_o(t_1)$ is found by entering the voltage transfer characteristic at $v_i = v_i(t_1)$ and the result is plotted at $t = t_1$ on the $v_o - t$ coordinate frame. Continuing this process, sufficient points can be generated to yield the complete $v_o(t)$ curve.

14.24 For the circuit of Fig. 14-9, let $V_{CC} = 15$ V, $R_B = 1$ kΩ, $R_C = 250$ Ω, $\beta = 30$, $R_{sat} = 5$ Ω, $I_{CEO} = 5$ μA, $R_F = 200$ Ω, $R_{CO} = 50$ kΩ, and $V_F = 0.65$ V. Sketch the asymptotic voltage transfer characteristic.

❚ The breakpoints are found by (2) and (3) of Problem 14.22:

$$v_{iC} = V_F + \frac{R_B + R_F}{\beta}I_{CEO} = 0.65 + \frac{1 \times 10^3 + 200}{30}(5 \times 10^{-6}) \approx 0.65 \text{ V}$$

$$v_{iS} = V_F + \frac{R_B + R_F}{\beta(R_C + R_{sat})}V_{CC} = 0.65 + \frac{1 \times 10^3 + 200}{30(250 + 5)}(15) \approx 3\text{V}$$

Cutoff region: By (1) of Problem 14.21,

$$v_o = V_{CC} - I_{CEO}R_C = 15 - (5 \times 10^{-6})(250) = 14.99, \qquad v_i < 0.65 \text{ V}$$

Active region: By (4) of Problem 14.21,

$$v_o = V_{CC} + \frac{\beta R_C}{R_B + R_F}V_F - \frac{\beta R_c}{R_B + R_F}v_i = 15 + \frac{30(250)}{1 \times 10^3 + 200}(0.65) - \frac{30(250)}{1 \times 10^3 + 200}v_i$$
$$= 19.063 - 6.25v_i, \qquad 0.65 \leq v_i \leq 3 \text{ V}$$

Saturation region: By (5) of Problem 14.21,

$$v_o = \frac{R_{sat}}{R_C}V_{CC} = \frac{5}{250}(15) = 0.3 \text{ V}, \qquad v_i > 3 \text{ V}$$

The resulting asymptotic voltage transfer characteristic is displayed by Fig. 14-12.

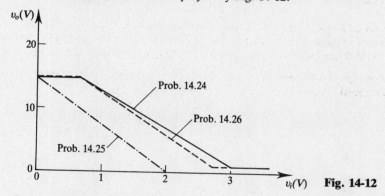

Fig. 14-12

14.25 Repeat Problem 14.24 if the BJT is ideal in that $V_F = I_{CEO} = R_F = R_{sat} = 0$. All other values are unchanged.

❚ By (2) and (3) of Problem 14.22 with the idealizing approximations implemented, the breakpoints are

$$v_{iC} = 0$$

$$v_{iS} = \frac{R_B}{\beta R_C}V_{CC} = \frac{1 \times 10^3}{30(250)}(15) = 2 \text{ V}$$

Cutoff region: By (1) of Problem 14.21,

$$v_o = V_{CC} = 15 \text{ V}, \qquad v_i < 0 \tag{1}$$

Active region: From (4) of Problem 14.21,

$$v_o = V_{CC} - \frac{\beta R_C}{R_B} v_i = 15 - \frac{30(250)}{1 \times 10^3} v_i = 15 - 7.5 v_i, \qquad 0 \le v_i \le 2 \text{ V} \tag{2}$$

Saturation region: By (5) of Problem 14.21,

$$v_o = 0, \qquad v_i > 2 \tag{3}$$

The sketch of (1)–(3) to give the idealized voltage transfer characteristic is added to Fig. 14-12. Comparing the results with those of Problem 14.25, significant difference exists between results obtained by the models of Figs. 14-6 and 14-7.

14.26 Repeat Problem 14.24 if the BJT is modeled by $I_{CEO} = R_F = 0$, $V_F = 0.7$, and all else is unchanged.

▮ Introducing the approximations in (2) and (3) of Problem 14.22, the breakpoints are

$$v_{iC} = V_F + \frac{R_B}{\beta} I_{CEO} = 0.7 + \frac{1 \times 10^3}{30}(5 \times 10^{-6}) \approx 0.7 \text{ V}$$

$$v_{iS} = V_F + \frac{R_B}{\beta(R_C + R_{sat})} V_{CC} = 0.7 + \frac{1 \times 10^3}{30(250 + 5)}(15) = 2.66 \text{ V}$$

Cutoff region: By (1) of Problem 14.21,

$$v_o = V_{CC} - I_{CEO} R_C = 15 - (5 \times 10^{-6})(250) = 14.99, \qquad v_i < 0.7 \text{ V}$$

Active region: By (4) of Problem 14.21,

$$v_o = V_{CC} + \frac{\beta R_C}{R_B} V_F - \frac{\beta R_C}{R_B} v_i = 15 + \frac{30(250)}{1 \times 10^3}(0.7) - \frac{30(250)}{1 \times 10^3} v_i = 20.25 - 4.87 v_i, \qquad 0.7 \le v_i \le 2.66 \text{ V}$$

Saturation region: By (5) of Problem 14.21,

$$v_o = \frac{R_{sat}}{R_C} V_{CC} = \frac{5}{250}(15) = 0.3 \text{ V}, \qquad v_i > 2.66 \text{ V}$$

The resulting voltage transfer characteristic is plotted in Fig. 14-12. Comparison with the results of Problem 14.24 shows that the above approximations ($I_{CEO} = R_F = 0$, $V_F = 0.7$ V) yield results that favorably compare with the more exact model of Fig. 14-6.

14.27 Several BJT switching circuits can be formed as special cases of the circuit in Fig. 14-13. If this circuit is to be operated in the switching mode, draw equivalent circuits that must be used to describe the circuit in the cutoff, active, and saturation regions. Use the practical BJT models of Fig. 14-6 with $I_{CEO} = R_F = 0$.

▮ See Fig. 14-14 where parts (a), (b) and (c) represent, respectively, the equivalent circuits for cutoff, active, and saturation region operation.

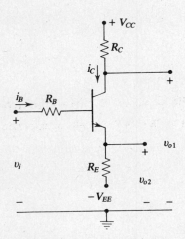

Fig. 14-13

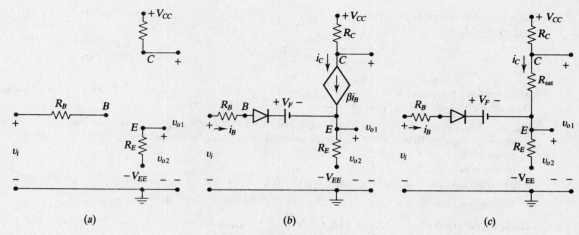

Fig. 14-14

14.28 Based on the equivalent circuits of Fig. 14-14, find the equations to describe the output voltage v_{o1} in terms of the input voltage v_i over the three regions of operation.

∎ Since the models are based on $I_{CEO} = 0$, the transition between cutoff and active region operation is characterized by $i_B = 0$. For active region operation, application of KVL around the base-emitter loop of Fig. 14-14(b) and use of $i_E = (\beta + 1)i_B$ yields

$$v_i = i_B R_B + V_F + (\beta + 1)i_B R_E - V_{EE}$$

Whence,
$$i_B = \frac{v_i + V_{EE} - V_F}{R_B + (\beta + 1)R_E} \tag{1}$$

Cutoff region: From (1) it is seen that, if $v_i < -V_{EE} + V_F$, i_B tends to have negative values and the BJT is cutoff so that $i_B = 0$. Hence, $i_C = \beta i_B = 0$, and from Fig. 14-14(a),

$$v_{o1} = V_{CC} \tag{2}$$

Active region: If $v_i \geq -V_{EE} + V_F$, it is seen from (1) that $i_B \geq 0$ and active region operation prevails as long as $v_{CE} > V_{CE\,sat}$. Applying KVL in Fig. 14-14(b) and recalling that $i_C = \beta i_B$ lead to

$$v_{o1} = V_{CC} - i_C R_C = V_{CC} - \beta i_B R_C \tag{3}$$

Substitution of (1) into (3) and rearrangement give

$$v_{o1} = V_{CC} - \frac{\beta R_C}{R_B + (\beta + 1)R_E}(V_{EE} - V_F) - \frac{\beta R_C}{R_B + (\beta + 1)R_E} v_i \tag{4}$$

Saturation region: Application of KVL around the collector-emitter loop of Fig. 14-14(c) leads to

$$i_C = I_{C\,sat} = \frac{V_{CC} + V_{EE}}{R_C + R_{sat} + R_E} \tag{5}$$

But
$$v_{o1} = V_{CC} - I_{C\,sat} R_C \tag{6}$$

Use (5) in (6) and rearrange to yield

$$v_{o1} = \frac{(R_{sat} + R_E)V_{CC} - R_C V_{EE}}{R_C + R_{sat} + R_E} \tag{7}$$

14.29 Determine the breakpoints v_{iC} and v_{iS} of the asymptotic voltage transfer characteristic, v_{o1}, vs. v_i, for the BJT switching circuit of Fig. 14-13.

∎ Rearrange (4) of Problem 14.28 to find

$$v_i = -(V_{EE} - V_F) + \frac{R_B + (\beta + 1)R_E}{\beta R_C}(V_{CC} - v_{o1}) \tag{1}$$

The breakpoint v_{iC}, boundary between cutoff and saturation region operation, follows from substitution of (2) from Problem 14.28 into (1) as

$$v_{iC} = -(V_{EE} - V_F) + \frac{R_B + (\beta+1)R_E}{\beta R_C}(V_{CC} - V_{CC}) = V_F - V_{EE} \tag{2}$$

The breakpoint v_{iS}, boundary between active and saturation region operation, is found by substitution of (7) from Problem 14.28 into (1):

$$v_{iS} = -(V_{EE} - V_F) + \frac{R_B + (\beta+1)R_E}{\beta R_C}\left[V_{CC} - \frac{(R_{sat}+R_E)V_{CC} - R_C V_{EE}}{R_C + R_{sat} + R_E}\right]$$

After simplification,

$$v_{iS} = V_F - V_{EE} + \frac{R_B + (\beta+1)R_E}{\beta(R_C + R_{sat} + R_E)}(V_{CC} + V_{EE}) \tag{3}$$

14.30 Utilizing the equivalent circuits of Fig. 14-14, find the equations to describe the output voltage v_{o2} in terms of the input voltage v_i over the three regions of operation.

Cutoff region: From (1) of Problem 14.28, if $v_i < -V_{EE} + V_F$, i_B attempts to take on negative values and the BJT is cutoff so that $i_B = 0$. Hence, $i_E = (\beta+1)i_B = 0$, and from Fig. 14-14(a)

$$v_{o2} = -V_{EE} \tag{1}$$

Active region: For $v_i \geq -V_{EE} + V_F$, it is seen from (1) of Problem 14.28 that $i_B \geq 0$ and active region operation exists as long as $v_{CE} > V_{CE\,sat}$. Applying KVL in Fig. 14-14(b) and recalling that $i_E = (\beta+1)i_B$ lead to

$$v_{o2} = -V_{EE} + i_E R_E + -V_{EE} + (\beta+1)i_B R_E \tag{2}$$

Substitution of (1) of Problem 14.28 into (2) and rearrangement give

$$v_{o2} = -V_{EE} + \frac{(\beta+1)R_E}{R_B + (\beta+1)R_E}(V_{EE} - V_F) + \frac{(\beta+1)R_E}{R_B + (\beta+1)R_E} v_i \tag{3}$$

Saturation region: Application of KVL to Fig. 14-14(c) results in

$$v_{o2} \approx -V_{EE} + I_{C\,sat} R_E \tag{4}$$

Use of (5) from Problem 14.28 in (4) and rearrangement yield

$$v_{o2} = \frac{-(R_C + R_{sat})V_{EE} + R_E V_{CC}}{R_C + R_{sat} + R_E} \tag{5}$$

14.31 Determine the breakpoints v_{iC} and v_{iS} of the asymptotic voltage transfer characteristic (v_{o2} vs. v_i) for the BJT switching circuit of Fig. 14-13.

Rearrange (3) of Problem 14.30 to find

$$v_i = -(V_{EE} - V_F) + \frac{R_B + (\beta+1)R_E}{(\beta+1)R_E}(V_{EE} + v_{o2}) \tag{1}$$

The breakpoint v_{iC} follows from substitution of (1) from Problem 14.30 into (1) as

$$v_{iC} = -(V_{EE} - V_F) + \frac{R_B + (\beta+1)R_E}{(\beta+1)R_E}(V_{EE} - V_{EE}) = V_F - V_{EE} \tag{2}$$

The breakpoint v_{iS} is found by substitution of (5) from Problem 14.30 into (1):

$$V_{iS} = -(v_{EE} - V_F) + \frac{R_B + (\beta+1)R_E}{(\beta+1)R_E}\left[V_{EE} + \frac{-(R_C+R_{sat})V_{EE} + R_E V_{CC}}{R_C + R_{sat} + R_E}\right]$$

After simplification,

$$v_{iS} = V_F - V_{EE} + \frac{R_B + (\beta+1)R_E}{(\beta+1)(R_C + R_{sat} + R_E)}(V_{EE} + V_{CC}) \tag{3}$$

14.32 For the BJT switching circuit of Fig. 14-13, let $V_{CC} = 10$ V, $V_{EE} = 0$, $R_C = R_E = 1$ kΩ, $R_B = 25$ kΩ, $\beta = 40$, $V_F = 0.7$ V, and $R_{sat} = 15$ Ω. Sketch the asymptotic voltage transfer characteristic v_{o1} vs. v_i.

The breakpoints are found by (2) and (3) of Problem 14.29 as

$$v_{iC} = 0.7 - 0 = 0.7 \text{ V}$$

$$V_{iS} = 0.7 - 0 + \frac{25 \times 10^3 + (40+1)(1 \times 10^3)}{40(1 \times 10^3 + 15 + 1 \times 10^3)}(10+0) = 8.89 \text{ V}$$

By (2), (4), and (7) of Problem 14.28, the respective expressions for v_{o1} over the cutoff, active, and saturation regions of operations are

$$v_{o1} = V_{CC} = 10 \text{ V}, \qquad v_i < 0.7 \text{ V} \tag{1}$$

$$v_{o1} = V_{CC} - \frac{\beta R_C}{R_B + (\beta+1)R_E}(V_{EE} - V_F) - \frac{\beta R_C}{R_B + (\beta+1)R_E}v_i$$

$$= 10 - \frac{40(1 \times 10^3)}{25 \times 10^3 + (40+1)(1 \times 10^3)}(0 - 0.7) - \frac{40(1 \times 10^3)}{25 \times 10^3 + (40+1)(1 \times 10^3)}v_i$$

$$= 10.424 - 0.606 v_i, \qquad 0.7 \leq v_i \leq 8.89 \text{ V} \tag{2}$$

$$v_{o1} = \frac{(R_{\text{sat}} + R_E)V_{CC} - R_C V_{EE}}{R_C + R_{\text{sat}} + R_E} = \frac{(15 + 1 \times 10^3)(10) - (1 \times 10^3)(0)}{1 \times 10^3 + 15 + 1 \times 10^3} = 5.04 \text{ V}, \qquad v_i > 8.89 \text{ V} \tag{3}$$

The voltage transfer characteristic is sketched in Fig. 14-15.

14.33 Repeat Problem 14.32 if $R_E = 0$ and all else is unchanged. Sketch the resulting voltage transfer characteristic on Fig. 14-15 to see the effect of R_E.

Proceeding as in Problem 14.32, the breakpoints are found by (2) and (3) of Problem 14.29 as

$$v_{iC} = 0.7 - 0 = 0.7 \text{ V}$$

$$v_{iS} = 0.7 - 0 + \frac{25 \times 10^3 + (40+1)(0)}{40(1 \times 10^3 + 5 + 0)}(10+0) = 6.86 \text{ V}$$

Using (2), (4), and (7) of Problem 14.28,

$$v_{o1} = V_{CC} = 10 \text{ V} \qquad v_i < 0.7 \text{ V} \tag{1}$$

$$v_{o1} = V_{CC} - \frac{\beta R_C}{R_B}(V_{EE} - V_F) - \frac{\beta R_C}{R_B}v_i = 10 - \frac{(40)(1 \times 10^3)}{25 \times 10^3}(0 - 0.7) - \frac{(40)(1 \times 10^3)}{25 \times 10^3}v_i$$

$$= 11.12 - 1.6 v_i, \qquad 0.7 \leq v_i \leq 6.86 \text{ V} \tag{2}$$

$$v_{o1} = \frac{R_{\text{sat}} V_{CC}}{R_C + R_{\text{sat}}} = \frac{(15)(10)}{1 \times 10^3 + 15} = 0.15 \text{ V}, \qquad v_i > 6.86 \text{ V} \tag{3}$$

The results are added to Fig. 14-15 where it is seen that $R_E > 0$ acts to significantly increase v_{o1} over the saturation region.

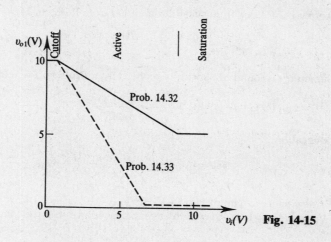

Fig. 14-15

14.34 Enhancement mode MOSFETs are used extensively in logic gate construction. However, unlike the BJT, the nonlinear nature of the MOSFET drain characteristic does not lend itself to simple mathematical analysis for the large active region signal excursion that occurs during switching operation. Hence, when needed for purposes of analysis, the voltage transfer characteristic is usually graphically determined. For the NMOS transistor switching circuit of Fig. 14-16, let $V_{DD} = 8$ V and $R_D = 1$ kΩ. If the MOSFET is described by the drain characteristic of Fig. 14-17, sketch the voltage transfer characteristic (v_o vs. v_i).

▌ The load line is constructed on the drain characteristic of Fig. 14-17 with ordinate intercept at $V_{DD}/R_D = 8/(1 \times 10^3) = 8$ mA and abscissa intercept at $V_D = 8$ V. For $v_i < V_T = 2$ V, $v_o = v_{DS} = 0$. As $v_i = v_{GS}$ is increased to values larger than 2 V, the point of operation moves up the load line, and values of $v_o = v_{DS}$ are read directly on the v_{DS} axis. For example, if $v_i = v_{GS} = 4$ V (point A), then $v_{DS} = 6$ V. The collection of points so determined are used to plot the resulting voltage transfer characteristic of Fig. 14-18.

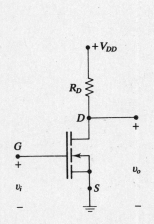

Fig. 14-16

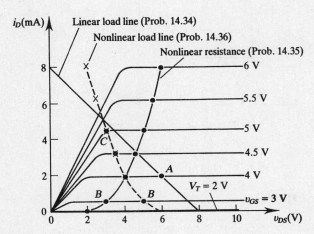

Fig. 14-17

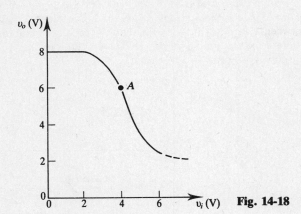

Fig. 14-18

14.35 When MOSFETs are fabricated in integrated circuits, another MOSFET is as easy to fabricate as a load resistor. Consequently, the load resistor R_D of the MOSFET switching circuit of Fig. 14-16 may be replaced by another MOS transistor such as shown in Fig. 14-19. Determine the nature of the load resistance formed by Q_1 if the drain characteristic of Fig. 14-17 is applicable.

▌ Since G_1 is connected to D_1, it is observed that $v_{GS1} = v_{DS1}$. The curve labeled nonlinear resistance on Fig. 14-17 connects all points for which $v_{GS1} = v_{DS1}$, and thus gives the $v_{DS1} - i_D$ or load characteristic of Q_1.

14.36 For the MOSFET switching circuit of Fig. 14-19, sketch the voltage transfer characteristic if both Q_1 and Q_2 are described by the drain characteristic of Fig. 14-17 and $V_{DD} = 8$ V.

▌ The load characteristic of Q_1 is determined in Problem 14.35 and is sketched on Fig. 14-17 (labeled nonlinear resistance). A load line is required to describe operation with the nonlinear resistance simulated by Q_1. By KVL,

$$v_{DS2} = V_{DD} - v_{DS1} \tag{1}$$

Further, Q_1 and Q_2 conduct the same current. Thus, (1) must be applied along lines of constant current i_D in Fig. 14-17. For example, if $i_D = 0.5$ mA, then $v_{DS1} = 3$ V (point B on the nonlinear resistance curve), and by (1) $v_{DS2} = 8 - 3 = 5$ V (point B on the nonlinear load line curve). The collection of points so determined plot out the complete nonlinear load line curve of Fig. 14-17. With the load line established, $v_i \geq 2$ V is allowed to change along the load line to determine corresponding values for $v_o = v_{DS2}$ from projection of the instantaneous operating point on the v_{DS} axis. The resulting voltage transfer characteristic is sketched in Fig. 14-20, where point C of Fig. 14-17 maps to point C of Fig. 14-20.

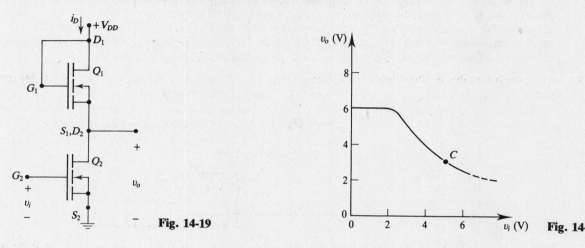

Fig. 14-19

Fig. 14-20

LOGIC FAMILIES

14.37 In practical electronic switches, the ON state consists of a range of output voltages $0 < v_o \leq V_L$ (called *low logic*), the OFF state consists of the range $V_L < V_H \leq v_o <$ power-supply voltage (called *high logic*); V_L is called the *low-logic limit*, and V_H the *high-logic threshold*. Subsequent circuitry must be capable of discriminating effectively between V_L and V_H; for good design, the indeterminate range between V_L and V_H is made as large as possible. Generate the truth table for the *diode logic* (DL) circuit of Fig. 14-21, showing that the circuit is an AND gate for positive logic.

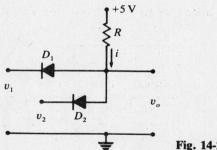

Fig. 14-21

▮ If either or both of v_1 and v_2 are logic low ($\leq V_L$), then v_o is logic low (≈ 0.7 V). If both v_1 and v_2 are logic high ($\geq V_H + 0.7$), then $v_o \geq V_H$. The truth table shows the circuit to be an AND gate.

v_1	v_2	v_o
0	0	0
0	1	0
1	0	0
1	1	1

14.38 Show that the DL circuit of Fig. 14-21 is an OR gate for negative logic.

If either or both v_1 and v_2 are equal V_L, then $v_o = V_L + 0.7 < V_H$. If both v_1 and v_2 are greater than or equal $V_H + 0.7$, then $v_o > V_H$. The truth table shows OR logic:

v_1		v_2		v_o	
actual	logic	actual	logic	actual	logic
V_L	1	V_L	1	$V_L + 0.7$	1
V_L	1	$V_H + 0.7$	0	$V_L + 0.7$	1
$V_H + 0.7$	0	V_L	1	$V_L + 0.7$	1
$V_H + 0.7$	0	$V_H + 0.7$	0	V_H	0

14.39 Set up the truth table for the diode logic circuit of Fig. 14-22, showing that the circuit is an OR gate for positive logic.

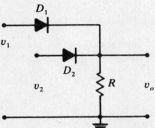

Fig. 14-22

▮ If either or both of v_1 and v_2 are logic high ($\geq V_H$), then $v_o \geq V_H - 0.7$. Otherwise, $v_o = 0$. The truth table shows OR logic.

v_1	v_2	v_o
0	0	0
0	1	1
1	0	1
1	1	1

14.40 Show that the diode logic circuit of Fig. 14-22 is an AND gate for negative logic.

▮ If either or both v_1 and v_2 are greater than V_H, then $v_o \geq V_H - 0.7$. Otherwise, $v_o = 0$. The resulting truth table shows AND operation for negative logic.

v_1		v_2		v_o	
actual	logic	actual	logic	actual	logic
V_L	1	V_L	1	0	1
V_L	1	V_H	0	$V_H - 0.7$	0
V_H	0	V_L	1	$V_H - 0.7$	0
V_H	0	V_H	0	$V_H - 0.7$	0

14.41 Through use of a truth table show that the basic *resistor transistor logic* (RTL) gate of Fig. 14-23 is a NOR gate for positive logic.

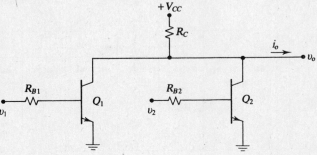

Fig. 14-23

▌ If $v_1 \geq v_H$, Q_1 is saturated and $v_o = V_{CE\,sat}$. Likewise, if $v_2 \geq V_H$, Q_2 is saturated and $v_o = V_{CE\,sat}$. Further, if Q_1 and Q_2 are both saturated, $v_o = V_{CE\,sat}$. However, if both v_1 and $v_2 < V_H$, both Q_1 and Q_2 are OFF and $v_o \approx V_{CC}$. The following truth table shows NOR logic.

v_1		v_2		v_o	
actual	logic	actual	logic	actual	logic
V_L	0	V_L	0	V_{CC}	1
V_L	0	V_H	1	$V_{CE\,sat}$	0
V_H	1	V_L	0	$V_{CE\,sat}$	0
V_H	1	V_H	1	$V_{CE\,sat}$	0

14.42 Show that the RTL gate of Fig. 14-23 is a NAND gate for negative logic.

▌ The circuit responses to actual voltages v_1 and v_2 are unchanged; however, the interpretation differs. The logic entries in the truth table of Problem 14.41 take on the logic complement values. When the resulting truth table is constructed, the NAND gate function is apparent.

v_1		v_2		v_o	
actual	logic	actual	logic	actual	logic
V_L	1	V_L	1	V_{CC}	0
V_L	1	V_H	0	$V_{CE\,sat}$	1
V_H	0	V_L	1	$V_{CE\,sat}$	1
V_H	0	V_H	0	$V_{CE\,sat}$	1

14.43 By use of a truth table, show that the circuit of Fig. 14-24 is a NOR gate if v_{o1} is taken as the output.

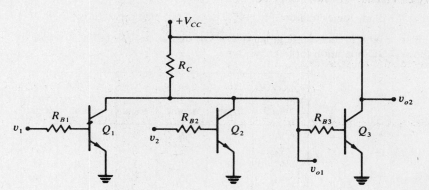

Fig. 14-24

▌ If $v_1 \geq V_H$, Q_1 is saturated and $v_{o1} = V_{CE\,sat}$. If $v_2 \geq V_H$, Q_2 is saturated and $v_{o1} = V_{CE\,sat}$. Also, if Q_1 and Q_2 are both saturated, $v_{o1} \approx V_{CE\,sat}$. Otherwise, $v_{o1} = V_{CC}$. A truth table shows NOR logic for v_{o1} as output.

v_1		v_2		v_{o1}		v_{o2}	
actual	logic	actual	logic	actual	logic	actual	logic
V_H	1	V_H	1	$V_{CE\,sat}$	0	V_{CC}	1
V_L	0	V_L	0	V_{CC}	1	$V_{CE\,sat}$	0
V_H	1	V_L	0	$V_{CE\,sat}$	0	V_{CC}	1
V_L	0	V_H	1	$V_{CE\,sat}$	0	V_{CC}	1

14.44 Show that the RTL circuit of Fig. 14-24 is an OR gate if v_{o2} is the output.

▮ Since $v_{o1} = V_{CE\,sat}$ is below the level that will render Q_3 conducting, and $v_{o1} = V_{CC}$ will drive Q_3 to saturation if R_{B3} is properly sized, the Q_3 stage is simply an inverter (a NOT gate). Thus, output v_{o2} is the *logic complement* (in which 0 s and 1 s are interchanged) of v_{o1}, as the truth table of Problem 14.42 shows, and the overall logic is that of an OR gate.

14.45 For the RTL circuit of Fig. 14-23, let $V_{CC} = 5$ V, $V_H = 3.5$ V, $R_C = 640\,\Omega$, $R_{B1} = R_{B2} = 450\,\Omega$, $V_{CE\,sat} = 0.2$ V, and $\beta = 50$. Determine the *fanout*, or number of similar gates that can be attached to v_o without risk of error in logic.

▮ For each successive gate that has a transistor in saturation, the current required is

$$I_{B\,sat} = \frac{I_{C\,sat}}{\beta} = \frac{V_{CC} - V_{CE\,sat}}{\beta R_C} = \frac{5 - 0.2}{50(640)} = 0.15 \text{ mA} \tag{1}$$

For n attached gates,

$$i_o = n I_{B\,sat} \tag{2}$$

To assure no logic error,

$$v_o = V_{CC} - i_o R_C \geq V_H = 3.5 \tag{3}$$

Substitute (2) into (3), rearrange, and use (1) to find

$$n \leq \frac{V_{CC} - 3.5}{R_C I_{B\,sat}} = \frac{5 - 3.5}{640(0.15 \times 10^{-3})} = 15.6$$

Thus, $n \leq 15$

14.46 Determine the power dissipated by the RTL, two-input NAND gate of Fig. 14-23 if the parameters of Problem 14.45 are pertinent.

▮ The worst-case condition is that both Q_1 and Q_2 are saturated due to high logic inputs. Power dissipation of the collector circuit only need be considered.

$$I_{C\,sat} = \frac{V_{CC} - V_{CE\,sat}}{R_C} = \frac{5 - 0.2}{640} = 7.5 \text{ mA}$$

Since there are two transistors,

$$\text{Total power} = 2 P_{CC} = 2 V_{CC} I_{C\,sat} = 2(5)(7.5 \times 10^{-3}) = 75 \text{ mW}$$

14.47 Use a truth table to show that the *diode transistor logic* (DTL) circuit of Fig. 14-25 is a NAND gate.

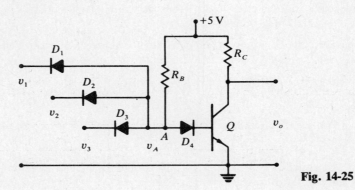

Fig. 14-25

▮ From Problem 14.37, we see that the signal at node A is the output of a three-input AND gate. Further, the transistor circuit to the right of node A forms an inverter [diode D_4 ensures that the transistor is not turned ON when $v_A \approx 0.7$ V (low logic)]. We extend the truth table of Problem 14.37 to include three inputs and the logic complement of v_A, showing the overall logic to be that of a NAND gate:

v_1	v_2	v_3	v_A	$v_o = \bar{v}_A$
0	0	0	0	1
0	0	1	0	1
0	1	0	0	1
0	1	1	0	1
1	0	0	0	1
1	0	1	0	1
1	1	0	0	1
1	1	1	1	0

14.48 Diode transistor logic (DTL) combines diodes and transistors to form logic gates. Use a truth table to show that the DTL circuit of Fig. 14-26 acts as a NOR gate.

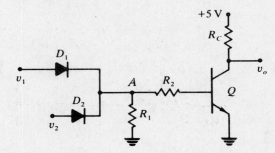

Fig. 14-26

▮ Recognize from Problem 14.39 that the signal at node A is the output of an OR gate. In addition, the transistor circuit to the right of node A forms an inverter. Thus, it is only necessary to form the logic complement for the output of the truth table of Problem 14.39 to see that the overall logic is that of a NOR gate.

v_1	v_2	v_A	$v_o = \bar{v}_A$
0	0	0	1
0	1	1	0
1	0	1	0
1	1	1	0

14.49 For design with practical diodes, D_4 in the DTL gate of Fig. 14-25 is usually replaced by two or more series-connected diodes to give a conservative design that avoids possible logic error when one or all of v_1, v_2, and v_3 are logic low. Justify this practice.

▮ Assume that only D_4 is present and v_3 drives a saturated transistor of another connected logic gate so that $v_3 = V_{CE\,sat} = 0.2$ V. If D_3 is a practical diode,

$$v_A = v_3 + v_{D3} = 0.2 + 0.7 = 0.9 \text{ V}$$

By KVL, $$v_{BE} = v_A - v_{D4} = 0.9 - 0.7 = 0.2 \text{ V}$$

This is not a sufficient margin of safety to assure that Q is OFF. Noise or a small increase in v_3 could lead to a false turn-on of Q.

14.50 For the DTL circuit of Fig. 14-25, let $R_B = 10$ kΩ and $R_C = 2$ kΩ. Determine the minimum value of $h_{FE\,min}$ (see Problem 14.18) for the transistor to assure saturation of Q if $v_1 = v_2 = v_3 = 5$ V.

▮ Assume that Q is in saturation and apply KVL to find

$$i_B = \frac{5 - v_{D4} - v_{BE}}{R_B} = \frac{5 - 0.7 - 0.7}{10 \times 10^3} = 0.36 \text{ mA}$$

Further use of KVL around the collector-emitter loop gives

$$i_C = \frac{5 - V_{CE\,sat}}{R_C} = \frac{5 - 0.2}{2 \times 10^3} = 2.4 \text{ mA}$$

Hence,
$$h_{FE\,min} = \frac{i_C}{i_B} = \frac{2.4 \times 10^{-3}}{0.36 \times 10^{-3}} = 6.67$$

14.51 Determine the power dissipated by the DTL gate of Fig. 14-25 if $v_1 = v_2 = v_3 = 5$ V.

▮ Using the results of Problem 14.50,

$$P_{CC} = (i_B + i_C)V_{CC} = (0.36 \times 10^{-3} + 2.4 \times 10^{-3})(5) = 13.8 \text{ mW}$$

14.52 The *transistor transistor logic* (TTL or T²L) circuit of Fig. 14-27 applies the input signal directly to transistor terminals (no current-limiting resistors). If either or both of v_1 and v_2 are logic low, Q_1 is driven to saturation. Set a truth table for this circuit showing that output v_{o1} produces AND logic and output v_{o2} produces NAND logic.

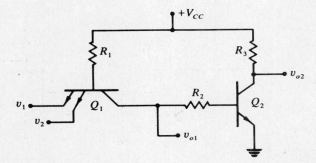

Fig. 14-27

▮ When Q_1 is saturated, v_{o1} is logic low; otherwise, v_{o1} is logic high. Further, if Q_1 is saturated so that v_{o1} is logic low, Q_2 is OFF and v_{o2} is logic high. Conversely, if Q_1 is OFF, v_{o2} is logic high, driving Q_2 to saturation, assuring that v_{o2} is logic low. Hence, v_{o2} is the logic complement of v_{o1}. The corresponding truth table follows.

v_1	v_2	v_{o1}	v_{o2}
1	1	1	0
1	0	0	1
0	1	0	1
0	0	0	1

14.53 TTL logic gates are usually applied as NAND gates. Other gate types are formed by appropriate use of inverters in conjunction with NAND gates. Construct a block diagram using the two-input TTL NAND gate of Fig. 14-27 in conjunction with inverters (NOT gates) to form an OR gate.

▮ See Fig. 14-28.

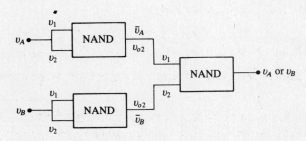

Fig. 14-28

14.54 Discuss the operation of the *complementary metal-oxide semiconductor* (CMOS) inverter of Fig. 14-29 and sketch its voltage transfer characteristic if $V_{DD} = 5$ V and the magnitudes of the threshold voltage (V_T) for the enhancement mode PMOS transistor (upper) and the NMOS transistor (lower) are both 2 V.

▌ Now $v_{GS1} = v_i$ and $v_{GS2} = v_i - V_{DD} = v_i - 5$.

If $0 \leq v_i \leq 2$ V, $\begin{cases} 0 \leq v_{GS1} \leq 2 & \text{and } Q_1 \text{ is OFF} \\ -5 \leq v_{GS2} \leq -3 & \text{and } Q_2 \text{ is ON} \end{cases}$

Thus, $v_o = 5$ V.

If $2 < v_i < 3$ V, $\begin{cases} 2 < v_{GS1} < 3 & \text{and } Q_1 \text{ is ON} \\ -3 < v_{GS2} < -2 & \text{and } Q_2 \text{ is ON} \end{cases}$

Thus, $v_{DS1} = v_{DS2}$ and $v_o = 5/2 = 2.5$ V.

If $3 \leq v_i \leq 5$, $\begin{cases} 3 \leq v_{GS1} \leq 5 & \text{and } Q_1 \text{ is ON} \\ -2 \leq v_{GS2} \leq 0 & \text{and } Q_2 \text{ is OFF} \end{cases}$

Whence, $v_o = 0$.

The resulting asymptotic voltage transfer characteristic is shown by Fig. 14-30.

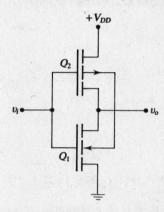

Fig. 14-29

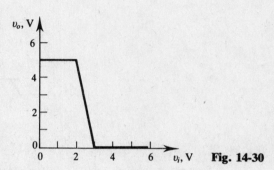

Fig. 14-30

14.55 Show that the NMOS circuit of Fig. 14-31 is a NOR gate for positive logic. Assume $V_T = 2$ V.

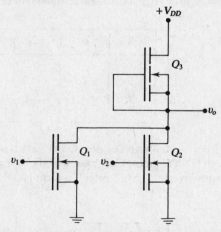

Fig. 14-31

▌ Q_3 is simply a nonlinear resistor. If either or both v_1 and v_2 are greater than $V_T = 2$ V, then either or both Q_1 and Q_2 are ON and $v_o = V_{D\,\text{on}}$. If $v_1 = v_2 \leq V_T$, $v_o = V_{DD}$. The resulting truth table shows NOR logic.

v_1		v_2		v_o	
actual	logic	actual	logic	actual	logic
$\leq V_T$	0	$\leq V_T$	0	V_{DD}	1
$\leq V_T$	0	$> V_T$	1	$V_{D\,\text{on}}$	0
$> V_T$	1	$\leq V_T$	0	$V_{D\,\text{on}}$	0
$> V_T$	1	$> V_T$	1	$V_{D\,\text{on}}$	0

14.56 Show that the NMOS circuit of Fig. 14-32 is a NAND gate for positive logic.

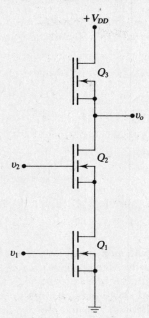

Fig. 14-32

▌ Q_3 acts as a nonlinear resistor. If $v_1 \geq V_T$ and $v_2 \geq V_{D\,on} + V_T$, $v_o = 2V_{D\,on}$. Otherwise, $v_o = V_{DD}$. The resulting truth table shows NAND logic.

v_1		v_2		v_o	
actual	logic	actual	logic	actual	logic
$\leq V_T$	0	$\leq V_{D\,on} + V_T$	0	V_{DD}	1
$\leq V_T$	0	$> V_{D\,on} + V_T$	1	V_{DD}	1
$> V_T$	1	$\leq V_{D\,on} + V_T$	0	V_{DD}	1
$> V_T$	1	$> V_{D\,on} + V_T$	1	$2V_{D\,on}$	0

MULTIVIBRATORS

14.57 The *bistable multivibrator* (BMV) or *flip-flop* (FF) is a switching circuit with two stable states. The circuit can be triggered (changed) from either state to the other by applying an input voltage via a suitable trigger circuit; it thus can be used as a basic memory element for digital logic. A simple BMV is shown in Fig. 14-33, where

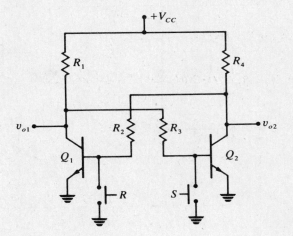

Fig. 14-33

either $v_{o1} \geq V_H$ ($=1$, or *high logic*) and $v_{o2} \leq V_L$ ($=0$, or *low logic*) or vice versa. The two momentary switches R and S set the values of v_{o1} and v_{o2} by *base triggering* (shorting the transistor base). Determine the voltages v_{o1} and v_{o2} if switch S is closed momentarily. Let $V_{CC} = 5$ V (a value that is commonly used in logic circuits), $R_1 = R_4 = 1$ kΩ, $R_2 = R_3 = 20$ kΩ, $h_{FE\,\min} = 50$ for both Q_1 and Q_2, and $V_{CE\,\text{sat}} = 0.2$ V for both Q_1 and Q_2. The transistors are Si devices.

▌ If S is closed, then $V_{BE2} \approx 0$ and Q_2 is OFF; however, Q_1 is forward-biased, and KVL requires that

$$I_{B1} = \frac{V_{CC} - V_{BE1}}{R_2 + R_4} = \frac{5 - 0.7}{21{,}000} \approx 205\ \mu\text{A}$$

Hence,
$$v_{o2} = V_{CC} - I_{B1}R_4 = 5 - (205 \times 10^{-6})(1000) \approx 4.8\text{ V}$$

and since Q_1 is driven into saturation,

$$v_{o1} = V_{CE\,\text{sat}} = 0.2\text{ V}$$

14.58 Determine the voltages v_{o1} and v_{o2} for the BMV circuit of Fig. 14-33 if switch R is closed momentarily.

▌ If R is closed, then by symmetry based on the work of Problem 14.57, $v_{o1} = 4.8$ V and $v_{o2} = 0.2$ V.

14.59 The BMV of Fig. 14-33 is shown with momentary switches R and S; however, in practical applications, triggering must be done electronically. One possibility would be to impress a short-duration low-logic signal on the upper contact connection of either R or S to trigger the switching of state. But if a negative voltage v_{BE} of short duration were placed across the base-emitter junction of the triggered transistor, the resulting reverse charge flow would accelerate the transistor turnoff process. Such a negative trigger voltage can be realized by use of an RC circuit (differentiator action) as shown in Fig. 14-34. Qualitatively discuss the reset process that occurs if Q_2 is operating in saturation when a rectangular pulse v_R arrives at terminal R in Fig. 14-34.

▌ First, let us examine the voltage v_1 ($\approx R_R C_1\, dv_R/dt$ if RC is small), assuming that negligible loading is presented by the base loop of Q_1. In that case, the arrival of a rectangular pulse at terminal R charges and discharges capacitor C_1, resulting in the output voltage v_1 depicted in Fig. 14-35 (provided the time constant $R_R C_1$ is much smaller than the duration τ of the incoming pulse).

Diode D_1 blocks the positive excursion of v_1, keeping it from the base-emitter junction of Q_1. However, the negative excursion of v_1 forward-biases D_1, whence $v_{BE1} < 0$ momentarily; this results in the turnoff of Q_1 and the BMV switches states as discussed in Problem 14.57.

It is apparent that this BMV is triggered by the trailing edge of pulse v_R; if response to the leading edge is desired, v_R can be inverted before it is impressed on terminal R. And if terminals R and S are connected together to form a terminal T, then the BMV can be made to *toggle*, or alternately switch states, by a train of pulses arriving at terminal T. (See Problem 14.60.)

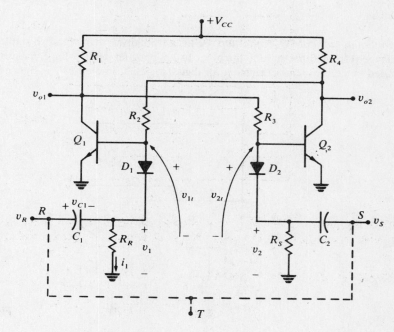

Fig. 14-34

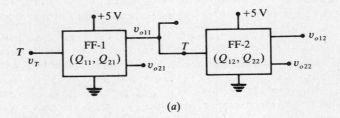

Fig. 14-35

14.60 Two flip-flops with toggle input T are cascaded as indicated in Fig. 14-36(a). Input v_T is a square pulse train (as from a clock) of frequency 100 kHz. Sketch the waveforms of v_{o11}, v_{o21}, v_{o12}, and v_{o22}, showing that the circuit is a *frequency divider* (the output waveform period is an integer multiple of the input waveform period).

▮ Waveform v_1 (or v_2) of Fig. 14-35 has no effect on Q_1 (or Q_2) of Fig. 14-34 if transistor Q_1 (or Q_2) is OFF, even with terminals R and S of Fig. 14-34 connected to form toggle T; thus, switching action must be triggered by the saturated transistor. A *timing diagram* shows the interrelationship of signals in a switching circuit as a function of time; that in Fig. 14-36(b) is based on the assumption that Q_{11} and Q_{12} (Q_1 of FF-1 and FF-2, respectively) are saturated at time $t=0$; then Q_{11} is switched to cutoff by the first trailing edge of v_T; and then Q_{12} is switched off by the first trailing edge of v_{o11}. Note the relationship among the waveform periods.

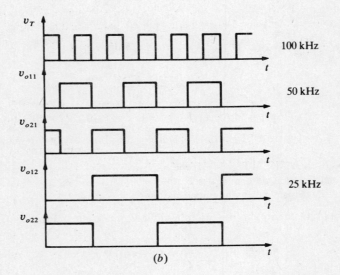

Fig. 14-36

14.61 For the BMV of Fig. 14-37, let $V_b = V_{CC} = 5$ V, $R_1 = R_4 = 2$ kΩ, $R_2 = R_3 = 20$ kΩ, and $R_5 = R_6 = 25$ kΩ. Also, for the Si transistors, $h_{FE\,min} = 50$ and $V_{CE\,sat} = 0.2$ V. Determine voltages v_{o1} and v_{o2} if switch S is closed momentarily.

▮ Upon closure of S, Q_{2A} is forward-biased with

$$I_{B2A} = \frac{V_b - V_{BE2A}}{R_5} = \frac{5 - 0.7}{25 \times 10^3} = 172\ \mu A$$

which is assumed to be adequate to drive Q_{2A} to saturation. Hence,

$$v_{o2} = V_{CE\,sat} = 0.2 \text{ V}$$

With S still closed, v_{o2} is not large enough to saturate Q_1; thus, Q_1 is in cutoff. Consequently, Q_2 is driven to saturation by

$$I_{B2} = \frac{V_{CC} - V_{BE2}}{R_2 + R_4} = \frac{5 - 0.7}{20 \times 10^3 + 2 \times 10^3} = 195.4 \text{ }\mu\text{A}$$

And

$$v_{o1} = V_{CC} - I_{B2}R_4 = 5 - (195.4 \times 10^{-6})(2 \times 10^3) = 4.61 \text{ V}$$

When S is released, Q_{2A} turns OFF, but the balance of the above analysis is unchanged.

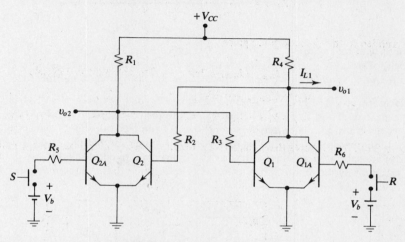

Fig. 14-37

14.62 For the BMV of Problem 14.61, determine voltages v_{o1} and v_{o2} if switch R is closed momentarily.

▌ Upon closure of R, Q_{1A} is forward-biased and driven to saturation with

$$I_{B1A} = \frac{V_b - V_{BE1A}}{R_6} = \frac{5 - 0.7}{25 \times 10^3} = 172 \text{ }\mu\text{A}$$

Hence,

$$v_{o1} = V_{CE\,sat} = 0.2 \text{ V}$$

With R still closed, v_{o1} is not large enough to maintain saturation of Q_2; thus, Q_2 is in cutoff. Consequently, Q_1 is driven to saturation by

$$I_{B1} = \frac{V_{CC} - V_{BE1}}{R_3 + R_1} = \frac{5 - 0.7}{20 \times 10^3 + 2 \times 10^3} = 195.4 \text{ }\mu\text{A}$$

whence,

$$v_{o2} = V_{CC} - I_{B1}R_1 = 5 - (195.4 \times 10^{-6})(2 \times 10^3) = 4.61 \text{ V}$$

When R is released, Q_{1A} turns OFF, but the balance of the above analysis remains valid.

14.63 For the transistors of the BMV in Fig. 14-37, the minimum value of base current to maintain saturation is 150 μA. With Q_2 saturated as determined in Problem 14.61, determine the maximum load current I_{L1} that can be drawn by subsequent logic gates while maintaining Q_2 in saturation.

▌ Let the subscript min denote the condition of minimum base current to maintain saturation. Then, by Ohm's law,

$$V_{R2\,min} = I_{B2\,min}R_2 = (150 \times 10^{-6})(20 \times 10^3) = 3 \text{ V}$$

By KVL,

$$V_{R4\,min} = V_{CC} - V_{R2\,min} - V_{BE2} = 5 - 3 - 0.7 = 1.3 \text{ V}$$

And, by Ohm's law,

$$I_{R4\,min} = \frac{V_{R4\,min}}{R_4} = \frac{1.3}{2 \times 10^3} = 0.65 \text{ mA}$$

KCL applied at the output node leads to

$$I_{L1} = I_{R4\,min} - I_{B2\,min} = 0.65 \times 10^{-3} - 150 \times 10^{-6} = 0.5 \text{ mA}$$

14.64 For the BMV of Fig. 14-38, let $V_{BB} = 1$ V, $V_{CC} = 5$ V, $R_{C1} = R_{C2} = 1$ kΩ, $R_{B1} = R_{B2} = 50$ kΩ, and $R_{B3} = R_{B4} = 15$ kΩ. The Si transistors are identical with $h_{FE\,min} = 50$ and $v_{CE\,sat} = 0.2$ V. Determine voltages v_{o1} and v_{o2} if switch S is closed momentarily.

▌ When S is closed, $V_{BE2} \approx 0$ and Q_2 is OFF; however, Q_1 is driven to saturation. The current through R_{B3} is found by KVL as

$$I_{RB3} = \frac{V_{CC} - V_{BE1}}{R_{C2} + R_{B3}} = \frac{5 - 0.7}{1 \times 10^3 + 15 \times 10^3} \approx 269\ \mu A$$

By KVL, $\quad I_{RB1} = \dfrac{V_{BE1} + V_{BB}}{R_{B1}} = \dfrac{0.7 + 1}{50 \times 10^3} = 34\ \mu A$

And KCL applied at the Q_1 base node gives

$$I_{B1} = I_{BR3} - I_{RB4} = 269 \times 10^{-6} - 34 \times 10^{-6} = 235\ \mu A$$

Hence, $\quad v_{o1} = V_{CE\,sat} = 0.2$ V

$$v_{o2} = V_{CC} - I_{BR3} R_{C2} = 5 - (269 \times 10^{-6})(1 \times 10^3) \approx 4.73\ \text{V}$$

When S is opened, V_{BB} assures that a negative voltage is applied to the base of Q_2 maintaining the transistor in cutoff.

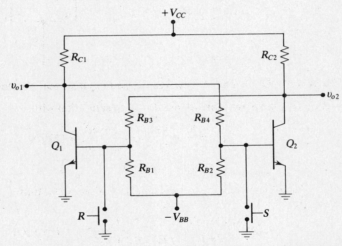

Fig. 14-38

14.65 Determine the voltage v_{o1} and v_{o2} for the two-source BMV of Problem 14.64 if switch R is closed momentarily.

▌ If R is closed, symmetry argument applied to the solution of Problem 14.64 leads to $v_{o1} = 4.73$ V and $v_{o2} = 0.2$ V.

Review Problems

14.66 Show that the circuit of Fig. 14-8 could be used as an inverter or NOT gate in a digital logic circuit.

▌ If $v_i\ (= V_H > 0)$ is large enough to saturate the transistor, then $v_o = v_{CE} = V_{CE\,sat} \approx 0$. If $v_i\ (= V_L < V_{BEQ} \approx 0)$ is small enough so that the transistor is cut off, then $v_o = v_{CE} \approx V_{CC}$. The truth table clearly shows NOT logic:

v_i		v_o	
actual	logic	actual	logic
V_H	1	0	0
V_L	0	V_{CC}	1

14.67 Rework Problem 14.18 using the model of Fig. 14-6 with $V_F = 0.5$ V, $R_{sat} = 15 \, \Omega$, and $R_F = 3$ kΩ.

■ The collector current is

$$I_{C\,sat} = \frac{V_{CC}}{R_{sat} + R_L} = \frac{15}{15 + 1 \times 10^3} = 14.78 \text{ mA}$$

The minimum base current is

$$I_{B\,sat} = \frac{I_{C\,sat}}{h_{FE\,min}} = \frac{14.78 \times 10^{-3}}{30} = 492.6 \, \mu\text{A}$$

14.68 Rework Problem 14.19 using the model of Fig. 14-6 with $V_F = 0.5$ V, $R_{sat} = 15 \, \Omega$, and $R_F = 3$ kΩ.

■ The collector current at saturation is

$$I_{C\,sat} = h_{FE\,min} I_{B\,sat} = 30(100 \times 10^{-6}) = 3 \text{ mA}$$

Whence,

$$R_L \leq \frac{V_{CC}}{I_{C\,sat}} - R_{sat} = \frac{15}{3 \times 10^{-3}} - 15 = 4.985 \text{ k}\Omega$$

14.69 Show that the collector power dissipation for BJTs operating in the switching mode is significantly less than that for the operation in the active region by calculating the collector power dissipated by the circuit of Fig. 14-8 for (a) cutoff, (b) saturation, and (c) $V_{CEQ} = 7.5$ V. Assume $R_L = 1$ kΩ, $V_{CC} = 15$ V, and that the transistor is characterized by $R_{sat} = 15 \, \Omega$, $V_{CE\,sat} = 0.2$ V, and $R_{CO} = 1$ MΩ.

■ (a) For cutoff, $I_B = 0$, and

$$I_C = \frac{V_{CC}}{R_L + R_{CO}} = \frac{15}{1 \times 10^3 + 1 \times 10^6} = 14.985 \, \mu\text{A}$$

By KVL,

$$V_{CE} = V_{CC} - I_C R_L = 15 - (14.985 \times 10^{-6})(1 \times 10^3) = 14.985 \text{ V}$$

Thus,

$$P_C = V_{CE} I_C = 14.985(14.985 \times 10^{-6}) = 0.225 \text{ mW}$$

(b) For the saturation conditions,

$$I_{C\,sat} = \frac{V_{CC}}{R_{sat} + R_L} = \frac{15}{15 + 1 \times 10^3} = 14.78 \text{ mA}$$

$$V_{CE\,sat} = V_{CC} - I_{C\,sat} R_L = 15 - (14.78 \times 10^{-3})(1 \times 10^3) = 0.22 \text{ V}$$

$$P_C = V_{CE\,sat} I_{C\,sat} = 0.22(14.78 \times 10^{-3}) = 3.25 \text{ mW}$$

(c) For the given active region condition,

$$I_C = \frac{V_{CC} - V_{CEQ}}{R_L} = \frac{15 - 7.5}{1 \times 10^3} = 7.5 \text{ mA}$$

$$P_C = V_{CEQ} I_C = 7.5(7.5 \times 10^{-3}) = 56.25 \text{ mW}$$

14.70 Use a truth table to show that the circuit of Fig. 14-39 is a NAND gate if v_{o1} is taken as the output.

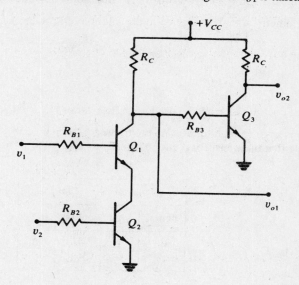

Fig. 14-39

SWITCHING CIRCUITS FOR DIGITAL LOGIC □ 449

▌ If $v_1 = v_2 \leq V_L$, $v_{o1} \approx V_{CC}$. If $v_1 (v_2) \geq V_H$, while $v_2 (v_1) \leq V_L$, $Q_1 (Q_2)$ is ON and $Q_2 (Q_1)$ is OFF and $v_{o1} \approx V_{CC}$. If $v_1 = v_2 \geq V_H$, both Q_1 and Q_2 are ON and $v_{o1} \approx 2V_{CE\,sat}$. The truth table shows NAND logic for v_{o1}.

v_1		v_2		v_{o1}		v_{o2}	
actual	logic	actual	logic	actual	logic	actual	logic
V_L	0	V_L	0	V_{CC}	1	$V_{CE\,sat}$	0
V_H	1	V_L	0	V_{CC}	1	$V_{CE\,sat}$	0
V_L	0	V_H	1	V_{CC}	1	$V_{CE\,sat}$	0
V_H	1	V_H	1	$2V_{CE\,sat}$	0	V_{CC}	1

14.71 Through use of a truth table, show that the circuit of Fig. 14-39 is an AND gate if v_{o2} is the output.

▌ For $v_{o1} = V_{CC}$, if R_{B3} is properly sized, Q_3 is saturated and $v_{o2} = V_{CE\,sat}$. However, if $v_{o1} = 2V_{CE\,sat}$, Q_3 remains OFF. Hence, v_{o2} is the logic complement of v_{o1}. When the result is added to the truth table of Problem 14.70, it is apparent that the overall logic is that of an AND gate.

14.72 If the BJT switching circuit of Fig. 14-13 is described by the parameters of Problem 14.32 except $R_C = 0$ and $R_B = 10\,k\Omega$, sketch the asymptotic voltage characteristic $v_{o2} - v_i$.

▌ The breakpoints are found by (2) and (3) of Problem 14.31 as

$$v_{iC} = 0.7 - 0 = 0.7\,V$$

$$v_{iS} = 0.7 - 0 + \frac{10 \times 10^3 + (40+1)(1 \times 10^3)}{(40+1)(0+15+1 \times 10^3)}(0+10) = 12.96\,V$$

By (1), (3), and (5) of Problem 14.30, the respective expressions for v_o over the cutoff, active, and saturation regions of operation are

$$v_{o2} = -V_{EE} = 0, \quad v_i < 0.7\,V \tag{1}$$

$$v_{o2} = -\frac{(\beta+1)R_E}{R_B + (\beta+1)R_E} V_F + \frac{(\beta+1)R_E}{R_B + (\beta+1)R_E} v_i$$

$$= -\frac{(40+1)(1 \times 10^3)}{10 \times 10^3 + (40+1)(1 \times 10^3)}(0.7) + \frac{(40+1)(1 \times 10^3)}{10 \times 10^3 + (40+1)(1 \times 10^3)} v_i$$

$$= -0.563 + 0.804 v_i, \quad 0.7 \leq v_i \leq 12.96\,V \tag{2}$$

$$v_{o2} = \frac{R_E V_{CC}}{R_{sat} + R_E} = \frac{(1 \times 10^3)(10)}{15 + 1 \times 10^3} = 9.85\,V, \quad v_i > 12.96\,V \tag{3}$$

The resulting voltage transfer characteristic is plotted in Fig. 14-40.

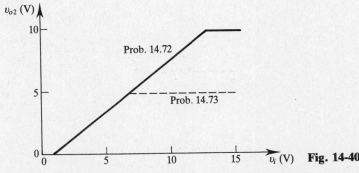

Fig. 14-40

14.73 Repeat Problem 14.72 if $R_C = 1\,k\Omega$ and all else is unchanged. Sketch the resulting voltage transfer characteristic on Fig. 14-40 for comparison with the result of Problem 14.72.

▌ Proceeding as in Problem 14.72, the breakpoints are determined by (2) and (3) of Problem 14.31:

$$v_{iC} = 0.7 - 0 = 0.7 \text{ V}$$

$$v_{iS} = 0.7 - 0 + \frac{10 \times 10^3 + (40+1)(1 \times 10^3)}{(40+1)(1 \times 10^3 + 15 + 1 \times 10^3)}(0+10) = 6.87 \text{ V}$$

By (1), (3) and (5) of Problem 14.30, the respective expression for v_{o2} over the cutoff, active, and saturation regions of operation are

$$v_{o2} = -V_{EE} = 0, \qquad v_i < 0.7 \text{ V} \tag{1}$$

$$v_{o2} = -\frac{(\beta+1)R_E}{R_B + (\beta+1)R_E}V_F + \frac{(\beta+1)R_E}{R_B + (\beta+1)R_E}v_i$$

$$= -\frac{(40+1)(1 \times 10^3)}{10 \times 10^3 + (40+1)(1 \times 10^3)}(0.7) + \frac{(40+1)(1 \times 10^3)}{10 \times 10^3 + (40+1)(1 \times 10^3)}v_i$$

$$= -0.563 + 0.804 v_i, \qquad 0.7 \leq v_i \leq 6.87 \text{ V} \tag{2}$$

$$v_{o2} = \frac{R_E V_{CC}}{R_C + R_{sat} + R_E} = \frac{(1 \times 10^3)(10)}{1 \times 10^3 + 15 + 1 \times 10^3} = 4.96, \qquad v_i > 6.87 \tag{3}$$

The resulting voltage transfer characteristic is added to Fig. 14-40 where it is seen that the addition of R_C significantly reduces the output voltage during the saturation region of operation. Also, saturation is reached at half the value of v_i for the case of $R_C = 0$.

14.74 For the BJT switching circuit of Fig. 14-13, let $V_{CC} = 0$, $V_{EE} = 5 \text{ V}$, $R_C = 1 \text{ k}\Omega$, $R_E = 0$, $R_B = 20 \text{ k}\Omega$, $\beta = 50$, $V_F = 0.7 \text{ V}$, and $R_{sat} = 10 \Omega$. Sketch the asymptotic $v_{o1} - v_i$ voltage transfer characteristic.

▌ The breakpoints are found by (2) and (3) of Problem 14.29 as

$$v_{iC} = 0.7 - 5 = -4.3 \text{ V}$$

$$v_{iS} = 0.7 - 5 + \frac{20 \times 10^3 + (50+1)(0)}{50(1 \times 10^3 + 10 + 0)}(5+0) = -2.32 \text{ V}$$

The respective expressions for v_{o1} over the cutoff, active, and saturation regions of operation are found by (2), (4), and (7) of Problem 14.28:

$$v_{o1} = V_{CC} = 0, \qquad v_i < -4.3 \text{ V} \tag{1}$$

$$v_{o1} = -\frac{\beta R_C}{R_B}(V_{EE} - V_F) - \frac{\beta R_C}{R_B}v_i = -\frac{50(1 \times 10^3)}{20 \times 10^3}(5 - 0.7) - \frac{50(1 \times 10^3)}{20 \times 10^3}v_i$$

$$= -10.75 - 2.5 v_i, \qquad -4.3 \leq v_i \leq -2.32 \text{ V} \tag{2}$$

$$v_{o1} = -\frac{R_C V_{EE}}{R_C + R_{sat}} = -\frac{(1 \times 10^3)(5)}{1 \times 10^3 + 10} = -4.95, \qquad v_i > -2.32 \text{ V} \tag{3}$$

The resulting voltage transfer characteristic is sketched on Fig. 14-41.

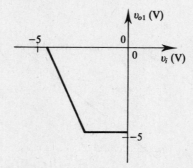

Fig. 14-41

14.75 For the BJT switching circuit of Problem 14.74, let $R_E = 1 \text{ k}\Omega$ and all else is unchanged. Sketch the asymptotic voltage transfer characteristic for v_{o2}.

The breakpoints are found by (2) and (3) of Problem 14.31 as

$$v_{iC} = 0.7 - 5 = -4.3 \text{ V}$$

$$v_{iS} = 0.7 - 5 + \frac{20 \times 10^3 + (50+1)(1 \times 10^3)}{(50+1)(1 \times 10^3 + 10 + 1 \times 10^3)}(5+0) = -0.836 \text{ V}$$

The respective expressions for v_{o2} over the cutoff, active, and saturation regions of operation follow from (1), (3), and (5) of Problem 14.30:

$$v_{o2} = -V_{EE} = -5 \text{ V}, \qquad v_i < -4.3 \text{ V} \tag{1}$$

$$v_{o2} = -V_{EE} + \frac{(\beta+1)R_E}{R_B + (\beta+1)R_E}(V_{EE} - V_F) + \frac{(\beta+1)R_E}{R_B + (\beta+1)R_E} v_i$$

$$= -5 + \frac{(50+1)(1 \times 10^3)}{20 \times 10^3 + (50+1)(1 \times 10^3)}(5 - 0.7) + \frac{(50+1)(1 \times 10^3)}{20 \times 10^3 + (50+1)(1 \times 10^3)} v_i$$

$$= -1.911 + 0.718 v_i, \qquad -4.3 \le v_i \le -0.836 \tag{2}$$

$$v_{o2} = -\frac{(R_C + R_{sat})V_{EE}}{R_C + R_{sat} + R_E} = -\frac{(1 \times 10^3 + 10)(5)}{1 \times 10^3 + 10 + 1 \times 10^3} = -2.51 \text{ V}, \qquad v_i > -0.836 \text{ V} \tag{3}$$

The voltage transfer characteristic is shown by Fig. 14-42.

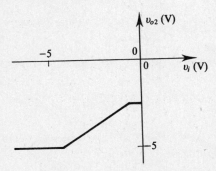

Fig. 14-42

14.76 For the BJT switching circuit of Fig. 14-13, let $V_{CC} = 10$ V, $V_{EE} = 5$ V, $R_C = R_E = 1$ kΩ, $R_B = 20$ kΩ, $\beta = 50$, $V_F = 0.7$, and $R_{sat} = 10$ Ω. Sketch the $v_{o1} - v_i$ voltage transfer characteristic.

The breakpoints are found by (2) and (3) of Problem 14.29 as

$$v_{iC} = 0.7 - 5 = -4.3 \text{ V}$$

$$v_{iS} = 0.7 - 5 + \frac{20 \times 10^3 + (50+1)(1 \times 10^3)}{50(1 \times 10^3 + 10 + 1 \times 10^3)}(10 + 5) = 6.3 \text{ V}$$

The respective expressions for v_{o1} over the cutoff, active, and saturation regions of operation are determined by (2), (4) and (7) of Problem 14.28:

$$v_{o1} = V_{CC} = 10 \text{ V}, \qquad v_i < -4.3 \text{ V} \tag{1}$$

$$v_{o1} = V_{CC} - \frac{\beta R_C}{R_B + (\beta+1)R_E}(V_{EE} - V_F) - \frac{\beta R_C}{R_B + (\beta+1)R_E} v_i$$

$$= 10 - \frac{50(1 \times 10^3)}{20 \times 10^3 + (50+1)(1 \times 10^3)}(5 - 0.7) - \frac{50(1 \times 10^3)}{20 \times 10^3 + (50+1)(1 \times 10^3)} v_i$$

$$= 6.97 - 7.04 v_i, \qquad -4.3 \le v_i \le 6.3 \text{ V} \tag{2}$$

$$v_{o1} = \frac{(R_{sat} + R_E)V_{CC} - R_C V_{EE}}{R_C + R_{sat} + R_E} = \frac{(10 + 1 \times 10^3)(10) - (1 \times 10^3)(5)}{1 \times 10^3 + 10 + 1 \times 10^3} = 2.54 \text{ V}, \qquad v_i > 6.3 \text{ V} \tag{3}$$

The asymptotic voltage transfer characteristic is depicted by Fig. 14-43.

452 □ CHAPTER 14

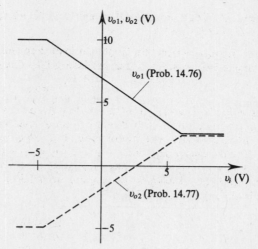

Fig. 14-43

14.77 Sketch the asymptotic $v_{o2} - v_i$ voltage transfer characteristic for the BJT switching circuit of Problem 14.76.

▎ The breakpoints follow from (2) and (3) of Problem 14.31:

$$v_{iC} = 0.7 - 5 = -4.3 \text{ V}$$

$$v_{iS} = 0.7 - 5 + \frac{20 \times 10^3 + (50+1)(1 \times 10^3)}{(50+1)(1 \times 10^3 + 10 + 1 \times 10^3)}(5 + 10) = 6.09 \text{ V}$$

The respective expressions for v_{o1} over the cutoff, active, and saturation regions of operation are determined by (1), (3), and (5) of Problem 14.30:

$$v_{o2} = -V_{EE} = -5 \text{ V}, \quad v_i < -4.3 \text{ V} \tag{1}$$

$$v_{o2} = -V_{EE} + \frac{(\beta+1)R_E}{R_B + (\beta+1)R_E}(V_{EE} - V_F) + \frac{(\beta+1)R_E}{R_B + (\beta+1)R_E}v_i$$

$$= -5 + \frac{(50+1)(1 \times 10^3)}{20 \times 10^3 + (50+1)(1 \times 10^3)}(5 - 0.7) + \frac{(50+1)(1 \times 10^3)}{20 \times 10^3 + (50+1)(1 \times 10^3)}v_i$$

$$= -1.911 + 0.718v_i \quad -4.3 \leq v_i \leq 6.09 \text{ V} \tag{2}$$

$$v_{o2} = -\frac{(R_C + R_{sat})V_{EE} + R_E V_{CC}}{R_C + R_{sat} + R_E} = -\frac{(1 \times 10^3 + 10)(5) + (1 \times 10^3)(10)}{1 \times 10^3 + 10 + 1 \times 10^3} = 2.46 \text{ V}, \quad v_i > 6.09 \text{ V} \tag{3}$$

The voltage transfer characteristic is added to Fig. 14-43.

14.78 The gate and source are connected for the depletion mode MOSFET Q_1 of Fig. 14-44 to establish a load resistor for Q_2. Sketch the voltage transfer characteristic for the MOSFET switching circuit if Q_1 is described by the

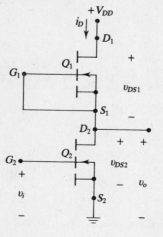

Fig. 14-44

drain characteristic of Fig. 14-45 and Q_2 is an enhancement mode MOS transistor described by the drain characteristic of Fig. 14-46. Let $V_{DD} = 8$ V.

▌ Since G_1 is shorted to S_1, $v_{GS1} = 0$ and the nonlinear load characteristic is the $v_{GS} = 0$ curve of Fig. 14-45. This nonlinear load characteristic must be used to establish a load line on Fig. 14-46. By KVL,

$$v_{DS2} = V_{DD} - V_{DS1} \qquad (1)$$

where the current through both Q_1 and Q_2 is identical. For example, if $i_D = 2$ mA (see point A of Fig. 14-45), $v_{DS1} = 7$ V. By (1), $v_{DS2} = 8 - 7 = 1$ V which is shown as point A on Fig. 14-46. In a similar manner, v_{DS1} for points B and C of Fig. 14-45 are found and the corresponding values of v_{D2} determined by (1) to give the points B and C of Fig. 14-46. The completed nonlinear load characteristic is shown by dashed line on Fig. 14-46. With the load line established, $v_i = v_{GS2} \geq 2$ V is allowed to change along the load line to determine the values of $v_o = v_{DS2}$ from the projection of the instantaneous operating point on the v_{DS} axis. The resulting voltage transfer characteristic is plotted as Fig. 14-47.

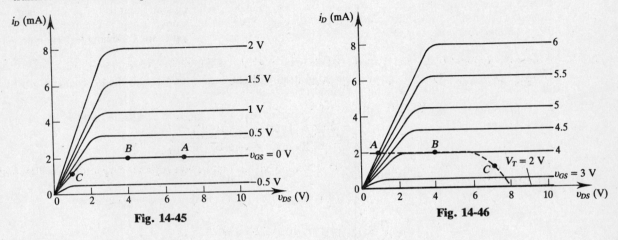

Fig. 14-45

Fig. 14-46

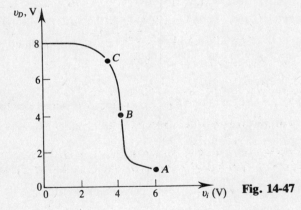

Fig. 14-47

CHAPTER 15
Boolean Algebra and Logic Gates

NUMBERING SYSTEMS

15.1 The radix (or *base*) r of a number system is defined as the quantity of different digits $(0, 1, 2, \ldots, r-1)$ that can possibly occur in each position of the number system. To allow expression of a number larger than the selected system base, the power series representation of a number N is introduced:

$$N = c_n r^n + c_{n-1} r^{n-1} + \cdots + c_1 r + c_0 r^0 + c_{-1} r^{-1} + \cdots + c_{-m} r^{-m} \qquad (1)$$

where the coefficients c_k range in value from 0 to $r-1$. Further, unless a statement to the contrary is made, the radix and its exponents are customarily recorded as base-10 numbers. Conventional expression of numbers uses the *positional notational system* where the coefficients of (1) are recorded in the following orderly manner:

$$N = (c_n c_{n-1} \cdots c_1 c_0 \cdot c_{-1} \cdots c_{-m})_r \qquad (2)$$

The subscript r is omitted when no likelihood of misunderstanding exists. Express the *decimal* (base-10) number $(2013)_{10}$, expressed in the positional notation of (2), in the power series form of (1).

I Comparing the given number with (2), it is apparent that $n = 3$ and $m = 0$. Also, $c_3 = 2$, $c_2 = 0$, $c_1 = 1$, and $c_0 = 3$. Hence, by (1),

$$(2013)_{10} = 2 \times 10^3 + 0 \times 10^2 + 1 \times 10 + 3$$

15.2 Express $(51.27)_{10}$ in the power series form of (1) from Problem 15.1.

I Comparing the given number with (2) of Problem 15.1, it is seen that $n = 1$, $m = 2$, $c_1 = 5$, $c_0 = 1$, $c_{-1} = 2$, and $c_{-2} = 7$. Whence, by (1) of Problem 15.1,

$$(51.27)_{10} = 5 \times 10 + 1 + 2 \times 10^{-1} + 7 \times 10^{-2}$$

15.3 Express the *quinary* (base-5) number $(1234.3)_5$ in the power series form of (1) from Problem 15.1.

I Comparison of the given number with (2) of Problem 15.1 shows that $n = 3$, $m = 1$, $c_3 = 1$, $c_2 = 2$, $c_1 = 3$, $c_0 = 4$, and $c_{-1} = 3$. Hence, by (1) of Problem 15.1,

$$(1234.3)_5 = 1 \times 5^3 + 2 \times 5^2 + 3 \times 5^1 + 4 \times 5^0 + 3 \times 5^{-1}$$

15.4 Write the *binary* (base-2) number $(1011.1)_2$ in the power series form of (1) from Problem 15.1.

I Comparison of the given number with (2) of Problem 15.1 shows that $n = 3$, $m = 1$, $c_3 = 1$, $c_2 = 0$, $c_1 = 1$, $c_0 = 1$, and $c_{-1} = 1$. Thus, by (1) of Problem 15.1,

$$(1011.1)_2 = 1 \times 2^3 + 0 \times 2^2 + 1 \times 2^1 + 1 \times 2^0 + 1 \times 2^{-1}$$

15.5 Determine the decimal equivalent of the binary number $(1011.1)_2$.

I The result is obtained directly by summation of the power series representation of Problem 15.4 as long as the radix and its exponents are expressed in decimal form:

$$(1011.1)_2 = 1 \times 2^3 + 0 \times 2^2 + 1 \times 2^1 + 1 \times 2^0 + 1 \times 2^{-1}$$
$$= 1 \times 8 + 0 \times 4 + 1 \times 2 + 1 \times 1 + 1 \times 0.5$$
$$= 8 + 0 + 2 + 1 + 0.5 = (11.5)_{10}$$

15.6 Find the decimal equivalent of the quinary number $(1234.3)_5$.

I From the power series representation of Problem 15.3,

$$(1234.3)_5 = 1 \times 5^3 + 2 \times 5^2 + 3 \times 5^1 + 4 \times 5^0 + 3 \times 5^{-1}$$
$$= 1 \times 125 + 2 \times 25 + 3 \times 5 + 4 \times 1 + 3 \times 0.2$$
$$= 125 + 50 + 15 + 4 + 0.6 = (194.6)_{10}$$

15.7 Convert the decimal number $(52)_{10}$ to an equivalent binary number.

▌ For integer numbers, the conversion is carried out through repeated integer division by the radix of the number system to which conversion is being made and with the elements of the converted number appearing as the discarded remainder of each operation, as illustrated below:

division	integer quotient	remainder	coefficient
52/2	26	0	c_0
26/2	13	0	c_1
13/2	6	1	c_2
6/2	3	0	c_3
3/2	1	1	c_4
1/2	0	1	c_5

The binary number is formed from the remainder terms as $(110100)_2$.

15.8 Convert the decimal number $(194)_{10}$ to a quinary number.

▌ Following the procedure of Problem 15.7,

division	integer quotient	remainder	coefficient
194/5	38	4	c_0
38/5	7	3	c_1
7/5	1	2	c_2
1/5	0	1	c_3

Hence, $(194)_{10} = (1234)_5$.

15.9 Convert the decimal fraction $(0.4375)_{10}$ to a binary fraction.

▌ For fractional numbers, the conversion is carried out through repeated multiplications by the radix of the number system to which conversion is being made and with the elements of the converted number appearing as the discarded integer of each operation. The process, as illustrated below, is repeated until either a zero fraction portion or the desired accuracy is obtained.

multiplication	integer portion	fraction portion	coefficient
0.4375×2	0	0.8750	c_{-1}
0.8750×2	1	0.7500	c_{-2}
0.7500×2	1	0.5000	c_{-3}
0.5000×2	1	0.0000	c_{-4}

The converted binary fraction is formed from the integer portion as $(0.0111)_2$.

15.10 Convert the decimal number $(10.75)_{10}$ to its binary equivalent.

▌ The integer portion and fraction portion are converted separately and combined to form the complete binary number. Following the procedure of Problem 15.7 for the integer portion gives

division	integer quotient	remainder	coefficient
10/2	5	0	c_0
5/2	2	1	c_1
2/2	1	0	c_2
1/2	0	1	c_3

Thus, $(10)_{10} = (1010)_2$.

Converting the fraction portion according to the procedure of Problem 15.9,

multiplication	integer portion	fraction portion	coefficient
0.75×2	1	0.500	c_{-1}
0.500×2	1	0.000	c_{-2}

Hence, $(0.75)_{10} = (0.11)_2$. The completed conversion is

$$(10.75)_{10} = (1010)_2 + (0.11)_2 = (1010.11)_2$$

15.11 Convert the decimal number $(194.6)_{10}$ to its equivalent quinary number.

▌ The integer portion was converted in Problem 15.8 giving $(194)_{10} = (1234)_5$. The fraction portion of the given number is converted to base-5 by the procedure of Problem 15.9:

multiplication	integer portion	fraction portion	coefficient
0.6×5	3	0.0	c_{-1}

Thus, $(0.6)_{10} = (0.3)_5$. And, the complete quinary number is

$$(194.6)_{10} = (1234)_5 + (0.3)_5 = (1234.3)_5$$

15.12 Addition of two binary numbers is performed in a manner similar to that of decimal addition. The complete binary addition table is

$$0 + 0 = 0$$
$$0 + 1 = 1$$
$$1 + 0 = 1$$
$$1 + 1 = 0 \quad \text{with a carry of 1}$$

where a carry is handled by addition to the next column to the left in the same manner as for decimal addition. Add the binary numbers $(101)_2$ and $(10)_2$.

▌ Using the above addition table,

$$\begin{array}{r} 101 \\ +10 \\ \hline (111)_2 \end{array}$$

15.13 Add the three binary numbers $1_2 + 1_2 + 1_2$.

▌ The process is most easily followed by use of associative addition, or

$$1 + 1 + 1 = (1 + 1) + 1 = 10 + 1 = (11)_2$$

15.14 Determine the sum of the two binary numbers $(101)_2$ and $(11)_2$.

▌ Using the addition table of Problem 15.12 and drawing an analogy for decimal addition to handle the resulting carries, we find

$$\begin{array}{r} 101 \\ +11 \\ \hline (1000)_2 \end{array}$$

15.15 Subtraction of positive binary numbers, if the minuend is larger than the subtrahend, is carried out by use of the following binary subtraction table:

$$0 - 0 = 0$$
$$1 - 0 = 1$$
$$1 - 1 = 0$$
$$0 - 1 = 1 \quad \text{with a borrow of 1}$$

Perform the binary subtraction $(111)_2 - (10)_2$.

▌ Applying the above subtraction table,

$$\begin{array}{r} 111 \\ -10 \\ \hline (101)_2 \end{array}$$

15.16 Carry out the binary subtraction $(101)_2 - (10)_2$.

▌ Applying the subtraction table of Problem 15.15,

$$\begin{array}{r} 101 \\ -10 \\ \hline (11)_2 \end{array}$$

where a borrow from the third column of 101 was necessary when handling the subtraction of the second column.

15.17 Implementation of subtraction in digital computers is simplified by use of *complements*. The *r's complement* of a positive number $P \neq 0$ in base r with an integer part of m digits is defined as $r^m - P$. For the digital number $(123)_{10}$, find the r's (10's) complement.

▌ Inspection of $(123)_{10}$ shows that $m = 3$; hence, the 10's complement is

$$r^m - P = 10^3 - 123 = 1000 - 123 = (877)_{10}$$

15.18 Determine the r's complement of $(41.75)_{10}$.

▌ Based on the definition in Problem 15.17,

$$r^m - P = 10^2 - 41.75 = 100 - 41.75 = (58.25)_{10}$$

15.19 Find the r's (2's) complement of $(101)_2$.

▌ Based on the definition in Problem 15.17,

$$r^m - P = (2^3)_{10} - (101)_2 = (1000)_2 - (101)_2 = (11)_2$$

15.20 Determine the 2's complement of $(0.101)_2$.

▌ Based on the definition in Problem 15.17,

$$r^m - P = (2^0)_{10} - (0.101)_2 = (1)_2 - (0.101)_2 = (0.011)_2$$

15.21 *Rule*: The r's complement of a number can be formed by subtracting each digit of the number from $r - 1$ and then adding 1 to the least significant digit of the resulting number. Verify this rule by finding the 10's complement of $(41.75)_{10}$ and comparing the answer with that of Problem 15.18.

▌ Since $r - 1 = 10 - 1 = 9$,

$$\begin{array}{r} 99.99 \\ -41.75 \\ \hline 58.24 \\ +1 \\ \hline (58.25)_{10} \end{array}$$

which checks with the answer to Problem 15.18.

15.22 Determine the 2's complement of $(101)_2$ by application of the rule of Problem 15.21.

Since $r - 1 = 2 - 1 = 1$,

$$\begin{array}{r} 111 \\ -101 \\ \hline 010 \\ +1 \\ \hline (011)_2 \end{array}$$

which agrees with the answer to Problem 15.19.

15.23 Demonstrate that the r's complement of a number returns the original number by use of the specific binary number $(101)_2$.

The 2's complement of $(101)_2$ has been determined as $(011)_2$ in Problem 15.22. All that remains is to form the 2's complement of $(011)_2$. Since $r - 1 = 2 - 1 = 1$,

$$\begin{array}{r} 111 \\ -011 \\ \hline 100 \\ +1 \\ \hline (101)_2 \end{array}$$

which is the original binary number.

15.24 The r's complement can be used for subtraction of two positive numbers, $M_r - N_r$, by application of the following two-step rule:
1. Add M_r and the r's complement of N_r.
2. The answer depends upon the existence of a carry from addition of the two most significant digits (called an *end-carry*):
   (a) If an end-carry results, $M_r - N_r$ is a positive number formed as the sum of part 1 with the end-carry discarded.
   (b) If there is no end-carry, $M_r - N_r$ is a negative number formed as the r's complement of the sum from part 1 preceded by a minus sign.

Using r's complement subtraction, find the difference $(50)_{10} - (41.75)_{10}$.

The 10's complement of $(41.75)_{10}$ was formed in Problem 15.21. Performing the addition of step 1,

$$\begin{array}{r} 50.00 \\ +58.25 \\ \hline (108.25)_{10} \end{array}$$

Discarding the one from the end-carry, the answer is $(8.25)_{10}$.

15.25 Use r's complement subtraction to find the difference $(50)_{10} - (75)_{10}$.

Forming the 10's complement of 75,

$$\begin{array}{r} 99 \\ -75 \\ \hline 24 \\ +1 \\ \hline (25)_{10} \end{array}$$

Then adding,

$$\begin{array}{r} 50 \\ +25 \\ \hline (75)_{10} \end{array}$$

Since no end-carry resulted from the addition step, form the 10's complement of $(75)_{10}$ and place a minus sign in front of the result to yield the answer:

$$
\begin{array}{r}
99 \\
-75 \\
\hline
24 \\
+1 \\
\hline
-(25)_{10}
\end{array}
$$

15.26 Use r's complement subtraction to find the difference $(111)_2 - (010)_2$.

▎ Forming the 2's complement of the subtrahend,

$$
\begin{array}{r}
111 \\
-010 \\
\hline
101 \\
+1 \\
\hline
(110)_2
\end{array}
$$

The resulting addition gives an end-carry which is disregarded to give the answer, or

$$
\begin{array}{r}
111 \\
+110 \\
\hline
(1101)_2
\end{array}
$$

Hence, $(111)_2 - (010)_2 = (101)_2$ which agrees with the answer of Problem 15.15.

15.27 By r's complement subtraction, determine $(010)_2 - (111)_2$.

▎ The 2's complement of the subtrahend is

$$
\begin{array}{r}
111 \\
-111 \\
\hline
000 \\
+1 \\
\hline
(001)_2
\end{array}
$$

Performing the addition,

$$
\begin{array}{r}
010 \\
+001 \\
\hline
(011)_2
\end{array}
$$

Since the addition produced no end-carry, the 2's complement of the sum and a leading negative sign are formed:

$$
\begin{array}{r}
111 \\
-011 \\
\hline
100 \\
+1 \\
\hline
-(101)_2
\end{array}
$$

15.28 The $(r-1)$'s complement of a positive base-r number $P = m \cdot n$ (an integer part of m digits and a fraction part of n digits) is defined as $r^m - r^{-n} - P$. For the digital number $(123)_{10}$, find the $(r-1)$'s or 9's complement.

▎ Inspection of $(123)_{10}$ shows that $m = 3$ and $n = 0$. Hence, the 9's complement is

$$r^m - r^{-n} - P = 10^3 - 10^{-0} - 123 = 1000 - 1 - 123 = (876)_{10}$$

15.29 Determine the $(r-1)$'s complement of $(41.75)_{10}$.

▎ Applying the definition of Problem 15.28.

$$r^m - r^{-n} - P = 10^2 - 10^{-2} - 41.75 = (58.24)_{10}$$

460 ☐ CHAPTER 15

15.30 Find the $(r-1)$'s or 1's complement of $(101)_2$.

▮ Based on the definition of Problem 15.28,

$$r^m - r^{-n} - P = (2^3)_{10} - (2^{-0})_{10} - (101)_2 = (1000)_2 - (001)_2 - (101)_2 = (010)_2$$

15.31 Find the 1's complement of $(0.101)_2$.

▮ Applying the definition of Problem 15.28,

$$r^m - r^{-n} - P = (2^0)_{10} - (2^{-3})_{10} - (0.101)_2 = (1.000)_2 - (0.001)_2 - (0.101)_2 = (0.010)_2$$

15.32 *Rule*: The $(r-1)$'s complement can be formed by subtracting each digit of the number from $r-1$. For the case of a binary number, this procedure is equivalent to replacing all 0's with 1's and all 1's with 0's. Verify this rule by finding the 9's complement of $(41.75)_{10}$ and comparing the answer with that of Problem 15.29.

▮ Since $r - 1 = 10 - 1 = 9$,

$$\begin{array}{r} 99.99 \\ -41.75 \\ \hline (58.24)_{10} \end{array}$$

which agrees with the answer to Problem 15.29.

15.33 Determine the 1's complement of $(101)_2$ by application of the rule of Problem 15.32.

▮ Since $r - 1 = 2 - 1 = 1$,

$$\begin{array}{r} 111 \\ -101 \\ \hline (010)_2 \end{array}$$

The same result could be obtained by interchange of 1's and 0's in $(101)_2$.

15.34 Demonstrate that the $(r-1)$'s complement of a number returns the original number by use of the binary number $(101)_2$.

▮ The 1's complement of $(101)_2$ has been determined as $(010)_2$ in Problem 15.33. All that remains is to form the 1's complement of $(010)_2$. Since $r - 1 = 1$,

$$\begin{array}{r} 111 \\ -010 \\ \hline (101)_2 \end{array}$$

which is the original binary number.

15.35 The $(r-1)$'s complement can be used for subtraction of two positive numbers, $M_r - N_r$, by application of the following two-step rule:
1. Add M_r to the $(r-1)$'s complement of N_r.
2. The answer depends upon the existence of an end-carry from the addition of part 1:
    (a) If there is an end-carry, $M_r - N_r$ is a positive number formed by discarding the end-carry and adding 1 to the least signficant digit of the sum of part 1 (called an *end-around carry*).
    (b) If there is no end-carry, $M_r - N_r$ is a negative number formed as the $(r-1)$'s complement of the sum from part 1 preceded by a minus sign.

Using the $(r-1)$'s complement subtraction, find the difference $(50)_{10} - (41.75)_{10}$.

▮ The 9's complement of $(41.75)_{10}$ was formed in Problem 15.32. Performing the addition of part 1 gives

$$\begin{array}{r} 50.00 \\ +58.24 \\ \hline (108.24)_{10} \end{array}$$

Since an end-carry resulted, implement an end-around carry to form the answer:

$$\begin{array}{r} 08.24 \\ +1 \\ \hline (8.25)_{10} \end{array}$$

15.36 Use $(r-1)$'s complement subtraction to find the difference $(50)_{10} - (75)_{10}$.

▌ Forming the 9's complement of 75,

$$\begin{array}{r} 99 \\ -75 \\ \hline (24)_{10} \end{array}$$

Then adding,

$$\begin{array}{r} 50 \\ +24 \\ \hline (74)_{10} \end{array}$$

Since there was no end-carry from the addition, form the 9's complement of $(74)_{10}$ and place a negative sign in front of the result to form the answer:

$$\begin{array}{r} 99 \\ -74 \\ \hline -(25)_{10} \end{array}$$

THEOREMS AND PROPERTIES OF BOOLEAN ALGEBRA

15.37 Digital circuits make use of ON-OFF devices to implement the operations of a system of logic called (two-valued) *Boolean algebra*. Its statements may take the form of algebraic expressions, logic block diagrams, or truth tables, as well as circuits. The theorems of Boolean algebra are listed in Table 15-1 in *algebraic*

TABLE 15-1 Boolean Algebra Theorems

number		theorem	name
1	(a)	$A + B = B + A$	Commutative law
	(b)	$A \cdot B = B \cdot A$	
2	(a)	$(A + B) + C = A + (B + C)$	Associative law
	(b)	$(A \cdot B) \cdot C = A \cdot (B \cdot C)$	
3	(a)	$A \cdot (B + C) = A \cdot B + A \cdot C$	Distributive law
	(b)	$A + (B \cdot C) = (A + B) \cdot (A + C)$	
4	(a)	$A + A = A$	Identity law
	(b)	$A \cdot A = A$	
5		$\bar{\bar{A}} = A$	Negation law
6	(a)	$A + A \cdot B = A$	Redundancy law
	(b)	$A \cdot (A + B) = A$	
7	(a)	$0 + A = A$	Boolean postulates
	(b)	$1 \cdot A = A$	
	(c)	$1 + A = 1$	
	(d)	$0 \cdot A = 0$	
8	(a)	$\bar{A} + A = 1$	
	(b)	$\bar{A} \cdot A = 0$	
9	(a)	$A + \bar{A} \cdot B = A + B$	
	(b)	$A \cdot (\bar{A} + B) = A \cdot B$	
10	(a)	$\overline{A + B} = \bar{A} \cdot \bar{B}$	DeMorgan's laws
	(b)	$\overline{A \cdot B} = \bar{A} \cdot \bar{B}$	

expression form

$A, B, C \equiv$ logic variables $\quad \cdot \equiv$ AND (intersection of sets)

$0 \equiv$ null (or zero) set $\quad + \equiv$ OR (union of sets)

$1 \equiv$ universal set $\quad \bar{A} \equiv$ negation (or complement) of A

By use of other theorems in Table 15-1, prove Theorem 4(*a*).

$$\begin{align} A + A &= 1 \cdot (A + A) && \text{by Theorem 7}(b) \\ &= (\bar{A} + A) \cdot (A + A) && \text{by Theorem 8}(a) \\ &= \bar{A} \cdot A + A \cdot A && \text{by Theorem 3}(a) \\ &= 0 + A && \text{by Theorems 8}(a) \text{ and 4}(b) \\ &= A && \text{by Theorem 7}(a) \end{align}$$

15.38 Use Table 15-1 to prove Theorem 4(*b*).

$$\begin{align} A \cdot A &= A \cdot A + 0 && \text{by Theorem 7}(a) \\ &= A \cdot A + \bar{A} \cdot A && \text{by Theorem 8}(b) \\ &= (A + \bar{A}) \cdot A && \text{by Theorem 3}(a) \\ &= 1 \cdot A && \text{by Theorem 8}(a) \\ &= A && \text{by Theorem 7}(b) \end{align}$$

15.39 Use Table 15-1 to prove Theorem 7(*c*).

$$\begin{align} 1 + A &= \bar{A} + A + A && \text{by Theorem 8}(a) \\ &= \bar{A} + A && \text{by Theorem 4}(a) \\ &= 1 && \text{by Theorem 8}(a) \end{align}$$

15.40 Use Table 15-1 to prove Theorem 7(*d*).

$$\begin{align} 0 \cdot A &= (\bar{A} \cdot A) \cdot A && \text{by Theorem 8}(b) \\ &= \bar{A} \cdot (A \cdot A) && \text{by Theorem 2}(b) \\ &= \bar{A} \cdot A && \text{by Theorem 4}(b) \\ &= 0 && \text{by Theorem 8}(b) \end{align}$$

15.41 Use Table 15-1 to prove Theorem 5.

$$\begin{align} \bar{\bar{A}} &= \overline{\overline{A \cdot A}} && \text{by Theorem 4}(b) \\ &= \overline{\bar{A} + \bar{A}} && \text{by Theorem 10}(b) \\ &= A \cdot A && \text{by Theorem 10}(a) \\ &= A && \text{by Theorem 4}(b) \end{align}$$

15.42 Prove Theorem 6(*a*) by using Table 15-1.

$$\begin{align} A + A \cdot B &= 1 \cdot A + A \cdot B && \text{by Theorem 7}(b) \\ &= 1 \cdot A + B \cdot A && \text{by Theorem 1}(b) \\ &= (1 + B) \cdot A && \text{by Theorem 3}(a) \\ &= 1 \cdot A && \text{by Theorem 7}(c) \\ &= A && \text{by Theorem 7}(b) \end{align}$$

15.43 Prove Theorem 6(b) by using Table 15-1.

$$A \cdot (A + B) = (0 + A) \cdot (A + B) \quad \text{by Theorem 7}(a)$$
$$= (A + 0) \cdot (A + B) \quad \text{by Theorem 1}(a)$$
$$= A + (0 \cdot B) \quad \text{by Theorem 3}(b)$$
$$= A + 0 \quad \text{by Theorem 7}(d)$$
$$= 0 + A \quad \text{by Theorem 1}(a)$$
$$= A \quad \text{by Theorem 7}(a)$$

15.44 The *Venn diagram* is a graphical illustration of the union and intersection of sets where the universal set is represented by a rectangular area and the elements of sets are represented by the interiors of correspondingly labeled circles. Show the validity of Theorem 7(b), $1 \cdot A = A$, from Table 15-1 by use of a Venn diagram.

▮ The Venn diagram is constructed in Fig. 15-1. First, $1 \cdot A$, or the intersection of A and the universal set is filled in with horizontal hatching. However, this hatched area is identically A; thus, $1 \cdot A = A$.

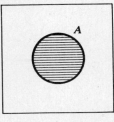

$1 \cdot A \equiv$ **Fig. 15-1**

15.45 Illustrate the validity of Theorem 7(c), $1 + A = 1$, from Table 15-1 through use of a Venn diagram.

▮ The Venn diagram is constructed in Fig. 15-2. First, the universal set is filled in with horizontal hatching. Then, A is filled in with vertical hatching. It is clearly seen that the union of the universal set and A is the universal set, or $A + 1 = 1$.

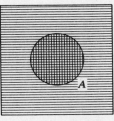

$1 \equiv \quad A \;|||$ **Fig. 15-2**

15.46 Validate Theorem 6(a), $A + A \cdot B = A$, of Table 15-1 by use of a Venn diagram.

▮ The Venn diagram is constructed in Fig. 15-3 where A has been filled by horizontal hatching. Then, the area that is common to both A and B is vertically hatched. The union of the vertical hatched and horizontal hatched areas is seen to be simply A, or $A + A \cdot B = A$.

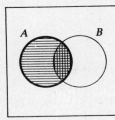

$A \equiv \quad A \cdot B \;|||$ **Fig. 15-3**

15.47 Illustrate Theorem 6(b), $A \cdot (A + B) = A$, of Table 15-1 by use of a Venn diagram.

▌ The Venn diagram is constructed in Fig. 15-4. First, A is filled by horizontal hatching. Then, the union of A and B is vertically hatched. The intersection of A and $A + B$ is the cross-hatched area, or $A \cdot (A + B) = A$.

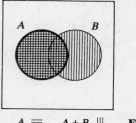

$A \equiv \quad A + B \;|||\quad$ **Fig. 15-4**

15.48 Show the validity of Theorem 9(a), $A + \bar{A} \cdot B = A + B$, from Table 15-1 through use of a Venn diagram.

▌ The Venn diagram is shown by Fig. 15-5 where A is filled by horizontal hatching. Also, the intersection of $\bar{A}$ and B is vertically hatched. The union of these two hatched areas is either A or B. Thus, $A + \bar{A} \cdot B = A + B$.

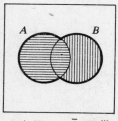

$A \equiv \quad \bar{A} \cdot B \;|||\quad$ **Fig. 15-5**

15.49 Illustrate Theorem 10(b), $\overline{A \cdot B} = \bar{A} + \bar{B}$, from Table 15-1 by use of a Venn diagram.

▌ As logic expressions increase in complexity, it is usually easier to follow the work if a Venn diagram is constructed for each side of the equation. First, the area that does not belong to the intersection of A and B is horizontally hatched in Fig. 15-6(a). Then, the area that is $\bar{A}$ is horizontally hatched in Fig. 15-6(b). Also, the area that is $\bar{B}$ is filled by vertical hatching. Now any area of Fig. 15-6(b) with hatching (vertical or horizontal) represents the union of $\bar{A}$ and $\bar{B}$. Since the areas with no hatching in both Figs. 15-6(a) and (b) are identical, we conclude that $\overline{A \cdot B} = \bar{A} + \bar{B}$.

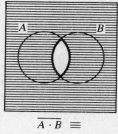 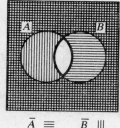

$\overline{A \cdot B} \equiv \qquad\qquad\qquad \bar{A} \equiv \quad \bar{B} \;|||\quad$ **Fig. 15-6**

(a) \qquad\qquad\qquad (b)

15.50 Illustrate Theorem 2(a), $(A + B) + C = A + (B + C)$, from Table 15.1 by use of a Venn diagram.

▌ First, the area representing the union of A and B is filled with horizontal hatching in Fig. 15-7(a). Then, the area belonging to C is vertically hatched. So it is seen that the union of $A + B$ and C is any area of Fig. 15-7(a) with hatching (vertical or horizontal). Next, the area belonging to A is filled by horizontal hatching in Fig. 15-7(b). Then, the union of B and C is vertically hatched. Hence, the union of A and $B + C$ is the area of Fig. 15-7(b) with any hatching. Since the areas of both Figs. 15-7(a) and 15-7(b) with any hatching are identical, then $(A + B) + C = A + (B + C)$.

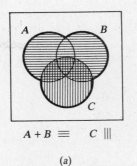
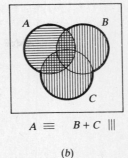

$A + B \equiv \quad C \;|||$ $\qquad A \equiv \quad B + C \;|||$

(a) $\qquad\qquad\qquad$ (b) $\qquad$ Fig. 15-7

15.51 Verify Theorem 3(a), $A \cdot (B + C) = A \cdot B + A \cdot C$, of Table 15-1 by use of a Venn diagram.

▌ First, the area belonging to A is filled by horizontal hatching in Fig. 15-8(a). Then, the area belonging to the union of B and C is vertically hatched. The intersection of A and $B + C$ is the area inside the heavy lines of Fig. 15-8(a). Next, the intersection of A and B is filled by horizontal hatching and the intersection of A and C is filled by vertical hatching in Fig. 15-8(b). The union of $A \cdot B$ with $A \cdot C$ is the area inside the heavy lines of Fig. 15-8(b). Since the areas enclosed by heavy lines in both Figs. 15-8(a) and 15-8(b) are identical, it is concluded that $A \cdot (B + C) = A \cdot B + A \cdot C$.

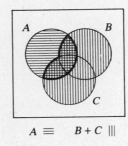
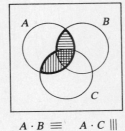

$A \equiv \quad B + C \;|||$ $\qquad A \cdot B \equiv \quad A \cdot C \;|||$

(a) $\qquad\qquad\qquad$ (b) $\qquad$ Fig. 15-8

15.52 Validate Theorem 3(b), $A + (B \cdot C) = (A + B) \cdot (A + C)$, of Table 15-1 through use of Venn diagrams.

▌ The area belonging to A is filled by horizontal hatching and the area for the intersection of B and C is filled by vertical hatching in Fig. 15-9(a); the union of A and $B \cdot C$ is enclosed by heavy lines. In Fig. 15-9(b), the union of A and B is horizontally hatched while the union of A and C is vertically hatched. The intersection of $A + B$ and $A + C$ is enclosed by heavy lines in Fig. 15-9(b). Since the areas enclosed by heavy lines in Fig. 15-9(a) and 15-9(b) are identical, we conclude that $A + (B \cdot C) = (A + B) \cdot (A + C)$.

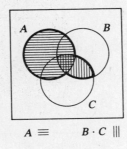
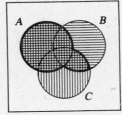

$A \equiv \quad B \cdot C \;|||$ $\qquad A + B \equiv \quad A + C \;|||$

(a) $\qquad\qquad\qquad$ (b) $\qquad$ Fig. 15-9

15.53 *Perfect induction* is a method of proof whereby a theorem is verified for all possible combinations of values that the variables may assume. For the case of binary variables, the truth table can be used to orderly present all possible combinations of variable values. Use a truth table to prove Theorem 6(a), $A + A \cdot B = A$, of Table 15-1 by perfect induction.

▌ The truth table for perfect induction can vary in number of columns, depending on the nature of the Boolean function to be implemented, and the number of rows, depending on the number of variables contained in the Boolean function. A sufficient table for Theorem 6(a), $A + A \cdot B = A$, is shown below:

A	B	$A \cdot B$	$A + A \cdot B$
0	0	0	0
0	1	0	0
1	0	0	1
1	1	1	1

Since the first column, representing A, and the last column, representing $A + A \cdot B$, are identical for all possible combinations of the variables A and B, Theorem 6(a) is thus proved by the method of perfect induction.

15.54 Prove Theorem 6(b), $A \cdot (A + B) = A$, of Table 15-1 by the method of perfect induction.

▮ A sufficient truth table for $A \cdot (A + B) = A$ is constructed as follows:

A	B	$A + B$	$A \cdot (A + B)$
0	0	0	0
0	1	1	0
1	0	1	1
1	1	1	1

Since the first column, A, and the last column, $A \cdot (A + B)$, are equal over all possible values of A and B, Theorem 6(b) is proved by perfect induction.

15.55 Prove Theorem 9(a), $A + \bar{A} \cdot B = A + B$, of Table 15-1 by the method of perfect induction.

▮ A sufficient truth table for $A + \bar{A} \cdot B = A + B$ is shown below:

A	B	$\bar{A}$	$\bar{A} \cdot B$	$A + \bar{A} \cdot B$	$A + B$
0	0	1	0	0	0
0	1	1	1	1	1
1	0	0	0	1	1
1	1	0	0	1	1

Since the last two columns are identical, Theorem 9(a) is proven by perfect induction.

15.56 Prove Theorem 9(b), $A \cdot (\bar{A} + B) = A \cdot B$, of Table 15-1 by the method of perfect induction.

▮ A sufficient truth table for Theorem 9(b), $A \cdot (\bar{A} + B) = A \cdot B$ is constructed as follows:

A	B	$\bar{A}$	$\bar{A} + B$	$A \cdot (\bar{A} + B)$	$A \cdot B$
0	0	1	1	0	0
0	1	1	1	0	0
1	0	0	0	0	0
1	1	0	1	1	1

Since the last two columns are equal over all possible values of A and B, the theorem is proven by perfect induction.

15.57 Prove Theorem 10(a), $\overline{A+B} = \bar{A} \cdot \bar{B}$, of Table 15-1 by the method of perfect induction.

▌ A sufficient truth table for $\overline{A+B} = \bar{A} \cdot \bar{B}$ is constructed below:

A	B	$\bar{A}$	$\bar{B}$	$A+B$	$\overline{A+B}$	$\bar{A} \cdot \bar{B}$
0	0	1	1	0	1	1
0	1	1	0	1	0	0
1	0	0	1	1	0	0
1	1	0	0	1	0	0

Since the last two columns are equal over all possible values of A and B, the theorem is proven by perfect induction.

15.58 Prove Theorem 10(b), $\overline{A \cdot B} = \bar{A} + \bar{B}$, of Table 15-1 by the method of perfect induction.

▌ A sufficient truth table for $\overline{A \cdot B} = \bar{A} + \bar{B}$, is shown below:

A	B	$\bar{A}$	$\bar{B}$	$A \cdot B$	$\overline{A \cdot B}$	$\bar{A} + \bar{B}$
0	0	1	1	0	1	1
0	1	1	0	0	1	1
1	0	0	1	0	1	1
1	1	0	0	1	0	0

Since the last two columns are equal, Theorem 10(b) is proven by the method of perfect induction.

15.59 Prove Theorem 2(a), $(A + B) + C = A + (B + C)$, of Table 15-1 by the method of perfect induction.

▌ A sufficient truth table for $(A+B)+C = A+(B+C)$ is constructed below:

A	B	C	$A+B$	$B+C$	$(A+B)+C$	$A+(B+C)$
0	0	0	0	0	0	0
0	0	1	0	1	1	1
0	1	0	1	1	1	1
0	1	1	1	1	1	1
1	0	0	1	0	1	1
1	0	1	1	1	1	1
1	1	0	1	1	1	1
1	1	1	1	1	1	1

Since the last two columns are equal over all possible values of A, B, and C, the theorem is proven by perfect induction.

GATE IMPLEMENTATION OF BOOLEAN FUNCTIONS

15.60 The elements of digital circuits are called *logic gates*, and their symbols are shown in Fig. 15-10. Circuit diagrams that include these symbols are called *logic block diagrams*. For the interconnected logic gates (logic block diagram) of Fig. 15-11, determine the Boolean function f generated.

▌ Referring to Fig. 15-11,

$$X = A \cdot B \qquad (1)$$

$$Y = \bar{C} \qquad (2)$$

$$f = X + Y \qquad (3)$$

Substitute (*1*) and (*2*) into (*3*) to find

$$f = A \cdot B + \bar{C} \tag{4}$$

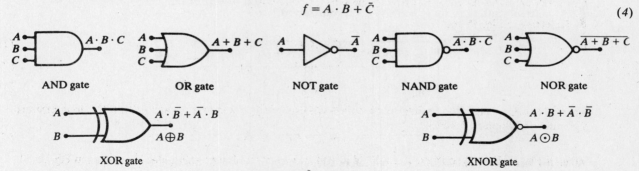

Fig. 15-10

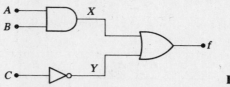

Fig. 15-11

15.61 Determine the Boolean function f simulated by the logic circuit of Fig. 15-12.

▌ The gate outputs are

$$X = A \cdot B \tag{1}$$
$$Y = \bar{A} \cdot \bar{B} \tag{2}$$
$$f = X + Y \tag{3}$$

Substitution of (*1*) and (*2*) into (*3*) yields

$$f = (A \cdot B) + (\bar{A} \cdot \bar{B}) \tag{4}$$

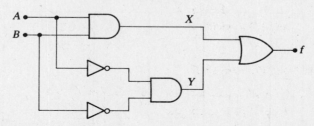

Fig. 15-12

15.62 Find the Boolean function f simulated by the logic circuit of Fig. 15-13.

▌ The gate outputs are as follows:

$$X = A + B \tag{1}$$
$$Y = \bar{A} + \bar{B} \tag{2}$$
$$f = X \cdot Y \tag{3}$$

Use of (*1*) and (*2*) in (*3*) gives

$$f = (A + B) \cdot (\bar{A} + \bar{B}) \tag{4}$$

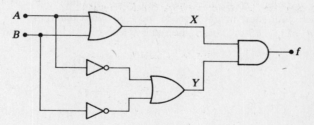

Fig. 15-13

15.63 In general, a logic circuit to simulate a particular Boolean function is not unique. Show that the logic circuit of Fig. 15-14 simulates the same Boolean function as the logic circuit of Fig. 15-12.

▌ Referring to Fig. 15-14,

$$X = \overline{A + B} = \bar{A} \cdot \bar{B} \tag{1}$$

$$Y = A \cdot B \tag{2}$$

$$f = X + Y \tag{3}$$

where DeMorgan's law was applied to the right-hand side of (1). Substitute (1) and (2) into (3) to determine the simulated Boolean function:

$$f = (\bar{A} \cdot \bar{B}) + (A \cdot B) \tag{4}$$

which is identical to (4) of Problem 15.61. Note that a smaller number of logic gates is required in Fig. 15-14 than in Fig. 15-12.

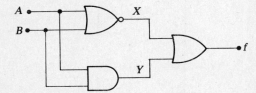

Fig. 15-14

15.64 Show that the logic circuit of Fig. 15-15 simulates the same Boolean function as the logic circuit of Fig. 15-13.

▌ Using DeMorgan's law as convenient,

$$X_1 = \overline{A + B} \tag{1}$$

$$X_2 = \overline{X_1 \cdot X_1} = \bar{X}_1 + \bar{X}_1 \tag{2}$$

Substitute (1) into (2) and use Theorems 5 and 4(a) of Table 15-1:

$$X_2 = \overline{\overline{A + B}} + \overline{\overline{A + B}} = A + B + A + B = A + B \tag{3}$$

From Fig. 15-15, $\qquad Y = \overline{A \cdot B} = \bar{A} + \bar{B} \tag{4}$

And, $\qquad f = X_2 \cdot Y \tag{5}$

Substitute (3) and (4) into (5) to yield

$$f = (A + B) \cdot (\bar{A} + \bar{B}) \tag{6}$$

which is identical to (4) of Problem 15.62.

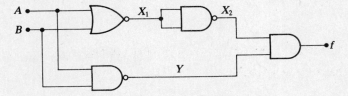

Fig. 15-15

15.65 Form a logic circuit to simulate the Boolean function $f = (\bar{A} \cdot \bar{B}) + (\bar{C} \cdot \bar{D})$.

▌ Applying DeMorgan's law to f, we see that

$$\bar{f} = \overline{(\bar{A} \cdot \bar{B}) + (\bar{C} \cdot \bar{D})} = (A + B) \cdot (C + D)$$

Now, $(A + B)$ and $(C + D)$ can each be generated by a 2-input OR gate and the results can be fed to a NAND gate to yield f. See Fig. 15-16 for the logic block diagram.

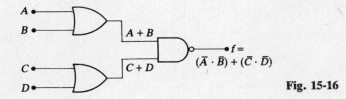

Fig. 15-16

15.66 Construct a logic circuit to simulate the Boolean function $f = A \cdot B \cdot \bar{C}$.

▮ See Fig. 15-17.

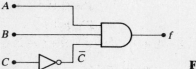

Fig. 15-17

15.67 Devise a logic circuit to simulate the Boolean function $f = A \cdot \bar{B} + C$.

▮ See Fig. 15-18 where $X = A \cdot \bar{B}$ and so $f = X + C = A \cdot \bar{B} + C$.

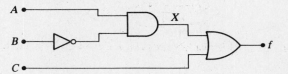

Fig. 15-18

15.68 Build up two different logic circuits to simulate the Boolean function $f = (A + B) \cdot \bar{C}$.

▮ In Fig. 15-19(a),
$$X = \overline{A + B} = \bar{A} \cdot \bar{B}$$
$$f = \overline{X + C} = \overline{(\bar{A} \cdot \bar{B}) + C} = (A + B) \cdot \bar{C}$$

In Fig. 15-19(b),
$$X = A + B$$
$$f = X \cdot \bar{C} = (A + B) \cdot \bar{C}$$

Thus, it is seen that both circuits of Fig. 15-19 simulate the required f.

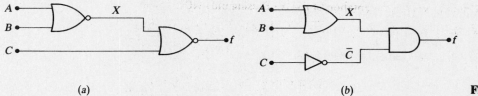

(a)   (b)   Fig. 15-19

15.69 Use only NAND gates to design a logic circuit to simulate the Boolean function $f = A \cdot (B + C)$.

▮ By Theorem 3(a) of Table 15-1,
$$f = A \cdot (B + C) = (A \cdot B) + (A \cdot C) \qquad (1)$$

Use Theorem 5 and then apply DeMorgan's law to (1), giving
$$f = \bar{\bar{f}} = \overline{\overline{(A \cdot B) + (A \cdot C)}} = \overline{(\bar{A} + \bar{B}) \cdot (\bar{A} + \bar{C})}$$

Now $X = \bar{A} + \bar{B}$ and $Y = \bar{A} + \bar{C}$ can each be simulated by a NAND gate. Then each output can be fed to a NAND gate to produce $f = \overline{X \cdot Y}$. See Fig. 15-20 for the completed logic circuit.

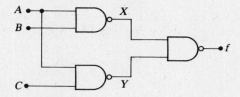

Fig. 15-20

15.70 Describe an implementation of the Boolean function $f = A \cdot B + C \cdot D + E \cdot F$ using only NAND gates.

▌ See Fig. 15-21 where
$$X = \overline{A \cdot B} = \bar{A} + \bar{B}$$
$$Y = \bar{C} + \bar{D}$$
$$Z = \bar{E} + \bar{F}$$

And,
$$f = \overline{X \cdot Y \cdot Z} = \bar{X} + \bar{Y} + \bar{Z} = \overline{\bar{A} + \bar{B}} + \overline{\bar{C} + \bar{D}} + \overline{\bar{E} + \bar{F}} = A \cdot B + C \cdot D + E \cdot F$$

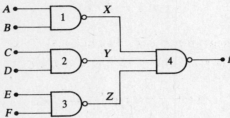

Fig. 15-21

15.71 If the logic circuit of Fig. 15-21 is available, describe a connection to implement the Boolean function $f = A \cdot B + C \cdot D + E$.

▌ Connect the two inputs of NAND gate 3 together to receive signal E and let all else remain unchanged. To verify, X and Y are unchanged from Problem 15.70; however,
$$Z = \bar{E} + \bar{E} = \bar{E}$$

Hence,
$$f = \overline{X \cdot Y \cdot Z} = \bar{X} + \bar{Y} + \bar{Z} = \overline{\bar{A} + \bar{B}} + \overline{\bar{C} + \bar{D}} + \bar{\bar{E}} = A \cdot B + C \cdot D + E$$

15.72 Describe the implementation of $f = A + B$ with only NOR gates.

▌ See Fig. 15-22 where $X = \overline{A + B} = \bar{A} \cdot \bar{B}$. Then,
$$f = \overline{X + X} = \bar{X} \cdot \bar{X} = \bar{X} = \overline{\bar{A} \cdot \bar{B}} = A + B$$

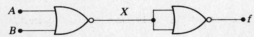

Fig. 15-22

15.73 Describe the implementation of $f = A \cdot B$ using only NOR gates.

▌ See Fig. 15-23 where $X = \overline{A + A} = \bar{A} \cdot \bar{A} = \bar{A}$ and $Y = \bar{B}$. Then,
$$f = \overline{\bar{A} + \bar{B}} = A \cdot B$$

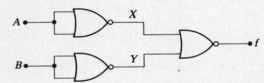

Fig. 15-23

15.74 Describe the implementation of $f = A \cdot B$ with only NAND gates.

▌ See Fig. 15-24 where $X = \overline{A \cdot B}$ and so
$$f = \overline{X \cdot X} = \bar{X} + \bar{X} = \bar{X} = \overline{\overline{A \cdot B}} = A \cdot B$$

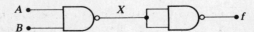

Fig. 15-24

15.75 Describe a logic circuit to implement $f = A \cdot B + C$.

▌ We see that the Boolean function above differs from f of Problem 15.60 in that $\bar{C}$ is replaced by C. Hence, the logic circuit of Fig. 15-11, with the inverter removed so that $Y = C$, will yield the required simulation. Verifying,
$$X = A \cdot B \qquad Y = C$$
and
$$f = X + Y = A \cdot B + C$$

15.76 Determine the output for the logic circuit of Fig. 15-12 if the two inverters are removed so that $Y = A \cdot B$.

▮ Based on (4) of Problem 15.61 and Theorem 4(b) of Table 15-1,

$$f = (A + B) \cdot (A + B) = A + B$$

15.77 Find the Boolean function simulated by the logic circuit of Fig. 15-14 if the AND gate is replaced by a NAND gate.

▮
$$X = \overline{A + B} = \bar{A} \cdot \bar{B} \qquad Y = \overline{A \cdot B} = \bar{A} + \bar{B}$$

and
$$f = X + Y = \bar{A} \cdot \bar{B} + \bar{A} + \bar{B} \qquad (1)$$

15.78 Show that if the AND gate of Fig. 15-14 were replaced by a NAND gate (see Problem 15.77), then the resulting circuit could be replaced by a single NAND gate.

▮ Using Theorem 7(a) of Table 15-1, (1) of Problem 15.77 can be rewritten as

$$f = \bar{A} \cdot \bar{B} + \bar{A} + \bar{B} = \bar{A} \cdot \bar{B} + 1 \cdot \bar{B} + \bar{A}$$

Applying the commutative law, the distributive law, and Theorems 7(c) and (b) of Table 15-1,

$$f = \bar{B} \cdot \bar{A} + \bar{B} \cdot 1 + \bar{A} = \bar{B} \cdot (\bar{A} + 1) + \bar{A} = \bar{B} \cdot 1 + \bar{A} = \bar{B} + \bar{A}$$

which is the output of a two-input NAND gate with A and B as the inputs.

15.79 Replace the logic circuit of Fig. 15-13 by a single logic gate.

▮ From Problem 15.62,

$$f = (A + B) \cdot (\bar{A} + \bar{B})$$

Construct a truth table to find the single logic gate that matches f:

A	B	$\bar{A}$	$\bar{B}$	$A+B$	$\bar{A}+\bar{B}$	$(A+B) \cdot (\bar{A}+\bar{B})$
0	0	1	1	0	1	0
0	1	1	0	1	1	1
1	0	0	1	1	1	1
1	1	0	0	1	0	0

It is seen that the last column, or f, is identical to the XOR logic (see Problem 14.10); thus, the circuit could be replaced by one XOR gate.

15.80 Replace the logic circuit of Fig. 15-12 by a single logic gate.

▮ From Problem 15.61,

$$f = (A \cdot B) + (\bar{A} \cdot \bar{B})$$

Construct a truth table to find the single logic gate that matches f:

A	B	$\bar{A}$	$\bar{B}$	$A \cdot B$	$\bar{A} \cdot \bar{B}$	$(A \cdot B) + (\bar{A} \cdot \bar{B})$
0	0	1	1	0	1	1
0	1	1	0	0	0	0
1	0	0	1	0	0	0
1	1	0	0	1	0	1

It is seen that the last column, or f, matches the logic of an XNOR gate (see Problem 14.11).

15.81 Design an XNOR gate with an AND, OR, and NOR gate.

▌ See Fig. 15-25 where
$$X = A \cdot B \qquad Y = \overline{A + B} = \bar{A} \cdot \bar{B}$$
Then,
$$f = X + Y = (A \cdot B) + (\bar{A} \cdot \bar{B})$$

which, from Problem 15.80, is known to be XNOR logic; hence, the logic circuit could be replaced by a single XNOR gate.

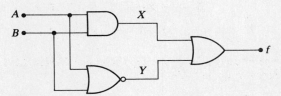

Fig. 15-25

15.82 Design an XOR gate using an AND, OR, and NAND gate.

▌ See Fig. 15-26 where
$$X = A + B \qquad Y = \overline{A \cdot B} = \bar{A} + \bar{B}$$
$$f = X \cdot Y = (A + B) \cdot (\bar{A} + \bar{B})$$

which, from Problem 15.79, is known to be XOR logic; thus, the circuit could be replaced by a single XOR gate.

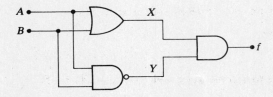

Fig. 15-26

15.83 Design a circuit to perform XOR logic using only NOR and NAND gates.

▌ See Fig. 15-27 where $X = \overline{A + B} = \bar{A} \cdot \bar{B}$. From Problem 15.74, $Y = A \cdot B$.
Thus,
$$f = \overline{X + Y} = \overline{(\bar{A} \cdot \bar{B}) + (A \cdot B)} = (A + B) \cdot (\bar{A} + \bar{B})$$

which from Problem 15.79 is known to be XOR logic.

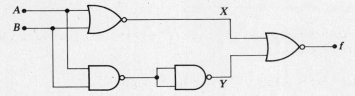

Fig. 15-27

15.84 Design a circuit to perform XOR logic using only NOR gates.

▌ See Fig. 15-28 where $X = \overline{A + B}$. Based on Problem 15.73, $Y = A \cdot B$. Hence,
$$f = \overline{X + Y} = \bar{X} \cdot \bar{Y} = \overline{\overline{(A + B)}} \cdot \overline{(A \cdot B)} = (A + B) \cdot (\bar{A} + \bar{B})$$

which from Problem 15.79 is known to be XOR logic.

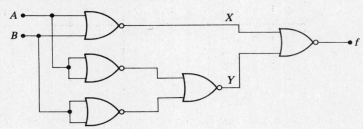

Fig. 15-28

15.85 Design a circuit to perform XOR logic using only NAND gates.

▮ See Fig. 15-29, where, based on Problem 15.74,
$$Y = A \cdot B$$
From the results of Problem 15.143,
$$X_1 = A + B$$
Now
$$X_2 = \overline{X_1 \cdot X_1} = \bar{X}_1 + \bar{X}_1 = \bar{X}_1 = \overline{A + B} = \bar{A} \cdot \bar{B}$$
Whence,
$$f = \overline{X_2 \cdot Y} = \bar{X}_2 + \bar{Y} = \overline{\bar{A} \cdot \bar{B}} + \overline{A \cdot B} = (A + B) \cdot (\bar{A} + \bar{B})$$
which from Problem 15.79 is known to be XOR logic.

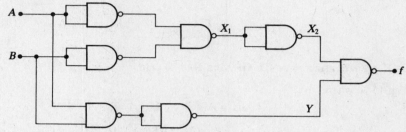

Fig. 15-29

15.86 Design a circuit to perform XNOR logic using only NOR and NAND gates.

▮ See Fig. 15-30 where
$$Y = \overline{A \cdot B} = \bar{A} + \bar{B}$$
Based on Problem 15.72,
$$X = A + B$$
Hence,
$$f = \overline{X \cdot Y} = \bar{X} + \bar{Y} = \overline{(A + B)} + \overline{(\bar{A} + \bar{B})} = (\bar{A} \cdot \bar{B}) + (A \cdot B)$$
which, from Problem 15.80, is known to be XNOR logic.

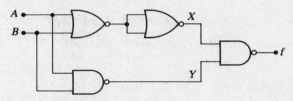

Fig. 15-30

15.87 Design a circuit to perform XNOR logic using only NOR gates.

▮ See Fig. 15-31 where
$$X = \overline{A + B} = \bar{A} \cdot \bar{B}$$
By the result of Problem 15.73,
$$Y = A \cdot B$$
From Problem 15.72,
$$f = X + Y = (\bar{A} \cdot \bar{B}) + (A \cdot B)$$
which, from Problem 15.80, is known to be XNOR logic.

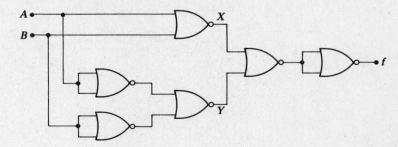

Fig. 15-31

BOOLEAN ALGEBRA AND LOGIC GATES □ 475

15.88 Find a circuit to perform XNOR logic using only NAND gates.

▌ See Fig. 15-32 where
$$X = \overline{A \cdot B}$$

By the result of Problem 15.143,
$$Y = \overline{A + B}$$

Hence,
$$f = \overline{X \cdot Y} = \bar{X} + \bar{Y} = \overline{\overline{(A \cdot B)}} + \overline{\overline{(A + B)}} = (A \cdot B) + (\bar{A} \cdot \bar{B})$$

which from Problem 15.80 is known to be XNOR logic.

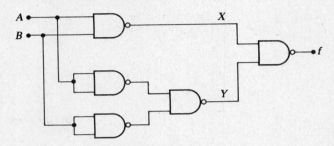

Fig. 15-32

15.89 Determine a single logic gate to replace the logic circuit of Fig. 15-33.

▌ The output is $f = \bar{A} \cdot \bar{B}$. Construct a truth table as follows:

A	B	$\bar{A}$	$\bar{B}$	$\bar{A} \cdot \bar{B}$
0	0	1	1	1
0	1	1	0	0
1	0	0	1	0
1	1	0	0	0

The last column of the truth table, or f, is the logic of a NOR gate (see Problem 14.9). Thus, the circuit could be replaced by a single two-input NOR gate.

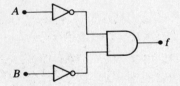

Fig. 15-33

15.90 Determine a single logic gate to replace the circuit of Fig. 15-34.

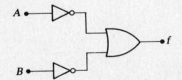

Fig. 15-34

▌ The output is $f = \bar{A} + \bar{B}$. Set up a truth table to examine all possible outputs:

A	B	$\bar{A}$	$\bar{B}$	$\bar{A} + \bar{B}$
0	0	1	1	1
0	1	1	0	1
1	0	0	1	1
1	1	0	0	0

The last column of the truth table, or f, is the logic of a NAND gate (see Problem 14.8). Thus, the circuit could be replaced by a two-input NAND gate.

SIMPLIFICATION OF BOOLEAN FUNCTIONS

15.91 Simplify by algebraic manipulation, that is, by use of the theorems of Table 15-1, the Boolean function $f = B \cdot (A + B)$.

$f = A \cdot B + B \cdot \bar{B}$	by Theorem 3(a)
$= A \cdot B$	by Theorem 8(b)

15.92 Algebraically simplify $f = A \cdot C + A \cdot \bar{B} \cdot C$.

$f = A \cdot C + A \cdot C \cdot \bar{B}$	by Theorem 1(b)
$= A \cdot C \cdot (1 + \bar{B})$	by Theorem 3(a)
$= A \cdot C$	by Theorem 7(c)

15.93 Algebraically simplify $f = \bar{A} + \bar{B} + A \cdot B \cdot \bar{C}$.

$f = \bar{A} \cdot (B\bar{C} + 1) + \bar{B} + A \cdot B \cdot \bar{C}$	by Theorem 7(c)
$= \bar{A} \cdot B \cdot \bar{C} + \bar{A} + \bar{B} + A \cdot B \cdot \bar{C}$	by Theorem 3(a)
$= (\bar{A} + A) \cdot B \cdot \bar{C} + \bar{A} + \bar{B}$	by Theorems 1(a) and 3(a)
$= B \cdot \bar{C} + \bar{A} + \bar{B}$	by Theorem 8(a)

15.94 Algebraically simplify $f = A + \bar{A} \cdot B$.

$f = A \cdot (B + 1) + \bar{A} \cdot B$	by Theorem 7(c)
$= A \cdot B + A + \bar{A} \cdot B$	by Theorem 3(a)
$= (A + \bar{A}) \cdot B + A$	by Theorems 1(a) and 3(a)
$= B + A$	by Theorem 8(a)

15.95 Algebraically simplify $f = A \cdot (A + B) + B \cdot (\bar{A} + B)$.

$f = A \cdot A + A \cdot B + \bar{A} \cdot B + B \cdot B$	by Theorem 3(a)
$= A + A \cdot B + \bar{A} \cdot B + B$	by Theorem 4(b)
$= A + (A + \bar{A}) \cdot B + B$	by Theorem 3(a)
$= A + B + B$	by Theorem 8(a)
$= A + B$	by Theorem 4(a)

15.96 Algebraically simplify $f = A + B + \bar{A}$.

$f = (A + \bar{A}) + B$	by Theorem 1(a)
$= 1 + B$	by Theorem 8(a)
$= 1$	by Theorem 7(c)

15.97 By algebraic manipulation, simplify $f = \bar{A} \cdot \bar{B} \cdot \bar{C} \cdot \bar{D} + \bar{A} \cdot \bar{B} \cdot \bar{C} \cdot D$.

$f = \bar{A} \cdot \bar{B} \cdot \bar{C} \cdot (\bar{D} + D)$	by Theorem 3(a)
$= \bar{A} \cdot \bar{B} \cdot \bar{C} \cdot (1)$	by Theorem 8(a)
$= \bar{A} \cdot \bar{B} \cdot \bar{C}$	by Theorem 7(b)

15.98 Use algebraic manipulation to simplify $f = \bar{A} \cdot \bar{B} \cdot \bar{C} \cdot \bar{D} + \bar{A} \cdot \bar{B} \cdot \bar{C} \cdot D + A \cdot \bar{B} \cdot \bar{C} \cdot D + A \cdot \bar{B} \cdot \bar{C} \cdot \bar{D}$.

▮ $\quad f = \bar{A} \cdot \bar{B} \cdot \bar{C} \cdot (\bar{D} + D) + A \cdot \bar{B} \cdot \bar{C} \cdot (\bar{D} + D) \quad$ by Theorem 3(a)

$\quad = \bar{A} \cdot \bar{B} \cdot \bar{C} \cdot (1) + A \cdot \bar{B} \cdot \bar{C} \cdot (1) \quad$ by Theorem 8(a)

$\quad = \bar{A} \cdot \bar{B} \cdot \bar{C} + A \cdot \bar{B} \cdot \bar{C} \quad$ by Theorem 7(b)

$\quad = (\bar{A} + A) \cdot \bar{B} \cdot \bar{C} \quad$ by Theorem 3(a)

$\quad = \bar{B} \cdot \bar{C} \quad$ by Theorems 8(a) and 7(b)

15.99 By algebraic manipulation, simplify $f = \bar{A} \cdot \bar{B} \cdot C + \bar{A} \cdot B \cdot C$.

▮ $\quad f = \bar{A} \cdot C \cdot \bar{B} + \bar{A} \cdot C \cdot B \quad$ by Theorem 1(b)

$\quad = \bar{A} \cdot C \cdot (\bar{B} + B) \quad$ by Theorem 3(a)

$\quad = \bar{A} \cdot C \cdot (1) \quad$ by Theorem 8(a)

$\quad = \bar{A} \cdot C \quad$ by Theorem 7(a)

15.100 Algebraically simplify $f = \bar{A} \cdot C + \bar{A} \cdot B + A \cdot \bar{B} \cdot C + B \cdot C$.

▮ $\quad f = \bar{A} \cdot B + (\bar{A} + A \cdot \bar{B} + B) \cdot C \quad$ by Theorems 1(a) and 3(a)

$\quad = \bar{A} \cdot B + [\bar{A} + A \cdot \bar{B} + (A + \bar{A}) \cdot B] \cdot C \quad$ by Theorem 8(a)

$\quad = \bar{A} \cdot B + [\bar{A} + A \cdot \bar{B} + A \cdot B + \bar{A} \cdot B] \cdot C \quad$ by Theorem 3(a)

$\quad = \bar{A} \cdot B + [\bar{A} + A \cdot (\bar{B} + B) + \bar{A} \cdot B] \cdot C \quad$ by Theorem 3(a)

$\quad = \bar{A} \cdot B + [\bar{A} + A + \bar{A} \cdot B] \cdot C \quad$ by Theorem 8(a)

$\quad = \bar{A} \cdot B + [1 + \bar{A} \cdot B] \cdot C \quad$ by Theorem 8(a)

$\quad = \bar{A} \cdot B + C \quad$ by Theorem 7(c)

15.101 A *Karnaugh map*, used in the simplification of Boolean expressions, is actually a truth table in another form. As such, the same number of possible combinations present in the truth table must be accounted for in the map; thus, for n variables, 2^n combinations exist. Simplification of a Boolean function, if possible, is accomplished by identification of the largest possible pattern of adjacent squares. Simplify $f = A + \bar{A} \cdot B$ by use of a Karnaugh map.

▮ A, the first term of f, spans both squares in the top row of the Karnaugh map; thus, the row is filled by 1's. The second term, $\bar{A} \cdot B$, spans only the lower right-hand square of the Karnaugh map and a 1 is so entered. 0's are entered in the balance of the map to give the completed Karnaugh map.

	$\bar{B}$	B
A	1	1
$\bar{A}$	0	1

When collecting patterns of adjacent squares, subareas of the map can be used more than once (justified by Theorem 4 of Table 15-1). For the above Karnaugh map, all 1's are used at least once by the two patterns consisting of the upper row (A) and the right column (B). Thus,

$$f = A + B$$

15.102 Simplify $f = A \cdot (A + B) + B \cdot (\bar{A} + B)$ by use of a Karnaugh map.

▮ First, f must be expressed in a *canonical form* (no associative terms) Boolean expression.

$$f = A + A \cdot B + \bar{A} \cdot B + B$$

The Karnaugh map is constructed:

	$\bar{B}$	B
A	1	1
$\bar{A}$	1	1

Since the map is filled, f represents the universal set, or $f = 1$.

15.103 Use a Karnaugh map to simplify $f = B \cdot (A + \bar{B})$.

▌ Expressing f as a canonical form Boolean function,

$$f = A \cdot B + B \cdot \bar{B}$$

The Karnaugh map is constructed:

	$\bar{B}$	B
A	0	1
$\bar{A}$	0	0

The term $B \cdot \bar{B} = 0$, and hence, occupies no space on the map. The simplified expression consists of the pattern expressible as $f = A \cdot B$.

15.104 Simplify $f = \bar{A} \cdot \bar{B} \cdot C + \bar{A} \cdot B \cdot C$ through use of a Karnaugh map.

▌ The three-variable map is constructed as follows:

	$\bar{B}$	$\bar{B}$	B	B
A	0	0	0	0
$\bar{A}$	0	1	1	0
	$\bar{C}$	C	C	$\bar{C}$

The pattern of two 1's is described by $f = \bar{A} \cdot C$.

15.105 Construct a Karnaugh map for $f = A \cdot C + A \cdot \bar{B} \cdot C$ and write a simplified expression for f.

▌ The Karnaugh map is

	$\bar{B}$	$\bar{B}$	B	B
A	0	1	1	0
$\bar{A}$	0	0	0	0
	$\bar{C}$	C	C	$\bar{C}$

Whence, the pattern of 1's is described by $f = A \cdot C$.

15.106 Use a Karnaugh map to simplify $f = \bar{A} + \bar{B} + A \cdot B \cdot \bar{C}$.

▌ Construct the map:

	$\bar{B}$	$\bar{B}$	B	B
A	1	1	0	1
$\bar{A}$	1	1	1	1
	$\bar{C}$	C	C	$\bar{C}$

From the pattern of 1's, $f = \bar{A} + \bar{B} + B \cdot \bar{C}$.

15.107 Simplify $f = \bar{A} \cdot C + \bar{A} \cdot B + A \cdot \bar{B} \cdot C + B \cdot C$ by use of a Karnaugh map.

▌ The Karnaugh map is as follows:

	$\bar{B}$	$\bar{B}$	B	B
A	0	1	1	0
$\bar{A}$	0	1	1	1
	$\bar{C}$	C	C	$\bar{C}$

Whence, $\qquad f = C + \bar{A} \cdot B$

15.108 Simplify $f = \bar{A} \cdot \bar{B} \cdot \bar{C} \cdot \bar{D} + \bar{A} \cdot \bar{B} \cdot \bar{C} \cdot D$ through use of a Karnaugh map.

▌ The four-variable map is constructed below:

	$\bar{B}$	$\bar{B}$	B	B	
A	0	0	0	0	$\bar{D}$
A	0	0	0	0	D
$\bar{A}$	1	0	0	0	D
$\bar{A}$	1	0	0	0	$\bar{D}$
	$\bar{C}$	C	C	$\bar{C}$	

The pattern of 1's is described by $f = \bar{A} \cdot \bar{B} \cdot \bar{C}$.

15.109 Construct a Karnaugh map and simplify $f = \bar{A} \cdot \bar{B} \cdot \bar{C} \cdot \bar{D} + \bar{A} \cdot \bar{B} \cdot \bar{C} \cdot D + A \cdot \bar{B} \cdot \bar{C} \cdot \bar{D} + A \cdot \bar{B} \cdot \bar{C} \cdot D$.

▌ The map is given as

	$\bar{B}$	$\bar{B}$	B	B	
A	1	0	0	0	$\bar{D}$
A	1	0	0	0	D
$\bar{A}$	1	0	0	0	D
$\bar{A}$	1	0	0	0	$\bar{D}$
	$\bar{C}$	C	C	$\bar{C}$	

The column of 1's is described by $f = \bar{B} \cdot \bar{C}$.

15.110 By use of a Karnaugh map, simplify $f = A \cdot \bar{B} \cdot C + \bar{A} \cdot B \cdot \bar{C} + \bar{A} \cdot B \cdot C + A \cdot \bar{B} \cdot \bar{C}$.

▌ The Karnaugh map is constructed below where it is seen, by grouping 1's, that $f = A \cdot \bar{B} + \bar{A} \cdot B$.

	$\bar{B}$	$\bar{B}$	B	B
A	1	1	0	0
$\bar{A}$	0	0	1	1
	$\bar{C}$	C	C	$\bar{C}$

15.111 Simplify the Boolean function $f = A \cdot B \cdot \bar{C} + \bar{A} \cdot \bar{B} \cdot \bar{C} + B \cdot \bar{C}$.

▌ The Karnaugh map constructed below shows that $f = \bar{A} \cdot \bar{C} + B \cdot \bar{C}$.

	$\bar{B}$	$\bar{B}$	B	B
A	0	0	0	1
$\bar{A}$	1	0	0	1
	$\bar{C}$	C	C	$\bar{C}$

15.112 Reduce the Boolean function $f = A \cdot B + A \cdot \bar{B} + \bar{A} \cdot C$ to a more simple expression through use of a Karnaugh map.

▌ The map, constructed below, shows that all 1's are encompassed at least once by $f = A + C$.

	$\bar{B}$	$\bar{B}$	B	B
A	1	1	1	1
$\bar{A}$	0	1	1	0
	$\bar{C}$	C	C	$\bar{C}$

15.113 Simplify the Boolean function $f = A \cdot B \cdot C \cdot D + A \cdot C \cdot D + A \cdot D$.

▌ Construct the Karnaugh map from which it is seen that the resulting row of 1's is described by $f = A \cdot D$.

	$\bar{B}$	$\bar{B}$	B	B	
A	0	0	0	0	$\bar{D}$
A	1	1	1	1	D
$\bar{A}$	0	0	0	0	D
$\bar{A}$	0	0	0	0	$\bar{D}$
	$\bar{C}$	C	C	$\bar{C}$	

15.114 By use of a Karnaugh map, show that $f_1 = f_2$ where

$$f_1 = A \cdot B \cdot C + A \cdot \bar{B} \qquad f_2 = A \cdot (\bar{B} + C) = A \cdot \bar{B} + A \cdot C$$

▌ Construct the map where it is seen that the mapping of f_1 and f_2 are identical.

	$\bar{B}$	$\bar{B}$	B	B
A	1	1	1	0
$\bar{A}$	0	0	0	0
	$\bar{C}$	C	C	$\bar{C}$

15.115 Simplify the Boolean function $f = A \cdot \bar{B} \cdot \bar{C} \cdot D + \bar{A} \cdot \bar{B} \cdot \bar{C} \cdot D + A \cdot \bar{B} \cdot \bar{C} \cdot \bar{D} + \bar{A} \cdot \bar{B} \cdot \bar{C} \cdot \bar{D} + A \cdot B \cdot \bar{C} \cdot D + \bar{A} \cdot B \cdot \bar{C} \cdot D$.

▮ From the Karnaugh map below it is seen that $f = \bar{B} \cdot \bar{C} + B \cdot \bar{C} \cdot D$.

	$\bar{B}$	$\bar{B}$	B	B	
A	1	0	0	0	$\bar{D}$
A	1	0	0	1	D
$\bar{A}$	1	0	0	1	D
$\bar{A}$	1	0	0	0	$\bar{D}$
	$\bar{C}$	C	C	$\bar{C}$	

15.116 Find a simpler expression for $f = A \cdot B \cdot C + A \cdot \bar{B} \cdot C + \bar{A} \cdot B \cdot C + \bar{A} \cdot \bar{B} \cdot C$.

▮ From the Karnaugh map, it is seen that simply $f = C$.

	$\bar{B}$	$\bar{B}$	B	B
A	0	1	1	0
$\bar{A}$	0	1	1	0
	$\bar{C}$	C	C	$\bar{C}$

15.117 Simplify $f = \bar{A} \cdot B \cdot C + \bar{A} \cdot B \cdot \bar{C} + \bar{A} \cdot \bar{B} \cdot C + \bar{A} \cdot \bar{B} \cdot \bar{C}$.

▮ After constructing the Karnaugh map, it is apparent that $f = \bar{A}$.

	$\bar{B}$	$\bar{B}$	B	B
A	0	0	0	0
$\bar{A}$	1	1	1	1
	$\bar{C}$	C	C	$\bar{C}$

15.118 Reduce the Boolean function $f = A \cdot B \cdot C \cdot D + A \cdot \bar{B} \cdot C \cdot D + \bar{A} \cdot B \cdot C \cdot D + \bar{A} \cdot \bar{B} \cdot C \cdot D$ to a simpler expression.

▮ From the Karnaugh map, the pattern of 1's can be represented by $f = C \cdot D$.

	$\bar{B}$	$\bar{B}$	B	B	
A	0	0	0	0	$\bar{D}$
A	0	1	1	0	D
$\bar{A}$	0	1	1	0	D
$\bar{A}$	0	0	0	0	$\bar{D}$
	$\bar{C}$	C	C	$\bar{C}$	

Review Problems

15.119 Use $(r-1)$'s complement subtraction to find the difference $(111)_2 - (010)_2$.

▮ Form the 1's complement of the subtrahend:

$$\begin{array}{r} 111 \\ -010 \\ \hline (101)_2 \end{array}$$

The resulting addition gives an end-carry; thus, the end-around carry operation is performed to form the answer:

$$\begin{array}{r} 111 \\ +101 \\ \hline (1100)_2 \end{array}$$

and

$$\begin{array}{r} 100 \\ +1 \\ \hline (101)_2 \end{array}$$

15.120 Apply $(r-1)$'s complement subtraction to find the difference $(010)_2 - (111)_2$.

▮ The 1's complement of the subtrahend is

$$\begin{array}{r} 111 \\ -111 \\ \hline (000)_2 \end{array}$$

Performing the addition,

$$\begin{array}{r} 010 \\ +000 \\ \hline (010)_2 \end{array}$$

Since no end-carry resulted in the addition, the answer is formed as the negative of the 1's complement of $(010)_2$, or

$$\begin{array}{r} 111 \\ -010 \\ \hline -(101)_2 \end{array}$$

15.121 Form the octal (base-8) equivalent of $(100)_{10}$.

▮ Following the procedure developed in Problem 15.7,

division	integer quotient	remainder	coefficient
100/8	12	4	c_0
12/8	1	4	c_1
1/8	0	1	c_2

The octal number is formed from the remainder terms as $(144)_8$.

15.122 Find the octal equivalent of $(100.250)_{10}$.

▮ Conversion of the integer portion was determined in Problem 15.121. The fraction portion is converted according to the procedure of Problem 15.9:

multiplication	integer portion	fraction portion	coefficient
0.250×8	2	0.000	c_{-1}

Hence, $(0.250)_{10} = (0.2)_8$, and the complete conversion is

$$(0.250)_{10} = (144)_8 + (0.2)_8 = (144.2)_8$$

15.123 Prove Theorem 2(*b*) of Table 15-1 by the method of perfect induction.

BOOLEAN ALGEBRA AND LOGIC GATES □ 483

▌ A sufficient truth table for Theorem 2(b), $(A \cdot B) \cdot C = A \cdot (B \cdot C)$, is constructed as follows:

A	B	C	A·B	B·C	(A·B)·C	A·(B·C)
0	0	0	0	0	0	0
0	0	1	0	0	0	0
0	1	0	0	0	0	0
0	1	1	0	1	0	0
1	0	0	0	0	0	0
1	0	1	0	0	0	0
1	1	0	1	0	0	0
1	1	1	1	1	1	1

Since the last two columns are equal over all possible values of A, B, and C, the theorem is proven by perfect induction.

15.124 Prove Theorem 3(a) of Table 15-1 by the method of perfect induction.

▌ A sufficient truth table for Theorem 3(a), $A \cdot (B + C) = A \cdot B + A \cdot C$, is constructed below:

A	B	C	B+C	A·B	A·C	A·(B+C)	A·B+A·C
0	0	0	0	0	0	0	0
0	0	1	1	0	0	0	0
0	1	0	1	0	0	0	0
0	1	1	1	0	0	0	0
1	0	0	0	0	0	0	0
1	0	1	1	0	1	1	1
1	1	0	1	1	0	1	1
1	1	1	1	1	1	1	1

Since the last two columns are equal over all possible values of A, B, and C, the theorem is proven by perfect induction.

15.125 Prove Theorem 3(b) of Table 15-1 by the method of perfect induction.

▌ A sufficient truth table for Theorem 3(b), $A + (B \cdot C) = (A + B) \cdot (A + C)$, is formed as follows:

A	B	C	B·C	A+B	A+C	A+(B·C)	(A+B)·(A+C)
0	0	0	0	0	0	0	0
0	0	1	0	0	1	0	0
0	1	0	0	1	0	0	0
0	1	1	1	1	1	1	1
1	0	0	0	1	1	1	1
1	0	1	0	1	1	1	1
1	1	0	0	1	1	1	1
1	1	1	1	1	1	1	1

Since the last two columns are equal over all possible values of the variables, the theorem is proven by perfect induction.

15.126 Algebraically simplify $f = A \cdot B \cdot (A \cdot \bar{B} \cdot \bar{C} + \bar{A} \cdot B \cdot \bar{C})$.

▌ Expand using the distributive and commutative laws to find

$$f = A \cdot A \cdot B \cdot \bar{B} \cdot \bar{C} + A \cdot \bar{A} \cdot B \cdot B \cdot \bar{C} = A \cdot (B \cdot \bar{B}) \cdot \bar{C} + (A \cdot \bar{A}) \cdot B \cdot \bar{C} = 0$$

484 □ CHAPTER 15

where, from Theorem 8(b),

$$A \cdot \bar{A} = B \cdot \bar{B} = 0$$

15.127 Design a logic circuit to implement the Boolean function $f = A \cdot B \cdot C \cdot D + A \cdot C \cdot D + A \cdot D$.

▌ The Boolean function has been simplified in Problem 15.113 to $f = A \cdot D$. Hence, inputs B and C have no bearing on the value of f. The logic circuit is simply an AND gate with A and D as inputs and f as output.

15.128 Design a logic circuit to implement the Boolean function $f = A \cdot B \cdot \bar{C} + \bar{A} \cdot \bar{B} \cdot \bar{C} + B \cdot \bar{C}$.

▌ The Boolean function has been simplified in Problem 15.111 to $f = \bar{A} \cdot \bar{C} + B \cdot \bar{C}$. See Fig. 15-35 for one logic circuit implementation.

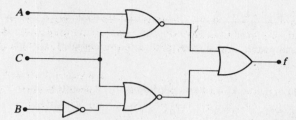

Fig. 15-35

15.129 Design a logic circuit to implement $f = \bar{A} \cdot \bar{B} \cdot \bar{C} \cdot (D + \bar{D})$.

▌ The Boolean function has simplified in Problem 15.97 to $f = \bar{A} \cdot \bar{B} \cdot \bar{C}$ which can be implemented by supplying A, B, and C to a three-input NOR gate, giving $f = \overline{A + B + C} = \bar{A} \cdot \bar{B} \cdot \bar{C}$.

15.130 Design a logic circuit to implement $f = \bar{A} \cdot \bar{B} \cdot \bar{C} + \bar{A} \cdot B \cdot C$.

▌ From Problem 15.99, $f = \bar{A} \cdot C$. The function can be implemented by the circuit of Fig. 15-36.

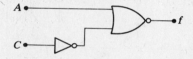

Fig. 15-36

15.131 Write an algebraic expression for X, the overall output of the *logic circuit* of Fig. 15-37.

▌ With the outputs of the intermediate logic gates as indicated in Fig. 15-37, the overall output must be

$$X = \overline{A \cdot B} + \overline{\bar{A} + B} \tag{1}$$

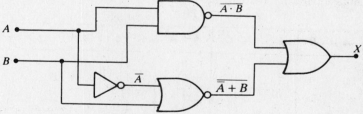

Fig. 15-37

15.132 Algebraically simplify (1) of Problem 15.131.

▌ Application of DeMorgan's laws and Theorems 7(c) and 4(a) yield

$$X = \bar{A} + (\bar{B} + \bar{A} \cdot \bar{B}) = \bar{A} + \bar{B}$$

which, according to DeMorgan's laws, can also be written as

$$X = \overline{A \cdot B} \tag{1}$$

15.133 It is apparent from (1) of Problem 15.132 that the entire logic network of Fig. 15-37 can be replaced with a single two-input NAND gate. Construct a truth table for the network to verify this observation.

▌ A truth table for Fig. 15-37 need only contain all pairs of values for the logic variables A and B and the resulting values of X; however, it is good practice to augment the basic truth table with columns showing intermediate values, as in the following table:

A	B	$\overline{A \cdot B}$	$\overline{A} + B$	X
0	0	1	0	1
0	1	1	0	1
1	0	1	1	1
1	1	0	0	0

Obviously, the output column is identical to the $\overline{A \cdot B}$ column and could be replaced by it.

15.134 A common procedure in digitial circuit design is to generate a logic statement or function describing the required circuit without regard to complexity, and then to manipulate the function using the theorems of Table 15-1 until an equivalent that requires a minimum combination of logic gates is found. A circuit is to be designed to realize the following logic:
1. If A, B, and C are all present (=1, true), then the process is correct (=1, true).
2. If A, B, and C are all absent (=0, false), then the process is correct (=1, true).
3. If B is present (=1, true), then the process is correct (=1, true).
4. Otherwise, the process is incorrect (=0, false).

Express the process state as a digital logic function.

▎ We simply form an OR expression encompassing all combinations that lead to a correct process (an output of 1). From statements 1 to 3,

$$X = A \cdot B \cdot C + \bar{A} \cdot \bar{B} \cdot \bar{C} + B \tag{1}$$

15.135 Algebraically simplify (1) of Problem 15.134.

▎ Operating on the second term of (1) from Problem 15.134 with the first of DeMorgan's laws gives

$$X = A \cdot B \cdot C + \overline{A + B + C} + B \tag{1}$$

Applying the redundancy law to (1) yields

$$X = (B + A \cdot B \cdot C) + \overline{A + B + C} = B + \overline{A + B + C} \tag{2}$$

15.136 Draw a logic circuit for the original and simplified system of Problem 15.134.

▎ A logic circuit for the original system, as described by (1) of Problem 15.134, is shown in Fig. 15-38(a), where six logic gates are required. Equation (2) of Problem 15.135 can be realized with only two logic gates, as illustrated by Fig. 15-38(b).

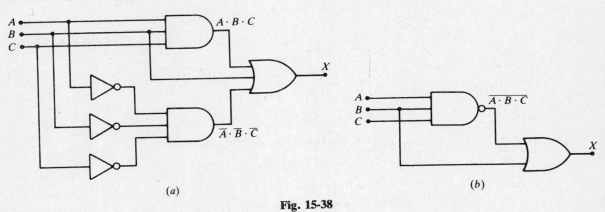

Fig. 15-38

15.137 A tedious but systematic approach to the construction of the logic function describing a process is to generate a table of all combinations of variables, identifying all *intersections* (AND combinations that are true). The logic function is then the *union* (OR combination) of all the intersections. Use this systematic approach to generate a logic function for the process of Problem 15.134.

■ All possible combinations of input variables, as well as the intersections that satisfy statements 1 to 3 of Problem 15.134, are shown in the following table:

A	B	C	intersections
0	0	0	$\bar{A}\cdot\bar{B}\cdot\bar{C}$
0	0	1	
0	1	0	$\bar{A}\cdot B\cdot\bar{C}$
0	1	1	$\bar{A}\cdot B\cdot C$
1	0	0	
1	0	1	
1	1	0	$A\cdot B\cdot\bar{C}$
1	1	1	$A\cdot B\cdot C$

The union of the intersections is then

$$X = \bar{A}\cdot\bar{B}\cdot\bar{C} + \bar{A}\cdot B\cdot\bar{C} + \bar{A}\cdot B\cdot C + A\cdot B\cdot\bar{C} + A\cdot B\cdot C \tag{1}$$

15.138 Draw a logic circuit for the process of Problem 15.134 as suggested by (1) of Problem 15.137.

■ The logic circuit suggested by (1) of Problem 15.137 is displayed in Fig. 15-39, where it is assumed that an OR gate with no more than four inputs is available (as a practical limit).

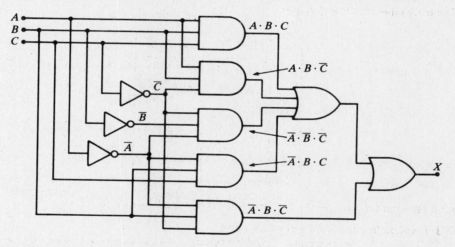

Fig. 15-39

15.139 Illustrate, for the case of three variables, that NAND gates can be interconnected to form the equivalent of an OR gate.

■ The desired logic function is

$$X = A + B + C \tag{1}$$

Applying the negation and the first of DeMorgan's laws to (1) yields

$$X = \overline{\overline{A+B+C}} = \overline{\bar{A}\cdot\bar{B}\cdot\bar{C}} \tag{2}$$

Now (2) could be synthesized by a NAND gate if the complements of the three variables were available as inputs. But by the identity law,

$$\overline{A\cdot A} = \bar{A} \tag{3}$$

An interpretation of (3) is that if A is the input to both gates of a two-input NAND gate, then its output is $\bar{A}$. Thus, (1) is realized by the logic circuit of Fig. 15-40.

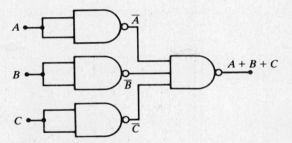

Fig. 15-40

15.140 Construct the truth table for the logic function $f = A \cdot \bar{B} + A \cdot B$.

▌ The truth table for f, showing the values of the individual terms, is:

A	B	$A \cdot \bar{B}$	$A \cdot B$	f
0	0	0	0	0
0	1	0	0	0
1	0	1	0	1
1	1	0	1	1

15.141 Constuct the truth table for $f = \bar{A} \cdot B \cdot C + \bar{A} \cdot \bar{B}$.

▌ The truth table for f is as follows:

A	B	C	$\bar{A} \cdot B \cdot C$	$\bar{A} \cdot \bar{B}$	f
0	0	0	0	1	1
0	0	1	0	1	1
0	1	0	0	0	0
0	1	1	1	0	1
1	0	0	0	0	0
1	0	1	0	0	0
1	1	0	0	0	0
1	1	1	0	0	0

15.142 Realize the logic function $f = A \cdot B + \bar{A} \cdot B$ using only NOR logic gates.

▌ The negation law and DeMorgan's laws give

$$\bar{f} = \overline{\overline{A \cdot B + \bar{A} \cdot B}} = \overline{\overline{(A \cdot B)} \cdot \overline{(\bar{A} \cdot B)}} = \overline{(\bar{A} + \bar{B}) \cdot (A + \bar{B})} = \overline{\overline{(\bar{A} + \bar{B})} + \overline{(A + \bar{B})}} \quad (1)$$

And, by the negation and identity laws,

$$f = \bar{\bar{f}} = \overline{\bar{f} + \bar{f}} \quad (2)$$

Use of (1) and (2) leads to the logic circuit of Fig. 15-41. Note, however, that $A \cdot B + \bar{A} \cdot B = B$, so f could be realized without any NOR gates (or other gates) by straight connection to signal B.

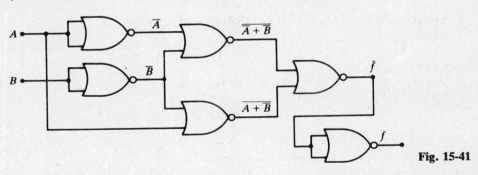

Fig. 15-41

15.143 Use DeMorgan's laws to find a combination of NAND gates that is identical in function to a two-input OR gate.

▌ The desired logic function is

$$f = A + B \tag{1}$$

Applying DeMorgan's law to (1) gives

$$f = \overline{\overline{A + B}} = \overline{\bar{A} \cdot \bar{B}} \tag{2}$$

Now (2) could be realized by a NAND gate if $\bar{A}$ and $\bar{B}$ were available. But, from Problem 15.139, the complement of a variable is available from a two-input NOR gate if the variable is fed to both inputs. Thus, (1) is realized by the logic circuit of Fig. 15-42.

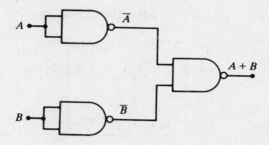

Fig. 15-42

15.144 Show that the logic function $X = \bar{A} \cdot \bar{B} + \bar{A} \cdot B + A \cdot B$ can be reduced to $\bar{A} + B$.

▌
$$\begin{aligned}
X &= \bar{A} \cdot (\bar{B} + B) + A \cdot B && \text{by Theorem 3}(a) \\
&= \bar{A} + A \cdot B && \text{by Theorem 8}(a) \\
&= \bar{A} + B && \text{by Theorem 9}(a)
\end{aligned}$$

15.145 If $X = \bar{A} \cdot B + A \cdot \bar{B}$, show that $\bar{X} = A \cdot B + \bar{A} \cdot \bar{B}$.

▌
$$\begin{aligned}
\bar{X} &= \overline{\bar{A} \cdot B + A \cdot \bar{B}} = \overline{(\bar{A} \cdot B)} \cdot \overline{(A \cdot \bar{B})} && \text{by Theorem 10}(a) \\
&= (A + \bar{B}) \cdot (\bar{A} + B) && \text{by Theorem 10}(a) \\
&= A \cdot \bar{A} + A \cdot B + \bar{A} \cdot \bar{B} + \bar{B} \cdot B && \text{by Theorem 3}(a) \\
&= A \cdot B + \bar{A} \cdot \bar{B} && \text{by Theorem 8}(b)
\end{aligned}$$

15.146 Show that $A \cdot B \cdot C + A \cdot B \cdot \bar{C} = A \cdot B$ by use of a truth table.

▌ Construct the truth table:

A	B	C	$\bar{C}$	$A \cdot B \cdot C$	$A \cdot B \cdot \bar{C}$	$A \cdot B$
0	0	0	1	0	0	0
0	0	1	0	0	0	0
0	1	0	1	0	0	0
0	1	1	0	0	0	0
1	0	0	1	0	0	0
1	0	1	0	0	0	0
1	1	0	1	0	1	1
1	1	1	0	1	0	1

Since the union of the second and third columns from the right equal the last column, the identity is established.

15.147 With the aid of a truth table, show that $(A + C) \cdot (\bar{A} + B) = A \cdot B + \bar{A} \cdot C$.

▌ By Theorems 3(a) and 8(b),

$$(A + C) \cdot (\bar{A} + B) = A \cdot \bar{A} + A \cdot B + \bar{A} \cdot C + B \cdot C = A \cdot B + \bar{A} \cdot C + B \cdot C \tag{1}$$

Comparing (1) with the given equation, it is sufficient to show that

$$A \cdot B + \bar{A} \cdot C + B \cdot C = A \cdot B + \bar{A} \cdot C$$

Construct a truth table as follows:

A	B	C	$\bar{A}$	$\bar{B}$	$\bar{A}\cdot B+\bar{A}\cdot C+B\cdot C$	$A\cdot B+\bar{A}\cdot C$
0	0	0	1	1	0	0
0	0	1	1	1	1	1
0	1	0	1	0	0	0
0	1	1	1	0	1	1
1	0	0	0	1	0	0
1	0	1	0	1	0	0
1	1	0	0	0	1	1
1	1	1	0	0	1	1

Since the last two columns are equal, the identity is established.

15.148 Use NAND gates only to construct a logic circuit with output $X = A + \bar{B}\cdot C$.

See Fig. 15-43.

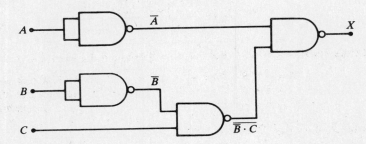

Fig. 15-43

15.149 Evaluate the Boolean expression $f = \bar{A}\cdot\bar{B} + A\cdot B + \bar{B} + C$ for $A = 1$, $B = 0$, and $C = 1$.

$$f = 0\cdot 1 + 1\cdot 0 + 1 + 1 = 1$$

15.150 Evaluate the Boolean expression for $f = A\cdot B\cdot\bar{C} + \bar{A}\cdot B + A\cdot\bar{B}$ for $A = 1$, $B = 0$, and $C = 1$.

$$f = 1\cdot 0\cdot 0 + 0\cdot 0 + 1\cdot 1 = 1$$

CHAPTER 16
Combinational and Sequential Logic

COMBINATIONAL LOGIC CIRCUITS

16.1 Logic circuits for which the output depends only upon the present input are called *combinational logic circuits*. One such logic circuit is the *half-adder*, a two-input, two-output circuit that is capable of implementing the binary addition table of Problem 15.12. Form the truth table for the half-adder.

▮ Let A and B be the half-adder (HA) inputs, S be the one binary digit (*bit*) sum, and C_o be the carry output that must be added to the next most significant column. Then, Table 16-1 describes the half-adder operation.

TABLE 16-1 Half-Adder

inputs		outputs	
A	B	S	C_o
0	0	0	0
0	1	1	0
1	0	1	0
1	1	0	1

16.2 Design a logic circuit to implement Table 16-1.

▮ Output S from Table 16-1 is seen to be XOR logic. Also, output C_o is AND logic. Thus, the half-adder can be simulated by the logic circuit of Fig. 16-1..

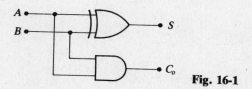

Fig. 16-1

16.3 If an XOR gate is not available, describe an alternate simulation for the half-adder of Fig. 16-1.

▮ The logic circuit of Fig. 15-13 is shown to exhibit XOR logic in Problem 15.79. Thus, the XOR gate of Fig. 16-1 could be replaced by the logic circuit of Fig. 15-13 to form a half-adder.

16.4 When adding binary numbers of more than one bit, the half-adder of Problem 16.2 does not provide a means to consider the possible carry from addition of the column of bits to the right of the present addition operation. A binary adder that can handle a carry from a previous stage of addition is called a *full-adder* (FA). Extend Table 16-1 to allow inputs A and B and a carry input C_i from the immediately previous stage, thus developing the full-adder truth table.

▮ As long as the carry input $C_i = 0$, the outputs are identical to the HA truth Table 16-1. However, with $C_i = 1$, any time A or B is not zero, a carry output must be generated. Table 16-2 describes the full-adder operation.

TABLE 16-2 Full-Adder

inputs			outputs	
A	B	C_i	S	C_o
0	0	0	0	0
0	1	0	1	0
1	0	0	1	0
1	1	0	0	1
0	0	1	1	0
0	1	1	0	1
1	0	1	0	1
1	1	1	1	1

16.5 Based on Table 16-2, write a Boolean function S that describes all conditions for sum $S = 1$.

▌ There are four conditions for $S = 1$ given by

$$S = \bar{A} \cdot B \cdot \bar{C}_i + A \cdot \bar{B} \cdot \bar{C}_i + \bar{A} \cdot \bar{B} \cdot C_i + A \cdot B \cdot C_i \tag{1}$$

Applying Theorem 3(a) of Table 15-1,

$$S = (\bar{A} \cdot B + A \cdot \bar{B}) \cdot \bar{C}_i + (\bar{A} \cdot \bar{B} + A \cdot B) \cdot C_i \tag{2}$$

We define $\bar{A} \cdot B + A \cdot \bar{B}$ as $A \oplus B$ and $\bar{A} \cdot \bar{B} + A \cdot B$ as $A \odot B$. However, XNOR logic is the complement of XOR logic, and (2) can be rewritten as

$$S = (A \oplus B) \cdot \bar{C}_i + \overline{(A \oplus B)} \cdot C_i \tag{3}$$

But, since $X \cdot \bar{Y} + \bar{X} \cdot Y = X \oplus Y$, (3) becomes

$$S = (A \oplus B) \oplus C_i \tag{4}$$

16.6 Using Table 16-2, write a Boolean function C_o that describes all conditions for output carry $C_o = 1$.

▌ The four conditions for $C_o = 1$ are given by

$$C_o = A \cdot B \cdot \bar{C}_i + \bar{A} \cdot B \cdot C_i + A \cdot \bar{B} \cdot C_i + A \cdot B \cdot C_i \tag{1}$$

By Theorems 1(a) and 3(a) of Table 15-1,

$$C_o = A \cdot B \cdot (\bar{C}_i + C_i) + (\bar{A} \cdot B + A \cdot \bar{B}) \cdot C_i \tag{2}$$

Using Theorem 8(a) of Table 15-1 and recognizing $\bar{A} \cdot B + A \cdot \bar{B}$ as XOR logic, (2) can be written as

$$C_o = A \cdot B + (A \oplus B) \cdot C_i \tag{3}$$

16.7 Based on (4) of Problem 16.5 and (3) of Problem 16.6, draw a logic circuit that simulates the full-adder.

▌ See Fig. 16-2.

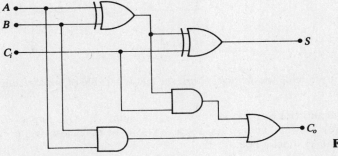

Fig. 16-2

16.8 Show that the two interconnected half-adders and the OR gate of Fig. 16-3 simulates the logic of a full-adder.

▌ Based on Table 16-1,

$$C_1 = A \cdot B \tag{1}$$

$$S_1 = A \oplus B \tag{2}$$

$$C_2 = S_1 \cdot C_i \tag{3}$$

$$S = S_1 \oplus C_i \tag{4}$$

Substituting (2) into (4) yields

$$S = (A \oplus B) \oplus C_i \tag{5}$$

which agrees with the full-adder sum of (4) from Problem 16.5. Now, from Fig. 16-3,

$$C_o = C_1 + C_2 \tag{6}$$

Using (1)–(3) in (6),

$$C_o = A \cdot B + (A \oplus B) \cdot C_i \tag{7}$$

Equation (7) is identical to the full-adder carry logic of (3) from Problem 16.6. Hence, the circuit of Fig. 16-3 simulates half-adder logic.

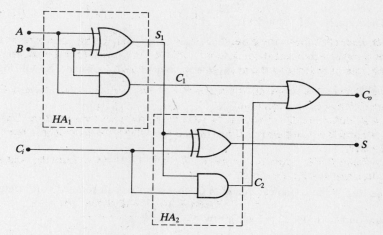

Fig. 16-3

16.9 Illustrate how to cascade-connect a half-adder and a full-adder to perform the parallel addition of two 2-bit binary numbers $A_2A_1 + B_2B_1 = C_oS_2S_1$. The resulting circuit is called a *ripple-carry adder*.

▌ See Fig. 16-4. The addition could also be implemented with two full-adders if C_i were set to zero on the least significant bit adder.

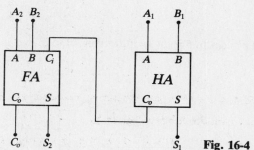

Fig. 16-4

16.10 Construct the truth table for the *half-subtractor* that is capable of implementing the binary subtraction table of Problem 15.14.

▌ Let A (minuend) and B (subtrahend) be the half-subtractor (HS) inputs, D be one binary bit difference, and B_o be the borrow output that must be subtracted from the next most significant column. Then, Table 16-3 describes the half-subtractor operation.

TABLE 16-3 Half-Subtractor

inputs		outputs	
A	B	D	B_o
0	0	0	0
0	1	1	1
1	0	1	0
1	1	0	0

16.11 Design a logic circuit to implement Table 16-3.

▌ Output D from Table 16-3 is seen to be XOR logic. Further, output $B_o = 1$ only if $A = 0$ and $B = 1$, or $B_o = \bar{A} \cdot B$. Thus, the half-subtractor is simulated by the logic circuit of Fig. 16-5.

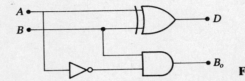

Fig. 16-5

16.12 The *full-subtractor* (FS) forms the difference between two bits, taking into account that a borrow may have been necessary by a lower significant stage. Extend Table 16-3 to account for the possibility of an immediately preceding borrow, thus forming the full-subtractor truth table.

▌ As long as the previous stage input borrow $B_i = 0$, the outputs are identical to Table 16-3. However, if $B_i = 1$, three difference cases can arise:
1. If $A = B$, difference $D = 1$ and an output borrow from the next most significant stage is necessary.
2. If $A = 0$ and $B = 1$, difference $D = 0$ and an output borrow $B_o = 1$ from the next most significant stage is necessary.
3. If $A = 1$ and $B = 0$, difference $D = 0$ and $B_o = 0$ as no borrow from the next stage is required.

Table 16-4 describes the FS operation.

TABLE 16-4 Full-Subtractor

inputs			outputs	
A	B	B_i	D	B_o
0	0	0	0	0
0	1	0	1	1
1	0	0	1	0
1	1	0	0	0
0	0	1	1	1
0	1	1	0	1
1	0	1	0	0
1	1	1	1	1

16.13 Based on Table 16-4, write a Boolean function D that describes all conditions for difference $D = 1$.

▌ There are four conditions for $D = 1$ given by

$$D = \bar{A} \cdot B \cdot \bar{B_i} + A \cdot \bar{B} \cdot \bar{B_i} + \bar{A} \cdot \bar{B} \cdot B_i + A \cdot B \cdot B_i \qquad (1)$$

Since (1) is identical to (1) of Problem 16.5 if C_i were replaced by B_i, symmetry with (4) of Problem 16.5 gives

$$D = (A \oplus B) \oplus B_i \qquad (2)$$

16.14 Based on Table 16-4, write a Boolean function B_o that describes all conditions for $B_o = 1$.

▌ The four conditions for $B_o = 1$ are given by

$$B_o = \bar{A} \cdot B \cdot \bar{B_i} + \bar{A} \cdot \bar{B} \cdot B_i + \bar{A} \cdot B \cdot B_i + A \cdot B \cdot B_i \qquad (1)$$

By use of Theorems 1(a) and 3(a) of Table 15-1, (1) can be rewritten as
$$B_o = \bar{A} \cdot B \cdot (\bar{B}_i + B_i) + (\bar{A} \cdot \bar{B} + A \cdot B) \cdot B_i \tag{2}$$
Using Theorem 8(a) of Table 15-1 and recognizing $\bar{A} \cdot \bar{B} + A \cdot B$ as XOR logic, (2) can be written as
$$B_o = \bar{A} \cdot B + (A \odot B) \cdot B_i \tag{3}$$
An alternate expression for (3) is formed by realizing that XOR logic is the complement of XNOR logic:
$$B_o = \bar{A} \cdot B + \overline{(A \oplus B)} \cdot B_i \tag{4}$$

16.15 Based on (2) of Problem 16.13 and (4) of Problem 16.14, draw a logic circuit that simulates the full-subtractor.

❙ See Fig. 16-6.

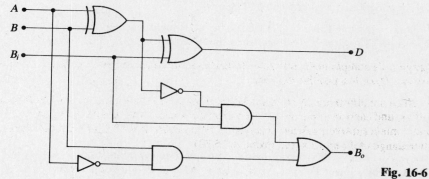

Fig. 16-6

16-16 Show that the two interconnected half-subtractors and the OR gate of Fig. 16-7 simulates the logic of a full-subtractor.

❙ Based on Table 16-3,
$$B_1 = \bar{A} \cdot B \tag{1}$$
$$D_1 = A \oplus B \tag{2}$$
$$B_2 = \bar{D}_1 \cdot B_i \tag{3}$$
$$D = D_1 \oplus B_i \tag{4}$$

Substituting (2) into (4) yields
$$D = (A \oplus B) \cdot B_i \tag{5}$$
which agrees with the full-subtractor difference of (2) from Problem 16.13. Now, from Fig. 16-7,
$$B_o = B_1 + B_2 \tag{6}$$
Using (1)–(3) in (6) leads to
$$B_o = \bar{A} \cdot B + \overline{(A \oplus B)} \cdot B_i \tag{7}$$

Equation (7) is identical to the full-subtractor output borrow logic of (4) from Problem 16.14. Hence, the logic circuit of Fig. 16-7 truly implements full-subtractor logic.

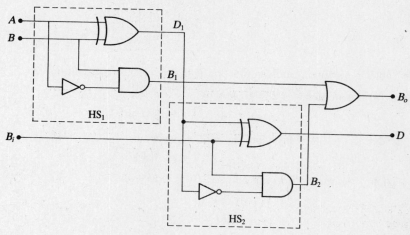

Fig. 16-7

16.17 Illustrate the cascaded connection of a full-subtractor and a half-subtractor to perform parallel subtraction of two 2-bit binary numbers $A_2A_1 - B_2B_1 = D_2D_1$. Assume $A_2A_1 > B_2B_1$.

▌ See Fig. 16-8. Since $A_2A_1 > B_2B_1$, $B_o = 0$ for the FS and is of no consequence.

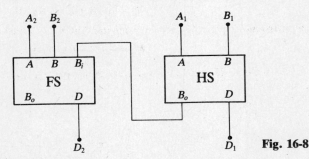

Fig. 16-8

16.18 Design a 1's complement subtractor to find the difference of two 2-bit binary numbers $A_2A_1 - B_2B_1 = D_2D_1$. Assume D_2D_1 is a positive number.

▌ Since the difference D_2D_1 is a positive number, the 1's complement of B_2B_1 is added to A_2A_1 and no end-around carry is implemented (see Problem 15.35). The two interconnected FA's of Fig. 16-9 perform the 1's complement subtraction where the inverters in the B_1 and B_2 lines form the 1's complement of B_2B_1 by interchange of 0's and 1's (see Problem 15.32).

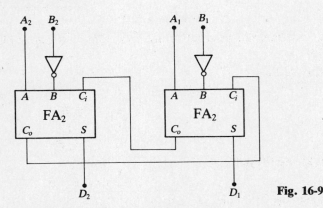

Fig. 16-9

16.19 Use full-adders to design a combinational logic circuit to generate the 2's complement of a 2-bit binary number.

▌ Comparing the rules of Problems 15.21 and 15.32, it is seen that the 2's complement of a number is no more than the 1's complement with 1 added to the least significant bit. Further, the 1's complement is formed by interchange of 0's and 1's. The logic circuit of Fig. 16-10 forms $B'_2B'_1$, the 2's complement of B_2B_1, where logic 1 is simply a connection to power supply voltage, and logic zero is obtained by connection to ground or signal common.

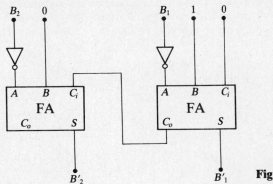

Fig. 16-10

16.20 Design a 2's complement subtractor to find the difference of two 2-bit binary numbers $A_2A_1 - B_2B_1 = D_2D_1$ if D_2D_1 is a positive number.

■ Based on Problem 15.24, the 2's complement of B_2B_1 should be added to A_2A_1. Further, since D_2D_1 is a positive number, the end-carry is to be discarded. Hence, the circuit of Fig. 16-11 performs the 2's complement subtraction where the 2's complement generator is detailed in Fig. 16-10.

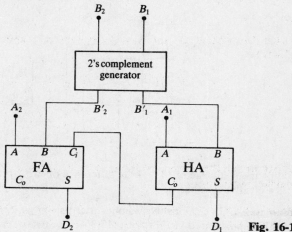

Fig. 16-11

16.21 *Decoders* are combinational logic circuits that produce an output signal in response to a specific signal pattern while ignoring all signals that do not match the proper pattern. Such a device allows use of a common set of lines to provide instructions to multiple devices. Design a decoder to be connected to a set of 4-bit address lines $A_3A_2A_1A_0$ that provides a logic high output signal if and only if the 4-bit word 0110 appears on the address lines.

■ Complete decoding (match on all lines) of the lines is implied. The output of the decoder should be logic high if and only if $\bar{A}_3 \cdot A_2 \cdot A_1 \cdot \bar{A}_0 = 1$. The required decoding is obviously accomplished by the logic circuit of Fig. 16-12.

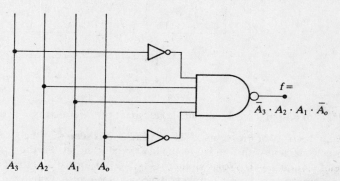

Fig. 16-12

16.22 The *multiplexer* (MUX) is a combinational logic circuit that connects one of multiple input lines $(D_{n-1}, \ldots, D_1, D_0)$ to an output line (f) according to a received *select code* ($S = S_{n-1} \cdots S_1 S_0$). Figure 16-13 illustrates the lowest order MUX possible (2-line to 1-line multiplexer). Explain how, through use of a 1-bit select code, this circuit is able to render f either the value of D_0 or D_1.

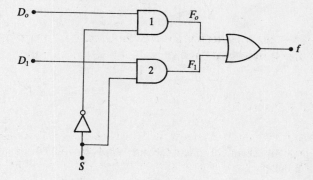

Fig. 16-13

▎ Referring to Fig. 16-13,

$$F_0 = D_0 \cdot \bar{S} \quad (1)$$

$$F_1 = D_1 \cdot S \quad (2)$$

$$f = F_0 + F_1 \quad (3)$$

A truth table is constructed realizing that S can be either 0 or 1, applying (1)–(3), and using Theorems 7(b) and 7(d) of Table 15-1.

D_0	D_1	S	$\bar{S}$	F_0	F_1	f
D_0	D_1	0	1	D_0	0	D_0
D_0	D_1	1	0	0	D_1	D_1

Whence it is apparent that if select code S is logic low, output line f matches input line D_0. Vice versa, if S is logic high, f matches input line D_1.

16.23 The *demultiplexer* (DEMUX) is a combinational logic circuit that connects a single-input line (D) to one of multiple output lines ($f_{n-1}, \ldots, f_1, f_0$) according to a received n-bit select code ($S = S_{n-1} \cdots S_1 S_0$). Figure 16-14 illustrates the lowest order DEMUX possible (1-line to 2-line demultiplexer). Explain how, through use of a 1-bit select code, this circuit is able to make only $f_1 = D$ or only $f_2 = D$.

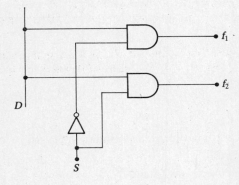

Fig. 16-14

▎ From Fig. 16-14,

$$f_1 = D \cdot \bar{S} \quad (1)$$

$$f_2 = D \cdot S \quad (2)$$

A truth table is constructed realizing that S can be either 0 or 1, applying (1) and (2), and using Theorems 7(b) and 7(d) of Table 15-1:

D	S	$\bar{S}$	f_1	f_2
D	0	1	D	0
D	1	0	0	D

Whence, it is apparent that if select code S is logic low, output $f_1 = D$ while output $f_2 = 0$. Conversely, if S is logic high, output $f_2 = D$ while output $f_1 = 0$.

SEQUENTIAL LOGIC CIRCUITS

16.24 Logic circuits for which the output at the present time depends on prior input values are called *sequential logic circuits*. The flip-flop (introduced in Chapter 14) can be built up from logic gates and is capable of storing (or remembering) 1 bit of binary data; hence, this memory device is one of the basic building blocks of sequential logic circuits. Two NOR gates can be connected as shown in Fig. 16-15 to form a *Reset-Set (RS) flip-flop*. Let Q_n be the value of Q before an RS switching event and Q_{n+1} be the value of Q after the switching event. Construct a truth table that characterizes the operation of this RS flip-flop.

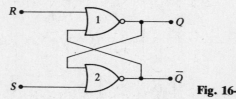

Fig. 16-15

▌ For a NOR gate with two inputs A and B, output $f = \overline{A + B} = \bar{A} \cdot \bar{B}$. Hence, if one input is 1, the output $f = 0$. Consider the logic circuit of Fig. 16-15 with $S = 1$ and $R = 0$. If $S = 1$, $\bar{Q} = 0$ since the NOR gate 2 has a logic high input. Consequently, $Q = 1$ since $R = \bar{Q} = 0$, or both inputs to NOR gate 1 are zero. Now if S becomes 0 and R remains 0, $\bar{Q}$ is unchanged from 0 since $Q = 1$ remains an input to NOR gate 2. Thus, the circuit displays memory of the values of Q and $\bar{Q}$ after the signal is removed.

By symmetry, if $S = 0$ and $R = 1$, $\bar{Q} = 1$ and $Q = 0$. Further, the values of Q and $\bar{Q}$ remain unchanged when R becomes 0 if S remains 0.

If $R = S = 1$, each NOR gate has a 1 input; thus, both Q and $\bar{Q}$ are necessarily 0. Such a condition contradicts the assumption of complementary outputs Q and $\bar{Q}$ implied by Fig. 16-15; consequently, this case is not allowed (N/A). The resulting truth table is constructed as in Table 16-5.

TABLE 16-5 RS Flip-Flop

S	R	Q_{n+1}	$\bar{Q}_{n+1}$
0	0	Q_n	$\bar{Q}_n$
0	1	0	1
1	0	1	0
1	1	N/A	N/A

16.25 NAND gates and inverters can be connected as shown in Fig. 16-16 to form an RS flip-flop. Construct a truth table that characterizes this flip-flop.

▌ For a NAND gate with two inputs (A and B), output $f = \overline{A \cdot B} = \bar{A} + \bar{B}$. Hence, if one input is 0, the output $f = 1$. The output $f = 0$ only if $A = B = 1$. Consider the logic circuit of Fig. 16-16 with $R = 1$ and $S = 0$. If $R = 1$, $\bar{R} = 0$ and $\bar{Q} = 1$ since the NAND gate 2 has a 0 input. Also, with $S = 0$, $\bar{S} = 1 = \bar{Q}$, and consequently, $Q = 0$. When R changes to zero, the output values are unchanged if S remains 0.

By symmetry, for $S = 1$ and $R = 0$, $\bar{Q} = 0$ and $Q = 1$. Further, Q and $\bar{Q}$ are unchanged when S goes to zero if R remains 0.

If $R = S = 1$, $\bar{R} = \bar{S} = 0$ and each NAND gate has a 0 input value requiring $Q = \bar{Q} = 1$. This case is not allowed (N/A) since the complementary output Q and $\bar{Q}$ is contradicted. The resulting truth table is identical to Table 16-5.

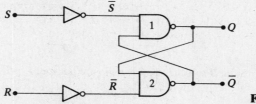

Fig. 16-16

16.26 The RS flip-flops of Problems 16.24 and 16.25, as presented, change states immediately upon a change in the RS switching pattern (called *asynchronous operation*). Frequently, logic circuits are allowed to change states only upon signal from a square-wave generator or clock (called *synchronous operation*). The NOR gate-based flip-flop of Fig. 16-15 is modified to form a *synchronous (or clocked) RS flip-flop* as shown by Fig. 16-17. Qualitatively describe the operation of this logic circuit and construct the truth table that reflects the synchronous operation.

COMBINATIONAL AND SEQUENTIAL LOGIC ☐ 499

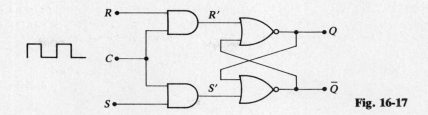

Fig. 16-17

▌ The values of Q and $\bar{Q}$ are controlled by the switching states of R' and S' as discussed for R and S in Problem 16.24. However, during the interval that the clock pulse has the value of low logic, $R' = S' = 0$ (due to the presence of the AND gates), regardless of the values of R and S. But, when the clock pulse has the value of high logic, $R' = R$ and $S' = S$; thus, if the RS input pattern has changed since the previous clock pulse, Q and $\bar{Q}$ can be changed. Letting a subscript n denote the value that existed at the end of the clock pulse interval of high logic, and subscript $n + 1$ denote the value that exists over the clock pulse interval of low logic, Table 16-6 will result.

TABLE 16-6 Clocked RS Flip-Flop

S_n	R_n	Q_{n+1}	$\bar{Q}_{n+1}$
0	0	Q_n	$\bar{Q}_n$
0	1	0	1
1	0	1	0
1	1	N/A	N/A

16.27 The D (*delay*) *flip-flop* has the characteristic that it delays the passage of unchanged data to its output ports (and consequently to subsequent elements of the system) until occurrence of a clock pulse; further, it maintains the data value over the low logic interval of the clock pulse. A D flip-flop is shown in Fig. 16-18. Explain its operation and construct a truth table to characterize its performance.

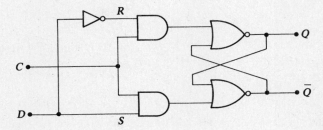

Fig. 16-18

▌ The circuit to the right of R and S is seen to be no more than a clocked RS flip-flop. Since $D = S$ and $R = \bar{D} = \bar{S}$, R can never equal S; thus, only the middle two rows of Table 16-6 are applicable. Table 16-7 is constructed based upon this observation:

TABLE 16-7 D Flip-Flop

D_n	S_n	R_n	Q_{n+1}	$\bar{Q}_{n+1}$
0	0	1	0	1
1	1	0	1	0

16.28 The *JK flip-flop* removes the indeterminate or not allowed state of the RS flip-flop. Discuss the operation of the JK flip-flop of Fig. 16-19 and construct its truth table.

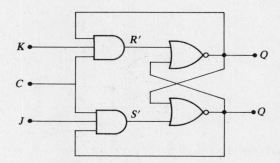

Fig. 16-19

▌ The circuit is seen to be that of a clocked RS flip-flop, except for the feedback paths from Q and $\bar{Q}$ to the three-input AND gates. Consequently, the analysis is similar to that of the clocked RS flip-flop, except for the case of $J = K = 1$. Since $R' = K \cdot Q \cdot C$, Q is set to 0 if $Q = 1$. Also, $S' = J \cdot \bar{Q} \cdot C$; thus, $\bar{Q}$ is set to 0 if $\bar{Q} = 1$. Thus, Q and $\bar{Q}$ change states or take on their complement values. The corresponding Table 16-8 will result.

TABLE 16-8 JK Flip-Flop

J_n	K_n	Q_{n+1}	$\bar{Q}_{n+1}$
0	0	Q_n	$\bar{Q}_n$
0	1	0	1
1	0	1	0
1	1	$\bar{Q}_n$	Q_n

16.29 The T (*toggle*) *flip-flop* is shown in Fig. 16-20. Describe its operation and construct its truth table.

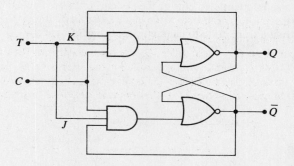

Fig. 16-20

▌ The logic circuit is seen to be a JK flip-flop with the inputs connected together. Thus, for $T = 0$, $J = K = 0$ and the first row of Table 16-8 is pertinent; if $T = 1$, $J = K = 1$ and the bottom row describes the operation. It is concluded that the T flip-flop does not change output states if $T = 0$; however, if $T = 1$, the output states are changed or toggled. The truth table is shown in Table 16-9.

**TABLE 16-9
T Flip-Flop**

T_n	Q_{n+1}
0	Q_n
1	$\bar{Q}_n$

16.30 In many practical applications, it is desired to *preset* (*PS*) a flip-flop so that $Q = 1$ or *clear* (*CR*) so that $Q = 0$. Such features can be added to the flip-flop by insertion of OR gates as illustrated for the RS flip-flop of Fig. 16-21. Explain the operation of the PS and CR features.

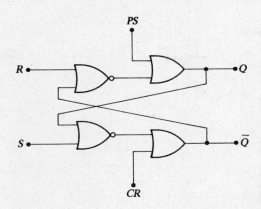

Fig. 16-21

▌ If PS = 1 and CR = 0, Q is set to a value of 1. For PS = 0 and CR = 1, $\bar{Q}$ is set to 1; thus, $Q = 0$. After the initializing preset or clear operation, PS = CR = 0 are maintained and the OR gates have no effect on the subsequent flip-flop operation. Although illustrated for the RS flip-flop, this technique is equally applicable to other flip-flop types.

16.31 The RS flip-flop of Fig. 16-17 (and other flip-flop types) is sensitive to the clock high logic duration in that Q (and $\bar{Q}$) can be altered by a change in the RS switching pattern until the clock pulse decreases to low logic. Typically, the clock high logic duration is 50 percent of a clock cycle. In a complex logic system, it may be difficult to avoid timing problems that lead to a logic error when the flip-flop can be changed by data that arrives at any time over the clock high logic duration. The *RS master-slave (M/S) flip-flop* of Fig. 16-22 avoids the data arrival timing ambiguity by setting the overall flip-flop (enclosed by dash lines) output according to the RS switching pattern that existed upon transition to low logic of the clock pulse. Explain the switching events of the RS M/S flip-flop of Fig. 16-22.

▌ Notice that clock input $C_2 = \bar{C}$, but $C_1 = C$. Thus, Q_1 and $\bar{Q}_1$ are set according to the instantaneous R and S switching pattern over the duration of $C_1 = C$ high logic. As $C_1 = C$ falls to low logic, $C_2 = \bar{C}$ makes the transition to high logic; thus, $Q = Q_2$ and $\bar{Q} = \bar{Q}_2$ are set according to the switching patterns of $S_2 = Q_1$ and $R_2 = \bar{Q}_1$. With $C_1 = C = 0$, Q_1 and $\bar{Q}_1$ change over the low logic duration of C and the values of Q and $\bar{Q}$ are maintained over this interval. In conclusion, Q and $\bar{Q}$ are set at the negative transition (trailing edge) of clock pulse C according to the values of R and S at that point in time and maintained at that value for one complete clock cycle.

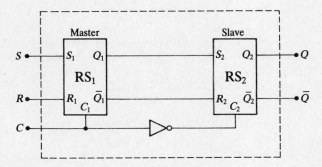

Fig. 16-22

16.32 Design a 2-bit *register* (storage cells for binary information that are insensitive to clock pulse duration) to *accumulate* data ($D = D_1 D_0$) from a previous operation and to *hold* D on output lines $B = B_1 B_0$ for later use by subsequent logic circuits.

▌ The D flip-flop is suited for use as a register. See Fig. 16-23 for the 2-bit register. The D flip-flops are master-slave (Problem 16.31) or leading edge triggered (Problem 16.44) so that the output data is not sensitive to clock pulse duration.

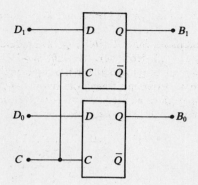

Fig. 16-23

16.33 Design a 2-bit *latch* (storage cells for binary information that are sensitive to clock pulse duration) to accumulate data from a previous operation for use by subsequent logic circuits.

▌ Since a latch is sensitive to clock pulse duration, its application is limited to use where data timing problems cannot arise; however, the device is a more economical choice in such cases. The two interconnected D flip-flops of Fig. 16-23 could satisfy the requirement if the flip-flops were the simple design of Fig. 16-18.

16.34 A register that is capable of transferring its binary information either to the left or right is known as a *shift register*. Using D flip-flops, design a 2-bit shift register that moves incoming data from left to right.

▌ See Fig. 16-24 for illustration of the 2-bit shift register. Each data bit is first entered as output of the left-hand M/S D flip-flop. Upon low transition of the clock, the data bit is stably available at output Q_1 which is also connected as the input of the right-hand M/S D flip-flop. Over the next clock pulse, the new arriving data bit at D_1 and the previous data bit at D_2 are set in the respective master flip-flops. As the clock pulse makes its transition to low logic, data bits are set in the respective slave flip-flops; the arriving data bit is set as output Q_1 while the previous data bit is set as the output Q_2. The shifting process continues as long as clock pulses are transmitted to the flip-flops.

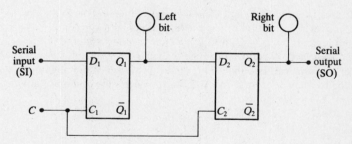

Fig. 16-24

16.35 The sequential logic circuit of Fig. 16-25 is a 2-bit *binary counter* where B_0 is the *least significant bit* (LSB) and B_1 is the *most significant bit* (MSB). Describe the output $B_1 B_0$ for the first five clock cycles if the counter were initially cleared ($B_0 = B_1 = 0$).

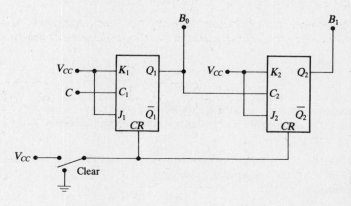

Fig. 16-25

▌ Assume M/S flip-flops are utilized. Since both J and K are connected to a voltage (V_{CC}) that produces high logic, $J = K = 1$ and from Table 16-8 it is realized that the fall of each arriving clock pulse results in replacement of Q by its complement. Since $C = C_1$, Q_1 will toggle each clock cycle so that $Q_1 = B_0$ produces the sequence $B_0 = (0, 1, 0, 1, 0, \ldots)$. However, the right-hand flip-flop is clocked by $Q_1 = C_2$; hence, Q_2 toggles upon transition to low logic of Q_1 producing the sequence $B_1 = (0, 0, 1, 1, 0, 0, 1, 1, \ldots)$. The first four clock cycles are tabulated (where $\downarrow$ denotes transition to low logic of clock pulse), from whence it is seen that the logic circuit outputs $B_1 B_0$ produce a binary count from $(00)_2$ to $(11)_2$, or 0_{10} to 3_{10}, and then resets to 0, repeating the sequence ad infinitum as long as C is present.

B_1	B_0	C
0	0	0
0	1	$\downarrow$
1	0	$\downarrow$
1	1	$\downarrow$
0	0	$\downarrow$

Since the fall to low logic of one flip-flop leads to setting of the successive flip-flop, the output settings propagate from left to right, leading to the name *ripple counter* for the presented configuration.

16.36 A counter that repeats a sequence of M (or M_{10}) binary numbers is called a *modulo-M counter*. Thus, the counter of Problem 16.35 is called a modulo-4 counter. Design a modulo-8 ripple counter.

▌ In order to count through a sequence of eight binary numbers, a 3-bit binary output is necessary. Logically, extension of the modulo-4 ripple counter of Problem 16.35 by cascade addition of a third M/S JK flip-flop with JK connected to toggle and its clock driven by $Q_2 = B_1 = (0, 0, 1, 1, 0, 0, 1, 1, \ldots)$ should yield the desired counter. It is noted that sequence B_1 makes a transition from high to low logic every fourth cycle of clock C. Then, output Q_3 toggles upon transition to low logic of Q_2 producing the sequence $Q_3 = B_3 = (0, 0, 0, 0, 1, 1, 1, 1, 0, 0, 0, 0, \ldots)$. When placed in tabular form, the repeated binary count from $(000)_2$ to $(111)_2$, or 0_{10} to 7_{10}, is apparent.

B_2	B_1	B_0	C
0	0	0	0
0	0	1	$\downarrow$
0	1	0	$\downarrow$
0	1	1	$\downarrow$
1	0	0	$\downarrow$
1	0	1	$\downarrow$
1	1	0	$\downarrow$
1	1	1	$\downarrow$
0	0	0	$\downarrow$

16.37 Convert the modulo-4 ripple counter of Problem 16.35 to a *synchronous counter* (a counter for which all bits are set at the same point in time by action of the system clock).

▌ For this 2-bit counter, the modification is relatively simple (see Fig. 16-26) in that the output Q_1 becomes the input to $J_2 K_2$. According to Table 16-8, the output $B_1 = Q_2$ is complemented each time that $Q_1 = B_0$ is 1 and is

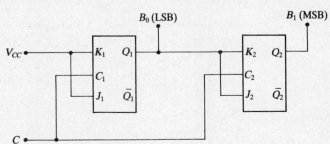

Fig. 16-26

unchanged each time that $Q_1 = B_0$ is 0. Outputs for the first four clock cycles are tabulated as follows:

B_1	B_0	C
0	0	0
0	1	↓
1	0	↓
1	1	↓
0	0	↓

16.38 The counters of Problems 16.35 to 16.37 have incremented from 0 to $M - 1$ before resetting (called *up counters*). A *down counter* decrements from $M - 1$ to 0 before resetting. Design a modulo-4 down counter.

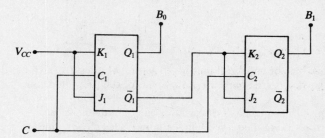

Fig. 16-27

▌ See Fig. 16-27 where the complementary output of the left-hand flip-flop is connected so that $J_2 = K_2 = \bar{Q}_1$. Assume the counter has been preset so that $Q_1 = B_0 = Q_2 = B_1 = 1$. Since $J_1 = K_1 = V_{CC}$, Q_1 will toggle each cycle of the clock. For the first clock cycle, $\bar{Q}_1 = K_2 = J_2 = 0$; thus, by Table 16-8, Q_2 is unchanged. For the second clock cycle, $\bar{Q}_1 = J_2 = K_2 = 1$; thus, Q_2 is toggled. During the third clock cycle, $\bar{Q}_1 = J_2 = K_2 = 0$ and Q_2 is unchanged. Q_2 is complemented over the fourth clock cycle since $\bar{Q}_1 = 1$. The down counter outputs are tabulated for the first four clock cycles.

B_1	B_0	C
1	1	0
1	0	↓
0	1	↓
0	0	↓
1	1	↓

16.39 Design a 2-bit *up-down counter*, a counter capable of incrementing from 0 to $M - 1$ and reset, or decrementing from $M - 1$ to 0 and reset based upon a direction command.

▌ Based on the work of Problems 16.37 and 16.38, we merely need a circuit to pass Q_1 to the successive flip-flop for up count and to pass $\bar{Q}_1$ to the successive flip-flop for down count. For the circuit of Fig. 16-28, $D = 1$ for up count and $D = 0$ for down count. If $D = 1$ ($\bar{D} = 0$), then $X = D \cdot Q_1 = Q_1$ and $Y = \bar{D} \cdot \bar{Q}_1 = 0$; hence, the logic circuit functions as the up counter of Fig. 16-26. For $D = 0$ ($\bar{D} = 1$), then $X = D \cdot Q_1 = 0$ and $Y = \bar{D} \cdot \bar{Q}_1 = \bar{Q}_1$, or the circuit functions as the down counter of Fig. 16-27.

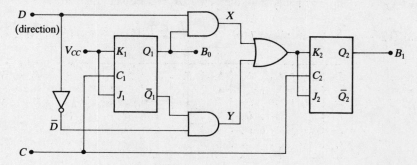

Fig. 16-28

Review Problems

16.40 Partial decoding (match on a subset of the address lines) is sometimes utilized. Design a decoder to connect to a set of 4-bit address lines $A_3A_2A_1A_0$ that provides a logic high output f only if either of the 4-bit words 1110 or 1111 appear on the address lines.

▌ Only the three most significant bits need to be decoded, and the logic is accomplished by the circuit of Fig. 16-29.

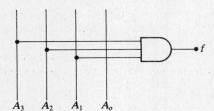

Fig. 16-29

16.41 Extend the MUX of Problem 16.22 to a 4- to 1-line multiplexer.

▌ The address code S must exhibit a unique value for each of the four data lines; hence, for four lines a 2-bit address code is necessary, or $S = S_2S_1$. Referring to Fig. 16-30,

$$F_1 = D_1 \cdot \bar{S}_1 \cdot \bar{S}_2 \tag{1}$$

$$F_2 = D_2 \cdot S_1 \cdot \bar{S}_2 \tag{2}$$

$$F_3 = D_3 \cdot \bar{S}_1 \cdot S_2 \tag{3}$$

$$F_4 = D_4 \cdot S_1 \cdot S_2 \tag{4}$$

$$f = F_1 + F_2 + F_3 + F_4 \tag{5}$$

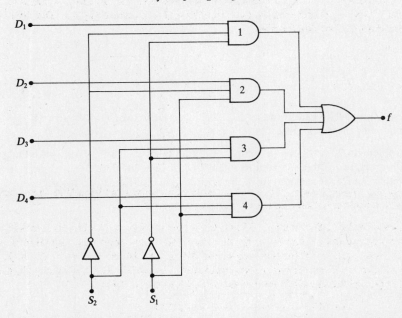

Fig. 16-30

Based on Theorems 7(b) and 7(d) of Table 15-1 and use of (1)–(5), the following truth table is constructed:

D_1	D_2	D_3	D_4	S_1	S_2	$\bar{S}_1$	$\bar{S}_2$	F_1	F_2	F_3	F_4	f
D_1	D_2	D_3	D_4	0	0	1	1	D_1	0	0	0	D_1
D_1	D_2	D_3	D_4	0	1	1	0	0	D_2	0	0	D_2
D_1	D_2	D_3	D_4	1	0	0	1	0	0	D_3	0	D_3
D_1	D_2	D_3	D_4	1	1	0	0	0	0	0	D_4	D_4

Whence, it is clear that one and only one of the data lines is connected to the output f for each value of the select code.

16.42 Explain the concept of designing a multiplexer-demultiplexer system to transmit eight lines of data between two points while using less than nine interconnecting lines.

▎ A 3-bit select code $S = S_2 S_1 S_0$ is necessary to give a unique value for each of the eight input data lines of the MUX and the eight output lines of the DEMUX. Four input AND gates are required in both the MUX and DEMUX—three select lines and one data line to each AND gate. Interconnection of the MUX and DEMUX requires one line to transmit data, three lines to transmit the select code, and a signal common or ground for a total of seven interconnecting lines. Without the use of multiplexing, a total of nine lines are necessary—eight data lines and a signal common line. The system is shown in Fig. 16-31.

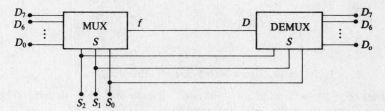

Fig. 16-31

16.43 Design a *comparator circuit* that gives an output of logic high if and only if two 2-bit binary numbers ($A = A_1 A_0$ and $B = B_1 B_0$) are equal.

▎ Two-input XNOR gate logic gives a logic high output if and only if the two inputs are equal. Thus, the bits of equal significance can be compared by XNOR logic and the results ANDed as illustrated by Fig. 16-32 where $f = \overline{(A_0 \oplus B_0)} \cdot \overline{(A_1 \oplus B_1)}$ which has a value of 1 only if $A_0 = B_0$ while $A_1 = B_1$.

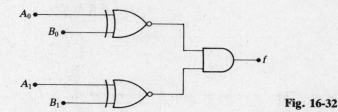

Fig. 16-32

16.44 Use an RS flip-flop, AND gates, and an RC differentiator network to design a JK flip-flop that is triggered on the leading edge of the clock pulse.

▎ From Problem 16.28, a JK flip-flop can be built up from an RS flip-flop by ANDing the Q and $\bar{Q}$ output with the K and J inputs, respectively. From Problem 14.59, we know that the clock pulse leading edge can be differentiated by an RC network to form the positive spike of Fig. 14-35 for use to set Q and $\bar{Q}$ according to the switching pattern of J and K at the time of arrival of the clock pulse leading edge. The resulting logic circuit is shown by Fig. 16-33.

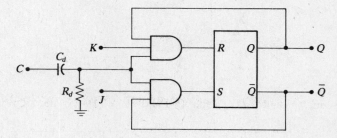

Fig. 16-33

16.45 Use two T flip-flops to design a circuit that receives the system clock at frequency f and generates two new lower frequency clock signals (C_1 and C_2) with characteristics $f_1 = f/2$ and $f_2 = f/4$.

▎ Use two cascade-connected T flip-flops as shown by Fig. 16-34. The input of the left-hand flip-flop is connected to clock input C. Output $Q_1 = C_1$ will toggle every other clock pulse; thus, its frequency is $f/2$. Output $Q_2 = C_2$ toggles every other cycle of C_1; thus, its frequency is $(f/2)/2 = f/4$.

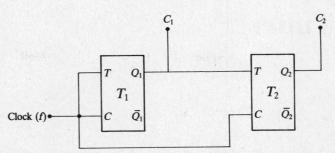

Fig. 16-34

16.46 Design a 2-bit shift register system that transfers data from shift register 1 to shift register 2, but retains the data in shift register 1.

▎ The shift register of Problem 16.34 loses the information in the right-hand bit after each clock cycle. However, if the right-hand bit were fed back to the input or left-hand bit as the serial input, the data is retained. This principle (called *recirculating*) is used in the proposed circuit of Fig. 16-35 to retain the data of shift register 1 while simultaneously transferring the data to shift register 2. For this set of 2-bit registers, two clock cycles are necessary to implement complete data transfer. The system clock is ANDed with a shift control (not illustrated) that is logic high for the duration of two clock cycles to effect the data transfer.

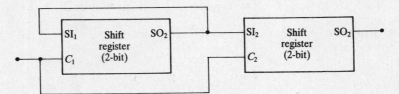

Fig. 16-35

16.47 Using JK flip-flops, design a 2-bit shift register that is initially parallel loaded with $D = D_1 D_0$.

▎ M/S JK flip-flops with preset (PS) feature are selected. The data can then be loaded through the PS terminals. The resulting shift register is shown in Fig. 16-36.

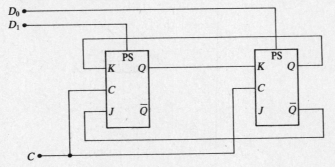

Fig. 16-36

16.48 Extend the modulo-4 synchronous counter of Problem 16.37 to a modulo-8 synchronous counter.

▎ Looking at the modulo-8 counting table of Problem 16.30, it is seen that bit B_2 should be complemented at the cycle following $B_0 = B_1 = 1$. This condition can be implemented by using an AND gate to set $J_3 = K_3 = B_0 \cdot B_1$. See Fig. 16-37 for the resulting modulo-8 synchronous counter.

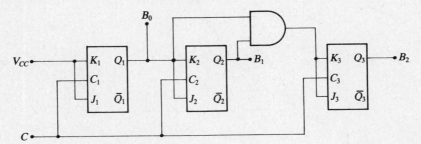

Fig. 16-37

CHAPTER 17
Vacuum Tubes

VACUUM DIODES

17.1 The symbol for the *vacuum diode* is shown in Fig. 17-1. Conduction is by electrons that escape the *cathode* and travel through the internal vacuum to be collected by the *anode*. The *filament* is an electric resistance heater used to elevate the cathode temperature to a level at which thermionic emission occurs. A typical i_P-versus-v_P characteristic curve (the *plate characteristic*) is depicted in Fig. 17-2(a). This characteristic is described by the *Childs-Langmuir three-halves-power law*:

$$i_P = \kappa v_P^{3/2} \tag{1}$$

where κ is the *perveance* (a constant that depends upon the mechanical design of the tube). Like the semiconductor diode, the vacuum diode is a near-unilateral conductor. However, the forward voltage drop of the vacuum diode is typically several volts, which renders the ideal-diode approximation unjustifiable. Most commonly, the vacuum diode is modeled in forward conduction as a resistance $R_P = V_P/I_P$, obtained as the slope of a straight-line approximation of the plate characteristic (Fig. 17-2). A reverse-biased vacuum diode acts like an open circuit.

A vacuum diode with perveance $\kappa = 0.1 \times 10^{-3}$ is operating with a plate voltage $V_P = 30$ V. Determine its approximate forward resistance.

▮ By (1),
$$I_P = \kappa V_P^{3/2} = (0.1 \times 10^{-3})(30)^{3/2} = 16.43 \text{ mA}$$

and
$$R_P = \frac{V_P}{I_P} = \frac{30}{16.43 \times 10^{-3}} = 1.826 \text{ k}\Omega$$

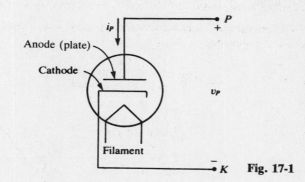

Fig. 17-1

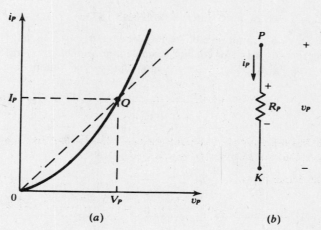

(a)   (b)   Fig. 17-2

17.2 The vacuum diode of Fig. 17-3 has plate characteristic $i_P = 0.08 v_P^{3/2}$ mA. If $E_{PP} = 50$ V and $R = 2.5$ kΩ, calculate the value of the plate current.

▌ By KVL,

$$i_P = \frac{E_{PP} - v_P}{R} \quad (1)$$

Equate (1) and the plate characteristic to find

$$0.08 \times 10^{-3} v_P^{3/2} = \frac{E_{PP} - v_P}{R} \quad (2)$$

Substitute given values, square both sides of (2), and rearrange to yield

$$v_P^3 - 25 v_P^2 + 2500 v_P - 62500 = 0 \quad (3)$$

By trial-and-error or use of a root formula, determine that (3) is satisfied by $v_P = 25$ V. Thus, from the plate characteristic,

$$i_P = 0.08 v_P^{3/2} = 0.08(25)^{3/2} = 10 \text{ mA}$$

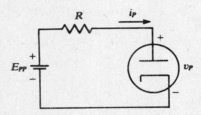

Fig. 17-3

17.3 On a common set of axes sketch the forward characteristics of a vacuum diode described by $i_P = 1.00 v_P^{3/2}$ mA and a semiconductor diode described by $i_D = 10^{-5} e^{40 v_D}$ mA for values of forward current from 0 to 100 mA. What major difference between the devices is apparent from your sketch?

▌ See Fig. 17-4 where the large forward voltage drop across the vacuum diode when compared with the semiconductor diode is apparent.

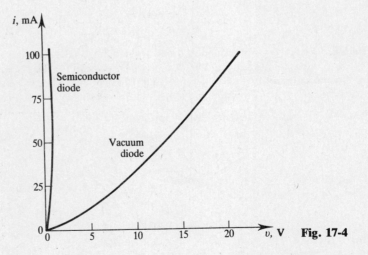

Fig. 17-4

17.4 If a vacuum diode has the static plate characteristic $i_P = 0.08 v_P^{3/2}$ mA and is operating with a quiescent plate current $I_P = 15$ mA, find the quiescent plate voltage V_P.

▌ From the plate characteristic,

$$V_P = \left(\frac{I_P}{8 \times 10^{-5}} \right)^{2/3} = \left(\frac{15 \times 10^{-3}}{8 \times 10^{-5}} \right)^{2/3} = 32.76 \text{ V}$$

17.5 Analogous to the dynamic resistance defined in Problem 2.50, the dynamic resistance of a vacuum diode is

$$r_P \equiv \left.\frac{dv_P}{di_P}\right|_Q = \frac{V_P}{I_P} \quad (1)$$

Determine the dynamic resistance for the diode of Problem 17.4.

▎ Differentiating the expression for the plate voltage from Problem 17.4,

$$r_P = \left.\frac{dv_P}{di_P}\right|_Q = \frac{2/3}{(8 \times 10^{-5})^{2/3}(I_P)^{1/3}}$$

$$= \frac{2/3}{(8 \times 10^{-5})^{2/3}(15 \times 10^{-3})^{1/3}} = 1.456 \text{ k}\Omega \quad (2)$$

17.6 Due to the nonlinear nature of the vacuum diode plate characteristic, the dynamic resistance (r_p) varies over a significant range for large signal excursion. Frequently, an average value of r_P is used for large signal excursion. If the diode of Problem 17.4 is to be applied with a plate voltage v_P that swings from 10 to 50 V, find an average value of dynamic resistance ($r_{p\text{ ave}}$) to use in analysis.

▎ The dynamic resistance is calculated for multiple operation points. Then, these resulting values are averaged. Arbitrarily decide to find r_P for v_P = 20, 30, and 40 V, a set of values that are spaced over the anticipated range of v_P. Using the plate characteristic of Problem 17.4 and (2) of Problem 17.5, the values of r_P are found as

20 V: $\quad i_P = 0.08 v_P^{3/2} = 0.08(20)^{3/2} = 7.16$ mA

$$r_P = \frac{2/3}{(8 \times 10^{-5})^{2/3}(i_P)^{1/3}} = \frac{2/3}{(8 \times 10^{-5})^{2/3}(7.16 \times 10^{-3})^{1/3}} = 1.861 \text{ k}\Omega$$

30 V: $\quad i_P = 0.08(30)^{3/2} = 13.14$ mA

$$r_P = \frac{2/3}{(8 \times 10^{-5})^{2/3}(13.14 \times 10^{-3})^{1/3}} = 1.52 \text{ k}\Omega$$

40 V: $\quad i_P = 0.08(40)^{3/2} = 20.24$ mA

$$r_P = \frac{2/3}{(8 \times 10^{-5})^{2/3}(20.24 \times 10^{-3})^{1/3}} = 1.316 \text{ k}\Omega$$

Whence, $\quad r_{p\text{ ave}} = \dfrac{1.861 + 1.52 + 1.316}{3} = 1.566$ kΩ

17.7 *Rectification efficiency* (η_R) is defined as the ratio of average output dc power supplied to a load to the average ac supplied to the rectifier diode and load. Find an expression for the rectification efficiency of the half-wave vacuum diode rectifier of Fig. 17-5. Assume that the vacuum diode can be modeled by a constant resistance R_F in the forward conduction direction and by an infinite resistance in the reverse direction.

▎ The half-wave rectified current can be expressed as

$$i_P = \begin{cases} I_{pm} \sin \omega t & 0 \le \omega t \le \pi \\ 0 & \pi < \omega t < 2\pi \end{cases} \quad (1)$$

Hence, the average value of load current is

$$I_{p0} = \frac{1}{2\pi} \int_0^\pi I_{pm} \sin \omega t\, d(\omega t) = \frac{I_{pm}}{\pi} \quad (2)$$

Also, the rms value of load current is

$$I_p = \left[\frac{1}{2\pi} \int_0^\pi I_{pm}^2 \sin^2 \omega t\, d(\omega t) \right]^{1/2} = \frac{I_{pm}}{2} \quad (3)$$

Thus, by the definition of rectification efficiency, and using (2) and (3),

$$\eta_R = \frac{P_{\text{dc}}}{P_{\text{ac}}}(100\%) = \frac{I_{p0}^2 R_L}{I_p^2(R_L + R_F)}(100\%) = \frac{(I_{pm}/\pi)^2 R_L}{(I_{pm}/2)^2(R_L + R_F)}(100\%) = \left(\frac{2}{\pi}\right)^2 \frac{R_L}{R_L + R_F} \quad (4)$$

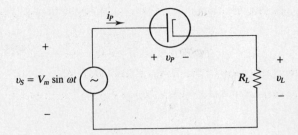

Fig. 17-5

17.8 Determine the upper limit of rectification efficiency for the half-wave rectifier of Problem 17.7.

▮ From (4) of Problem 17.7, it is seen that the upper limit would occur if $R_L \gg R_F$, giving

$$\eta_{R\,max} = (2/\pi)^2 (100\%) = 40.53\%$$

17.9 For the half-wave rectifier of Fig. 17-5, the diode is that of Problem 17.4. If $R_L = 25$ kΩ and $v_S = 50 \sin \omega t$ V, evaluate the rectification efficiency.

▮ Based on the results of Problem 17.6, we note that R_L is significantly larger than $r_p = R_F$. Consequently, the rectification efficiency is not sensitive to a small error in r_p and the value of $r_{p\,ave}$ of Problem 17.6 will be used for R_F. Then, by (4) of Problem 17.7,

$$\eta_R = \left(\frac{2}{\pi}\right)^2 \frac{R_L}{R_L + r_{p\,ave}}(100\%) = \left(\frac{2}{\pi}\right)^2 \frac{25 \times 10^3}{25 \times 10^3 + 1.566 \times 10^3}(100\%) = 38.1\%$$

17.10 The vacuum diodes in the rectifier circuit of Fig. 17-6(a) are identical and can be modeled by $R_F = 400$ Ω in the forward direction and by an infinite resistance in the reverse direction. If $C = 0$, $R_L = 5$ kΩ, $V_{L0} = 110$ V, and $v_S = 120\sqrt{2} \sin 120\pi t$ V, calculate the turns ratio of the ideal transformer.

▮ Owing to the symmetry of the problem, it is necessary to analyze only one leg of the rectifier circuit, as indicated in Fig. 17-6(b). Since the output is a rectified sine wave, for which (from Problem 3.51) $V_{L0} = 0.637 V_{Lm}$, during the positive half cycle of v_S the load voltage can be written as

$$v_L(t) = \frac{110}{0.636} \sin 120\pi t = 172.96 \sin 120\pi t \text{ V}$$

[$v_L(t)$ is zero during the negative cycle of v_S]. By Ohm's law,

$$i_L(t) = \frac{v_L(t)}{R_L} = \frac{172.96}{5000} \sin 120\pi t \times 0.03459 \sin 120\pi t \text{ A}$$

Also,

$$v_{D1}(t) = R_F i_L(t) = 400(0.03459 \sin 120\pi t) = 13.84 \sin 120\pi t \text{ V}$$

so that

$$V_{S1}(t) = v_{D1}(t) + v_L(t) = 186.8 \sin 120\pi t \text{ V}$$

and the turns ratio is given by

$$a = \frac{v_S}{v_{S1}} = \frac{120\sqrt{2}}{186.8} = 0.908$$

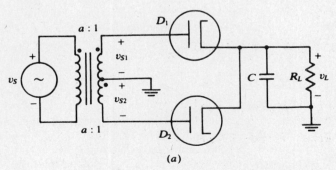
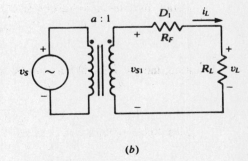

Fig. 17-6

17.11 Find the rectification efficiency for the full-wave vacuum diode rectifier of Fig. 17-6.

▌ Based on the results of Problem 17.10,

$$i_L = \begin{cases} 0.03459 \sin 120\pi t \text{ A} & 0 \leq \omega t < \pi \\ -0.03459 \sin 120\pi t \text{ A} & \pi \leq \omega t < 2\pi \end{cases}$$

The average value of load current follows as

$$I_{L0} = \frac{1}{\pi} \int_0^\pi \frac{I_{Lm} \sin 120\pi t}{R_L + R_F} d(120\pi t) = \frac{2}{\pi} I_{Lm} \qquad (1)$$

The effective value of load current is

$$I_L = \left[\frac{1}{\pi} \int_0^\pi I_{Lm}^2 \sin^2(120\pi t) d(120\pi t) \right]^{1/2} = \frac{I_{Lm}}{\sqrt{2}} \qquad (2)$$

By the definition of rectification efficiency from Problem 17.7 and using (1) and (2),

$$\eta_r = \frac{P_{dc}}{P_{ac}}(100\%) = \frac{I_{Lo}^2 R_L}{I_L^2 (R_L + R_F)}(100\%) = \frac{[(2/\pi)I_{Lm}]^2 R_L}{(I_{Lm}/\sqrt{2})^2 (R_L + R_F)}(100\%)$$

$$= \frac{8}{\pi^2} \frac{R_L}{R_L + R_F}(100\%) = \frac{8}{\pi^2} \frac{5 \times 10^3}{5 \times 10^3 + 400}(100\%) = 75.05\%$$

VACUUM TRIODES

17.12 The voltage-current characteristics of the vacuum triode (shown physically and schematically by Fig. 17-7) are experimentally determined with the cathode sharing a common connection with the input and output ports. If plate voltage v_P and grid voltage v_G are taken as independent variables, and grid current i_G as the dependent variable, then the *input characteristics* (or *grid characteristics*) have the form

$$i_G = f_1(v_P, v_G) \qquad (1)$$

of which Fig. 17-8(a) is a typical experimentally determined plot. Similarly, with v_P and v_G as independent variables, the plate current i_P becomes the dependent variable of the *output characteristics* (or *plate characteristics*)

$$i_P = f_2(v_P, v_G) \qquad (2)$$

of which a typical plot is displayed in Fig. 17-8(b).

The triode input characteristics of Fig. 17-8(a) show that operation with a positive grid voltage results in flow of grid current; however, with a negative grid voltage (the common application), negligible grid current flows and the plate characteristics are reasonably approximated by a three-halves-power relationship involving a linear combination of plate and grid voltages:

$$i_P = \kappa(v_P + \mu v_G)^{3/2} \qquad (3)$$

where κ again denotes the perveance and μ is the *amplification factor*.

To establish a range of triode operation favorable to the signal to be amplified, a quiescent point must be determined by dc bias circuitry. The basic triode amplifier of Fig. 17-9 has a grid power supply V_{GG} of such polarity as to maintain v_G negative (the more common mode of operation). Determine a set of constraints placed on the grid and plate characteristics to establish the quiescent point of operation.

▌ From Fig. 17-9 with no input signal ($v_S = 0$), application of KVL around the grid loop yields the equation of the *grid bias line*,

$$i_G = -\frac{V_{GG}}{R_G} - \frac{v_G}{R_G} \qquad (4)$$

which can be solved simultaneously with (1) or plotted as indicated on Fig. 17-8(a) to determine the quiescent values I_{GQ} and V_{GQ}. If V_{GG} is of the polarity indicated in Fig. 17-9, the grid is negatively biased, giving the Q point labeled Q_n. At that point, $I_{GQ} \approx 0$ and $V_{GQ} \approx -V_{GG}$; these approximate solutions suffice in the case of negative grid bias. However, if the polarity of V_{GG} were reversed, the grid would have a positive bias, and the quiescent point Q_p would give $I_{GQ} > 0$ and $V_{GQ} < V_{GG}$.

Voltage summation around the plate circuit of Fig. 17-9 leads to the equation of the *dc load line*

$$i_P = \frac{V_{PP}}{R_L} - \frac{v_P}{R_L} \qquad (5)$$

which, when plotted on the plate characteristics of Fig. 17-8(*b*), yields the quiescent values V_{PQ} and I_{PQ} at its intersection with the curve $v_G = V_{GQ}$.

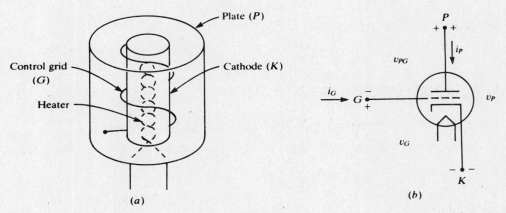

Fig. 17-7

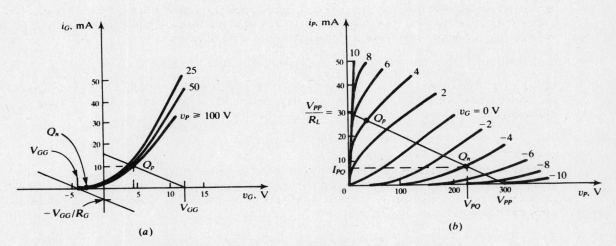

Fig. 17-8 (*a*) Grid characteristics, (*b*) Plate characteristics.

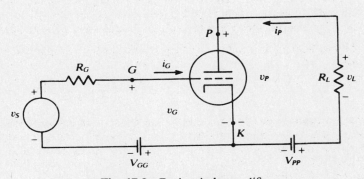

Fig. 17-9 Basic triode amplifier.

17.13 In the triode amplifier of Fig. 17-9, $V_{GG} = 4$ V, $V_{PP} = 300$ V, $R_L = 10$ kΩ, and $R_G = 2$ kΩ. The plate characteristics for the triode are given by Fig. 17-8(b). Draw the dc load line.

▌ For the given values, the dc load line, given by (5) of Problem 17.12, has the i_P intercept

$$\frac{V_{PP}}{R_L} = \frac{300}{10 \times 10^3} = 30 \text{ mA}$$

and the v_P intercept $V_{PP} = 300$ V. These intercepts have been utilized to draw the dc load line on the plate characteristics of Fig. 17-8(b).

17.14 Determine the quiescent grid current I_{GQ} for the triode amplifier of Problem 17.13.

▌ Since the polarity of V_{GG} is such that v_G is negative, negligible grid current will flow ($I_{GQ} \approx 0$).

17.15 Find the quiescent grid voltage for the amplifier of Problem 17.13.

▌ For negligible grid current, (5) of Problem 17.12 evaluated at the Q point yields $V_{GQ} = -V_{GG} = -4$ V.

17.16 Determine the value of quiescent plate current for the amplifier of Problem 17.13.

▌ The quiescent plate current is read as the projection of Q_n onto the i_p axis of Fig. 17-8(b) and is $I_{PQ} = 8$ mA.

17.17 Evaluate the quiescent plate voltage for the triode amplifier of Problem 17.13.

▌ Projection of Q_n onto the v_P axis of Fig. 17-8(b) gives $V_{PQ} = 220$ V.

17.18 For the triode amplifier of Fig. 17-9, let $V_{PP} = 300$ V, $V_{GG} = -12$ V, $R_L = 10$ kΩ, and $R_G = 750$ Ω. If the triode is described by the grid and plate characteristics of Fig. 17-8, draw the dc load line.

▌ The dc load line is identical to the load line determined in Problem 17.13.

17.19 Find the quiescent grid current for the amplifier of Problem 17.18.

▌ For the given values, the grid bias line, given by (4) of Problem 17.12, has the i_G intercept

$$-\frac{V_{GG}}{R_G} = -\frac{-12}{750} = 16 \text{ mA}$$

and v_G intercept $-V_{GG} = 12$ V. These intercepts are used to draw the grid bias line on the grid characteristic of Fig. 17-8(a). From the Q_p projection back on the i_G axis, $I_{GQ} \approx 10$ mA.

17.20 Determine the quiescent grid voltage for the amplifier of Problem 17.18.

▌ In Fig. 17-8(a), projection of Q_p on the v_G axis gives $V_{GQ} \approx 4$ V.

17.21 Find the value of quiescent plate current for the amplifier of Problem 17.18.

▌ The value of $V_{GS} = 4$ from Problem 17.20 allows us to identify the quiescent point of Q_p on the dc load of Fig. 17-8(b). Projection of Q_p to the i_P axis gives $I_{PQ} \approx 27$ mA.

17.22 Evaluate the quiescent plate voltage for the amplifier of Problem 17.18.

▌ Projection from Q_p to the v_P axis of Fig. 17-8(b) gives $V_{PQ} \approx 30$ V.

17.23 Determine the plate power dissipation at quiescent conditions for the triode amplifier of Problem 17.18.

▌ Based on results of Problems 17.21 and 17.22,

$$P_P = V_{PQ}I_{PQ} = 30(27 \times 10^{-3}) = 810 \text{ mW}$$

17.24 Since the amplifier of Problem 17.18 is biased with a positive quiescent grid voltage, the grid draws current, and this also dissipates power. Determine the grid power dissipated at quiescent conditions.

▌ Based on the results of Problems 17.19 and 17.20,

$$P_G = V_{GQ}I_{GQ} = 4(10 \times 10^{-3}) = 40 \text{ mW}$$

17.25 Determine the total power supplied to the amplifier of Problem 17.18.

I Based on given values and the results of Problems 17.19 and 17.20,
$$P_{BB} = V_{GG}I_{GQ} + V_{PP}I_{PQ} = 12(10 \times 10^{-3}) + 300(27 \times 10^{-3}) = 8.22 \text{ W}$$

17.26 The triode amplifier of Fig. 17-9 has V_{GG}, V_{PP}, R_G, and R_L as given in Problem 17.13. If the plate characteristics of the triode are given by Fig. 17-10 and $v_S = 2 \sin \omega t$ V, graphically find v_P and i_P.

I The dc load line, with the same intercepts as in Problem 17.13, is superimposed on the characteristics of Fig. 17-10; however, because the plate characteristics are different from those of Problem 17.13, the quiescent values are now $I_{PQ} = 11.3$ mA and $V_{PQ} = 186$ V. The time axis on which to plot $v_G = -4 + 2 \sin \omega t$ V is constructed perpendicular to the dc load line at the Q point. Time axis for i_P and v_P are also constructed as shown, and values of i_P and v_P corresponding to particular values of $v_G(t)$ are found by projecting through the dc load line, for one cycle of v_G. The result, in Fig. 17-10, shows that v_P varies from 152 to 218 V and i_P ranges from 8.1 to 14.7 mA.

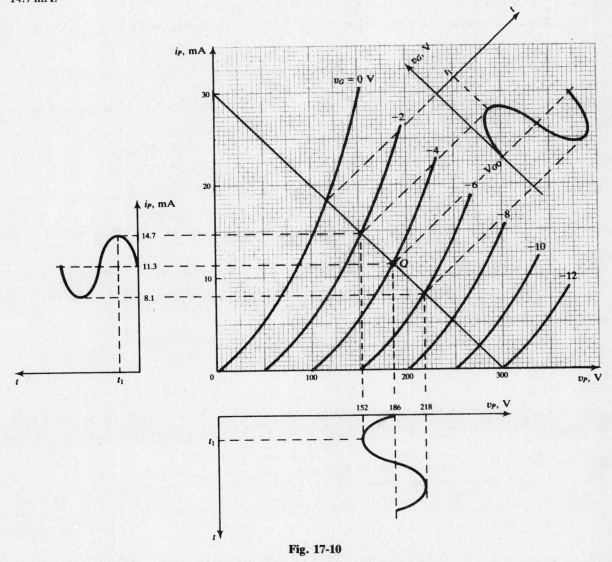

Fig. 17-10

17.27 For a triode with plate characteristics given by Fig. 17-10, find the perveance κ.

I The perveance can be evaluated at any point on the $v_G = 0$ curve. Choosing the point with coordinates $i_P = 15$ mA and $v_P = 100$ V, we have, from (3) of Problem 17.12,
$$\kappa = \frac{i_P}{v_P^{3/2}} = \frac{15 \times 10^{-3}}{100^{3/2}} = 15 \ \mu\text{A/V}^{3/2}$$

17.28 For a triode with plate characteristics of Fig. 17.10, find the amplification factor μ.

▮ The amplification factor is most easily evaluated along the v_P axis. From (3) of Problem 17.12, for the point $i_P = 0$, $v_P = 100$ V, $v_G = -4$ V, we obtain

$$\mu = -\frac{v_P}{v_G} = -\frac{100}{-4} = 25$$

17.29 The amplifier of Problem 17.13 has plate current

$$i_P = I_P + i_p = 8 + \cos \omega t \text{ mA}$$

Determine the power delivered by the plate supply voltage V_{PP}.

▮ The power supplied by the source V_{PP} is found by integration over a period of the ac waveform:

$$P_{PP} = \frac{1}{T}\int_0^T V_{PP} i_P \, dt = V_{PP} I_P = 300(8 \times 10^{-3}) = 2.4 \text{ W}$$

17.30 Find the average value of power delivered to the load R_L for the amplifier of Problem 17.13 with i_P as described by Problem 17.29.

▮
$$P_L = \frac{1}{T}\int_0^T i_P^2 R_L \, dt = R_L(I_P^2 + I_p^2) = 10 \times 10^3 \left[(8 \times 10^{-3})^2 + \left(\frac{1 \times 10^{-3}}{\sqrt{2}}\right)^2\right] = 0.645 \text{ W}$$

17.31 Determine the average value of power dissipated by the plate for the triode in the amplifier of Problem 17.13 with i_P as given in Problem 17.29.

▮ Based on the results of Problems 17.29 and 17.30, the average power dissipated by the plate is

$$P_P = P_{PP} - P_L = 2.4 - 0.645 = 1.755 \text{ W}$$

17.32 If the tube for the amplifier of Problem 17.13, with plate current as given by Problem 17.29, has a plate rating of 2 W, is it being properly applied?

▮ The tube is not properly applied. If the signal is removed (so that $i_p = 0$), then the plate dissipation increases to $P_P = P_{PP} = 2.4$ W, which exceeds the power rating.

17.33 Determine a small-signal equivalent circuit model for the vacuum triode valid for negligible grid current flow.

▮ For the case of negligible grid current, (1) of Problem 17.12 degenerates to $i_G = 0$ and the grid acts as an open circuit. For small excursions (ac signals) about the Q point, $\Delta i_P = i_p$ and an application of the chain rule to (2) of Problem 17.12 leads to

$$i_p = \Delta i_P \approx d i_P = \frac{1}{r_p} v_p + g_m v_g \tag{1}$$

where we have defined

$$\text{Plate resistance} \quad r_p \equiv \left.\frac{\partial v_P}{\partial i_P}\right|_Q \approx \left.\frac{\Delta v_P}{\Delta i_P}\right|_Q \tag{2}$$

$$\text{Transconductance} \quad g_m \equiv \left.\frac{\partial i_P}{\partial v_G}\right|_Q \approx \left.\frac{\Delta i_P}{\Delta v_G}\right|_Q \tag{3}$$

Under the condition $i_G = 0$, (1) is simulated by the current-source equivalent circuit in Fig. 17-11(a). The frequently used voltage-source model of Fig. 17-11(b) is developed in Problem 17.34.

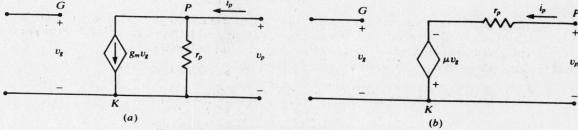

Fig. 17-11 Triode small-signal equivalent circuits.

17.34 Use the current-source small-signal triode model of Fig. 17-11(a) to derive the voltage-source model of Fig. 17-11(b).

▮ We need to find the Thévenin equivalent for the circuit to the left of the output terminals in Fig. 17-11(a). If the independent source is deactivated, then $v_g = 0$; thus, $g_m v_g = 0$, and the dependent current source acts as an open circuit. The Thévenin resistance is then $R_{Th} = r_p$. The open-circuit voltage appearing at the output terminals is

$$v_{Th} = -g_m v_g r_p \equiv -\mu v_G$$

where $\mu \equiv g_m r_p$ is the *amplification factor*. Proper series arrangement of v_{Th} and R_{Th} gives the circuit of Fig. 17-11(b).

17.35 For the amplifier of Problem 17.26, use (2) of Problem 17.33 to evaluate the plate resistance.

▮
$$r_p \approx \left.\frac{\Delta v_P}{\Delta i_P}\right|_{v_G=-4} = \frac{218 - 152}{(14.7 - 8.1) \times 10^{-3}} = 10 \text{ k}\Omega$$

17.36 Use (3) of Problem 17.33 to evaluate the transconductance of the triode of the amplifier in Problem 17.26.

▮
$$g_m \approx \left.\frac{\Delta i_P}{\Delta v_G}\right|_{v_P=186} = \frac{(14.7 - 8.1) \times 10^{-3}}{-2 - (-6)} = 1.65 \text{ mS}$$

17.37 Find an expression for the voltage gain $A_v = v_p/v_g$ of the basic triode amplifier of Fig. 17-9, using an ac equivalent circuit.

▮ The equivalent circuit of Fig. 17-11(b) is applicable if R_L is connected from P to K. Then, by voltage division in the plate circuit,

$$v_p = \frac{R_L}{R_L + r_p}(-\mu v_g) \quad \text{so} \quad A_v = \frac{v_p}{v_g} = \frac{-\mu R_L}{R_L + r_p}$$

17.38 The *plate efficiency* of a vacuum-tube amplifier is defined as the ratio of ac signal power delivered to the load to plate supply power, or $P_{L\,ac}/P_{PP}$. Calculate the plate efficiency of the amplifier of Problem 17.29.

▮
$$\eta = \frac{P_{L\,ac}}{P_{PP}}(100\%) = \frac{I_p^2 R_L}{V_{PP} I_P}(100\%) = \frac{(10^{-3}/\sqrt{2})^2 (10 \times 10^3)}{2.4}(100\%) = 0.208\%$$

17.39 What is the maximum possible plate efficiency for the amplifier of Problem 17.29 without changing the Q point or clipping the signal?

▮ Ideally, the input signal could be increased until i_p swings ± 8 mA; thus, based on the result of Problem 17.38,

$$\eta_{\max} = \left(\frac{8}{1}\right)^2 (0.208\%) = 13.31\%$$

17.40 The triode amplifier of Fig. 17-12 utilizes *cathode bias* to eliminate the need for a grid power supply. The very large resistance R_G provides a path to ground for stray charge collected by the grid; this current is so small, however, that the voltage drop across R_G is negligible. It follows that the grid is maintained at a negative bias, so

$$v_G = -R_K i_P \tag{1}$$

A plot of (1) on the plate characteristics is called the *grid bias line*, and its intersection with the dc load line determines the Q point. Let $R_L = 11.6$ kΩ, $R_K = 400$ Ω, $R_G = 1$ MΩ, and $V_{PP} = 300$ V. If the plate characteristics of the triode are given by Fig. 17-13, draw the dc load line.

▮ The dc load line has horizontal intercept $V_{PP} = 300$ V and vertical intercept

$$\frac{V_{PP}}{R_{dc}} = \frac{V_{PP}}{R_L + R_K} = \frac{300}{(11.6 + 0.4) \times 10^3} = 25 \text{ mA}$$

as shown on the plate characteristics of Fig. 17-13.

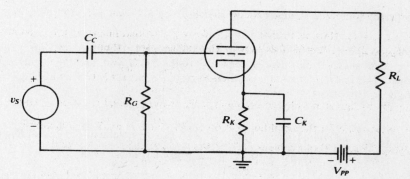

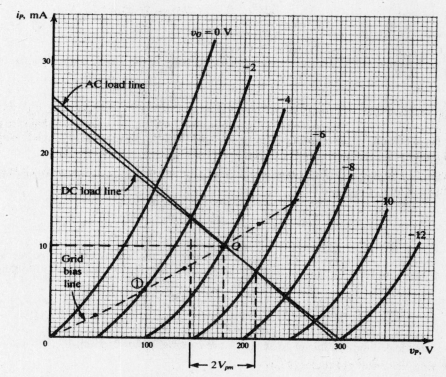

Fig. 17-12

Fig. 17-13

17.41 Sketch the grid bias line for the amplifier of Problem 17.40.

▮ Points for the plot of (1) of Problem 17.40 are found by selecting values of i_P and calculating the corresponding values of v_G. For example, if $i_P = 5$ mA, then $v_G = -400(5 \times 10^{-3}) = -2$ V, which plots as point 1 of the dashed grid bias line in Fig. 17-13. Note that this is not a straight line.

17.42 Determine the Q-point quantities for the amplifier of Problem 17.40.

▮ From the intersection of the grid bias line with the dc load line on Fig. 17-13, $I_{PQ} = 10$ mA, $V_{PQ} = 180$ V, and $V_{GQ} = -4$ V.

17.43 In the amplifier of Problem 17.40, let $v_S = 2 \cos \omega t$ V. Draw the ac load line on Fig. 17-13.

▮ If capacitor C_K appears as a short circuit to ac signals, then application of KVL around the plate circuit of Fig. 17-12 gives, as the equation of the ac load line, $V_{PP} + V_{GQ} = i_P R_L + v_P$. Thus, the ac load line has vertical and horizontal intercepts

$$\frac{V_{PP} + V_{GQ}}{R_L} = \frac{300 - 4}{11.6 \times 10^3} = 25.5 \text{ mA} \quad \text{and} \quad V_{PP} + V_{GQ} = 296 \text{ V}$$

as shown on Fig. 17-13.

17.44 Graphically determine the voltage gain for the amplifier of Problem 17.43.

▌ We have $v_g = v_S$; thus, as v_g swings ± 2 V along the ac load line from the Q point in Fig. 17-13, v_P swings a total of $2V_{pm} = 213 - 145 = 68$ V as shown. The voltage gain is then

$$A_v = -\frac{2V_{pm}}{2V_{gm}} = -\frac{68}{4} = -17$$

where the minus sign is included to account for the phase reversal between v_p and v_g.

17.45 Find the voltage gain for the amplifier of Problem 17.43 using small-signal analysis.

▌ Applying (2) and (3) of Problem 17.33 at the Q point of Fig. 17-13 yields

$$r_p = \frac{\Delta v_P}{\Delta i_P}\bigg|_{v_G = -4} = \frac{202 - 168}{(15 - 8) \times 10^{-3}} = 4.86 \text{ k}\Omega$$

$$g_m = \frac{\Delta i_P}{\Delta v_G}\bigg|_{v_p = 180} = \frac{(15.5 - 6.5) \times 10^{-3}}{-3 - (-5)} = 4.5 \text{ mS}$$

Then, $\mu \equiv g_m r_p = 21.87$, and Problem 17.37 yields

$$A_v = -\frac{\mu R_L}{R_L + r_p} = -\frac{(21.87)(11.6 \times 10^3)}{(11.6 + 4.86) \times 10^3} = -15.41$$

17.46 The input admittance to a triode modeled by the small-signal equivalent circuit of Fig. 17-11(b) is obviously zero; however, there are interelectrode capacitances that must be considered for high-frequency operation. Add these interelectrode capacitances (grid-cathode capacitance C_{gk}; plate-grid, C_{pg}; and plate-cathode, C_{pk}) to the small-signal equivalent circuit of Fig. 17-11(b) and find the input admittance Y_{in}.

▌ With the interelectrode capacitances in position, the small-signal equivalent circuit is given by Fig. 17-14. The input admittance is

$$Y_{in} = \frac{I_S}{V_S} = \frac{I_1 + I_2}{V_S} \tag{1}$$

But

$$I_1 = \frac{V_S}{1/sC_{gk}} = sC_{gk}V_S \tag{2}$$

and

$$I_2 = \frac{V_S - V_o}{1/sC_{pg}} = sC_{pg}(V_S - V_o) \tag{3}$$

Substituting (2) and (3) into (1) and rearranging give

$$Y_{in} = s\left[C_{gk} + \left(1 + \frac{V_o}{V_S}\right)C_{pg}\right] \tag{4}$$

Now, from the result of Problem 17.37,

$$\frac{V_o}{V_S} = -\frac{\mu R_L}{R_L + r_p}$$

so (4) becomes

$$Y_{in} = s\left[C_{gk} + \left(1 + \frac{\mu R_L}{R_L + r_p}\right)C_{pg}\right] \tag{5}$$

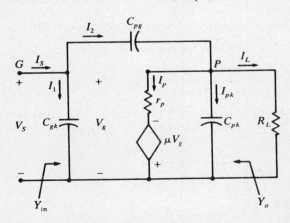

Fig. 17-14

17.47 Find an expression for the output admittance Y_o of the triode small-signal equivalent circuit of Fig. 17-14.

▮ The output admittance is

$$Y_o = -\frac{I_L}{V_o} = -\frac{I_2 - I_p - I_{pk}}{V_o} \tag{1}$$

and

$$I_{pk} = sC_{pk}V_o \tag{2}$$

Let Y_o' be the output admittance that would exist if the capacitances were negligible; then

$$I_p = Y_o'V_o \tag{3}$$

so that

$$Y_o = s\left[\left(1 + \frac{R_L + r_p}{\mu R_L}\right)C_{pg} + C_{pk}\right] + Y_o' \tag{4}$$

17.48 Using the results of Problems 17.46 and 17.47, develop a high-frequency model for the triode.

▮ From (5) of Problem 17.46 and (4) of Problem 17.47, we see that high-frequency triode operation can be modeled by Fig. 17-11(b) with a capacitor $C_{in} = C_{gk} + [1 + R_L/(R_L + r_p)]C_{pg}$ connected from the grid to the cathode, and a capacitor $C_o = [1 + (R_L + r_p)/\mu R_L]C_{pg} + C_{pk}$ connected from the plate to the cathode.

17.49 The circuit of Fig. 17-15 is a *cathode follower*, so called because v_o is in phase with v_S and nearly equal to it in magnitude. Find a voltage-source equivalent circuit of the form of Fig. 17-11(b) that models the cathode follower.

▮ By direct analogy with the work of Problem 8.32, the equivalent circuit of Fig. 17-16 follows.

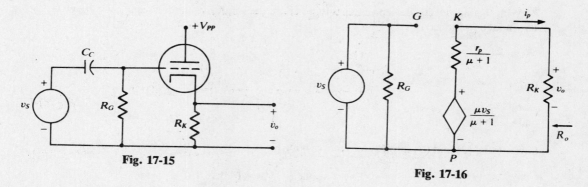

Fig. 17-15

Fig. 17-16

17.50 For the cathode follower of Fig. 17-15, $r_p = 5$ kΩ, $\mu = 25$, and $R_K = 15$ kΩ. Use the equivalent circuit of Fig. 17-16 to find a formula for the voltage gain.

▮ By voltage division applied to the output network of Fig. 17-16,

$$v_o = \left[\frac{R_K}{R_K + r_p/(\mu + 1)}\right]\frac{\mu v_S}{\mu + 1} \tag{1}$$

The voltage gain results from rearrangement of (1):

$$A_v = \frac{v_o}{v_S} = \frac{\mu R_K}{r_p + (\mu + 1)R_K} \tag{2}$$

17.51 Evaluate the voltage gain for the cathode follower of Problem 17.50.

▮ Using the given values and (2) of Problem 17.50,

$$A_v = \frac{25(15 \times 10^3)}{5 \times 10^3 + (25 + 1)(15 \times 10^3)} = 0.95$$

17.52 The cathode follower is frequently used as a final-stage amplifier to effect an impedance match with a low-impedance load for maximum power transfer. In such a case, the load (resistor R_L) is capacitor-coupled to the right of R_K in Fig. 17-16. Find an expression for the internal impedance (output impedance) of the cathode follower as seen by the load.

With v_S deactivated (shorted) in Fig. 17-16,

$$R_o = R_K \left\| \left(\frac{r_p}{\mu + 1} \right) = \frac{r_p R_K}{r_p + (\mu + 1)R_K} \right. \tag{1}$$

17.53 Evaluate the output impedance for the cathode follower of Problem 17.50.

Using the given values and (1) of Problem 17.52,

$$R_o = \frac{(50 \times 10^3)(15 \times 10^3)}{5 \times 10^3 + (25 + 1)(15 \times 10^3)} \approx 190 \ \Omega$$

17.54 The amplifier of Fig. 17-17 is a *common-grid* amplifier. By finding a Thévenin equivalent for the network to the right of G, K and another for the network to the left of R_P, the small-signal circuit of Fig. 17-18 can be shown to be valid. Find an expression for the voltage gain.

By voltage division applied to the output circuit of Fig. 17-18,

$$v_o = \left[\frac{R_P}{R_P + r_p + (\mu + 1)R_K} \right] (\mu + 1) v_s$$

After rearrangement,

$$A_v = \frac{v_o}{v_S} = \frac{(\mu + 1)R_P}{R_P + r_p + (\mu + 1)R_K} \tag{1}$$

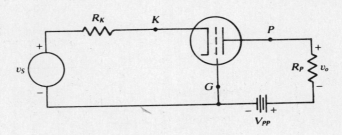

Fig. 17-17

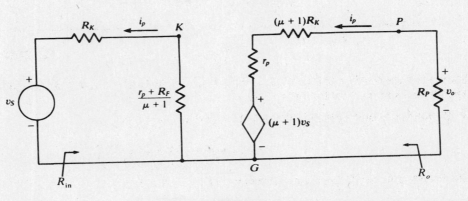

Fig. 17-18

17.55 Evaluate the voltage gain for the common grid amplifier using the typical values $\mu = 20$, $r_p = 5 \ \text{k}\Omega$, $R_K = 1 \ \text{k}\Omega$, and $R_P = 15 \ \text{k}\Omega$.

By (1) of Problem 17.54,

$$A_v = \frac{(20 + 1)(15 \times 10^3)}{15 \times 10^3 + 5 \times 10^3 + (20 + 1)(1 \times 10^3)} = 7.7$$

17.56 Determine the input resistance R_{in} of the common grid amplifier for the typical values given in Problem 17.55.

The input resistance is

$$R_{in} = R_K + \frac{r_p + R_K}{\mu + 1} = 1 \times 10^3 + \frac{5 \times 10^3 + 1 \times 10^3}{20 + 1} = 1.95 \ \text{k}\Omega$$

17.57 Determine the output resistance R_o of the common grid amplifier for the typical values of Problem 17.55.

▮ With the independent source v_S deactivated (shorted), the controlled source $(\mu + 1)v_S = 0$. Hence,

$$R_o = r_p + (\mu + 1)R_K = 5 \times 10^3 + (20 + 1)(1 \times 10^3) = 26 \text{ k}\Omega$$

VACUUM PENTODES

17.58 Interelectrode capacitances limit the usefulness of the triode at high frequencies; the plate-grid capacitance C_{pg} leads to the greatest detriment in performance because its effect is magnified by tube parameters and because it constitutes undesired coupling between input and output, leading to a reduction in control of the output. A marked reduction in C_{pg} results if a *screen grid* is inserted between the control grid and the plate to form a *tetrode* (four-element *tube*). The screen grid is maintained at a positive potential v_{G2} with respect to the cathode. However, when the plate voltage is less than the screen-grid voltage ($v_P < v_{G2}$), the plate characteristics exhibit an erratic nonlinearity due to electrons being dislodged from the plate by incident electrons arriving from the cathode (*secondary emission*) and being attracted to the screen grid. This undesirable nonlinearity can be eliminated by adding a third grid (a *suppressor grid*) between the screen grid and the plate to form a *pentode* (five-element *tube*). The suppressor grid is maintained at the same potential as the cathode by electrical connection, as indicated by the schematic of the vacuum pentode in Fig. 17-19.

The grid characteristics of a pentode are similar to those of a triode [see Fig. 17-8(a)]. The experimentally determined plate characteristics are typically like those in Fig. 17-20. (A similarity to JFET drain characteristics is apparent.)

In the pentode amplifier of Fig. 17-19, assume that the control grid is biased for $V_{G1Q} = -2.5$ V and the screen grid is biased for $V_{G2Q} = 150$ V. Let $V_{PP} = 300$ V and $R_L = 15$ kΩ. The pentode is characterized by Fig. 17-20. Draw the dc load line.

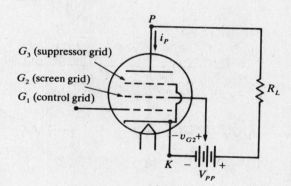

Fig. 17-19

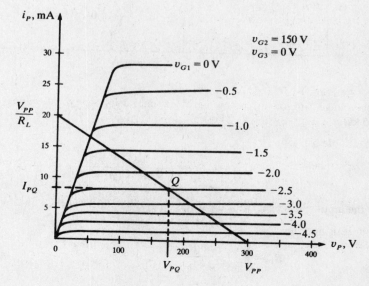

Fig. 17-20

VACUUM TUBES ◻ 523

▌ For the given values the dc load line is described by (5) of Problem 7.12 and has i_P intercept

$$\frac{V_{PP}}{R_L} = \frac{300}{15 \times 10^3} = 20 \text{ mA}$$

and v_P intercept $V_{PP} = 300$ V. These intercepts have been used to draw the dc load line of Fig. 17-20.

17.59 Determine the quiescent control grid current for the pentode amplifier of Problem 17.58.

▌ If $v_{G1} < 0$, then the control grid draws negligible current and $I_{G1Q} \approx 0$.

17.60 Find the quiescent value of plate current for the pentode amplifier of Problem 17.58.

▌ The quiescent plate current is read, from the projection of Q onto the i_P axis of Fig. 17-20, as $I_{PQ} = 8.13$ mA.

17.61 Determine the quiescent value of plate voltage for the pentode amplifier of Problem 17.58.

▌ The quiescent plate voltage is read from the projection of Q onto the v_P axis of Fig. 17-20 and is $V_{PQ} = 175$ V.

17.62 Since the plate characteristics of a pentode can be described by (2) of Problem 17.12, the small-signal equivalent circuits of Fig. 17-11, with parameters determined by (2) and (3) of Problem 17.33, are applicable. For the pentode amplifier of Problem 17.58, find the parameters of the small-signal equivalent circuit.

▌ Evaluating (2) and (3) of Problem 17.33 at the Q point of Fig. 17-20 gives

$$r_p \approx \left.\frac{\Delta v_P}{\Delta i_P}\right|_{v_{G1}=-2.5} = \frac{250 - 100}{(8.44 - 8.125) \times 10^{-3}} = 476.2 \text{ k}\Omega$$

and

$$g_m \approx \left.\frac{\Delta i_P}{\Delta v_{G1}}\right|_{v_P=175} = \frac{(10.94 - 5.94) \times 10^{-3}}{-2 - (-3)} = 5 \text{ mS}$$

17.63 Find the voltage gain for the pentode amplifier of Problem 17.58 by small-signal analysis.

▌ Using the small-signal equivalent circuit of Fig. 17-11(a) and the parameters determined in Problem 17.61, we have

$$A_v = \frac{v_p}{v_g} = -g_m(r_p \| R_L) = \frac{-g_m r_p R_L}{r_p + R_L} = \frac{-(5 \times 10^{-3})(476.2 \times 10^3)(15 \times 10^3)}{(476.2 + 15) \times 10^3} = -72.71$$

Review Problems

17.64 Suppose the amplifier of Problem 17.29 has plate resistance $r_p = 20$ kΩ and $v_S = 1 \cos \omega t$ V. Find its amplification factor μ using the small-signal voltage-source model of Fig. 17-11(b).

▌ Applying KVL around the plate circuit with $v_g = v_S$,

$$\mu v_g = \mu v_S = i_p(r_p + R_L)$$

Solving for μ and using $V_{Sm} = 1$ V and $I_{pm} = 1$ mA,

$$\mu = \frac{i_p(r_p + R_L)}{v_S} = \frac{I_{pm}(r_p + R_L)}{V_{Sm}} = \frac{(1 \times 10^{-3})(20 \times 10^3 + 10 \times 10^3)}{1} = 30$$

17.65 Suppose the bypass capacitor C_K is removed from the amplifier of Fig. 17-12. Find an expression for the voltage gain.

▌ With C_K removed, the equivalent circuit is shown by Fig. 17-21. KVL applied around the input loop gives

$$v_g = v_S - R_K i_p \quad (1)$$

KVL around the plate circuit yields

$$\mu v_g = (R_L + r_p + R_K) i_p \quad (2)$$

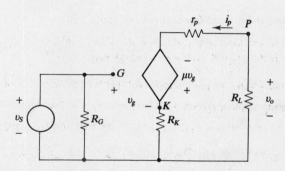

Fig. 17-21

Substitute (1) into (2) and rearrange to find

$$i_p = \frac{\mu v_S}{R_L + r_p + (\mu + 1)R_K} \tag{3}$$

But,
$$v_o = -i_p R_L \tag{4}$$

Substitute (3) into (4) and solve for the voltage gain.

$$A_v = -\frac{\mu R_L}{R_L + r_p + (\mu + 1)R_K} \tag{5}$$

17.66 Determine the percentage voltage gain deviation for the amplifier of Problem 17.45 if the bypass capacitor C_K were removed.

▌ Evaluating (5) of Problem 17.65 to find the gain with C_K removed,

$$A_v = -\frac{21.87(11.6 \times 10^3)}{11.6 \times 10^3 + 4.86 \times 10^3 + (21.87 + 1) + 400} = -9.91$$

The gain is seen to decrease upon removal of C_K:

$$\text{percent decrease} = \frac{-15.41 - (-9.91)}{-15.41}(100\%) = 35.7\%$$

17.67 The plate characteristics for the transformer coupled triode power amplifier of Fig. 17-22 are shown by Fig. 17-23. Maximum plate power dissipation rating for the triode is 20 W. $R_L = 8\,\Omega$ and the transformer is ideal. Determine R_K to place the Q point at $I_{PQ} = 75$ mA and $V_{GQ} = -40$ V.

▌ Since the transistor is ideal ($R = 0$), then R_K is dictated by the Q-point condition, or

$$R_K = -\frac{V_{GQ}}{I_{PQ}} = -\frac{-40}{75 \times 10^{-3}} = 533\,\Omega$$

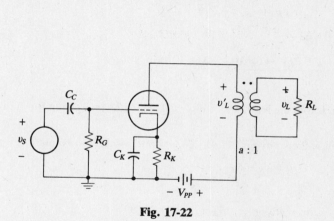

Fig. 17-22

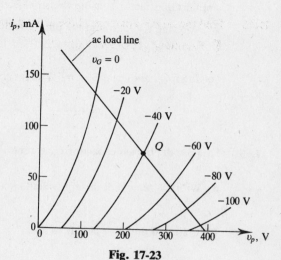

Fig. 17-23

17.68 Find the value of V_{PQ} for the amplifier of Problem 17.67.

▌ Enter the plate characteristic of Fig. 17-23 with $I_{PQ} = 75$ mA and $V_{GSQ} = -40$ V to establish the quiescent point marked Q. Projecting Q onto the v_P axis gives $V_{PQ} = 240$ V.

17.69 Determine the necessary value of plate supply voltage for the amplifier of Problem 17.67.

▌ Apply KVL around the plate circuit and use the results of Problems 17.67 and 17.68 to find
$$V_{PP} = V_{PQ} + I_{PQ}R_K = 240 + (750 \times 10^{-3})(533) = 280 \text{ V}$$

17.70 Is the tube of Problem 17.67 operating with rated plate power at quiescent conditions?

▌
$$P_P = V_{PQ}I_{PQ} = 240(75 \times 10^{-3}) = 18 \text{ W}$$

Since 18 W < 20 W, the tube is not exceeding rated plate power.

17.71 If the ideal transformer turns ratio $a = 16$, construct the ac load line for the amplifier of Problem 17.67 on the plate characteristic of Fig. 17-23.

▌ The ac resistance of the plate circuit is given by
$$R_{ac} = R'_L = a^2 R_L = (16)^2(8) = 2048 \text{ }\Omega$$

By extension of (2) from Problem 4.72,
$$v_{P\,max} = V_{PQ} + I_{PQ}R_{ac} = 240 + (75 \times 10^{-3})(2048) \approx 394 \text{ V}$$

The ac load line is drawn on Fig. 17-23 passing through the Q point with v_P-axis intercept of $v_{P\,max} = 394$ V.

17.72 If a signal $v_S = v_g = 40 \sin \omega t$ V is applied, find the power delivered to R_L for the amplifier of Problem 17.67.

▌ Moving ± 40 V from the Q point along the ac load line of Fig. 17-23, it is seen that v_P swings from approximately 130 V to approximately 350 V. Hence, the swing is nearly symmetric, and
$$v_p \approx 110 \sin \omega t = V'_{Lm} \sin \omega t \text{ V}$$

Whence,
$$P_L = \frac{(V'_L)^2}{R'_L} = \frac{(V'_{Lm}/\sqrt{2})^2}{R'_L} = \frac{(110/\sqrt{2})^2}{2048} = 2.95 \text{ W}$$

17.73 Find the plate supply power delivered to the amplifier of Problem 17.67.

▌
$$P_{PP} = V_{PP}I_{PQ} = 280(75 \times 10^{-3}) = 21 \text{ W}$$

17.74 Determine the efficiency of the amplifier of Problem 17.67.

▌
$$\eta = \frac{P_L}{P_{PP}}(100\%) = \frac{2.95}{21}(100\%) \approx 14\%$$

17.75 Find the i_P-axis intercept of the ac load line on Fig. 17-23.

▌ Extending (1) of Problem 4.72,
$$i_{P\,max} = \frac{V_{PQ}}{R'_L} + I_{PQ} = \frac{240}{2048} + 75 \times 10^{-3} = 192 \text{ mA}$$

Index

AC load lines, 4.71, 4.76, 4.97, 4.103, 17.43
Active mode of transistor operation, 4.86, 4.136
Active region, 14.24–14.26, 14.28, 14.30
Adjustable-output voltage regulator, 13.57–13.58
Admittance (or y) parameters (*see y parameters*)
Alpha (α) in BJT, 4.7, 4.10
Amplification factor, 8.3, 17.34
Amplifiers:
 buffer, 13.24
 cascaded, 9.15–9.23, 9.27–9.29, 9.59–9.71, 9.90–9.92, 12.26–12.28
 classification of, 4.108–4.112
 common-base (CB), 7.63–7.92, 7.126–7.129, 7.145–7.148, 7.159–7.160
 common-collector (CC), 4.73, 7.5, 7.93–7.123
 common-drain (CD), 8.31–8.58, 8.69, 8.81–8.88
 common-emitter (CE), 7.21–7.62, 7.132–7.144, 7.149–7.158
 common-gate (CG), 8.59–8.68
 common-grid, 17.54–17.57
 common-source (CS), 8.9–8.30, 8.89–8.96
 complementary-symmetry, 11.65–11.76
 current-feedback-biased, 6.67
 current-series, 12.56–12.83
 current-shunt, 12.100–12.111
 Darlington-pair, 9.72–9.81, 9.88–9.89
 difference, 9.25
 differential, 5.58, 5.107–5.110, 9.26, 13.21
 direct-coupled, 11.1–11.20
 drain-feedback biased, 5.31, 8.22
 efficiency of, 11.2, 11.9–11.11, 11.19–11.20, 11.39
 emitter-follower (EF), 4.73, 7.5, 7.93–7.123
 exponential, 13.65–13.64, 13.119
 frequency effects (*see* Frequency effects in amplifiers)
 inductor-coupled, 11.21–11.30, 11.100–11.102
 inverse log, 13.63
 inverting, 13.3, 13.5–13.9
 inverting summer (adder), 13.18–13.19
 logarithmic, 13.61–13.62, 13.64, 13.117–13.119

Amplifiers (*Cont.*):
 n-channel MOSFET (NMOS), 5.86–5.90, 5.98
 noninverting, 13.10–13.12, 13.15–13.17, 13.125
 pentode, 17.58–17.63
 push-pull, 11.54–11.62
 transformer-coupled, 11.31–11.51, 11.109–11.111
 triode, 17.12–17.26
 two-stage, 9.1–9.9
 unity-follower, 13.23–13.24
 voltage-series, 12.10–12.55
 voltage-shunt, 12.84–12.99, 13.42
 (*See also* Feedback amplifiers; Multiple-transistor amplifiers; Small-signal frequency-independent BJT amplifiers; Small-signal frequency-independent FET amplifiers; Operational amplifiers; Power amplifiers)
Amplitude ratio, 10.4
Analog multiplier, 13.81–13.82, 13.127
AND gate, 14.5–14.6
Anode, 17.1
Associative law, Boolean algebra, 15.37
Avalanche breakdown voltage, 11.93
Average power, 1.105–1.106, 1.154, 2.145
Average value, 1.98–1.99, 1.101, 1.103, 1.105–1.106, 3.2, 3.6
Avogardo's law, 2.3

β cutoff frequency, 10.63–10.64
Band-pass filter, 13.101–13.102
Band-reject filter, 13.103–13.104
Base, in BJT, 4.1
Base-emitter-junction resistance, 10.60
Base ohmic resistance, 10.60
Base triggering, 14.57
Beta (β) in BJT, 4.8–4.9, 4.11, 4.74, 4.122, 6.1–6.12, 6.26
Bias:
 β-independent, 4.40, 4.142, 4.150, 6.4, 6.12, 11.47
 β-uncertainty in, 6.1–6.12, 6.26, 6.91, 6.102, 6.105, 6.111–6.112
 in bipolar junction transistors, 4.38–4.40
 cathode, 17.40
 considerations in transistors, 6.1–6.130
 constant-base-current, 4.61, 6.1–6.3, 6.18, 6.24

Bias (*Cont.*):
 constant-emitter-current, 6.2
 in field-effect transistors, 5.23–5.24, 5.75
 fixed, 6.78
 Q-point-bounded, 6.77–6.89
 self-, 5.33, 5.39, 5.41
 shunt-feedback, 6.5–6.7, 6.19, 6.57, 6.111–6.112
 source, 5.75
 stability factor analysis in, 6.36–6.76, 6.93–6.95, 6.99–6.100, 6.103–6.104, 6.107–6.110, 6.112
 temperature effects in, 6.15–6.23, 6.96–6.97, 6.127
 voltage-divider, 5.23, 6.119–6.120
BJT amplifiers (*see* Small-signal frequency-independent BJT amplifiers)
Binary counter, 16.35
Binary functions, 14.1–14.15
Binary (base-2) numbers, 15.4–15.5, 15.7, 15.12–15.16
Bipolar junction transistors (BJT), 4.1–4.151
 ac load lines in, 4.71, 4.76, 4.97, 4.103
 alpha (α) in, 4.7
 beta (β) in, 4.8–4.9, 4.11, 4.74, 4.122
 bias and, 4.38–4.40
 collector-base leakage in, 4.7, 4.127
 collector-emitter leakage in, 4.9
 common-base (CB), 4.29
 common-emitter (CE), 4.32
 constant-base-current bias in, 4.61, 6.1–6.3, 6.24, 6.63, 6.102
 constant-emitter-current bias in, 6.4, 6.105
 dc load lines in, 4.41–4.42, 4.56, 4.73, 4.96, 4.104
 Ebers–Moll model in, 4.19–4.23, 4.114
 high-frequency models of, 10.60–10.73
 in multiple-transistor amplifiers, 9.1–9.34
 nonlinear-element stabilization for, 6.27–6.35
 operating modes of, 4.86–4.112, 4.140–4.151
 operating principles of, 4.1–4.28, 4.113–4.129
Bistable multivibrator, 14.57–14.65
Bit, 16.1
Block diagrams:
 feedback amplifiers, 12.1

527

Block diagrams (*Cont.*):
  functional, 12.1
  logic, 15.60
  mathematical, 12.1
Bode plots, 10.5–10.21
  asymptotic, 10.10, 10.12, 10.15,
    10.21, 10.27, 10.31, 10.35,
    13.94, 13.96, 13.102
Boolean algebra, 15.1–15.150
  gate implementation of, 15.60–
    15.90
  simplification of functions, 15.91–
    15.118, 15.132, 15.135
  theorems and properties of, 15.37–
    15.59
Boolean postulates, 15.37
Break frequency, 10.8
Bridge rectifier circuit, 3.56
Buffer amplifier, 13.23
Buffer circuit, 14.1
Bypass capacitors, 10.22–10.25,
  10.32, 10.57

CE hybrid parameters, 7.1
Capacitors:
  bypass, 10.22–10.25, 10.32, 10.57
  coupling, 10.28, 10.42–10.46,
    10.54
  filter, 3.46, 3.48, 3.54
Cascaded amplifiers, 9.15–9.23,
  9.27–9.29, 9.59–9.71, 9.90–9.92,
  10.38–10.40, 12.26–12.28
Cathode, 17.1
Cathode bias, 17.40
Cathode follower circuit, 17.49–
  17.53
Childs–Langmuir three-halves-power
  law, 17.1
Circuit(s):
  bridge rectifier, 3.56
  buffer, 14.1
  cathode follower, 17.49–17.53
  clamping, 2.58
  clipping, 2.63–2.64
  combinational logic, 16.1–16.23
  comparative, 16.43
  decoder, 16.21, 16.40
  demultiplexer (DEMUX), 16.23
  diode logic (DL), 14.37–14.40
  flip-flop (FF), 14.57, 14.60, 16.24–
    16.31, 16.44–16.45
  full-adder (FA), 16.4–16.8
  full-subtractor (FS), 16.12, 16.15–
    16.17
  function-generator, 2.73
  half-adder (HA), 16.1–16.3
  half-subtractor (HS), 16.10–16.11
  inverter, 14.1
  limiter, 13.68
  multiplexer (MUX), 16.22, 16.41
  negative clipping, 2.63–2.64
  phase-inverter, 11.60
  phase-splitter, 8.111
  power supply, 3.43–3.56, 3.75–3.84
  precision zero limiter, 13.76–13.77
  regulator, 2.89

Circuit(s) (*Cont.*):
  reset/set (RS) flip-flop, 16.24–
    16.26, 16.31, 16.44
  ripple-carry adder, 16.9
  Schmitt trigger, 13.78–13.80
  sequential logic, 16.24–16.39
  tee-equivalent, 7.11–7.12, 7.103,
    7.127
  zero limiter, 13.75–13.77
Circuit analysis (port-based),
  1.1–1.154
Clamping circuit, 2.58
Clipping circuit, 2.63–2.64
Clocked RS flip-flop, 16.26
Collector, in BJT, 4.1
Collector-base leakage, 4.7, 4.127
Collector characteristics:
  of common-base (CB) connection,
    4.29
  of common-emitter (CE)
    connection, 4.32
Collector-emitter leakage, 4.9
Collector power dissipation, 11.1,
  11.7, 11.15, 11.56, 11.63
Combinational logic circuits,
  16.1–16.23
  decoders, 16.21, 16.40
  demultiplexer (DEMUX), 16.23
  full-adder (FA), 16.4–16.8
  full-subtractor (FS), 16.12,
    16.15–16.17
  half-adder (HA), 16.1–16.3
  half-subtractor (HS), 16.10–16.11
  multiplexer (MUX), 16.22, 16.41
  ripple-carry adder, 16.9
Common-base (CB) amplifiers,
  7.63–7.92, 7.126–7.129, 7.145–
  7.148, 7.159–7.160
Common-base (CB) connection, 4.29
Common-base forward short circuit
  current gain, 4.22
Common-base reverse short circuit
  current gain, 4.21
Common-collector (CC) amplifier,
  4.73, 7.5, 7.93–7.123
Common-drain (CD) amplifier,
  8.31–8.58, 8.69, 8.81–8.88
Common-emitter (CE) amplifier,
  7.21–7.62, 7.133–7.144, 7.149–
  7.158
Common-emitter (CE) connection,
  4.32
Common-gate (CG) amplifier,
  8.59–8.68
Common-grid amplifier, 17.54, 17.57
Common-mode gain, 13.14
Common-mode rejection, 13.21
Common-mode rejection ratio
  (CMMR), 13.14–13.16
Common-source (CS) amplifier,
  8.9–8.30, 8.89–8.96
Common-source (CS) connection, in
  JFET, 5.5
Commutative law, Boolean algebra,
  15.37
Comparative circuit, 16.43

Complement, 14.6, 15.17–15.36
Complementary metal-oxide
  semiconductor (CMOS)
  inverter, 14.54
Complementary-symmetry amplifier,
  11.65–11.76
Conductivity, 2.4, 2.8
Constant-base-current bias, 4.61,
  6.1–6.3, 6.18, 6.24, 6.63, 6.102
Constant-emitter-current bias, 6.4,
  6.105
Controlled voltage sources, 1.69
Converters:
  Current-to-voltage, 13.37
  Negative-impedance (NIC), 13.24
  Voltage-to-current, 13.38
Corner frequency, 10.8
Counters, 16.24–16.39
  binary, 16.35
  down, 16.38
  modulo-M, 16.36
  ripple, 16.35
  synchronous, 16.37
  up, 16.38
  up-down, 16.39
Coupling capacitor, 10.28,
  10.42–10.46, 10.54
Crossover distortion, 11.106
Current-series feedback amplifiers,
  12.56–12.83
Current-shunt feedback amplifiers,
  12.100–12.111
Cutoff, 4.86
Cutoff region, 14.24–14.26, 14.28,
  14.30

DC load line:
  in bipolar junction transistors,
    4.41–4.42, 4.56, 4.73, 4.96,
    4.104
  in diodes, 2.46
  in field-effect transistors, 5.25
  in vacuum tiodes, 17.13, 17.18
Darlington-pair emitter-follower,
  9.30–9.34, 9.72–9.81, 9.88–9.89
Darlington transistor pair, 4.79–4.81
Decimal (base-10) numbers, 15.1
Decoders, 16.21, 16.40
Delay (D) flip-flops, 16.27, 16.32
DeMorgan's laws, Boolean algebra,
  15.37
Demultiplexer (DEMUX) circuit,
  16.23
Depletion-enhancement-mode
  MOSFET, 5.22, 5.125
Difference amplifier, 9.25
Differential amplifier, 5.58,
  5.107–5.110, 9.26, 13.21
Digital logic (*see* Switching circuits
  for digital logic)
Diode(s):
  characteristic, 2.44
  ideal, 2.38–2.39
  light-emitting (LED), 2.99–2.103
  power level applications of,
    3.1–3.84

INDEX

Diode(s) (*Cont.*):
  rectifier, 2.38–2.82
  signal level application of, 2.1–2.152
  vacuum, 17.1–17.11
  varactor, 2.104–2.105
  Zener, 2.83–2.98
    (*See also* Semiconductor diodes; Vacuum diodes)
Diode logic (DL) circuit, 14.37–14.40
Diode transistor logic (DTL), 14.47–14.51
Dissipation derating curve, 11.94
Distortion:
  considerations in power amplifiers, 11.77–11.92
  crossover, 11.106
  turn-off, 11.87
  turn-on, 11.87, 11.103
Distributive law, Boolean algebra, 15.37
Down counter, 16.38
Drain characteristics:
  of JFET, 5.5
  of MOSFET, 5.22
Drain-feedback bias line, 5.31
Drain-feedback biased amplifier, 5.31, 8.22
Drift velocity, 2.5
Driving-point impedance, 1.44, 1.149
Dynamic load line, 2.45
Dynamic resistance of diode, 2.50

Ebers–Moll transistor model, 4.19–4.23, 4.114
Efficiency of amplifier, 11.2, 11.9–11.11, 11.19–11.20, 11.39
Emitter, in BJT, 4.1
Emitter-follower (EF) amplifier, 4.73, 7.5, 7.93–7.123
Enhancement-mode operation (MOSFET), 5.15, 5.36, 5.46, 5.49, 5.71, 5.126, 14.34
Exclusive-NOR (XNOR) gate, 14.11
Exclusive-OR (XOR) gate, 14.10
Exponential amplifier, 13.63–13.64, 13.119

FET amplifiers (*see* Small-signal frequency-independent FET amplifiers)
Feedback amplifiers, 12.1–12.133
  concepts in, 12.1–12.9
  current-series, 12.56–12.83
  current-shunt, 12.100–12.111
  voltage-series, 12.10–12.55
  voltage-shunt, 12.84–12.99, 13.42
Feedback resistor, 13.3
Feedback transfer ratio, 12.1
Field-effect transistors (FET), 5.1–5.127
  bias in, 5.23–5.24, 5.75
  dc load line in, 5.25
  junction (*see* Junction field-effect transistors)

Field-effect transistors (FET) (*Cont.*):
  high-frequency models, 10.74–10.84
  metal-oxide semiconductor (*see* Metal-oxide-semiconductor field-effect transistor)
  in multiple transistor amplifiers, 9.35–9.65
  operating principles of, 5.1–5.22
  Q-point-bounded bias for, 6.77–6.89
Filament, 17.1
Filter(s):
  band-pass, 13.101–13.102
  band-reject, 13.103–13.104
  in diode applications, 3.46
  high-pass, 10.24, 10.87–10.89, 10.92, 13.93, 13.99–13.101
  low-pass, 13.84–13.88, 13.97–13.98, 13.105–13.107, 13.121
  in operational amplifiers, 13.84–13.104
Filter capacitor, 3.46, 3.48, 3.54
Fixed bias, 6.78
Flip-flop (FF) circuit, 14.57, 14.60, 16.24–16.31, 16.44–16.45
  delay (D), 16.27, 16.32
  JK, 16.28
  RS master-slave (M/S), 16.31
  reset–set (RS), 16.24–16.26, 16.44
  toggle (T), 16.29, 16.45
Forced response, 3.7
Forward current gain:
  in common-base (CB) connection, 7.3
  in common-collector (CC) connection, 7.7
  in common-emitter (CE) connection, 7.1
Forward transfer ratio, 12.1
Frequency, break or corner, 10.8
Frequency effects in amplifiers, 10.1–10.110
  Bode plots, 10.5–10.21
  of bypass and coupling capacitors, low-frequency, 10.22–10.59
  frequency response, 10.1–10.4
  high-frequency BJT models, 10.60–10.73
  high-frequency FET models, 10.74–10.84
Frequency modulation, 2.104
Frequency transfer function, 10.4
Full-adder (FA) circuit, 16.4–16.8
Full-subtractor (FS) circuit, 16.12, 16.15–16.17
Full-wave rectifier, 3.50
Function-generator circuit, 2.73
Functional block diagram, 12.1

g parameters, 1.93–1.94, 1.97, 1.151
Gain-bandwidth product (GBW), 12.8, 12.112
Gain ratio, 10.4

Gate implementation, (*see* Logic gates)
Gyrator, 13.114–13.115

h (or hybrid) parameters, 1.76–1.81, 1.144, 1.146–1.147, 7.15, 7.20, 7.39, 7.44, 7.52, 7.63–7.65, 12.49–12.50
  CB, 7.9
  CC, 7.14, 7.126
  CE, 7.1, 7.14
Half-adder (HA) circuit, 16.1–16.3
Half-cycle average value, 1.99
Half-subtractor (HS) circuit, 16.10–16.11
Half-wave rectifier, 3.1
High-frequency cutoff point, 10.60, 10.77, 10.107, 10.110
High-frequency range, 10.1
High-frequency region, 10.60
High-frequency voltage-gain ratio, 10.62, 10.66, 10.76, 10.81, 10.84
High logic, 14.37
High-logic threshold, 14.37
High-pass filter, 10.24, 10.87–10.89, 10.92, 13.93, 13.99–13.101
Hybrid (or h) parameters (*see* h parameters)
Hybrid-π BJT model, high-frequency, 10.60–10.61, 10.63, 10.65

Ideal diode, 2.38–2.39
Identity law, Boolean algebra, 15.37
Impedance:
  driving-point, 1.44
  input, 7.24, 7.59–7.60, 7.66, 7.73–7.75, 7.96, 7.125, 8.25–8.26, 9.5
  output, 7.25, 7.37, 7.62, 7.97, 8.27–8.28, 9.9, 9.85–9.86, 10.49
  reflection rule, FET amplifier, 8.67
Inductor-coupled amplifiers, 11.21–11.30, 11.100–11.102
Input borrow, 16.12
Input characteristics:
  of common-base (CB) connection, 4.29
  of common-emitter (CE) connection, 4.32
  of vacuum triodes, 17.12
Input impedance, 7.24, 7.59–7.60, 7.66, 7.73–7.75, 7.96, 7.125, 8.25–8.26, 9.5
Input resistance:
  in common-base (CB) connection, 7.3
  in common-collector (CC) connection, 7.7
  in common-emitter (CE) connection, 7.1
Instantaneous power, 1.5–1.6, 1.108, 1.118, 1.153
Interelectrode capacitance, 17.58
Intersection, 15.137

Inverse log amplifier, 13.63
Inverse mode of transistor operation, 4.86
Inverter circuit, 14.1
Inverting amplifier, 13.3, 13.5–13.8
Inverting differentiator, 13.29
Inverting input, 13.1
Inverting integrator, 13.30
Inverting summer (adder) amplifier, 13.18–13.19

JK flip-flop, 16.28, 16.47
Junction field-effect transistor (JFET), 5.1–5.14, 5.33–5.36, 5.39–5.45, 5.53–5.68, 5.72–5.85, 5.91, 5.94, 5.99–5.102, 5.105–5.123
 in common-source (CS) connection, 5.5
 drain characteristics of, 5.5
 n-channel, 5.1–5.2
 output characteristics of, 5.5
 p-channel, 5.1
 self-bias in, 5.33, 5.39, 5.41
 shorted-gate parameters of, 5.5, 5.40, 5.42
 transfer characteristics of, 5.5–5.6, 5.40

Karnaugh map, 15.101–15.118

Lag–lead network, 13.90
Laplace-domain voltage-gain ratio, 10.1
Laplace transforms, 3.22, 3.26, 3.58, 3.70–3.71
Latch, 16.33
Lead network, 10.7
Lead–lag network, 13.90
Leakage current, temperature dependence, 6.16, 6.18–6.20
Least significant bit (LSB), 16.35
Level clamp, 13.67, 13.108
Light-emitting diode (LED), 2.99–2.103
Limiter circuit, 13.68
Linear function operations of op amps, 13.18–13.60
Linear mode of transistor operation, 4.86
Load line:
 ac, 4.71, 4.76, 4.97, 4.103, 17.43
 dc, in BJT, 4.41–4.42, 4.56, 4.73, 4.96, 4.104
 dc, in diodes, 2.46
 dc, in FET, 5.25
 dc, in vacuum triodes, 17.13, 17.18
 dynamic, 2.45
Load power of amplifier, 11.1, 11.5, 11.14, 11.23, 11.38, 11.66
Logarithmic amplifier, 13.61–13.62, 13.64, 13.117–13.119
Logic block diagrams, 15.60
Logic complement, 14.44
Logic families, 14.37–14.56

Logic gates, 14.4–14.15, 15.60–15.90
 AND, 14.5–14.6, 14.12, 14.71
 DTL (diode transistor logic), 14.47–14.51
 NAND, 14.8, 14.70, 15.69–15.71, 15.74, 15.77–15.78, 15.148
 NOR, 14.9, 15.72–15.73, 15.142
 OR, 14.7
 RTL (resistor transistor logic), 14.41–14.56
 TTL (transistor transistor logic), 14.52–14.53
 XNOR (exclusive-NOR), 14.11
 XOR (exclusive-OR), 14.10
Loop-sampling, loop-comparison feedback amplifier, 12.56
Low-frequency cutoff point, 10.24, 10.31, 10.35, 10.41, 10.91–10.92
Low-frequency range, 10.1
Low-frequency region, 10.24
Low-frequency voltage-gain ratio, 10.24, 10.37
Low logic, 14.37
Low-logic limit, 14.37
Low-pass filters, 13.84–13.88, 13.97–13.98, 13.105–13.107, 13.121
Luminous intensity, 2.99–2.101, 2.103

Majority carrier (hole), 2.16–2.17
Mass-action law, 2.14
Matched-complement pair transistors, 11.71
Mathematical block diagram, 12.1
Maximum-dissipation hyperbola, 11.18
Maximum symmetrical swing, 4.90, 4.92–4.93, 4.102, 4.151, 11.26–11.29, 11.36
Metal-oxide-semiconductor field-effect transistor (MOSFET), 5.15–5.22, 5.31–5.32, 5.36–5.38, 5.46–5.52, 5.69–5.71, 5.93, 5.103–5.104, 5.124, 14.34–14.36
 depletion-mode operation of, 5.22, 5.125
 drain characteristics of, 5.15
 enhancement-mode operation of, 5.15, 5.36, 5.46, 5.49, 5.71, 5.126, 14.34
 saturation current of, 5.20
 transfer characteristics of, 5.15, 5.18
Midfrequency range, 7.1, 10.1
Midfrequency voltage-gain ratio, 10.25, 10.30, 10.70
Miller admittance, 10.82–10.83
Miller capacitance, 10.79–10.80
Minority carrier (electron) 2.16–2.17
Mobility of electrons, 2.4
Modulo-M counter, 16.36
Most significant bit (MSB), 16.35
Multiple-transistor amplifiers:
 BJTs in, 9.1–9.34

Multiple-transistor amplifiers (Cont.):
 cascaded construction of, 9.15–9.23, 9.27–9.29, 9.59–9.71, 9.90–9.92, 12.26–12.28
 FETs in, 9.35–9.65
 two-stage construction of, 9.1–9.9
Multiplexer (MUX) circuit, 16.22, 16.41
Multivibrators, 14.57–14.65
 bistable (BMV), 14.57

N-channel MOS, 5.86–5.90, 5.98, 14.55–14.56
NAND gate, 14.8, 15.69–15.71, 15.74, 15.77–15.78, 15.148
Natural response, 3.7, 3.41
Negation law, Boolean algebra, 15.37
Negative clipping circuit, 2.63–2.64
Negative feedback, 12.2
Negative-impedance converter (NIC), 13.34–13.35
Negative logic, 14.6
Network(s):
 first-order, 10.4–10.7
 lag–lead, 13.90
 lead, 10.7
 lead–lag, 13.90
 one-port, 1.1–1.25, 1.108–1.118, 1.153–1.154
 theorems, 1.26–1.66
 two-port, 1.67–1.97, 1.144–1.152
Network impedances, generalized, 1.16–1.18, 1.113
Network phase angle, 10.2
Node-sampling, loop comparison feedback amplifier, 12.10
Node-sampling, node comparison feedback amplifier, 12.84
Noninverting amplifier, 13.10–13.12, 13.15–13.17, 13.125
Noninverting input, 13.1
Nonlinear diode compensation, 6.33
Nonlinear-element stabilization, in BJT, 6.27–6.35
Nonlinear function operations of op amps, 13.61–13.83
NOR gate, 14.9, 15.72–15.73, 15.142
Norton's theorem, 1.46, 1.48, 1.50–1.51, 1.54–1.55, 1.57, 1.59, 1.125, 1.133
Numbering systems, 15.1–15.36
 binary (base-2), 15.4–15.5, 15.7, 15.12–15.16
 complements in, 15.17–15.36
 decimal (base-10), 15.1
 octal (base-8), 15.121–15.122
 power series representations of, 15.1–15.6
 quinary (base-5), 15.3, 15.6, 15.8
 radix of, 15.1

Octal (base-8) numbers, 15.121–15.122

Ohmic region, 5.5
Open-circuit impedance (or z) parameters (*see* z parameters)
Operational amplifiers, 13.1–13.127
　filter applications in, 13.84–13.104
　fundamentals of, 13.1–13.17
　ideal, 13.3–13.4, 13.9
　linear function operations of, 13.18–13.60
　nonlinear function operations of, 13.61–13.83
　practical, 13.3, 13.7–13.9
OR gate, 14.7
Output admittance:
　in common-base (CB) connection, 7.3
　in common-collector (CC) connection, 7.7
　in common-emitter (CE) connection, 7.1
Output borrow, 16.12
Output characteristics:
　of common-base (CB) connection, 4.29
　of common-emitter (CE) connection, 4.32
　of JFET, 5.5
　of vacuum triodes, 17.12
Output impedance, 7.25, 7.37, 7.62, 7.97, 8.27–8.28, 9.9, 9.85–9.86, 10.49

Parallel-output, parallel-input feedback amplifier, 12.100
Parallel-output, series-input feedback amplifier, 12.10
Parameters:
　admittance (or y), 1.84–1.92, 1.95–1.96, 12.91
　g, 1.93–1.94, 1.97
　hybrid (or h), 1.76–1.81, 7.9, 7.15, 7.20, 7.39, 7.44, 7.52, 7.63–7.65, 12.49–12.50
　open-circuit impedance (or z), 1.67–1.75, 1.80, 1.90–1.92, 7.9, 7.150, 12.62
　r, 7.10, 7.69
Passive sign convention, 1.106, 1.122
Pentode amplifier, 17.58–17.63
Perfect induction, 15.53–15.59, 15.123–15.125
Perveance, 17.1
Phase-inverter circuit, 11.60
Phase-splitter circuit, 8.111
Pinchoff, 5.5
Pinchoff voltage, 5.5, 5.13–5.14
Plate characteristic:
　of vacuum diodes, 17.1
　of vacuum triodes, 17.12
Plate efficiency, vacuum-tube amplifier, 17.38–17.39
Plate resistance, 17.33
Port, 1.1
Positional notation system, 15.1
Positive logic, 14.5–14.6

Power:
　average, 1.105–1.106, 1.154
　instantaneous, 1.5–1.6, 1.108, 1.110, 1.118, 1.153
　single-ended, supply, 11.71
Power amplifiers, 11.1–11.111
　complementary-symmetry, 11.65–11.76
　direct-coupled, 11.1–11.20
　distortion considerations in, 11.77–11.92
　inductor-coupled, 11.21–11.30, 11.100–102
　push–pull, 11.52–11.64
　thermal considerations in, 11.93–11.97, 11.107–11.108
　transformer-coupled, 11.31–11.51, 11.109–11.111
Power-dissipating capability, 11.18
Power supply circuits, 3.43–3.56, 3.75–3.84
Power system representations, 15.1–15.6
Practical integrator, 13.32
Precision zero limiter circuit, 13.76–13.77
Push–pull amplifiers, 11.52–11.64

Q-point (*see* Quiescent point)
Q-point-bounded bias (FET), 6.77–6.89
Quiescent point (Q-point):
　of bipolar junction transistors, 4.38, 4.101
　in field-effect transistors, 5.26, 5.94, 5.101
　of semiconductor diodes, 2.46
　stability, 6.36
Quinary (base-5) numbers, 15.3, 15.6, 15.8

r parameter, 7.10
r-parameter model, 7.11, 7.13, 7.27, 7.31, 7.69, 7.119
r's complement, 15.17–15.23
r's complement subtraction, 15.24–15.27
(r-1)'s complement, 15.28–15.34
(r-1)'s complement subtraction, 15.35–15.36, 15.119–15.120
Radix of numbering system, 15.1
Rectification efficiency, 17.7–17.9, 17.11
Rectifier diodes, 2.38–2.82, 2.136–2.143
Redundancy law, Boolean algebra, 15.37
Register, 16.32
Regulator circuit, 2.89
Reset–set (RS) flip-flop circuits, 16.24–16.26, 16.44
　master-slave (M/S), 16.31
Residue theory, 3.71
Resistance:
　base-emitter-junction, 10.60

Resistance (*Cont.*):
　base ohmic, 10.60
　thermal, in power amplifiers, 11.94
Resistor–transistor logic (RTL), 14.41–14.46
Reverse saturation current, 2.30, 2.35, 2.37, 2.121, 2.126
Reverse voltage ratio:
　in common-base (CB) connection, 7.3
　in common-collector (CC) connection, 7.7
　in common-emitter (CE) connection, 7.1
Ripple-carry adder circuit, 16.9
Ripple counter, 16.35
Ripple factor, 3.45–3.46, 3.52–3.53
Rott-mean-square (rms) values, 1.100, 1.102, 1.104–1.105, 1.107, 2.78, 2.144, 3.6

Saturation, 4.86, 4.137
Saturation current, 2.18–2.19, 5.10–5.11, 5.20
Saturation region, 14.24–14.26, 14.28, 14.30
Saturation voltage, 13.1
Schmitt trigger circuit, 13.78–13.80
Screen grid, 17.58
Second-harmonic distortion factor, 11.79, 11.81, 11.83, 11.85–11.86
Select code, 16.22
Self-bias, 5.33, 5.39, 5.41
Semiconductor diodes:
　in general circuits, power-level, 3.1–3.42, 3.57–3.74
　ideal, 2.38–2.39
　light-emitting (LED), 2.99–2.103
　material properties of, 2.1–2.17, 2.106–2.119
　in power supply circuits, 3.43–3.56, 3.75–3.84
　in rectifier circuit analysis, 2.38–2.82, 2.136–2.143
　terminal characteristics of, 2.18–2.37, 2.121–2.132
　varactor, 2.104–2.105
　Zener, 2.83–2.98
　(*See also* Diodes)
Sensitivity factor, 6.36–6.76, 6.94–6.95, 6.101, 6.113–6.118, 6.126, 6.128
Sequential logic circuits, 16.24–16.39
　counters, 16.35–16.39
　flip-flops, 16.21–16.34
Series-input, series-output feedback amplifier, 12.56
Shift register, 16.34, 16.46–16.47
Short-circuit admittance parameters, 12.91
Shunt-feedback bias, 6.5–6.7, 6.19, 6.57, 6.111–6.112
Single-ended power supply, 11.71
Small-signal equivalent circuit models, 7.1

Small-signal frequency-independent BJT amplifiers, 7.1–7.165
  common-base (CB) analysis, 7.63–7.92, 7.126–7.129, 7.145–7.148, 7.159–7.160
  common-collector (CC) analysis, 7.5, 7.93–7.123
  common-emitter (CE) analysis, 7.21–7.62, 7.132–7.144, 7.149–7.158
  equivalent circuit models of, 7.1–7.20
  tee-equivalent circuits in, 7.11–7.12, 7.103, 7.127
Small-signal frequency-independent FET amplifiers, 8.1–8.113
  common-drain (CD) analysis, 8.31–8.58, 8.69, 8.81–8.88
  common-gate (CG) analysis, 8.59–8.68
  common-source (CS) analysis, 8.9–8.30, 8.89–8.96
  equivalent circuit models of, 8.1–8.8, 8.71, 8.80, 8.104, 8.108
  impedance and voltage reflection rules for, 8.67
Source-drain resistance, 8.1
Stability factor analysis, in transistor bias, 6.36–6.76, 6.93, 6.99–6.101, 6.103–6.104, 6.107–6.110, 6.121
Steady-state response, 10.3
Subtractor amplifier, 13.21
Summing integrator, 13.48
Superposition theorem, 1.26–1.38, 1.119, 1.121, 1.125
Supplied power of amplifier, 11.1, 11.3–11.4, 11.13, 11.37, 11.44, 11.55
Suppressor grid, 17.58
Switching circuits for digital logic, 14.1–14.78
  binary functions in, 14.1–14.15
  logic families in, 14.37–14.56
  logic gates in, 14.4–14.15
  multivibrators in, 14.57–14.65
  transistor switching concepts in, 14.16–14.36
Synchronous counter, 16.37, 16.48

Tee-equivalent circuit, 7.11–7.12, 7.103, 7.127

Terminal characteristics, semiconductor diodes, 2.18–2.37, 2.121–2.132
Theorem(s):
  Boolean algebra, 15.37–15.59
  network, 1.26–1.66
  Norton's, 1.46
  superposition, 1.26–1.38
  Thevenin's, 1.44–1.45
Thermal resistance, in power amplifiers, 11.94
Thevenin's theorem, 1.44–1.45, 1.47, 1.49–1.50, 1.52–1.53, 1.56, 1.58, 1.60–1.61, 1.63–1.66, 1.82–1.83, 1.126–1.127, 1.130–1.132, 1.134–1.135, 1.137–1.139, 1.141, 1.143
Three-halves-power law, 17.1
Threshold voltage, 5.15
Time constant, 10.4
Timing diagram, 14.60
Toggle (T) flip-flop, 16.29, 16.45
Transcendental equation, 3.8, 3.31
Transconductance, 8.1, 17.33, 17.36
Transconductance ratio, 12.1
Transfer bias line, 5.24
Transfer characteristics:
  of common-base (CB) connection, 4.29
  of common-emitter (CE) connection, 4.32
  of diode, 2.56, 2.61, 2.66–2.68
  of JFET, 5.5–5.6
  of MOSFET, 5.15, 5.18, 5.22
  of voltage, 14.20–14.26, 14.32–14.34, 14.72–14.78
Transfer function, 10.1
Transformer-coupled amplifiers, 11.31–11.51, 11.109–11.111
Transistors:
  bias considerations in, 6.1–6.130
  switching concepts in circuits for digital logic, 14.16–14.36
  (See also Bipolar junction transistors; Field effect transistors)
Transistor transistor logic (TTL, or T^2L), 14.52–14.53
Transresistance ratio, 12.1
Triodes, vacuum, 17.12–17.57
Truth table, 14.1–14.3, 14.5–14.11, 14.37–14.43, 14.47–14.48, 14.70, 15.53–15.59, 15.123–15.125
Two-stage amplifier, 9.1–9.9

Tubes, vacuum (see Vacuum tubes)
Turn-off distortion, 11.87
Turn-on distortion, 11.87, 11.103

Unions, 15.137
Unity-follower amplifier, 13.23–13.24
Unity-gain bandwidth, 13.107
Up counter, 16.38
Up-down counter, 16.39

Vacuum pentodes (see Vacuum tubes)
Vacuum triodes (see Vacuum tubes)
Vacuum tubes, 17.1–17.75
  cathode follower circuits, 17.49–17.53
  diodes, 17.1–17.11
  pentodes, 17.58–17.63
  triode amplifiers, 17.12–17.26
  triodes, 17.12–17.57
Varactor diode, 2.104–2.105
Venn diagram, 15.44–15.52
Virtual ground, 13.5
Voltage:
  avalanche-breakdown, 11.86
  pinchoff, 5.5, 5.13–5.14
  saturation, 13.1
  threshold, 5.15
Voltage-divider bias, 5.23, 6.119–6.120
Voltage reflection rule, FET amplifiers, 8.67
Voltage regulation, in diodes, 3.44
Voltage-series feedback amplifiers, 12.10–12.55
Voltage-shunt feedback amplifiers, 12.84–12.99, 13.42
Voltage transfer characteristic, 14.20–14.26, 14.32–14.34, 14.72–14.78

XNOR (exclusive-NOR) gate, 14.11
XOR (exclusive-OR) gate, 14.10

y (or admittance) parameters, 1.84–1.92, 1.95–1.96, 12.91

z (or open-circuit impedance) parameters, 1.67–1.75, 1.80, 1.90–1.92, 1.144–1.145, 1.148, 1.150, 7.9, 7.150, 12.62
Zener diode, 2.83–2.98
Zero limiter circuit, 13.75–13.77